INTRODUCTION TO MODERN CIRCUIT ANALYSIS

Introduction to Modern Circuit Analysis

DONALD A. CALAHAN
ALAN B. MACNEE
E. LAWRENCE McMAHON

Department of Electrical and Computer Engineering

UNIVERSITY OF MICHIGAN

HOLT, RINEHART AND WINSTON, INC.

New York, Chicago, San Francisco, Atlanta, Dallas, Montreal, Toronto, London, Sydney

Calahan, Donald Albert, 1935–
Introduction to modern circuit analysis.

Includes bibliographical references.
1. Electric circuits. 2. Electronic data processing—Electric networks. I. Macnee, Alan, 1920– joint author. II. McMahon, Edward Lawrence, 1931– joint author. III. Title.
TK454.C34 621.319′2 73-21900
ISBN 0-03-080269-5

4 5 6 7 8 071 9 8 7 6 5 4 3 2 1

To Martha, Lois, and Marge

PREFACE

Introduction

An introductory circuits course serves as a cornerstone of most electrical engineering curricula. Frequently, it is the first point of the contact between the engineering student and his chosen field of specialization, and it usually is scheduled for the end of the sophomore or the beginning of the junior year.

Traditionally, the objectives of a first circuit course have been twofold:

1. To develop the student's background in the use of continuous mathematics by its use in a representative application area.
2. To provide a foundation in the analysis of lumped, linear circuits.

Although some elements of electronic circuits analysis (such as controlled sources) have found their way into typical first circuit courses, the emphasis has traditionally remained on the analysis of relatively simple *RLC* circuits.

In recent years there has been a considerable expansion and shifting of electrical engineering curricula which has stimulated reassessment of the objectives and content of this first course. It is only a slight oversimplification to say that this reexamination is a response to the computer revolution. On the one hand, this revolution has given rise to the rapid appearance and expansion of a new group of professionals whose principal interests are mathematically oriented in the area of computer hardware and software, and who view circuits from a purely functional level. Equally important, on the other hand, has been a step function change in the computing power which now normally is available for engineers, even if their principal interests continue to lie in more traditional areas. In a very real sense, today's well-trained young engineers must view digital computation as their predecessors did slide rule calculations.

As might be expected, the response of electrical engineering educators to the computer and computer-related developments has been far from uniform. At one extreme, faculties are experimenting with a first course which attempts to give an equal emphasis on both discrete and continuous systems, usually with a considerable attenuation of the electrical aspects of such systems. At the other extreme, we find no change in the introductory courses leaving a study of the computer and its usage for subsequent introduction in junior and senior electives.

Our General Philosophy

In this text, we have attempted to identify both (1) a body of knowledge and (2) an approach to its study that strikes a balance between these divergent points of view.

1. Topically, we feel that the student should at the first opportunity get an exposure to the modelling-analysis phase of the design process. Focusing on electronic circuits, this involves exposure to nonlinear device models, the inclusion of these models in simple circuits, and the

analysis of these circuits. Further, because circuits can now be viewed functionally for most purposes, it is important to provide a study continuum that includes a "how-it-works" level of discussion involving highly simplified models where appropriate. This inevitably requires a reduced discussion of some of the more esoteric aspects of circuit theory that have made their way into even elementary courses.

2. Our approach is guided by the belief that every electrical engineer, whether his ultimate professional activity is mathematically or physically oriented, should feel at ease with digital computation. It is our further observation that many students will sense the artificiality of computer problems that are merely grafted onto (traditional) circuit analysis topics. We feel that the computer use should be integrated into the network analysis presentation.

Our attempts to implement these viewpoints in the first circuits course were tempered by the spirited comments of our teaching colleagues, who were forced to mesh our written ideas with their classroom experiences. The recurring theme that emerged from these experiments was the following: Although instructors were tolerant or enthusiastic of our efforts to extend circuit analysis into the realm of nonlinear resistive circuits using the digital computer, they opposed dilution or significant alteration of the treatment of linear dynamic circuits. Sinusoidal steady-state and pole-zero analysis were particularly hallowed ground, the feeling being that these concepts were best developed by traditional approaches. Consequently, in this text, we do not interrupt treatment of these topics with computer involvement, except as clearly optional material at the end of certain chapters. We feel that this contributes to the utility of this text as a replacement for other circuit texts in terms of its coverage of what most instructors would feel are the "vital" topics.

To make room for the expanded topical coverage, we have tried to keep the presentation of the various subjects as concise as possible while still being reasonably complete. Based on our classroom experiences with this material, most students would like to have more examples illustrating the analysis methods presented and showing typical engineering circuit applications. We believe that the individual instructor is in the best position to select such examples based on personal experiences and interests. If a laboratory is part of, or accompanies, the course which employs this text, circuits used in the experiments afford natural examples which will evoke high student interest.

The Role of the Computer in the Text

The use of the digital computer is essential to the motivational and pedagogical development of much of the text material. In particular, use of the computer:

1. Allows the analysis of familiar electronic (transistor) circuits, using simple nonlinear device models.
2. Motivates discussion of linearization and discretization concepts, which are posed as computer-oriented counterparts to the transform methods that form the basis of analytical solution procedures.
3. Develops an early appreciation for the insight versus calculation choices that influence hand- and computer-oriented design.

Only in a few instances is the digital computer used in problem solving without linkage between the algorithmic and analytic approach. For example, the iteration methods of Chapters 3 and 5 are related in Chapter 6 to small-signal analysis of resistive networks; small- and large-signal analysis are again related for dynamic circuits in Chapter 14.

In general, where choices exist between simplicity and the generality of algorithms used, we chose the simplest algorithms that are *representative* of more powerful state-of-the-art techniques. Thus, in Chapter 13 Euler's method is used rather than Runge-Kutta's method to solve simultaneous first-order differential equations, because the former is far easier to understand. More efficient techniques are discussed in some of the problems. Inevitably, in an introductory text we cannot explore computational approaches in depth. For example, the very basic concept of numerical stability is only alluded to in Chapter 14. Similarly, modification of the Newton iteration for diodes is not discussed in Chapters 3 or 5. The many sophisticated numerical techniques (as well as the "tricks") that make state-of-the-art CAD programs work efficiently are well beyond the reach of this text.

Concerning the use of canned circuit analysis programs in an introductory course, we appreciate that there is a sharp division of opinion even among instructors enthusiastic for the inclusion of computer-oriented material. However, dc analysis is necessarily a prelude to small-signal analysis of nonlinear circuits, a topic covered in detail in Chapter 6 and used throughout the remainder of the text in the dynamic analysis of amplifiers and other nearly linear circuits. Clearly, it is wasteful of student time and computational resources to reprogram the dc analysis phase of such problems. Therefore, we included a simple yet sufficiently general nonlinear dc analysis program, DCAP, in the Instructor's Manual. Some instructors, after becoming acquainted with this program, are inclined to use a similar program for (linear) dynamic analysis. In such cases, we have found useful the CORNAP program (reference 11.5), which computes pole-zero, frequency, and transient responses as well as the state equations.

A few comments are in order about a novel pedagogy used in this text to develop computer-aided problem solving. The general analysis of nonlinear resistive networks in Chapter 5 is based on the use of models of the nonlinear elements which we have called *companion models*. This approach leads the student into the powerful Newton method of analysis in a natural, simple fashion. A further advantage of this approach is that it leads naturally in Chapter 6 into linear incremental modeling and analysis which is useful for the treatment of many electronic circuits. Finally, in Chapter 14, after the treatment of linear dynamic circuits in Chapters 8 and 13, the use of companion models is extended to the numerical integration of both linear and nonlinear dynamic circuits.

Fitting in the Electronics Curriculum

Although we draw rather freely from electronic circuits for sample problems, the large- and small-signal analysis methods that distinguish the text are based on analytical device models; in contrast, much of traditional electronic circuit design has been performed with graphical and piecewise-linear device models. It is not expected that a course using this text will replace the usual first electronic circuits course, but rather that it will serve as a motivating introduction to such courses.

We observe that the familiarization which the student gains from working problems involving transistor circuits allows us to move rapidly in a following electronics course. We have chosen to use this additional available time to introduce topics in packaged electronics, so that a student can, in three circuits/electronics courses, achieve a modest awareness of modern electronic circuit design.

For students outside of electrical engineering it is feasible to consider the text as offering a two-course compaction of the common three-course circuits-electronics sequence. This is possible because of the considerable involvement of large- and small-signal models of electronic circuits in the problems of the text, both on a functional and a detailed analysis level. For example, the student could be expected to develop familiarity with the form of simple amplifier and digital computer circuits and to verify even their large signal dynamic detailed performance. He would learn little of the physical principles of operation of the electronic devices, however, and would be unfamiliar with many of the elements of electronic circuit design. This involvement may be considered adequate for certain students (for example, those interested in computer engineering).

Connections with a Laboratory

A laboratory is often associated with the first electrical engineering course to enhance the motivation and "real world" experience of the student. We feel that the introduction given to the analysis of simple digital and amplifier circuits in Chapters 3, 5, and 14 can be the basis for a number of significant "wet-lab" and "dry-lab" (simulation-based) experiments. One could, for example, develop the functional uses of digital circuits at greater length in a laboratory-lecture format. Of course, the rather standard treatment of dynamic circuits offers the usual laboratory possibilities for observing frequency- and time-domain phenomena.

Course Planning

An elementary circuits text is likely to be used in a variety of curriculum formats ranging from an intense one-semester 4-hour/week course to two 3-hour two-semester or three-quarter courses. Most parts of this text have been used in a two-course sequence of 3 trimester hours each at the University of Michigan.

In our course sequence, the first 3-hour course concentrates on the static analysis of both linear and nonlinear circuits (Chapters 1 through 6) and covers some dynamic circuit analysis (usually through Chapter 8). This course is taken by both electrical and computer engineering (CE) students. The second course, not required in the CE program, offers further continued coverage of dynamic circuits, which are treated in Chapters 9 through 15. This course is accompanied in our curriculum by a 3-trimester-hour course which presents much of the mathematics that often has been taught in introductory circuits courses: ordinary differential equations, Fourier series, complex variables, and Laplace transform theory. Chapter 11 and Appendix C are provided to allow for coverage of such material by instructors who wish to present the mathematics in the circuits course.

To assist in planning other course sequences, it is well to point out some of those topics that we view as essential to an orderly development of the material in this text and those we regard as optional. Elective topics are those not later used in the development of other topics, but may be referred to in the problems.

Required (Chapters 2 to 6)

i-v characteristics (2)
controlled sources (2)
diode and (bipolar) transistor models (3)
amplifier fundamentals (3)
digital circuit fundamentals (3)
matrices, determinants (4)
Gauss elimination (5)
companion model (5)
DNET program (5)
complete transistor models (5)
use of DCAP (5)
incremental and small-signal analysis (6)

Elective

topological analysis (4)
logic & memory circuits (5)
mutual inductance (7)
convolution (9)
load-flow analysis (10)
small-signal dynamic analysis (11)
state variable analysis (13)
computer-oriented transient analysis (15)
two-port analysis (15)
transformers (16)
analogs (16)
Fourier analysis (Appendix C)
fundamentals of physical electronics (Appendix C)

In planning the extent of computer use for the course, it is important to realize that there is no need to assign a computer program wherever there is one given in the text. Often it is the descretization or linearization concept that is important; only occasionally will an instructor dwell on a concept sufficiently to warrant a programming assignment. For example, we will typically assign a digital or amplifier circuit problem in Chapter 3, a DCAP problem in Chapter 5 (which we may subsequently use in small-signal analysis in Chapter 6) and a nonlinear transient analysis problem in Chapter 14 (possibly building on a program from Chapter 5). Spread over a 78-lecture-hour course, this involvement seems reasonable yet modest.

It should be noted that the ordering of the topics presented after Chapter 7 can be modified to some extent. For example, the treatment of transformers given in the first half of Chapter 16 has been written so that it can be covered immediately after Chapter 7, which introduces mutual inductance. Chapters 8 and 9 and Chapters 10, 11, and 12 form natural blocks of traditional linear circuit analysis material. However, it is quite reasonable to present the computer methods of dynamic circuit analysis in Chapter 14 at any time after the introduction of the energy storage elements in Chapter 7. We have found such realignment can be useful in avoiding end-of-term loading of computational facilities.

To assist the instructor teaching this material for the first time, several teaching aids have been provided in conjunction with the computer-oriented material in Chapters 3, 5, and 14:

1. At the end of these chapters, a number of simple problems are given that illustrate the solution principles *without* using a digital computer.
2. Over two dozen Fortran programs are given in the instructors' manual as solutions to problems described in the text. This allows the instructor to adjust to the methodologies presented without risking assignment of unworkable or numerically ill-conditioned problems.

Program Availability

The programs displayed in the text and the DCAP program in the Instructor's Manual are available to instructors from the authors either on punched cards or user-supplied magnetic tape in OS compatible, labeled, 800 BPI, 9-track format.

D. A. CALAHAN
A. B. MACNEE
E. L. MC MAHON

CONTENTS

1

Introduction to Systems and Circuit Analysis

1.1 Engineering System Analysis

An engineering system is a collection of physical components that together serve a useful function. Electrical engineers have a vital role to play in the design of such systems, because the components are often based on electrical phenomena. This is illustrated in Figure 1.1, where a number of engineering systems associated with the construction of a possible next-generation automobile are shown. Tire pressure, engine condition, and emission products could all be electronically monitored in such a car. One might even consider a radar-guided, anti-collision device which would contain (1) a radar transmitter/receiver, (2) an electrical system to control the fuel and braking systems, and (3) a computer to decide when a collision is imminent. Systems such as these are all well within the domain of expertise of the modern electrical engineer.

The success of such systems depends to a great extent on the engineering judgment and analytical abilities of the designer. The judgment is based mostly on experience and knowledge of the state-of-the-art within a particular field of design. Analytical facility, on the other hand, is gained by practice, using appropriate mathematics in the analysis of "typical" systems. Such mathematical studies, applied to the solution of electrical circuit problems, are the subject of this text.

1.2 Electrical Circuit Analysis

The study of the behavior of electrical components and circuits serves a number of useful functions.

First, as indicated in Figure 1.1, nearly all electrical systems require circuit components. The designer of the radar, control system, or computer system who does not have a basic understanding of circuit components is not well prepared to assume responsibility for a subsystem design.

Second, the cornerstone of the engineering design process is systematic analysis, which

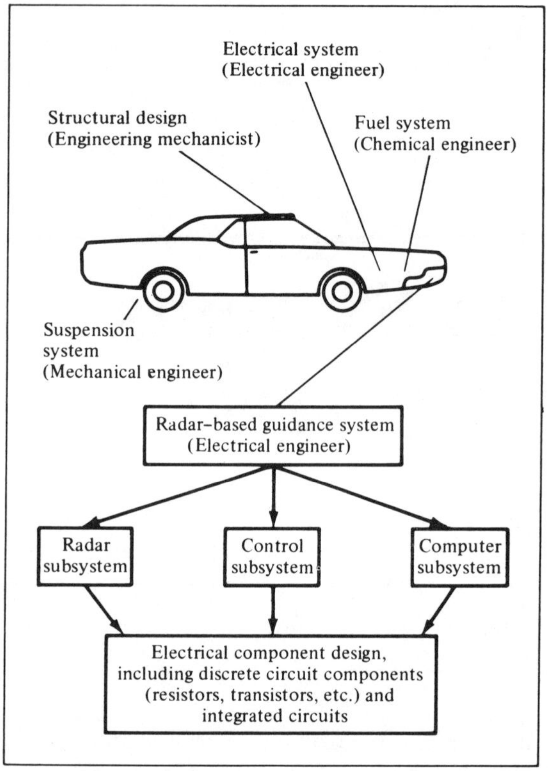

FIGURE 1.1 Example of an engineering system: radar-guided automobile.

can be viewed as consisting of the following steps:

1. An analysis of the design objectives to determine whether they are consistent with the physical characteristics of available components.
2. A choice of potentially useful components and an initial design approach.
3. Development of mathematical models of components.
4. Analysis of the interconnection of components making up the initial design to yield a set of mathematical equations which model the complete system.
5. Solution of these mathematical equations and a comparison between the performance of the system model and the design objectives (if necessary, revise design).
6. Comparison of the system model performance with measurements of a prototype physical system.
7. If necessary, revision of component models to correct deficiencies in the system model.
8. Optimization of the system performance.

Although we will not have time to become familiar with all of these steps in design, we will find Steps 3 to 5 are well-illustrated by the material of this text.

Finally, many readers will have at least casual acquaintence with some of the circuits and circuit functions we will study. In a very real sense, then, we can relate the results of our analysis to the performance of circuits that we may have observed.

1.3 Circuit Functions

The variety of functions performed by electrical circuits contributes to their usefulness for study. Many classifications of these functions are possible. The following list was chosen as indicative of the more familiar functions.

1. *Amplification.* A circuit that magnifies an input signal to produce a larger output signal with the same shape is termed an amplifier. (See Figure 1.2.) The most popular example of this function is the audio amplifier used in stereo and public address systems.

2. *Energy conversion.* We are probably all familiar with the need to convert the alternating current (ac) power source at an electrical outlet to a direct current (dc) form; this dc power source is necessary for the operation of radios, televisions, battery chargers, and so on. However, the reverse process, dc to ac conversion, is also vital to other electrical systems such as car radios.

3. *Information processing.* Signals are commonly classified as either *analog* or *digital.* The former are continuous functions of time, as would occur from measurement of a physical phenomenon; digital signals have the form of a sequence of pulses representing, for example,

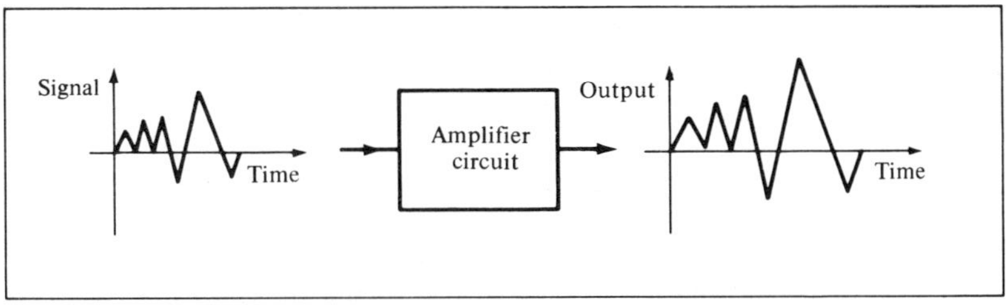

FIGURE 1.2 An amplifier circuit.

successive digits of a number. The usefulness of either type signal is specified by its *information* content. This information can be altered by electrical circuits. As shown in Figure 1.3, an analog signal buried in a noisy background can be *filtered* to produce an essentially pure signal. We are all familiar with the role of computer circuits in operating on digital input information to produce new information in the form of output data.

4. *Memory*. Circuits that are designed to hold information for later processing perform a *memory* function. Although memories exist to store analog signals, the most common examples are electronic circuits used in digital computer memories. It is quite common to find millions of such memory elements in a modern computer (Figure 1.4).

As we encounter a variety of circuits in this text, the reader may find it useful to view them as belonging to one of the above types.

1.4 A Fundamental Classification

As we will learn, there are a number of types or classifications of circuits based on certain characteristics of the circuit components. From the standpoint of the organization of this text, however, there is one dominant characteristic that will determine when the circuit will be studied.

To illustrate this characteristic, consider the

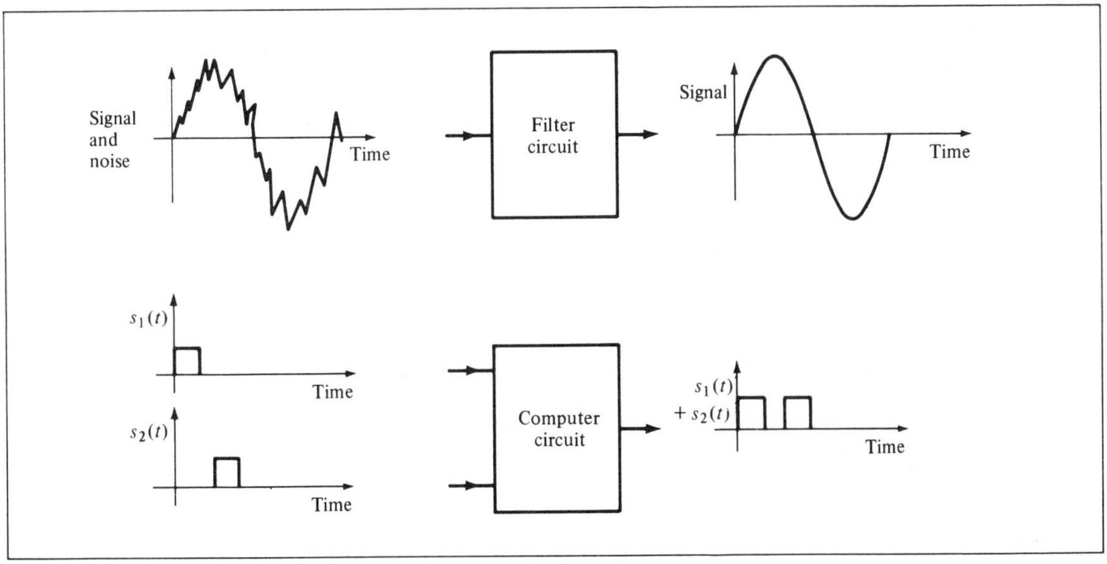

FIGURE 1.3 Analog and digital information processing.

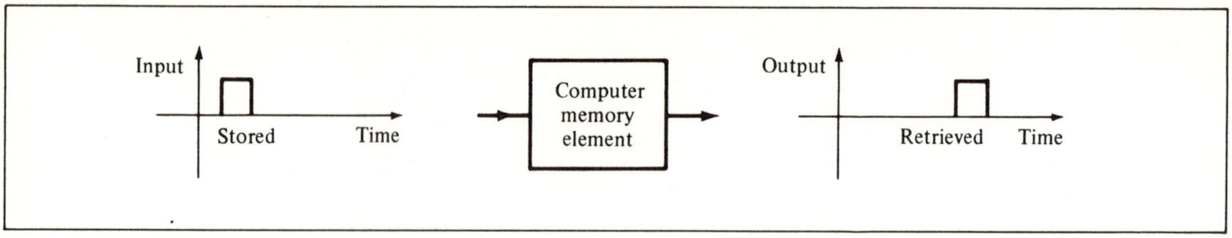

FIGURE 1.4 Digital memory function.

simple lighting system shown in Figure 1.5(a). If we close the switch S at an arbitrary time, which can be designated as $t = 0$, the lamp "appears" to light instantaneously. We know from basic physics that this cannot be so; at the very least there must be a time interval equal to the time it takes light to travel 100 feet (ft) before the lamp can possibly light. That time is

$$t = \frac{\text{distance}}{\text{velocity of light}} = \frac{100 \times 12 \times 2.54}{3 \times 10^{10}}$$

$$= 1.016 \times 10^{-7} \text{ sec}$$

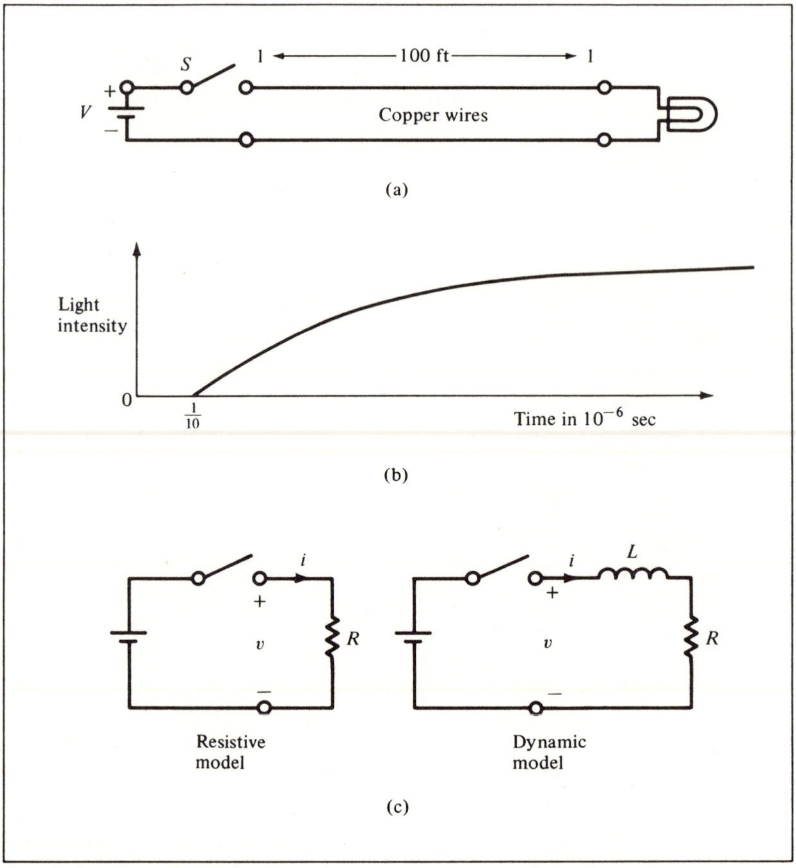

FIGURE 1.5 A simple lighting circuit, its transient response, and its models.

or about 0.1 of a microsecond (μsec), and standing at the switch we could not see the light for another 0.1 μsec! Two-tenths μsec may seem to be a very short time. It is if one is interested in a lighting circuit, but it represents about 1/16 in. on the screen of a 23-in. home television set, and modern computers can easily add two numbers to a precision of ten decimal digits in that time. This issue of whether the transmission time is important to our study profoundly effects the method of analysis.

If we instrumented the circuit of Figure 1.5(a) so that the lamp intensity could be observed accurately, we might observe the waveform sketched in Figure 1.5(b), depending on such things as the resistances of the lamp, source, and lines, and the geometric configuration of the lines. In every case, after a transient period, the intensity settles down to a steady value.

If we are interested only in the steady value, we could describe the lighting circuit by *algebraic* equations, in this case Ohm's law (Chapter 2),

$$v = Ri$$

We will discuss this law more formally later; for our purposes we observe only that v and i will follow each other *instantaneously*. This is known as a *resistive*, or *static*, model of the lighting circuit shown in Figure 1.5(c).

In order to calculate waveforms which are functions of time such as shown in Figure 1.5(b), we must use a *dynamic* model described by some form of *differential* equation. As shown in Figure 1.5(c), a crude dynamic model could use an inductor (Chapter 7) to represent certain storage characteristics of the 100-ft connecting line; the resulting differential equation would be of the form

$$v = L\frac{di}{dt} + Ri$$

In the text, the nondynamic (algebraic) characteristics of networks are studied in Chapters 2 through 6; the dynamic properties are investigated in Chapters 7 through 14. Because the solution of algebraic equations is easier to describe than the solution of differential equations, the reader will find the first part of the text devoted to developing network concepts and simple mathematical solution techniques; the second part will be more challenging mathematically and will heavily rely on the prior or concurrent experience of the reader in a course on differential equations.

2

Introduction to Network Analysis

2.1 Physical Basis of Circuit Behavior

Charge

The basic electrical quantity is the property of matter known as charge. Because electrical circuits utilize charge to achieve their results, it may seem natural to preface the subject of circuit analysis with a discussion of "what charge is." This is the approach taken in elementary physics courses, where the reader probably studied forces between charges near other charges, in magnetic fields, and so on. These topics are not germain to the study of circuit analysis, however; we are usually interested only in the flow of charges in wires, resistors, batteries, and so on, which we view as obeying rather simple laws.

The only physical principle of immediate use to us is Coulomb's[1] law

$$F \sim \frac{q_1 q_2}{r^2} \tag{2.1}$$

[1] Charles Augustin Coulomb (1736–1806) was a French physicist. Between 1786 and 1789 he published memoirs describing his measurements of electric repulsion and attraction and formulating the law which now bears his name.

where F is a force and r the distance between two charges of value q_1 and q_2.

There are two kinds of charge: positive and negative. Like charges repel one another and unlike charges attract; therefore, if F in Equation (2.1) is taken as a repulsive force, the force is negative when the two charges have opposite signs and is an attractive force.

Current

In the study of electric circuits, our main interest is in moving charges. Suppose we have a wire along which charge is moving; if 1 coulomb per second is moving past some point in the wire, then the wire is carrying a current i of 1 ampere.[2] That is, 1 ampere (amp) = 1 coulomb/sec.

To say that 1 coulomb/sec is passing a given point is an incomplete description, because

[2] Andre Marie Ampere (1775–1836) pioneered in electromagnetism while serving as a professor of mathematics at the Ecole Polytechnique in Paris. On September 11, 1820, he heard of H. C. Oersted's discovery that a magnetic needle is influenced by a current, and 7 days later he presented a paper to the Academy explaining that and similar phenomena.

neither the kind of charge (positive or negative) nor its direction have been stated. We can take care of both these questions by adopting a convention. We assume that current results from the motion of positive charge, and we take the positive direction of current to be the direction in which the positive charge is moving. When we say that there is a current of 1 amp flowing from left to right in a wire, what we are really saying is that either positive charge is moving from left to right at the rate of 1 coulomb/sec or that negative charge is moving from right to left at the rate of 1 coulomb/sec.

Current is measured by a device known as an *ammeter*, a sketch of which is shown in Figure 2.1. The ammeter is provided with a directional reference by marking one terminal with a "+" sign (or by coloring it red) and the other with a "−" sign (or by coloring it black). This establishes a reference direction *from* the plus (red) terminal *toward* the minus (black) terminal through the ammeter. If the meter indicates a positive value, by a clockwise deflection, this means that the actual current direction is the same as the reference direction, while a negative value means that the actual current direction is opposite the reference direction.

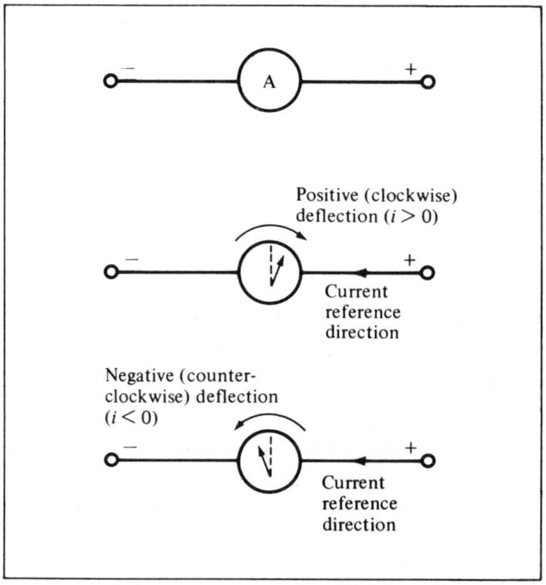

FIGURE 2.1 Ammeter representation.

Example 2.1

Suppose we are given the simple circuit shown in Figure 2.2(a). The first step in its analysis is the assignment of reference *directions* for currents (and voltages, as we shall see). As shown in Figure 2.2(b), these assignments are quite arbitrary, and require no a priori knowledge of the circuit behavior. As a result of a mathematical analysis of the circuit based on the methods of this text, suppose that we determine the currents to have the numerical values of Figure 2.2(c). As we see, some of the values such as i_1 may be negative; this implies that, in the *physical* circuit, current (positive charge) flows opposite to our reference direction. This is indicated by the direction of the ammeter deflection.

Voltage

In general, there is energy (work) associated with the movement of charge. If w joules of energy are required to move a positive charge of strength q from point a to point b in a circuit, then there is a voltage[3] $v = w/q$ between a and b. We say that "b is positive with respect to a." If, on the other hand, the circuit imparts energy to the charge as it moves from a to b, a is positive with respect to b.

The voltage between any two points is independent of the path taken by the charge. Furthermore, there can be a voltage between two points even if no charge is actually moving from one to the other, because voltage is a measure of how much energy *would be* required *if* a charge were moved.

Reference polarities are required for voltage, just as reference directions are required for cur-

[3] Alessandro Volta (1745–1827), an Italian physicist, is credited with inventing the electroscope and the voltaic pile, the predecessor of the storage battery.

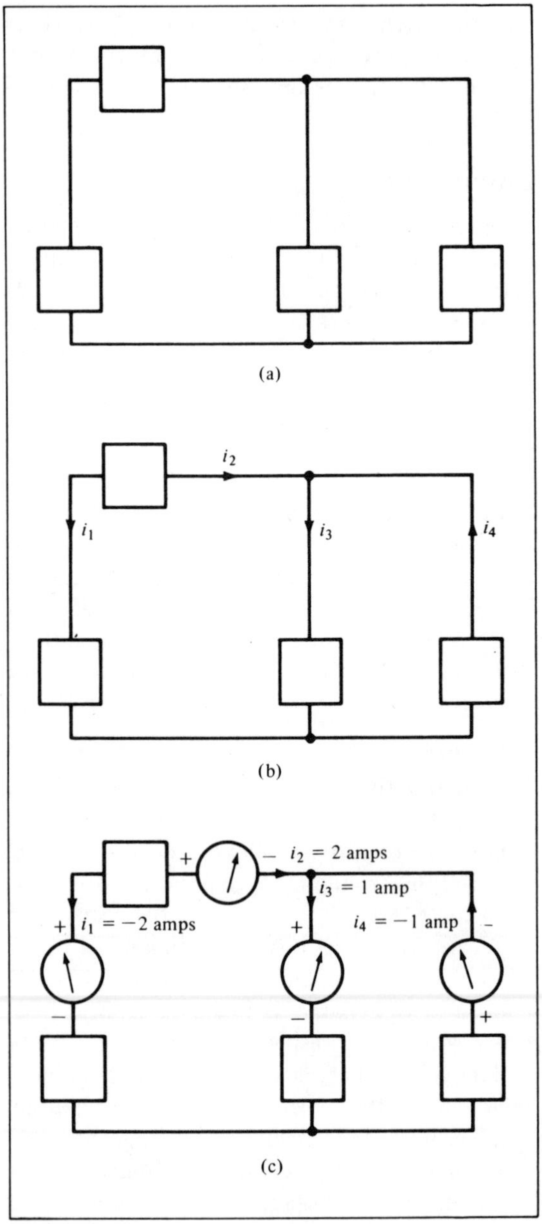

FIGURE 2.2 Use of references (boxes represent resistors, batteries, and so on). (a) Given circuit; (b) reference assignment; (c) result of analysis.

rent. These can be provided by marking each pair of points with a "+" sign on one point and a "−" sign on the other. A positive voltage is then one with the same polarity as the reference, and a negative voltage, one with the opposite polarity. In practice, the need for these marks is often eliminated by adopting the convention of putting the plus reference at the tail of the current reference arrow and the minus reference at the head of the arrow. Thus, both current and voltage references are automatically defined if either current arrows or voltage "+" or "−" signs are given.

Voltage is measured by a device known as a *voltmeter*, shown in Figure 2.3. References are important when using the voltmeter. If the meter indicates a positive value by a positive (clockwise) deflection, this means that a positive amount of work is required to move a positive charge from the "−" to the "+" terminal.

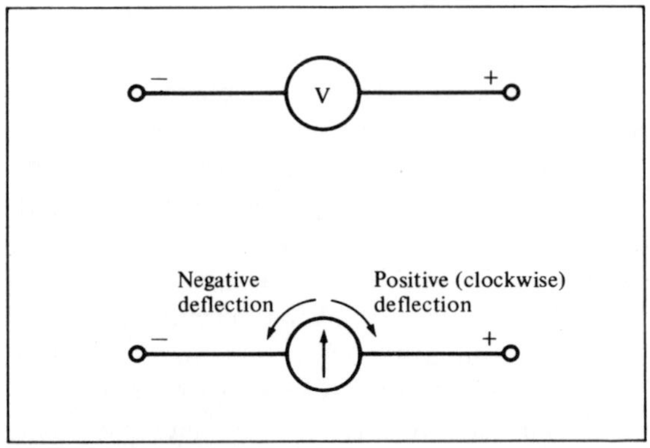

FIGURE 2.3 Voltmeter representation.

Example 2.2

Let the circuit of Figure 2.4 (a) be assigned the voltage references shown. After an analysis, it is found that the voltages have the numerical values shown in Figure 2.4(b). As with currents, these may have both positive and negative values. If voltmeters are connected as shown, the deflections will reflect the signs of the calculated results.

Power

Let us consider a typical circuit element with current and voltage references as shown

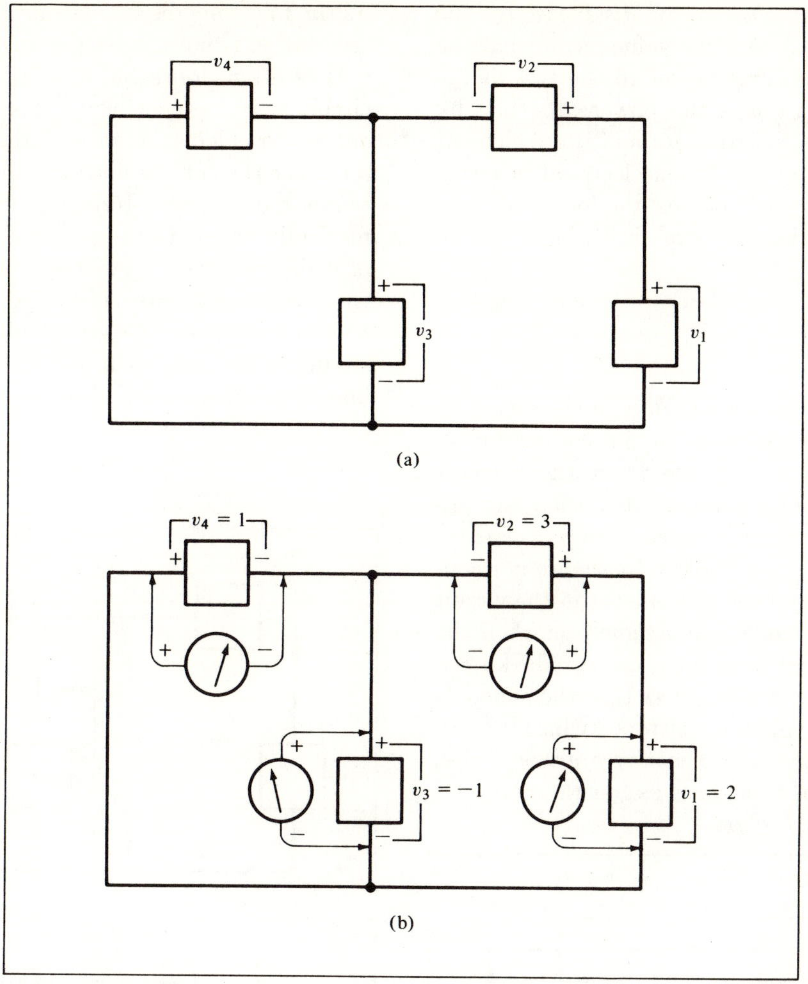

FIGURE 2.4 Illustration of voltage references. (a) Assignment of references; (b) results of analysis.

in Figure 2.5. Because v is the number of joules required to move 1 coulomb from b to a, then if i is the number of coulombs/sec moving from a to b, it is easy to see that their product is the number of joules/sec, or watts, received by the element. That is,

$$vi = \frac{\text{joules}}{\text{coulomb}} \times \frac{\text{coulombs}}{\text{second}}$$

$$= \frac{\text{joules}}{\text{second}} = \text{watts} = p \qquad (2.2)$$

If the convention of putting the + voltage reference at the tail of the current reference is followed, then a positive value of p means that

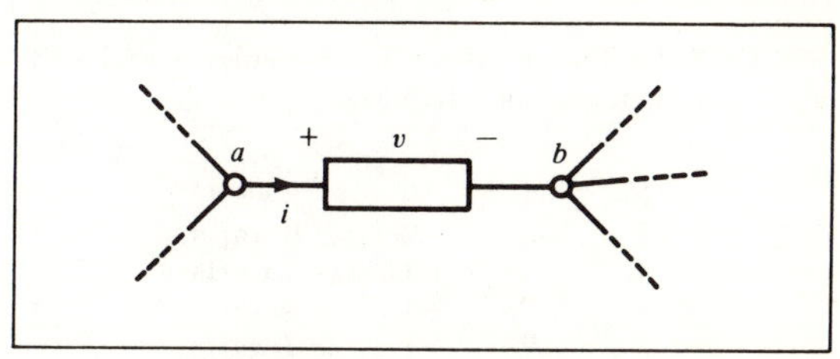

FIGURE 2.5 Typical element.

power is being *absorbed* or *dissipated* by the element, while a negative value means that the element is supplying power to the rest of the circuit. That is, p is the power absorbed by the element; a negative p implies the element is generating power. When electrical power is absorbed, it is, of course, not lost, but only converted to another form (heat, light, and so on).

Example 2.3

In Figure 2.6(a) current and voltage references are chosen consistent with the convention of Figure 2.5. If an analysis gives the currents and voltages of Figure 2.6(b), then the absorbed powers p_j are formed from $p_j = v_j i_j$.

The reader should note that we are not concerned with the type of elements in the circuit when the references are assigned or when the p_j are calculated from the i_j and v_j. The knowledgeable reader might guess that the elements with positive p_j are resistors (which absorb power) and those with negative p_j are batteries, which might be thought of as generating power.

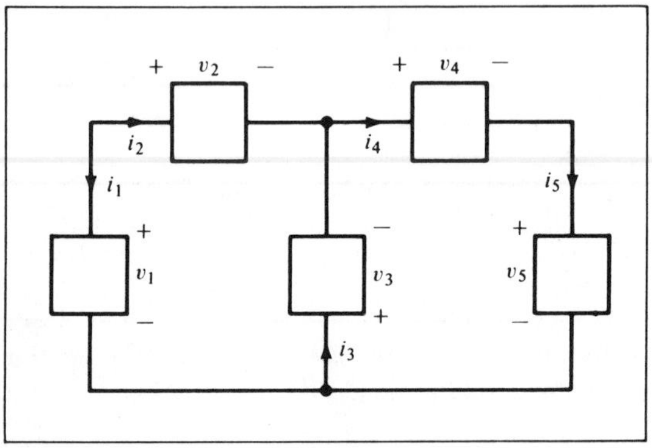

FIGURE 2.6 Power calculation for current and voltage. (a) Assignment of references:

j	v_j	i_j	p_j
1	1	2	2 watts
2	−3	−2	6 watts
3	−4	1	−4 watts
4	−3	−1	3 watts
5	7	−1	−7 watts

(b) results of analysis.

As the following discussion shows, this physical association is not always possible.

It is often desireable to consider parts of a circuit as a single element. This need not alter our power calculation procedure. So long as we reference the current flowing *into* the positively referenced voltage terminal, we will automatically calculate the power *absorbed* by the part of the circuit viewed as a single element. In Figure 2.7, for example, the product $p = v(\pm i)$ gives the power *absorbed* by the circuit to the right or to the left of the dashed line, depending on whether the *reference* direction of i at the positive reference of v points right or left. From Example 2.3 $v = -v_3 = 4$

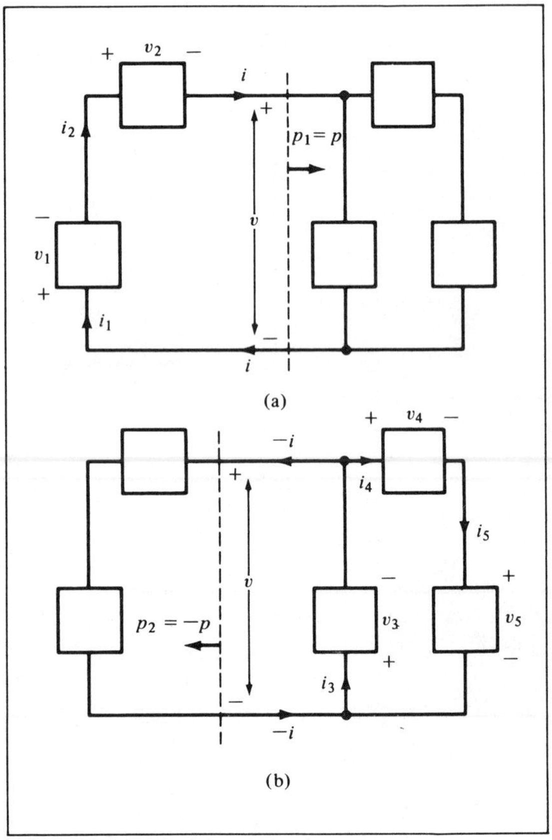

FIGURE 2.7 Generalized power calculation for example of Figure 2.6. (a) vi = power absorbed by circuit to right of dashed line; (b) $v(-i)$ = power absorbed by circuit to left of dashed line.

TABLE 2.1

MKS Units Used in this Text

Quantity	Symbol	MKS Units
Charge	q	coulombs
Current	i	amperes
Voltage	v	volts
Energy	w	joules
Power	p	watts

volts and $i = i_2 = -2$ amps. The power absorbed by the circuit to the right is therefore -8 watts; this part of the circuit is therefore a generator of 8 watts even though one of the elements ($j = 4$) absorbs power.

The example hints at one of the more powerful theorems associated with circuit analysis, namely, that the algebraic sum of the absorbed powers in a circuit is zero. This will be an excellent check on calculation procedures (see Problem 2.2).

Units and Prefixes

We have discussed the three most common quantities of electrical measurement: current, voltage, and power. Table 2.1 summarizes the mks units we will use to measure these quantities. In addition, the prefixes of Table 2.2 are often used with these units to indicate multiplication or division by a large power of ten. Thus, 5 times 10^{-6} sec is 5 microseconds (μsec), 2 times 10^3 watts is 2 kilowatts (kw), and 5 times 10^{-3} amps is 5 milliamps (ma). In the future, we will use these abbreviations frequently, because they are part of the engineering jargon related to circuit analysis.

TABLE 2.2

Unit Prefixes

Multiple or Submultiple	Prefix	Symbol
10^9	giga	G
10^6	mega	M
10^3	kilo	k
10^{-3}	milli	m
10^{-6}	micro	μ
10^{-9}	nano	n
10^{-12}	pico	p

2.2 Description of Resistive Circuit Elements

In the first six chapters of this text we will restrict our attention to circuits containing only elements related by algebraic equations. This will permit the development of basic network concepts without being burdened by the more complex mathematics required for the analysis of circuits described by differential equations.

Resistance

A *resistor* is an element with two terminals (points at which other elements may be connected), characterized by

$$i = f(v) \qquad \text{or} \qquad v = g(i) \tag{2.3}$$

The important property of a resistor is that the voltage (or current) at any instant of time is completely determined by the current (or voltage) at that instant of time. A resistor is therefore a *memoryless* element in the sense that previous values have no effect on present values of current or voltage.

A very important type of resistor is the *linear resistor*, described by

$$v = iR \tag{2.4}$$

$$i = Gv \tag{2.5}$$

where

$$G = \frac{1}{R} \tag{2.6}$$

Equation (2.4) is known as Ohm's law.[4] The parameter R is called the *resistance*, expressed in ohms (Ω); its reciprocal G is the *conductance*,

[4] Georg Simon Ohm (1787–1854) was a German physicist. He developed the law, which bears his name, in a pamphlet published in 1827.

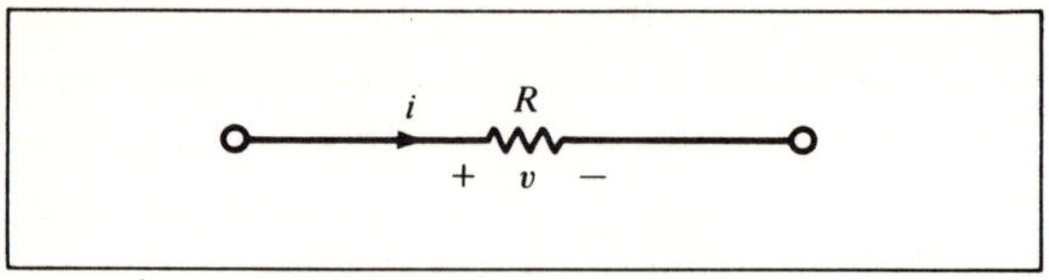

FIGURE 2.8 Resistor symbol.

expressed in mhos (℧). The linear resistor is so common that when the unqualified term "resistor" is used, what is ordinarily meant is a linear resistor; any other resistor is usually referred to as a nonlinear resistor. The symbol for a linear resistor is shown in Figure 2.8.

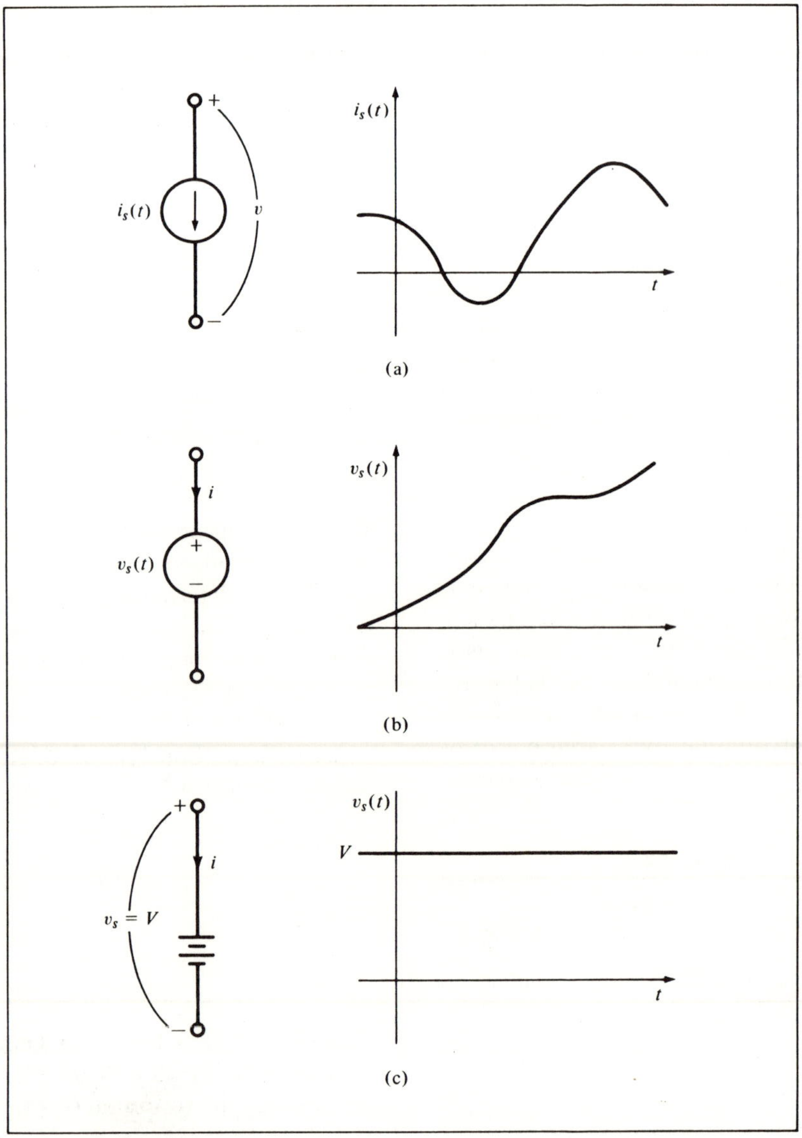

FIGURE 2.9 Characterization of sources. (a) Current source representation; (b) voltage source representation; (c) alternative constant voltage source representation (battery).

It follows directly from Equations (2.2) and (2.4) that for a linear resistor the power is given by

$$p = i^2R = v^2G \tag{2.7}$$

Therefore, if R is positive, the power is always positive; the resistor can only absorb energy, converting it to heat or some other energy form.

Example 2.4

A 2000 Ω resistor is carrying a current of 5 times 10^{-3} amps. The voltage, from Equation (2.4), is

$$v = iR = 5 \times 10^{-3} \times 2 \times 10^3 = 10 \text{ volts}$$

The power being absorbed by the resistor can be found from Equation (2.2)

$$p = vi = 10(5 \times 10^{-3}) = 50 \times 10^{-3} \text{ watts}$$
$$= 50 \text{ mw}$$

or directly from Equation (2.7)

$$p = i^2R = (5 \times 10^{-3})^2(2 \times 10^3)$$
$$= 25 \times 10^{-6} \times 2 \times 10^3 = 50 \text{ mw}$$

Sources

In contrast to resistors, sources have the property that their voltage and current are unrelated to each other. A voltage source has two fundamental properties: (1) its voltage is some function of time $v(t)$ (possibly a constant), quite independent of the current through it; (2) its current is arbitrary and is determined by the circuit to which the source is connected. A current source, on the other hand, has a prescribed current $i(t)$ which is independent of the voltage across it, and the voltage is arbitrary. These elements are exemplified in Figure 2.9(a)–(c). We are all familiar with devices that behave approximately as voltage sources: batteries, electric generators, radio and television power supplies; but current sources may be less familiar. Devices such as photocells and transistors approach such current source behavior.

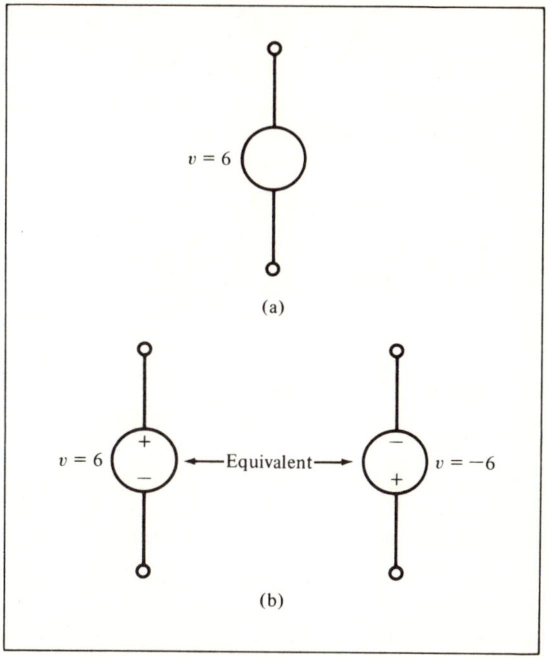

FIGURE 2.10 Voltage source definitions.

The reference indications for current and voltage sources are essential to either definition. For example, the voltage source of Figure 2.10(a) is not completely defined because no reference direction for v is shown. In Figure 2.10(b), a reversal of reference symbols is shown to achieve the same effect as changing the sign of the source value.

Sources can be further classified as independent and dependent (controlled) sources. In an independent source, the voltage or current is a specified function of time. In a dependent source, the voltage or current is a function of some other voltage or current in the circuit. Dependent sources will be encountered in representations of transistors and other electronic devices in later chapters; we will study them in more detail later in this chapter.

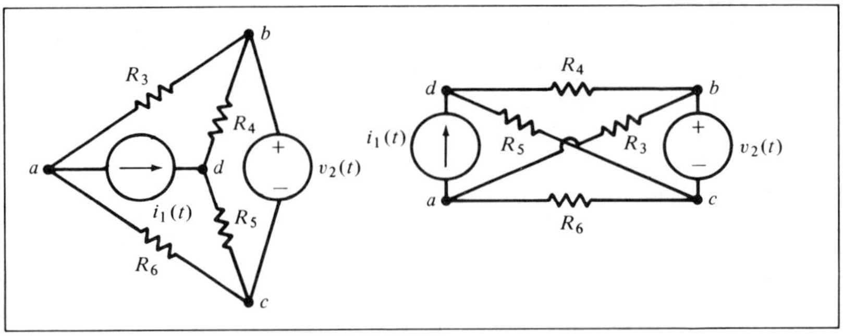

FIGURE 2.11 Equivalent networks.

2.3 Laws of Networks

Networks

Thus far, we have considered the behavior of individual circuit elements. We now turn our attention to the way in which elements behave when connected together. We begin with some definitions.

Any collection of elements connected at their terminals (nodes) constitutes a *network*. Only the manner in which the elements are connected affects the electrical behavior of the network; the position of the elements and vertices in space is of no consequence. Thus, the two networks shown in Figure 2.11 are the same, because they are structurally equivalent.

Kirchhoff's Laws

The two basic laws of network behavior are Kirchhoff's current law and Kirchhoff's voltage law.[5] Kirchhoff's current law (KCL) states that the net current entering any node is zero. This follows from the observation that charge flowing into a node cannot be created or destroyed. For example, in Figure 2.12(a), we can write

$$i_1 + i_2 + i_3 + i_4 + i_5 = 0 \qquad (2.8)$$

This equation implies that some of the currents must necessarily have a zero or negative value. An alternative statement of the law is that the total current entering any node is equal to the total current leaving that node. For the node of Figure 2.12(b) we would write

$$i_1 + i_2 + i_3 = i_4 + i_5 \qquad (2.9)$$

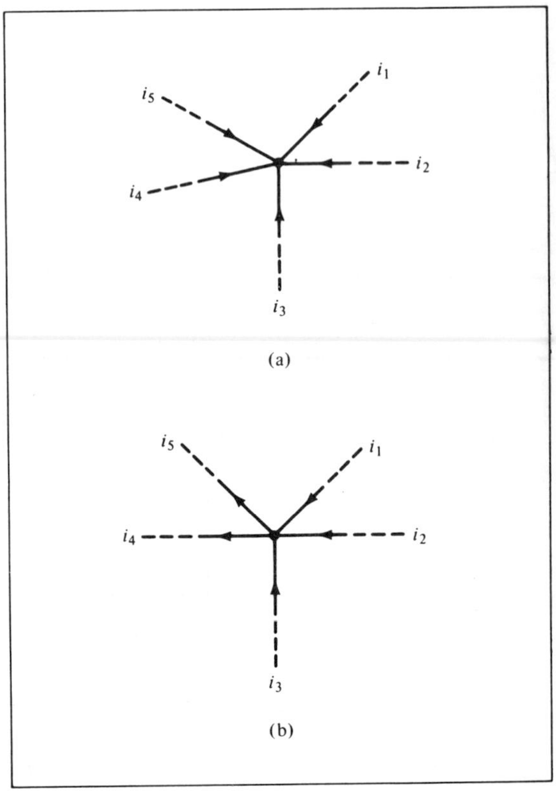

FIGURE 2.12 Kirchhoff's current law.

[5] Gustav Robert Kirchhoff (1824–1887), a German physicist, generalized Georg S. Ohm's theory for a linear conductor to the case of conductors in three dimensions and developed a theorem which gives the distribution of currents in a network.

This does not mean that the currents are necessarily all positive; it only means that the reference directions of i_1, i_2, and i_3 are toward the node, while the reference directions of i_4 and i_5 are away from the node. We could just as well write for Figure 2.12(b) the equation

$$i_1 + i_2 + i_3 - i_4 - i_5 = 0 \qquad (2.10)$$

where the minus signs result from the fact that we are summing currents *entering* the node, while i_4 and i_5 have reference directions *away* from the node; a positive current leaving the node is the same as a negative current entering the node.

In writing KCL equations, there are always two sets of signs to contend with. The first are the signs in the equation, which are determined by the reference directions of the currents relative to the node. The other signs are those due to the actual current direction relative to the reference direction. For example, in Figure 2.12(b), let us assume that the current values are

$$i_1 = 2 \qquad i_2 = -4, \qquad i_3 = 1 \qquad i_4 = 3$$
$$i_5 = -4 \qquad (2.11)$$

In numerical form, Equation (2.10) then becomes

$$(2) + (-4) + (1) - (3) - (-4) = 0 \qquad (2.12)$$

where the signs outside the parentheses are those of the equation while the signs inside are those of the numerical values of the currents.

Kirchhoff's voltage law (KVL) states that the net voltage around any closed path is zero or, equivalently, that the voltage between any two points is independent of the path taken between them. Because voltage is proportional to the work to move a unit charge, this law requires that the work to move a charge from one point to another is independent of the path over which the charge is moved. For example, summing voltages around the closed path in Figure 2.13(a), we have

$$v_1 + v_2 + v_3 + v_4 + v_5 = 0 \qquad (2.13)$$

while in Figure 2.13(b) we have

$$v_1 + v_2 + v_3 - v_4 - v_5 = 0 \qquad (2.14)$$

The minus signs in this equation arise from the fact that the reference directions (and therefore reference polarities) of elements 4 and 5 are opposite the direction of the closed path. In Equations (2.13) and (2.14), the voltages themselves have signs which depend on the relation between the actual and reference polarities, and these signs must be included in the equation.

Equation (2.14) can be written in a way which emphasizes the alternative statement of KVL, namely,

$$v_1 + v_2 + v_3 = v_4 + v_5 \qquad (2.15)$$

where both sides of the equation are expressions for the voltage from point a to point b.

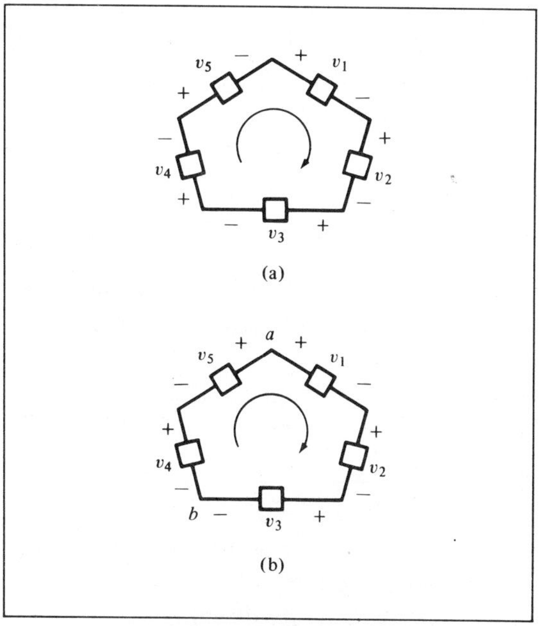

FIGURE 2.13 Kirchhoff's voltage law.

2.4 Analysis of Simple Networks

Series and Parallel Networks

To show the application of KCL and KVL, let us consider the simple networks shown in Figure 2.14.

The series network is analyzed as follows. Applying KVL, we have

$$v_s = v_1 + v_2 + v_3 \tag{2.16}$$

From Ohm's law we find

$$v_1 = R_1 i_1 \qquad v_2 = R_2 i_2 \qquad v_3 = R_3 i_3 \tag{2.17}$$

while from KCL, applied at each node, it follows that

$$i_1 = i_2 = i_3 = i \tag{2.18}$$

Combining these three equations we get

$$v_s = (R_1 + R_2 + R_3)i \tag{2.19}$$

$$i = \frac{v_s}{R_1 + R_2 + R_3} \tag{2.20}$$

Substituting this result back into Equations (2.17) gives us

$$v_1 = \frac{R_1}{R_1 + R_2 + R_3} v_s \tag{2.21}$$

$$v_2 = \frac{R_2}{R_1 + R_2 + R_3} v_s \tag{2.22}$$

$$v_3 = \frac{R_3}{R_1 + R_2 + R_3} v_s \tag{2.23}$$

These results show a number of important properties of series networks. Comparing Equation (2.20) with Ohm's law, Equation (2.5), we see that we would get exactly the same current if the source v_s were connected to a single resistor given by

$$R_{eq} = R_1 + R_2 + R_3 \tag{2.24}$$

In other words, so far as the behavior at terminals a–b is concerned, the series connection of R_1, R_2, and R_3 is *equivalent* to R_{eq}. It is easy to see that this result is general: A series combination of resistors is equivalent to a single resistor of value equal to the sum of all the resistances.

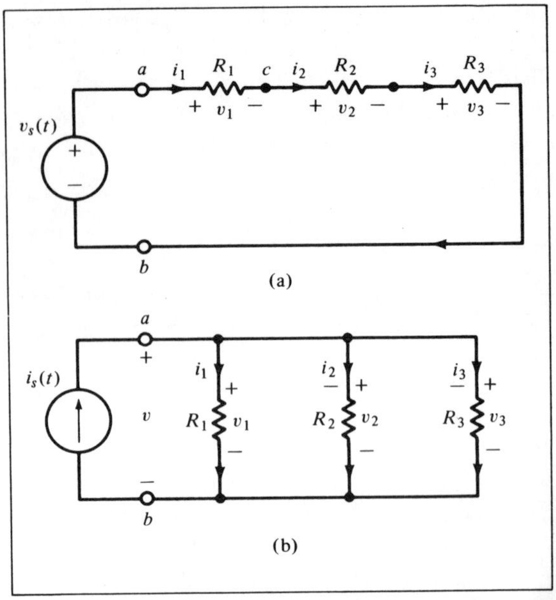

FIGURE 2.14 Series and parallel networks.

We can now write Equation (2.21) in the form

$$v_1 = \frac{R_1}{R_{eq}} v_s \tag{2.25}$$

with similar expressions for v_2 and v_3. This is known as the *voltage divider* relation, whose general expression is: In any series combination of resistors, the fraction of the total voltage which appears across any one resistor is the same as the fraction of the total resistance represented by the resistor.

Example 2.5

For the network of Figure 2.15,

$$R_{eq} = R_1 + R_2 + R_3 = 5 + 10 + 15 = 30\ \Omega$$

The current is therefore

$$i = \frac{v_s}{R_{eq}} = \frac{90 \text{ volts}}{30\ \Omega} = 3 \text{ amps}$$

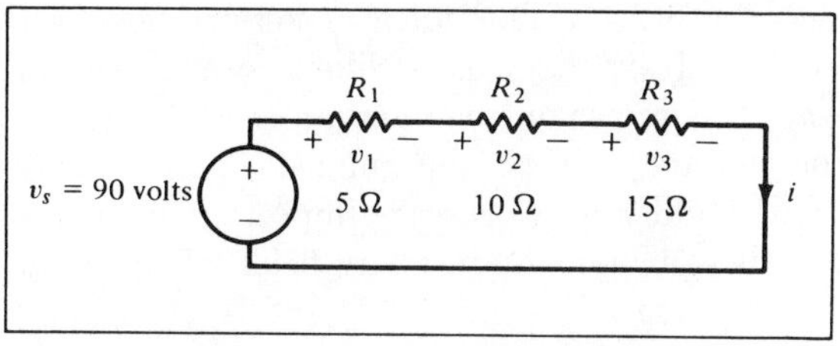

FIGURE 2.15 Series network example.

The voltages are given by

$$v_1 = \frac{5}{30}90 = 15 \qquad v_2 = \frac{10}{30}90 = 30$$

$$v_3 = \frac{15}{30}90 = 45 \text{ volts}$$

These three voltages, of course, add to 90 volts, the source voltage.

An analogous analysis can be carried out for the parallel network of Figure 2.14(b). To analyze this network, we first write KCL at node a as

$$i_s = i_1 + i_2 + i_3 \tag{2.26}$$

From Ohm's law we have

$$i_1 = G_1v_1 \qquad i_2 = G_2v_2 \qquad i_3 = G_3v_3 \text{ amps} \tag{2.27}$$

while from KVL

$$v_1 = v_2 = v_3 = v \tag{2.28}$$

Combining Equations (2.26), (2.27), and (2.28), we get

$$i_s = (G_1 + G_2 + G_3)v \tag{2.29}$$

$$v = \frac{i_s}{G_1 + G_2 + G_3} \tag{2.30}$$

Putting this result back into Equation (2.27), we obtain

$$i_1 = \frac{G_1}{G_1 + G_2 + G_3} i_s \tag{2.31}$$

and similar expressions for i_2 and i_3.

Comparing Equation (2.30) with Ohm's law, we see that the same voltage v would result if we replaced the parallel combination of G_1, G_2, and G_3 with a single resistor of value

$$G_{eq} = G_1 + G_2 + G_3 \tag{2.32}$$

This result is easily generalized: A parallel combination of resistors is equivalent to a single resistor of conductance value equal to the sum of the conductances.

Using this result, we can now write Equation (2.31) in the form

$$i_1 = \frac{G_1}{G_{eq}} i_s \tag{2.33}$$

We can also generalize this result, the current divider relation, in the following statement: In a parallel combination of resistors, the fraction of the total current which flows in any one resistor is equal to the fraction of the total conductance represented by that resistor.

Example 2.6

The equivalent conductance across terminals a–b in Figure 2.16 is

$$G_{eq} = G_1 + G_2 + G_3 = \tfrac{1}{2} + \tfrac{1}{3} + \tfrac{1}{6} = 1 \text{ ℧}$$

(Note that the values given on the diagram are resistances, not conductances.) Then from Equation (2.33) we can write

$$i_1 = \frac{\frac{1}{2}}{1}2 = 1 \qquad i_2 = \frac{\frac{1}{3}}{1}2 = \tfrac{2}{3}$$

$$i_3 = \frac{\frac{1}{6}}{1}2 = \tfrac{1}{3} \text{ amp}$$

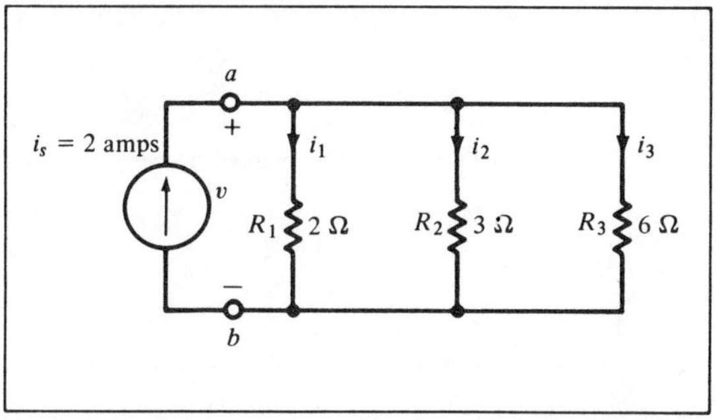

FIGURE 2.16 Parallel network example.

The sum of the three currents is, of course, 2 amps.

There is a very obvious similarity between the behavior of the series network and the parallel network. In fact, if we interchange voltage and current, series and parallel, resistance and conductance, the two analysis procedures are exactly the same. This is an illustration of the principal of *duality*: A network and its dual network can be analyzed similarly when the above interchanges are made. We will study this subject in more detail in Section 4.5.

A complicating factor is that resistors are commonly expressed in terms of their resistance rather than their conductance. Therefore, the formula for resistances in parallel might better be written as

$$R_{\text{eq}} = \frac{1}{1/R_1 + 1/R_2 + \cdots + 1/R_n} \qquad (2.34)$$

For two resistances, this can be expressed in the commonly used form

$$R_{\text{eq}} = \frac{R_1 R_2}{R_1 + R_2} \qquad (2.35)$$

Expressions similar to Equation (2.35) can be developed for three or more resistances in parallel, but they are too complicated to be of practical value.

Ladder Network Analysis

A more complicated example of the application of KCL and KVL is the *ladder* network shown in Figure 2.17. This form of network is characterized by its "ladderlike" form with an *input* source at one end and an *output* (current or voltage) to be determined at the other end. Ladder networks are important in representing phenomena in many fields of electrical engineering.

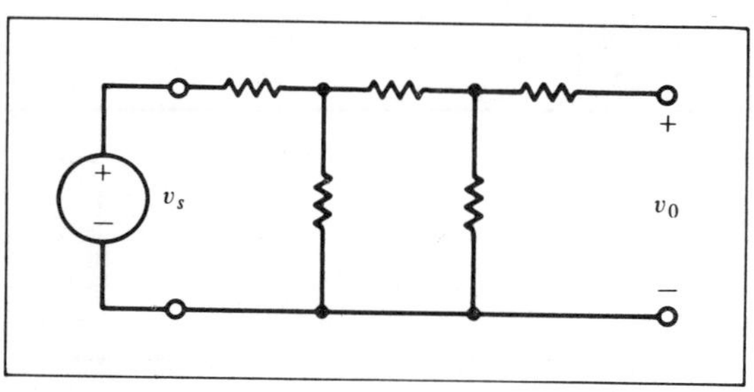

FIGURE 2.17 Form of ladder network.

One approach to the analysis is via the concept of equivalent resistance using the series and parallel resistor, current divider, and voltage divider relations already developed. This can be illustrated for the ladder network of Figure 2.18(a), as follows.

The geometry of the ladder structure allows any element to separate the network into two parts by making the element an open, or short, circuit; this assures that there are a number of sections through the network where a portion of the network can be replaced by a single *equivalent*, resistor. Five such sections (1)–(1) through (5)–(5) are marked in Figure 2.18(a). Starting at (5)–(5) and looking to the right, we see two 3Ω resistors in series. Therefore, as far as the rest of the network is concerned, they can be replaced by a single equivalent resistor of value

$$R_{\text{eq}_5} = 3 + 3 = 6\ \Omega$$

as indicated in Figure 2.18(b). Now at (4)–(4) we see two parallel resistors to the right which can be replaced by the single resistor of value

$$G_{\text{eq}_4} = \frac{1}{2} + \frac{1}{6} = \frac{2}{3}\ \mho$$

as shown in Figure 2.18(c).

This process is continued

$$R_{\text{eq}_3} = 2 + \frac{3}{2} = \frac{7}{2}\ \Omega$$

$$G_{\text{eq}_2} = 1 + \frac{2}{7} = \frac{9}{7}\ \mho$$

until finally we can calculate the value of the equivalent resistor seen across the voltage source terminals

$$R_{\text{eq}_1} = 1 + \frac{7}{9} = \frac{16}{9}\ \Omega$$

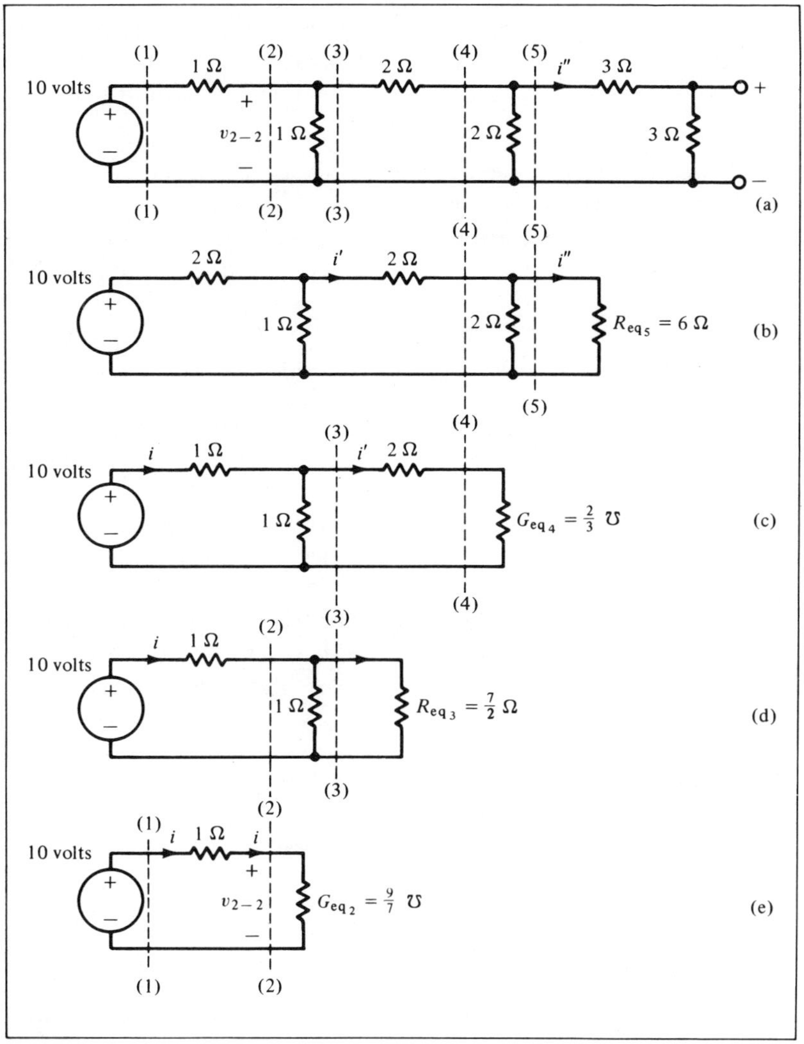

FIGURE 2.18 Systematic analysis of a ladder network using equivalent resistances.

Therefore, the current flow from the source is

$$i = \frac{10}{(16/9)} = \frac{45}{8} \text{ amps}$$

All that is involved in calculating the equivalents above is that in a series connection resistances add, and in a parallel connection conductances add.

Once we have found the input current, it is a simple process to go back through the network calculating the current through, voltage across, and power dissipated in every resistor in the original structure. From Figure 2.18(e), by Ohm's law the voltage across the first 1Ω resistor is 45/8 volts and across G_{eq_2} is

$$v_{2-2} = \frac{45}{8} \times \frac{7}{9} = \frac{35}{8} \text{ volts}$$

Now, moving up to Figure 2.18(d), the current through the shunt 1Ω resistance is

$$i_{1\,\Omega} = \frac{35}{8} \text{ amps}$$

and through R_{eq_3} is

$$i' = v_{2-2} \times G_{eq_3} = \frac{35}{8} \times \frac{2}{7} = \frac{5}{4} \text{ amps}$$

The continuation of this process through the balance of the ladder is straightforward, and the reader can easily verify that the voltage across the final 3Ω resistor is 15/16 volt.

It should be noted that, if we wanted to find only this voltage, the current divider relation could be used to advantage. Thus, to the right of (2)–(2) the input current divides between a 1℧ path and a 2/7℧ path; therefore, the current divider equation gives directly

$$i' = i \times \frac{\frac{2}{7}}{1 + \frac{2}{7}} = \frac{2}{9} i$$

To the right of (4)–(4), i' divides between a $\frac{1}{2}$℧ and a $\frac{1}{6}$℧ path; therefore, we have

$$i'' = i' \times \frac{\frac{1}{6}}{\frac{1}{2} + \frac{1}{6}} = \frac{1}{4} i'$$

Combining these two equations, we can write the current through the final 3Ω resistor directly

$$i'' = \frac{1}{4} \times \frac{2}{9} i = \frac{1}{18} \times \frac{45}{8} = \frac{5}{16} \text{ amp}$$

and therefore, the output voltage is

$$v_{3\Omega} = \frac{5}{16} \times 3 = \frac{15}{16} \text{ volt}$$

This example provides an opportunity to apply the generalized procedures for calculating power illustrated in Figure 2.7. The power absorbed by the network to the right of (2)–(2) is, from Figure 2.18(e),

$$p_{2-2} = (v_{2-2})(i) = \left(\frac{35}{8}\right)\left(\frac{45}{8}\right)$$

$$= \frac{1575}{64} \text{ watts}$$

By calculating the powers absorbed by every resistor to the right of (2)–(2), the reader can show that these sum to p_{2-2}.

2.5 Port Analysis

Introduction

As we witnessed in Figure 2.7, we can often treat a *group of elements* as a single element for the purposes of analysis. Reducing the analysis of a complicated network to the analysis of a small number of groups of elements, we can often reduce the overall calculation to be performed as well as gain insight into the behavior of the element groups. The latter is important if, for example, the group represents a physical device such as a transistor. In the following section, we initiate the formal study of the characterization of such element groups.

Port Representation

The basis for our discussion is the concept of a *port.*

Definition

A *port* is a pair of accessible nodes to which we associate a voltage v and a current i. The same current i (which may be zero) must flow in one terminal and out the other. These nodes can be used for measurement purposes or for connecting an external source or another network. However, a set of voltage and current references *must* be defined for a port; furthermore, it will be our custom to reference the current positive flowing *inward* at the positive voltage reference terminal as shown in Figure 2.19(b)–(c).

A port results either from (a) a *soldering-iron* or (b) a *pliers* type of entry into a network. These are illustrated in Figure 2.19(b)–(c). In the first case, a pair of wires is simply attached to a pair of nodes in the network to define the port. In a sense, we are looking *across* part of the network. In the "pliers"

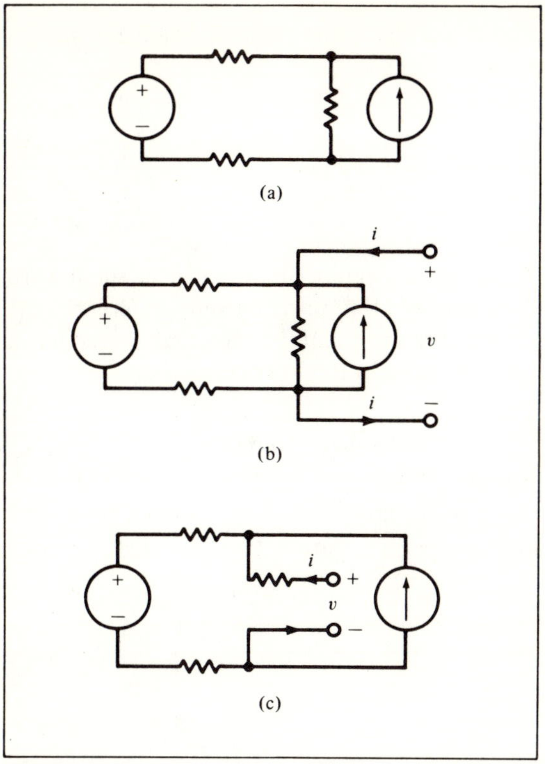

FIGURE 2.19 Distinguishing a port. (a) A network; (b) a one-port (soldering iron entry); (c) a one-port (pliers entry).

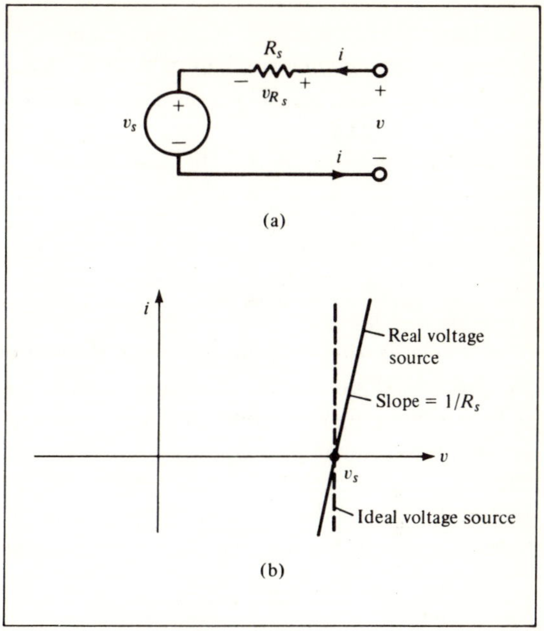

FIGURE 2.20 Models of voltage source. (a) Network model; (b) $i - v$ characteristic.

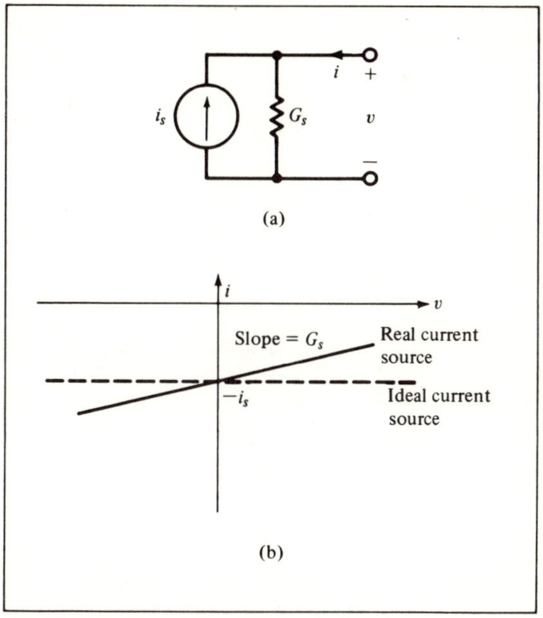

FIGURE 2.21 Models of a current source. (a) Network model; (b) $i - v$ characteristic.

entry, a cut is made in one of the connecting wires, and we look *into* the network at that point.

A Simple One-port: The Nonideal Source

In our discussion of sources, we said that a source has the property that its current and voltage are independent of one another. This property is never found in physical devices (such as batteries) that are intended to behave as sources. *Real* or *nonideal* sources always have internal resistances associated with them as shown in Figure 2.20(a) for a voltage source. We can write i in terms of v for such a voltage source model as

$$v = v_s + R_s i \tag{2.36}$$

An alternative representation of this internal resistance is shown by plotting i versus v. The resultant *i-v characteristic* for such a nonideal source is distinguished in Figure 2.20(b) by a large but finite slope. When the port current i is zero, the voltage v_{R_s} is zero and, from Equation (2.36), $v = v_s$. When the port current i increases positively in the direction shown, the voltage v_{R_s} increases, resulting in an increasingly larger

$$v = v_s + v_{R_s}$$

This effect is depicted in Figure 2.20(b) by the positive slope, where v increases with i.

In a similar way, we find that ideal current sources never occur physically; inevitably, real current sources have internal effects represented by a parallel resistor having a large resistance value R_s (or small G_s) as shown in Figure 2.21(a). The port equation for such a source is

$$i = -i_s + G_s v \tag{2.37}$$

When $G_s = 0$, we have

$$i = -i_s \tag{2.38}$$

The negative sign occurs because of the inward direction of the port current reference. As shown in Figure 2.21(b), the real source has a positive slope in its i-v characteristic.

Equivalent Source Models

Because Equations (2.36) and (2.37) both describe straight lines, it is possible, by a suitable choice of parameters, to describe *any*

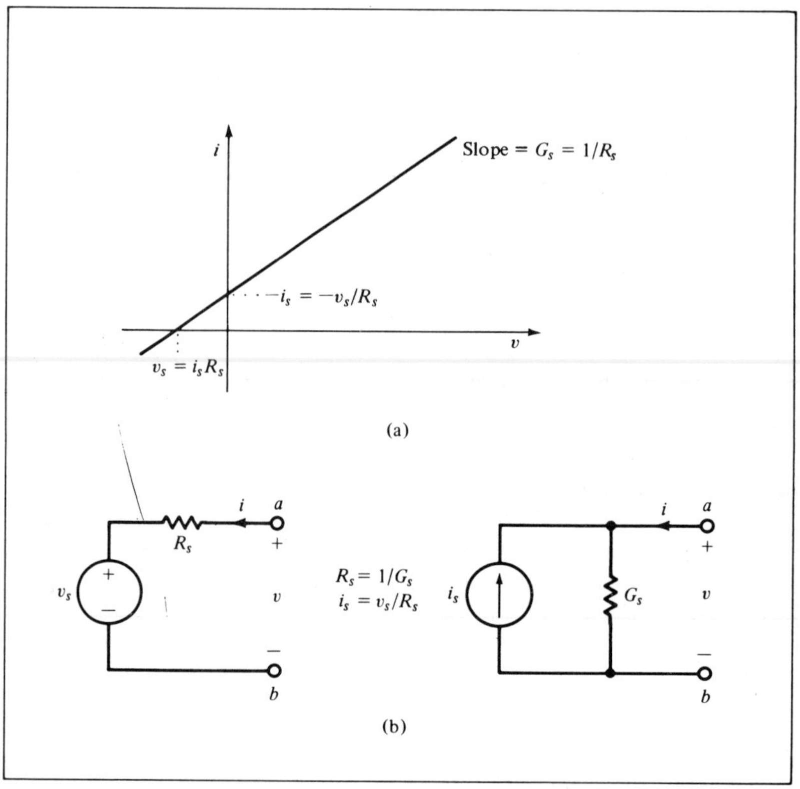

FIGURE 2.22 Equivalent models of nonideal current or voltage source. (a) General linear $i - v$ characteristic; (b) equivalent models.

straight line *i-v* characteristic with *either* equation. Solving Equation (2.37) for v and setting the result equal to v in Equation (2.36) gives

$$v = \frac{i + i_s}{G_s} = v_s + R_s i \tag{2.39}$$

To satisfy this equation, we must have

$$R_s = \frac{1}{G_s}$$

$$v_s = \frac{i_s}{G_s} = i_s R_s$$

If these two equations are satisfied, then the two models in Figure 2.22 are equivalent in the sense that no change is observed at terminals *a–b* when one model is substituted for the other in any network. Thus, we can use *either model* to represent *either type of nonideal source.*

Example 2.7

The current source equivalent of the storage battery model of Figure 2.23(a) is found from

$$i_s = \frac{v_s}{R_s} = \frac{12.6 \text{ volts}}{2\ \Omega} = 6.3 \text{ amps}$$

$$G_s = \frac{1}{R_s} = 0.5\ \mho$$

We can have, then, a current source-resistor model of what is physically a nonideal voltage source!

More on i–v Characteristics

The manipulation of the simple nonideal source models can be extended with surprising ease to much more complicated networks. Before we do this, however, let us formalize the concept of *equivalence* that permitted us to say that the networks of Figure 2.22 behaved identically.

If excitations are applied to a one-port, the response at that port is uniquely represented by its *i-v* characteristic. This must be true because v and i are the only variables common to the one-port and any network connected to it. We can gain a further appreciation of this statement by examining a one-port with a variety of networks connected.

Let a voltage source v be applied at the port and plot the resultant port current i that flows [Figure 2.24(a)–(b)]. Now apply a current source i and plot the resultant voltage v. These *i-v* characteristics must be identical for the following reason. Inside the one-port network, the network always reacts so as to satisfy KVL and KCL. It would be inconsistent to claim that the network of Figure 2.24(a), with 5 volts across nodes 1–2 and 2 amps flowing into and out of these nodes, satisfies KVL and KCL while the same network of Figure 2.24(b) does not. Indeed, we could attach any form of network [Figure 2.24(c)] across the pair of terminals and the one-port will react similarly, that is, according to its *i-v* characteristic. Note that we do not claim that the *i-v* characteristic is continuous, single-valued, or otherwise well-behaved. We say only that it represents the

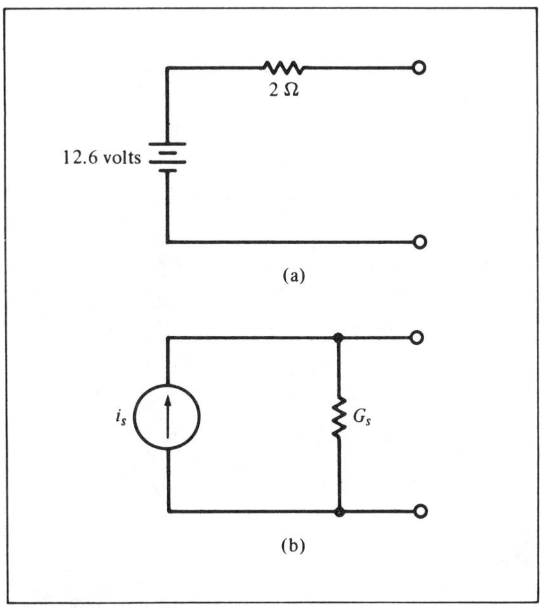

FIGURE 2.23 Current source equivalent of the storage battery.

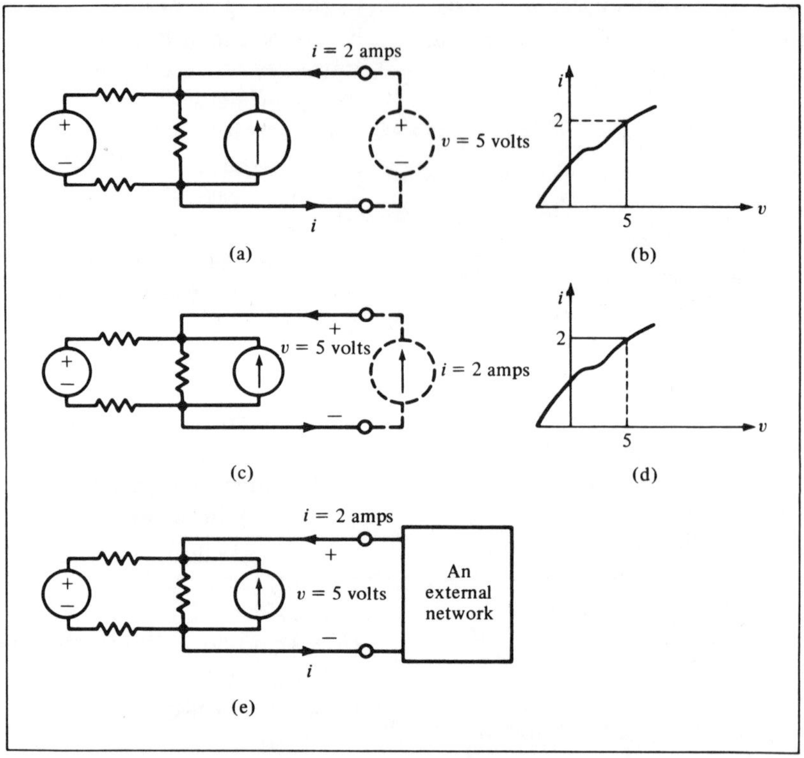

FIGURE 2.24 One-port measurements.

behavior of the one-port when connected to an external network.

Because the *i-v* characteristic describes all possible forms of interaction between the one-port and other networks, we conclude that *all one-port resistive networks that have the same current-voltage characteristic are equivalent.*

A one-port can thus be described by either (1) its detailed network model, as in Figure 2.24(a), or (2) an equation or a graph relating i and v, as in Figure 2.24(b).

The graphical model will be an especially effective way of displaying certain properties of one-port networks.

Example 2.8

The significance of the *i-v* characteristic in establishing equivalence is displayed in Figure 2.25(a). Here, we are asked to find a simplified equivalent of a current and voltage source in parallel.

This is a perplexing problem if we try to resort to intuition. However, we have the answer immediately if we establish the *i-v* characteristic; we need only observe that $v = 3$ volts regardless of i, because the port terminals are across a voltage source. The *i-v* characteristic is therefore a vertical line, which can be obtained more simply with a 3-volt source [Figure 2.25(c)]. The 2-amp current source is invisible from the port terminals. Internally, the current source supplies a constant 2-amp current through the voltage source as shown in Figure 2.25(a).

A similar problem is illustrated in Problem 2.24, where a current and voltage source are in series.

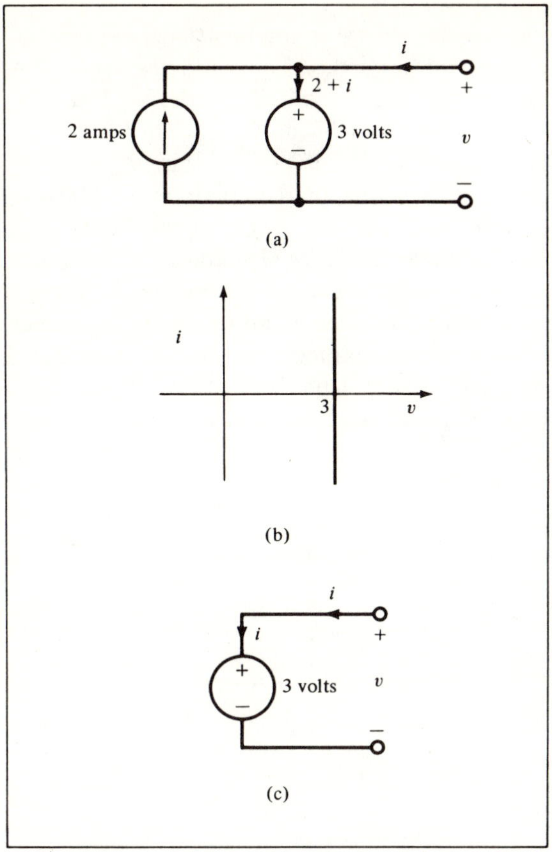

FIGURE 2.25 Deducing a source equivalent.

General One-port Representations

The nonideal source models of Figure 2.22 are of far greater utility than indicated above. Their general forms are shown again in Figure 2.26, with notation we will carry throughout. We will call the series voltage-source-resistor combination [Figure 2.26(a)] a *Thevenin* configuration or equivalent; a parallel current-source-resistance combination will be termed a *Norton* configuration or equivalent.[6] Not only are they equivalent to each other; *any network of sources and linear resistors* can be reduced to these forms. We will prove this general statement in Chapter 6; for the present, let us become familiar with the reduction operation itself.

The reduction is usually performed in one of three ways. The most general will be discussed in Chapter 6; the two simpler methods are now considered in detail.

Method 1. *Determination from network reduction.* This method begins by combining any series or parallel resistors and any series voltage sources or parallel current sources. This reduces the complexity of the network. Next, we single out a part of the network which we recognize as either a Thevenin or Norton configuration. If it is a Norton configuration, we use the source transformation of Figure 2.26 to obtain a Thevenin equivalent and vice versa. This step will usually result in parallel or series resistors or sources. We combine such elements, thus reducing complexity further, and search for more Thevenin-Norton configurations. In many cases, this process may be continued until a simple Thevenin or Norton configuration remains.

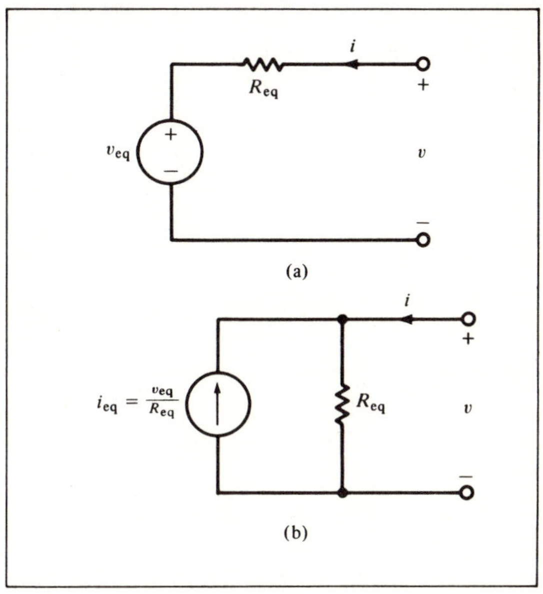

FIGURE 2.26 Model terminology.

[6] This equivalence was first proposed in 1883 by Jean C. Thevenin (1857–1926), a French telegraph engineer. The dual form was developed some fifty years later by E. C. Norton (1898–), an engineer at the Bell Telephone Laboratories.

This reduction operation is demonstrated for a voltage divider in Figure 2.27(a). The final result, shown in Figure 2.27(e), is of the predicted simplified form.

We need not restrict our attention to networks with only one source. A series of simple reduction steps similarly converts the network of Figure 2.28(a) with three sources to a simple Thevenin equivalent. Again, all networks of Figure 2.28 are electrically indistinguishable from the port.

Method 2. Determination from i-v characteristics. Method 1 depends on our being able to locate source-resistor combinations and cannot always be applied. Also it requires that the network model be available; however, if a one-port is uniquely defined by its i-v characteristic, then we should be able to obtain R_{eq} and v_{eq} (or i_{eq}) from this characteristic only. Method 2 describes such a procedure which can be applied to *any* network of sources and resistors.

If a one-port has a Thevenin or Norton equivalent, then from our discussion of nonideal sources, its i-v characteristic must be a straight line. Any two independent measurements of v and i are sufficient to determine the line and

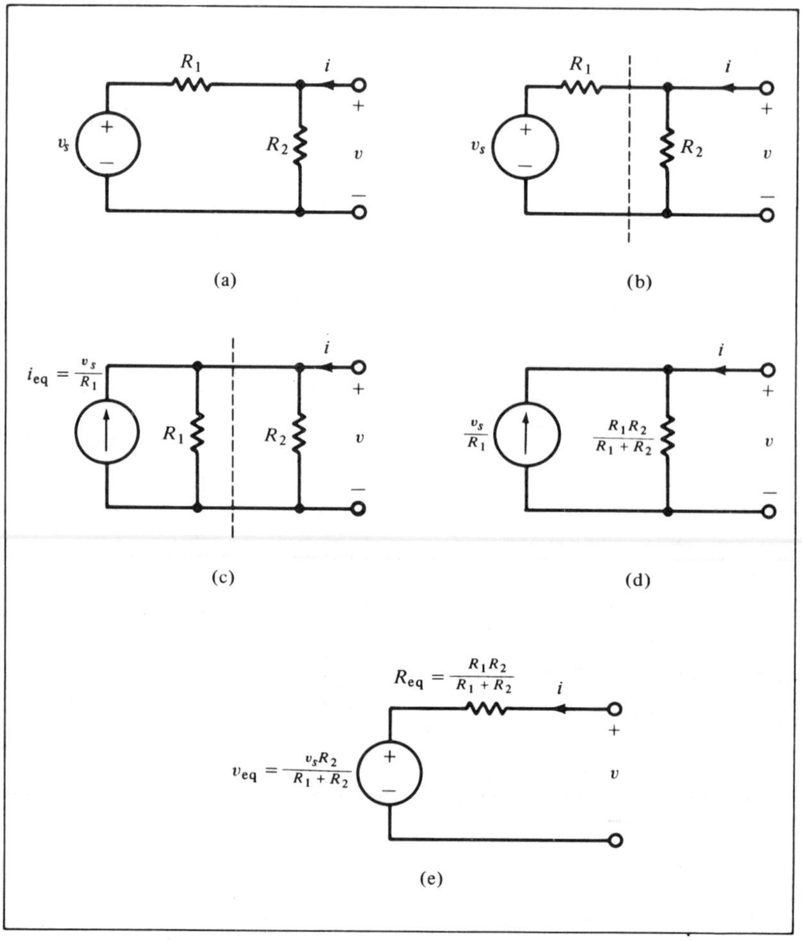

FIGURE 2.27 Reduction of voltage divider to a Thevenin equivalent. (a) Voltage divider; (b) recognize a Thevenin configuration; (c) convert to Norton equivalent; (d) combine parallel resistors; (e) convert to Thevenin equivalent.

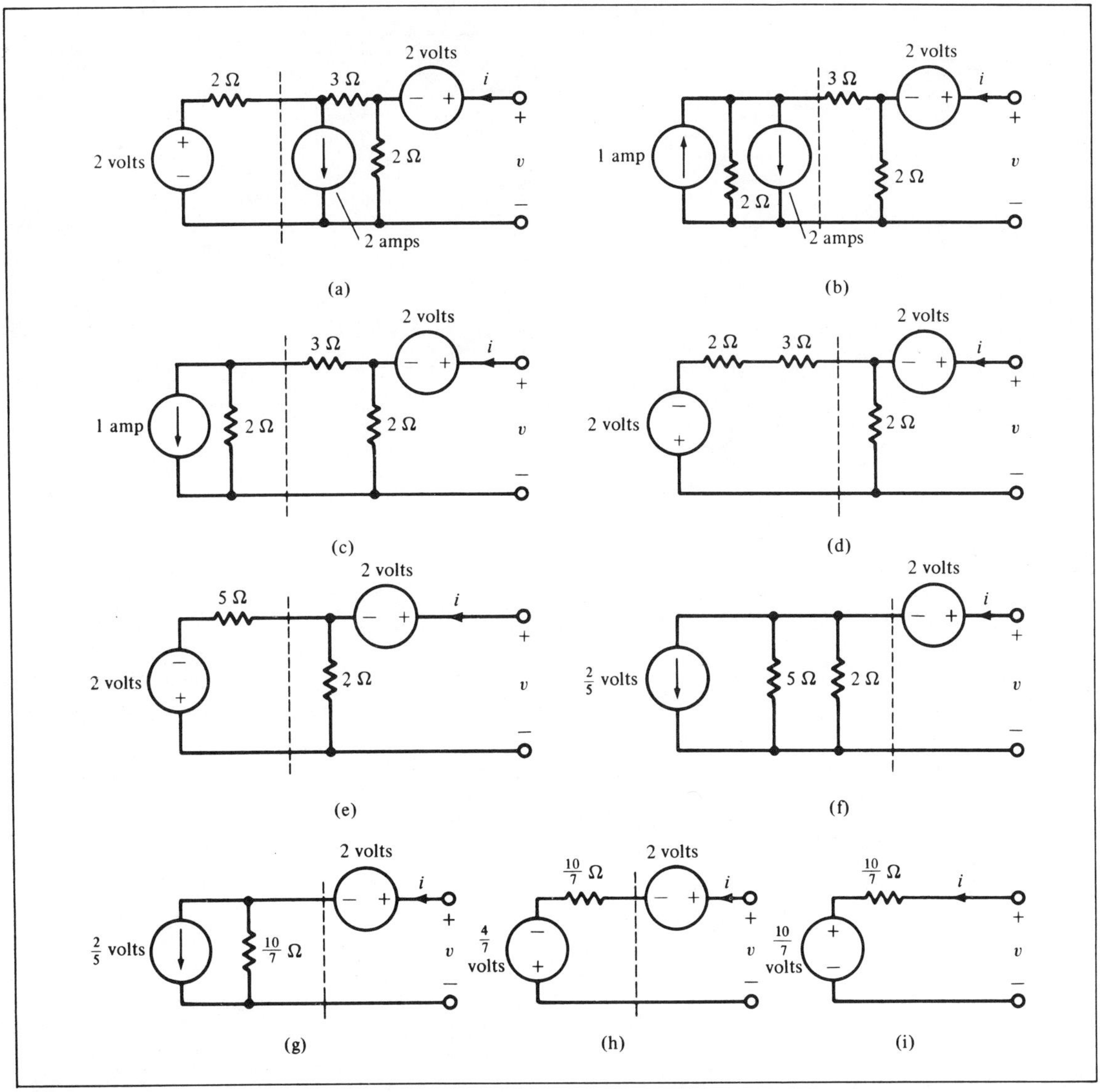

FIGURE 2.28 Complicated reduction example. (a) Thevenin configuration recognized; (b) conversion to Norton equivalent; (c) parallel current sources combined: (d) Norton-Thevenin conversion; (e) series resistors combined; (f) Thevenin-Norton conversion; (g) parallel resistors combined; (h) Norton-Thevenin conversion; (i) series voltage sources combined.

thus determine the parameters v_{eq} and R_{eq}. Conceptually, the easiest measurements to make are shown in Figure 2.29.

When the terminals of the network (or its Thevenin equivalent) are open-circuited ($i = 0$), we have from Figure 2.29(a),

$$v_{oc} = v_{eq} \tag{2.40}$$

If the terminals are shorted ($v = 0$), we have

$$i_{sc} = \frac{v_{eq}}{R_{eq}} \tag{2.41}$$

yielding

$$R_{eq} = \frac{v_{eq}}{i_{sc}} = \frac{v_{oc}}{i_{sc}} \tag{2.42}$$

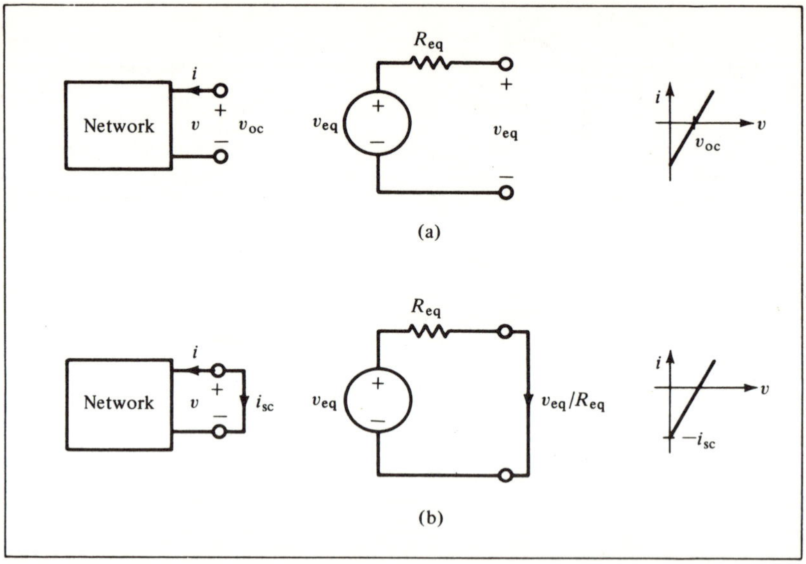

FIGURE 2.29 Determining v_{eq} and R_{eq} of a network. (a) Open-circuit measurement; (b) short-circuit measurement.

Equations (2.40) and (2.42) permit calculation of v_{eq} and R_{eq} directly from measurements v_{oc} and i_{sc}.

Although most useful when port *measurements* of v_{oc} and i_{sc} can be made, this method can occasionally be used when the one-port is represented as a simple network so that v_{oc} and i_{sc} can be obtained by inspection.

Example 2.9

The open-circuit voltage of the voltage divider of Figure 2.30(a) is

$$v_{oc} = \frac{R_2}{R_1 + R_2} v_s$$

With terminals *a–b* shorted, the current in the short is easily seen to be

$$i_{sc} = \frac{v_s}{R_1}$$

because all the current will flow through the short rather than R_2. Therefore, the Thevenin equivalent resistance is

$$R_{eq} = \frac{v_{oc}}{i_{sc}} = \frac{R_1R_2}{R_1 + R_2}$$

Variations of this open-circuit, short-circuit procedure are possible. Any two measurements will allow us to establish the equation for the linear *i-v* characteristic.

Example 2.10

A storage battery is found to have a rated open-circuit voltage of $v_{oc} = 12.6$ volts. When a 1Ω headlight is connected by closing the switch shown in Figure 2.31, the battery terminal voltage v_t decreases to 12.5 volts.

To find R_{eq}, we note that with the switch closed, the voltage v_R across R_{eq} is $v_{eq} - v_t = 12.6 - 12.5 = 0.1$ volt. The current through R_{eq} is identical to that through R_L; that is, 12.5 volts/1 Ω = 12.5 amps. Therefore,

$$R_{eq} = \frac{(0.1)}{(12.5)} = 0.008\ \Omega$$

2.6 Controlled Sources

Definition

A number of common circuit functions cannot be represented with only independent sources and resistors. One example is amplification: A

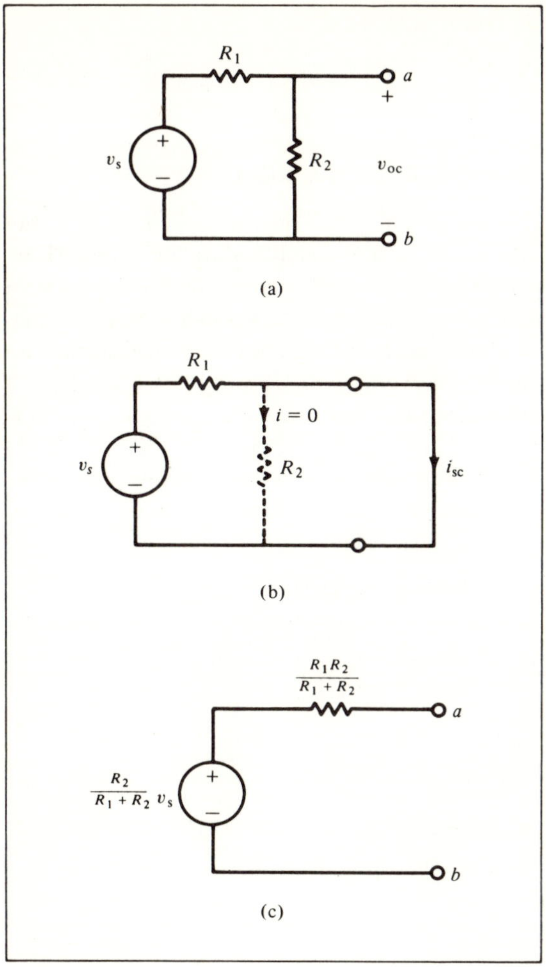

FIGURE 2.30 A network and its equivalent. (a) Open-circuit calculation; (b) short-circuit calculation; (c) Thevenin equivalent.

network of resistors cannot amplify voltage, current, or power. Yet it will be vital to our study in later chapters that we be able to model behavior such as amplification.

A (linear) *controlled* or *dependent* source has the property that the strength of the source is *proportional* to a current or voltage elsewhere in the circuit. If the strength of this source is large enough, we can represent amplification. For instance, the dependent voltage source v_2 of Figure 2.32 is a simple representation of a voltage amplification of 1000.

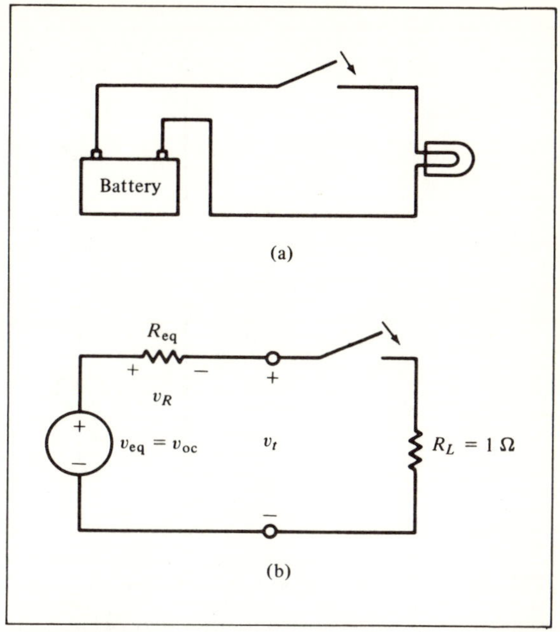

FIGURE 2.31 Thevenin equivalent of a storage battery.

There are four types of controlled sources, corresponding to the control of a current or voltage source by another current or voltage. These are shown in Figure 2.33. Note that controlled sources are represented by diamond-shaped symbols, rather than the circles used for independent sources.

Network Analysis with Controlled Sources

The following two examples show how controlled sources are manipulated in the determination of Thevenin and Norton equivalents.

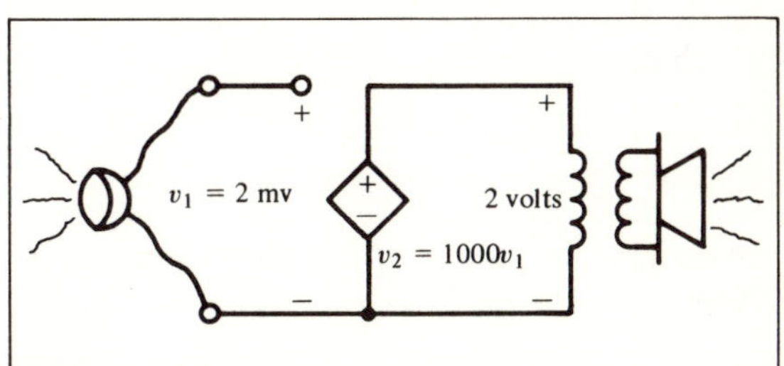

FIGURE 2.32 An amplifier model.

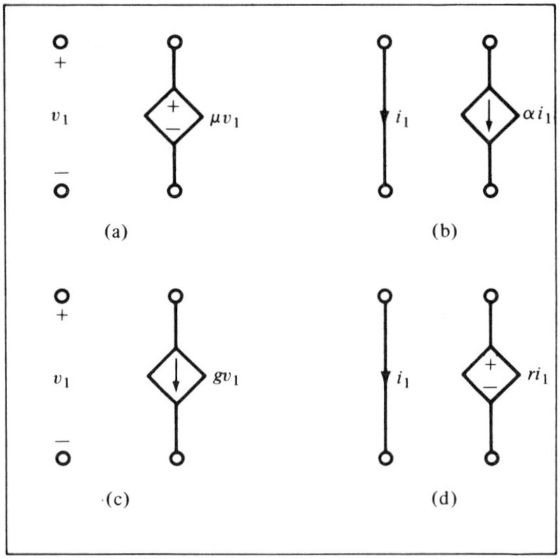

FIGURE 2.33 Controlled source symbols. (a) Voltage-controlled voltage source (VCVS); (b) current-controlled current source (CCCS); (c) voltage-controlled current source (VCCS); (d) current-controlled voltage source (CCVS).

Example 2.11 (Method 2)

The network of Figure 2.34 is sufficiently simple to consider the calculation of v_{oc} and i_{sc}. We can write, for the open-circuit model,

$$\begin{aligned} v_{eq} &= v_{oc} = v_s + Ri + \mu v_1 \\ &= v_s + R(0) + \mu(0) \\ &= v_s \end{aligned}$$

For the short-circuited model we have

$$v_s + v_1 + \mu v_1 = 0$$

which gives

$$v_1 = \frac{-v_s}{1+\mu}$$

The current through the shorted terminals is then

$$i_{sc} = \frac{-v_1}{R} = \frac{v_s}{(1+\mu)R}$$

Therefore, we determine

$$R_{eq} = \frac{v_{oc}}{i_{sc}} = (1+\mu)R$$

Example 2.12 (Method 1)

The reduction of the network of Figure 2.35(a) represents a more challenging problem. Here, we initially view the controlled source as an independent source and proceed with the reduction to Figure 2.35(d). The only precaution we must observe is *not to eliminate the variable* (i) *controlling the dependent source* during the network reduction. The network of Figure 2.35(d) is not in final Thevenin form, however, because the voltage source is dependent on i. We can write at this stage

$$\begin{aligned} v &= (R_1 + R_2 + R_3)i \\ &\qquad + v_s - \alpha R_2 i + i_s(R_1 + R_2) \\ &= (R_1 + R_3 + (1-\alpha)R_2)i \\ &\qquad + (v_s + i_s(R_1 + R_2)) \\ &= (R_{eq})i + (v_{eq}) \end{aligned}$$

which yields the equivalent of Figure 2.35(e). The reader may note that by making $\alpha > 1$

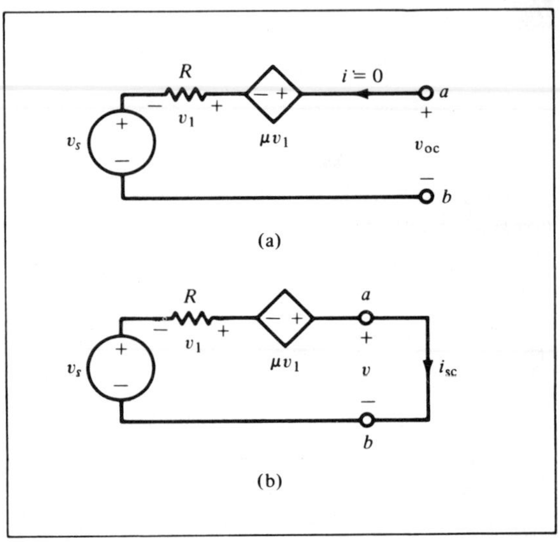

FIGURE 2.34 Controlled source example.

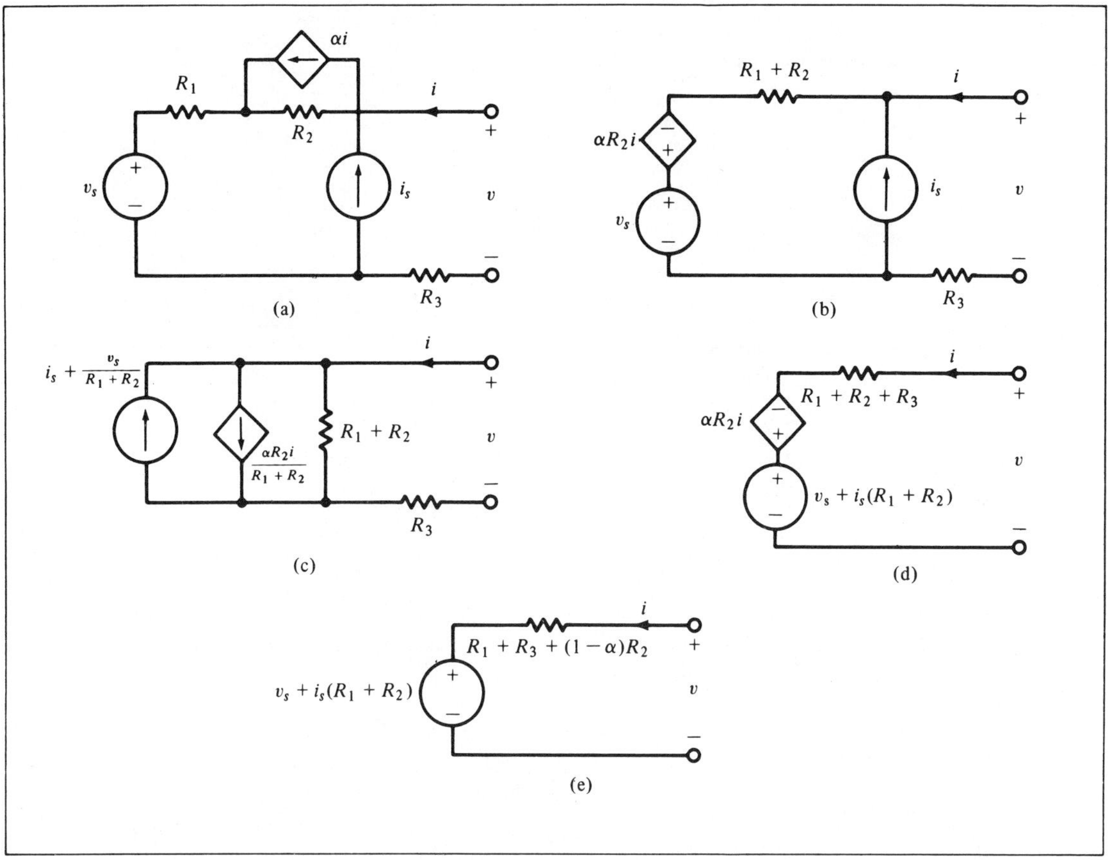

FIGURE 2.35 Network reduction with controlled source.

and $R_1 = R_3 = 0$, we can obtain an effective negative resistance of value $(1 - \alpha)R_2$!

Operational Amplifiers

A highly useful circuit involving the controlled source concept is the *differential* amplifier. Although the amplifier itself may consist of 20 or more transistors, we can often represent the functional behavior of the circuit by a single controlled source model.

The symbol for this device is shown in Figure 2.36(a). It produces a voltage v_c at a pair of "output" terminals that is A times the difference between two "input" voltages. It is easily seen that the VCVS of Figure 2.36(b) models this behavior. In practice, the positively marked input terminal is often connected to the reference (ground) node as shown in Figure 2.36(c); this reduces the amplifier model to one having a simple output voltage/input voltage (v_0/v_a) ratio of $-A$. Such a voltage ratio is often referred to as the "gain" of the amplifier.

The most common form of differential amplifier is the *operational* amplifier. This circuit has such a high gain ($A > 50{,}000$) that for many purposes we can consider $A = \infty$.

For practical reasons we cannot discuss, the operational amplifier cannot operate with such a large voltage gain. Instead, external resistors are added as shown in Figure 2.37 to control the gain. The nature of this control is determined by analysis of the model of Figure

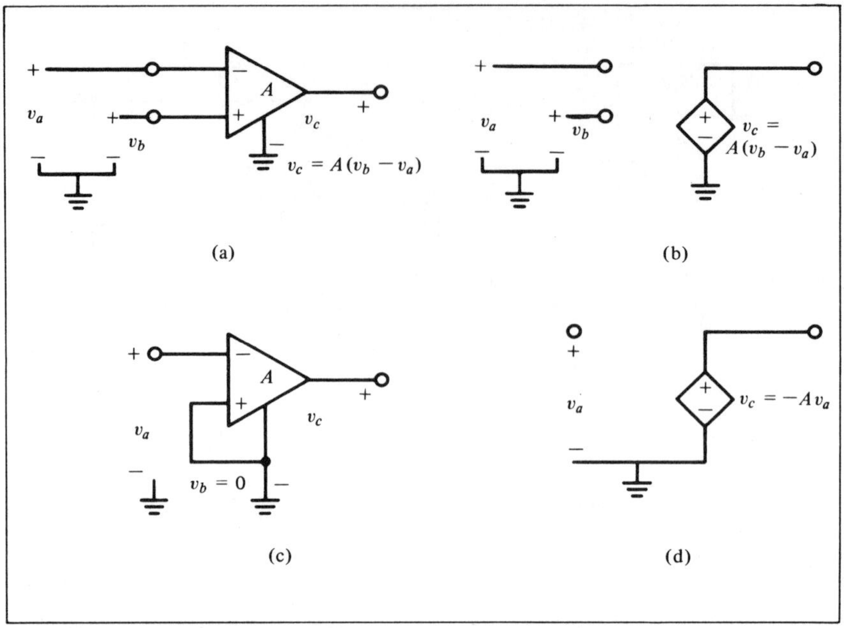

FIGURE 2.36 Differential amplifiers and their models. (a) and (b) General case; (c) and (d) grounded input.

2.37(b). First, the current through the resistors is found from Ohm's law

$$i = \frac{\text{voltage across resistor}}{\text{resistance}}$$

giving

$$i_s = \frac{v_s - v_a}{R_s} \qquad i_f = \frac{-Av_a - v_a}{R_f}$$

Solving for v_a, we obtain

$$v_a = \frac{R_f}{(A+1)R_s + R_f} v_s$$

The output voltage v_o is then determined from

$$v_o = -Av_a$$

$$= \frac{-AR_f}{(A+1)R_s + R_f} v_s$$

When A is large, the gain approaches

$$\lim_{A\to\infty} \frac{-AR_f}{(A+1)R_s + R_f} = \frac{-R_f}{R_s}$$

Therefore, the gain can be made -1, -10, -100, or any value smaller (in magnitude) than A by proper choice of R_f and R_s.

At present, operational amplifiers are mass produced, resulting in a retail cost considerably less than \$0.50 per unit. This cost, plus the option to obtain any voltage gain merely by addition of two resistors has resulted in the operational amplifier being among the most common types of electronic circuits. Other aspects of their performance are studied in Problems 2.38 to 2.39.

2.7 Linearity

The terminology "linear element" or "linear network" has been and will be used throughout this text to describe a rather special property. We must be able to identify networks that have this important property.

Consider again the property of a voltage divider to split an applied voltage according

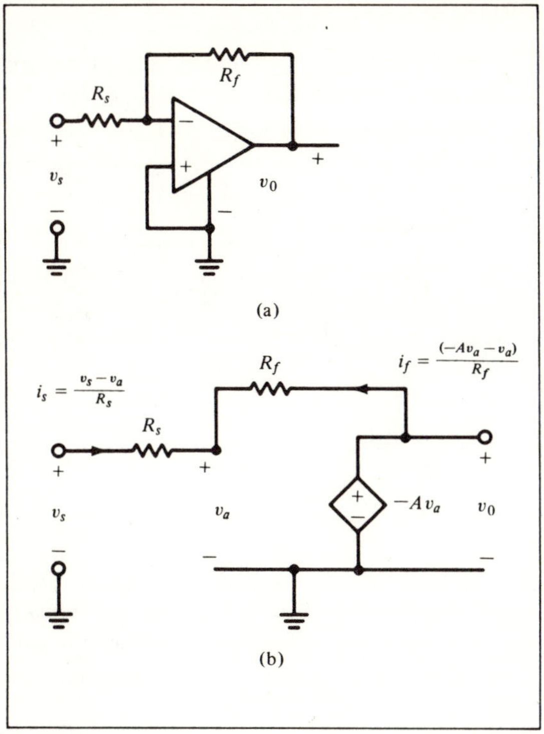

FIGURE 2.37 Operational amplifier with external resistors.

to the ratio of its resistances, that is, for the network of Figure 2.38.

$$v_o = \frac{R_2 v_s}{R_1 + R_2}$$

If we regard the voltage v_s as an *input* (excitation) to the voltage divider and v_o as its *output*

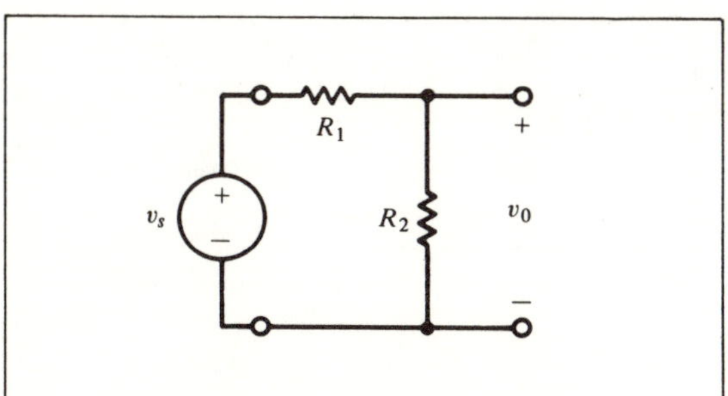

FIGURE 2.38 Voltage divider revisited.

(response), then the output is proportional to the input. That is, if α times the input is applied, α times the output will result. A network with such a proportional excitation-response characteristic is termed a *linear* network. It is therefore consistent to term an element described by

$$v = Ri$$

as a *linear* resistor because the response v is proportional to the excitation i.

In general, let us define a response current or voltage as y and an excitation current or voltage as x. Then, if $y = f(x)$, the network or element is linear if

$$\alpha y = f(\alpha x) \tag{2.43}$$

that is, y is proportional to x.

It may come as a surprise that an independent source such as a battery is *not* a linear element, because the battery voltage and current are independent of one another! However, a controlled source is a linear element because

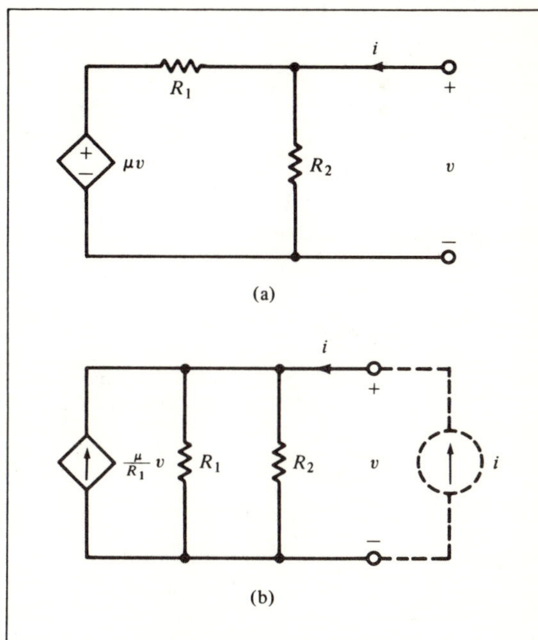

FIGURE 2.39 A linear one-port.

the controlled variable (the response) is proportional to the controlling variable (the excitation).

We will prove in Chapter 6 the quite reasonable assertion that any network of linear elements—resistors and controlled sources—is a linear network. The following is an example.

Example 2.13

A one-port network is linear if the port current or voltage response is proportional to a port voltage or current excitation, respectively. Applying a current source to the one-port of Figure 2.39(a) we can write after the source transformation of Figure 2.39(b),

$$i = -\mu v G_1 + v(G_1 + G_2)$$
$$= [(1 - \mu)G_1 + G_2]v \qquad (2.44)$$

Because v is proportional to i, the one-port is linear, as expected.

PROBLEMS

2.1 (a) Calculate i_4 for the network structure of Figure P 2.1(a). (b) Calculate v_5 for the network structure of Figure P 2.1(b).

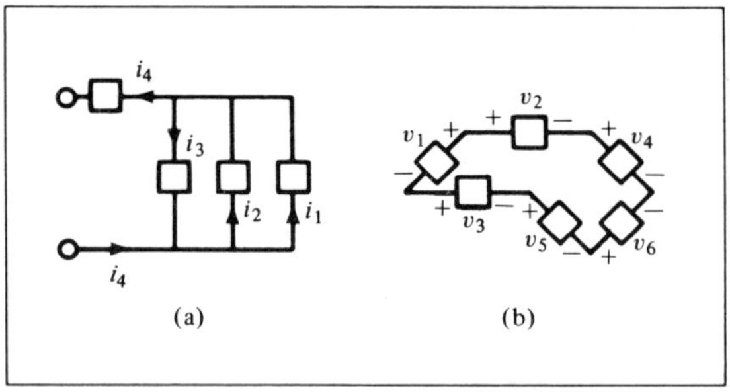

FIGURE P 2.1 (a) $i_1 = 6$ amps; $i_2 = 7$ amps; $i_3 = -6$ amps. (b) v_1, $v_3 = 2$ volts; v_2, $v_4 = 3$ volts; $v_6 = 7$ volts.

2.2 In each of the networks of Figure P 2.2, you are asked to find the voltage across and the current through each branch of the network. Check answers by showing that the sum

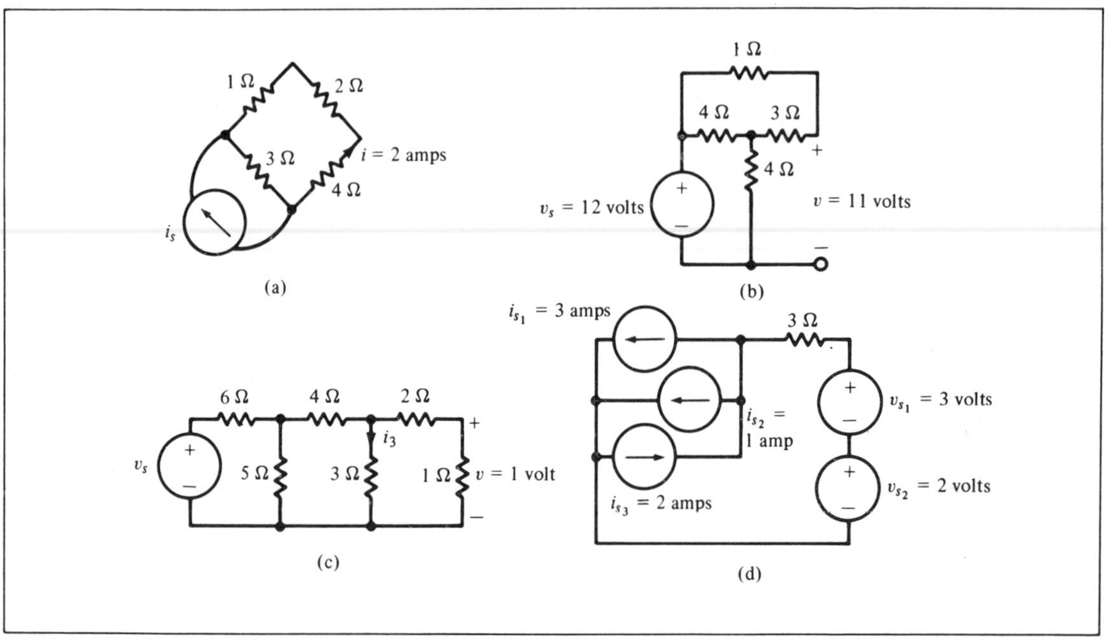

FIGURE P 2.2 (a) i_s must be found; (c) v_s must be found. [*Hint*: Show first that $i_3 = 1$ amp.]

of the powers absorbed in the network elements is zero (Tellegen's theorem).

2.3 In Figure P 2.3, use KVL to find the voltages across the resistors; then use Ohm's law and KCL to find currents in all the elements.

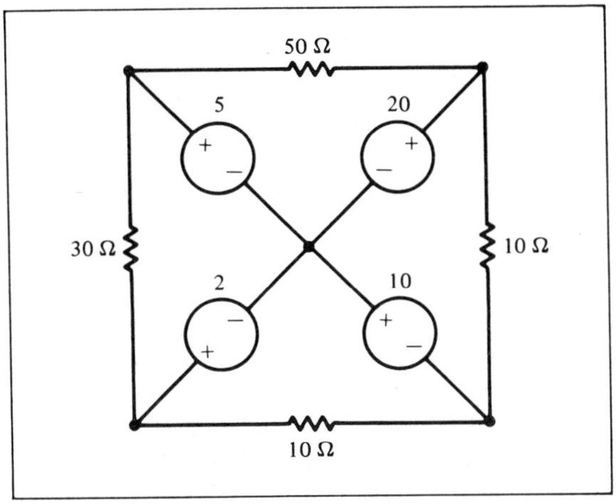

FIGURE P 2.3 Voltage source (values in volts).

2.4 In Figure P 2.4, use KCL to find the resistor currents; then use Ohm's law and KVL to find the voltages in all elements.

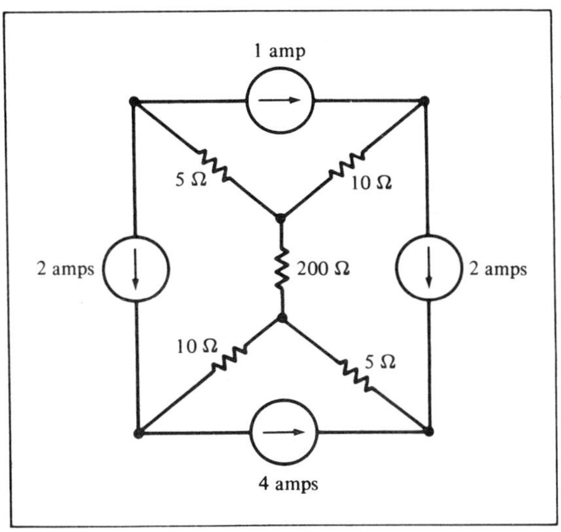

FIGURE P 2.4

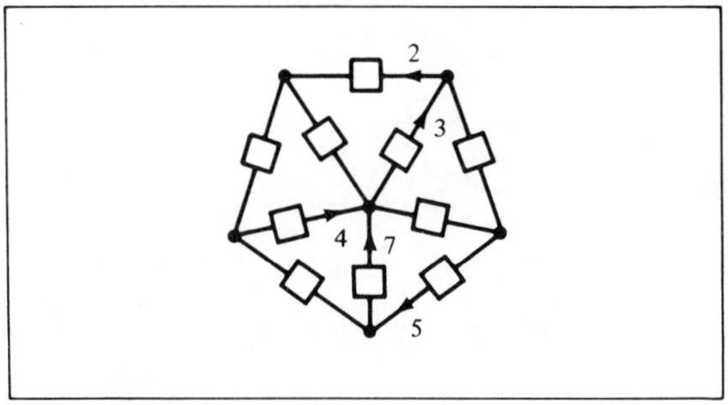

FIGURE P 2.5 All values shown in milliamps (ma).

2.5 Find all the currents (in amps) through elements shown in Figure P 2.5.

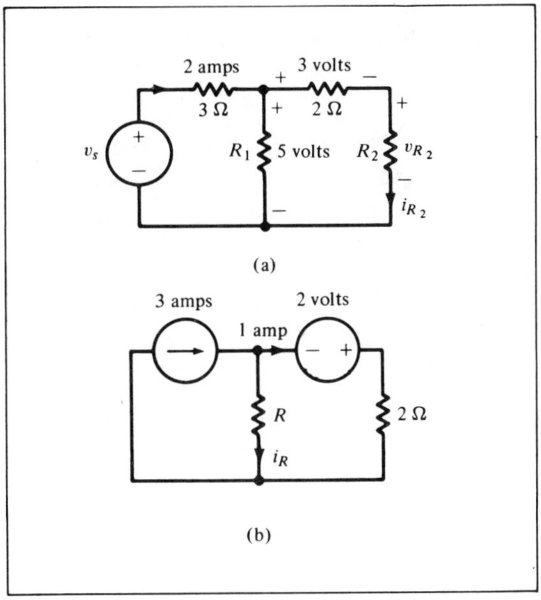

FIGURE P 2.6

2.6 Calculate v_{R_2}, i_{R_2}, R_2, and R_1 in Figure P 2.6(a).

2.7 Calculate R in Figure P 2.6(b).

2.8 By combining series and parallel resistances, verify the R's of Figure P 2.7.

2.9 Find the equivalent resistance at the terminals of each of the networks of Figure P 2.8.

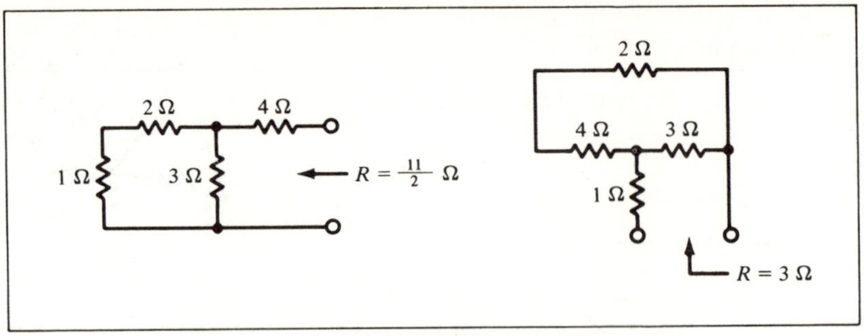

FIGURE P 2.7

2.10 Calculate v_{R_1} and v_{R_2} in Figure P 2.9(a); then determine $\hat{v}_{R_1}$ in Figure P 2.9(b); finally, show by differentiation that in this case

$$\frac{\partial v_{R_2}}{\partial G_1} = v_{R_1}\hat{v}_{R_1}$$

[*Comment:* In general, networks have the interesting property that differentiation with respect to R's and G's can be performed without knowing the symbolic expression for voltages and currents. Instead, two related networks are analyzed and the derivative formed as the product of an appropriate voltage from each network, as above; see Reference 2.4.]

2.11 Calculate v_{R_2} and v_{R_4} in Figure P 2.10, using the voltage divider relationship. Check by showing that

$$v_{12} = \frac{(R_2R_3 - R_1R_4)}{(R_1 + R_2)(R_3 + R_4)} v$$

2.12 Use the voltage divider relationship to show that $v_1 = 4$ volts and $v_2 = 8/3$ volts in Figure P 2.11.

2.13 Use the current divider relationship to show that $i = 2/9$ amp and $v = 15/9$ volts for the network of Figure P 2.12.

2.14 For the network of Figure P 2.13: (a) Write the KCL equation for the current flowing into each node, using the node ordering

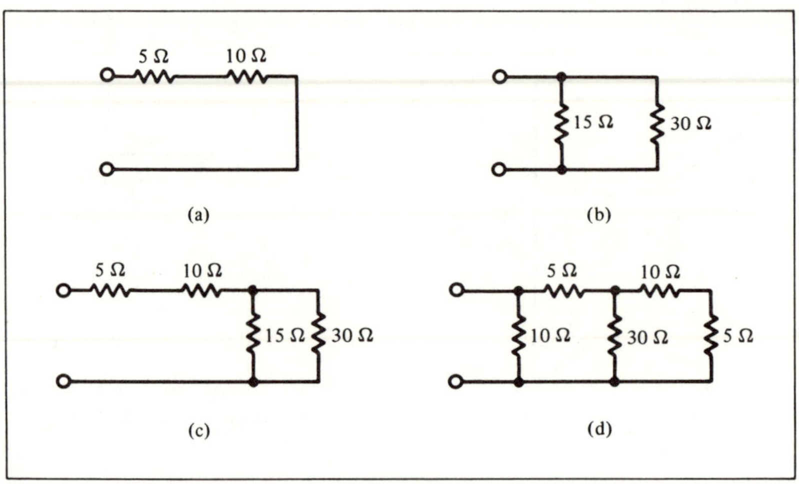

FIGURE P 2.8

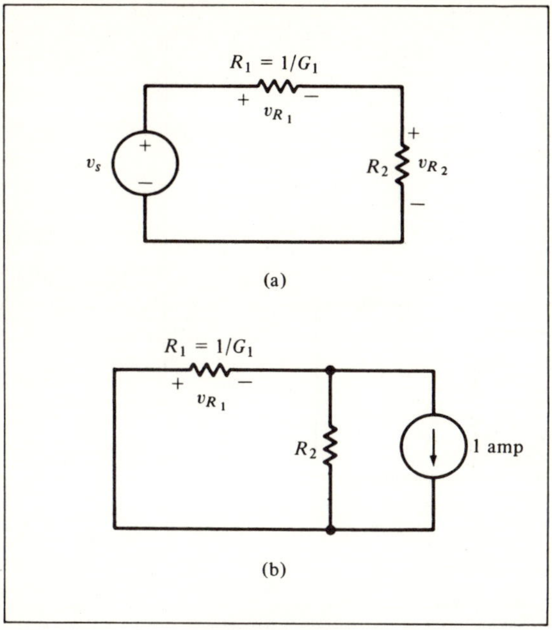

FIGURE P 2.9

shown. (b) Show that the equation formed from the sum of the first two KCL equations represents the current flowing across the cut shown in Figure P 2.13. (c) Show the cut described by the sum of the first three equations. (d) How many times does each current appear in the original KCL equations? What sign property do these terms possess?

2.15 For a certain light bulb, the current is found to depend on the voltage as

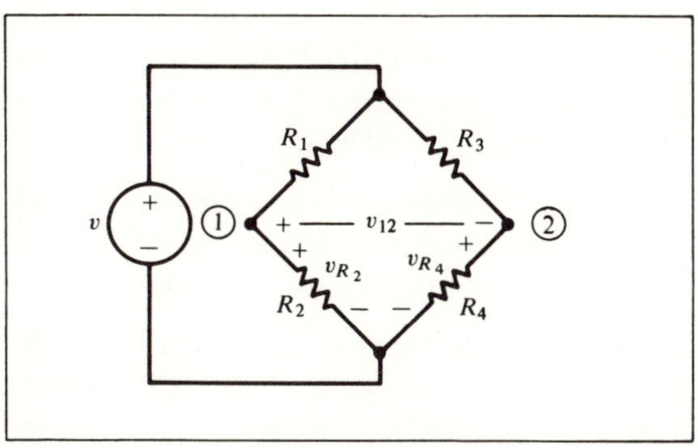

FIGURE P 2.10

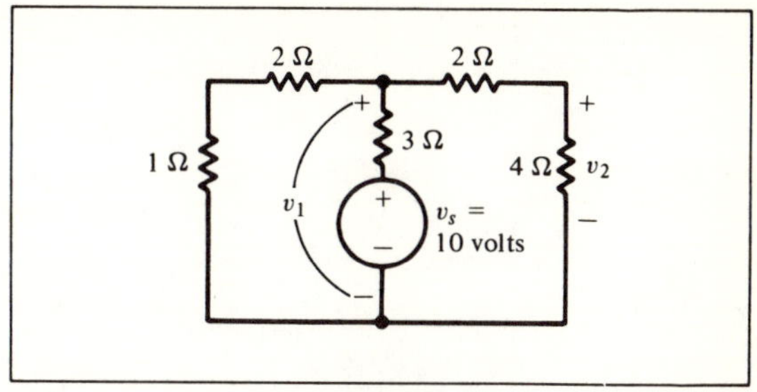

FIGURE P 2.11

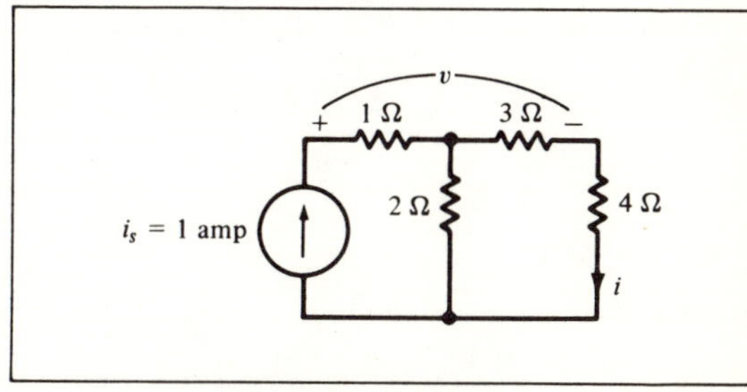

FIGURE P 2.12

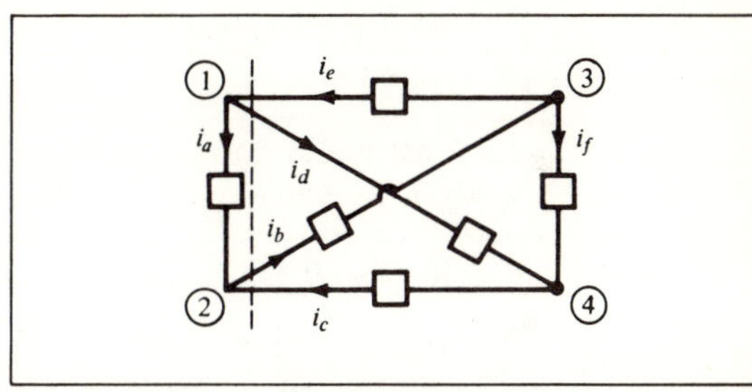

FIGURE P 2.13

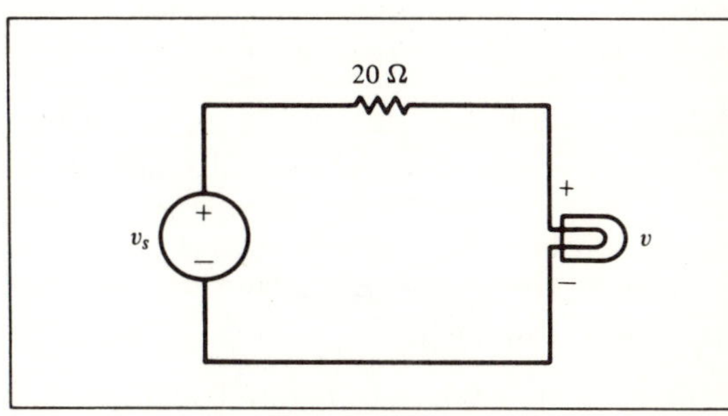

FIGURE P 2.14

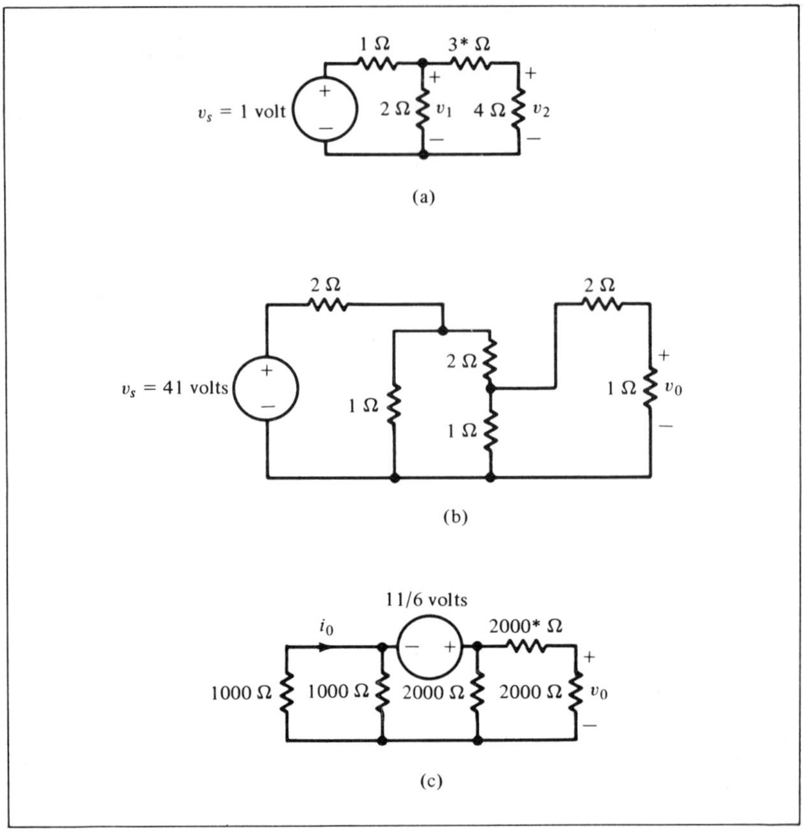

FIGURE P 2.15 (a) $v_1 = 14/23$ volt, $v_2 = 8/23$ volt; (b) $v_0 = 1$ volt; (c) $i_0 = \frac{1}{2}$ ma, $v_1 = \frac{2}{3}$ volt.

$$i = 0.1305 \times 10^{-1}v - 0.329 \times 10^{-6}v^3$$

What source voltage is required to make $v = 115$ volts? What power is dissipated in the lamp? (Figure P 2.14.)

2.16 Verify the voltages and currents shown in the ladder networks of Figure P 2.15.

2.17 (a)–(c) Solve the problems of Figure P 2.15 with asterisked values multiplied by k, where k is furnished by the instructor.

2.18 Solve the networks of Figure P 2.15 with asterisked values doubled.

2.19 Show that the i's of Figure P 2.16 are the same.

2.20 Show that the v's of Figure P 2.17 are the same.

2.21 For the network of Figure P 2.18 determine v_0.

2.22 (a)–(i) Verify the Thevenin and Norton equivalents of Figure P 2.19. Voltages are in volts, currents in amperes.

2.23 (a)–(i) Solve the problems of Figure P 2.19 with asterisked values multiplied by k, where k is furnished by the instructor.

2.24 Determine the Norton equivalent of the networks shown in Figure P 2.20 by successive Thevenin-Norton transformations. Voltages are in volts, currents in amperes.

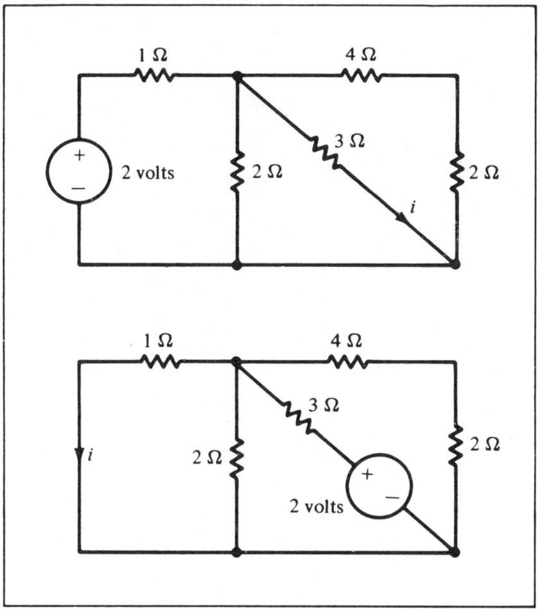

FIGURE P 2.16

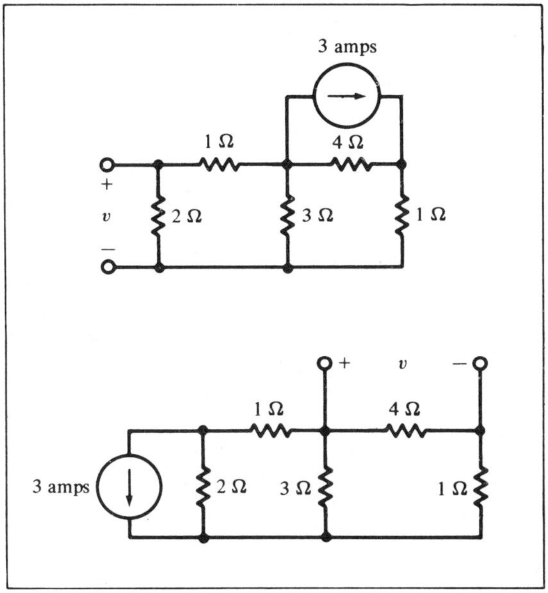

FIGURE P 2.17

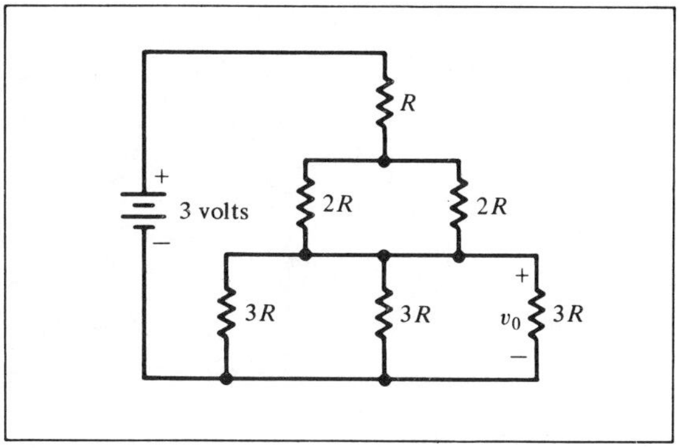

FIGURE P 2.18

2.25 A 10Ω resistor is connected to the network N of Figure P 2.21. Without the 10Ω resistor connected (that is, $i_R = 0$), $v_R = 10$ volts. With the resistor connected, $v_R = 7$ volts. Find the Norton equivalent of N.

2.26 Networks A and B of Figure P 2.22 are connected as shown. Find the value of v_s which results in $i_s = 0$.

2.27 Networks A and B are connected as shown in Figure P 2.23. Find the Thevenin equivalent of network A and then join the networks and find v_0.

2.28 An engineer has a single calibrated 1000Ω resistor and a box of 101 *identical* resistors of unknown value. He also has an ammeter (uncalibrated) and a battery (potential unknown). He connects the network of Figure P 2.24 and observes the direction of current flow in the ammeter connected to different points along the chain of 100 resistors. The current reverses direction as the connection is shifted across the fortieth resistor. What can be established concerning the value of the 101 unknown resistors?

2.29 (a) As R is varied in Figure P 2.25 what are the minimum and maximum values possible for the current $|i|$? (b) Plot $|i|$ versus v as R is varied over the range $(0, \infty)$. (c) Plot the power absorbed in R as R is varied in (b). When does R absorb maximum power?

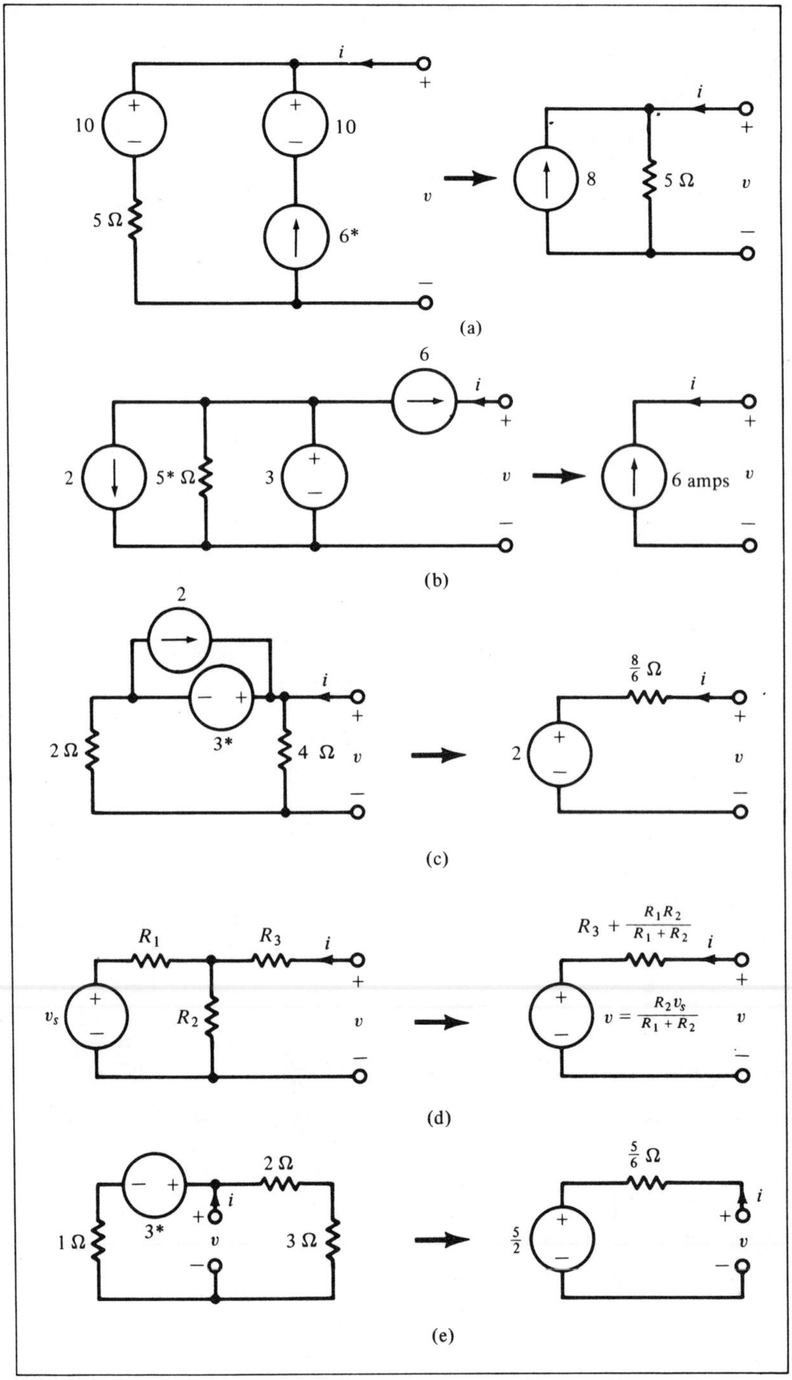
i
+
10
10
5 Ω
6*
v
−
8
5 Ω
(a)
6
2
5* Ω
3
6 amps
(b)
2
3*
2 Ω
4 Ω
8/6 Ω
(c)
R1
R3
R2
vs
R3 + R1R2/(R1 + R2)
v = R2vs/(R1 + R2)
(d)
2 Ω
1 Ω
3 Ω
5/6 Ω
5/2
(e)

FIGURE P 2.19

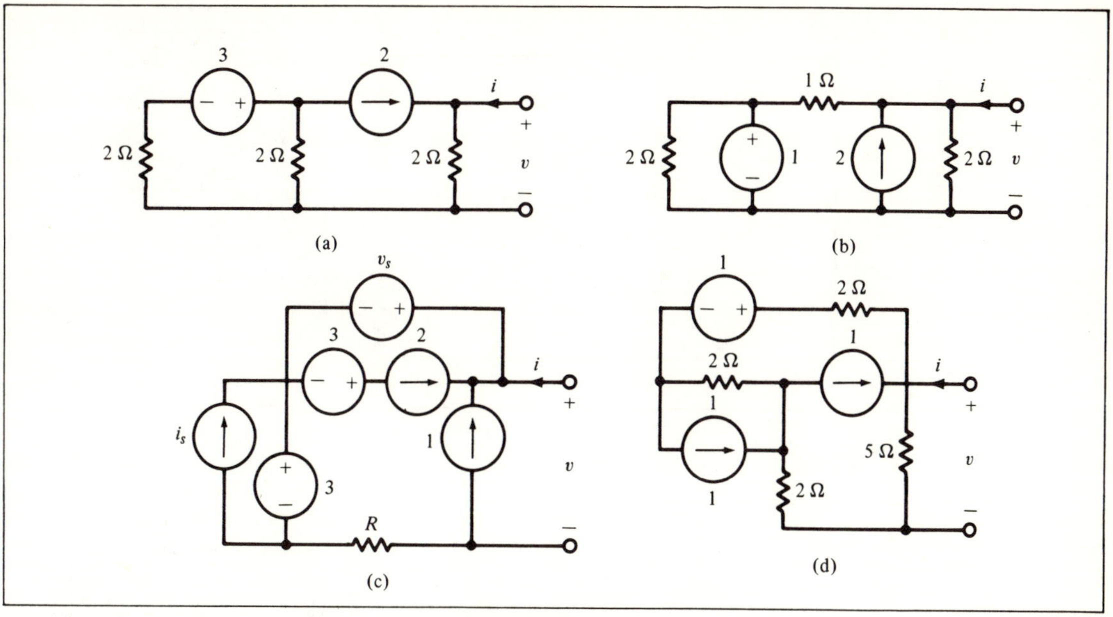

FIGURE P 2.20

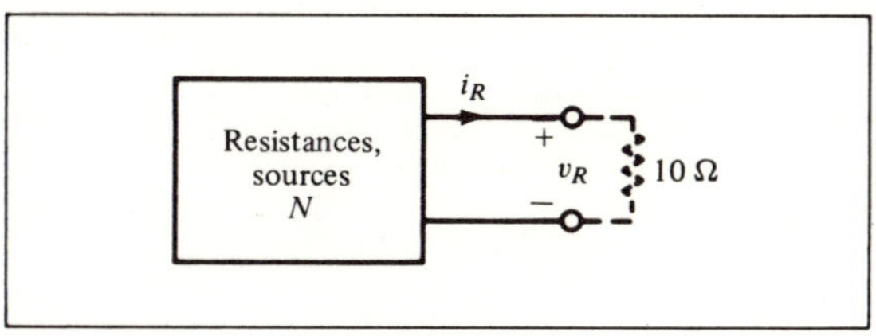

FIGURE P 2.21

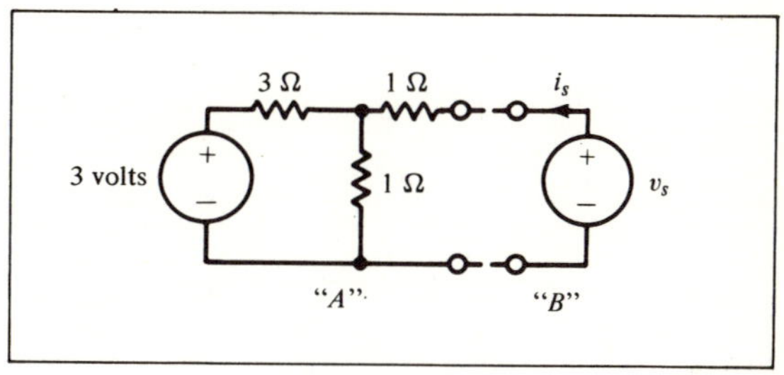

FIGURE P 2.22

2.30 Show by calculation that the sum of the three branch currents flowing across the network cut shown in Figure P 2.26 is zero. Why is this true?

2.31 Show that $v_s = 6.002$ volts in Figure P 2.27.

2.32 If $v_s = -19.5$ volts and $v_1 = 1$ volt, show that $R = 0\ \Omega$ for the network of Figure P 2.28.

2.33 The following problems show the ability of controlled sources to change the character of a network to which they are connected. (a) Show that $R = 1/R_L$ for the network of Figure P 2.29(*a*). (b) Show that $R = -R_L$ for

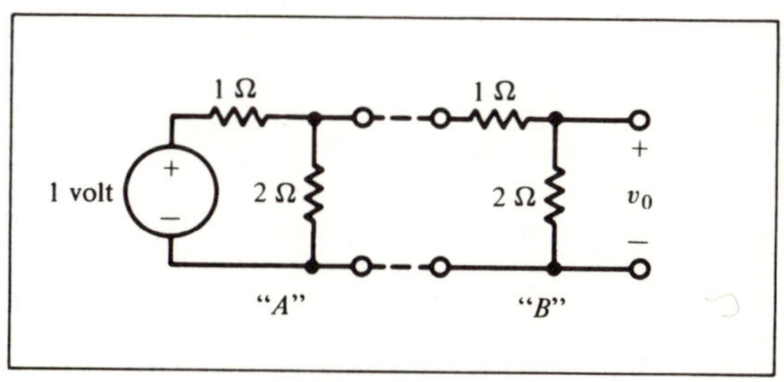

FIGURE P 2.23

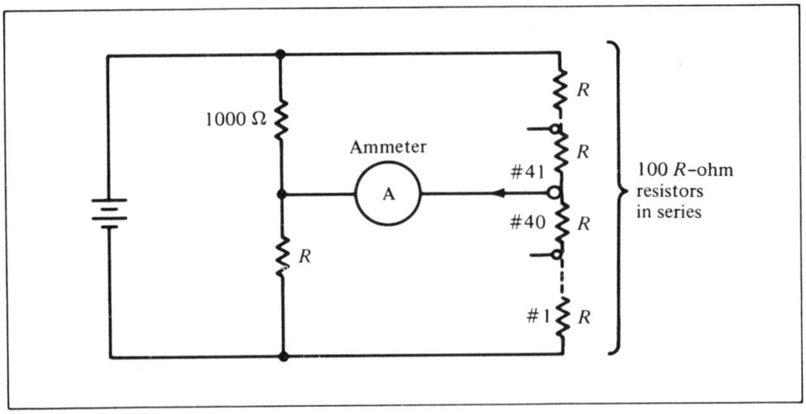

FIGURE P 2.24

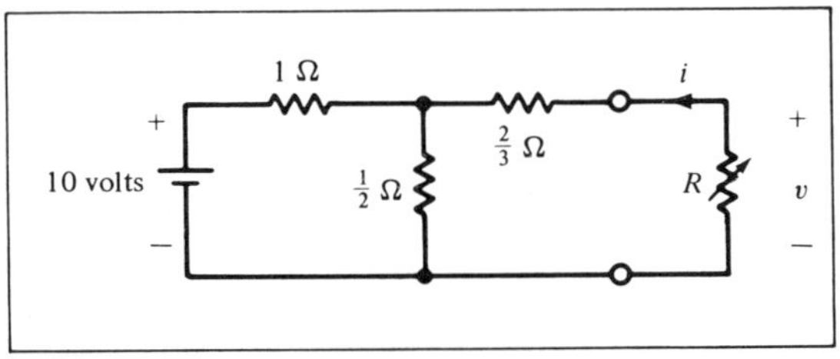

FIGURE P 2.25

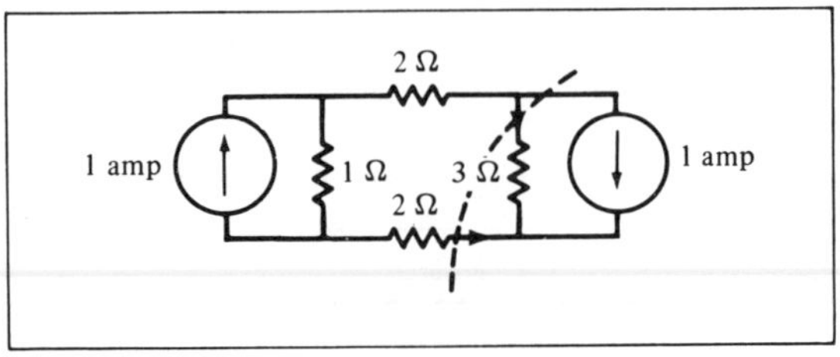

FIGURE P 2.26

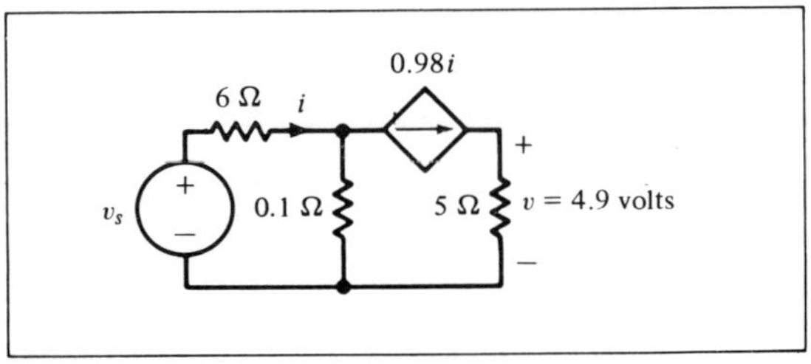

FIGURE P 2.27

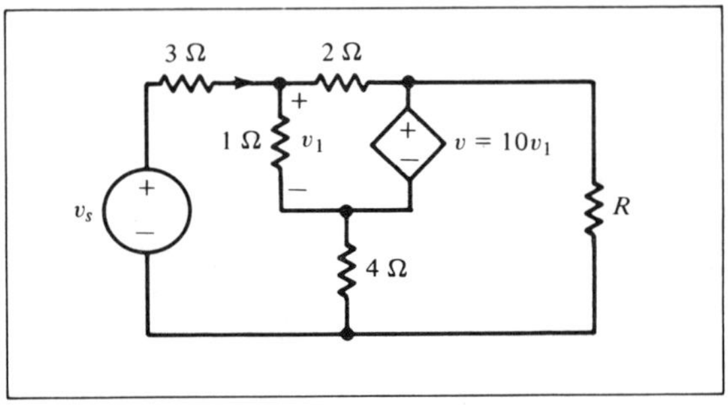

FIGURE P 2.28

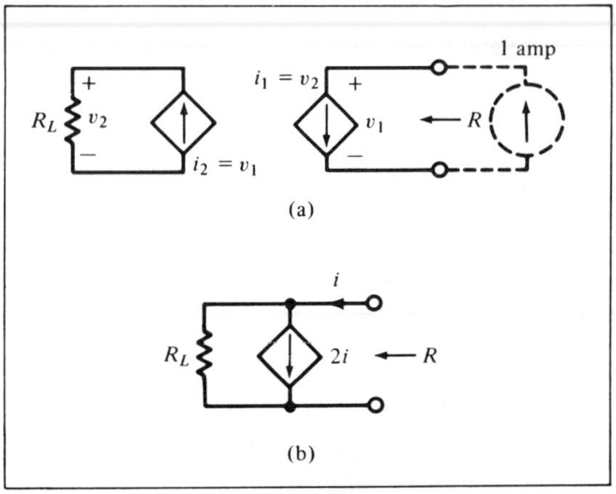

FIGURE P 2.29

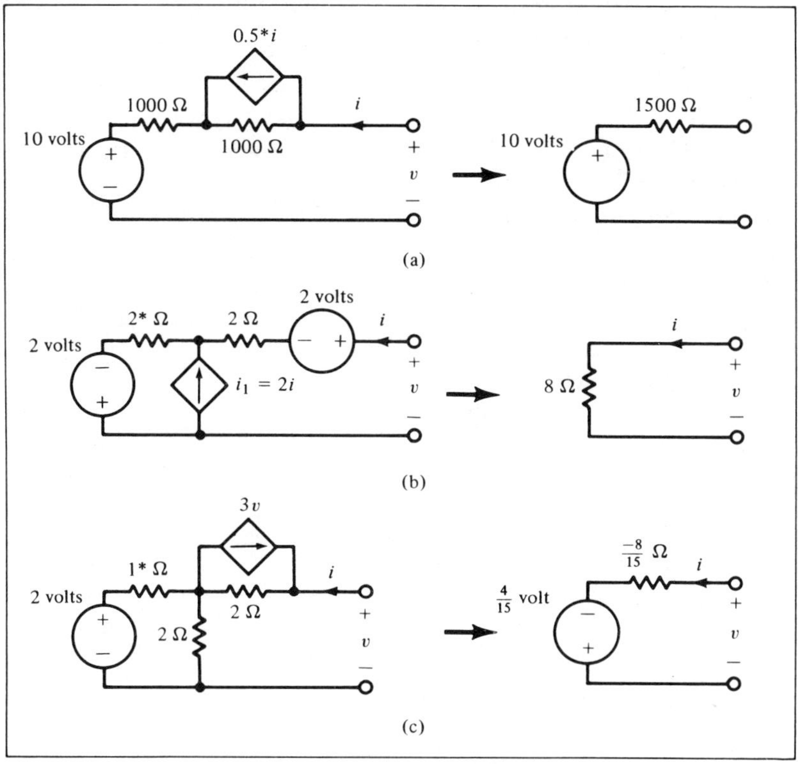

FIGURE P 2.30

the network of Figure P 2.29(b). (c) Determine the power absorbed in the controlled sources in Figure P 2.29(a) if a current source of 1 amp is applied as shown.

2.34 Verify the Thevenin and Norton equivalents shown in Figure P 2.30.

2.35 (a)–(c) Solve the problems of Figure 2.30 with asterisked values multiplied by k, where k is furnished by the instructor.

2.36 Show that the "input resistance" R_i of

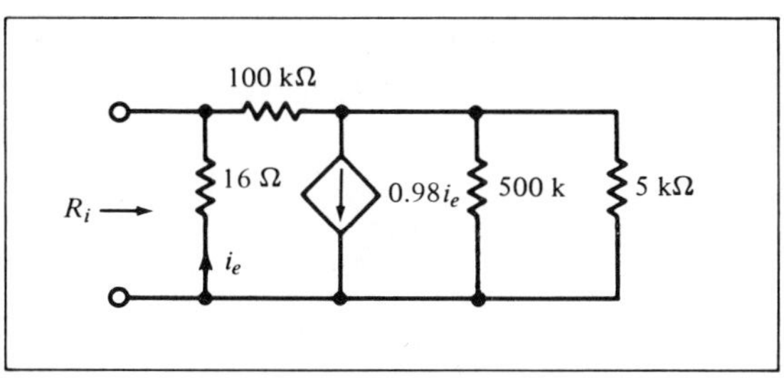

FIGURE P 2.31

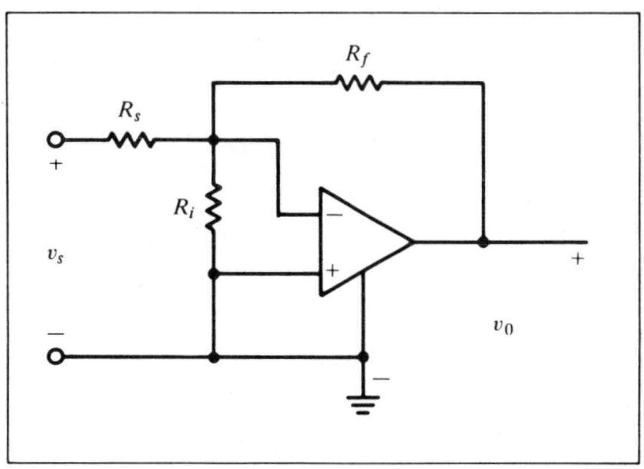

FIGURE P 2.32

the circuit shown in Figure P 2.31 is 16.8 Ω. (R_i is the Thevenin equivalent resistance looking into the port.) How can this be larger than 16 Ω?

2.37 What properties must the i–v characteristic of a linear network possess?

2.38 The operational amplifier circuit of Figure P 2.32 is to be studied as a one-port as follows (assume $R_i = \infty$). (a) Apply a voltage source v_s at the input terminal. Find the Thevenin equivalent at the output terminals. (b) With the output left open-circuited, determine the resistance looking into the input terminals, both with A finite and $A = \infty$.

2.39 Solve Problem 2.38 with a finite R_i.

REFERENCES AND ADDITIONAL READING

2.1. W. H. HAYT, JR., AND J. E. KEMMERLY, *Engineering Circuit Analysis*. New York: McGraw-Hill Book Company, Inc., 1963, Chapter 2.

2.2. W. H. HAYT, JR., AND G. W. HUGHES, *Introduction to Electrical Engineering*. New York: McGraw-Hill Book Company, Inc., 1968, Chapter 3.

2.3. NORMAN BALABANIAN, *Fundamentals of Circuit Theory*. Boston: Allyn and Bacon, Inc., 1964, Chapter 2.

2.4. D. A. CALAHAN, *Computer-Aided Network Design*. New York: McGraw-Hill Book Company, Inc., 1972, Chapter 5.

3

Nonlinear Network Analysis

3.1 Introduction

In Chapter 2, we have studied in a rather informal way the principles of elementary linear network analysis. At this point, we could proceed directly into generalizing these techniques (Chapter 4). The text is written to provide this option. However, we can also extend these methods into the realm of simple *nonlinear* networks. This offers the opportunity to observe at an early stage how familiar "real" circuits operate and at the same time to develop more facility with the linear circuit analysis methods of Chapter 2.

In this chapter, we will study the operation of (1) simple digital computer circuits and (2) single-transistor amplifiers. We will observe how both of these kinds of circuits function, and conduct a detailed analysis of each using the principles of network analysis contained in Chapter 2.

3.2 A Simple Nonlinear Element

The *diode* is the most common of the resistive electronic components that are nonlinear in their normal range of operation. The *pn* junction diode (we will term it simply a *diode*) is described by the algebraic equation[1]

$$i_d = I_s(e^{\lambda v_d} - 1) \tag{3.1}$$

Note that this *i-v* characteristic does *not* have the form $i \sim v$ as required of a linear element. Typical values of I_s (10^{-9} amp) and λ (40 volts^{-1}) result in the characteristics in Figure 3.1(a). It is easy to show by calculation that for

$$v_d > \epsilon, \qquad i_d \cong I_s e^{\lambda v_d}$$

$$v_d < -\epsilon, \qquad i_d \cong -I_s$$

where ϵ is in the range of 0.1 volt. We observe that only a small negative current ($-I_s$) can flow when v_d is negative, but current increases sharply with positive v_d.

The above diode equation is a reasonably accurate description of a real diode's behavior and is useful for a detailed evaluation of circuit

[1] Descriptions of the physical principles on which the equation is based are given in Appendix D and Reference 3.1.

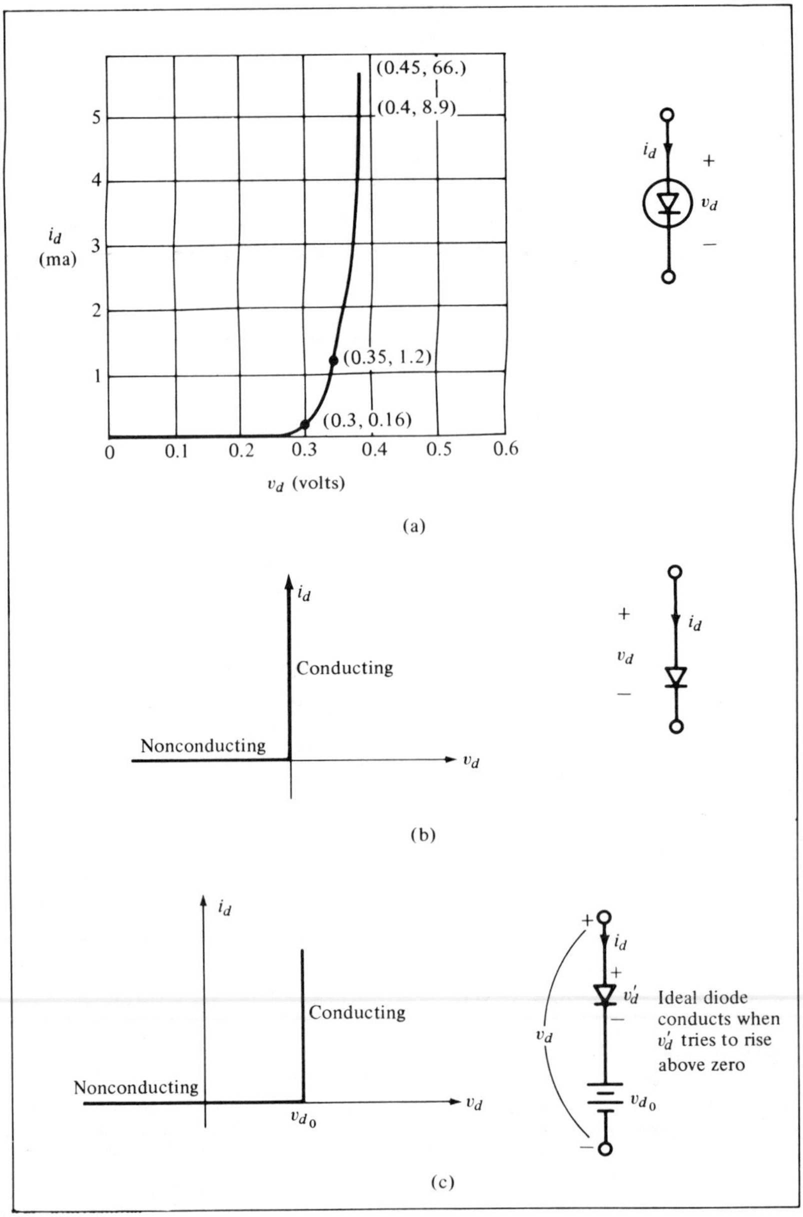

FIGURE 3.1 Diode characteristics and symbols: (a) *pn* junction diode; (b) ideal diode; (c) a model of the junction diode [$v_{d0} \cong 0.4$ volt for $I_s = 10^{-9}$ amp., $\lambda = 40$ volt^{-1} of part (a)].

performance. However, it is too complicated if we want to know only the function of a circuit ("how it works"). For this purpose, there are a number of idealizations of the graph.

The simplest is shown in Figure 3.1(b), described by

$$
\begin{aligned}
i &= 0 \qquad (v < 0) \qquad (\text{"nonconducting," "off"}) \\
v &= 0 \qquad (i > 0) \qquad (\text{"conducting," "on"})
\end{aligned}
\tag{3.2}
$$

This is termed an *ideal diode* and is represented by the special symbol shown in Figure 3.1(b). This diode model behaves as an open circuit if $v < 0$ and a short circuit if $i > 0$. This simple "open-short" model idealizes for many purposes the nearly vertical and horizontal portions of the real diode characteristic. However, it leaves something to be desired in modeling the displacement of the vertical portion v_{d_0} volts to the right of the vertical axis. As shown in Figure 3.1(c), the *i-v* characteristic can be shifted by adding a voltage source (battery) in series with the ideal diode. As v_d is increased from a zero value, the ideal diode will not conduct until v_d attempts to rise above v_{d_0} volts. The voltage v_d will then be held at v_{d_0} volts, because the diode will become a short, and the complete model will appear to be simply the voltage source.

In the rest of this chapter, we will have occasion to use all three diode models at different stages in our analysis as we proceed from a macroscopic to a microscopic evaluation of circuit behavior.

3.3 Elementary Digital Computer Circuits

Introduction

Design of computer systems begins by establishing the mathematical function that each circuit is to perform and then developing the appropriate circuit. There are, fortunately, only a few types of such circuits that are used thousands (or hundreds of thousands) of times throughout computers. In the following section, we will examine some of these elementary building blocks and the simple mathematical functions they perform. We will conduct two levels of analysis on these: one using ideal diode models to indicate how the mathematical function is realized by the circuit and then a detailed circuit analysis to detect departures from this idealized performance.

Logical Signals

The functions performed by digital circuits are described by the operations of *discrete* mathematics. In this mathematics, constants and variables can take on only *integer* values. If we specialize this to the case when only two integers, **0** and **1**, are allowed, we can view these numbers as representing *logical* (FALSE-TRUE) quantities. The circuits that we will describe process such logical information to produce new logical information; they are therefore called *logic circuits.*

This does not imply that voltages and currents in a logic circuit must have the values 0 and **1** volt. For example, it is common to represent a logical **0** (FALSE) by a voltage level of 0 volts and a logical **1** (TRUE) by a signal level of 5 volts. However, the opposite assignment could also be made, as Table 3.1 shows.

TABLE 3.1

Typical Logical-Voltage Level Assignments

Logical	Positive Logic	Negative Logic
0	0 volts	5 volts
1	5 volts	0 volts

In general, *negative* (positive) logic denotes the assignment of a logical **1** to the more negative (positive) voltage. In theory, we must be told which type of logic, positive or negative, is assumed before we can determine the function of a logical device; in practice, *positive logic is assumed unless otherwise specified.*

Logic Circuit Examples

The two logical *elements* or *operations* we consider in this chapter are the AND and the OR elements. These are defined in Figure 3.2 by *truth tables.* For example, the AND element accepts two logical inputs (A, B) and produces one logical output. From the truth table, the value of the output is a logical **1** if and only if both A *and* B are logical **1**s. Similarly, the OR

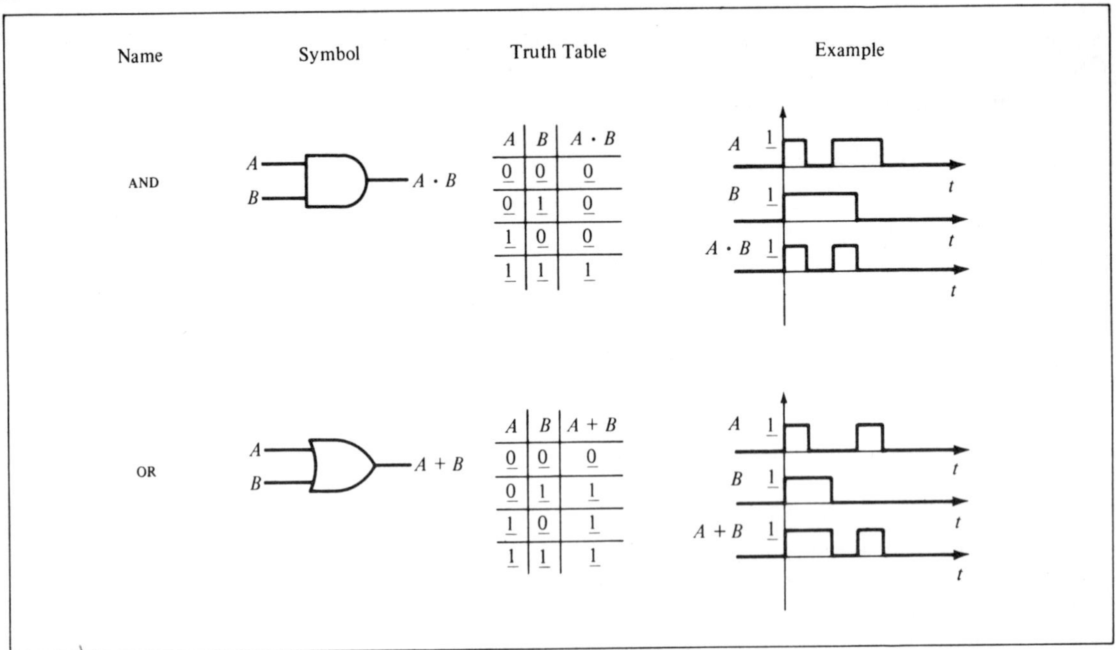

FIGURE 3.2 AND and OR logic elements.

element produces a logical **1** if either the A *or* B input is a logical **1**.

An Electronic AND Circuit

We launch our study of digital circuit analysis by showing that the circuit of Figure 3.3(a) realizes an AND function. We adopt the ideal diode model, which allows us to regard the diodes as either open or short circuits.

We define the logical states of the circuit as

$$\text{State of } A \text{ or } B \text{ or } C: \quad \begin{array}{l} \mathbf{0} \text{ if } 0 \text{ volts} \\ \mathbf{1} \text{ if } 5 \text{ volts} \end{array} \tag{3.3}$$

In deducing how the circuit will react to a **0** state of A and B, assume for the moment that both diodes D_A and D_B are nonconducting, so that they behave as open circuits. Clearly, from Figure 3.4, the voltage at point C (v_0) will try to rise to 10 volts, the value of the applied source. But the diodes will conduct and become short circuits when v_0 attempts to become positive. Therefore, the only voltage at C consistent with both the diode characteristics and the rest of the circuit is $v_0 = 0$ volts, with both diodes conducting. The current flowing through R is then 10/1000 amp; each diode will (ideally) conduct $\frac{1}{2}$ of this current, or 5 ma.

If input B is raised to 5 volts, v_0 still cannot rise above 0 volts, because diode D_A will again conduct. The total current through D_A is now 10 ma. A similar argument applies if A is at 5 volts and B at 0 volts.

When both inputs are at 5 volts, then the output v_0 will again try to rise above 0 volts (Figure 3.5). Now, however, if $v_0 = 0$, both diodes will remain nonconducting because the negative side of each diode is maintained at 5 volts. Voltage v_0 can therefore continue to rise. Only when v_0 attempts to rise above 5 volts will each diode voltage become positive; again, the only consistent condition is $v_0 = 5$ volts; each diode then ideally conducts $(\frac{1}{2})(10 - 5)/1000 = 0.0025$ amp.

An Electronic OR Circuit

Although it seems quite reasonable to define 0 volts as a logical **0**, this is not at all necessary. Let us see what happens to Table 3.2 if we

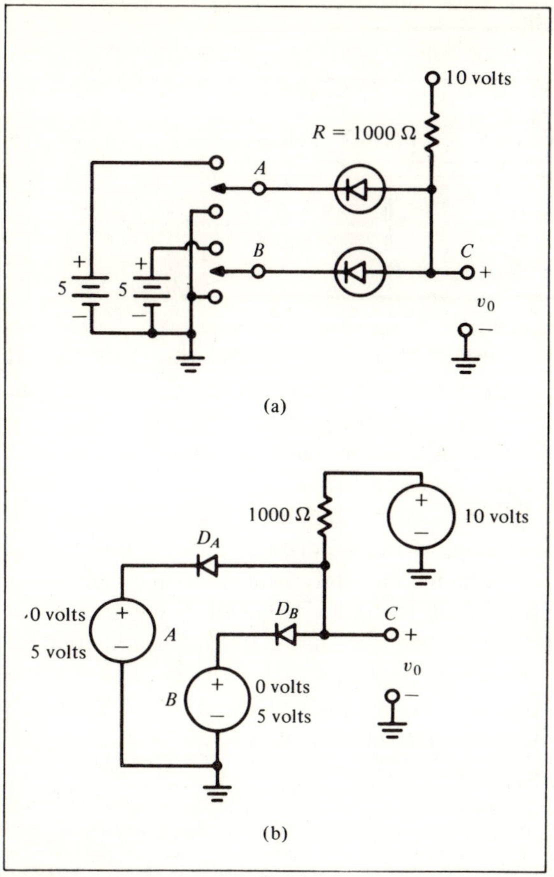

FIGURE 3.3 Electronic AND circuit and its network model. (a) Circuit; (b) network model.

use the negative logic assignments

$$\text{state of } A \text{ or } B \text{ or } C: \quad \begin{matrix} \mathbf{0} \text{ if 5 volts} \\ \mathbf{1} \text{ if 0 volts} \end{matrix} \tag{3.4}$$

The same four possible arrangements of input and output voltages are now interpreted differently, as shown in Table 3.3. A logical **1** is now produced if either A *or* B is in a logical **1** state. The same circuit that performed an AND function with positive logic now performs an OR function with negative logic.

Diode OR circuits can be designed to operate with positive logic; an example is given in Problem 3.3, which the reader is invited to study.

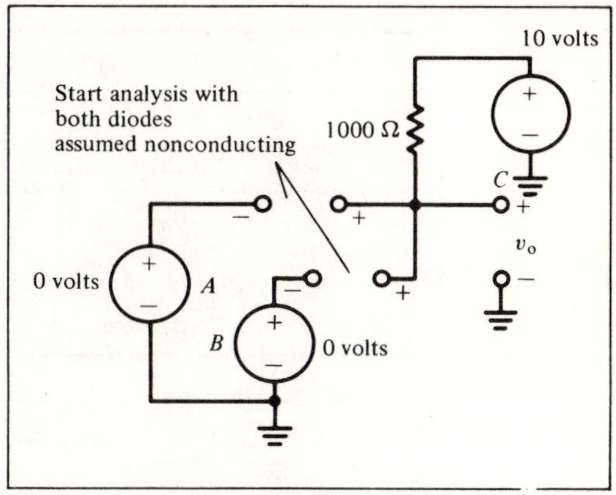

FIGURE 3.4 Both inputs in 0 state. v_0 tries to rise to 10 volts; stopped at 0 volt.

Neither an AND nor an OR circuit accomplishes a significant processing function in itself. However, these elements can, in combination, realize the addition and other operations we associate with a digital computer. These topics are considered in courses on digital design.

Detailed Analysis of Simple Logic Circuits

If we interconnect many simple logic circuits to perform a complex function such as addition, we would predict from the above idealized model an exact electrical realization of the mathematical functions. Unfortunately, a circuit using real diodes could behave far differ-

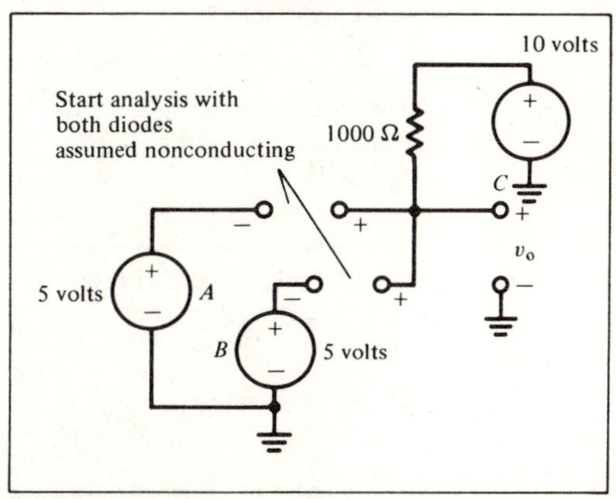

FIGURE 3.5 Both inputs in 1 state. v_0 tries to rise to 10 volts; stopped at 5 volts.

TABLE 3.2

Logic Table for AND Circuit

	A		*B*		*C*	
	State	Voltage	State	Voltage	State	Voltage
	0	0	**0**	0	**0**	0
⟶	**1**	5	**0**	0	**0**	0
	0	0	**1**	5	**0**	0
	1	5	**1**	5	**1**	5

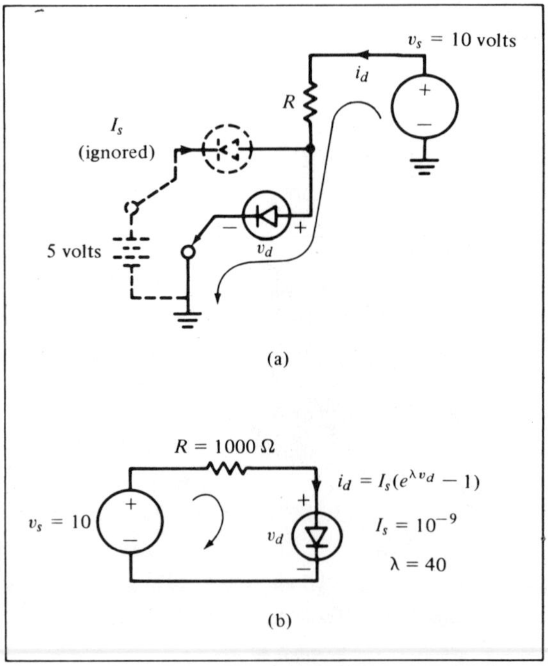

FIGURE 3.6 AND circuit and network model.

ently. A more detailed analysis then would be necessary, using either the diode model of Figure 3.1(c) or, better, the exponential characteristic of Figure 3.1(a), In this section, we will construct such an analysis of the AND circuit of Figure 3.6(a) using exponential diode characteristics. We will later extend the methodology in Chapter 5 to determine the behavior of transistorized logic circuits.

Consider the analysis of the AND circuit when the input A has a value of 5 volts and the input B has a value of 0 volts (indicated by the arrow in Table 3.2). We can describe the model of Figure 3.6(a) by writing a KVL equation around the loop shown.

$$-v_s + i_d R + v_d = 0 \tag{3.5}$$

Adding the equation for the diode

$$i_d = I_s(e^{\lambda v_d} - 1) \tag{3.6}$$

yields two equations in two unknowns. Although the first equation is linear in i_d and v_d, the second is not. No substitution of one equation into the other will yield an analytic solution.

To solve problems similar to the above, we propose two solution methods, namely:

Table 3.3

Logic Table for OR Circuit

A		*B*		*C*	
State	Voltage	State	Voltage	State	Voltage
0	5	**0**	5	**0**	5
1	0	**0**	5	**1**	0
0	5	**1**	0	**1**	0
1	0	**1**	0	**1**	0

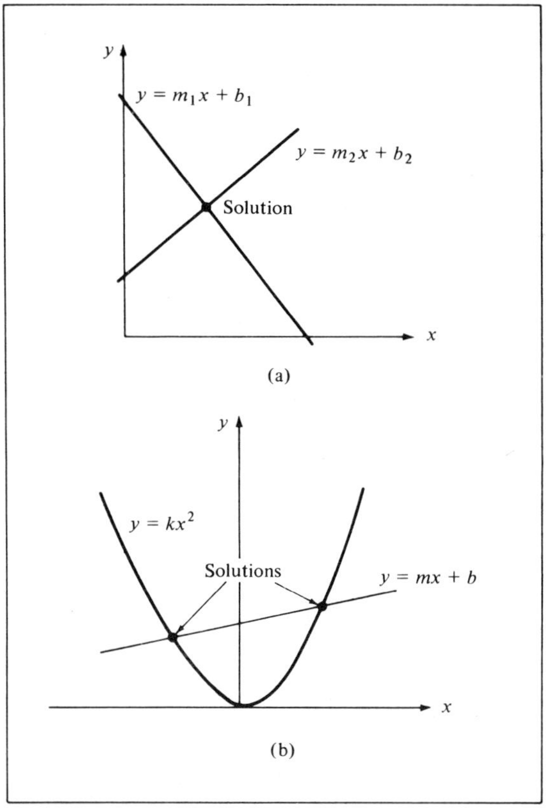

FIGURE 3.7 Graphical solution of simultaneous equations. (a) Linear; (b) nonlinear.

1. Graphical analysis, where the *i-v* characteristics used in explaining Thevenin and Norton equivalents assume a new importance as a means of solution.

2. Iterative solution with the digital computer.

Graphical Analysis

The reader is probably familiar with graphical methods for solution of simultaneous equations. If the equations are linear, then no more than one solution can exist. This is illustrated in Figure 3.7(a), where the solution is at the intersection point shown. When one of the equations is nonlinear, then multiple solutions can exist as indicated in Figure 3.7(b).

If we regard Equations (3.5) and (3.6) each as representing i_d as a function of v_d, then plotting these equations together yields the solution at the intersection of the graphs. The plot of Equation (3.5) is known as a *load line,* the resistor being viewed as a "load" on the diode. A sample construction is shown in Figure 3.8. It is useful to note that the voltage-axis intercept of the load line is equal to the value of the voltage source v_s, whereas the current-axis intercept is v_s/R. This makes its construction particularly easy.

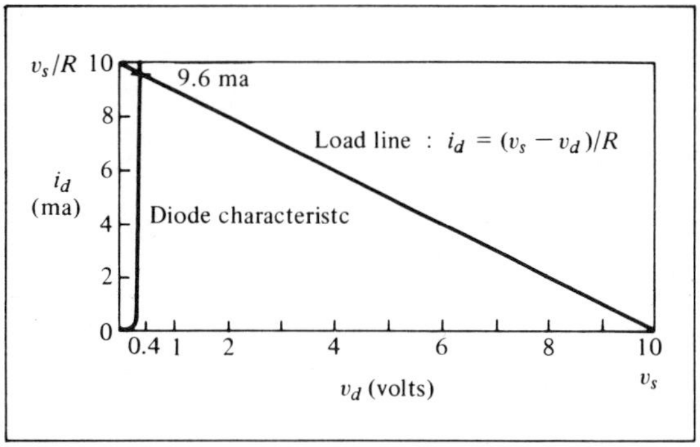

FIGURE 3.8 Graphical analysis of logic circuit.

This graphical analysis shows the solution of the analysis problem to be at $i_d = 9.6$ ma and $v_d \cong 0.4$ volt. Hence, the voltage v_0 is not zero as indicated in Table 3.2, but approximately 0.4 volt. This difference between 0 and 0.4 volt could be important if the AND circuit drives other logic circuits at C.

Computer Analysis

Graphical techniques have the serious limitation that they cannot be easily extended to circuits containing more than one nonlinearity. In contrast, the computer-related methods we now propose will serve as the basis for more general techniques for analysis of large nonlinear networks in Chapter 5.

We again return to the simple diode-resistor model of Figure 3.6(b) for development of a computer algorithm. First, we are faced with the need to solve Equations (3.5) and (3.6),

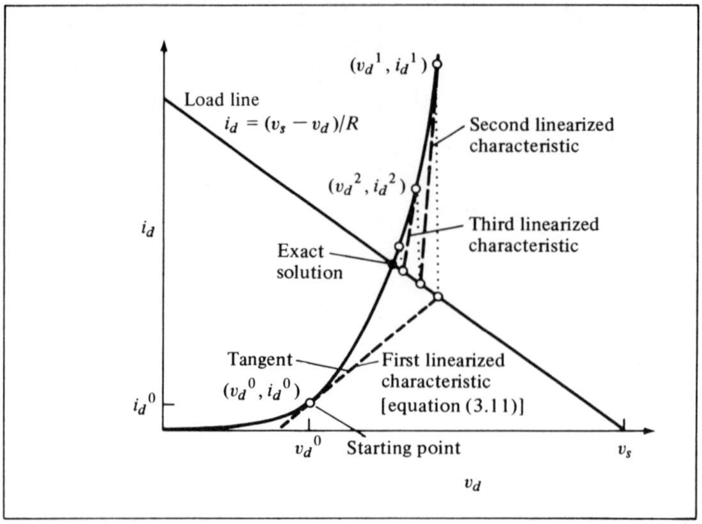

FIGURE 3.9 Iterative solution of resistor-diode network.

which we rewrite

$$i_d = \frac{(v_s - v_d)}{R} \tag{3.7}$$

$$i_d = I_s(e^{\lambda v_d} - 1) \tag{3.8}$$

The basis for development of our computer-aided solution is the concept of *iteration*. The procedure is first to make an *initial guess* of the solution, and then to *refine* successively this guess until the v_d and i_d satisfy both equations within a certain error.

To explain, at the mth step in the process, assume that we have an estimate of (v_d, i_d). Let this guess be denoted (v_d^m, i_d^m), $m = 0, 1, \ldots$. For example, the initial guess is designated (v_d^0, i_d^0) in Figure 3.9. We then approximate the nonlinear characteristic in the vicinity of this point by a straight line which passes through the starting point and has the same slope. Combining this linearized characteristic with the linear equation (3.7), we can solve for a new value v_d^{m+1}. For example, when $m = 0$, the linearized characteristic has the form

$$i_d = i_d^0 + \left.\frac{\partial i_d}{\partial v_d}\right|_{v_d = v_d^0} (v_d - v_d^0) \tag{3.9}$$

$$= I_s(e^{\lambda v_d^0} - 1) + \lambda I_s e^{\lambda v_d^0}(v_d - v_d^0) \tag{3.10}$$

Equation (3.9) has the form of a Taylor's series expansion of i_d about v_d^0, but with terms beyond the linear term ignored. Clearly, this expression (1) is linear in i_d and v_d, (2) passes through $(i_d, v_d) = (i_d^0, v_d^0)$, and (3) has the same slope $\partial i_d/\partial v_d$ as the nonlinear diode characteristic at the tangent point.

Solving for v_d in Equation (3.7) (which becomes v_d^1 using our notation) yields

$$v_d^1 = -i_d^1 R + v_s$$

$$= -\left[i_d^0 + \left.\frac{\partial i_d}{\partial v_d}\right|_{v_d = v_d^0} (v_d^1 - v_d^0)\right] R + v_s$$

or

$$v_d^1 = \frac{-\left[i_d^0 - \left.\frac{\partial i_d}{\partial v_d}\right|_{v_d = v_d^0} v_d^0\right] R + v_s}{1 + R \left.\frac{\partial i_d}{\partial v_d}\right|_{v_d = v_d^0}}$$

Note that $\partial i_d/\partial v_d$ has the dimension of current/voltage or *mhos*. If we define the conductance $G_d^0 \triangleq (\partial i_d/\partial v_d)_{v_d=v_d^0}$, then we can write

$$v_d^1 = \frac{-(i_d^0 - G_d^0 v_d^0)R + v_s}{1 + RG_d^0}$$

From v_d^1, we can calculate

$$i_d^1 = I_s(e^{\lambda v_d^1} - 1)$$

from Equation (3.8). The point (v_d^1, i_d^1) is on the diode characteristic as shown in Figure 3.9, and a new linearized (tangent) characteristic may be defined. This process of successive linearization and updating of the nonlinear function is continued; it is known as the Newton-Raphson, or simply Newton,[2] iteration. The general form of the iteration equation is

$$v_d^{m+1} = \frac{-i_d^m R + G_d^m v_d^m R + v_s}{1 + RG_d^m} \tag{3.11}$$

where $G_d^m \triangleq (\partial i_d/\partial v_d)_{v_d=v_d^m}$.

Convergence of the Iteration

The Newton iteration described above can be observed in Figure 3.9 to approach the solution at the intersection of the diode characteristic and the load line. We cannot precisely determine $|v_d^m - v_d|$, that is, the difference between the mth iterant and the exact solution value v_d, because the latter is unknown. However, we can monitor the progress made at the $(m+1)$st iteration by forming

$$|v_d^{m+1} - v_d^m|$$

Near the solution value, this quantity necessarily becomes smaller. Therefore, if

$$|v_d^{m+1} - v_d^m| < \epsilon \tag{3.12}$$

where ϵ is a small number (say 10^{-3}), then we can usually conclude that progress is too small to warrant further iteration because we are near the solution. In fact, it can be shown (Problem 5.30) that near the solution,

$$|v_d^{m+1} - v_d| = K|v_d^m - v_d|^2 \tag{3.13}$$

where K is a constant dependent on the characteristics of the solution point. Thus, the error

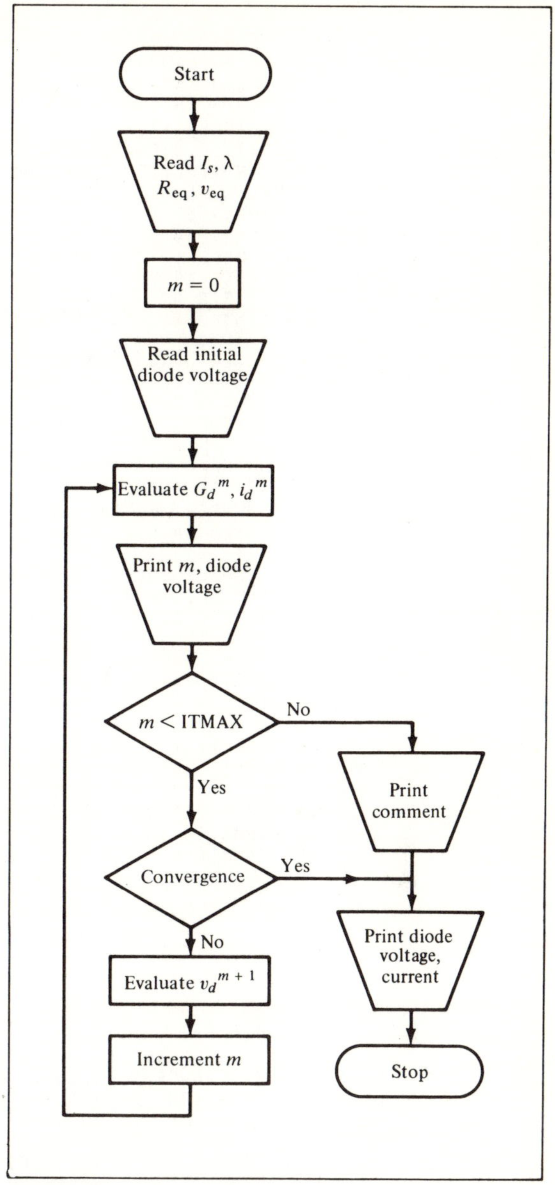

FIGURE 3.10 Flow chart of program of figure 3.11(a).

[2] Isaac Newton (1642–1727), an English mathematician and experimental philosopher, made major contributions to astronomy, the theory of calculus, and formulated the laws of gravitation.

```
C****  LOGIC CIRCUIT EXAMPLE (LNET)
      REAL*4 LAMDA
C****  READ SATURATION CURRENT IN NANOAMPS
      READ(5,15)CSAT,LAMDA,REQ,VEQ
      CSAT=CSAT*1.E-9
      WRITE(6,8)CSAT,LAMDA,REQ,VEQ
C****  INITIALIZE ITERATION PARAMETERS
      M=0
      ITMAX=15
      DVD=1.E60
C****  GUESS INITIAL DIODE VOLTAGE
      READ(5,15)VDM
      WRITE(6,9)
C****  ENTER NEWTON RAPHSON ITERATION
      GO TO 1
C****  CALCULATE (M+1)ST DIODE VOLTAGE
    3 VDMP1=(-(CURM-GDM*VDM)*REQ+VEQ)/(1.+GDM*REQ)
      DVD=VDMP1-VDM
      VDM=VDMP1
      M=M+1
C****  UPDATE DIODE-RELATED TERMS
    1 EXP1=EXP(LAMDA*VDM)
      CURM=CSAT*(EXP1-1.)
      GDM=LAMDA*CSAT*EXP1
      WRITE(6,16)M,VDM
C****  CONVERGENCE CRITERION
      IF(M.LE.ITMAX)GO TO 2
      WRITE(6,11)M
      GO TO 12
    2 IF(ABS(DVD).GT.1.E-6)GO TO 3
   12 WRITE(6,4)VDM,CURM
      STOP
    9 FORMAT(2X,'ITERATION',4X,'VOLTAGE')
    4 FORMAT(//2X,'VOLTAGE =',E12.4,2X,'CURRENT =',E12.4)
    8 FORMAT(2X,'CSAT CURRENT =',E12.4,2X,'LAMDA =',E12.4/
     12X,'RESISTANCE =',E12.4,2X,'SOURCE =',E12.4)
   11 FORMAT(2X,'NO CONVERGENCE AFTER ',I2,'ITERATIONS')
   15 FORMAT(4F10.0)
   16 FORMAT(I6,E18.4)
      END
```

(a)

```
      1.        40.       1000.      10.
      .6
```

(b)

```
 CSAT CURRENT =   0.1000E-08  LAMDA =   0.4000E 02
 RESISTANCE =   0.1000E 04  SOURCE =   0.1000E 02
 ITERATION     VOLTAGE
      0        0.6000E 00
      1        0.5750E 00
      2        0.5500E 00
      3        0.5251E 00
      4        0.5003E 00
      5        0.4758E 00
      6        0.4521E 00
      7        0.4304E 00
      8        0.4134E 00
      9        0.4042E 00
     10        0.4020E 00
     11        0.4019E 00
     12        0.4019E 00
 VOLTAGE =   0.4019E 00   CURRENT =   0.9598E-02
```

(c)

FIGURE 3.11 Program, input, and output for solution of logic circuit of Figure 3.6(a).

in the solution is *squared* at each iteration. An error $|v_d{}^m - v_d| = 10^{-3}$ would become an error $|v_d{}^{m+1} - v_d| = 10^{-6}$ at the next iteration. The selection of ϵ is not critical, because the additional iteration required to check the convergence with Inequality (3.12) puts the result ($v_d{}^{m+1}$) considerably closer to the solution v_d than ϵ; that is,

$$|v_d{}^{m+1} - v_d| \ll |v_d{}^{m+1} - v_d{}^m| < \epsilon$$

Each of the following network examples has only one solution value for v_d. The reader should be reminded from Figure 3.7(b) that, because the equations are nonlinear, multiple solutions may exist. Such multiple solutions imply that the circuit which the equations describe can operate with a number of different sets of voltages and currents. We reserve the study of such circuits for Chapter 5.

A Solution Program

Equation (3.11) is a simple formula to represent in program form. Figure 3.10 shows the flow chart for such a program. The addition of input-output statements, updating expressions, and convergence tests results in the Fortran program of Figure 3.11(a). This program accepts as input data the quantities I_s, λ, R, v_s, and $v_d{}^0$ of Figure 3.6(b) and prints as output data (1) the diode voltage at each iteration and (2) the diode voltage and current after convergence.

For the parameter values represented in Figure 3.6(b), the output data of Figure 3.11(c) is obtained. The convergence is observed to be relatively slow initially, the correction applied to $v_d{}^0$ being only $0.025/0.198 \cong 1/8$ of the total correction necessary. Near the solution, this ratio increases; at the ninth iteration, 0.0022/0.0023 of the remaining correction is applied.

Both the final diode voltage and current previously obtained from graphical analysis are within 1 percent of the precise computer-calculated values. This is well within the accuracy with which I_s, λ, R, and v_s could be measured. In this case, then, graphical analysis would be adequate.

3.4 Introduction to Transistor Amplifier Circuits

Introduction

We are all familiar with the use of electronic circuits to amplify signals. As mentioned in Chapter 2, circuits with positive-valued resistors cannot amplify; the presence of a controlled source is essential to achieving this function with linear elements. The transistor is the circuit component most commonly used to obtain the effect of a controlled source.

In the following section, we will briefly discuss the physical processes that occur within the transistor and develop from them a simple model in terms of a diode and a controlled source. We then study several simple transistor amplifiers to show how the amplification occurs, and how the circuit performance can be predicted.

Transistor Models

The basic transistor operation and models are displayed in Figure 3.12(a)–(d). We can crudely describe the transistor as consisting of two diodes constructed in close physical proximity (for example, within 0.0001 in.), as shown in Figure 3.12(b).

In normal operation, the emitter diode is conducting or forward-biased,[3] and the collector diode nonconducting or reverse-biased.[3] We would therefore expect the collector diode current to be small ($\cong -I_s$). However, because of the small separation between the two diode junctions, the internal motion of charges is not described by the simple exponential relationships of two diodes. In particular, an appreciable fraction (α) of the emitter diode charge expected to flow through the base is diverted to the collector (hence, the words "emitter"

[3] The terms reverse- and forward-biased are traditionally preferred for exponential models of diodes.

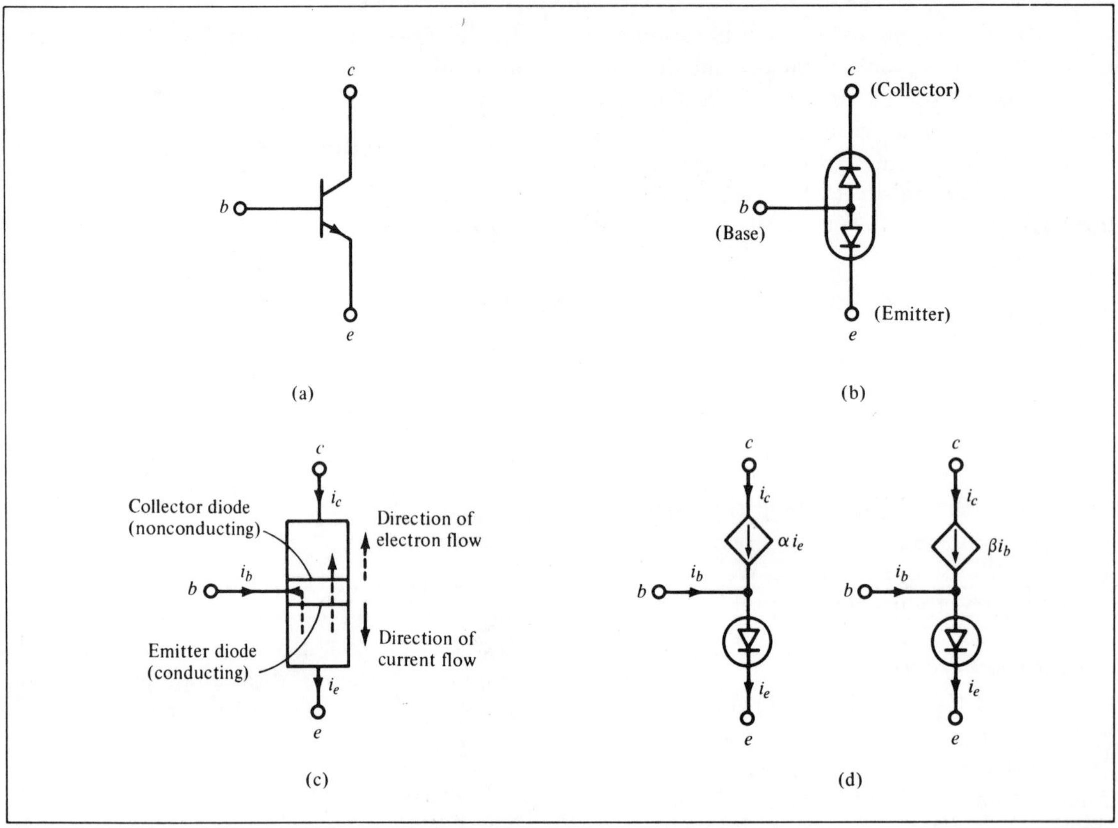

FIGURE 3.12 The *npn* transistor and its models. (a) Schematic; (b) elementary representation; (c) internal physical processes; (d) elementary models [$\beta = \alpha/(1 - \alpha)$].

and "collector"). Thus, the collector, although reverse-biased, can pick up charges injected by the emitter into the base and a large collector current can flow! We can describe this charge flow by

$$i_c = \alpha i_e \tag{3.14}$$

where α is commonly between 0.98 and 0.999. Because the algebraic sum of the currents flowing into the transistor must be zero from KCL, we can also write

$$i_e = i_b + i_c \tag{3.15}$$

Two network models of the transistor are shown in Figure 3.12(d). These models satisfy Equations (3.14) and (3.15) and additionally account for the presence of the emitter diode. The collector diode, assumed reverse-biased, is ignored in the studies of this chapter; its effects will be considered in Chapter 5. It suffices to say at this time that the transistor operation is highly nonlinear when the collector diode conducts and the device fails to have any amplifying properties.

The right-hand equivalent transistor model in Figure 3.12(d) is obtained by eliminating i_e from (3.14) and (3.15) as

$$i_c = \alpha(i_b + i_c)$$

$$i_c = \left(\frac{\alpha}{1 - \alpha}\right) i_b$$

$$= \beta i_b$$

It is important to note the presence of the controlled current source in these models. As we are about to show, it is this component in the model that is primarily responsible for the amplifying capabilities of the transistor.

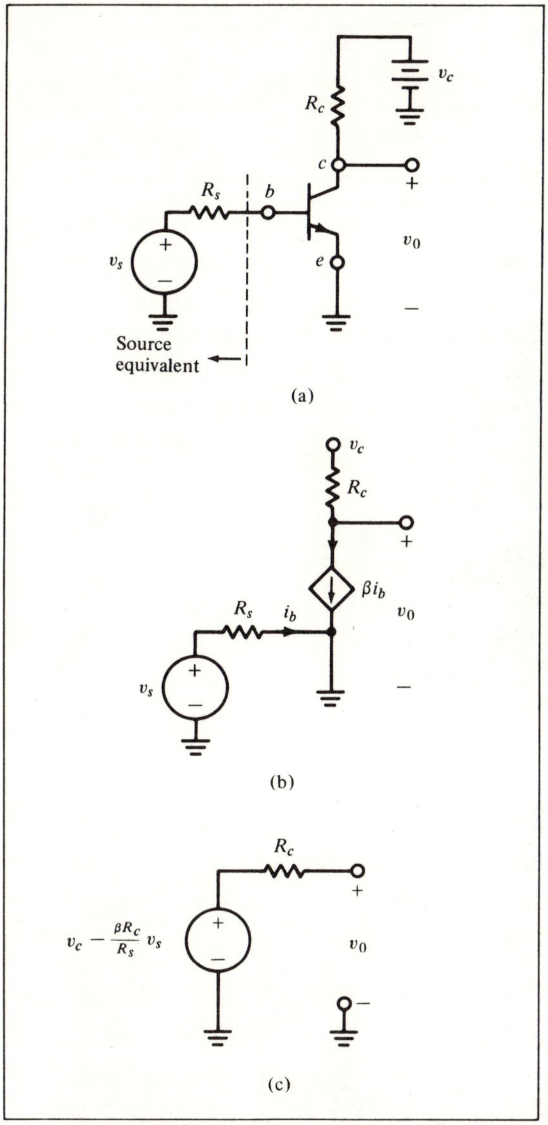

FIGURE 3.13 A transistor amplifier.

Amplification

The amplifying capacity of the transistor can be illustrated by the simple amplifier circuit of Figure 3.13(a). Here, v_s is regarded as the input and v_0 as the output; v_c is a constant (supply) source. The circuit is considered an amplifier if changes in v_s produce larger changes in v_0.

The analysis of this network begins by determining the base current

$$i_b = \frac{v_s}{R_s}$$

The collector current follows as

$$i_c = \beta i_b = \frac{\beta}{R_s} v_s$$

and the output voltage is

$$\begin{aligned} v_0 &= v_c - i_c R_c \\ &= v_c - \beta \frac{R_c}{R_s} v_s \end{aligned} \tag{3.16}$$

This is the open-circuit output voltage and therefore is equal to the Thevenin equivalent voltage v_{eq} as observed across the output of the amplifier. The complete Thevenin equivalent across the output has the form of Figure 3.13(c). The Thevenin resistance $R_{eq} = R_c$ is known as the output resistance of the amplifier.

To study the amplification function, we apply a *step signal* as shown in Figure 3.14(a) and observe its effect on the output voltage v_0. With no signal, the output voltage has a constant or *dc* (*d*irect *c*urrent) value $v_0 = v_c$, from Equation (3.16). An input change v_{s0} is shown to produce a change in the output.

The voltage *gain* of the amplifier normally is defined as the ratio of the change in the output voltage to the change of the input voltage. In Figure 3.14, an input or signal change v_{s0} produces an output change $-\beta R_c v_{s0}/R_s$. Therefore, we have

$$\text{Voltage gain} = -\beta \frac{R_c}{R_s}$$

If $R_c = R_s$, then the gain is $-\beta$, a large negative number. This negative sign implies that a *positive* input change produces a *negative* change in the output. We can thus achieve a large

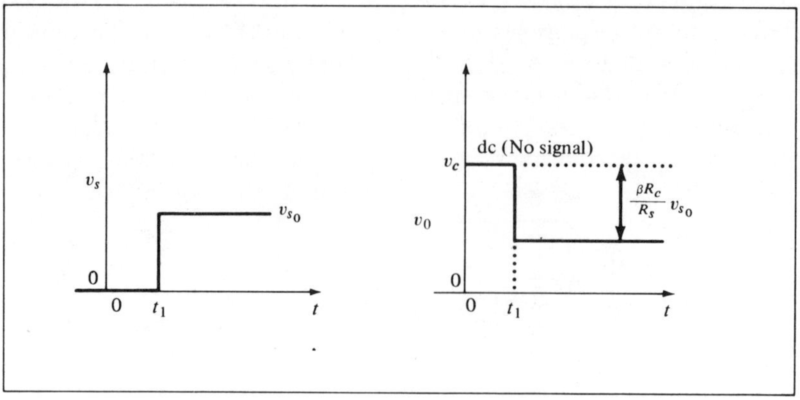

FIGURE 3.14 Amplifier input and output.

voltage gain (say, -50 to -100) with a single-transistor amplifier!

As an exercise in applying the principles of Chapter 2, the reader is urged to verify that Figure 3.13 (c) is a correct Thevenin equivalent for the amplifier as viewed from its output port

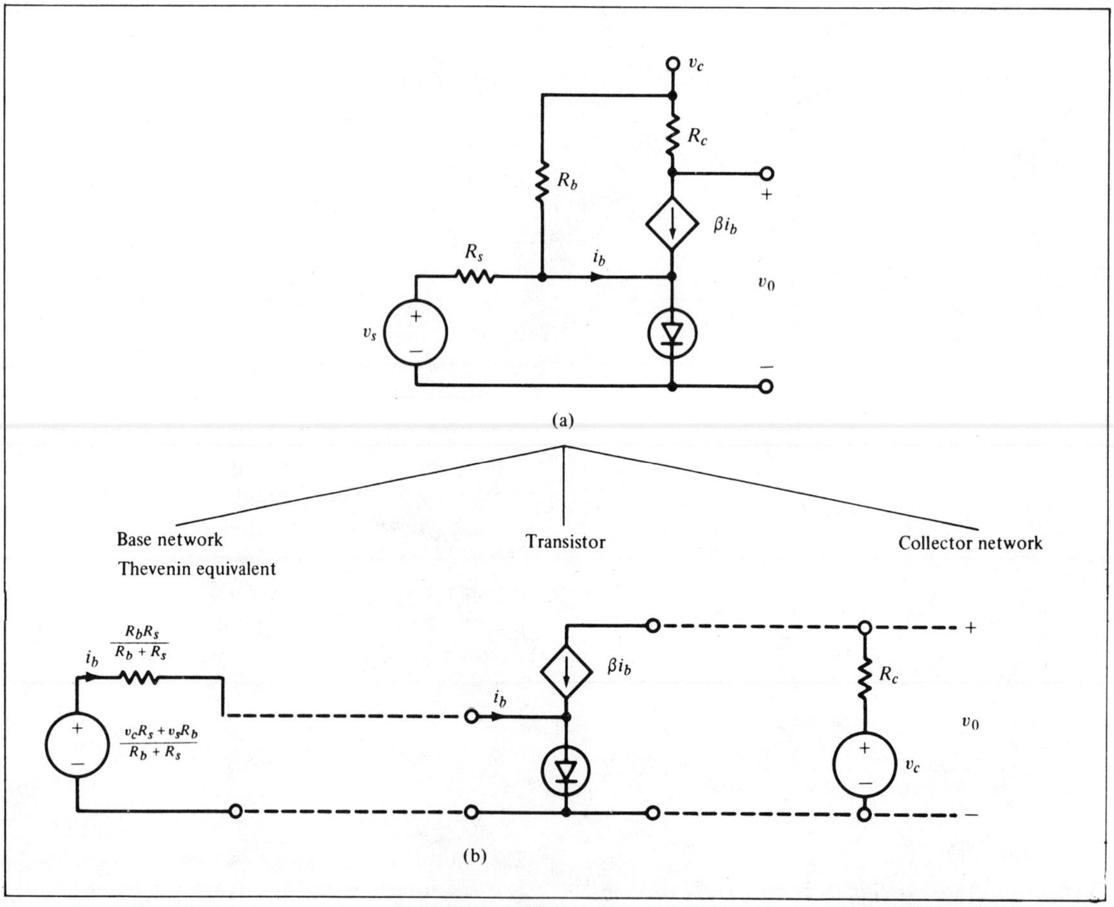

FIGURE 3.15 Partitioned study of a transistor amplifier. (a) Nonlinear model; (c) simplified model; (d) refined

(the v_0 terminals); the use of Method 2 is suggested. The Thevenin resistance $R_{eq} = R_c$ determines how changes in v_0 are reduced when an external load resistor is connected across the output terminals.

3.5 Detailed Analysis of Amplifiers

Introduction

The circuit of Figure 3.13 is the simplest of amplifiers. Even so, we simplified its analysis by modeling the emitter diode by a short circuit and by ignoring any effects of the collector diode. In the following sections we examine more complicated amplifier circuits involving progressively more realistic transistor models. These investigations culminate in a computer analysis of a typical single transistor amplifier using the nonlinear transistor model of Figure 3.12(d).

Load-Line Analysis

We begin with the somewhat more complicated example of Figure 3.15(a), where a resistor has been added to the amplifier of Figure 3.13 connecting the base of the transistor to the constant voltage source v_c. Although we could write and solve the network equations of this circuit, we reserve that approach for the next section. Instead, we introduce the notion of an *approximate* analysis based on certain justifiable, simplifying assumptions. This procedure gives us further insight into the operation (and even the design) of transistor amplifiers as

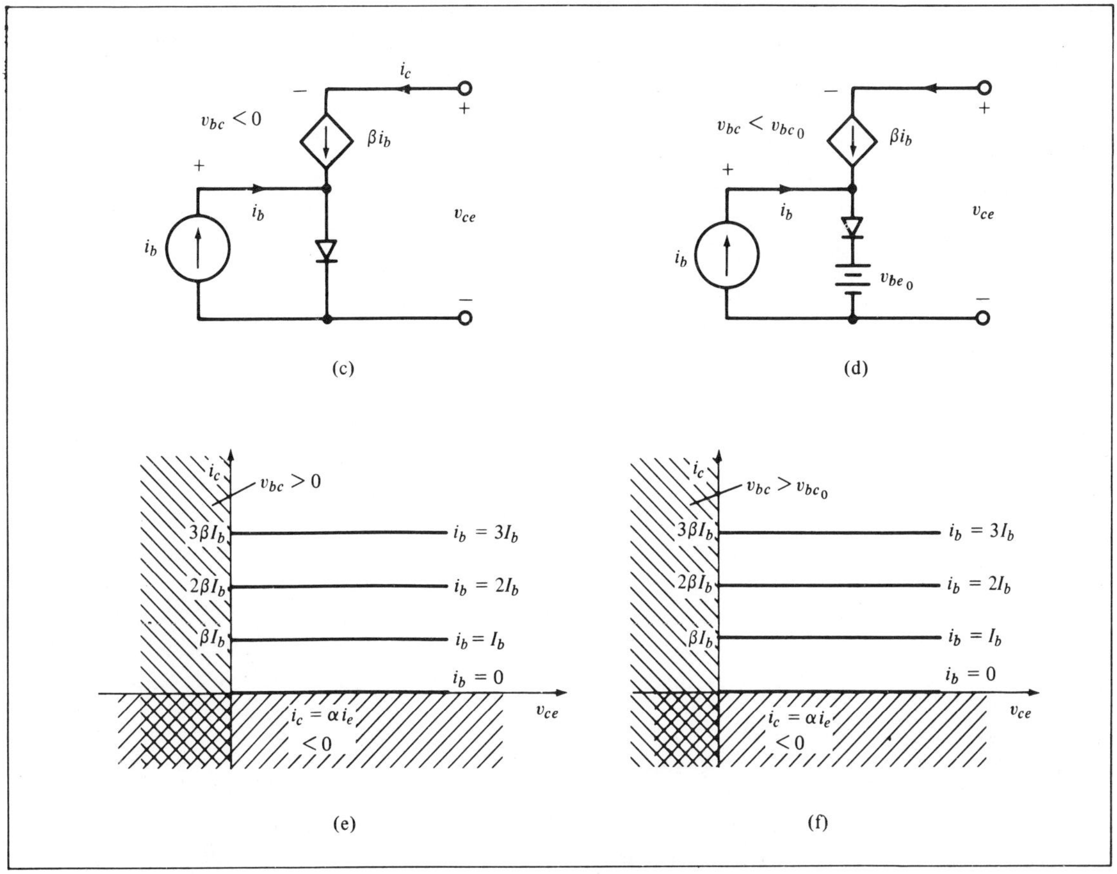

model; (e) output characteristics from (c); (f) output characteristics from (d).

well as illustrating an approach (partitioning) that engineers frequently use when faced with the analysis of large circuits and systems.

The approximate analysis proceeds in this case by breaking up the network into three parts. The (v_s, R_s, v_c, R_b) elements are viewed from the transistor input (base-emitter terminals) as a one-port and are represented by their Thevenin equivalent; the (v_c, R_c) voltage source and resistor combination is seen from the transistor output (collector-emitter terminals) as a one-port that is already in the Thevenin form; and the transistor has both an input and output port. This two-port configuration seems to be beyond our present analysis abilities. Let us study the transistor operation and model in more detail, however, to see how a few key approximations will allow us to piece together these three apparently interdependent network partitions.

Concentrating on the transistor model as an isolated network, let us apply a constant current source $i_b = I_b > 0$ to the input base terminal, as shown in Figure 3.15(c). The output of the transistor (the collector-emitter terminals) can then be considered a *one-port* described by an $i_c - v_{ce}$ characteristic. Since

$$i_c = \beta i_b = \beta I_b \tag{3.17}$$

the transistor behaves from this output port as a current source of value βI_b. Its $i_c - v_{ce}$ characteristic is therefore a horizontal line. For each different value of I_b, a new horizontal $i_c - v_{ce}$ characteristic is defined. The resultant set of *output characteristics* is shown in Figure 3.15(e). They show in compact pictorial form how the output current and voltage are affected by different input currents.

To complete this graphical representation of the current model of Figure 3.15(c), observe that Equation (3.17) is not valid when $i_c < 0$. To see this, recall that an ideal model for the emitter diode requires that $i_e \geqq 0$; since $i_b = (1 - \alpha)i_e$, this necessitates $i_b \geqq 0$ and $i_c \geqq 0$. The output characteristics do not, then, extend below the $i_c = 0$ axis. Further, since we are studying the case where the collector is not conducting ($v_{bc} \leqq 0$, or $v_{cb} \geqq 0$), the output voltage is limited by

$$v_{ce} = v_{cb} + v_{be} \geqq v_{be} \geqq 0 \tag{3.18}$$

Therefore, the region to the left of $v_{ce} = 0$ axis is excluded from present consideration. As we mentioned, transistor operation is highly nonlinear in this region. These forbidden regions are shown shaded in Figure 3.15(e).

Although usually adequate from the output port, the model of Figure 3.15(c) does not represent the input port with sufficient accuracy. The simple ideal emitter diode model does not include the constant voltage v_{beo} across the conductor emitter diode [$\cong$0.4 volt in Figure 3.1(c)]. This voltage v_{beo} is included in the improved model of Figure 3.15(d). Further, we can, from Figure 3.1(a), assume that the collector diode does not conduct appreciable current if v_{bc} is below the conduction voltage v_{bco}, that is, $v_{bc} < v_{bco}$ or $v_{cb} > -v_{bco}$. This restricts the output voltage to

$$\begin{aligned} v_{ce} &= v_{cb} + v_{be} \\ &\geqq -v_{bco} + v_{beo} \\ &\geqq 0 \end{aligned} \tag{3.19}$$

provided that both diodes have the same conduction voltage ($v_{beo} = v_{bco}$). This inequality for the improved model is identical to Equation (3.18) for the simplified model. The regions of validity of the output characteristics are therefore identical, as shown in Figure 3.15. The difference is that the emitter model now has a constant voltage source of value v_{beo}.

This transistor model is now adequate for our purposes; we can begin reconnecting the base and collector circuits. The base current that flows when the base network is attached to the transistor can be determined as follows. From the base network diagram of Figure 3.15(b) we can write

$$-\frac{v_sR_b + v_cR_s}{R_b + R_s} + i_b\frac{R_bR_s}{R_b + R_s} + v_{beo} = 0$$

or

$$i_b = \left[\frac{v_c R_s - v_{beo}(R_b + R_s)}{R_b R_s}\right] + \frac{v_s}{R_s} \qquad (3.20)$$

This base current will flow *regardless* of the currents or voltages elsewhere in the circuit, provided only that the emitter diode is conducting.

From Equation (3.20), we can observe the purpose of adding R_b, the resistor from the base to v_c. When $R_b = \infty$, the base current with no signal ($v_s = 0$) is determined solely by v_{beo}, over which we have no direct control. With R_b added, the dc value of base current can be adjusted to nearly any value we desire by changing R_b.

Focusing next on the collector circuit, we will discuss two ways of determining the output voltage v_0

Analytic solution. The output voltage is calculated as in Section 3.4, with

$$\begin{aligned} v_0 &= v_c - i_c R_c \\ &= v_c - \beta i_b R_c \\ &= v_c - \frac{\beta R_c}{R_b R_s} (v_s R_b + v_c R_s \\ &\qquad - v_{beo}(R_b + R_s)) \qquad (3.21) \\ &= \left[v_c - \frac{\beta R_c}{R_b R_s} (v_c R_s - v_{beo}(R_b + R_s))\right] \\ &\qquad - \frac{\beta R_c}{R_s} v_s \end{aligned}$$

The bracketed term is a constant, determined by the constant (dc) sources, resistor values, and value assumed for v_{beo}. The other term is the signal component of the output and is identical to the signal output found in Section 3.4. Thus, R_b does not affect the signal gain.

Example 3.1

(a) Let

$v_c = 10$ $\quad R_s = 20{,}000\ \Omega$ $\quad R_b = 240{,}000\ \Omega$

$R_c = 5000\ \Omega$ $\quad \alpha = 0.98\ (\beta = 49)$ $\quad v_{beo} = 0.4$

for the network of Figure 3.15(a). The voltage gain is to be found.

Analytic solution. We find from Equation (3.20) that

$$i_b = \left[\frac{200{,}000 - v_{beo}(240{,}000 + 20{,}000)}{(20{,}000)(240{,}000)}\right] + \frac{v_s}{20{,}000}$$

$$= (0.04166 - 0.0543 v_{beo} + 0.05 v_s) \times 10^{-3}$$

For $v_{beo} = 0.4$ volts, we have

$$i_b = (0.02 + 0.05 v_s) \text{ ma}$$

The value of $v_{ce} = v_0$ is found from Equation (3.21) as

$$\begin{aligned} v_0 &= v_c - \beta i_b R_c \\ &= 10 - (49)(0.04166 - 0.0543 v_{beo} + 0.05 v_s) \\ &\quad \times (10^{-3})(5000) \\ &= -0.22 + (13.30 v_{beo} - 12.25 v_s) \\ &= 5.1 - 12.25 v_s \end{aligned}$$

for $v_{beo} = 0.4$. From this equation, we see that the gain is -12.25, so that the circuit does function as an amplifier.

To insure that we are operating in the region of the transistor model validity depicted in Figure 3.15(f), we set $v_s = 0$; we note that under this condition:

1. The base current is positive as required.
2. The base-collector voltage is

$$\begin{aligned} v_{bc} &= -v_{cb} \\ &= -(v_{ce} - v_{be}) = -(v_0 - v_{be}) \\ &= -(5.1 - 0.4) \\ &= -4.7 \end{aligned}$$

which is negative, so that the base-collector junction is nonconducting as required in Equation (3.17).

Graphical solution. The base current of Equa-

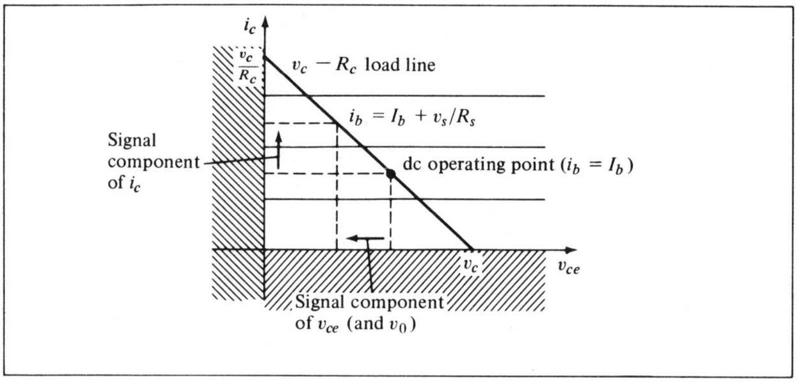

FIGURE 3.16 Graphical representation of amplifier analysis.

tion (3.20) can be written

$$i_b = I_b + \frac{v_s}{R_s}$$

The constant value I_b defines a single horizontal i-v curve on the transistor output characteristics as shown in Figure 3.16. If the $v_c - R_c$ load line is constructed on the same graph, the intersection of the load line with the $i_b = I_b$ characteristic determines the dc value of i_c and v_{ce} as shown. When the signal changes i_b to a value $I_b + v_s/R_s$, then the operating point of the transistor moves to a new output characteristic at $i_b = I_b + v_s/R_s$. This point *remains on the same load line*, because v_c and R_c are the same. As the signal increases positively or negatively, we can represent the resultant change directly on the transistor output characteristic, as shown in Figure 3.16, by excursions along the load line around the dc operating point. The changes in i_c and v_{ce} are then the *signal components* of each. From knowledge of the signal component of $v_0 = v_{ce}$, we obtain the gain of the amplifier at the dc operating point.

The entire process of computing the amplifier gain is summarized by the following relationship.

$$\left.\begin{matrix} v_s(t) \\ v_c \end{matrix}\right\} \xrightarrow[\text{}]{\text{Thevenin equivalent of base circuit}} I_b + \frac{v_s(t)}{R_s} \xrightarrow{\text{Transistor output characteristics}} v_{ce}(t) = v_0(t)$$

This procedure "works" because v_{be} is assumed constant when the emitter diode conducts; if v_{be} were strongly dependent on the emitter current i_e, and therefore on the collector current i_c, the base circuit could not be separated from the collector or output circuit and the method would fail.

(b) The dc operating point and the output signal of the amplifier of Figure 3.15(a) are to be studied using the transistor output characteristic to determine the signal levels that will cause the transistor to operate beyond the normal operating region that we have defined.

Figure 3.17 presents an accurate display of the transistor characteristics and the $v_c - R_c$ load line. We will study its operation at points (a), (b), and (c), the latter two being the limits of normal operation.

1. When $v_s(t) = 0$, the dc value of base current is, from Example 3.1(a), 0.02 ma. The dc value of i_c is βi_b, or $(49)(0.02) = 0.98$ ma, resulting in a v_{ce} of $v_c - R_c i_c = 10 - (5000)(0.98) \times 10^{-3} = 5.1$ volts.

2. The maximum value of collector current occurs at point (b), when the collector diode threatens to conduct. The v_{ce} at this point is 0 volts, so that $i_{c_{max}} = (v_c - v_{ce})/R_c = 10/5000 = 2$ ma. The base current at this collector current is

$$i_{b_{max}} = i_{c_{max}}/\beta = 2/49 = 0.0408 \text{ ma}$$

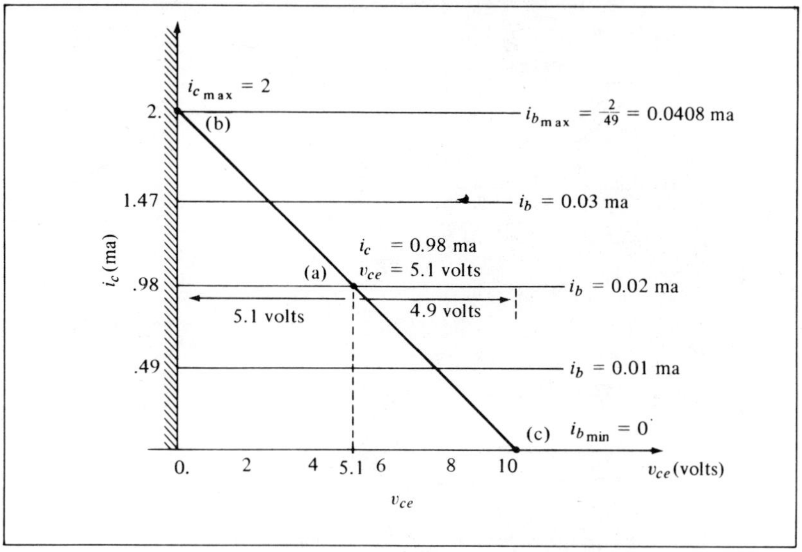

FIGURE 3.17 Graphical analysis of amplifier.

Then

$$i_{b_{max}} = 0.02 + 0.05v_{s_{max}} = 0.0408 \text{ ma}$$

from which

$$v_{s_{max}} = 0.416 \text{ volt}$$

3. The minimum value of collector current is zero, at point (c), resulting in

$$i_{b_{min}} = \frac{i_{c_{min}}}{\beta} = 0$$

Therefore,

$$i_{b_{min}} = 0.02 + 0.05v_{s_{min}} = 0$$

$$v_{s_{min}} = -0.4 \text{ volt}$$

If the signal has equal positive and negative peak values $\pm V_s$ as will be the case in our next study, this peak is limited by

$$V_s \leqq |v_{s_{min}}| = 0.4 \text{ volt} \qquad (3.22)$$

We see from the above that graphical analysis does not give any new quantitative results. It does, however, give considerable insight into qualitative behavior. For example:

1. The dc and signal components of v_{ce} (or v_0) are quite evident.
2. The limitations on i_b, i_c, and v_{ce} that result in either reverse-biasing of the emitter diode or forward-biasing of the collector diode are all displayed.

Also, we might ask *design* questions about the effects on the output signal of changing v_c, R_c, or the dc value (I_b) of the base current. These aspects are illustrated in Figure 3.18 and discussed in Problem 3.13.

Analysis by Digital Computer

We have progressed from showing feasibility of amplification (due to a controlled source) to the analytical and graphical analysis of a simple amplifier. However, the latter methods have shortcomings:

1. When more complicated amplifier networks are to be analyzed.
2. When the validity of simplifying assumptions that make possible partitioned analysis of the network, such as a constant v_{beo}, are to be evaluated.
3. When the effects of nonlinearities, such as the emitter diode characteristic, must be determined.

In this chapter, we continue to examine only simple circuits, saving larger network analysis

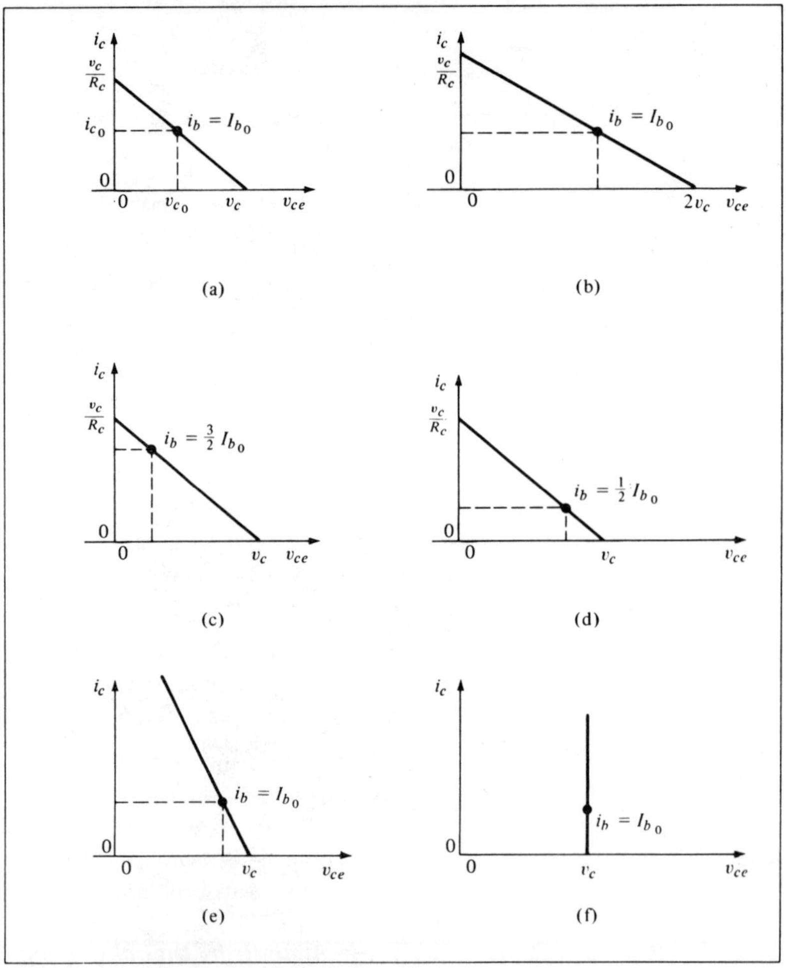

FIGURE 3.18 Effects of varying design parameters. (a) $v_{c0} = v_c/2$, $i_{c0} = v_c/2R_c$; (b) v_c, R_c doubled; (c) dc base current increased; (d) dc base current decreased; (e) R_c halved; (f) $R_c = 0$.

problems for Chapter 5. We shall, however, carry out a precise computer analysis of small amplifier circuits to evaluate the extent to which these circuits, because of the emitter diode nonlinearity, distort the signals they amplify.

An alternative resistor arrangement used to obtain a dc operating point is shown in Figure 3.19. Here, the resistor R_b of the amplifier we studied in Figure 3.15 is connected to the collector rather than to v_c. We are not in a position to comment on the merits of this amplifier circuit versus the one from which it was derived. This is a topic that might be considered in a later electronics course, where circuit *design* rather than simply circuit analysis would be stressed.

For analysis purposes, however, note that this new connection makes it difficult to separate the input (base) network from the output (collector) network, because R_{bc} now connects both networks. This is strong motivation to turn to the digital computer for assistance.

The extension of the graphical and computer analysis procedures used in logic circuit analysis

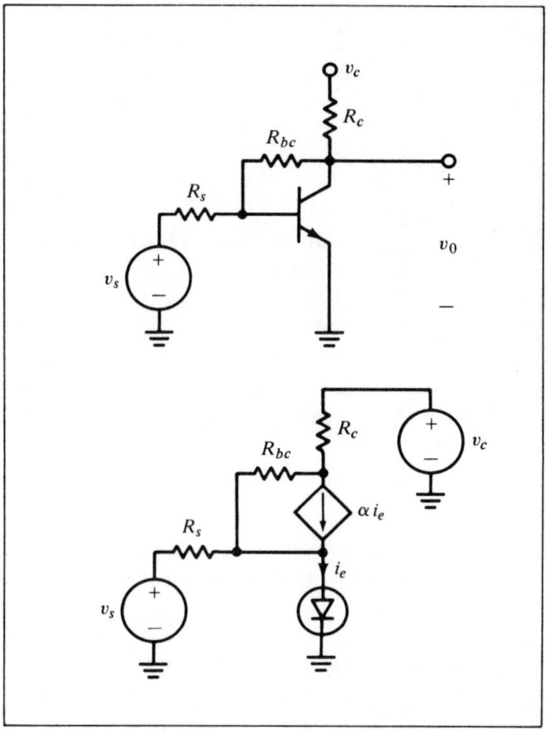

FIGURE 3.19 An alternative amplifier circuit.

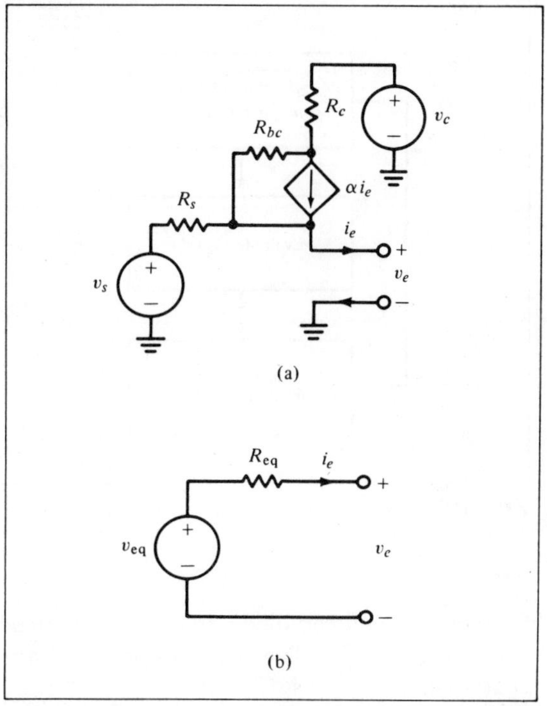

FIGURE 3.21 Model after withdrawing emitter diode. (b) $R_{eq} = [R_s(R_c + R_{bc}) - \alpha R_{bc}R_s]/(R_s + R_c + R_{bc})$; $v_{eq} = [v_s(R_c + R_{bc}) + v_cR_s]/(R_s + R_c + R_{bc})$

to transistor amplifier analysis is straightforward. The process begins by separating the emitter diode from the rest of the amplifier model, leaving a network (N) containing only resistors and sources. Then N may be represented by a Thevenin equivalent at the diode port. The resultant model [Figure 3.20(b)] is of the same form encountered in the logic circuit analysis; it can therefore be solved in a similar manner.

To illustrate this process, the emitter diode is withdrawn from the network to form a one-port, as depicted in Figure 3.21. The Thevenin equivalent is found using the methods of Chapter 2, with the result shown. The connec-

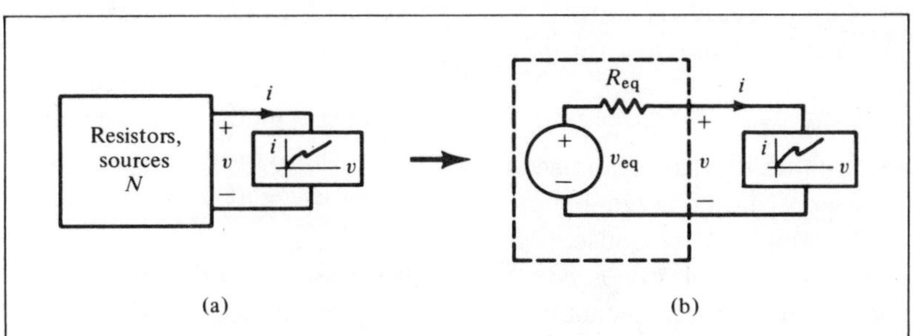

FIGURE 3.20 Separation of single nonlinearity.

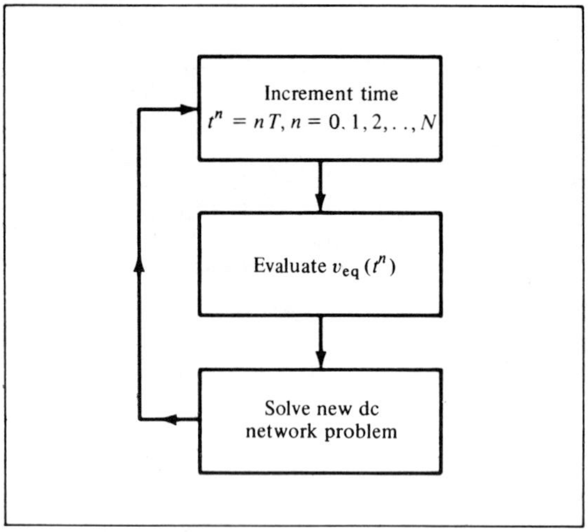

FIGURE 3.22 Flow chart for transient analysis.

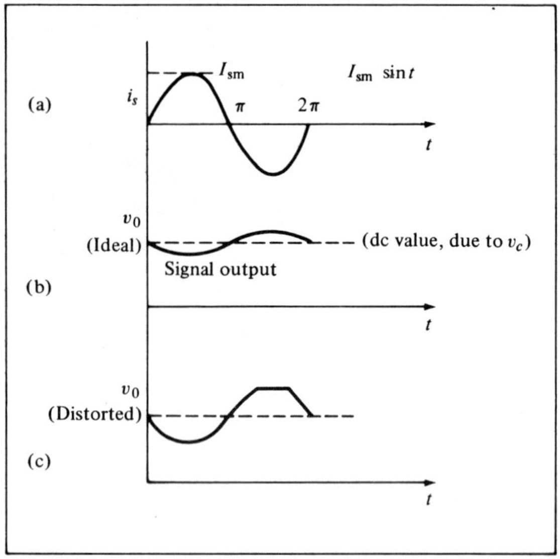

FIGURE 3.23 Input and output amplifier waveforms.

tion of the emitter diode to this equivalent gives us the familiar resistor-diode-voltage source model.

We can make this problem somewhat more interesting by letting v_s be a function of time $v_s(t)$. To solve for $v_0(t)$, we fix time at $t^n = nT$, $n = 0, 1, 2, \ldots, N$. This is known as *discretization* of the time variable, that is, representation of the continuous variable t by certain of its sample values. Then $v_{eq}(t^n)$ defines a fixed Thevenin equivalent voltage source

$$v_{eq}(t^n) = \frac{v_c R_s}{(R_s + R_c + R_{bc})} + \frac{v_s(t^n)(R_c + R_{bc})}{(R_s + R_c + R_{bc})}$$

The nonlinear network at $t = t^n$ can therefore be viewed as posing a dc network analysis problem. For $n = 0, 1, 2, \ldots, N$, a total of $N + 1$ dc network analyses problems are defined. (See Figure 3.22.)

The signal chosen in this case is $v_s(t) = V_{sm} \sin \omega t$, termed a sinusoid. It is often used as a test signal to observe the behavior of an amplifier. Ideally, the amplifier response [v_0 of Figure 3.19] should consist of (1) a constant (dc) component due to the constant source v_c and (2) a sinusoidal component due to v_s. If this latter signal term is not sinusoidal, then the amplifier is said to distort the signal. Examples of $v_0(t)$ are shown in Figure 3.23(b) and (c).

Figure 3.24 contains an *incomplete* solution program for determination of the emitter current $i_e(t^n)$ and the output voltage $v_0(t^n)$ at 20 time intervals during the application of the sinusoid of Figure 3.23(a), that is, every $2\pi/20$ sec.

Using a value $V_{sm} = 0.7$ and the other element values used in Example 3.1(a), the output data of Figure 3.25(b) is obtained. A sketch of this data (Figure 3.26) shows that there is indeed distortion of the signal component of the output voltage, because the peaks of the sinusoidal component are not equally displaced about the dc value of 7.13 volts. The origin of this distortion is principally the emitter diode becoming nearly reverse-biased at $t = 4.7124$ sec as shown in Figure 3.25.

It is possible to develop a number of mathematical measures of this distortion with a sinusoidal signal (see Reference 3.2). One

```
C****  AMPLIFIER EXAMPLE (ANET)
       REAL*4 LAMDA/40./,CSAT/1.E-9/
       READ(5,15)VC,VSM,ALF,RC,RBC,RS
       WRITE(6,8)VC,VSM,ALF,RC,RBC,RS
C****  INITIALIZE ITERATION PARAMETERS
       M=0
       ITMAX=15
       TMAX=6.2832
       T=TMAX/20.
       TIME=0.
       REQ=RS*(RC+RBC)/(RS+RC+RBC)-ALF*RBC*RS/(RS+RC+RBC)
       WRITE(6,9)
C****  GUESS INITIAL DIODE VOLTAGE
       VDM=.6
     7 DVD=1.E60
C****  UPDATE VEQ
       VS=VSM*SIN(TIME)
       VEQ=
C****  ENTER NEWTON RAPHSON ITERATION
       GO TO 1
C****  CALCULATE (M+1)ST DIODE VOLTAGE
     3 VDMP1=
       DVD=VDMP1-VDM
       VDM=VDMP1
       M=M+1
C****  UPDATE DIODE-RELATED TERMS
     1 EXP1=EXP(LAMDA*VDM)
       CURM=CSAT*(EXP1-1.)
       GDM=LAMDA*CSAT*EXP1
C****  CONVERGENCE CRITERION
       IF(M.LE.ITMAX)GO TO 2
       WRITE(6,11)M
       GO TO 12
     2 IF(ABS(DVD).GT.1.E-6)GO TO 3
    12 COLCUR=CURM-(VS-VDM)/RS
       VOUT=VC-COLCUR*RC
       IF(VOUT.LT..1E0)WRITE(6,16)
       WRITE(6,4)TIME,VS,CURM,VOUT
       IF(TIME.GE.TMAX)GO TO 5
       M=0
       TIME=
       GO TO 7
     5 STOP
     9 FORMAT(2X,'TIME         SIGNAL       IE            VOUT')
     4 FORMAT(4(1PE12.4))
     8 FORMAT(2X,'VC =',1PE10.3,2X,'VSM =',1PE10.3,2X,'ALF =',1PE10.3/
      12X,'RC =',1PE10.3,2X,'RBC =',1PE10.3,2X,'RS =',1PE10.3)
    11 FORMAT(2X,'NO CONVERGENCE AFTER ',I2,'ITERATIONS')
    15 FORMAT(6F10.0)
    16 FORMAT(' COLLECTOR DIODE FORWARD BIASED')
       END
```

FIGURE 3.24 Program for transient analysis of amplifier.

simple form is the expression

$$\text{Distortion } (\%) = 100 \frac{|\text{Average of peak signal values} - \text{dc value}|}{|\text{Peak signal value}|}$$

$$= 100 \frac{\left| \dfrac{v_0(\pi/2) + v_0(3\pi/2)}{2} - v_0(0) \right|}{|v_0(\pi/2) - v_0(0)|} \tag{3.23}$$

10.	.7	.98	5000.	240000.	20000.

(a)

```
VC = 1.000E 01  VSM = 7.000E-01  ALF = 9.800E-01
RC = 5.000E 03  RBC = 2.400E 05  RS = 2.000E 04
```

TIME	SIGNAL	IE	VOUT
0.0	0.0	5.7229E-04	7.0557E 00
3.1416E-01	2.1631E-01	8.3011E-04	5.8184E 00
6.2832E-01	4.1145E-01	1.0656E-03	4.6882E 00
9.4248E-01	5.6631E-01	1.2537E-03	3.7855E 00
1.2566E 00	6.6574E-01	1.3748E-03	3.2039E 00
1.5708E 00	7.0000E-01	1.4166E-03	3.0033E 00
1.8850E 00	6.6574E-01	1.3748E-03	3.2039E 00
2.1991E 00	5.6631E-01	1.2537E-03	3.7855E 00
2.5133E 00	4.1145E-01	1.0656E-03	4.6882E 00
2.8274E 00	2.1631E-01	8.3010E-04	5.8184E 00
3.1416E 00	-2.2308E-06	5.7229E-04	7.0557E 00
3.4558E 00	-2.1631E-01	3.2140E-04	8.2597E 00
3.7699E 00	-4.1145E-01	1.1285E-04	9.2601E 00
4.0841E 00	-5.6631E-01	8.1419E-06	9.7614E 00
4.3982E 00	-6.6574E-01	2.5913E-07	9.7975E 00
4.7124E 00	-7.0000E-01	7.2681E-08	9.7978E 00
5.0266E 00	-6.6574E-01	2.5915E-07	9.7975E 00
5.3407E 00	-5.6631E-01	8.1427E-06	9.7614E 00
5.6549E 00	-4.1145E-01	1.1286E-04	9.2601E 00
5.9690E 00	-2.1631E-01	3.2140E-04	8.2597E 00
6.2832E 00	3.1265E-06	5.7229E-04	7.0557E 00
6.5973E 00	2.1631E-01	8.3011E-04	5.8184E 00

(b)

FIGURE 3.25 Input and output data for ANET program.

The distortion would be zero if the diode were a linear element; for then the peaks will be equally displaced about the dc value and their average will be precisely the dc value (see Problem 3.9). For the example at hand, this distortion is calculated as

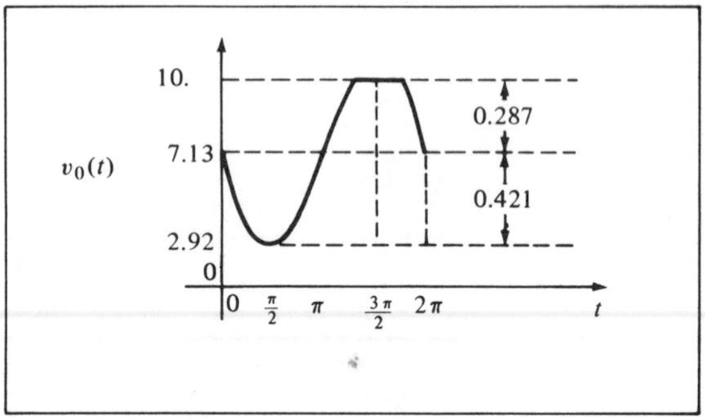

FIGURE 3.26 Output voltage of amplifier.

$$\text{Distortion} = 100 \times \left| \frac{\dfrac{2.92 + 10.}{2} - 7.13}{2.92 - 7.13} \right|$$

$$= 100 \times \frac{0.67}{4.21}$$

$$= 15.9\%$$

PROBLEMS

3.1 (a)–(d) It is occasionally necessary to represent the diode by a model more complicated than that shown in Figure 3.1(c), but still composed of an ideal diode with other simple elements such as resistors and sources. Some examples are shown in Figure P 3.1. The i_d–v_d characteristics of these models are known as piecewise-linear characteristics, because they are composed of the normal straight-line port characteristics of source-resistor networks (studied in Chapter 2), but the slopes change abruptly when the ideal diode conducts. Plot the i_d–v_d characteristics of these models, label-

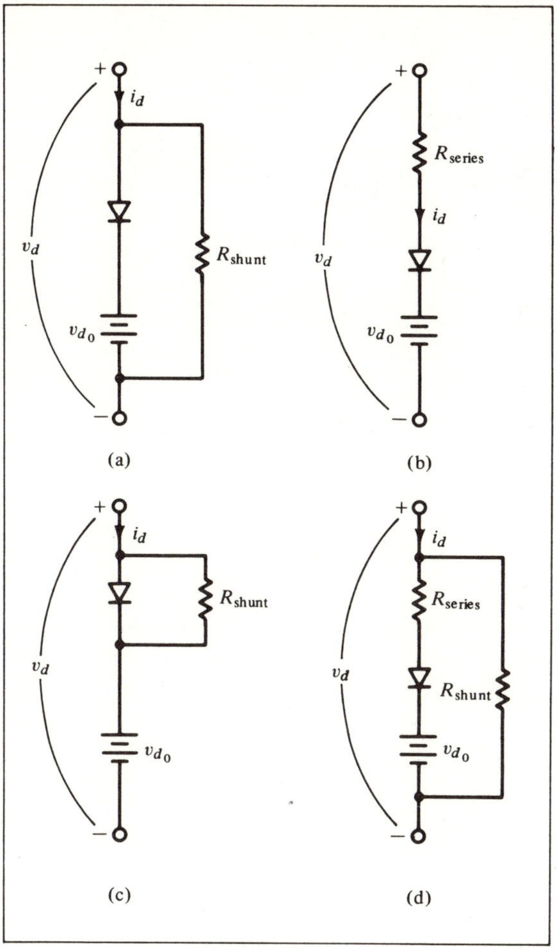

FIGURE P3.1

ing all slopes and points where the slope changes. To resemble the original diode characteristic, assume that R_{series} is small and R_{shunt} is large.

3.2 (a)–(d) The diodes of Figure P 3.2 are described by

$$i_{d_1} = I_{s_1}(e^{\lambda_1 v_{d_1}} - 1)$$
$$i_{d_2} = I_{s_2}(e^{\lambda_2 v_{d_2}} - 1)$$

where the current and voltage references of Figure 3.1 are used. Assume $\lambda_2 > \lambda_1$, $I_{s_1} > I_{s_2}$. Construct the i_d–v_d characteristic analytically. Show graphically how the two diode characteristics add to produce the overall i_d–v_d characteristic.

3.3 When point A is at 0 volts and point B is at +1 volts, what is the voltage at point C in Figure P 3.3? (a) Is this an OR circuit or an AND circuit if +1 volts represents a **1** and 0 volts a **0** (positive logic)? (b) Construct a truth table for this circuit.

3.4 When point A is at 0 volts and point B at −1 volts, what is the voltage at point C in Figure P 3.4? (a) Is this an OR circuit or an AND circuit if −1 volts represents a **1** and 0 volts a **0** (negative logic)? (b) Construct a truth table for this circuit.

3.5 Another input is added to the circuit of Figure P 3.3. If v_A, v_B, and v_C are functions of time as displayed in Figure P 3.5, calculate $v_0(t)$.

3.6 For each of the circuits given in Figure P 3.6, calculate the current through each diode and the voltage across each diode. Use graphical analysis. In every case, the diode characteristic is given by

$$i_d = 10^{-6}(e^{40v_d} - 1)$$

3.7 (a) The network of Figure P 3.7(a) is intended to produce a voltage of 5 volts at terminals ①, ②, and ③ when $v_B = 5$ volts and $v_A = 0$ volts. Explain the state of each diode under these conditions. (Note that ideal diode models are shown.) (b) Explain the condition of each diode if $v_A = 5$ volts and $v_B = 0$ volts. What voltages appear at the terminals ①–③ under these conditions? This circuit is known as an encoder, because it converts a logical **1** (positive logic) at ⓐ or ⓑ to a logical **1** at ①, ②, and/or ③, depending on how the diodes are connected. (c) Add another input (v_C) and connect a set of diodes so as to produce a logical **1** at ① and ② (but not at ③) when ⓒ is a logical **1**. (d) If each voltage source has an internal resistance of 10 Ω, calculate the voltage that will appear at terminals ①, ②, and ③ when $v_A = 0$ volts, $v_B = 5$ volts.

3.8 The neon bulb of Figure P 3.7(b) has the approximate i–v characteristic shown. Show on

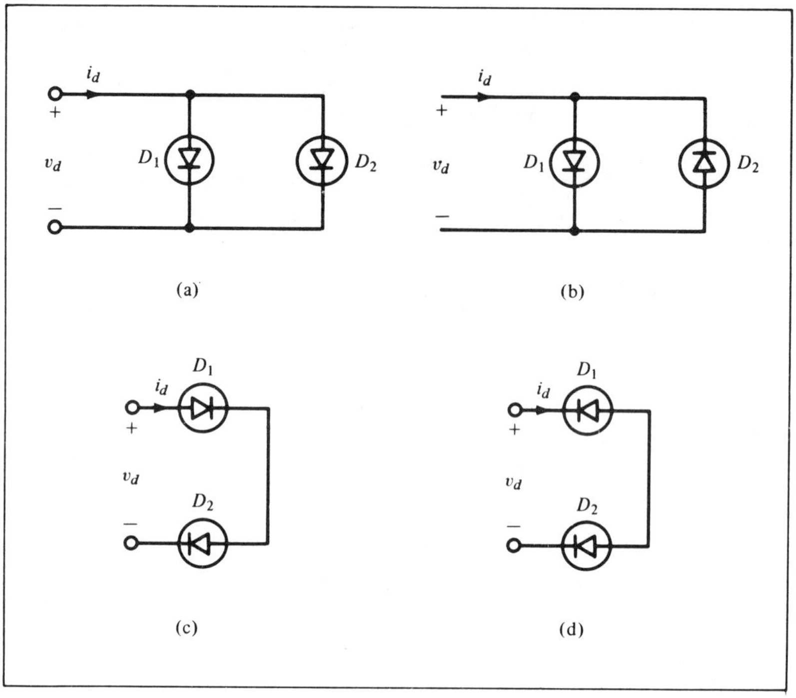

FIGURE P 3.2

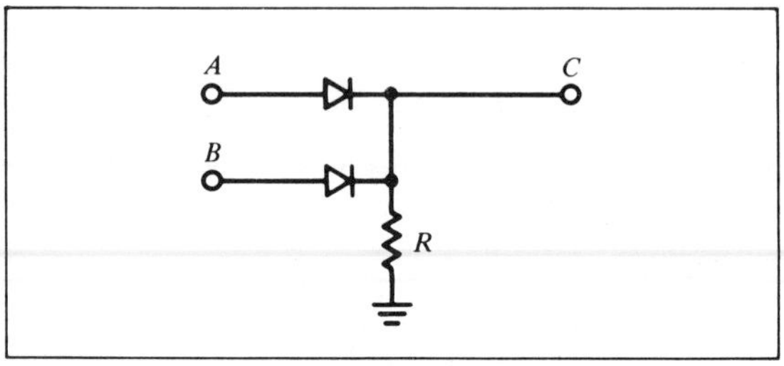

FIGURE P 3.3

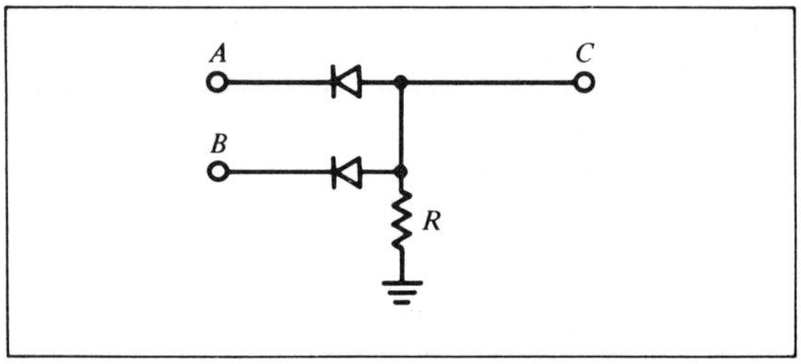

FIGURE P 3.4

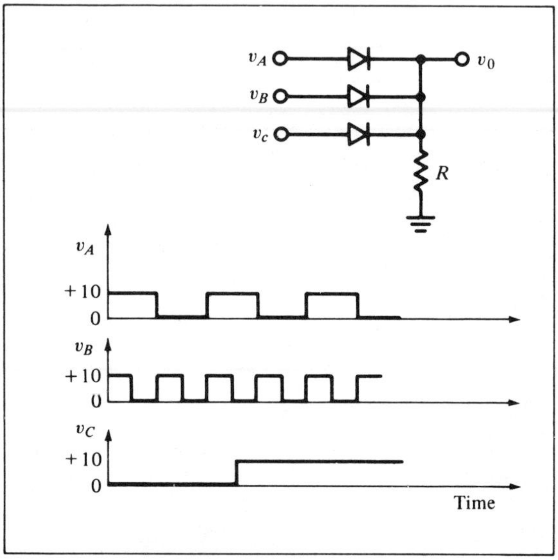

FIGURE P 3.5 Calculate $v_0(t)$.

the characteristic the positions of two load lines that have (a) one and (b) multiple dc solutions. Assume that $v_s > 0$ and $R > 0$.

3.9 The network of Figure P 3.8(a) is connected to the diode configuration of Figure P 3.2(c). Find the resultant v_d and i_d by computer analysis. Check this result by graphical analysis. Let $I_{s_1} = I_{s_2} = 10^{-6}$, $\lambda_1 = \lambda_2 = 40$.

3.10 Repeat Problem 3.9 if the same pair of diodes are connected to the network of Figure P 3.8(b).

3.11 The i–v characteristics of a *tunnel diode* are shown in Figure P 3.9(b)–(c). Denote on the graph all the solutions of the resistor-diode circuit of Figure P 3.9(a). Which solution would a Newton-Raphson iteration find be-

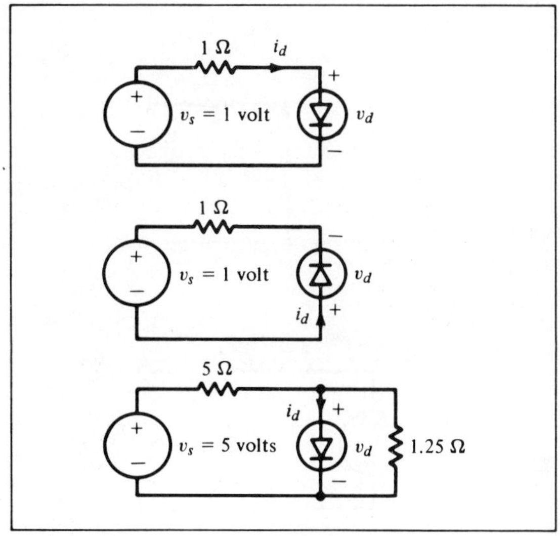

FIGURE P 3.6

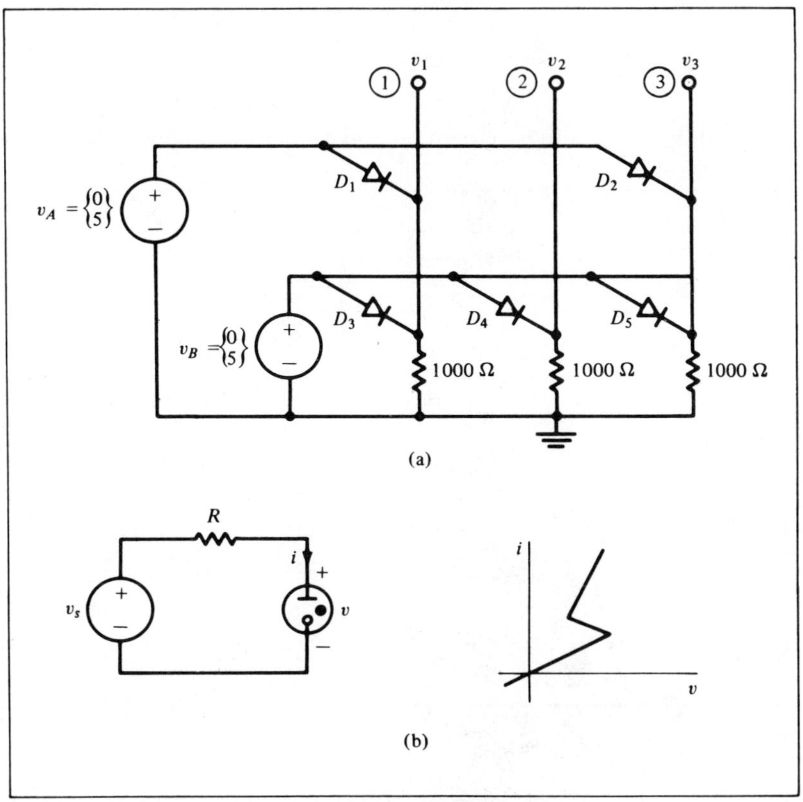

FIGURE P 3.7

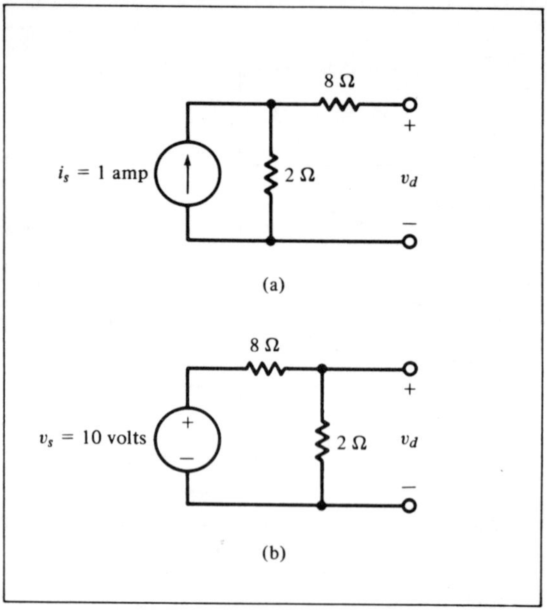

FIGURE P 3.8

ginning at the point (v_d^0, i_d^0) shown. Show the iteration graphically.

3.12 Give the input data for the logic circuit analysis program of Figure 3.10 to simulate the condition when both inputs are a logical **0.** A logical **1.** Assume identical diodes.

3.13 The following questions concern the effect of changing element values in the circuit of Example 3.1(a). In each case, calculate the new i_b, i_c, v_0, and signal gain at the *dc* operating point. Also, depict the solution on the output characteristics and determine the largest positive and negative signal peaks allowed to maintain transistor operation in the region shown in Figure 3.15(f). (a) v_{be_0} changes to 0.41 volt. (b) v_c changes to 11 volts. (c) R_s changes to 15,000 Ω. (d) R_c changes to 4000 Ω. (e) R_b changes to 200,000 Ω. (f) β changes to 60.

3.14 The output characteristics, load line, and dc operating point of a transistor are shown in Figure 3.18(a). The dc base current is at a level such that the dc operating point is at the middle of the load line. Therefore, the same magnitudes of positive and negative signal levels are permitted within the normal region of transistor operation. (a) If the signal com-

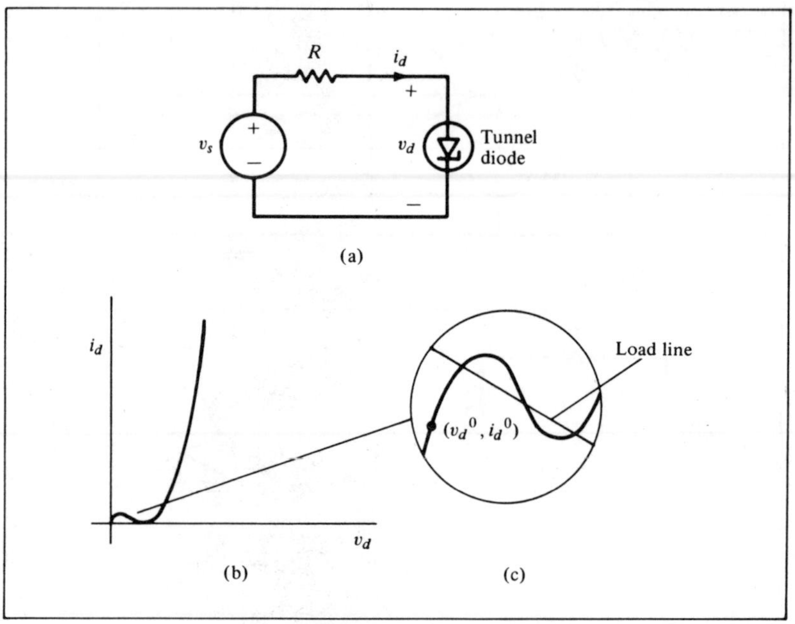

FIGURE P 3.9

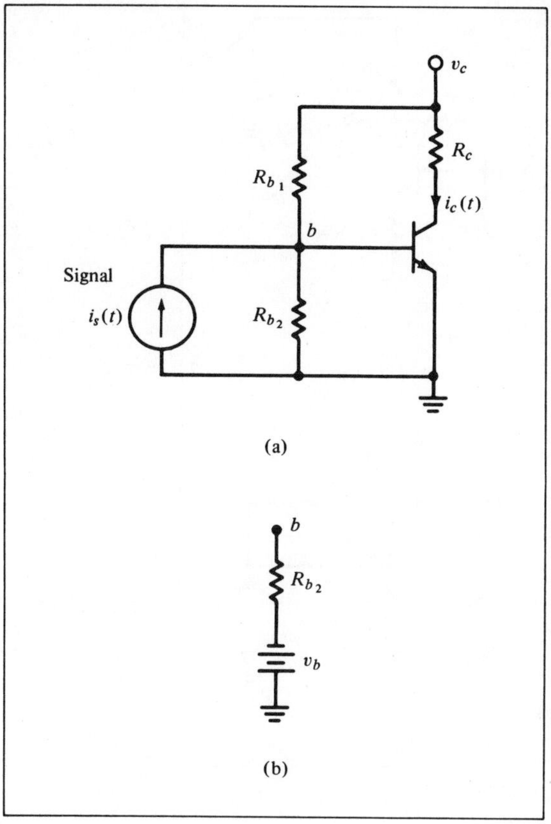

FIGURE P 3.10

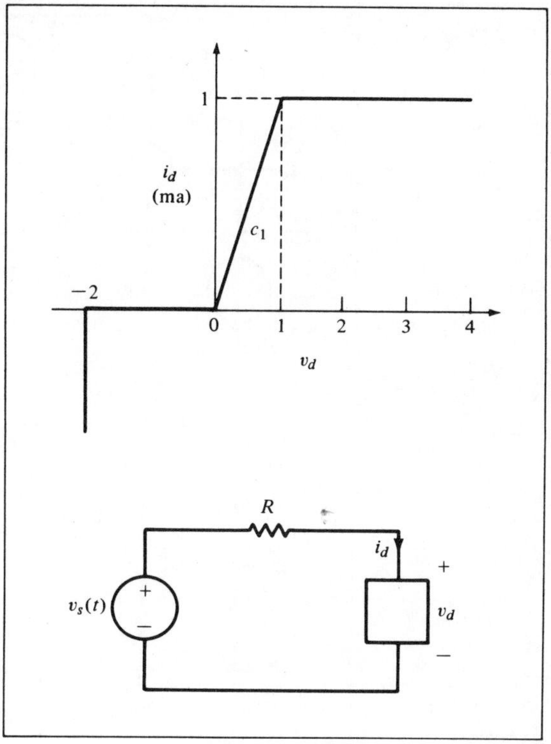

FIGURE P 3.11

ponent of the base current is $v_s(t)/R_s$ as in Example 3.1(a), determine the maximum and minimum allowable signal levels in terms of v_c, R_c, R_s, and β. (b) If v_c and R_c are both doubled, the v_{ce}-axis intercept of the load line is doubled, the slope is halved, but the i_c-axis intercept is left unchanged [Figure 3.18(b)] if R_b is changed so that the transistor still operates on the same output characteristic $i_b = I_{b_0}$ as in Figure 3.18(a). How is the signal gain altered? How large a positive signal $v_s(t)$ is permitted? How large a negative signal is permitted? (c) If the dc base current of Figure 3.18(a) is increased by 50 percent as in Figure 3.18(c) by changing R_b, answer the three questions of part (b) above. (d) If the dc base current of Figure 3.18(a) is halved as in Figure 3.18(d) by changing R_b, answer the three questions of (b) above. (e) If the value of R_c of Figure 3.18(a) is halved, the slope of the load line doubles as in Figure 3.18(e). Answer the three questions of (b) above under this condition. Note that the transistor still operates on the same dc output characteristic $i_b = I_{b_0}$. (f) If R_c is set to zero as in Figure 3.18(f), answer the same three questions as in (b) above.

3.15 For some purposes, the current gain illustrated by $\Delta i_c/\Delta i_s$ in Figure P 3.10(a) is a measure of amplifier performance. (a) Model the base network of this amplifier by a Thevenin equivalent, as in Figure 3.15. (b) Connect tbis equivalent to the transistor and determine $i_b(v_{beo} \neq 0)$. (c) Calculate the signal component of $i_c(t)$ and hence the current gain. (d) Represent the dc and signal components of $i_c(t)$

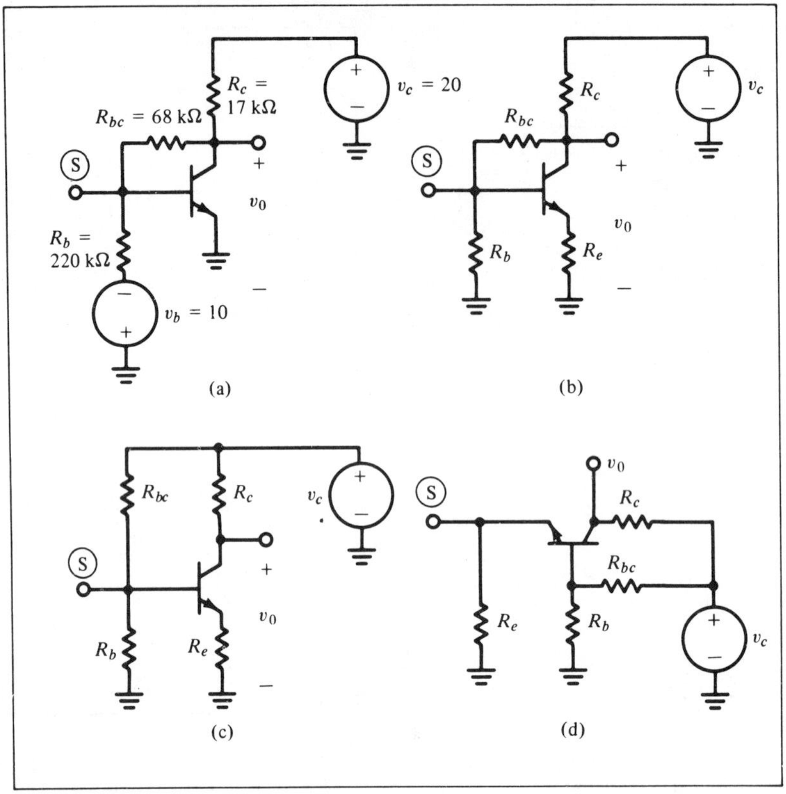

FIGURE P 3.12 Use values $\alpha = 0.98$, $i_s = 10^{-9}$, $\lambda = 40$.

on the transistor output characteristic. (e) If $i_s(t) = I_{\text{sm}} \cos t$, how large can I_{sm} be without exceeding the range of transistor validity shown in Figure 3.15?

3.16 Consider a nonlinear device that has the current-voltage relationship of Figure P 3.11. Plot the current $i_d(t)$ in each of the following cases, using a graphical construction that involves a moving load line. (a) $R = 1000\ \Omega$, $v_s(t) = 1 + \sin t$. (b) $R = 1000\ \Omega$, $v_s(t) = 2 + \sin t$. (c) $R = 1000\ \Omega$, $v_s(t) = 1 + 2 \sin t$. (d) $R = 0$, $v_s = \sin t$. (e) $R = 1000\ \Omega$, $v_s = -2 + \sin t$.

3.17 Replace the c_1 part of the characteristic of Figure P 3.11 by the curve

$$i_d = 10^{-3} v_d^2$$

For $R = 1000$, $v_s(t) = 1 + \sin t$: (a) Find $i_d(t)$ for $t = n\pi/10$ for $n = 0, 1, \cdots, 20$ using a computer. (b) Plot $i_d(t)$ using a graphical construction that involves a moving load line.

3.18 The following problems apply to the amplifier of Figure P 3.12(a). Use $\lambda = 40$, $I_s = 10^{-9}$, $\alpha = 0.98$. ANET may be used as a model program. (a) Derive the Thevenin equivalent seen from the emitter diode; in this example, $v_{\text{eq}} = -0.2786 v_b + 0.7213 v_c$, and $R_{\text{eq}} = 13{,}243\ \Omega$. (b) Compute i_e by graphical analysis; find v_0 from i_e. (c) Let a current source

$$i_s(t) = I_{\text{sm}} \sin t$$

be applied at Ⓢ. Construct a Thevenin equivalent to include $i_s(t)$. Compute $i_e(t^n)$ and $v_0(t^n)$ at intervals of $2\pi/20$ sec for $0 \leqq t \leqq 2\pi$ sec. Let $I_{\text{sm}} = 50\ \mu$amps. (d) Compute $i_c(t^n)$ and

$v_{ce}(t^n)$ in (e). Plot some of these values on the transistor output characteristic and show that they lie in a straight line. Extend this line to predict those value of $i_s(t)$ that cause operation at one of the limits of the normal region of transistor operation.

3.19 Repeat Problem 3.18 for amplifier of Figure P 3.12(b). In this example, $v_{eq} = R_b v_c/(R_b + R_{bc} + R_c)$, $R_{eq} = R_e + (R_b R_c + (1-\alpha) R_b R_{bc})/(R_b + R_{bc} + R_c)$. Use $v_c = 10$, $R_c = 5000\ \Omega$, $R_{bc} = 18\ \text{k}\Omega$, $R_b = 12.8\ \text{k}\Omega$, $R_e = 750\ \Omega$.

3.20 Repeat Problem 3.18 for the amplifier of Figure P 3.12(c). Use $v_c = 10$, $R_c = 5\ \text{k}\Omega$, $R_e = 1\ \text{k}\Omega$, $R_{bc} = 30\ \text{k}\Omega$, $R_b = 4\ \text{k}\Omega$.

3.21 Repeat Problem 3.18 for the amplifier of Figure P 3.12(d). Use $v_c = 11$, $R_c = 5\ \text{k}\Omega$, $R_{bc} = 167\ \text{k}\Omega$, $R_b = 20\ \text{k}\Omega$, $R_e = 1\ \text{k}\Omega$.

3.22 In Equation (3.23), show that if the sinusoidal peaks are equally displaced about the dc value $v_0(0)$, then there is no distortion.

3.23 In Problem 3.29, after computing the value of $i_s(t)$ that results in $i_c = 0$, apply an $i_s(t)$ 5 percent larger and determine $v_0(t^n)$ by computer analysis. Find the percent distortion in the resultant waveform.

3.24 An operational amplifier is connected as shown in Figure P 3.13(a). The amplifier has a large negative gain, but the output is necessarily limited to some finite magnitude A_0 as shown. The amplifier is said to be *saturated* when operating at this limit. (a) Remove the operational amplifier and find g where

$$v_i = g(v_0, v_s)$$

The resultant equation is similar to the load-line equation of the text of the form $v_d = h(i_d, v_s)$. (b) Indicate the graphical solution on the $v_0 - v_i$ characteristic, assuming that the amplifier is not saturated. Determine the intercepts of the load line with the two axes, and calculate $v_0 = g_0(v_s)$. (c) Repeat (b) when the operational amplifier is saturated at $v_0 = A_0$. (d) If the operational amplifier v_0–v_i characteristic is described by the general function $v_0 = f(v_i)$, construct an iteration equation for v_i^{m+1} in terms of v_i^m and the circuit parameters.

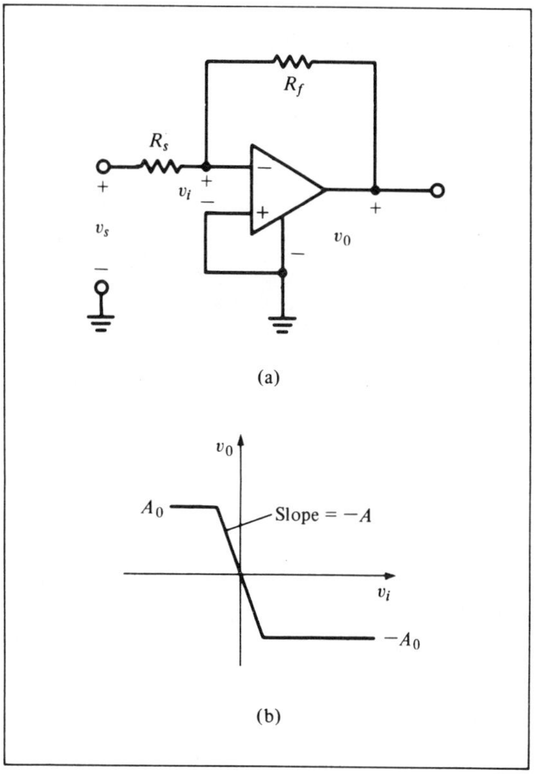

FIGURE P 3.13

REFERENCES AND ADDITIONAL READING

3.1. M. S. GHAUSI, *Electronic Circuits.* New York: Van Nostrand Reinhold Co., 1971, Chapter 3.

3.2. M. S. GHAUSI, *Electronic Circuits.* New York: Van Nostrand Reinhold Co., 1971, Section 5.5.

3.3. D. L. SCHILLING AND C. BELOVE, *Electronic Circuits: Discrete and Integrated.* New York: McGraw-Hill Book Company, Inc., 1968, Chapter 2.

3.4. E. J. ANGELO, JR., *Electronic Circuits.* New York: McGraw-Hill Book Company, Inc., 1964, Chapter 2.

4

General Methods of Linear Network Analysis

4.1 Introduction

The methods of linear network analysis studied in Chapter 2 were largely intuitive, demanding that the reader recognize certain configurations of elements that allowed simplification. These techniques cannot always be used; for example, consider the problem of finding v_0 for the network of Figure 4.1. This network contains no series or parallel resistors and the current source cannot be transformed to an equivalent voltage source. If R_5 were replaced with a diode, the problem would become even more imponderable.

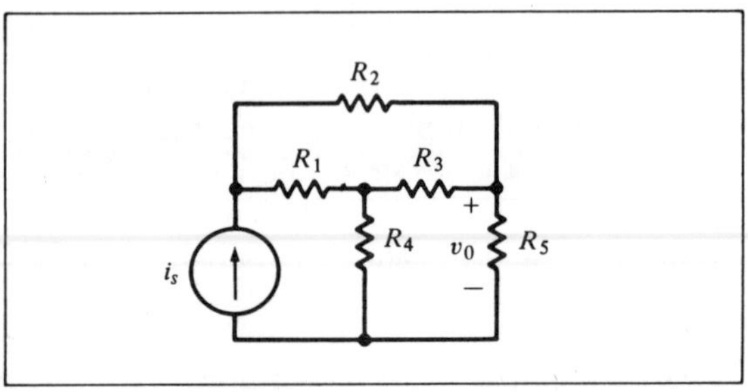

FIGURE 4.1 A small network defying analysis.

To avoid such an impasse, we must develop a more formal approach to the solution. The general analysis methods which we will develop have two distinctive phases:

1. The equation formulation phase, where we decide upon a set of network variables (currents or voltages) as unknowns and, using KVL and KCL, we write a set of independent equations involving these variables.
2. The equation solution phase, where, depending on the network complexity, we obtain a hand or computer-aided numerical solution.

In contrast to the methods of previous chapters, which depended on continuous reference to the network model that was reduced to simpler form, we will use the network only to write a set of descriptive equations. The solution of these equations will then be found using standard mathematical techniques, without reference to the network. We may lose some insight by abandoning the network model and using this more formal mathematical approach; this, however, is the price of generality.

We will examine several general methods of analysis, some involving network voltages as unknowns and some involving network currents. The reader may well ask why, if one method of analysis can handle all network structures, we do not learn just that one. The answer is that, as engineers, we are also very much interested in the "cost" of accomplishing our goals, because in the long run it is the limiting factor for most engineering tasks. One analysis method may be capable of handling all problems, but for a particular network it may be possible to perform a hand analysis using one method while another method would require a computer analysis because more unknowns are involved. Also, more arithmetic operations may be required in a computer or hand solution even though the number of unknowns is the same.

4.2 Elementary Node Analysis

Node Voltage Variables

For most networks a convenient choice of solution variables are the node voltages. These are defined by selecting a node in the network to be a *reference* node (also called the ground node) and then using as variables the voltages of all other nodes relative to the reference node. This is illustrated in Figure 4.2, where node ④ has been selected as the reference and the node voltages are v_1, v_2, and v_3 as indicated.

In defining node voltages, the references shown in Figure 4.2 are most commonly used; the node voltages are all selected as positive relative to the reference node. It is most important to bear in mind that the polarities shown are only *reference* polarities. When one solves for the actual voltages present in a particular network, any of these voltages can turn out to be negative; that is, they have the opposite polarity from the selected reference polarities.

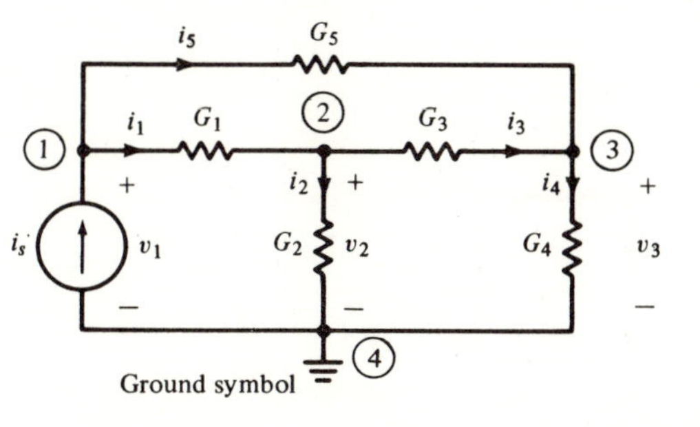

FIGURE 4.2 Network example for node analysis.

Writing the Node Equations

A systematic method of carrying out a node analysis of a network of n nodes is the following:

Step 1. Select a reference node and label every other node voltage positive relative to this reference. This defines $n - 1$ voltages.

Step 2. Write KCL at the $n - 1$ nonreference nodes; some of the currents will be specified sources.[1]

Step 3. Using Ohm's law write all of the unknown currents in terms of the node voltages.

Step 4. Solve the resulting $n - 1$ equations for the $n - 1$ voltage variables.

Step 5. Find every other voltage in the network as a difference of two of the node voltages and then find every current by Ohm's law.

These steps can be illustrated with the network shown in Figure 4.2. The node at the bottom of this network will be chosen as the reference. With node voltages as indicated, the branch current directions are arbitrarily assigned as shown.

Now KCL at nodes ①, ②, and ③ is

$$\begin{aligned} i_1 \qquad\qquad + i_5 &= i_s \\ -i_1 + i_2 + i_3 \qquad\qquad &= 0 \\ -i_3 + i_4 - i_5 &= 0 \end{aligned} \qquad (4.1)$$

Next, Ohm's law for each conductance is

$$\begin{aligned} G_1(v_1 - v_2) &= i_1 \\ G_2 v_2 &= i_2 \\ G_3(v_2 - v_3) &= i_3 \\ G_4 v_3 &= i_4 \\ G_5(v_1 - v_3) &= i_5 \end{aligned} \qquad (4.2)$$

[1] For the moment we will assume that only current sources are present; we will remove this restriction shortly.

Substituting these relations into Equations (4.1) and collecting terms gives the node equations

$$(G_1 + G_5)v_1 - G_1v_2 - G_5v_3 = i_s \quad (4.3)$$

$$-G_1v_1 + (G_1 + G_2 + G_3)v_2 - G_3v_3 = 0 \quad (4.4)$$

$$-G_5v_1 - G_3v_2 + (G_3 + G_4 + G_5)v_3 = 0 \quad (4.5)$$

Node Equations by Inspection

It is instructive to reconsider writing the node equations (4.3) to (4.5) more directly. We omit the writing of branch currents separately as in Equation (4.1) and simply write these currents directly in terms of the node voltages. Considering the currents at any node in Figure 4.2, two possibilities exist: Either the current flows between the node and the reference (or ground) node, or it flows between two non-reference nodes. If the current flows in a conductance, it can be written as the product of the conductance value and the voltage across that conductance. Otherwise, it is a source current that can be written immediately on the right-hand side of the equations. For Figure 4.2, the currents leaving node ① are

The current in $G_5 \rightarrow G_5(v_1 - v_3)$

The current in $G_1 \rightarrow G_1(v_1 - v_2)$

The current source $i_s \rightarrow -i_s$

$$(G_1 + G_5)v_1 - G_1v_2 - G_5v_3 - i_s$$

which by KCL add to zero. This is Equation (4.3). Similarly, currents leaving node ② are

Current in $G_1 \rightarrow G_1(v_2 - v_1)$

Current in $G_2 \rightarrow G_2(v_2 - 0)$

Current in $G_3 \rightarrow G_3(v_2 - v_3)$

$$-G_1v_1 + (G_1 + G_2 + G_3)v_2 - G_3v_3$$

which by KCL add to zero. This is Equation (4.4).

The pattern of this process is now easily identified. The currents *leaving* the jth node can be written as a sum of:

1. v_j times the sum of all conductances connected directly to node j minus
2. Every other node voltage times the value of the conductance between node j and the other node.

These currents should equal the sum of all source currents *entering* the node.

Example 4.1

As a numerical example consider the network of Figure 4.3. There are five nodes. One of these is taken as the ground node; the remaining are designated ①, ②, ③, and ④. The sum of the conductances connected directly to ① is $0.1 + 1.0 + 0.1 = 1.2$℧; further, the source

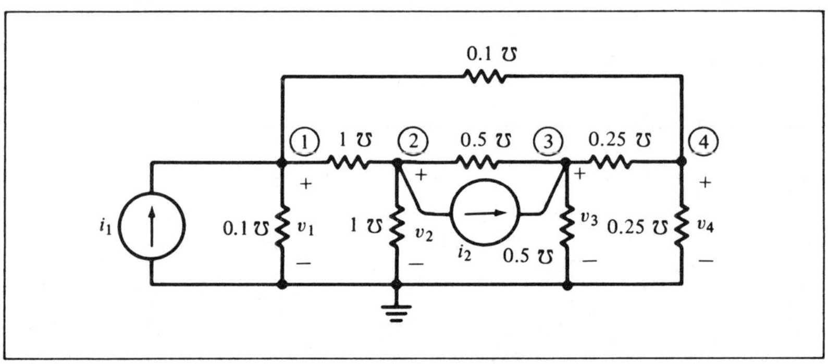

FIGURE 4.3 Large example for node analysis.

current enters this node, so the first node equation is

$$1.2v_1 - 1.0v_2 - 0.1v_4 = i_1 \qquad (4.6)$$

Notice that a v_3 term does not appear in this equation because there is no conductance between nodes 1 and 3. Similarly, at nodes ②, ③, and ④ we obtain

$$-1.0v_1 + 2.5v_2 - 0.5v_3 = -i_2 \qquad (4.7)$$

$$-0.5v_2 + 1.25v_3 - 0.25v_4 = i_2 \qquad (4.8)$$

$$-0.1v_1 \quad -0.25v_3 + 0.6v_4 = 0 \qquad (4.9)$$

4.3 Methods of Solution

Solution with Cramer's Rule

Equations (4.3) to (4.5) and (4.6) to (4.9) are simultaneous linear equations in the unknowns v_i. One method of solution of such equations is with Cramer's rule (see Appendix A). For example, the unknown v_3 can be written

$$v_3 = \frac{\begin{vmatrix} G_1 + G_5 & -G_1 & i_s \\ -G_1 & G_1 + G_2 + G_3 & 0 \\ -G_5 & -G_3 & 0 \end{vmatrix}}{\begin{vmatrix} G_1 + G_5 & -G_1 & -G_5 \\ -G_1 & G_1 + G_2 + G_3 & -G_3 \\ -G_5 & -G_3 & G_3 + G_4 + G_5 \end{vmatrix}} \qquad (4.10)$$

where the denominator is the determinant of the entire node matrix and the numerator is the determinant of a similar matrix, but with the third column replaced by the right-hand side current vector. These determinants are routinely evaluated as

$$v_3 = \frac{(G_1G_3 + G_5(G_1 + G_2 + G_3))i_s}{(G_1 + G_5)(G_1 + G_2 + G_3)(G_3 + G_4 + G_5) - 2G_1G_3G_5 - G_5^2(G_1 + G_2 + G_3) - G_3^2(G_1 + G_5) - G_1^2(G_3 + G_4 + G_5)}$$

$$= \frac{(G_1G_3 + G_1G_5 + G_2G_5 + G_3G_5)i_s}{(G_1 + G_5)(G_3(G_2 + G_4) + G_2G_4) + G_1G_5(G_2 + G_4)} \qquad (4.11)$$

In general, we can determine *any* node voltage v_r by evaluating the determinant formed when the rth column of the node matrix is replaced by the current vector. The denominator remains the same regardless of r.

Solution from the Matrix Inverse

Cramer's rule has several disadvantages as a solution method. First, it is a relatively inefficient method from the standpoint of computation, requiring more arithmetic operations than the methods we shall discuss in Chapter 5. Secondly, the solution must always be represented in terms of two determinants; seemingly, the solution of a 100-node network would be represented as the ratio of two 99×99 determinants! We would prefer a shorthand representation.

An alternative representation of the Cramer's rule solution is obtained from a matrix representation of the node equation. For example, Equations (4.3) to (4.5) could be written

$$\begin{bmatrix} G_1 + G_5 & -G_1 & -G_5 \\ -G_1 & G_1 + G_2 + G_3 & -G_3 \\ -G_5 & -G_3 & G_3 + G_4 + G_5 \end{bmatrix} \begin{bmatrix} v_1 \\ v_2 \\ v_3 \end{bmatrix} = \begin{bmatrix} i_s \\ 0 \\ 0 \end{bmatrix} \qquad (4.12)$$

or more succinctly as

$$\mathbf{Gv} = \mathbf{i}$$

where boldface denotes a matrix or vector (a

matrix with one column). Then the solution vector $\mathbf{v}$ is given by

$$\mathbf{v} = \mathbf{G}^{-1}\mathbf{i} \tag{4.13}$$

As described in Appendix A, the inverse in this case has the form

$$\begin{bmatrix} v_1 \\ v_2 \\ v_3 \end{bmatrix} = \begin{bmatrix} \frac{\Delta_{11}}{\Delta} & \frac{\Delta_{21}}{\Delta} & \frac{\Delta_{31}}{\Delta} \\ \frac{\Delta_{12}}{\Delta} & \frac{\Delta_{22}}{\Delta} & \frac{\Delta_{32}}{\Delta} \\ \frac{\Delta_{13}}{\Delta} & \frac{\Delta_{23}}{\Delta} & \frac{\Delta_{33}}{\Delta} \end{bmatrix} \begin{bmatrix} i_s \\ 0 \\ 0 \end{bmatrix} = \begin{bmatrix} \frac{\Delta_{11}}{\Delta} i_s \\ \frac{\Delta_{12}}{\Delta} i_s \\ \frac{\Delta_{13}}{\Delta} i_s \end{bmatrix} \tag{4.14}$$

Here, Δ is the determinant evaluated in Equation (4.11) and Δ_{ij} is a cofactor, evaluated as $(-1)^{i+j} \times$ (determinant left after eliminating ith row and jth column). Solving for v_3 from Equation (4.14), we find that

$$v_3 = \frac{\Delta_{13}}{\Delta} i_s \tag{4.15}$$

$$= \frac{(+1)\begin{vmatrix} -G_1 & (G_1 + G_2 + G_3) \\ -G_5 & -G_3 \end{vmatrix} i_s}{\Delta}$$

$$= \frac{(G_1G_3 + G_1G_5 + G_2G_5 + G_3G_5)}{\Delta} i_s \tag{4.16}$$

which agrees with Equation (4.11).

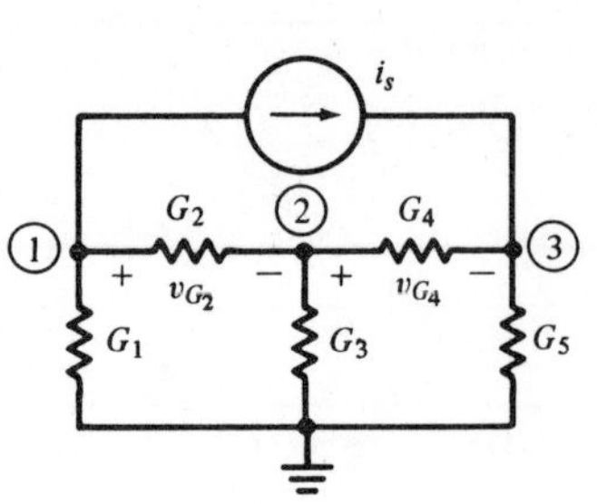

FIGURE 4.4

As a further illustration of this important solution procedure, Example 4.2 is given.

Example 4.2

Solve for node and resistor voltages in the network of Figure 4.4.

$$\begin{bmatrix} G_1 + G_2 & -G_2 & 0 \\ -G_2 & G_2 + G_3 + G_4 & -G_4 \\ 0 & -G_4 & G_4 + G_5 \end{bmatrix} \begin{bmatrix} v_1 \\ v_2 \\ v_3 \end{bmatrix} = \begin{bmatrix} -i_s \\ 0 \\ i_s \end{bmatrix}$$

$$\begin{bmatrix} v_1 \\ v_2 \\ v_3 \end{bmatrix} = \frac{1}{\Delta} \begin{bmatrix} \Delta_{11} & \Delta_{21} & \Delta_{31} \\ \Delta_{12} & \Delta_{22} & \Delta_{32} \\ \Delta_{13} & \Delta_{23} & \Delta_{33} \end{bmatrix} \begin{bmatrix} -i_s \\ 0 \\ i_s \end{bmatrix} \tag{4.17}$$

Node voltages:

$$v_1 = \frac{1}{\Delta}(-\Delta_{11} + \Delta_{31})i_s$$

$$= \frac{1}{\Delta}(-G_2G_5 - G_3G_4 - G_3G_5 - G_4G_5)i_s$$

$$v_2 = \frac{1}{\Delta}(-\Delta_{12} + \Delta_{32})i_s$$

$$= \frac{1}{\Delta}(-G_2G_5 + G_1G_4)i_s$$

$$v_3 = \frac{1}{\Delta}(-\Delta_{13} + \Delta_{33})i_s$$

$$= \frac{1}{\Delta}(G_1G_2 + G_1G_3 + G_1G_4 + G_2G_3)i_s$$

Resistor voltages (sample calculation):

$$
\begin{aligned}
v_{G_2} &= v_1 - v_2 \\
&= \frac{1}{\Delta}(-\Delta_{11} + \Delta_{31} + \Delta_{12} - \Delta_{32})i_s \\
&= \frac{1}{\Delta}(-G_3G_4 - G_3G_5 - G_4G_5 - G_1G_4)i_s \\
v_{G_4} &= v_2 - v_3 \\
&= \frac{1}{\Delta}(-\Delta_{12} + \Delta_{32} + \Delta_{13} - \Delta_{33})i_s \\
&= \frac{1}{\Delta}(-G_2G_5 - G_1G_2 - G_1G_3 - G_2G_3)i_s
\end{aligned}
$$

4.4 Node Analysis Extended

The procedure of Section 4.2 for writing node equations is easily remembered and rarely causes difficulties. The origin of most difficulties in applying node analysis is the handling of voltage sources and the four types of controlled sources. We will consider several of these cases in the following sections. In particular, our discussion will be restricted to networks that contain resistors, independent current sources, independent voltage sources, and voltage-controlled current sources (VCCS's); such networks can be described by node equations with slight adjustments of our present procedures.

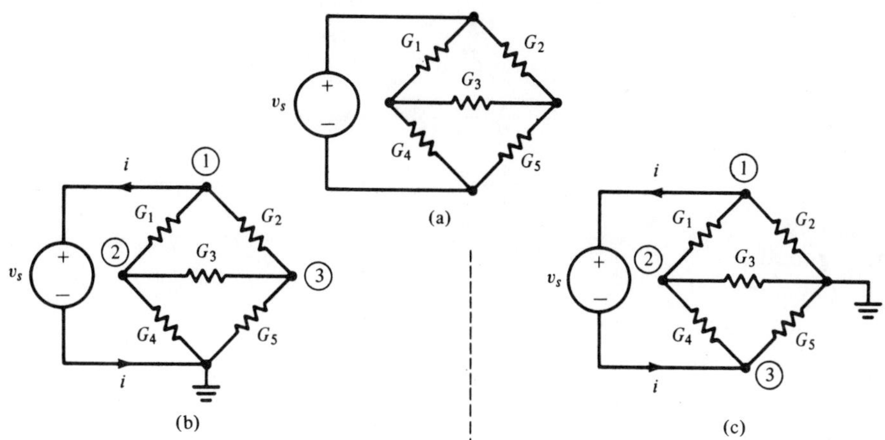

(1) Node analysis

$$
\begin{bmatrix} G_1 + G_2 & -G_1 & -G_2 \\ -G_1 & G_1 + G_3 + G_4 & -G_3 \\ -G_2 & -G_3 & G_2 + G_3 + G_5 \end{bmatrix}
\begin{bmatrix} v_1 \\ v_2 \\ v_3 \end{bmatrix} = \begin{bmatrix} -i \\ 0 \\ 0 \end{bmatrix}
$$

(2) $v_1 = v_s$; first equation can be eliminated.

$$
\begin{bmatrix} G_1 + G_3 + G_4 & -G_3 \\ -G_3 & G_2 + G_3 + G_5 \end{bmatrix}
\begin{bmatrix} v_2 \\ v_3 \end{bmatrix} = \begin{bmatrix} v_sG_1 \\ v_sG_2 \end{bmatrix}
$$

(d)

(1) Node analysis

$$
\begin{bmatrix} G_1 + G_2 & -G_1 & 0 \\ -G_1 & G_1 + G_3 + G_4 & -G_4 \\ 0 & -G_4 & G_4 + G_5 \end{bmatrix}
\begin{bmatrix} v_1 \\ v_2 \\ v_3 \end{bmatrix} = \begin{bmatrix} -i \\ 0 \\ i \end{bmatrix}
$$

(2) i is unknown; move to left-hand side; $v_1 - v_3 = v_s$; form a new equation.

$$
\begin{bmatrix} G_1 + G_2 & -G_1 & 0 & 1 \\ -G_1 & G_1 + G_3 + G_4 & -G_4 & 0 \\ 0 & -G_4 & G_4 + G_5 & -1 \\ 1 & 0 & -1 & 0 \end{bmatrix}
\begin{bmatrix} v_1 \\ v_2 \\ v_3 \\ i \end{bmatrix} = \begin{bmatrix} 0 \\ 0 \\ 0 \\ v_s \end{bmatrix}
$$

(e)

FIGURE 4.5 Node analysis with a voltage source.

Independent Voltage Sources

If a voltage source is in series with a resistor, the configuration can be converted to a Norton equivalent and previous methods for handling current sources used. When the voltage source is not in series with a resistor, as in Figure 4.5 (a), either of two procedures can be followed:

1. If possible, one of the terminals of the voltage source is selected as the reference node. The node voltage at the remaining source terminal [node ① in Figure 4.5 (b)] is then known. The corresponding node equation can be eliminated and the known node voltage transferred to the right-hand side of the equations, as shown in Figure 4.5 (d). This *reduces* the dimension of the node matrix **G** by one. An alternative view of this process is illustrated in Problem 4.21.

2. The first procedure will fail if, for example, two voltage sources are not connected to a common node, for then both sources cannot be simultaneously grounded. The unknown current through the ungrounded source [i in Figure 4.5 (c)] can then formally be recognized as an unknown and brought to the left-hand side of the equations. In addition, an auxiliary equation may be written which relates the source voltage to the unknown node voltages. We will call the resulting equations *node equations,* even though they involve a current as unknown. This *increases* the dimension of the node matrix **G** by one, as shown in Figure 4.5(e).

The expanded size of the node matrix resulting from Step 2 does not necessarily result in more calculation. For example, the pattern of ± 1's in Figure 4.5 (e) can be exploited in calculation of the node determinant.

Illustration

Adding coumn 3 to column 1 and row 3 to row 1 in Figure 4.5(e) gives

$$\Delta = \begin{vmatrix} G_1+G_2+G_4+G_5 & -G_1-G_4 & G_4+G_5 & 0 \\ -G_1-G_4 & G_1+G_3+G_4 & -G_4 & 0 \\ G_4+G_5 & -G_4 & G_4+G_5 & -1 \\ 0 & 0 & -1 & 0 \end{vmatrix}$$

$$= (-1)(-1)\begin{vmatrix} G_1+G_2+G_4+G_5 & -G_1-G_4 & 0 \\ -G_1-G_4 & G_1+G_3+G_4 & 0 \\ G_4+G_5 & -G_4 & -1 \end{vmatrix}$$

$$= (-1)(-1)(-1)\begin{vmatrix} G_1+G_2+G_4+G_5 & -G_1-G_4 \\ -G_1-G_4 & G_1+G_3+G_4 \end{vmatrix}$$

$$= (-1)\{[(G_1+G_4)+(G_2+G_5)][(G_1+G_4)+G_3]-(G_1+G_4)^2\}$$

$$= -[(G_1+G_4)(G_2+G_3+G_5)+G_3(G_2+G_5)]$$

Analysis with a VCCS

If a controlled source other than a voltage-controlled current source (VCCS) appears in a network, it is often possible to manipulate the controlled or controlling variables using elementary network analysis to obtain a VCCS. Two examples are shown in Figure 4.6. Deferring to Chapter 5 those cases when such a direct conversion is impossible, we will now examine how the presence of a VCCS modifies the node equation formulation.

Our previous encounter with controlled sources occurred in Chapter 2 in the reduction of a one-port to a Thevenin or Norton equivalent. In the initial part of this reduction, we treated the controlled source as an independent source. That is, this source could be involved in source transformations to reduce network complexity, *so long as the controlling variable* was not lost in the reduction process.

A similar philosophy can be adopted here. The node equations of the network of Figure 4.7, for example, are initially written as if the VCCS i_1 is an independent source. Therefore, i_1 will appear on the right-hand side of the node equations as

$$\begin{bmatrix} 1/R_1 + 1/R_2 & -1/R_2 & 0 \\ -1/R_2 & 1/R_2 + 1/R_3 & 0 \\ 0 & 0 & 1/R_4 \end{bmatrix} \begin{bmatrix} v_1 \\ v_2 \\ v_3 \end{bmatrix} = \begin{bmatrix} i_s \\ i_1 \\ -i_1 \end{bmatrix} \tag{4.18}$$

$$= \begin{bmatrix} i_s \\ g_m(v_1 - v_2) \\ -g_m(v_1 - v_2) \end{bmatrix} \tag{4.19}$$

Obviously, Equation (4.18) cannot be solved because i_1 is dependent on node voltages v_1 and v_2 as shown in Equation (4.19). Therefore, the equations are put in final form by transferring the expressions involving v_1 and v_2 to the left-hand side as

$$\begin{bmatrix} 1/R_1 + 1/R_2 & -1/R_2 & 0 \\ -1/R_2 - g_m & 1/R_2 + 1/R_3 + g_m & 0 \\ g_m & -g_m & 1/R_4 \end{bmatrix} \begin{bmatrix} v_1 \\ v_2 \\ v_3 \end{bmatrix} = \begin{bmatrix} i_s \\ 0 \\ 0 \end{bmatrix}$$

The solution using determinants now proceeds as before, namely,

$$v_2 = \frac{\Delta_{12}}{\Delta} i_s$$

$$= \frac{\left(\dfrac{1}{R_2} + g_m\right) i_s}{\dfrac{1}{R_1}\left(\dfrac{1}{R_2} + \dfrac{1}{R_3} + g_m\right) + \dfrac{1}{R_2}\dfrac{1}{R_3}}$$

We see that the controlled source disturbs the symmetry of the node matrix. Like any resistor, however, the strength of the VCCS (g_m) appears in four locations of the node matrix, unless one of the four nodes involved is a reference node.

4.5 Topological Considerations

Introduction

At this point we depart from our procedure-oriented approach for a brief formal study of the principles upon which many of our general analysis methods are based. Although we cannot devote the effort to rigorously justifying many of these procedures, the discussion will strongly hint at the existence of such proofs, which can be found in various references (4.1

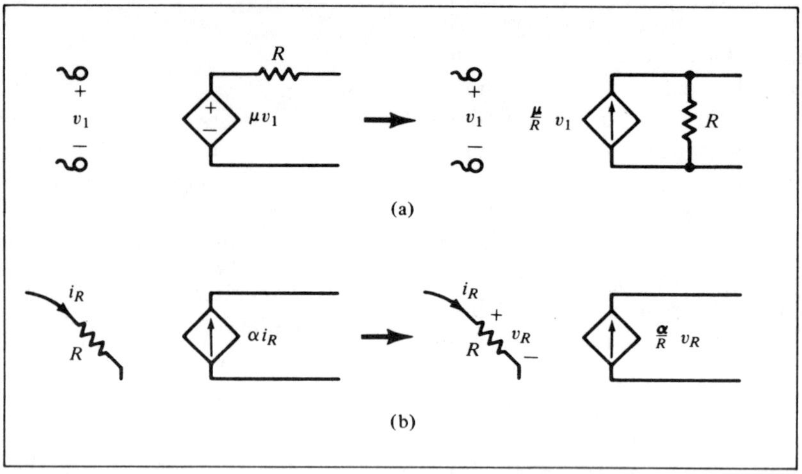

FIGURE 4.6 VCCS representations of other controlled sources. (a) VCVS to VCCS using Norton equivalent; (b) CCCS to VCCS.

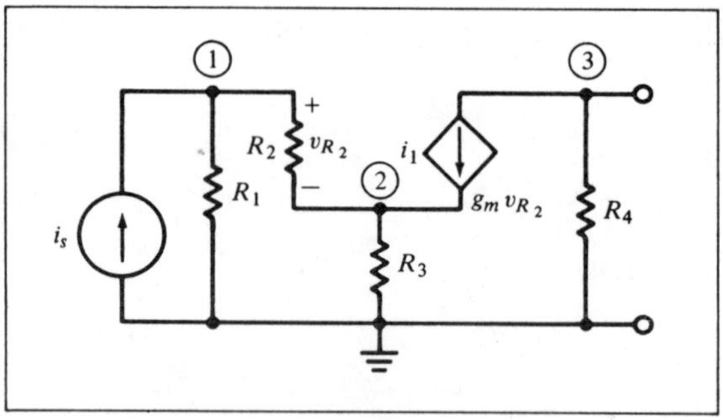

FIGURE 4.7 An analysis example.

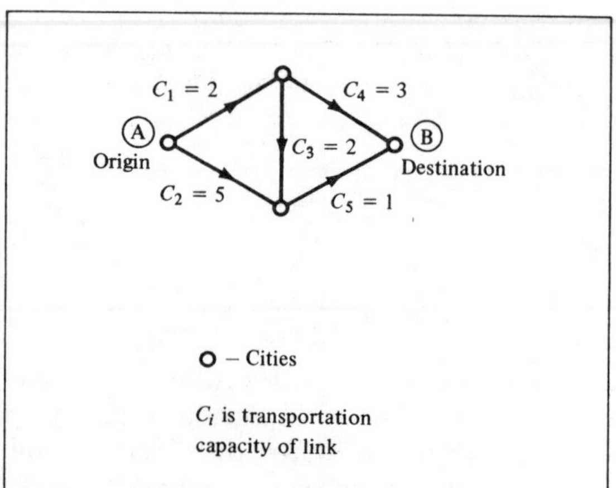

FIGURE 4.8 Network flow example.

and 4.2). At a minimum, we will come to appreciate the generality of the network concept.

Connection laws similar to KCL and KVL are observed throughout natural, social, and economic systems. For example, the problem of transporting goods and services between destinations is termed a *network flow* problem. As illustrated in Figure 4.8, we are given certain capacities for the connections between the "nodes" of the transportation problem. These capacities could represent trucking space limitations or perhaps the ability of a cable television system to transmit information. These problems are not, however, analogous to the electrical network problem. Although KCL must be satisfied at each node (the same goods or information must flow in and out), there are no quantities analogous to voltage. Most social and economic network models have this property.

Natural systems (to which electrical networks belong) typically share both KCL- and KVL-like connection rules. Water and gas distribution systems are an interesting and important example (see Problem 16.15). Other examples are considered in Chapter 16. The connection equations are known generally as

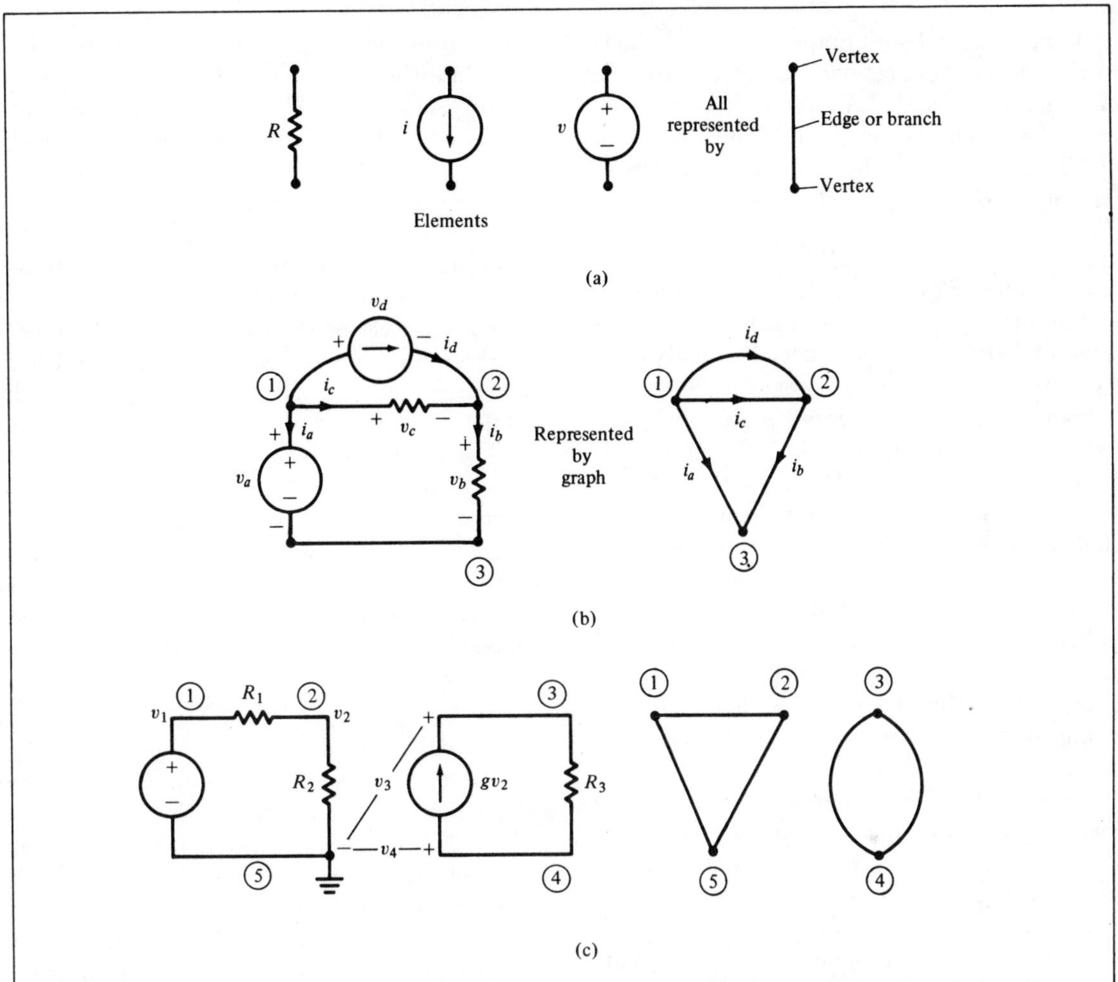

FIGURE 4.9 Topological representation of networks. (a) Elemental representation; (b) network representation; (c) unconnected graph.

topological relationships and describe the *structure* of the problem. Thus, many natural systems satisfy very similar topological laws. The equations for the elements which relate current- and voltage-like variables (such as Ohm's law) are called *constitutive* relationships. These too are similar for natural systems, at least for linear system components.

Topological Concepts

The similarity of topological relationships in different network systems has given rise to a number of structure-related network concepts, described as follows.

Network elements (resistors, independent or controlled sources) are represented topologically by a solid line called an *edge* or a *branch*. The nodes to which the element is connected are termed *vertices*, as shown in Figure 4.9(a). A network of elements is represented by a collection of such edges connected to one another only at their vertices. This figure is known as the network *graph*, and is illustrated in Figure 4.9(b).

Because the network graph displays the network element connections, we can write the network KCL and KVL equations directly from the graph. For example, from the graph of Figure 4.9(b), we observe that, *regardless of the types of elements present*, at node ①

$$i_a + i_c + i_d = 0$$

Concentration on the network graph assists in the identification of certain important network properties that can affect the analysis of the network. One of these properties is termed *connectivity*. A *connected graph* is one in which a path exists along the edges of the graph between any two vertices. The network of Figure 4.9(c) is not connected topologically, even though the two parts of the network are electrically related through the controlled source. Note that we cannot solve for all the node voltages of such a network. In this case, we cannot solve for v_3 and v_4, because the right-hand part of the network can "float" at any voltage difference with respect to the left-hand part. A nodal analysis of the network would therefore fail! (The reader may want to formulate the nodal equations and see how this failure is manifested.) We can, however, solve for the voltage difference $v_3 - v_4 = gR_3v_2$. Also, we can attain a solution by coalescing vertices ④ and ⑤ (that is, grounding node ④). This does not affect v_1, v_2, or $v_3 - v_4$, because no current will flow in this connection. For practical purposes, we can make all graphs connected to avoid an impasse such as the above. At the same time, we should appreciate that we are being something less than elegant in failing to accomodate the more general problem posed by unconnected graphs!

Trees

A key topological concept that we will put to frequent use is the notion of a *tree*. A tree is a collection of edges that connects all the vertices without forming any closed paths.

Two trees of the network graph of Figure 4.10(b) are shown in Figure 4.10(c)–(d). The set of branches that make up a tree are called tree branches, and for most graphs there are many such sets. The sets shown here are $\{d, e, g\}$ and $\{b, e, f\}$. The reader can easily identify other trees such as $\{b, f, g\}$ and $\{b, d, g\}$.

Although there are many trees for a particular graph, it will be noted that *the number of branches (edges) in the trees* is a fixed number (3 in the example). One way of constructing a tree is to select one of the branches in the network graph as a tree branch and then add to it new tree branches one at a time until no additional branches can be added without creating a closed path. Each new tree branch added is chosen to have one vertex in common with the partial tree already formed; the other vertex must then be a new vertex to avoid closing a path. The first branch, however, connects two vertices, so that the number of tree branches n_t is equal to the number of vertices minus one:

$$n_t = n - 1 \qquad (4.20)$$

The addition of any branch connected between two vertices i and j to a set of tree branches must complete a closed path, because a path already existed between i and j through the tree branches. Accordingly, the nontree branches are called *links*. Clearly, a branch cannot be classified as a link until a particular tree has been chosen. (How many links are in a graph of n nodes and b branches?)

Network Equations and Solution Variables

In a practical sense, the only requirement of a general network solution method is that it yield every network current or voltage. This requires at a minimum that we have the same number of network equations as solution variables. For example, in node analysis, there are $n - 1$ of each.

In our more formal study, we shall establish, apart from the *number* of equations and unknowns, that:

1. The *equations* are *independent*; that is, no equation can be obtained from the other equa-

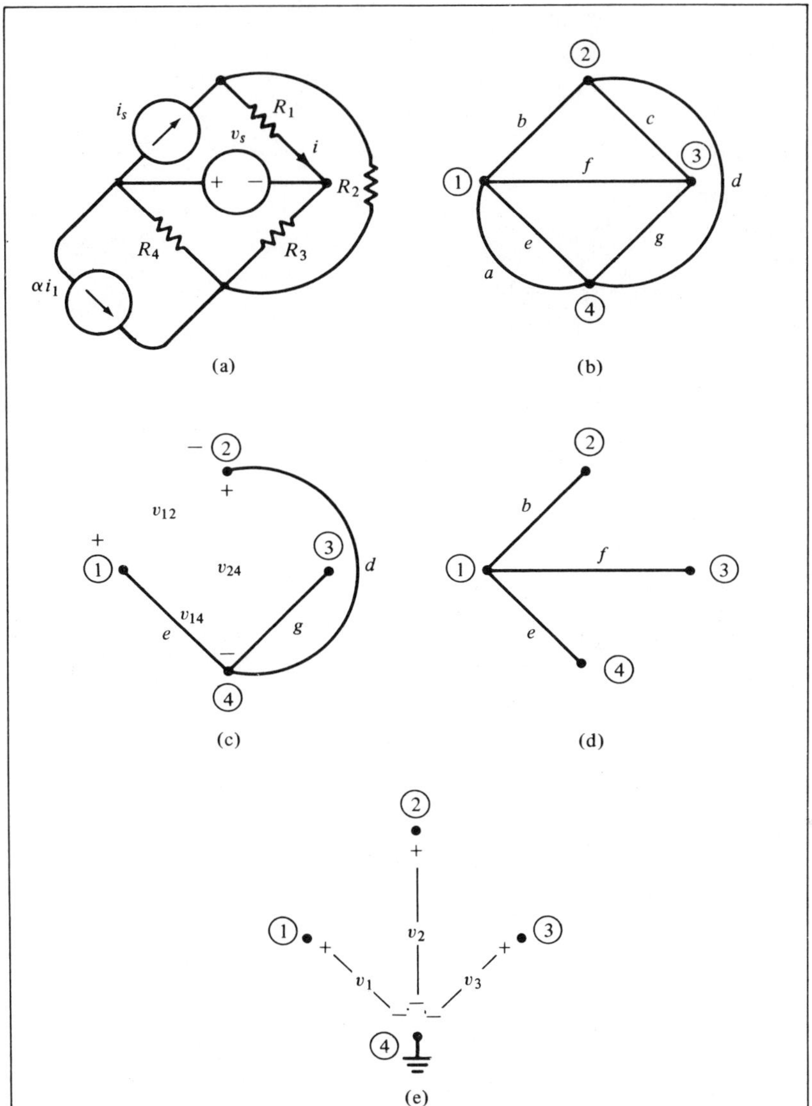

FIGURE 4.10 Network graph and two trees. (a) Electrical network; (b) network graph; (c) tree 1; (d) tree 2; (e) node to datum voltages.

tions; this directly affects our ability to obtain a unique solution of the equations.

2. The solution *variables* are *independent*; as in part 1, this is necessary to obtain a unique solution.

3. The solution variables are a *complete* set, so that once we solve for them, every current and voltage in the network can be trivially obtained using, for example, Ohm's law; clearly, the node voltages are such a set.

We will not attempt to prove that node analysis satisfies the above requirements. Rather, we will propose several different methods of analysis, for which the above properties are more readily established. These new procedures will not be as convenient to use as node analysis, however.

Solution with Voltage Variables

The *voltages across the tree branches* have a special role in establishing the independence of network voltage equations in general. It is easy to establish that these $n-1$ tree branch voltages are a *complete* and *independent* set of variables, as follows.

Because every node is connected to every other node via a path through the tree, the voltage between any two nodes can be written as the algebraic sum of the tree branch voltages along this path. The tree branch voltages are then a *complete* set of variables. For example, tree branch voltages for Figure 4.10(c) might be chosen to be v_{14}, v_{24}, and v_{34}. Then the voltages across the remaining network branches are:

$$\text{Branch } b\text{: } v_{12} = v_{14} - v_{24}$$

$$\text{Branch } f\text{: } v_{13} = v_{14} - v_{34}$$

$$\text{Branch } c\text{: } v_{23} = v_{24} - v_{34}$$

$$\text{Branch } e\text{: } v_{14} = v_{14}$$

The *independence* of tree branch voltages as solution variables follows from the observation that, because a tree contains no closed paths, tree branch voltages cannot be related by KVL equations containing *only* tree branch voltages. That is, each tree branch voltage can have a value independent of the values of the other tree branch voltages. This concept of independence is illustrated in Figure 4.11, where the voltage v_3 of Figure 4.11(b) is determined by the rest of the network branches (not shown) rather than by only voltages v_1 and v_2.

Cut-Set Equations

Having chosen $n-1$ tree-branch voltage variables for an analysis, how do we obtain $n-1$ independent equations in these unknowns? Our approach will depend on the topological notion of a *cut-set*.

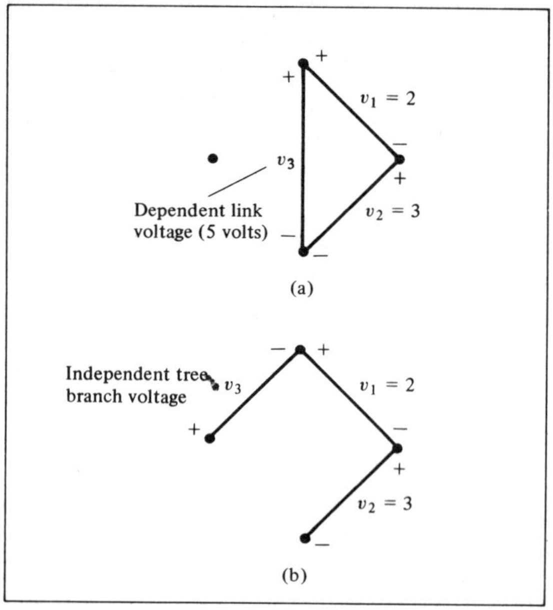

FIGURE 4.11 Dependent and independent variables; v_1 and v_2 are tree branch voltages.

To explain this concept, consider the network of Figure 4.12(a), together with the graph of Figure 4.12(c) representing its structure. A *cut* of the graph is an incision made across the branches that *divides the vertices of the graph into two groups*. For example, cut 4 separates nodes ② and ④ from nodes ①, ③, and ⑤. The branches cut in this process form a *cut-set*, which are the branches representing G_1, G_4, G_5, G_6, and G_2.

The total current flowing across such a cut must sum to zero. For example, the current flowing from bottom to top across cut 4 is

$$i_{G_1} + i_{G_4} + i_{G_6} + i_{G_5} - i_{G_2} = 0$$

Our reason for this claim is similar to that for justifying KCL in Chapter 2: Neither partition involved in a cut can store current. In fact, as Figure 4.12(b) shows, if the cuts are made around a single node, the above current summation becomes the familiar KCL current summation at a node.

If we choose a sequence of cuts properly and write the corresponding current equations, we can easily show that the equations have the independence for which we are looking. The process begins by choosing any tree of the network graph [two are shown in Figure 4.12(b)–(c)]. We define a sequence of cuts such that *each cut-set contains one and only one tree branch.* This results in $n - 1$ cut-sets and $n - 1$ cut-set current equations. In Figure 4.12(b), the tree is carefully chosen so that the current equations are simply the KCL equations at nodes ①, ②, ③, and ⑤. We know from Section 4.2 that these equations are the basis of a nodal analysis, with node ④ selected as the reference node. We could also base the cut-set equations on the tree of Figure 4.12(c). The resulting cut-set equations are

1. $\mathbf{i}_{G_7} + i_{G_4} + i_{G_1} - i_s = 0$
2. $-i_{G_1} + \mathbf{i}_{G_3} + i_{G_2} = 0$
3. $-i_{G_2} + i_{G_5} + \mathbf{i}_{G_8} = 0$
4. $i_{G_1} + i_{G_4} + \mathbf{i}_{G_6} + i_{G_5} - i_{G_2} = 0$

where the tree branch currents are shown boldfaced. Again, there will be $n - 1$ such equations—one for each tree branch. Most important, *none of these equations can be derived from the remaining because each contains a unique tree branch current.* In terms of the tree branch voltages, the current equations become

1. $v_{G_7}G_7 + (v_{G_7} + v_{G_6})G_4 + (v_{G_7} + v_{G_6} - v_{G_3})G_1 = i_s$
2. $(v_{G_7} + v_{G_6} - v_{G_3})G_1 + G_3 v_{G_3} + (v_{G_3} - v_{G_6} - v_{G_8})G_2 = 0$
3. $-(v_{G_3} - v_{G_6} - v_{G_8})G_2 + (v_{G_8} + v_{G_6})G_5 + G_8 v_{G_8} = 0$
4. $(v_{G_7} + v_{G_6} - v_{G_3})G_1 + (v_{G_7} + v_{G_6})G_4 + v_{G_6}G_6 + (v_{G_8} + v_{G_6})G_5 - (v_{G_3} - v_{G_6} + v_{G_8})G_2 = 0$

These equations can be solved for the tree branch voltages v_{G_7}, v_{G_6}, v_{G_3}, and v_{G_8} which, being a complete solution set, can be used to solve for all other voltages and currents.

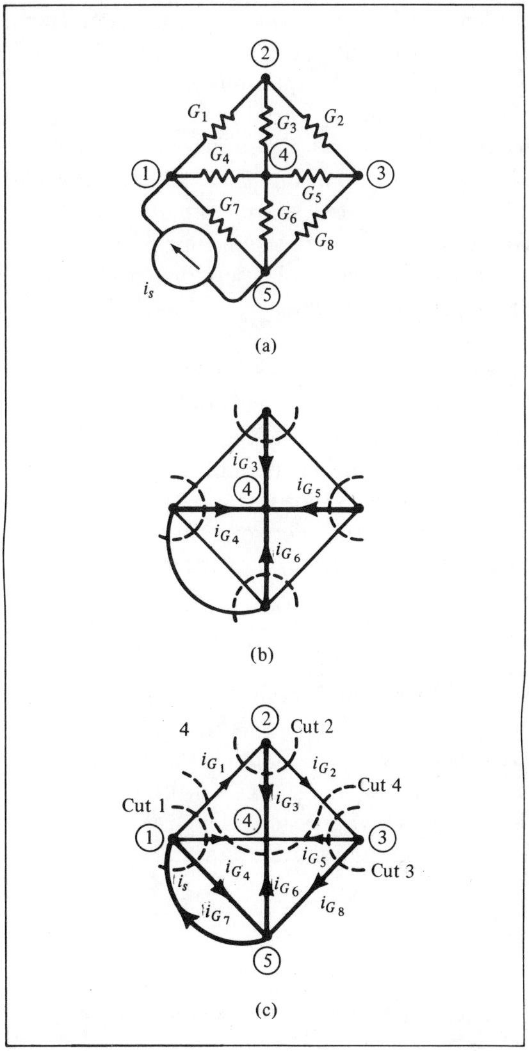

FIGURE 4.12 Illustration of node- and tree-branch cuts.

The above is not intended as a convenient method for hand analysis, because the cut-set equations are not as readily written by inspection as the node equations. However, we can now perhaps appreciate the topological foundation on which network equations are based. For example, we obtain the same number of cut-set equations as node voltage equations: $(n-1)$. Indeed, there are at most $(n-1)$ independent KCL equations and $(n-1)$ independent voltage variables in any network, regardless of how they are chosen. This is established by showing that all voltages can be uniquely related to any set of tree branch voltages. We have not developed or assumed the mathematical sophistication to rigorously justify this statement.

Selection of Current Solution Variables

In the last section we considered one systematic method of analysis which used the tree branch voltages as solution variables. We can also choose currents as solution variables. For example, consider the network in Figure 4.13. Clearly, the currents i_1, i_2, i_3, i_4 are all equal to

$$i = \frac{v_s}{R_1 + R_2 + R_3 + R_4} \tag{4.21}$$

We *could* write and solve three node equations in terms of the node voltages v_2, v_3, and v_4 (because the first node is connected to the reference by the voltage source v_s), but this would be a senseless approach. The reason is seen from the graph: i flows around a single closed path (or *loop*) and therefore a KVL equation around the loop gives the result of Equation (4.21) in one equation. The current i is known as a *loop* current.

We now intend to show that these notions of a loop and a loop current can be generalized to more complicated networks. Consider, for example, the network of Figure 4.14(a) and its graph in 4.14(b). There are two obvious

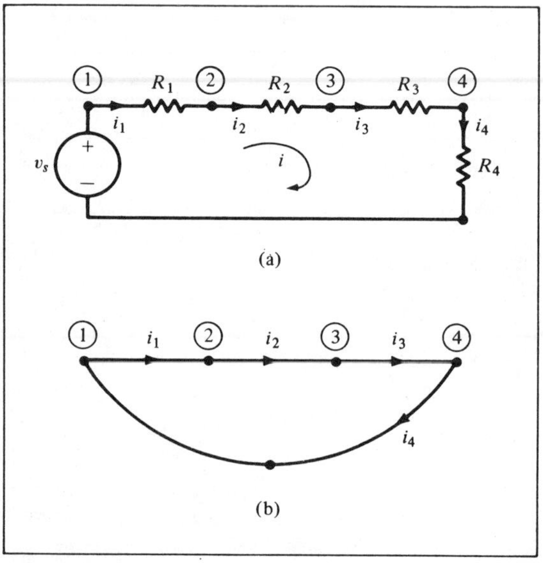

FIGURE 4.13. A sample network and its graph.

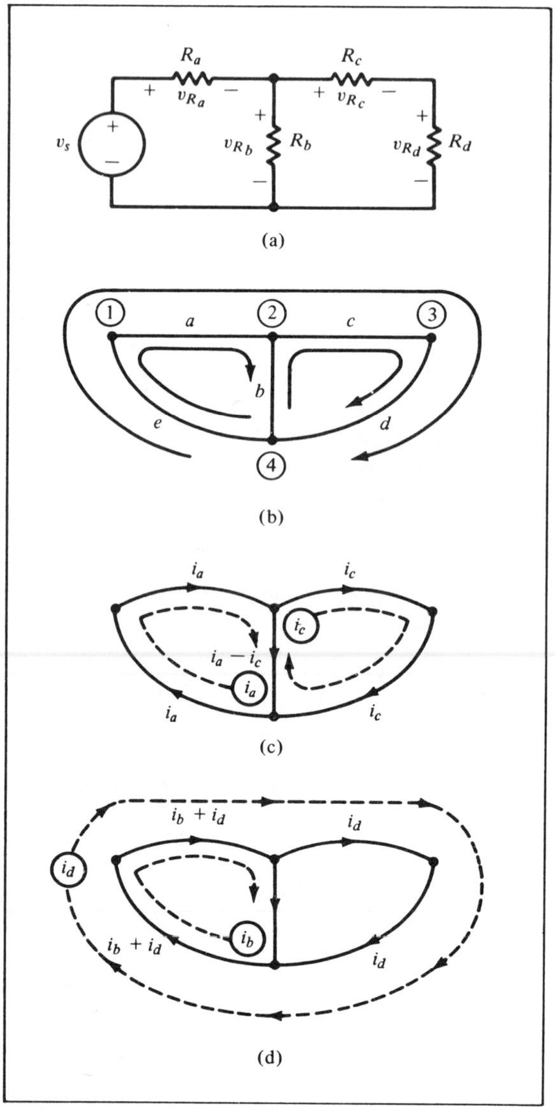

FIGURE 4.14 A two-loop network and graphs illustrating some possible choices of current variables.

loops: the paths through branches e, a, b, and through branches b, c, d. However, there are other loops as well; an example is the exterior loop through branches e, a, c, d. Let us define for each loop a loop current that flows around the loop, just as we did for the single loop case. It is clear from Figure 4.14(c)–(d) that any two of these three-loop currents *form a complete set of solution variables*. That is, in either figure, all branch currents can be determined once the loop currents are found.

Loop Equations

To show that these loop currents can in fact be determined, we propose the following systematic procedure for constructing equations involving loop currents for solution variables.

1. Starting with the graph of the network, remove the minimum number of branches necessary to open circuit all closed paths in the network. The residual graph will be a tree.
2. As each of the removed branches (they are *link* branches) is reconnected, a new loop will be formed. Assign an unknown loop current to each such loop.
3. Finally, write KVL around each loop. This will yield as many equations as there are loop currents.

For example, the graph of the network of Figure 4.15(a) is given in Figure 4.15(b). The selection of the tree $a\,c\,e$ leaves branches $b\,d$ as links. We associate one loop current with each link as shown. We can then write KVL around these loops as

$$v_{R_b} + v_{R_a} = v_s \qquad (4.22)$$

$$v_{R_d} + v_{R_a} + v_{R_c} = v_s \qquad (4.23)$$

These are necessarily *independent equations*, because each contains a link voltage not present in the other equation. Next, we can use Ohm's law to write these voltages in terms of the loop currents, namely,

$$R_b i_b + R_a(i_b + i_d) = v_s \qquad (4.24)$$

$$R_d i_d + R_c i_d + R_a(i_b + i_d) = v_s \qquad (4.25)$$

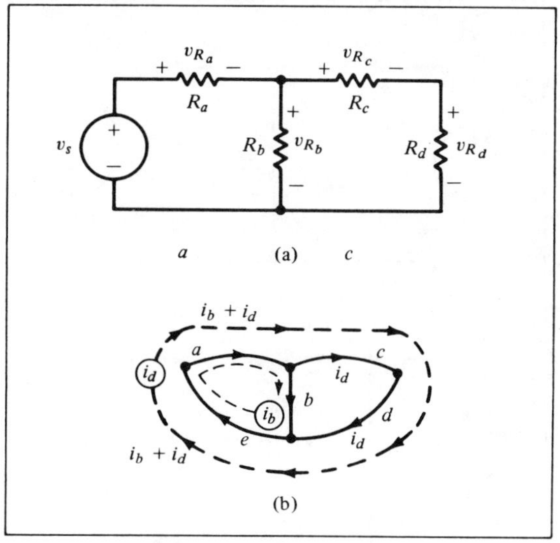

FIGURE 4.15 Loop analysis example.

These two equations in two unknowns can now be solved for the unknowns i_b and i_d.

The number of loop equations is equal to the number of links. If there are l loops, b branches, and n nodes, then

$$l = b - n + 1 \qquad (4.26)$$

loop equations and loop currents are associated with each tree.

We have not proved the independence of the loop currents. This is left as a problem (Problem 4.19).

Dual Networks

Our study of network principles concludes with a consideration of *dual* networks. Two networks are duals if the current equations describing one are identical to the voltage equations describing the other. For example, the network of Figure 4.16(a) is described by the loop equation

$$3i_1 - 2i_2 = 4$$

$$-2i_1 + 5i_2 = 2$$

whereas the network of Figure 4.16(b) has the nodal equation

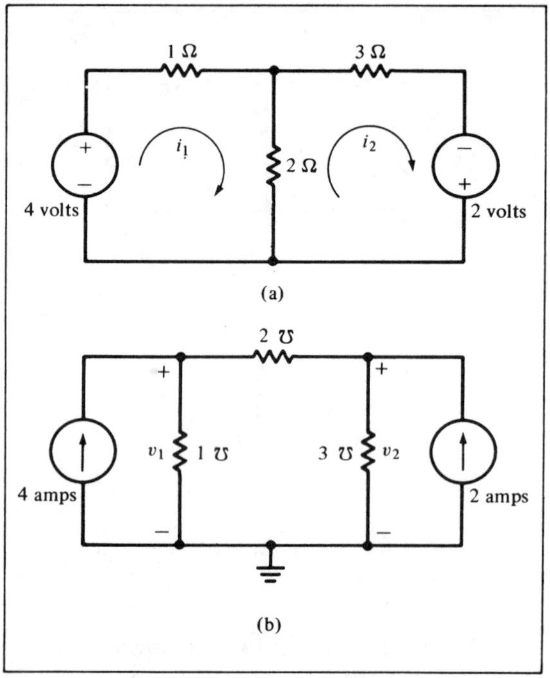

FIGURE 4.16 A network and its dual.

$$3v_1 - 2v_2 = 4$$
$$-2v_1 + 5v_2 = 2$$

These equations have the same numerical solution; more important, the equations show a correspondence between what might seem to be unrelated concepts. For example, it is interesting to note that an equation is generated in loop analysis by a link; the tree branch performs an identical function in cut-set analysis. Other relationships are shown in Table 4.1.

These analogies suggest how we might construct such dual networks *without writing the network equations*. This is achieved in part by producing graphs such that one tree branch exists in one network for every link branch in the other network. For networks that are *planar* (see below) a simple graphical construction will produce such dual graphs automatically. The interested reader should refer to Problem 4.20.

4.6 Mesh Analysis

Although the loop analysis technique of the preceding section applies to any network, a less general form of analysis involving current variables is more common. This is termed *mesh analysis*. It can be used only when the network is *planar*, that is, when the network can be drawn on a flat surface *without crossing branches*. For example, it would apply to the network of Figure 4.17(a) constructed with solid lines, which can be redrawn as shown in Figure 4.17(b) to prevent branch crossing. However, if the four resistors that are shown dashed are added in Figure 4.17(a), then the branches cannot be uncrossed and the more general loop analysis procedure of the previous section must be used.

Formally, a *mesh* is a closed path (loop) that contains no branches within it. Pictorially, meshes can be identified as "windows" between the network branches and nodes as illustrated in Figure 4.17(b).

Mesh analysis is based on the definition of

TABLE 4.1

Dual Quantities		
	Loop Analysis	Cut-set Analysis
Solution variables	Link currents	Tree voltages
One equation per	Link branch	Tree branch
Element replacement	Resistance Voltage source VCCS	Conductance Current sourse CCVS

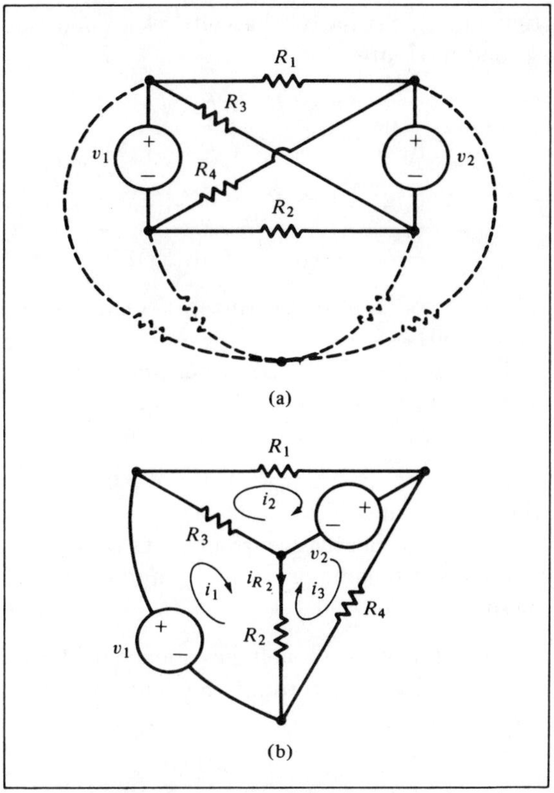

FIGURE 4.17 Mesh indentification.

tion for every mesh in the network. Let the number of meshes be l.

Step 2. Write KVL around the l meshes; some of the voltages may be voltage sources.

Step 3. Using Ohm's law, write all of the unknown voltages in terms of the mesh currents.

Step 4. Solve the resulting l equations for the mesh currents.

Step 5. Find the current in every branch as the sum or difference of mesh currents. Then find every voltage by using Ohm's law.

These steps are illustrated with the network of Figure 4.18. The two meshes are the loops

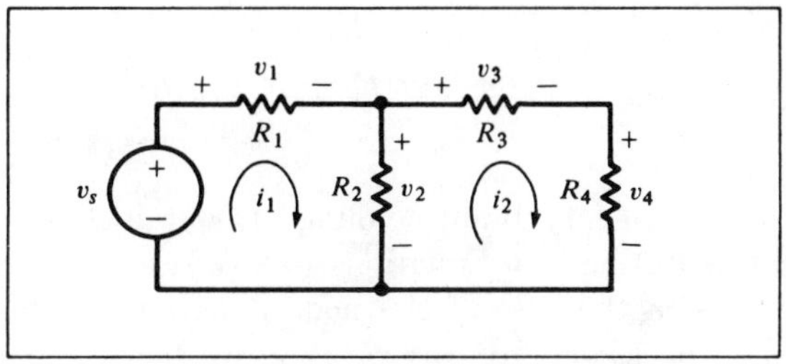

FIGURE 4.18 Mesh analysis example.

containing elements v_s, R_1, R_2 and R_2, R_3, R_4. The mesh currents i_1 and i_2 are assigned positive directions, clockwise in each mesh. Application of KVL around these meshes gives

$$\begin{aligned} v_1 + v_2 &= v_s \\ -v_2 + v_3 + v_4 &= 0 \end{aligned} \tag{4.27}$$

Next, Ohm's law for each resistor is

$$\begin{aligned} R_1 i_1 &= v_1 \\ R_2(i_1 - i_2) &= v_2 \\ R_3 i_2 &= v_3 \\ R_4 i_2 &= v_4 \end{aligned} \tag{4.28}$$

Substituting these relations in Equation (4.27) and collecting terms gives the mesh equations

$$(R_1 + R_2)i_1 - R_2 i_2 = v_s \tag{4.29}$$

$$-R_2 i_1 + (R_2 + R_3 + R_4)i_2 = 0 \tag{4.30}$$

mesh currents, which are depicted as flowing through the branches around the boundaries of each mesh. An example is displayed in Figure 4.17(b), where three currents are defined to correspond to the three meshes. Unlike node voltages, these currents are not natural quantities for measurement. For example, an ammeter inserted in any branch will not necessarily read a single mesh current but rather a sum or difference of mesh currents, namely,

$$i_{R_2} = i_1 - i_3$$

This loss of physical interpretation can best be compensated by examining the following simple analysis procedure which involves these mesh currents as unknowns.

Step 1. Assign a positive mesh current direc-

These can be written in matrix form

$$\begin{bmatrix} R_1 + R_2 & -R_2 \\ -R_2 & R_2 + R_3 + R_4 \end{bmatrix} \begin{bmatrix} i_1 \\ i_2 \end{bmatrix} = \begin{bmatrix} v_s \\ 0 \end{bmatrix}$$

or compactly,

$$\mathbf{Ri} = \mathbf{v}$$

Solving these equations using determinants gives the mesh currents

$$i_1 = \frac{\Delta_{11} v_s}{\Delta} = \frac{(R_2 + R_3 + R_4)v_s}{(R_1 + R_2)(R_3 + R_4) + R_1 R_2} \tag{4.31}$$

and

$$i_2 = \frac{\Delta_{12} v_s}{\Delta} = \frac{R_2 v_s}{(R_1 + R_2)(R_3 + R_4) + R_1 R_2} \tag{4.32}$$

Finally, every branch voltage is obtained by substituting into Equations (4.28).

As in the case of the node equations, it is not necessary to carry out Step 1 to 3 in full detail once the pattern is recognized. Step 1 is essential; but if we adopt the convention of always assigning a single positive mesh current direction—clockwise, for example—Equations (4.29) and (4.30) can be written by inspection. Considering the voltages that are to be summed around each mesh, two possibilities exist: Either the voltage is common to two meshes, or it appears in only one mesh. If the voltage is across a resistor, it will be equal to the resistance value times the net current flow through it. If the voltage is across a source, the source value can be written down directly. Thus, in Figure 4.18 the voltage drops around the first mesh are

Voltage across $R_1 \rightarrow R_1 i_1$

Voltage across $R_2 \rightarrow R_2(i_1 - i_2)$

Source voltage $\rightarrow -v_s$

$(R_1 + R_2)i_1 - R_2 i_2 - v_s$

which by KVL must sum to zero. This is Equation (4.29). Similarly, the voltages around the second mesh are

Voltage across $R_2 \rightarrow R_2(i_2 - i_1)$

Voltage across $R_3 \rightarrow R_3 i_2$

Voltage across $R_4 \rightarrow R_4 i_2$

$-R_2 i_1 + (R_2 + R_3 + R_4)i_2$

which by KVL also add to zero; this is Equation (4.30).

The pattern of these equations can be seen. The voltages encountered moving clockwise around the jth mesh can be written as a sum of:

1. i_j times the sum of all the resistances in the mesh minus
2. Every other mesh current times the values of the resistor common to mesh j and the other mesh.

The sum of these voltages must equal the sum of voltages across all the sources in the mesh.

Example 4.3

The mesh equations of the network of Figure 4.19 are to be found.

There are three meshes, and currents i_1, i_2, and i_3 are assigned to these meshes as shown. Now, for the first mesh the total resistance is $R_3 + R_2$; the resistance common to meshes 1

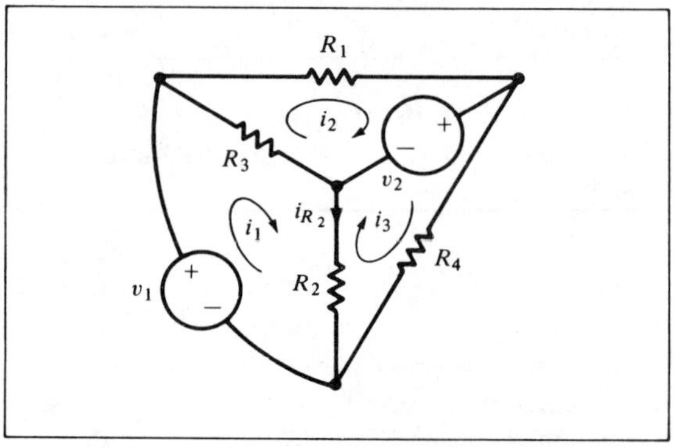

FIGURE 4.19 Three-mesh analysis example.

and 2 is R_3; and the resistance common to meshes 1 and 3 is R_2. The only voltage source in mesh 1 is v_1; therefore, the mesh 1 equation is

$$(R_2 + R_3)i_1 - R_3 i_2 - R_2 i_3 = v_1 \quad (4.33)$$

In the second mesh the total resistance is $R_1 + R_3$; also, R_3 is common to meshes 1 and 2, and the voltage source v_2 is contained in the mesh. The mesh 2 equation is therefore

$$-R_3 i_1 + (R_1 + R_3)i_2 = -v_2 \quad (4.34)$$

Similarly, for mesh 3 we find

$$-R_2 i_1 + (R_2 + R_4)i_3 = v_2 \quad (4.35)$$

In matrix form, the mesh equations for this example are

$$\begin{bmatrix} (R_2 + R_3) & -R_3 & -R_2 \\ -R_3 & (R_1 + R_3) & 0 \\ -R_2 & 0 & (R_2 + R_4) \end{bmatrix} \begin{bmatrix} i_1 \\ i_2 \\ i_3 \end{bmatrix} = \begin{bmatrix} v_1 \\ -v_2 \\ v_2 \end{bmatrix} \quad (4.36)$$

In these examples of mesh analysis it has been tacitly assumed that the sources were all independent voltage sources. In the following sections, mesh analysis with other types of controlled and independent sources is examined.

Mesh Analysis with Controlled Voltage Sources

The approach to formulating mesh equations for networks with controlled *voltage* sources is identical to that of Section 4.3 for writing node equations to include controlled *current* sources. The following example illustrates the procedure.

Example 4.4

The input resistance R_{eq} of the one-port of Figure 4.20(a) is to be found.

We could apply a current source i_s at nodes ① and ④, and calculate the resultant voltage v_{14} by node analysis. This would require three node equations and the evaluation of a 3 by 3 determinant. We choose to apply a voltage source v_s and determine the current i_1 through the source by mesh analysis. After the transformation of source i_c [Figure 4.20(b)], only two mesh currents need be defined. Initially regarding v_c as an independent source, we can write

$$\begin{bmatrix} 125 & -100 \\ -100 & 110{,}100 \end{bmatrix} \begin{bmatrix} i_1 \\ i_2 \end{bmatrix} = \begin{bmatrix} v_s \\ v_c \end{bmatrix} = \begin{bmatrix} v_s \\ 99{,}000 i_1 \end{bmatrix}$$

Transferring the term involving i_1 to the left-hand side gives the mesh equations in final form as

$$\begin{bmatrix} 125 & -100 \\ -99{,}100 & 110{,}100 \end{bmatrix} \begin{bmatrix} i_1 \\ i_2 \end{bmatrix} = \begin{bmatrix} v_s \\ 0 \end{bmatrix}$$

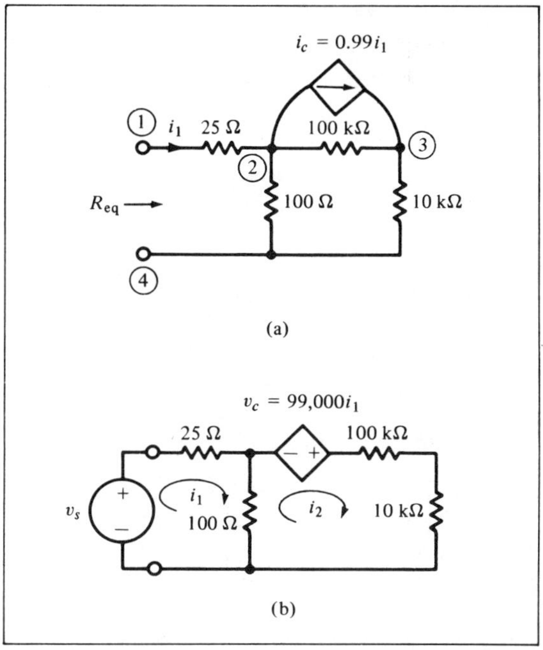

FIGURE 4.20 Calculation of equivalent resistance.

We solve for R_{eq} as

$$R_{eq} = \frac{v_s}{i_1} = \frac{v_s}{\frac{\Delta_{11} v_s}{\Delta}} = \frac{\Delta}{\Delta_{11}}$$

$$= \frac{3.84 \times 10^6}{110{,}100}$$

$$= 35.1\ \Omega$$

Mesh Analysis with Current Sources

Although rarely encountered in analysis of common electronic currents, it is interesting to consider mesh analysis of networks with a current source that cannot be removed by Norton-Thevenin source transformations.

An example is shown in Figure 4.21 (a). One approach is to write a mesh equation in the normal manner and then recognize that each current source reduces the number of unknowns by one. This is closely analogous to the node analysis procedure with a voltage source present, as demonstrated in Figure 4.5. An equivalent method which yields the same final mesh equations involves replacement of each current source by a string of current sources distributed around the mesh to be eliminated. For example, it is clear that the source currents flowing into nodes ①, ②, ③ of Figure 4.21 (a) are identical to those of Figure 4.21 (b). If each of these current sources can then be converted to a Thevenin equivalent as in Figure 4.21 (c), a reduced number of mesh equations can then be written. The analogous situation with voltage sources is illustrated in Problem 4.21.

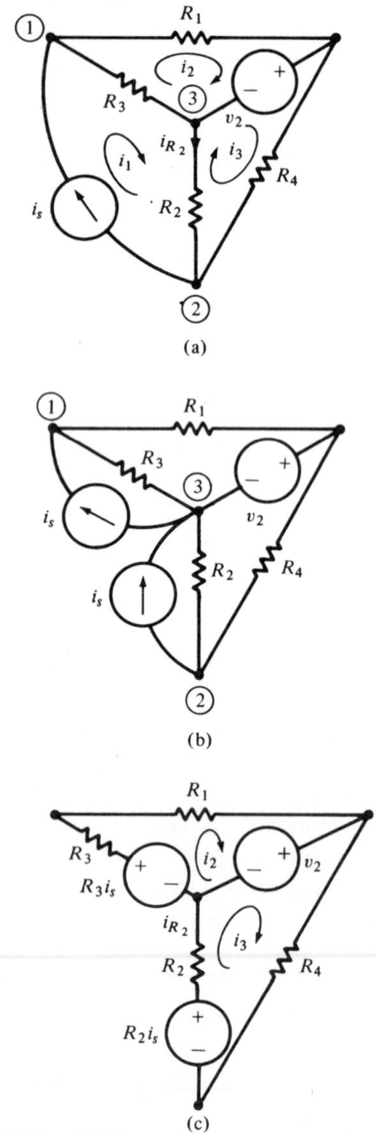

FIGURE 4.21 Source transformation to reduce number of meshes.

PROBLEMS

4.1 (a)–(d) Write the node equations for the networks of Figure P 4.1. One source transformation is allowed.

4.2 (a)-(d) Determine the voltages v_0 in Problem 4.1 in terms of determinants and cofactors of the node matrix.

4.3 (a) If

$$\begin{bmatrix} a_{11} & a_{12} & a_{13} \\ a_{21} & a_{22} & a_{23} \\ a_{31} & a_{32} & a_{33} \end{bmatrix} \begin{bmatrix} x_1 \\ x_2 \\ x_3 \end{bmatrix} = \begin{bmatrix} b_1 \\ 0 \\ b_3 \end{bmatrix}$$

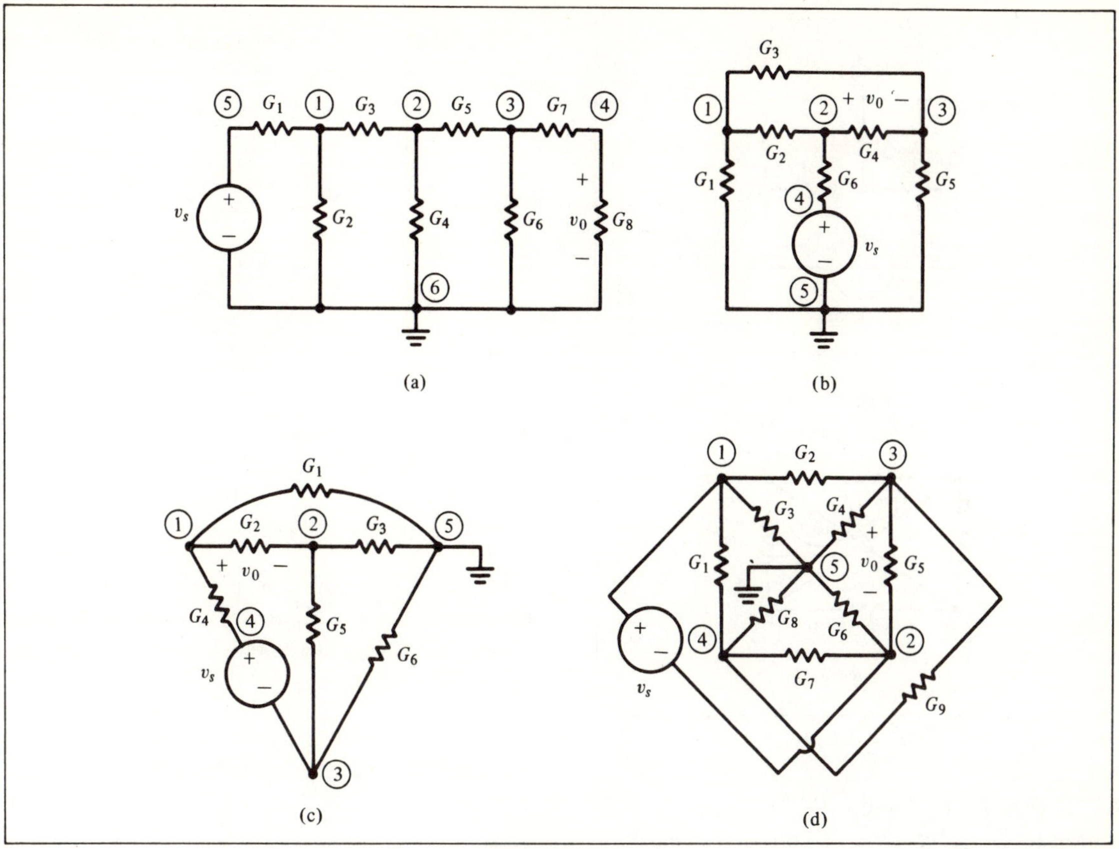

FIGURE P 4.1

fill in the following Δ_{ij} subscripts:

$$x_1 = \frac{\Delta_{??}b_1 + \Delta_{??}b_3}{\Delta}$$

(b) Solve the following simultaneous equations for x_3. First express x_1, x_2, and x_3 as the matrix of cofactors and a determinant.

$$\begin{bmatrix} 7 & 3 & 1 \\ -1 & -1 & 0 \\ 2 & 5 & 3 \end{bmatrix} \begin{bmatrix} x_1 \\ x_2 \\ x_3 \end{bmatrix} = \begin{bmatrix} 12 \\ -2 \\ 13 \end{bmatrix}$$

4.4 (a)–(b) Find v_0 in each of the networks of Figure P 4.2.

4.5 The node voltage equations for a network are

$$1.6v_1 - 0.5v_2 - v_3 = 1$$

$$-0.5v_1 + 1.6v_2 - 0.1v_3 = 0$$

$$-v_1 - 0.1v_2 + 3.1v_3 = 0$$

(a) Draw the network. (b) Show that the current flowing directly between nodes ② and ③ is zero, without solving for v_2 and v_3 separately.

4.6 (a) Using node equations, solve for v_0 of Figure P 4.3. (b) Suppose the magnitude and direction of the left-hand current source can be changed at will. Is it possible to find a value for this source which will reduce v_0 to zero? What value is required?

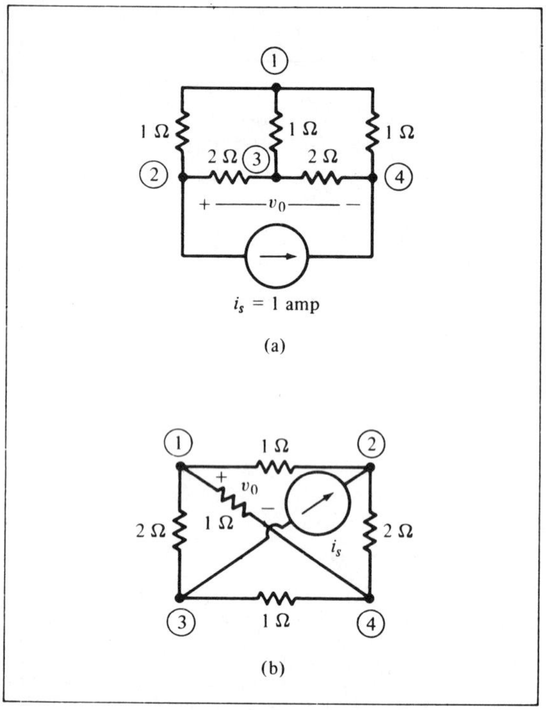

FIGURE P 4.2

4.7 Repeat Problem 4.6 with the 2-amp current source replaced by a 2-volt voltage source.

4.8 (a)–(d) Represent v_0 in terms of appropriate cofactors and determinants for the networks of Figure P 4.4, using node analysis.

4.9 (a)–(d) Evaluate the determinants of Problem 4.8 to find v_0.

4.10 Show that the ratio v_0/i_s for the network of Figure P 4.5 is -3.1×10^4.
[*Hint:* Use Thevenin and Norton transformations to reduce left and right halves of the network as much as possible.]

4.11 For the network of Figure 4.2, show that

$$\frac{v_3}{v_2} = \frac{\begin{vmatrix} -G_1 & (G_1 + G_2 + G_3) \\ -G_5 & -G_3 \end{vmatrix}}{\begin{vmatrix} G_1 & -G_3 \\ G_5 & (G_3 + G_4 + G_5) \end{vmatrix}}$$

$$= \frac{G_1G_3 + G_5(G_1 + G_2 + G_3)}{G_1(G_3 + G_4 + G_5) + G_3G_5}$$

4.12 (a)–(d) Construct a graph for each of the networks of Figure P 4.1. Find three trees for each of these graphs. Identify the links associated with each tree.

4.13 (a)–(b) Write the cut-set equations for the networks of Figure P 4.2. In each case, choose the tree that connects nodes ①, ②, ③, ④ in that order.

4.14 Explain why the node voltages are both a *complete* and an *independent* set of solution variables.

4.15 (a)–(d) For each of the networks of Figure P 4.1, define loop currents and write the loop equations. In each case, display the tree and links on which the loops are based.

4.16 Find the networks for which the loop equations are

(a)
$$\begin{aligned} (R_1 + R_2)i_1 - R_2i_2 &= v_s \\ -R_2i_1 + (R_2 + R_3 + R_4)i_2 - R_4i_3 &= 0 \\ -R_4i_2 + (R_4 + R_5)i_3 &= 0 \end{aligned}$$

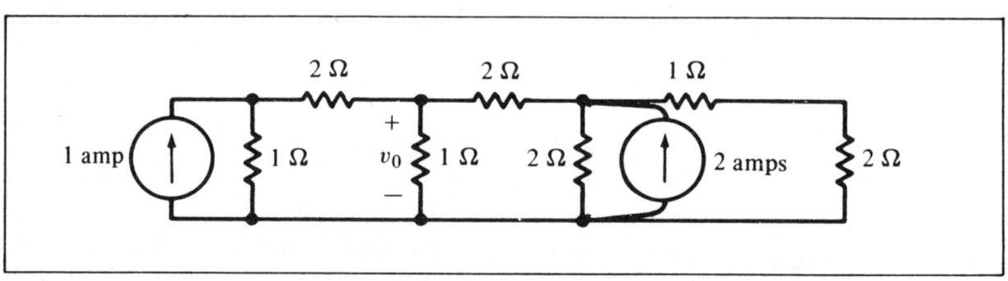

FIGURE P 4.3

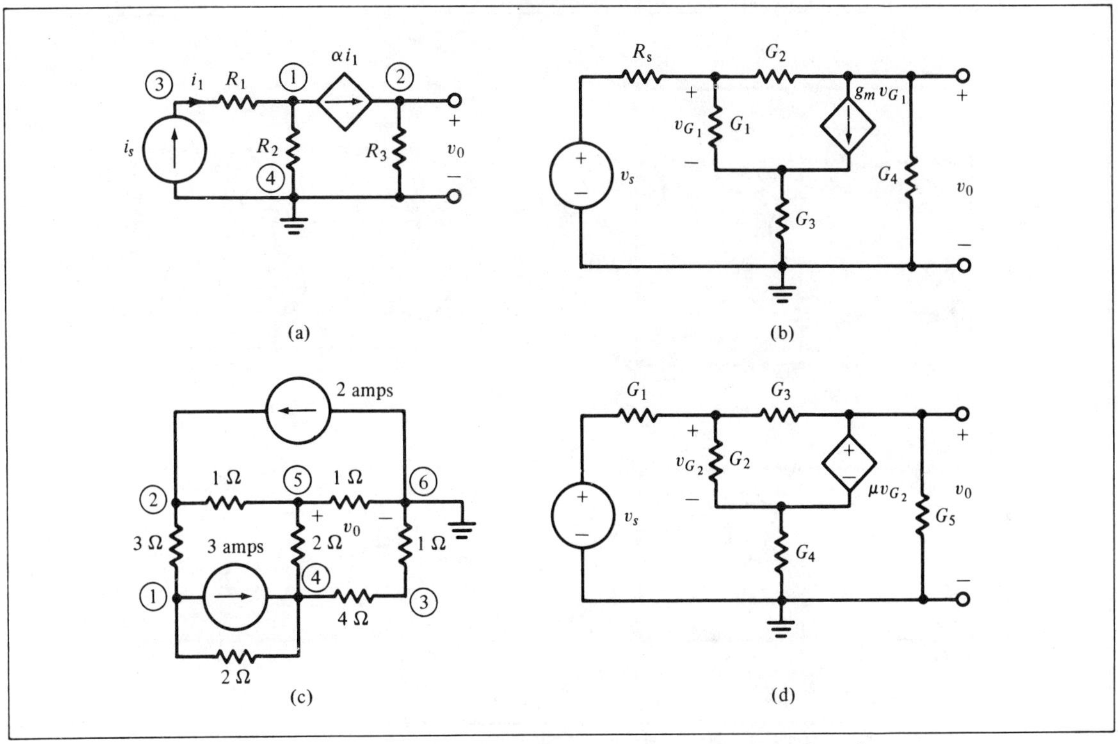

FIGURE P 4.4 (d) One source transformation allowed.

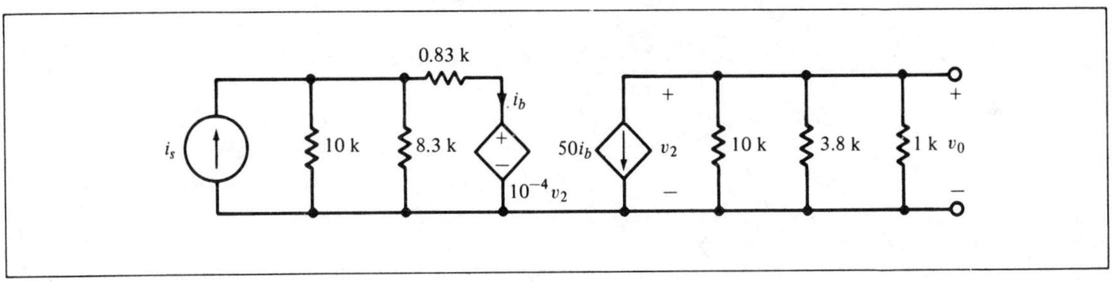

FIGURE P 4.5

(b) $(R_1 + R_3 + R_4)i_1 + (R_3 + R_4)i_2$

$$+ R_4 i_3 = v_0$$

$$(R_3 + R_4)i_1 + (R_2 + R_3 + R_4)i_2$$

$$+ R_4 i_3 = 0$$

$$R_4 i_1 + R_4 i_2 + (R_4 + R_5)i_3 = v$$

4.17 How many node and loop equations are required to describe the network of Figure P 4.6?

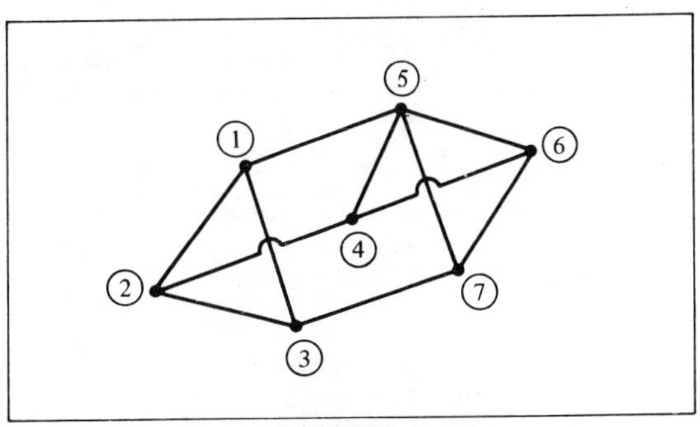

FIGURE P 4.6

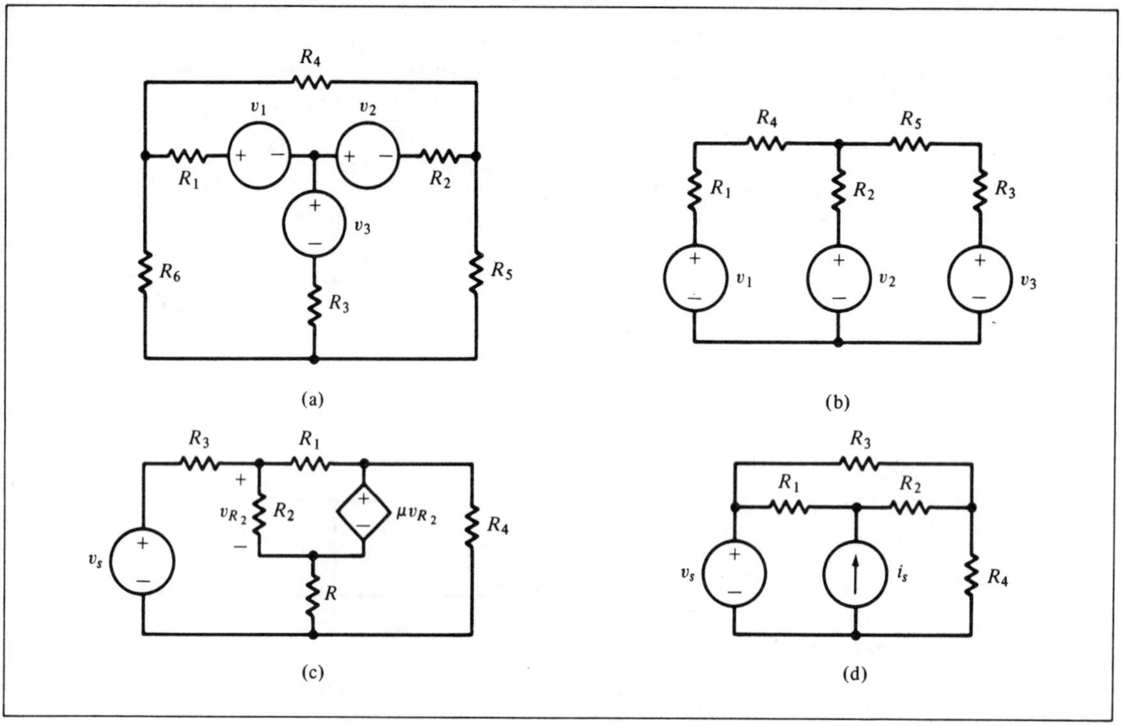

FIGURE P 4.7

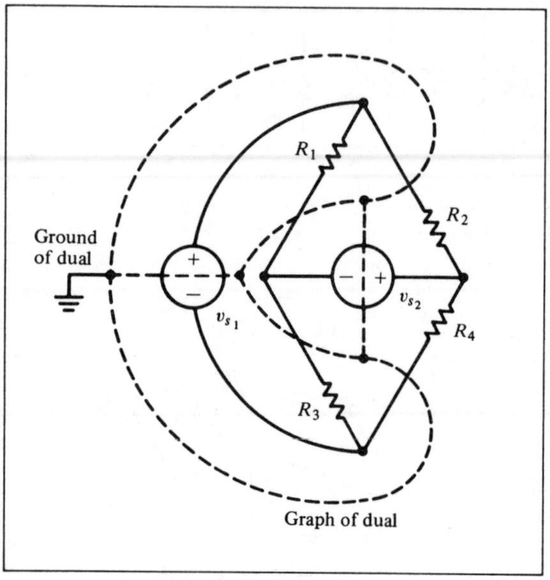

FIGURE P 4.8

4.18 (a)–(d) Write mesh equations for the networks of Figure P 4.7.

4.19 Show that the loop currents are independent variables. You may want to review the argument used to show the independence of tree branch voltages.

4.20 One method of constructing a dual network proceeds as follows. First, dots are inserted within the loops (meshes), and one dot is placed outside the network; the connection of these dots as illustrated in Figure P 4.8 then determines the graph of the dual network. The components of the dual network are taken from Table 4.1, where the dual element is used to replace the element it crosses. (a) Construct the dual of the network of Figure P 4.8. (b) Verify the duality by showing that the node equations of the the dual are identical to the mesh equations of the original network.

4.21 A voltage source (independent or controlled) can be passed through a node as shown in Figure P 4.9 without affecting the behavior of any network connected to terminals *a*, *b*, or *c*. A Thevenin-Norton transformation then reduces the number of nodes by one (from four to three in this case). Show that this is equivalent to the procedure illustrated in Figure 4.5(a).

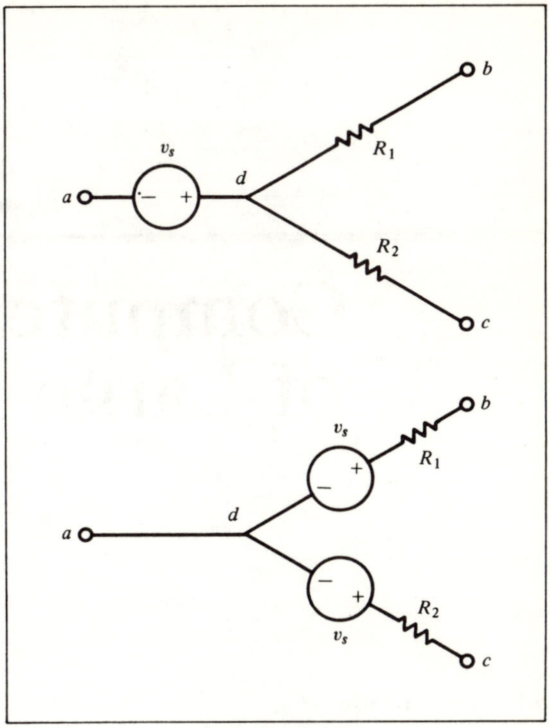

FIGURE P 4.9

REFERENCES AND ADDITIONAL READING

4.1. W. H. HAYT, JR., AND J. E. KEMMERLY, *Engineering Circuit Analysis*. New York: McGraw-Hill Book Company, Inc., 1963, Chapter 18.

4.2. B. J. LEON AND P. A. WINTZ, *Basic Linear Networks for Electrical and Electronics Engineers*. New York: Holt, Rinehart and Winston, Inc., 1970, Chapter 3 and Appendix II.

4.3. NORMAN BALABANIAN, *Fundamentals of Circuit Theory*. Boston: Allyn and Bacon, Inc., 1964, Chapters 2 and 3.

5

Computer Analysis of Large Networks

5.1 Introduction

Beyond Hand Computation

Prior to widespread use of the digital computer in network design, the student could hope to leave an introductory network analysis course with knowledge of mesh and node formulation methods and numerical solution methods such as Cramer's rule that were developed for hand computation. After practice in application of these techniques, the student was later asked to solve in an electronics course network models that were carefully tailored to match the limitations of hand computation.

With a modest investment in the study of computational algorithms, we can leave this course with far more powerful tools. In this chapter, we will develop algorithms for computer analysis of reasonably general forms of linear and nonlinear resistive networks. The implementation of these algorithms in program form will be demonstrated at two levels of generality:

1. Programs to solve *particular* networks but written so that new networks can be accommodated with modest reprogramming effort; these programs require that the equations be formulated by the reader for the particular network of interest, but the numerical solution of these equations is performed automatically by computer.
2. A program to solve rather arbitrary forms of resistive networks from a list of elements and their node connections in the circuit; both the equation formulation and numerical solution are performed automatically.

This latter program is representative of much more powerful general analysis programs encountered in all fields of engineering problem solving.

The above analysis capability will allow us to examine amplifier and digital circuit performance more completely than we could in Chapter 3, where we were restricted to circuits with a single diode nonlinearity. We will be able to study digital circuits operating beyond the normal region of transistor operation, as well as amplifiers involving one or more transistors. Finally, the interested reader can become acquainted with the operation of a different form of transistor known as the field-effect transistor; we will use this device in some of the problems at the end of the chapter to provide an exercise in applying our general analysis techniques to new forms of circuits.

Equation Formulation and Numerical Solution

We have already developed most of the concepts and techniques necessary for our computer-aided analysis. The linearization and iteration procedures of Chapter 3 remain the conceptual basis for the solution process. The equations will be formulated using node analysis. Unlike mesh or loop analysis, node analysis is both general and easily programmed. The only new technique we must learn involves the numerical solution of simultaneous equations. The principal solution method used in Chapter 4, determinant evaluation, is conceptually compact but numerically inefficient. For example, evaluation of the determinant of a full n by n matrix using the common row-column expansions of Appendix A involves more than $n!$ multiplications. In contrast, the Gauss elimination procedure we examine next can yield the node voltage vector in approximately $n^3/3$ operations. For $n > 2$, the latter becomes increasingly effective.

5.2 Solution of Simultaneous Linear Equations

The Gauss Elimination Procedure

Consider the problem of solving sets of simultaneous linear equations expressed in the equivalent forms:

$$\begin{array}{l} a_{11}x_1 + a_{12}x_2 + \cdots + a_{1n}x_n = b_1 \\ a_{21}x_1 + a_{22}x_2 + \cdots + a_{2n}x_n = b_2 \\ \cdot \quad \cdot \quad \cdot \quad \cdot \\ \cdot \quad \cdot \quad \cdot \quad \cdot \\ \cdot \quad \cdot \quad \cdot \quad \cdot \\ a_{n1}x_1 + a_{n2}x_2 + \cdots + a_{nn}x_n = b_n \end{array}$$

or

$$\begin{bmatrix} a_{11} & a_{12} \ldots a_{1n} \\ a_{11} & a_{22} \ldots a_{2n} \\ \cdot & \cdot \quad\quad \cdot \\ \cdot & \cdot \quad\quad \cdot \\ \cdot & \cdot \quad\quad \cdot \\ a_{n1} & a_{n2} \ldots a_{nn} \end{bmatrix} \begin{bmatrix} x_1 \\ x_2 \\ \cdot \\ \cdot \\ \cdot \\ x_n \end{bmatrix} = \begin{bmatrix} b_1 \\ b_2 \\ \cdot \\ \cdot \\ \cdot \\ b_n \end{bmatrix}$$

or in the representational form

$$\mathbf{Ax} = \mathbf{b}$$

where the a_{ij} and b_i are given and the x_i are to be found. For example, the equation

$$\begin{bmatrix} 4 & -2 & -1 \\ -2 & 17/2 & -2 \\ -1 & -2 & 11 \end{bmatrix} \begin{bmatrix} x_1 \\ x_2 \\ x_3 \end{bmatrix} = \begin{bmatrix} 2 \\ 0 \\ -2 \end{bmatrix} \tag{5.1}$$

has the $\mathbf{x}$ solution

$$\mathbf{x} = \frac{1}{85} \begin{bmatrix} 44 \\ 8 \\ -10 \end{bmatrix}$$

In this section we will discuss a simple yet often effective method of finding $\mathbf{x}$ from $\mathbf{A}$ and $\mathbf{b}$.

To motivate our study, suppose we were given the pair of equations

$$2x_1 + x_2 = 4$$

$$4x_1 - 3x_2 = -2$$

to solve for x_1 and x_2. The most obvious solution method is to solve the first equation for x_1, namely,

$$x_1 = \frac{4 - x_2}{2} = 2 - 0.5x_2 \tag{5.2}$$

and then substitute the result in the second, as

$$4(2 - 0.5x_2) - 3x_2 = -2$$

$$5x_2 = 10$$

$$x_2 = 2$$

Substituting this value of x_2 into Equation (5.2) gives

$$x_1 = 1$$

which completes the solution. This is a form of *elimination* process, that is, the variable x_1 was eliminated from the first equation. In the following extension to the case of more than two equations and unknowns, we will systematize this process so that one unknown is eliminated *at each step* until we are left with one equation and one unknown, as above. This method is called the Gauss[1] elimination procedure.

The Gauss procedure begins by attempting to eliminate x_1 from all the equations. If we divide the first equation by a_{11}, we obtain

$$x_1 + \left(\frac{a_{12}}{a_{11}}\right) x_2 + \cdots + \left(\frac{a_{1n}}{a_{11}}\right) x_n = \frac{b_1}{a_{11}}$$

This equation is now used as a reference in the following procedure:

1. Multiply this reference equation by a_{21} and subtract the result from the second equation, term-by-term:

$$\begin{array}{lll}
\text{Second equation} & \{ & a_{21}x_1 + a_{22}x_2 + \cdots + a_{2n}x_n = b_2 \\
-a_{21}\text{ times reference equation} & \{ & -a_{21}x_1 - a_{21}(a_{12}/a_{11})x_2 + \cdots - a_{21}(a_{1n}/a_{11})x_n = -a_{21}b_1/a_{11} \\
\hline
\text{Sum} & \{ & (0)x_1 + (a_{22} - a_{21}a_{12}/a_{11})x_2 + \cdots + (a_{2n} - a_{21}a_{1n}/a_{11})x_n = b_2 - a_{21}b_1/a_{11}
\end{array}$$

Note that x_1 has been made to vanish in the new equation.

2. In general, to eliminate the first unknown from the ith equation, $(i > 1)$ multiply the reference equation by a_{i1} and subtract the result from the ith equation.

The process is more readily followed in matrix form. After all the a_{i1} have been reduced to zero the first column is zero, (except the diagonal position). We then have

$$\begin{bmatrix} 1 & a_{12}/a_{11} & \cdots & a_{1n}/a_{11} \\ 0 & a_{22} - a_{21}a_{12}/a_{11} & \cdots & a_{2n} - a_{21}a_{1n}/a_{11} \\ \cdot & \cdot & \cdots & \cdot \\ \cdot & \cdot & \cdots & \cdot \\ \cdot & \cdot & \cdots & \cdot \\ 0 & a_{n2} - a_{n1}a_{12}/a_{11} & \cdots & a_{nn} - a_{n1}a_{1n}/a_{11} \end{bmatrix} \begin{bmatrix} x_1 \\ x_2 \\ \cdot \\ \cdot \\ \cdot \\ x_n \end{bmatrix} = \begin{bmatrix} b_1/a_{11} \\ b_2 - a_{21}b_1/a_{11} \\ \cdot \\ \cdot \\ \cdot \\ b_n - a_{n1}b_1/a_{11} \end{bmatrix} \tag{5.3}$$

The example of Equation (5.1) would be of the form

$$\begin{bmatrix} 1 & -1/2 & -1/4 \\ 0 & 15/2 & -5/2 \\ 0 & -5/2 & 43/4 \end{bmatrix} \begin{bmatrix} x_1 \\ x_2 \\ x_3 \end{bmatrix} = \begin{bmatrix} 1/2 \\ 1 \\ -3/2 \end{bmatrix} \tag{5.4}$$

at this stage of solution.

If we examine Equation (5.3) or (5.4), we observe that the lower right partition shown involves $n - 1$ equations in $n - 1$ unknowns (x_2 and x_3). We can therefore repeat the same algorithm to eliminate x_2, where the reference equation is now the second one. For the example of Equation (5.4) the second equation is divided by 15/2, multiplied by $-5/2$, and then subtracted from the third equation to yield

$$\begin{bmatrix} 1 & -1/2 & -1/4 \\ 0 & 1 & -1/3 \\ 0 & 0 & \frac{119}{12} \end{bmatrix} \begin{bmatrix} x_1 \\ x_2 \\ x_3 \end{bmatrix} = \begin{bmatrix} 1/2 \\ 2/15 \\ -7/6 \end{bmatrix} \tag{5.5}$$

[1] Karl Friedrich Gauss (1777–1855), a German mathematician and astronomer, made major contributions to geometry, to the theory of numbers, and to mathematical physics.

This process of eliminating matrix entries below the diagonal continues until the matrix is in *upper triangular* form as in Equation (5.5). The final equation is then divided by the diagonal element,

$$\begin{bmatrix} 1 & -1/2 & -1/4 \\ 0 & 1 & -1/3 \\ 0 & 0 & 1 \end{bmatrix} \begin{bmatrix} x_1 \\ x_2 \\ x_2 \end{bmatrix} = \begin{bmatrix} 1/2 \\ 2/15 \\ -2/17 \end{bmatrix} \quad (5.6)$$

leaving the matrix in the general form

$$\begin{bmatrix} 1 & a'_{12} & \cdot & \cdot & \cdot & a'_{1,n-1} & a'_{1n} \\ 0 & 1 & \cdot & \cdot & \cdot & a'_{2,n-1} & a'_{2n} \\ \cdot & \cdot & \cdot & & & \cdot & \cdot \\ \cdot & \cdot & & \cdot & & \cdot & \cdot \\ \cdot & \cdot & & & \cdot & \cdot & \cdot \\ 0 & 0 & \cdot & \cdot & \cdot & 1 & a'_{n-1,n} \\ 0 & 0 & \cdot & \cdot & \cdot & 0 & 1 \end{bmatrix} \begin{bmatrix} x_1 \\ x_2 \\ \cdot \\ \cdot \\ \cdot \\ x_{n-1} \\ x_n \end{bmatrix} = \begin{bmatrix} b'_1 \\ b'_2 \\ \cdot \\ \cdot \\ \cdot \\ b'_{n-1} \\ b'_n \end{bmatrix} \quad (5.7)$$

This completes the *forward reduction* of the matrix.

We can immediately solve for x_n in Equation (5.7), because the last equation requires that $x_n = b_n'$. The next to last equation has the form

$$x_{n-1} = b'_{n-1} - a'_{n-1,n}x_n$$

from which x_{n-1} can be determined. This *back substitution* continues up the $\mathbf{x}$ vector, solving successively for x_n, x_{n-1}, x_{n-2}, ..., x_1 by the formula

$$x_r = b_r' - \sum_{j=r+1}^{n} a'_{rj}x_j$$

$$(r = n-1,\quad n-2, \ldots, 1)$$

For the example of Equation (5.1), we find from Equation (5.5) that

$$x_3 = b_3' = -2/17$$

$$x_2 = b_2' - a'_{23}x_3 = 2/15 - (-1/3)(-2/17)$$

$$= 8/85$$

$$x_1 = b_1' - a'_{12}x_2 - a'_{13}x_3$$

$$= 1/2 - (-1/2)(8/85)$$

$$- (-1/4)(-2/17) = 44/85$$

Example 5.1

We are to solve

$$\begin{bmatrix} 2 & 1 & 0 \\ 1 & 3 & -1 \\ 4 & 0 & 6 \end{bmatrix} \begin{bmatrix} x_1 \\ x_2 \\ x_3 \end{bmatrix} = \begin{bmatrix} 3 \\ 3 \\ 10 \end{bmatrix}$$

for the x_i.

The solution proceeds as follows:

---

1. Divide the first equation by a_{11}.

$$\begin{bmatrix} 1 & 0.5 & 0 \\ 1 & 3 & -1 \\ 4 & 0 & 6 \end{bmatrix} \begin{bmatrix} x_1 \\ x_2 \\ x_3 \end{bmatrix} = \begin{bmatrix} 1.5 \\ 3 \\ 10 \end{bmatrix}$$

2. Eliminate x_1 from the last two equations by multiplying the first equation by a_{i1} and subtracting from the ith equation.

$$\begin{bmatrix} 1 & 0.5 & 0 \\ 0 & 2.5 & -1 \\ 0 & -2 & 6 \end{bmatrix} \begin{bmatrix} x_1 \\ x_2 \\ x_3 \end{bmatrix} = \begin{bmatrix} 1.5 \\ 1.5 \\ 4 \end{bmatrix}$$

3. Divide the second equation by the value in (2, 2) diagonal position.

$$\begin{bmatrix} 1 & 0.5 & 6 \\ 0 & 1 & -0.4 \\ 0 & -2 & 6 \end{bmatrix} \begin{bmatrix} x_1 \\ x_2 \\ x_3 \end{bmatrix} = \begin{bmatrix} 1.5 \\ 0.6 \\ 4 \end{bmatrix}$$

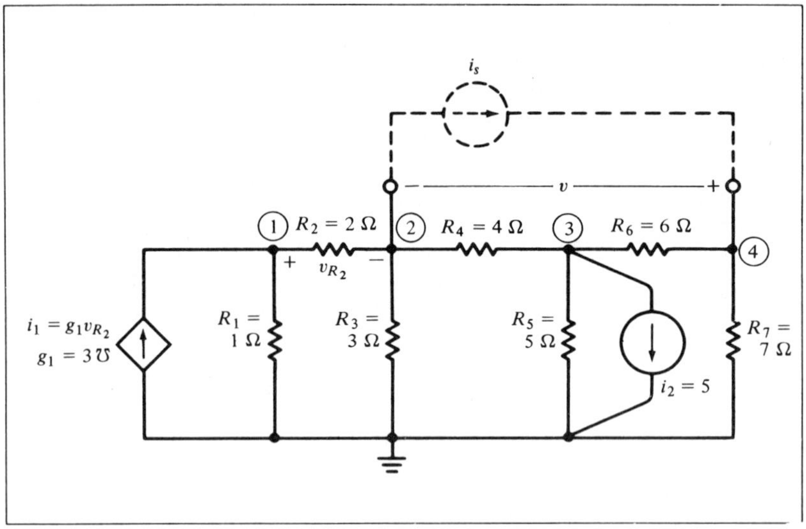

FIGURE 5.1 Network solved by program of Figure 5.2(a). Resistance values in ohms.

4. Eliminate x_2 from the last equation.

$$\begin{bmatrix} 1 & 0.5 & 0 \\ 0 & 1 & -0.4 \\ 0 & 0 & 5.2 \end{bmatrix} \begin{bmatrix} x_1 \\ x_2 \\ x_3 \end{bmatrix} = \begin{bmatrix} 1.5 \\ 0.6 \\ 5.2 \end{bmatrix}$$

5. Divide the last equation by the value in (3, 3) position; back substitute to obtain x_i.

$$x_3 = \frac{5.2}{5.2} = 1$$

$$x_2 = 0.6 + (0.4)(1) = 1$$

$$x_1 = 1.5 - 0.5(1) = 1$$

5.3 Examples of Computer Analysis of Linear Networks

With node analysis selected as a formulation method and Gauss elimination as a solution method, we are in a position to analyze networks well beyond the range of hand computation. As mentioned in the introduction, we will discuss several programs of increasing complexity and generality. All programs will be organized in four phases:

1. Data input phase (subroutine READ).
2. Node equation formulation phase (subroutine NODEQ).
3. Node equation solution phase using Gauss elimination (subroutine REDUCE).
4. Data output phase (subroutine OUTPUT).

The common structure will assist the reader in progressing from the special to the general program (Section 5.8). For the reader interested in the details of only the simpler programs, this common structure will provide insight into the organization of the general analysis program without study of the program itself.

Example 5.2

We are to determine a point on the *i-v* characteristic of the one-port network of Figure 5.1. Specifically, we will calculate the voltage v resulting from the application of a current $i_s = -1.5$ amps.

The node equations of the network can be determined using the methods of Chapter 4.

These have the final form

$$\begin{bmatrix} 1/R_1 + 1/R_2 - g_1 & -1/R_2 + g_1 & 0 & 0 \\ -1/R_2 & 1/R_2 + 1/R_3 + 1/R_4 & -1/R_4 & 0 \\ 0 & -1/R_4 & 1/R_4 + 1/R_5 + 1/R_6 & -1/R_6 \\ 0 & 0 & -1/R_6 & 1/R_6 + 1/R_7 \end{bmatrix} \begin{bmatrix} v_1 \\ v_2 \\ v_3 \\ v_4 \end{bmatrix} = \begin{bmatrix} 0 \\ -i_s \\ -i_2 \\ i_s \end{bmatrix} \tag{5.8}$$

```
C****  CALCULATION OF POINT ON I-V CHARACTERISTIC (RNET)
      IMPLICIT REAL*8(A-H,O-Z)
      REAL*8 G(10,10),I(10),V(10)
      DIMENSION NODOUT(2,5)
      NDIM=10
C****  READ INPUT DATA
    1CALL READ(NN,NODOUT,NOUT)
C****  SET UP NODE EQUATIONS
    3 CALL NODEQ(G,I,NN,NDIM)
C****  SOLVE NODE EQUATIONS
      CALL REDUCE(G,NN,I,NDIM)
      DO 2 J=1,NN
    2 V(J)=I(J)
C****  CALCULATE AND PRINT OUTPUT VOLTAGES
      CALL OUTPUT(NODOUT,NOUT,V)
      STOP
      END
      SUBROUTINE READ(NN,NODOUT,NOUT)
      IMPLICIT REAL*8(A-H,O-Z)
      REAL*8 I2,IS
      COMMON G1,I2,R1,R2,R3,R4,R5,R6,R7,IS
      DIMENSION NODOUT(2,1)
C****  NN = NUMBER OF NODES
C****  NOUT = NUMBER OF OUTPUTS
C****  NODOUT CONTAINS OUTPUT NODE NUMBERS
      READ(5,3)G1,I2,R1,R2,R3,R4,R5,R6,R7,IS
      READ(5,4)NN,NOUT,(NODOUT(1,J),NODOUT(2,J),J=1,NOUT)
      WRITE(6,2)G1,I2,R1,R2,R3,R4,R5,R6,R7,IS
      RETURN
    2 FORMAT(' G1 =',D12.4,' I2 =',D12.4,' R1 =',D12.4/
     1' R2 =',D12.4,' R3 =',D12.4,' R4 =',D12.4/
     2' R5 =',D12.4,' R6 =',D12.4,' R7 =',D12.4/
     3' IS =',D12.4)
    3 FORMAT(6F10.0)
    4 FORMAT(20I2)
      END
      SUBROUTINE NODEQ(G,I,NN,NDIM)
      IMPLICIT REAL*8(A-H,O-Z)
      REAL*8 G(NDIM,1),I(1),I2,IS
      COMMON G1,I2,R1,R2,R3,R4,R5,R6,R7,IS
C****  ZERO NODE MATRIX G AND CURRENT VECTOR I
      DO 1 K=1,NN
      I(K)=0.D0
      DO 1 J=1,NN
    1 G(J,K)=0.D0
C****  LOAD G AND I
      G(1,1)=1.D0/R1+1.D0/R2-G1
```

FIGURE 5.2 (a) Program to solve network of Figure 5.1; (b) computer input; (c) computer output.

```
      G(1,2)=-1.D0/R2+G1
      G(2,1)=-1.D0/R2
      G(2,2)=1.D0/R2+1.D0/R3+1.D0/R4
      G(2,3)=-1.D0/R4
      G(3,2)=G(2,3)
      G(3,3)=1.D0/R4+1.D0/R5+1.D0/R6
      G(3,4)=-1.D0/R6
      G(4,3)=G(3,4)
      G(4,4)=1.D0/R6+1.D0/R7
      I(2)=-IS
      I(3)=-I2
      I(4)=IS
      RETURN
      END
      SUBROUTINE OUTPUT(NODOUT,NOUT,V)
      IMPLICIT REAL*8 (A-H,O-Z)
      DIMENSION NODOUT(2,1),V(1)
      DO 2 I=1,NOUT
      VOUT=0.D0
      N1=NODOUT(1,I)
      N2=NODOUT(2,I)
      IF(N1.NE.0)VOUT=VOUT+V(N1)
      IF(N2.NE.0)VOUT=VOUT-V(N2)
      WRITE(6,3)N1,N2,VOUT
    2 CONTINUE
      RETURN
    3 FORMAT('OUTPUT NODES =',2I3,'  OUTPUT VOLTAGE =',F15.7)
      END
      SUBROUTINE REDUCE(A,N,B,NDIM)
      IMPLICIT REAL*8(A-H,O-Z)
      DIMENSION A(NDIM,1),B(1)
C****  THIS SUBROUTINE SOLVES A*X=B BY GAUSS ELIMINATION
C****  N IS THE MATRIX SIZE; X IS RETURNED IN B; A IS DESTROYED
C****  CONVERT A TO UPPER TRIANGULAR FORM
      NM1=N-1
      IF(N.EQ.1)GO TO 8
      DO 1 J=1,NM1
      IJ=J+1
      DO 4 K=IJ,N
    4 A(J,K)=A(J,K)/A(J,J)
      B(J)=B(J)/A(J,J)
      DO 1 I=IJ,N
      ALPHA=A(I,J)
      DO 2 K=IJ,N
    2 A(I,K)=A(I,K)-ALPHA*A(J,K)
    1 B(I)=B(I)-ALPHA*B(J)
    8 B(N)=B(N)/A(N,N)
C****  PERFORM BACK SUBSTITUTION
      IF(N.LT.2)GO TO 7
      DO 3 J=2,N
      LAST=N-J+1
      LASTP1=LAST+1
      DO 3 I=LASTP1,N
    3 B(LAST)=B(LAST)-A(LAST,I)*B(I)
    7 RETURN
      END
```

(a)

FIGURE 5.2 (Continued)

A program (RNET) to solve these simultaneous equations is given in Figure 5.2(a). The values of the seven resistors, the two independent sources (i_s and i_2), and the strength of the dependent source (g_1) are entered in the READ subroutine. The node matrix is formed in the G array and the right-hand side current vector in the I array, in subroutine NODEQ. The resultant matrix equation is solved in the REDUCE subroutine, the solution being returned in the same I array. [The reader may wish to reexamine the backsubstitution process to in-

```
3.        5.        1.        2.        3.        4.
5.        6.        7.        -1.5
 4 1 4 2
```

(b)

```
G1 =  0.3000D 01 I2 =  0.5000D 01 R1 =  0.1000D 01
R2 =  0.2000D 01 R3 =  0.3000D 01 R4 =  0.4000D 01
R5 =  0.5000D 01 R6 =  0.6000D 01 R7 =  0.7000D 01
IS = -0.1500D 01
OUTPUT NODES =  4  2  OUTPUT VOLTAGE =    -3.6666667
```

(c)

FIGURE 5.2 (Continued)

sure that the **x** and **b** vectors can be stored in the same array.] The voltage v is then determined from node voltages $(v_4 - v_2)$ and printed in the OUTPUT subroutine.

The input data of Figure 5.2(b) result in the output of Figure 5.2(c). We conclude from this result that the point $(v, i) = (-3\frac{2}{3}, -\frac{3}{2})$ is a point on the one-port i-v characteristic.

Example 5.3

The network of Figure 5.3 represents a linear model for a two transistor amplifier (the origin of this model will be studied in Problem 6.14). A signal source v_s is represented by its Norton equivalent. We desire to determine v_0/v_s, the ratio of output signal voltage v_0 to input signal voltage v_s.

The node equations are routinely written for this network, except that the current-controlled current sources (i_1 and i_2) must be converted to VCCS's. This is achieved by writing

$$i_1 = \alpha_1 i_{R_2}$$
$$= \left(\frac{\alpha_1}{R_2}\right)(v_1 - v_2)$$

and

$$i_2 = \alpha_2 i_{R_7}$$
$$= -\left(\frac{\alpha_2}{R_7}\right) v_4 \tag{5.9}$$

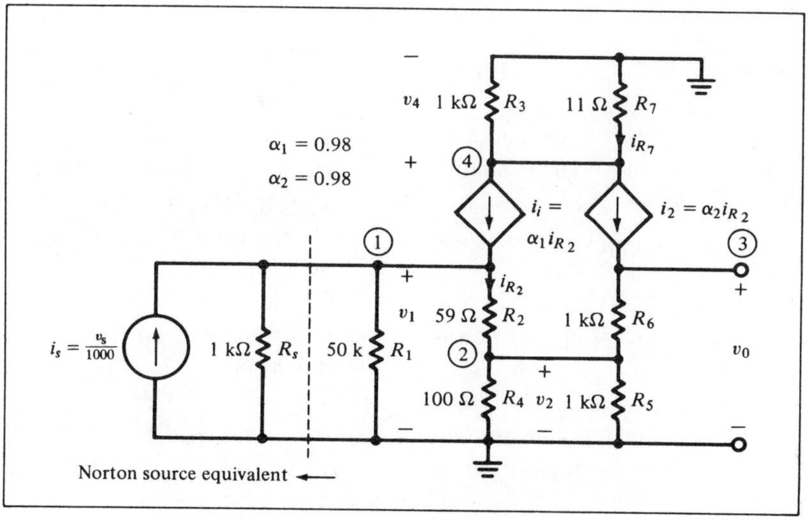

FIGURE 5.3 Amplifier analysis example.

A significant portion of the program of Figure 5.2(a), written to solve the network of Figure 5.1, can be used to analyze this amplifier. The names of elements are similar in both networks so that the READ subroutine can be used intact; only the subroutine NODEQ need be changed to accommodate the different node equations.

The ratio v_0/v_s is found as follows. Because only resistors and controlled sources are present, the network is linear, and the response v_0 is proportional to the excitation v_s (or i_s). This implies that the ratio v_0/v_s is a constant (call it μ) and independent of v_s. Therefore, we can apply any voltage v_s, and the resultant v_0 permits calculation of $\mu = v_0/v_s$. In particular, choosing $v_s = 1$ ($i_s = 0.001$) gives the result $\mu = v_0$.

It is left as an exercise to write the NODEQ subroutine. As a check, with $v_s = 1$, we obtain the signal gain

$$\mu = \frac{v_0}{v_s} = v_0 = 11.127 \qquad (5.10)$$

5.4 Analysis of Nonlinear Resistive Networks

Introduction

In the introduction to Chapter 4, we mentioned that the formal two-step formulation-solution process could result in a loss of insight into network performance when compared with the methods of Chapter 2 where the network model itself was reduced. In the analysis of nonlinear networks, a similar insight-versus-formality choice exists:

1. We can adopt a formal approach, whereby the network equations, which are not necessarily linear in the solution variables, are written and solved using an extension of the Newton iteration of Chapter 3.

2. We can construct, *element by element*, a network model of this Newton iteration process, without explicitly writing the complete set of nonlinear network equations.

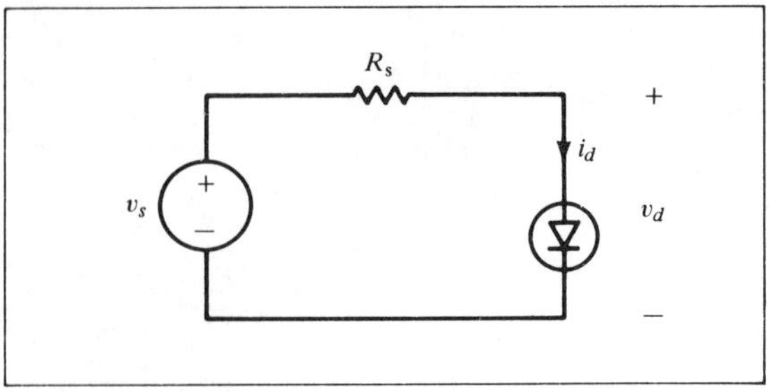

FIGURE 5.4 Simple diode example revisited.

The second method provides considerably more insight into network performance, a feature we will explore further in Chapter 6. Also, this method is easily programmed in a form that allows us to solve rather general nonlinear networks. On the other hand, the more formal approach is applicable to solution of other kinds of engineering problems involving nonlinear algebraic equations. We will consider both procedures, with emphasis on the network model approach.

Companion Network Models

In the iterative solution of the simple diode circuit of Figure 5.4 by Newton's method, we solved at each iteration the equation

$$v_d^{m+1} = \frac{-R_s i_d^m + (R_s \lambda I_s e^{\lambda v_d^m}) v_d^m + v_s}{1 + R_s \lambda I_s e^{v_d^m}} \qquad (5.11)$$

It was not the goal of our discussion in Chapter 3 to present an organized approach to analysis of larger nonlinear networks. However, now that we are familiar with general methods for solving *linear* resistive networks, as with RNET, it is reasonable to ask whether similar methods can be developed to solve *nonlinear* networks.

The key to more general methods is the equation describing the linearization of the nonlinear characteristic. For the diode, we have

$$i_d^{m+1} = i_d^m + \left[\left. \frac{\partial i_d}{\partial v_d} \right|_{v_d = v_d^m} \right] (v_d^{m+1} - v_d^m) \qquad (5.12)$$

$$= I_s(e^{\lambda v_d^m} - 1) + (\lambda I_s e^{\lambda v_d^m})(v_d^{m+1} - v_d^m) \quad (5.13)$$

We observed in Chapter 3 that the slope of the tangent line was

$$G_d^m \triangleq \left.\frac{\partial i_d}{\partial v_d}\right|_{v_d = v_d^m}$$

Equation (5.12) then can be written more compactly as

$$i_d^{m+1} = i_d^m + G_d^m(v_d^{m+1} - v_d^m) \quad (5.14)$$

Now consider the simple network of Figure 5.5(b). If we write KCL at node ①, we have

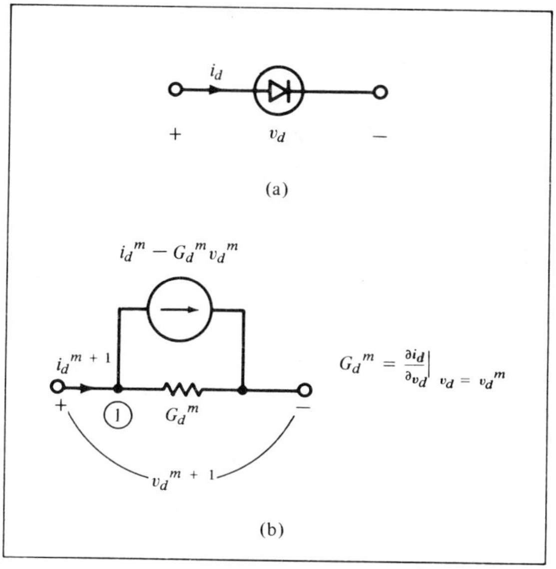

FIGURE 5.5 Companion model of diode.

$$i_d^{m+1} = \overbrace{(i_d^m - G_d^m v_d^m)}^{\text{Due to current source}} + \overbrace{v_d^{m+1} G_d^m}^{\text{Due to conductance}}$$

$$= i_d^m + G_d^m(v_d^{m+1} - v_d^m)$$

the same as Equation (5.14)!

If this branch model is inserted into the complete network (Figure 5.4) in place of the diode, we observe in Figure 5.6(a) that the new network is now linear and therefore solvable by mesh or node analysis. We will term the linearized network a "companion" network model.

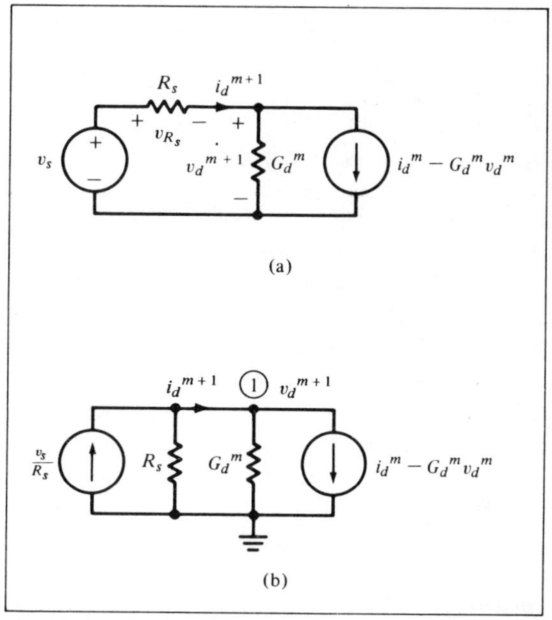

FIGURE 5.6 Application of Norton transformation to companion network.

The companion network has two important properties which permit us to substitute it for the original nonlinear network.

Property 1. Branch currents and voltages satisfy the same KCL and KVL (topological) laws as the original network, because the structure of both networks is the same.

Property 2. The current and voltage (constitutive) relationships within a branch *approximate* those in the original network; for example, the diode current and voltage relationship is a linearized approximation to the exponential branch relationship.

Of more importance is the meaning of the companion network model as a description of the iterative solution process. If we are at the mth iteration, then v_d^m is known (or guessed if m is zero). Therefore,

1. $i_d^m = I_s(e^{\lambda v_d^m} - 1)$

is known, and

2. $G_d{}^m = \lambda I_s e^{\lambda v_d{}^m}$

is known and is the slope of the $i_d - v_d$ characteristic at $(v_d{}^m, i_d{}^m)$. Numerical values for both the current source $i_d{}^m - G_d{}^m v_d{}^m$ and the linearized conductance $G_d{}^m$ of the companion network can thus be calculated. All the branches of the companion network are then determined, and we can solve for $v_d{}^{m+1}$ of Figure 5.6(a) by mesh or node analysis. From $v_d{}^{m+1}$, both $i_d{}^{m+1}$ and $G_d{}^{m+1}$ are formed and the iteration continues. If the iteration converges, branch currents and voltages of the linear companion network must *approach* those of the nonlinear network (for example, $i_d{}^{m+1} \to i_d$, $v_d{}^{m+1} \to v_d$). Thus, Property 2 becomes an exact relationship after convergence.

It may appear that we have gained only pedagogically because the solution of the network of Figure 5.6(a) for $i_d{}^{m+1}$ and $v_d{}^{m+1}$ yields the same results as solving two simultaneous linear equations, the solution method we adopted in Chapter 3. The advantage of the network model is that any of the general network analysis methods or network theorems applicable to linear resistive networks may be invoked in the solution, *after the companion network is formed.* For example, we could perform the source transformation of Figure 5.6(b). The single node equation at node ① then becomes

$$\left(\frac{1}{R_s} + G_d{}^m\right) v_d{}^{m+1} = -i_d{}^m + G_d{}^m v_d{}^m + \frac{v_s}{R_s}$$

which we solve to yield

$$v_d{}^{m+1} = \frac{-R_s i_d{}^m + R_s G_d{}^m v_d{}^m + v_s}{1 + R_s G_d{}^m}$$

This iteration formula agrees with Equation (5.11).

Analysis of a Large Network

The power of the above network model approach is more evident from consideraion of a more complicated problem. We will perform a dc analysis of the amplifier of Figure 5.7(a) using the companion network model of the entire circuit.

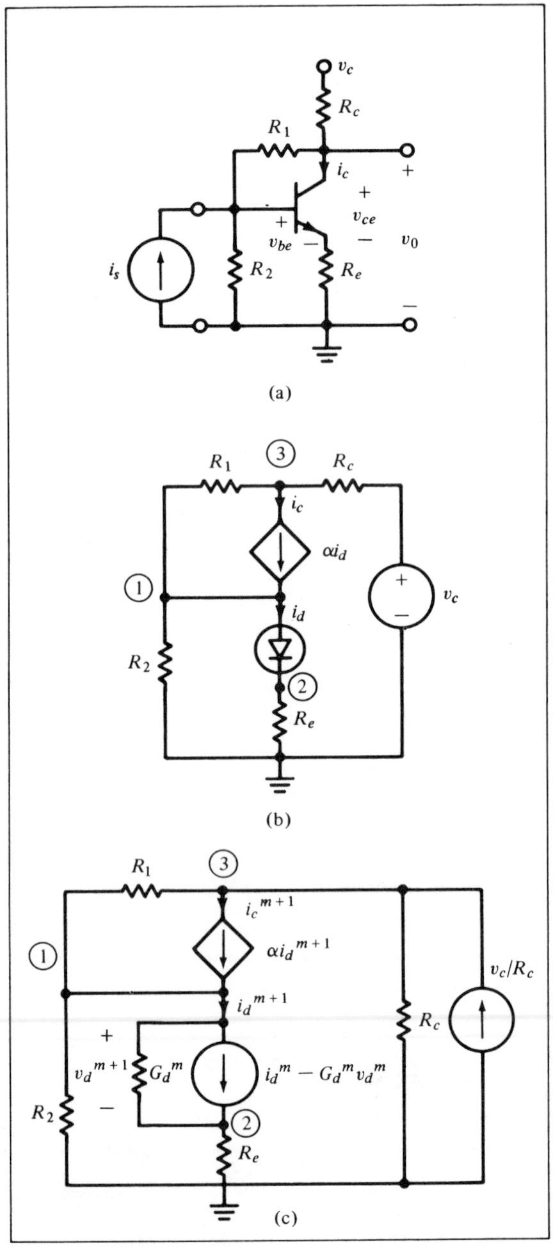

FIGURE 5.7 Evolution of companion model of transistor amplifier. (a) Transistor amplifier schematic; (b) network model; (c) companion network model ($R_1 = 20{,}000\ \Omega$, $R_2 = 3000\ \Omega$, $R_e = 300\ \Omega$, $R_c = 5000\ \Omega$, $v_c = 10$ volts, $\alpha = 0.98$).

The analysis of the companion network model of Figure 5.7(c) can be regarded as a two-step process. The node equations are first written with all current sources on the right-hand side, namely,

$$\begin{bmatrix} \frac{1}{R_1}+\frac{1}{R_2} & 0 & -\frac{1}{R_1} \\ 0 & \frac{1}{R_e} & 0 \\ -\frac{1}{R_1} & 0 & \frac{1}{R_c}+\frac{1}{R_1} \end{bmatrix} \begin{bmatrix} v_1^{m+1} \\ v_2^{m+1} \\ v_3^{m+1} \end{bmatrix} = \begin{bmatrix} -i_d^{m+1}+\alpha i_d^{m+1} \\ i_d^{m+1} \\ -\alpha i_d^{m+1}+\frac{v_c}{R_c} \end{bmatrix} \tag{5.15}$$

The (m) and $(m+1)$ superscripts are critical to the correct interpretation of this equation, and it is most important that *they be maintained throughout the solution process.* For example, i_d^{m+1} is an unknown; however, it can be written from Equation (5.14) as

$$i_d^{m+1} = (i_d^m - G_d^m v_d^m) + G_d^m(v_1^{m+1} - v_2^{m+1}) \tag{5.16}$$

Although $v_d^m = v_1^m - v_2^m$ is known from the previous iteration, v_1^{m+1} and v_2^{m+1} are unknowns in the present $(m+1)$st iteration. Like a controlled source, then, i_d^{m+1} cannot remain on the right-hand side of Equation (5.15). Substituting Equation (5.16) into Equation (5.15) and transferring expressions involving v_i^{m+1} to the left-hand side, we finally have

$$\begin{bmatrix} \frac{1}{R_1}+\frac{1}{R_2}+(-\alpha+1)G_d^m & -(1-\alpha)G_d^m & -\frac{1}{R_1} \\ -G_d^m & \frac{1}{R_e}+G_d^m & 0 \\ -\frac{1}{R_1}+\alpha G_d^m & -\alpha G_d^m & \frac{1}{R_c}+\frac{1}{R_1} \end{bmatrix} \begin{bmatrix} v_1^{m+1} \\ v_2^{m+1} \\ v_3^{m+1} \end{bmatrix} = \begin{bmatrix} (\alpha-1)(i_d^m - G_d^m v_d^m) \\ i_d^m - G_d^m v_d^m \\ -\alpha(i_d^m - G_d^m v_d^m) + v_c/R_c \end{bmatrix} \tag{5.17}$$

The iteration process begins by setting $m = 0$ and guessing a value for v_d^0, or, equivalently, for $v_1^0 - v_2^0$. From v_d^0, we calculate $i_d^0 = I_s(e^{\lambda v_d^0} - 1)$ and $G_d^0 = I_s\lambda e^{\lambda v_d^0}$. This calculation establishes both sides of Equation (5.17) and we may solve for v_i^1. Then we set $m = 1$ and repeat the process.

A program (DNET) to solve successively this equation is given in Figure 5.8(a). The main program is similar to the linear resistive network analysis program except that the convergence criterion

$$\text{ERR} = \sum_{j=1}^{3} (v_j^{m+1} - v_j^m)^2$$

is tested after every iteration. If ERR is less than some tolerance (chosen 10^{-6}), iteration is stopped.

The result of executing this program with the input data of Figure 5.8 (b) is shown in Figure 5.8(c). The iteration is seen to converge in six steps.

```
C****  DC ANALYSIS OF TRANSISTOR AMPLIFIER (DNET)
      IMPLICIT REAL*8(A-H,O-Z)
      REAL*8 G(10,10),I(10),V(10)
      DIMENSION NODOUT(2,5)
      NDIM=10
C****  READ INPUT DATA
    1 M=0
      CALL READ(NN,NODOUT,NOUT)
      WRITE(6,15)
      DO 4 J=1,NN
    4 V(J)=0.D0
C****  SET UP NODE EQUATIONS
    3 CALL NODEQ(G,V,I,NN,NDIM)
C****  SOLVE NODE EQUATIONS
      CALL REDUCE(G,NN,I,NDIM)
C****  V(J)-I(J) IS THE INCREMENT IN THE JTH NODE VOLTAGE
      ERR=0.D0
      DO 2 J=1,NN
      ERR=ERR+(V(J)-I(J))**2
    2 V(J)=I(J)
C****  CALCULATE AND PRINT OUTPUT VOLTAGE
      CALL OUTPUT(NODOUT,NOUT,V,M)
      M=M+1
C****  TEST CONVERGENCE CRITERION
      IF(ERR.GT.1.D-6.AND.M.LT.15)GO TO 3
      IF(M.EQ.15)WRITE(6,16)
   15 FORMAT(//'  DC ITERATION'//)
   16 FORMAT(' NO CONVERGENCE AFTER 15 ITERATIONS'/)
      STOP
      END
      SUBROUTINE READ(NN,NODOUT,NOUT)
      IMPLICIT REAL*8(A-H,O-Z)
      DIMENSION NODOUT(2,1)
      COMMON R1,R2,RE,RC,VC,ALF,VDI
      READ(5,3)R1,R2,RE,RC,VC,ALF,VDI
      READ(5,4)NN,NOUT,(NODOUT(1,J),NODOUT(2,J),J=1,NOUT)
      WRITE(6,2)R1,R2,RE,RC,VC,ALF
      RETURN
    2 FORMAT(' R1 =',1PD11.4,' R2 =',1PD11.4,' RE =',1PD11.4/
     1' RC =',1PD11.4,' VC =',1PD11.4,' ALF =',1PD11.4)
    3 FORMAT(6F10.0)
    4 FORMAT(20I2)
      END
      SUBROUTINE NODEQ(G,V,I,NN,NDIM)
      IMPLICIT REAL*8 (A-H,O-Z)
      REAL*8 G(NDIM,1),I(1),V(1),LAMDA
      COMMON R1,R2,RE,RC,VC,ALF,VDI
      DATA INIT/0/
      DATA CSAT/1.D-9/,LAMDA/40.D0/
C****  ZERO NODE MATRIX AND CURRENT VECTOR
      DO 1K=1,NN
      I(K)=0.D0
      DO 1 J=1,NN
    1 G(J,K)=0.D0
C****  CALCULATE DIODE PARAMETERS
      VDIODE=V(1)-V(2)
C****  APPLY INITIAL DIODE VOLTAGE IF ON FIRST ITERATION
```

FIGURE 5.8 (a) Diagram to perform dc analysis of amplifier of Figure 5.7; (b) input data for DNET; (c) output data for DNET.

5.5 Network Scaling

The analysis of a large network sometimes involves numbers either too large or too small for modern computers to handle. For an IBM 370 series machine, for example, numbers must lie between 10^{75} and 10^{-75} in magnitude. For this reason, it is often necessary to scale problems prior to computation.

```
      IF(INIT.EQ.0)VDIODE=VDI
      DEX=DEXP(LAMDA*VDIODE)
      GD=LAMDA*CSAT*DEX
      CUR=CSAT*(DEX-1.D0)-GD*VDIODE
C****  LOAD NODE MATRIX
      G(1,2)=-GD*(1.D0-ALF)
      G(1,3)=-1.D0/R1
      G(1,1)=-G(1,2)-G(1,3)+1.D0/R2
      G(2,1)=-GD
      G(3,1)=ALF*GD-1.D0/R1
      G(2,2)=GD+1.D0/RE
      G(3,2)=-ALF*GD
      G(3,3)=1.D0/RC+1.D0/R1
C****  LOAD CURRENT VECTOR
      I(1)=(ALF-1.D0)*CUR
      I(2)=CUR
      I(3)=-ALF*CUR+VC/RC
      INIT=1
      RETURN
      END
      SUBROUTINE OUTPUT(NODOUT,NOUT,V,M)
      IMPLICIT REAL*8 (A-H,O-Z)
      DIMENSION NODOUT(2,1),V(1),VOUT(5)
      IF(M.EQ.0)WRITE(6,4)(NODOUT(1,J),NODOUT(2,J),J=1,NOUT)
    3 DO 2 I=1,NOUT
      VOUT(I)=0.D0
      N1=NODOUT(1,I)
      N2=NODOUT(2,I)
      IF(N1.NE.0)VOUT(I)=VOUT(I)+V(N1)
      IF(N2.NE.0)VOUT(I)=VOUT(I)-V(N2)
    2 CONTINUE
      WRITE(6,1)M,(VOUT(I),I=1,NOUT)
      RETURN
    1 FORMAT(4X,I4,4X,5(1PD12.4))
    4 FORMAT('  ITERATION  ',5(I2,' -',I2,' NODES'))
      END
      SUBROUTINE REDUCE(A,N,B,NDIM)
      IMPLICIT REAL*8(A-H,O-Z)
      DIMENSION A(NDIM,1),B(1)
C****  THIS SUBROUTINE SOLVES A*X=B BY GAUSS ELIMINATION
C****  N IS THE MATRIX SIZE; X IS RETURNED IN B; A IS DESTROYED
C****  CONVERT A TO UPPER TRIANGULAR FORM
      NM1=N-1
      IF(N.EQ.1)GO TO 8
      DO 1 J=1,NM1
      IJ=J+1
      DO 4 K=IJ,N
    4 A(J,K)=A(J,K)/A(J,J)
      B(J)=B(J)/A(J,J)
      DO 1 I=IJ,N
      ALPHA=A(I,J)
      DO 2 K=IJ,N
    2 A(I,K)=A(I,K)-ALPHA*A(J,K)
    1 B(I)=B(I)-ALPHA*B(J)
    8 B(N)=B(N)/A(N,N)
C****  PERFORM BACK SUBSTITUTION
      IF(N.LT.2)GO TO 7
      DO 3 J=2,N
      LAST=N-J+1
      LASTP1=LAST+1
      DO 3 I=LASTP1,N
    3 B(LAST)=B(LAST)-A(LAST,I)*B(I)
    7 RETURN
      END
```

(a)

FIGURE 5.8 (Continued)

```
20000.    3000.    300.    5000.    10.    .98
.4
 3 2 3 1 1 2
```

(b)

```
R1 = 2.0000D 04 R2 = 3.0000D 03 RE = 3.0000D 02
RC = 5.0000D 03 VC = 1.0000D 01 ALF = 9.8000D-01
 DC ITERATION
 ITERATION    3 - 1 NODES 1 - 2 NODES
       0        4.4112D 00  3.7722D-01
       1        4.3352D 00  3.5791D-01
       2        4.2861D 00  3.4543D-01
       3        4.2693D 00  3.4117D-01
       4        4.2678D 00  3.4078D-01
       5        4.2678D 00  3.4078D-01
```

(c)

FIGURE 5.8 (Continued)

In an electrical network analysis problem, either all currents or all voltages can be scaled in magnitude (in some rare cases, both must be scaled). We call such scaling *current* and *voltage* scaling.

To exemplify, consider the network illustrated in Figure 5.9 containing a diode and a resistor described by

$$i_d = I_s(e^{\lambda v_d} - 1) \qquad i_R = \frac{v_R}{R}$$

Suppose that we scale currents by a factor b;

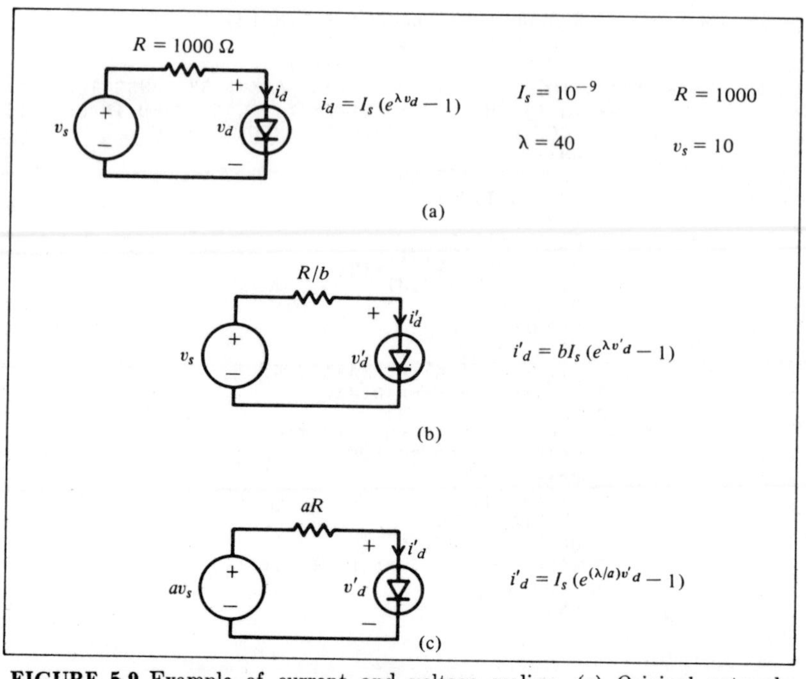

FIGURE 5.9 Example of current and voltage scaling. (a) Original network ($i_d = 9.6$ ma, $v_d = 0.402$); (b) current scaled network ($b = 1000$, $i_d' = 9.6$ amps, $v_d' = 0.402$); (c) voltage scaled network ($a = 10$, $i_d' = 9.6$ ma, $v_d' = 4.02$).

that is, every current of value i is to be replaced by one of value $i' = bi$. Then the above equations become

$$i_d' = bi_d = bI_s(e^{\lambda v_d'} - 1)$$

$$i_R' = bi_R = \frac{bv_R'}{R} = \frac{v_R'}{(R/b)}$$

The *voltages* are identical before and after scaling, so that $v = v'$ for each element. Therefore, the scaled network contains a resistor of value (R/b) and a diode with a saturation current of value (bI_s). The value of i_R' *after solution* of the scaled network must then be divided by b to obtain the correct unscaled value. For example, if $I_s = 10^{-9}$ amps and $R = 1000\Omega$ before scaling, then a scaling factor of $b = 1000$ results in $I_s = 10^{-6}$ amps and $R = 1\Omega$. If a current somewhere in the scaled network is calculated to be 9.6 amps, then the true value is $i = i'/b = 9.6/1000 = 0.0096$ amp or 9.6 ma.

Voltage scaling is similar. If we replace every voltage v by a voltage of value $v' = av$, the original diode and resistor equations then become

$$i_d' = I_s(e^{(\lambda/a)v_d'} - 1) \qquad i_R' = \frac{v_R'}{aR}$$

Note that the effect on the linear resistance value is similar to that observed from current scaling, but the diode equation, which represents a nonlinear resistance, is scaled quite differently.

5.6 Generalized Newton Iteration

The iteration formula of Equation (5.17) can also be obtained without the companion model from linearization of the equations that describe the nonlinear network. For example, the application of KCL at the nodes of the network model of Figure 5.7(b) yield the node equations

$$\left(\frac{1}{R_1} + \frac{1}{R_2}\right)v_1 - \frac{1}{R_1}v_3 + (1 - \alpha)i_d = 0 \tag{5.18}$$

$$\frac{1}{R_e}v_2 - i_d = 0 \tag{5.19}$$

$$-\frac{1}{R_1}v_1 + \left(\frac{1}{R_1} + \frac{1}{R_c}\right)v_3 + \alpha i_d - \frac{v_c}{R_c}v_c = 0 \tag{5.20}$$

where

$$i_d = I_s(e^{\lambda(v_1 - v_2)} - 1)$$

These equations are of the general form

$$F_1(v_1, v_2, \ldots, v_n) = 0 \tag{5.21}$$

$$F_2(v_1, v_2, \ldots, v_n) = 0 \tag{5.22}$$

$$\vdots$$

$$F_n(v_1, v_2, \ldots, v_n) = 0 \tag{5.23}$$

where $n = 3$. Our objective is to find a set of node voltages v_i that result in the satisfaction of Equations (5.21) to (5.23).

We denote the value of the v_i at the mth iteration as v_i^m and the value of F_j as

$$F_j^m \triangleq F_j(v_1^m, v_2^m, \ldots, v_n^m)$$

We assume that the v_i^m are known so that F_j^m can be evaluated. If all F_j^m are zero, we have satisfied Equations (5.21) to (5.23) and the v_i^m are the desired solution voltages. If not, then the value of F_j^{m+1} can be approximated by the first term in a Taylor's series expansion of F_j in the n variables $v_1, \ldots, v_n$ about the point $v_1^m, \ldots, v_n^m$. This is analogous to the expansion of i_d^{m+1} about v_d^m in Equation (5.12) and has the form

$$F_j^{m+1} = F_j^m + \sum_{i=1}^{n} \left(\frac{\partial F_j}{\partial v_i}\right)^m (v_i^{m+1} - v_i^m)$$

$$= 0 \qquad (j = 1, 2, \ldots, n)$$

The iteration algorithm is established by setting all $F_j^{m+1} = 0$, as required by KCL, leaving

$$\sum_{i=1}^{n} \left(\frac{\partial F_j}{\partial v_i}\right)^m (v_i^{m+1} - v_i^m) = -F_j^m \tag{5.24}$$

In component form, this becomes

$$\begin{bmatrix} \frac{\partial F_1}{\partial v_1} & \frac{\partial F_1}{\partial v_2} & \cdots & \frac{\partial F_1}{\partial v_n} \\ \frac{\partial F_2}{\partial v_1} & \frac{\partial F_2}{\partial v_2} & \cdots & \frac{\partial F_2}{\partial v_n} \\ \cdot & \cdot & & \cdot \\ \cdot & \cdot & & \cdot \\ \cdot & \cdot & & \cdot \\ \frac{\partial F_n}{\partial v_1} & \frac{\partial F_n}{\partial v_2} & \cdots & \frac{\partial F_n}{\partial v_n} \end{bmatrix}^m \begin{bmatrix} \Delta v_1^m \\ \Delta v_2^m \\ \cdot \\ \cdot \\ \cdot \\ \Delta v_n^m \end{bmatrix} = \begin{bmatrix} -F_1^m \\ -F_2^m \\ \cdot \\ \cdot \\ \cdot \\ -F_n^m \end{bmatrix} \tag{5.25}$$

where $\Delta v_i^m = v_i^{m+1} - v_i^m$. When $F_j = 0$ for all i at the solution, then the right-hand side vector is zero, all $\Delta v_i^m = 0$, and the iteration has converged. This iteration process is known as the generalized Newton iteration.

Applied to the F_j of Equations (5.18) to (5.20), the matrix equation of (5.25) has the form

$$\begin{bmatrix} \frac{1}{R_1} + \frac{1}{R_2} + (-\alpha + 1)G_d{}^m & -(1-\alpha)G_d{}^m & -\frac{1}{R_1} \\ -G_d{}^m & \frac{1}{R_e} + G_d{}^m & 0 \\ -\frac{1}{R_1} + \alpha G_d{}^m & -\alpha G_d{}^m & \frac{1}{R_c} + \frac{1}{R_1} \end{bmatrix} \begin{bmatrix} \Delta v_1^m \\ \Delta v_2^m \\ \Delta v_3^m \end{bmatrix}$$

$$= \begin{bmatrix} -\left(\frac{1}{R_1} + \frac{1}{R_2}\right) v_1^m + \frac{1}{R_1} v_3^m - (1-\alpha) i_d{}^m \\ -\frac{1}{R_2} v_2^m + i_d{}^m \\ \frac{1}{R_1} v_1^m - \left(\frac{1}{R_1} + \frac{1}{R_c}\right) v_3^m - \alpha i_d{}^m + \frac{v_c}{R_c} \end{bmatrix}$$

where $\Delta v_i^m = v_i^{m+1} - v_i^m$. In this equation the matrix is identical to that derived in (5.17) which resulted from the network model formulation. However, the *changes* in the v_i are determined above. The updated voltages $v_i^{m+1} = v_i^m + \Delta v_i^m$ are identical in either case.

5.7 Transistor Models

Introduction

The ability to analyze nonlinear networks of greater complexity permits us to investigate the functional and detailed circuit performance of a large class of electronic circuits. One important class is the transistor switching circuit, used in digital computers, which is described later in this chapter. The transistors in these circuits do not operate in the linear region. Therefore, the transistor model discussed in Chapter 3 [Figure 5.10(a)] is inadequate for detailed analysis of such circuits.

Complete Transistor Models[2]

The first addition usually made to the simple familiar model is a representation of the collector diode, which appears in Figure 5.10(b) between base and collector. Next, just as the

[2] In teaching this material, instructors may find it useful to precede this modeling discussion with a descriptive presentation of transistor physics, as given in Appendix D.

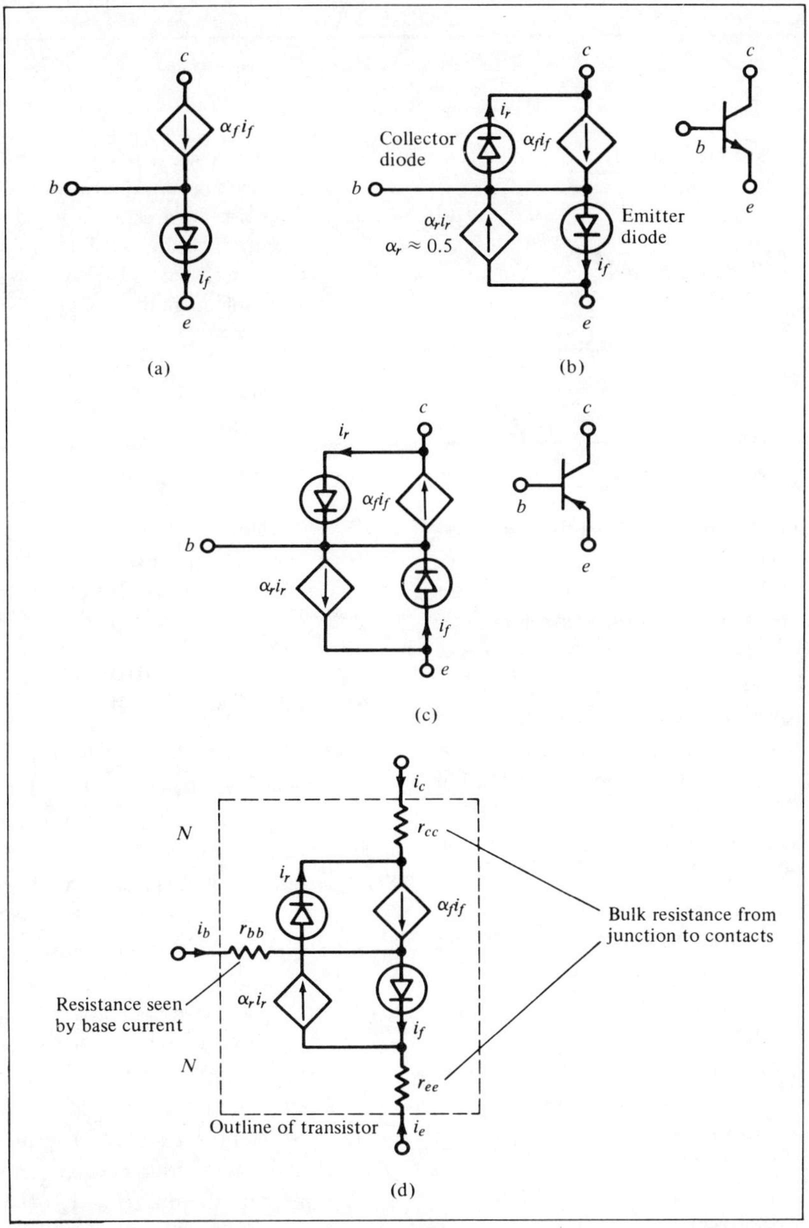

FIGURE 5.10 Transistor models [the *pnp* model of (c) represents a transistor fabricated so that the diodes and controlled sources of the *npn* model are effectively reversed]. (a) Elementary model (*npn*); (b) complete model (*npn*); (c) complete model (*pnp*); (d) *npn* transistor model for computer analysis.

emitter generates charges that flow to the collector diode, so too can the collector produce charges that travel toward the emitter. Due to the internal construction of the transistor, only a small fraction ($\alpha_r \cong 0.5$) reach the emitter. This fraction is indicated in the model of Figure 5.10 (b) by a CCCS of value α_r, which is analogous to the α_f generator of the simpler model of Figure 5.10 (a).

The transistor, with the collector diode

TABLE 5.1

Transistor States

State	Emitter Diode	Collector Diode	Comment
Forward	Forward-biased	Reverse-biased	Linear forward region; $i_c = \beta_f i_b \cong 50 i_b$ (typical); high amplification possible.
Reverse	Reverse-biased	Forward-biased	Linear reverse region; $i_e = \beta_r i_b \cong i_b$ (typical); amplification possible, usually low.
Saturated[a]	Forward-biased	Forward-biased	Transistor modeled by voltage sources; collector current unresponsive to base current changes.
Cutoff[a]	Reverse-biased	Reverse-biased	Transistor modeled by open circuits; collector current unresponsive to base voltage changes.

[a] See explanation of Section 5.9.

added, can operate in any of four states or regions, as described in Table 5.1. We have encountered the forward and cutoff states before in Chapter 3. The reverse state has little practical use because $\beta_r = \alpha_r/(1 - \alpha_r)$ is usually small (although it is interesting to observe that the collector and emitter are usually interchangeable without disasterous results). The saturated state will be examined in detail in the discussion in Section 5.9.

For high current applications, even the models of Figures 5.10(b)–(c) are inadequate. The emitter, collector, and base portions of the transistor all have a linear resistance associated with them as depicted in Figure 5.10(d). These resistances can be made to vary considerably (by changing the transistor fabrication process) to match the needs of an application. For example, we will shortly demonstrate a need for digital circuits to have a small collector resistance r_{cc}.

Field-Effect Transistors

The transistor we have studied to this point is the so-called "bipolar" transistor, the name alluding to the fact that both positive and negative charge carriers take part in device operation. In contrast, a more recently discovered device known as the *field-effect transistor* (FET) operates by the flow of only one type of carrier. The reader is invited to study its operation and model, which are described in Appendix D. A number of problems throughout the text involve the use of the FET.

5.8 A General Nonlinear Computer-Aided Analysis Program

The Need for a General Program

The technique of generating linearized models of each network branch seemingly provides the opportunity to analyze nonlinear networks of any size. However, the manipulation of the node equations necessary to describe controlled sources, voltage sources, and variables such as i_d^{m+1} of Figure 5.7(c) which depend on unknown node voltages is a bookkeeping problem beyond human reliability and endurance for any but the smallest useful networks.

When transistor models such as the above are used in the analysis of amplifiers, it becomes worthwhile to develop an automatic procedure for equation formulation by the computer. A general program to carry out both the formulation and numerical solution of nonlinear resistive networks is described in detail in the Instructor's Manual accompanying the text. Feeling that many readers will not wish to become involved in the rather complicated details of its construction, but will find it useful in evaluating circuit performance, we consider only certain aspects of its use in the next section.

(a) A name card; Cols. 1–60 will be printed on output sheet.
(b) Cards describing branches of the network (cards may be in any order). Cols. 1–4: Branch name (left justified) beginning with

V—voltage source
I—current source
R—resistance
D—exponential diode
J—FET VCCS

Col. 5: Type of control (*V* or *I*) if controlled source; otherwise blank. Cols. 6–9: Name of controlling branch (left justified) if controlled source; otherwise blank. Cols. 11–12 and 14–15: Node numbers establishing the following polarity (for all branches, including sources): The numbering must be consecutive, beginning with 00, but ordering is arbitrary. Col. 17: Polarity of output; blank for plus, – for minus. Col. 18: If branch voltage is to be considered an output quantity, use symbol V. Cols. 21–30: Value of element or strength of source, or initial diode voltages, or I_{dss}. Cols. 31–40, 41–50, . . . , 71–80: Values of stepped elements or stepped strengths of stepped controlled sources; all stepped elements must be stepped the same number of times; stepped element values may *not* be continued on succeeding cards. The last stepped value of an element may not be zero.

(c) Constant card (default values assigned if this card is omitted) Cols. 1–4: KON. Cols. 21–30: I_s – diode saturation current in nanoamps (default, 1). Cols. 31–40: λ—diode exponential constant (default, 40). Cols. 41–50: V_p—FET pinchoff constant (default, –4).

(d) Last card. Cols. 1–4: AMEN

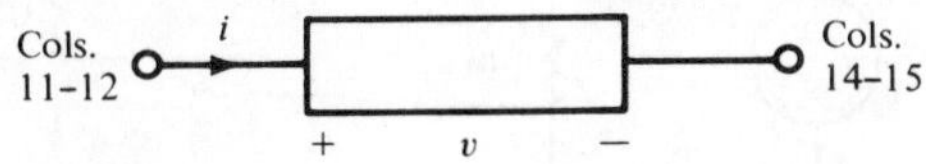

FIGURE 5.12 Input data format for DCAP.

A General dc Analysis Program (DCAP)

General engineering analysis programs are written for the convenience of the user. This implies that the "language" which the engineer uses to describe his problem to the computer should be a natural one. In the case of electrical networks, the input language must contain all the topological and constitutive information that characterize a network. The most common procedure is to describe each branch by its type (resistor, source, and so on), node numbers, and value, as shown in Figure 5.11.

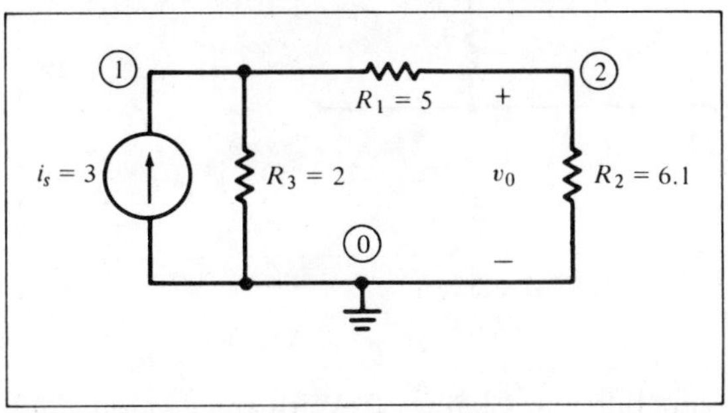

Type (v, R, or i)	Node Numbers	Output	Value
i_s	0 1		3.
R_1	1 2		5.
R_2	2 0	v	6.1
R_3	1 0		2.

FIGURE 5.11 Network analysis input language.

An extended form of this language is used in the DCAP program and is described in Figure 5.12. This program has the ability to perform a dc analysis of networks of (1) resistors, (2) independent current and voltage sources, (3) VCCS's and CCCS's, (4) diodes, and (5) FET nonlinear VCCS's. The use of the program is illustrated by the example.

Example 5.4

The output voltage v_0 of the two-transistor amplifier of Figure 5.13(a) is to be found with a signal change of 0.05 volt applied abruptly at $t = 0$. The complete transistor models of Figure 5.10(b) and (c) will be used.

The problem is solved by carrying out two analyses of the dc network model of Figure 5.13(b), one for each value of the input signal. The input data to describe this problem to DCAP are shown in Figure 5.14(a). The results of Figure 5.14(b) can be summarized as follows:

$$v_{s_1} = 0 \qquad v_{0_1} = 4.5760$$
$$v_{s_2} = 0.05 \qquad v_{0_2} = 5.1336$$

The signal change of 0.05 volt therefore produces an output change of 0.5576 volt. We have determined the amplifier has a voltage gain of

$$\mu = \frac{v_{0_2} - v_{0_1}}{v_{s_2} - v_{s_1}} = \frac{0.5576}{0.05} = 11.15$$

a value we recognize from Section 5.2, Example 5.3! We will explain this in Problem 6.14.

5.9 Digital Circuit Analysis

Introduction

In Chapter 4, our introduction to digital circuits was limited to simple diode AND and OR

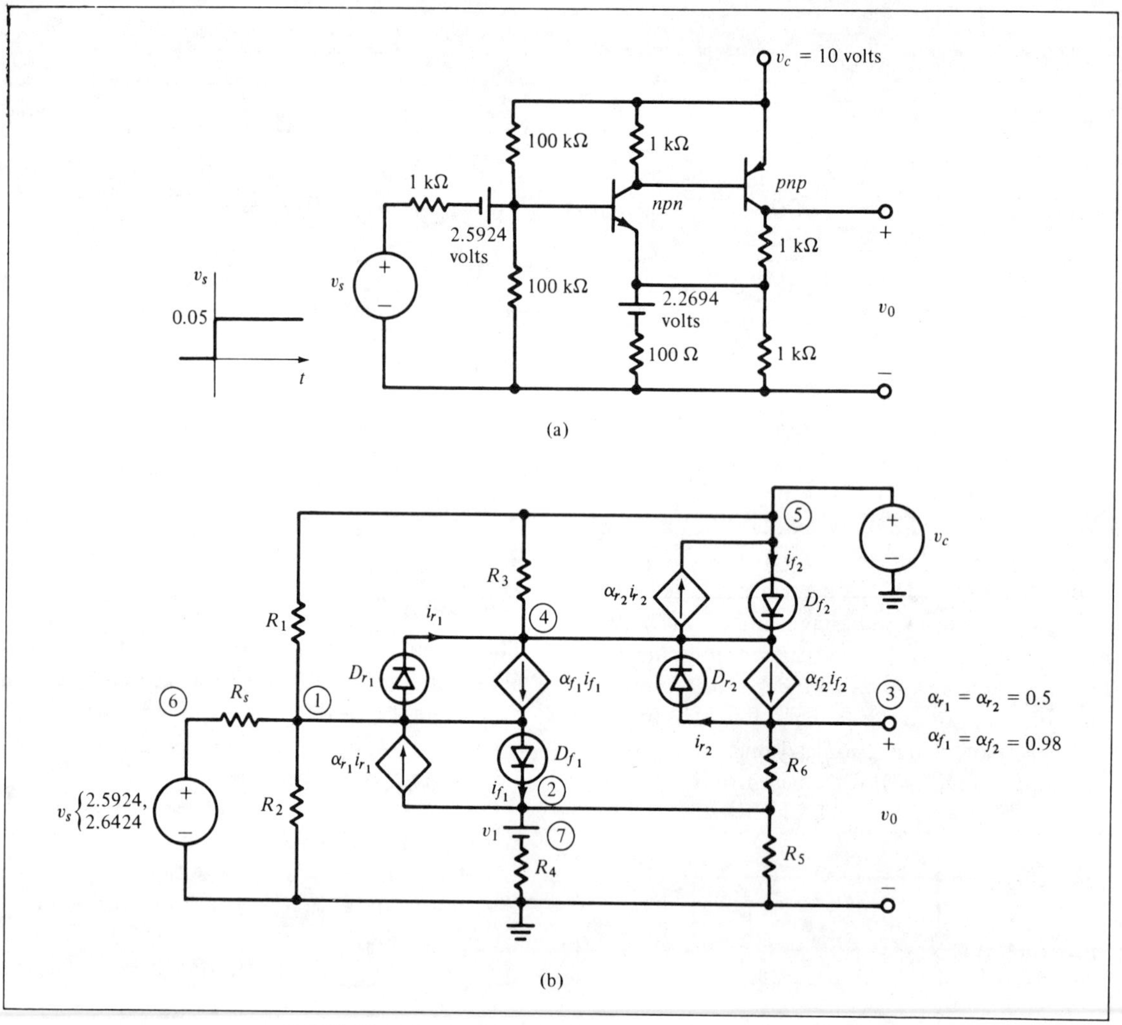

FIGURE 5.13 Transistor amplifier for DCAP analysis.

gates. Clearly, transistors should have a vital role to play in the design of digital circuits.

One reason for deferring study of such transistorized digital circuits was that we did not have available at that time the complete nonlinear transistor models of Figure 5.10, nor the extended analysis capability offered by the companion model or the DCAP program to correlate simple functional and detailed circuit analyses of a circuit. Also, many useful digital circuits require *two* transistors for operation, posing a circuit analysis problem beyond our capability in Chapter 3. With the general computer-aided analysis techniques of this chapter, however, we can study the behavior of such circuits in reasonable detail.

Inverter

One of the simplest transistor logic circuits is the inverter or NOT circuit of Figure 5.15. Partly because it consists of resistors and a transistor, it belongs to a class known as RTL (resistor-transistor-logic) digital logic. The logical function performed by this circuit is

```
TWO-STAGE TRANSISTOR AMPLIFIER
VS          06 00       2.5924      2.6424
RS          06 01       1000.
R1          05 01       100000.
R2          01 00       100000.
DR1         01 04        0.
IR1  IDR1   02 01        .5
DF1         01 02         .4
IF1  IDF1   04 01        .98
V1          02 07         2.2694
R4          07 00        100.
R3          04 05          1000.
R5          02 00          1000.
R6          03 02          1000.
DR2         03 04          0.
IR2  IDR2   04 05          .5
DF2         05 04          .4
IF2  IDF2   04 03          .98
VC          05 00          10.
IOUT        03 00   V      0.
AMEN
```

(a)

```
**********TWO-STAGE TRANSISTOR AMPLIFIER
BRANCH NODE      ELEMENT     BRANCH CON-
 NAME    NOS.     VALUE      NAME    TROL
VS     6   0     2.592D 00
RS     6   1     1.000D 03
R1     5   1     1.000D 05
R2     1   0     1.000D 05
IR1    2   1     5.000D-01   DR1      I
DF1    1   2     4.000D-01
IF1    4   1     9.800D-01   DF1      I
V1     2   7     2.269D 00
R4     7   0     1.000D 02
R3     4   5     1.000D 03
R5     2   0     1.000D 03
R6     3   2     1.000D 03
IR2    4   5     5.000D-01   DR2      I
DF2    5   4     4.000D-01
IF2    4   3     9.800D-01   DF2      I
VC     5   0     1.000D 01
    DC ITERATION
  ITERATION     3 - 0 NODES
       0        3.9467D 00
       1        4.2178D 00
       2        4.4280D 00
       3        4.5442D 00
       4        4.5743D 00
       5        4.5760D 00
       6        4.5760D 00
   DIODE INFORMATION AT DC SOLUTION
         DC INFO. (CHAPTER V)      INCREMENTAL INFO. (CHAPTER VI)
DIODE     VOLTAGE      CURRENT         CONDUCTANCE RESISTANCE
DR1    -7.002D 00   -1.000D-09        1.000D-60    1.000D 60
DF1     3.238D-01    4.211D-04        1.684D-02    5.937D 01
DR2    -5.058D 00   -1.000D-09        1.000D-60    1.000D 60
DF2     3.664D-01    2.315D-03        9.259D-02    1.080D 01
BRANCH NODE      ELEMENT     BRANCH CON-
 NAME    NOS.     VALUE      NAME    TROL
VS     0   8     2.642D 00
  ITERATION     3 - 0 NODES
       0        5.1324D 00
       1        5.1336D 00
       2        5.1336D 00
   DIODE INFORMATION AT DC SOLUTION
         DC INFO. (CHAPTER V)      INCREMENTAL INFO. (CHAPTER VI)
DIODE     VOLTAGE      CURRENT         CONDUCTANCE RESISTANCE
DR1    -6.948D 00   -1.000D-09        1.000D-60    1.000D 60
DF1     3.247D-01    4.369D-04        1.747D-02    5.722D 01
DR2    -4.495D 00   -1.000D-09        1.000D-60    1.000D 60
DF2     3.714D-01    2.835D-03        1.134D-01    8.818D 00
```

(b)

FIGURE 5.14 (a) DCAP input data for amplifier of Figure 5.13(b); (b) DCAP computer output.

described by the truth table of Figure 5.15 (b). A logical **0** (0 volts) is converted to a logical **1** (v_c volts) and vice versa.

The circuit function is easy to verify because the circuit configuration is identical to that of the basic amplifier studied in Chapter 3. However, the transistor operation is not restricted to its normal (linear) region; rather, the input signal voltage change (and the resultant change in base current) is sufficiently large that the transistor operates at the two limits of its linear region.

The one limit corresponds to saturation of the transistor when $v_s = v_c$ (a logical **1**). Both diodes conduct as shown in Figure 5.16 (a) and (b). We recall from Chapter 3 that a conducting diode can be simply modeled as a battery of value v_{d0}. The emitter and collector diodes can then be modeled for our analysis by voltage sources of values v_{be0} and v_{bc0} as shown. If we calculate the input and output voltages of the saturated transistor model, we find that $v_{be_{sat}} = v_{be0}$ and $v_{ce_{sat}} = v_{be0} - v_{bc0}$. The alternative transistor representation of Figure 15.16 (b)

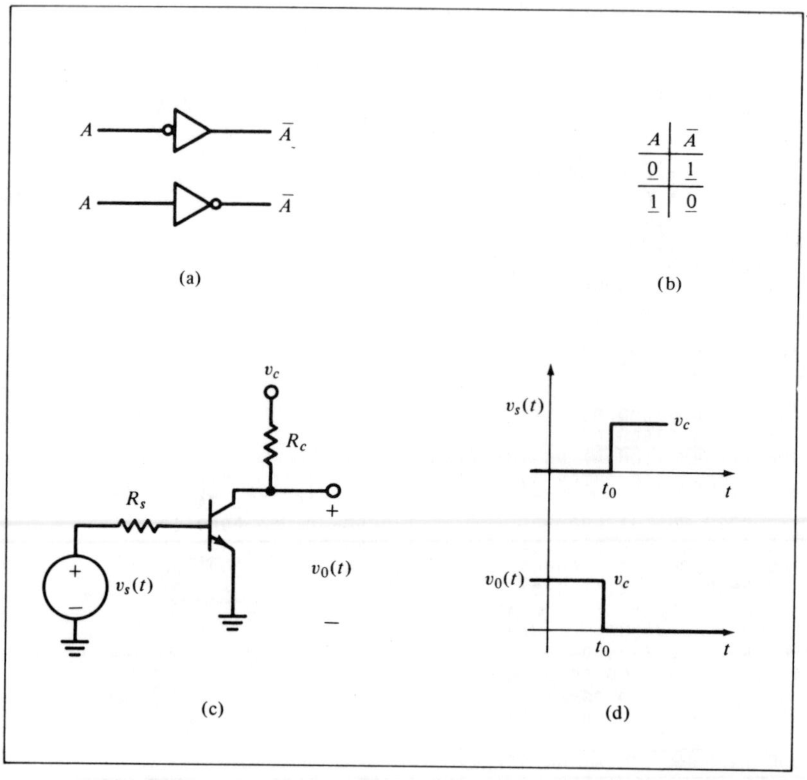

FIGURE 5.15 The inverter. (a) Symbol; (b) truth table; (c) circuit and signals (ideal).

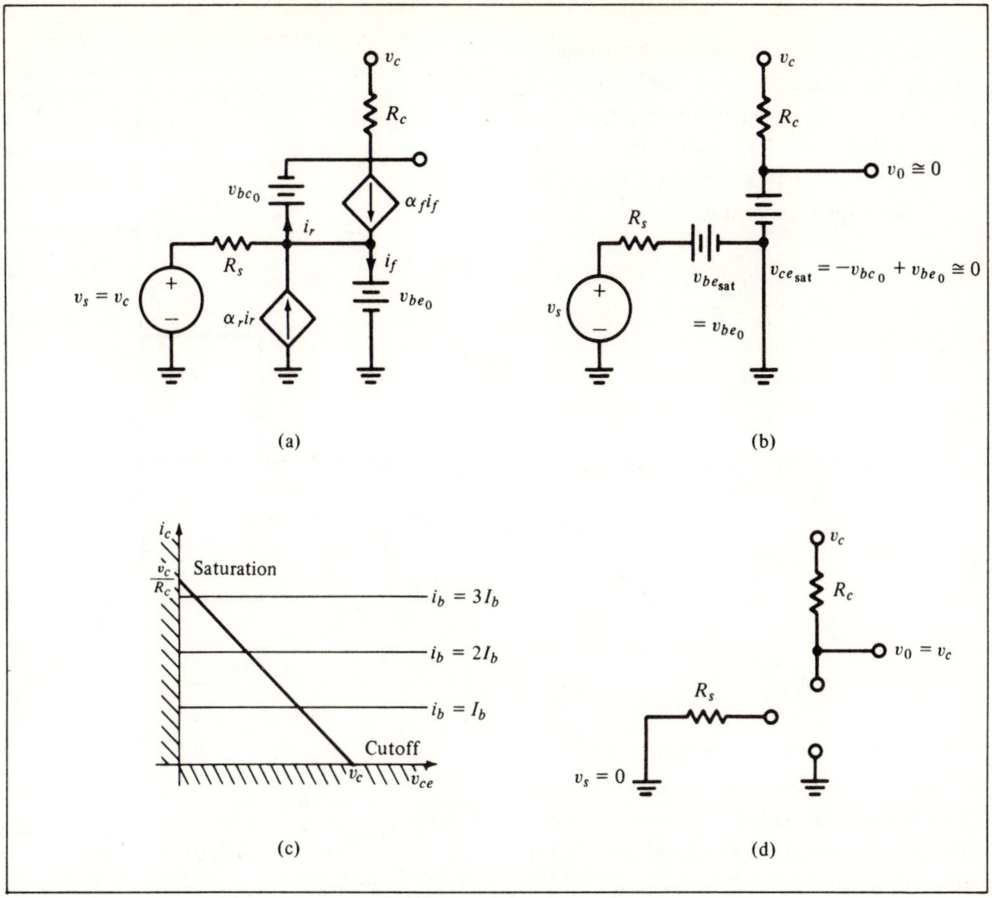

FIGURE 5.16 Inverter analysis. (a) and (b) Idealized saturated models; (c) output characteristic; (d) idealized cutoff model.

with constant voltage sources of value $v_{be_{sat}}$ and $v_{ce_{sat}}$ across the input and output behaves identically to that of Figure 15.16(a). Clearly, if both conducting diodes have nearly equal voltages, so that $v_{bc0} \cong v_{be0}$, then $v_{ce_{sat}} \cong 0$. With this assumption, we can come to the conclusion that a logical **0** *output* is provided for a logical **1** *input*.

With a logical **0** (0 volts) input, the base current is zero, both emitter and collector diodes are reverse-biased and the transistor operates in the cutoff state as indicated on the load line of Figure 5.16(c). The output voltage is $v_0 = v_c$, a logical **1**. Thus, with these simple transistor models, we conclude that the circuit does indeed operate as an inverter or NOT circuit.

It is worth noting that the linearity of the transistor characteristics between the saturation and cutoff states evidently is not critical to operation of the inverter, because the transistor is not operated in this region except during the *transitions* between these states. However, the linear model *is* useful in estimating the transistor β_f required to perform the switching. From Chapter 3, Equation (3.16), in the linear region of operation a signal change Δv_s produces a change in v_0 of

$$\Delta v_0 = \frac{-\beta_f R_c}{R_s} \Delta v_s$$

where $\beta_f = \alpha_f/(1 - \alpha_f)$. This output voltage change must be equal to $-v_c$ in order for $v_0 = v_{ce}$ to change from cutoff to saturation. Because the input signal at saturation has a value $v_s = v_c$ to represent a logical **1** input, to affect the desired output change $\Delta v_0 = -v_c$ we require

$$-v_c = \frac{-\beta_f R_c}{R_s} v_s$$

$$= \frac{-\beta_f R_c}{R_s} v_c$$

Therefore, saturation will be achieved when

$$\beta_f = \frac{R_s}{R_c} \tag{5.26}$$

This value of β_f is the *minimum* value necessary. If

$$\beta_f > \frac{R_s}{R_c} \tag{5.27}$$

then the controlled source α_f will simply force more current through the emitter and collector diodes; the transistor will, however, continue to appear as a constant voltage source from both its input and output terminals.

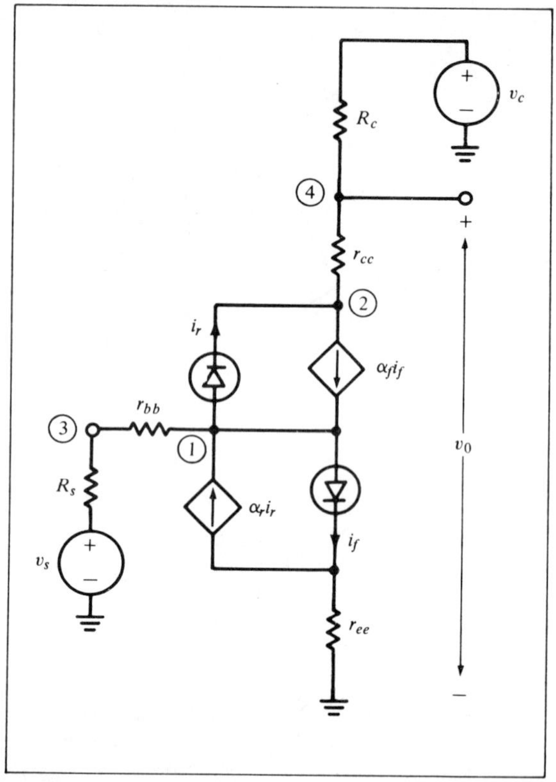

FIGURE 5.17 A typical inverter circuit. v_c = 5 volts, v_s = 0.5 volts, R_c = 500 Ω, R_s = 500 Ω, r_{bb} = 15 Ω, r_{ee} = 0.5 Ω (ignored), r_{cc} = 10 Ω, α_b = 0.98, α_r = 0.5, $I_s = 10^{-9}$ amps, $\lambda = 40$ volts^{-1}.

Computer Analysis of Inverter

Unfortunately, the preceding amplifier analysis omits certain key properties of the inverter which affect its connection to other digital circuits. We can use the complete transistor model of Figure 5.10(d) to obtain a representative analysis of a real inverter circuit. This leads us naturally to use of the computer-aided analysis methods developed earlier in the chapter.

The complete model of a typical inverter is illustrated in Figure 5.17. As v_s is varied from 0 to 5 volts—a logical **0** and **1** respectively—the transistor can be expected to range from cutoff to saturation. The latter is reasonable in this case, because $\beta_f \gg R_s/R_c$ so that inequality (5.27) is satisfied.

It is often important to know the manner in which this transistion between logical states is made. The dc *input-output* or *transfer* characteristic of Figure 5.18(a) is useful for this purpose. This plot shows the output voltage v_0 for every level of input voltage between 0 and 5 volts. We can determine from this graph what levels of input voltage produce logical levels of **0** and **1** at the output. For example, an input voltage $0 \leqq v_s \leqq 0.3$ produces a 5-volt output; a voltage $v_s \geqq 0.6$ produces an output voltage less than 0.18 volt. If we were to assign the logical levels

Logical **0** for $v \leqq 0.1$ volt

Logical **1** for $v \geqq 4.9$ volts

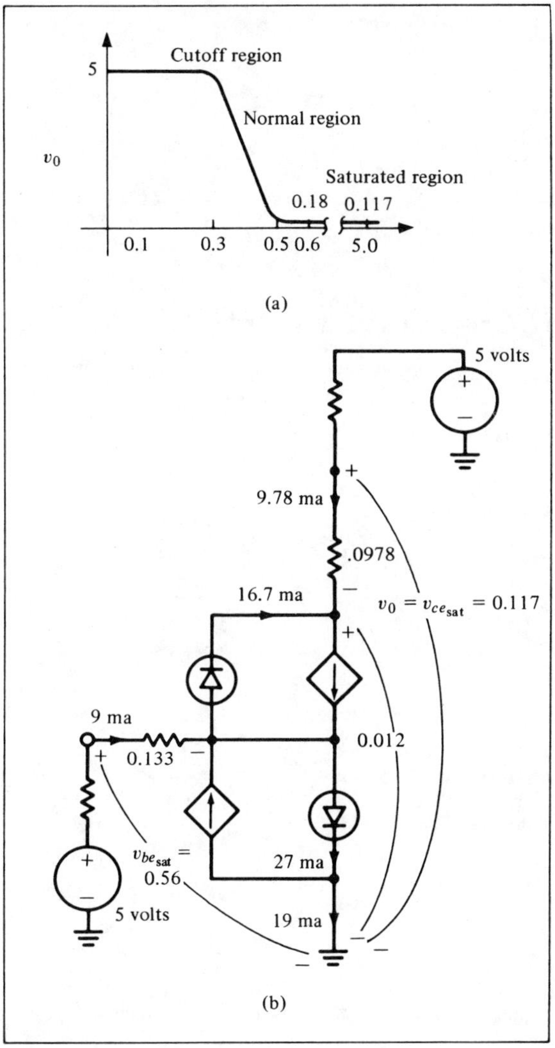

FIGURE 5.18 Results of computer evaluation of inverter.

then the circuit would *not* function as an inverter because the output would never achieve a logical **0** state. If, instead, we made the assignment

Logical **0** for $v \leqq 1$ volt

then an input of 0.6 volt (logical **0**) would produce an output of 0.18 volt (logical **0**); again the circuit would fail to function as an inverter. A little study will show that an acceptable assignment would be

Logical **0** for $v \leqq 0.4$ volt

Having made this assignment, we would then have to insure it was consistent with logic level assignments elsewhere in the digital system in which the inverter is used.

The transfer characteristic can be determined in several ways using the digital computer:

1. With the DCAP program, by stepping the input v_s over the entire input range $0 \leqq v_s \leqq 5$ volts in a coarse step (say 1 volt) and then choosing a smaller step size over the transition range ($0.3 \leqq v_s \leqq 0.6$ in this case).
2. With the INVRT program of Figure 5.19; this program has the same flow chart as the DNET program of Figure 5.8, except that the input voltage is stepped to generate the transfer characteristic.

It is interesting to observe the distribution of voltages and currents in the transistor at saturation, to compare with the simplified saturation model of Figure 5.16(a). In Figure 5.18(b), it is clear that the principal contribution to the output voltage $v_{ce_{sat}}$ is the voltage across r_{cc}, the bulk collector resistance. It is therefore quite important that transistors for such applications be fabricated to have a small r_{cc}. Also, in the saturated state, the base and collector currents are no longer related by β_f; both currents are determined largely by the values of R_s and R_c, because

$$i_b = \frac{v_s - v_{be_{sat}}}{R_s} \cong \frac{4.5}{500} = 9 \text{ ma}$$

$$i_c = \frac{v_c - v_{ce_{sat}}}{R_c} \cong \frac{4.9}{500} \cong 10 \text{ ma}$$

Note that i_c remains essentially constant, independent of small changes in $v_{ce_{sat}}$. But from Figure 5.18(b), $v_{ce_{sat}} \cong i_c r_{cc}$, which must then also be nearly constant. The conclusion is that

```
C****  DC ANALYSIS OF TRANSISTOR INVERTER (INVRT)
       IMPLICIT REAL*8(A-H,O-Z)
       REAL*8 G(10,10),I(10),V(10)
       DIMENSION NODOUT(2,5)
       NDIM=10
C****  READ INPUT DATA
       DO 4 J=1,NDIM
     4 V(J)=0.D0
     1 M=0
       CALL READ(NN,NODOUT,NOUT)
C****  SET UP NODE EQUATIONS
     3 CALL NODEQ(G,V,I,NN,NDIM)
C****  SOLVE NODE EQUATIONS
       CALL REDUCE(G,NN,I,NDIM)
C****  V(J)-I(J) IS THE INCREMENT IN THE JTH NODE VOLTAGE
       ERR=0.D0
       DO 2 J=1,NN
       ERR=ERR+(V(J)-I(J))**2
     2 V(J)=I(J)
C****  CALCULATE AND PRINT OUTPUT VOLTAGE
       CALL OUTPUT(NODOUT,NOUT,V,M)
       M=M+1
C****  TEST CONVERGENCE CRITERION
       IF(ERR.GT.1.D-6.AND.M.LT.50)GO TO 3
       IF(M.EQ.50)WRITE(6,16)
    16 FORMAT(' NO CONVERGENCE AFTER 50 ITERATIONS'/)
       GO TO 1
       END
       SUBROUTINE READ(NN,NODOUT,NOUT)
       IMPLICIT REAL*8(A-H,O-Z)
       DIMENSION NODOUT(2,1)
       DATA INIT/0/
       COMMON RCC,RBB,RS,RC,VC,ALFF,ALFR,VS,DVS
       INIT=INIT+1
       IF(INIT.GT.1)GO TO 5
       READ(5,1)RCC,RBB,RS,RC,VC,ALFF,ALFR,VS,DVS
       READ(5,2)NN,NOUT,(NODOUT(1,J),NODOUT(2,J),J=1,NOUT)
       WRITE(6,3)RCC,RBB,RS,RC,VC,ALFF,ALFR,VS,DVS
     6 RETURN
     5 VS=VS+DVS
       IF(INIT.EQ.7)STOP
       WRITE(6,7)VS
     7 FORMAT(/,' VS =',1PD11.4,/)
       GO TO 6
     1 FORMAT(6F10.0)
```

FIGURE 5.19 (a) Program for inverter circuit analysis; (b) input data for INVRT program; (c) summary of output data from INVRT program.

$v_{ce_{sat}}$ will be constant once the transistor achieves saturation.

Other Transistor Logic Gates

A common digital circuit arrangement is shown in Figure 5.20(a). Two transistor inverters have their emitters and collectors connected together. If *either* v_A or v_B is raised to v_c, the corresponding transistor is saturated and the voltage v_0 is forced to zero. This behavior is summarized in the truth table shown, or symbolically, by OR and NOT logical elements, assuming positive logic. The composite logical function is called a NOR element. Any number of inputs can be added to the NOR logical element by connecting additional inverters across the output. Saturation of any transistor will then result in a zero-valued output. An application of this circuit is given in Problem 5.21.

The most common logic circuit building block is the NAND gate, illustrated in Figure

```
    2 FORMAT(20I2)
    3 FORMAT(' RCC =',1PD11.4,' RBB =',1PD11.4,' RS =',1PD11.4,
     &/' RC =',1PD11.4,' VC =',1PD11.4,' ALFF =',1PD11.4,
     &/' ALFR =',1PD11.4,' VS =',1PD11.4,' DVS =',1PD11.4)
      END
      SUBROUTINE NODEQ(G,V,I,NN,NDIM)
      IMPLICIT REAL*8 (A-H,O-Z)
      REAL*8 G(NDIM,1),I(1),V(1),LAMDA
      COMMON RCC,RBB,RS,RC,VC,ALFF,ALFR,VS,DVS
      DATA CSAT/1.D-9/,LAMDA/40.D0/
C****  ZERO NODE MATRIX AND CURRENT VECTOR
      DO 1 K=1,NN
      I(K)=0.D0
      DO 1 J=1,NN
    1 G(J,K)=0.D0
      GS=1.D0/RS
      GBB=1.D0/RBB
      GC=1.D0/RC
      GCC=1.D0/RCC
      VDF=DMIN1(V(1),.6D0)
      VDR=DMIN1(V(1)-V(2),.6D0)
      DEXF=CSAT*DEXP(LAMDA*VDF)
      DEXR=CSAT*DEXP(LAMDA*VDR)
      GDF=LAMDA*DEXF
      GDR=LAMDA*DEXR
      CURF=DEXF-CSAT-GDF*VDF
      CURR=DEXR-CSAT-GDR*VDR
      G(1,1)=GBB+GDF+GDR-ALFF*GDF-ALFR*GDR
      G(1,2)=-GDR+ALFR*GDR
      G(1,3)=-GBB
      G(2,1)=-GDR+ALFF*GDF
      G(2,2)=GCC+GDR
      G(2,4)=-GCC
      G(3,1)=G(1,3)
      G(3,3)=GS+GBB
      G(4,2)=G(2,4)
      G(4,4)=GC+GCC
      I(1)=ALFR*CURR+ALFF*CURF-CURF-CURR
      I(2)=CURR-ALFF*CURF
      I(3)=VS*GS
      I(4)=VC*GC
      RETURN
      END
      SUBROUTINE OUTPUT(NODOUT,NOUT,V,M)
      IMPLICIT REAL*8 (A-H,O-Z)
      DIMENSION NODOUT(2,1),V(1),VOUT(5)
      IF(M.EQ.0)WRITE(6,4)(NODOUT(1,J),NODOUT(2,J),J=1,NOUT)
    3 DO 2 I=1,NOUT
      VOUT(I)=0.D0
      N1=NODOUT(1,I)
      N2=NODOUT(2,I)
      IF(N1.NE.0)VOUT(I)=VOUT(I)+V(N1)
      IF(N2.NE.0)VOUT(I)=VOUT(I)-V(N2)
    2 CONTINUE
      WRITE(6,1)M,(VOUT(I),I=1,NOUT)
```

FIGURE 5.19 (Continued)

5.21. The circuit realization of Figure 5.21(b) is an example of a DTL (diode-transistor-logic) circuit. Summarizing its operation, the diodes perform the AND function as we studied in Chapter 3 and the transistor performs a logical inversion of the result, restoring the level of the logical signal degraded by the voltage across the diodes during conduction. The circuit in Figure 5.21(c) is an example of a TTL (transistor-transistor-logic) realization.

```
      RETURN
    1 FORMAT(4X,I4,4X,5(1PD12.4))
    4 FORMAT('  ITERATION  ',5(I2,' -',I2,' NODES'))
      END
      SUBROUTINE REDUCE(A,N,B,NDIM)
      IMPLICIT REAL*8(A-H,O-Z)
      DIMENSION A(NDIM,1),B(1)
C****  THIS SUBROUTINE SOLVES A*X=B BY GAUSS ELIMINATION
C****  N IS THE MATRIX SIZE; X IS RETURNED IN B; A IS DESTROYED
C****  CONVERT A TO UPPER TRIANGULAR FORM
      NM1=N-1
      IF(N.EQ.1)GO TO 8
      DO 1 J=1,NM1
      IJ=J+1
      DO 4 K=IJ,N
    4 A(J,K)=A(J,K)/A(J,J)
      B(J)=B(J)/A(J,J)
      DO 1 I=IJ,N
      ALPHA=A(I,J)
      DO 2 K=IJ,N
    2 A(I,K)=A(I,K)-ALPHA*A(J,K)
    1 B(I)=B(I)-ALPHA*B(J)
    8 B(N)=B(N)/A(N,N)
C****  PERFORM BACK SUBSTITUTION
      IF(N.LT.2)GO TO 7
      DO 3 J=2,N
      LAST=N-J+1
      LASTP1=LAST+1
      DO 3 I=LASTP1,N
    3 B(LAST)=B(LAST)-A(LAST,I)*B(I)
    7 RETURN
      END
```

(a)

```
 10.        15.        500.       500.       5.         .98
 .5         .2         .08
  4 1 4 0
```

(b)

VS	VO
0	5.000
.2	4.998
.28	4.965
.36	4.447
.44	2.457
.52	.221
.6	.180
5.	.117

(c)

FIGURE 5.19 (Continued)

The diodes of the DTL circuit are replaced by emitter diodes of the transistors to improve certain characteristics of the circuit. A detailed analysis of these circuits is proposed in Problem 5.27.

Memory Circuits

A circuit is said to have memory if its state depends on the history of the inputs to the circuit, *after the inputs have been removed.* None of the circuits that we have studied have this property; for example, amplifier and inverter circuits will produce an output voltage that is dependent on the input signal. When the input is removed, the output *must* return to the value existing before the input was applied, *because the circuit equations have only a single solution with no signal applied.*

The equations of a memory circuit must have *more than one solution* with the input removed, so that the output is *not* uniquely determined by the input. The feasibility of a circuit with multiple solutions is established by the example of Figure 5.22.

Here, a device known as a *tunnel diode* is depicted by its *i-v* characteristic which has a region with a *negative slope.* A load line can then be found which passes through three points of the characteristic. All of these points represent solutions to the circuit equations, and we would conclude from this analysis that the circuit could exist in any of these three states. This circuit therefore satisfies the essential requirement of a memory element that it have multiple solutions.

The most common memory circuit used in

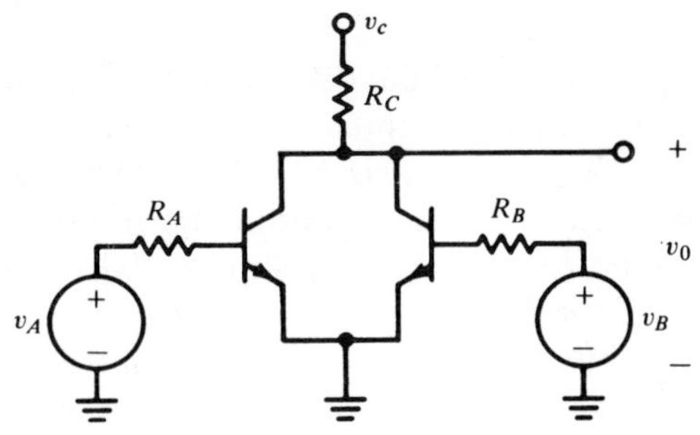

(a)

A	B	C
0	0	1
1	0	0
0	1	0
1	1	0

(b)

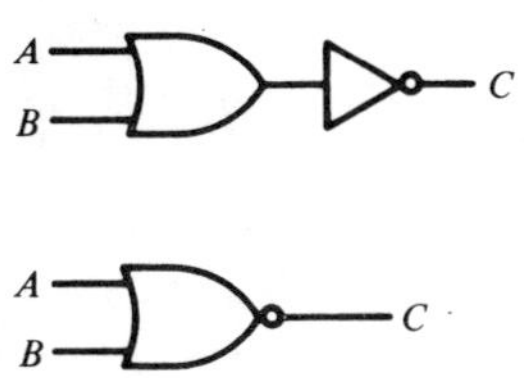

(c)

FIGURE 5.20 NOR gate circuit.

A	B	C
0	0	1
1	0	1
0	1	1
1	1	0

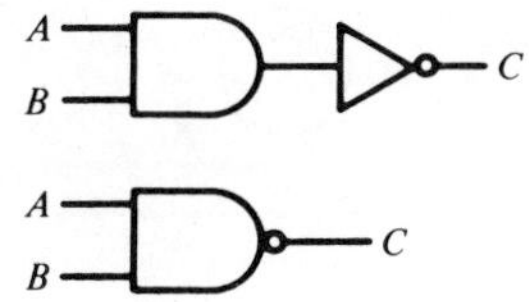

(a)

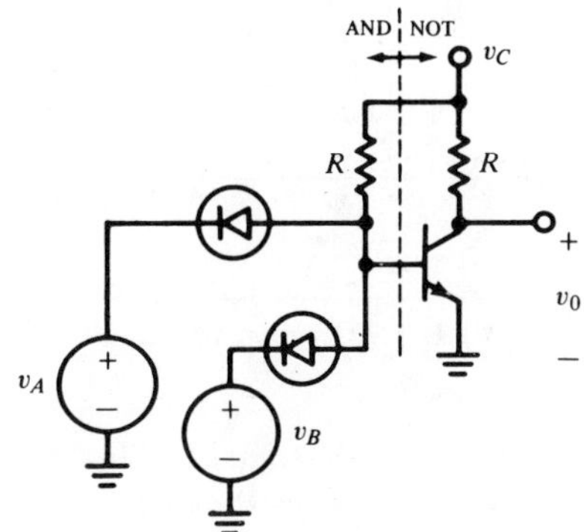

(b)

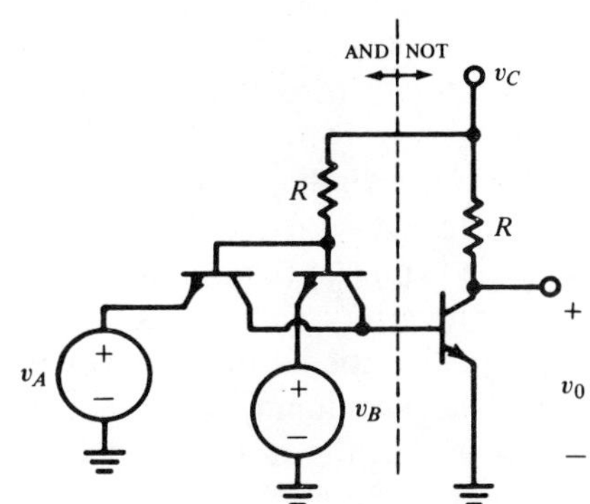

(c)

FIGURE 5.21 NAND gate circuits.

digital computers is the *flip-flop*, illustrated by the simple circuit of Figure 5.23. It is shown symbolically in Figure 5.23(a) as two inverters cross-coupled with the output of one being the input to the other. Two possible states of the flip-flop exist, each consistent with the truth table of the inverters from which it is formed. If we define the state of the element from the state of output Q, we can say that the flip-flop is either in the **1** state or in the **0** state, as shown.

A circuit realization of this function can take

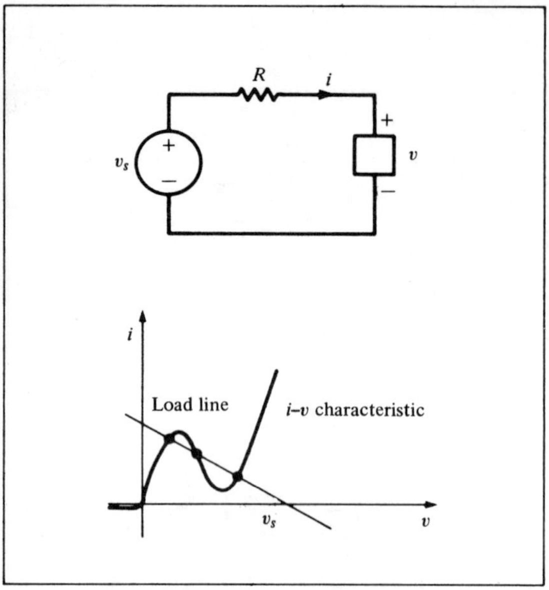

FIGURE 5.22 A memory circuit.

the form of two single-transistor inverter circuits, cross-coupled just as the logical elements. Such a circuit, in Figure 5.23 (b), will ideally operate in either of the two dc states shown, assuming that the transistor β_f's are sufficiently large so that a transistor saturates when a logical **1** is applied to the R_s attached to its base.

The state of the flip-flop (**0** or **1**) is set by an input which forces it to the desired state. In terms of logic symbols, this is affected by converting the inverters to NOR gates [Figure 5.23 (c)]. Two additional inputs are provided as shown. The SET input, when equal to a logical **1**, results in the state $Q = \mathbf{1}$ being the only stable state. Similarly, when RESET $= \mathbf{1}$, the only stable state is $Q = \mathbf{0}$. These possibilities are shown in Figure 5.23 (d). The flip-flop will remain in the state to which it was forced by its most recent input (a SET or a RESET). Once the flip-flop is in either state, it will remain in that state indefinitely, after the SET and RESET inputs have been removed. The result: a logical memory element.

A practical flip-flop would not be constructed as shown in Figure 5.23 for a variety of reasons. Modern technology makes the cost of adding transistors to a circuit design so small that real flip-flops may have 5 to 20 transistors to achieve the temperature stability, speed, and other characteristics required for digital circuit applications.

The actual switching between the two states when an input is applied is a complicated process. This process is governed by the internal motion of charge within the transistor, and must be described by differential rather than algebraic equations. Its study, then, is reserved for later in the text; the interested reader may achieve some insight into the switching process by the study of the switching of an inverter circuit, as illustrated in Section 14.4.

PROBLEMS

5.1 Solve the following simultaneous equations by Gauss elimination. Check solutions by substitution in the original equations.

(a)
$$\begin{bmatrix} 1 & 2 & 3 \\ 3 & 1 & 2 \\ 2 & 3 & 1 \end{bmatrix} \begin{bmatrix} x_1 \\ x_2 \\ x_3 \end{bmatrix} = \begin{bmatrix} 14 \\ 11 \\ 11 \end{bmatrix}$$

(b)
$$\begin{bmatrix} 2 & 1 & 0 \\ 0 & 3 & 1 \\ 3 & 2 & 1 \end{bmatrix} \begin{bmatrix} x_1 \\ x_2 \\ x_3 \end{bmatrix} = \begin{bmatrix} 1 \\ -3 \\ 1 \end{bmatrix}$$

(c)
$$\begin{bmatrix} -1 & 2 & 3 \\ 2 & 0 & 0 \\ 3 & 0 & -1 \end{bmatrix} \begin{bmatrix} x_1 \\ x_2 \\ x_3 \end{bmatrix} = \begin{bmatrix} -4 \\ -2 \\ 0 \end{bmatrix}$$

5.2 Count the number of multiplications and divisions involved in solving Problem 5.1 (a) by: (a) Gauss elimination, and (b) Cramer's

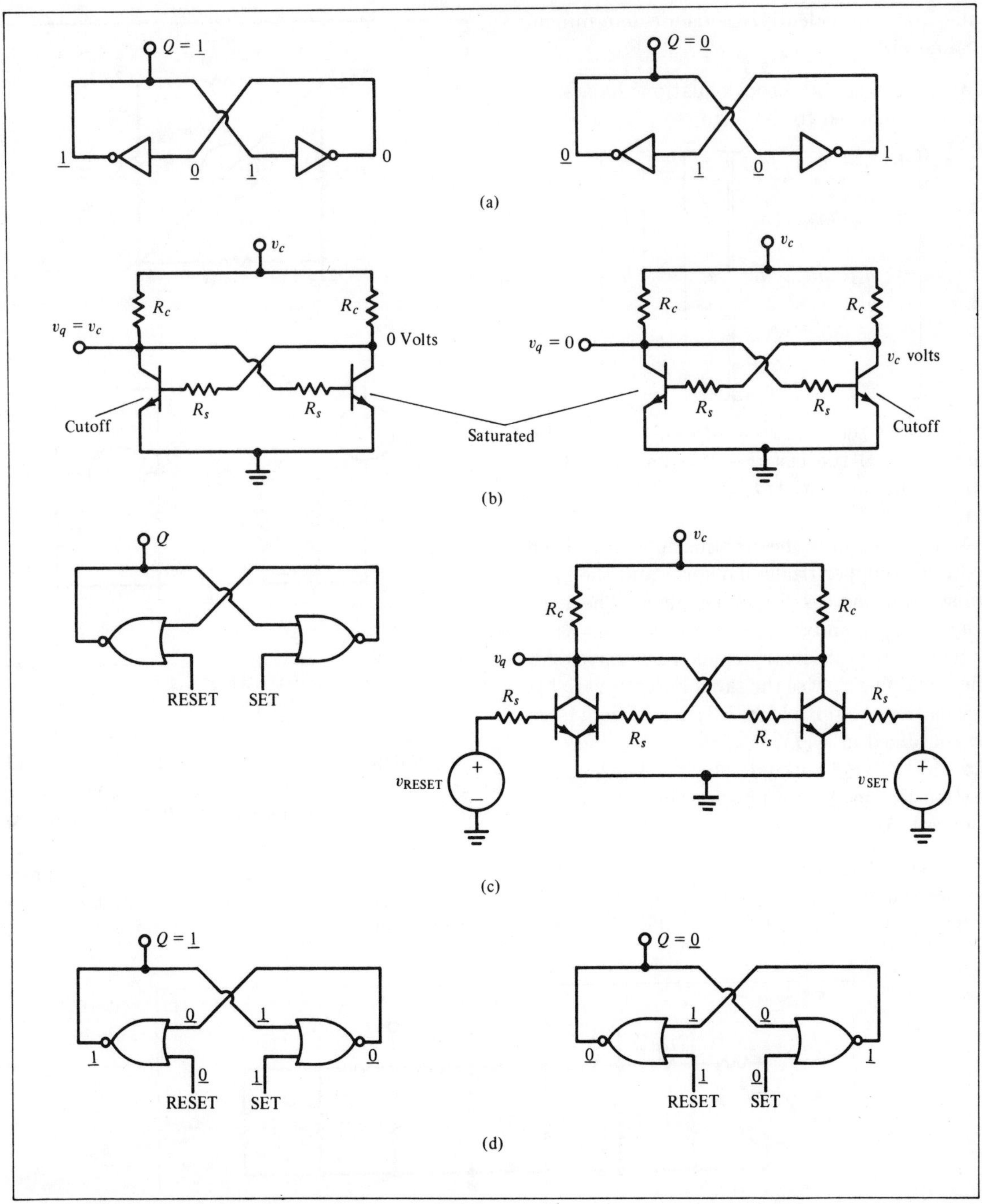

FIGURE 5.23 The flip-flop. (a) Symbolic representation and stable states; (b) circuit realization; (c) flip-flop with inputs; (d) stable states with SET and RESET equal to logical **1**.

rule (or equivalently, cofactor-determinant evaluation).

5.3 A set of simultaneous equations have been partially reduced to the form

$$\begin{bmatrix} 1 & a_{12} & a_{13} \ldots a_{1n} \\ 0 & a_{22} & a_{23} \ldots a_{2n} \\ 0 & a_{32} & a_{33} \ldots a_{3n} \\ \cdot & \cdot & \cdot\ \cdot \\ \cdot & \cdot & \cdot\quad \cdot \\ \cdot & \cdot & \cdot \qquad \cdot \\ 0 & a_{n2} & a_{n3} \quad a_{nn} \end{bmatrix} \begin{bmatrix} x_1 \\ x_2 \\ x_3 \\ \cdot \\ \cdot \\ \cdot \\ x_n \end{bmatrix} = \begin{bmatrix} b_1 \\ b_2 \\ b_3 \\ \cdot \\ \cdot \\ \cdot \\ b_n \end{bmatrix}$$

Construct a set of Fortran statements to eliminate x_2 from the last $n - 2$ equations. Here, a_{ij} is contained in $A(I, J)$, b_i in $B(I)$, and n in N.

5.4 A set of simultaneous equations have been reduced to upper triangular form and the back substitution process is taking place. The variables x_n, x_{n-1}, and x_{n-2} have been found. Write a set of Fortran instructions that solves for the remaining x_i. Use the same Fortran variable associations as in Problem 5.3; in addition, let x_i be contained in $X(I)$.

5.5 (a)–(b) Set up and solve by Gauss elimination the node equations for the networks of Figure P 5.1.

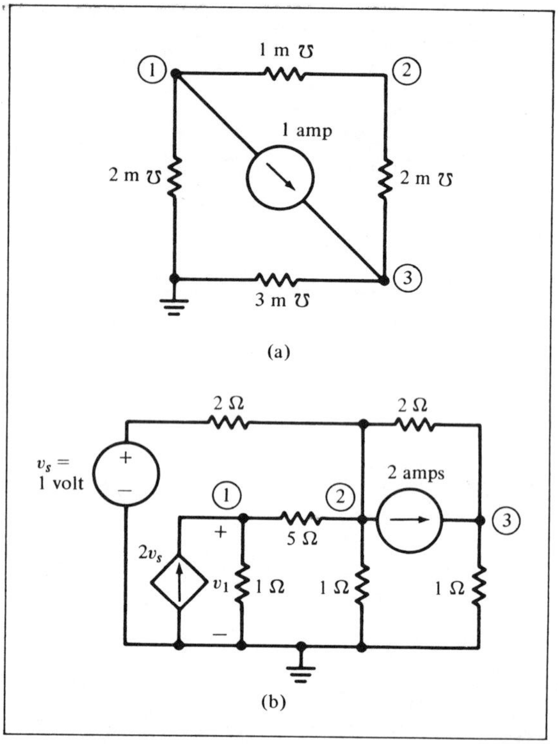

FIGURE P 5.1

5.6 The ladder network of Figure P 5.2 is termed an attenuator and is used to divide voltage v_s in a specified manner. The Thevenin equivalent between any of the 6 taps and ground has the same R_{eq}, but the v_{eq} is a specified fraction of v_s. Determine R_{eq} and v_{eq} across each of the 6 taps. Use RNET as a model program. [Attenuators can be designed to have any desired Thevenin resistance and any desired attenuation at each tap (Reference 5.5).]

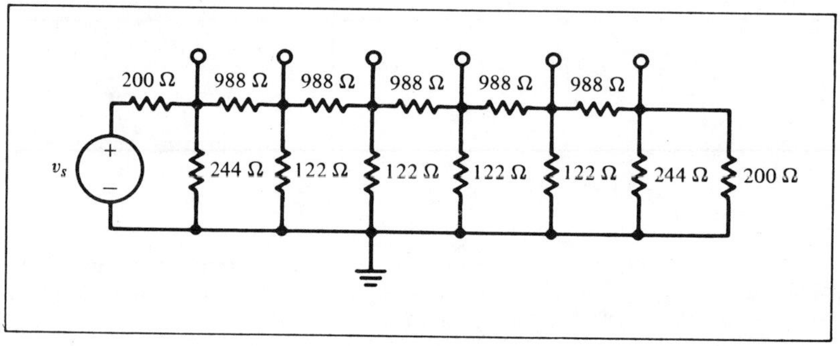

FIGURE P 5.2

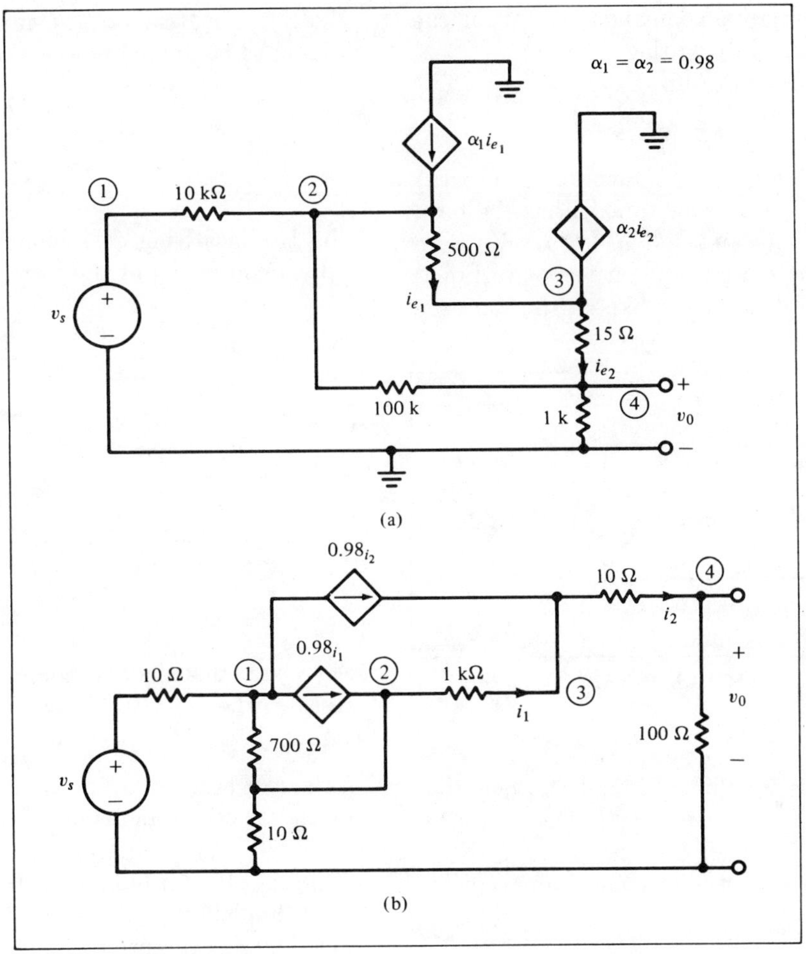

FIGURE P 5.3 (b) v_0/v_s is intentionally small.

5.7 Find the Thevenin equivalent across v_0 in Figure P 5.3(a). Also find v_0/v_s. Use RNET as a model.

5.8 Repeat Problem 5.7 for the network of Figure P 5.3(b).

5.9 Obtain a companion model for the lamp of Problem 2.15.

5.10 Sketch the i_d–v_d characteristic and develop a companion model for the Zener diode of Figure P 5.4. (Special iteration methods are usually necessary with this element because of the steep vertical slope of the i_d–v_d characteristic.)

$$v_d \geqq -0.1: \qquad i_d = I_s(e^{\lambda v_d} - 1)$$

$$v_d < -0.1: \qquad i_d = \frac{-I_s}{1 - (v_d/6)^6}$$

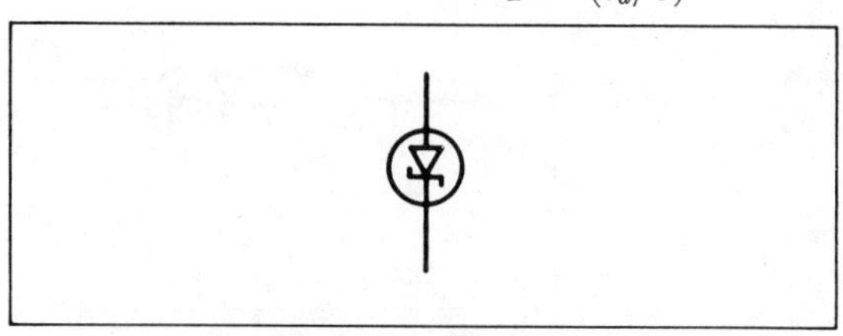

FIGURE P 5.4 Zener diode.

5.11 An alternative companion model for the diode is obtained by writing

$$v_d = \frac{1}{\lambda} \ln \left(\frac{i_d}{I_s} + 1 \right)$$

and expanding v_d^{m+1} as a function of i_d^m. (a) Show that the companion model has the form of Figure P 5.5, and find R_d^m. (b) Write an iteration equation for i_d^{m+1} analogous to Equation (5.12) for v_d^{m+1}. (c) Sketch the iteration path on the i_d–v_d characteristic, similar to Figure 3.9. From this path and the fact that a forward-biased diode has a steep vertical characteristic, which iteration (in i_d^{m+1} or v_d^{m+1}) do you expect to converge more rapidly?

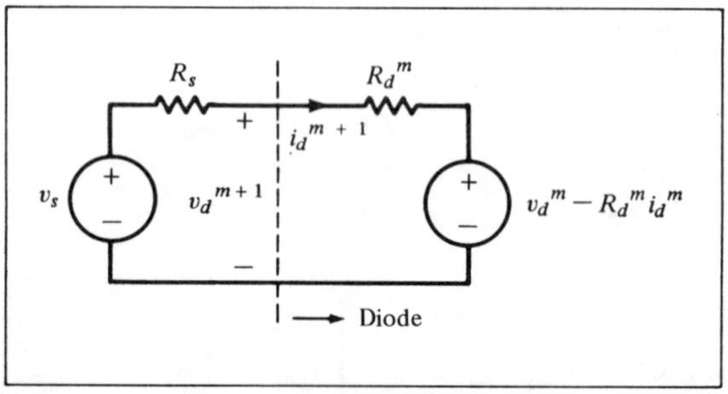

FIGURE P 5.5

5.12 The resistance of many materials rises linearly with temperature

$$R = R_0 + \alpha(T - T_0)$$

where R_0 is the resistance at room temperature T_0(°K). The temperature in turn is linearly related to dissipated power, that is,

$$T - T_0 = \beta(i^2R)$$

(a) Write

$$v = f_1(i)$$

(b) By linearizing v^{m+1} about $f_1(i^m)$, obtain a companion model of the form of Figure P 5.6.

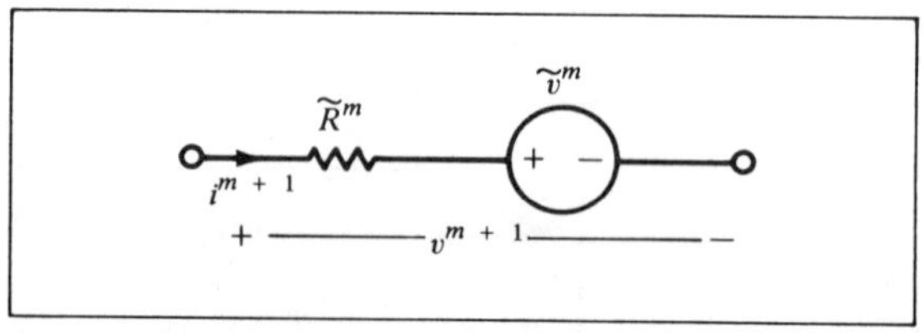

FIGURE P 5.6 $\tilde{R}^m = f_2(R_0, \alpha, \beta, i^m)$; $\tilde{v}^m = f_3(R_0, \alpha, \beta, i^m)$.

Show that this model becomes that of a simple resistor when $\alpha = 0$ or $\beta = 0$.

5.13 A real diode is often represented with a series resistance R as shown in Figure P 5.7(a) Complete the companion model shown in Figure P 5.7(b); that is, find the function f. Then find the formula for obtaining v_d^{m+1} from $\tilde{v}_d^{m+1}$. Explain why this latter formula is necessary for the computing algorithm.

5.14 The companion model for the FET described in Appendix D is incorporated in DCAP. We explore its development here

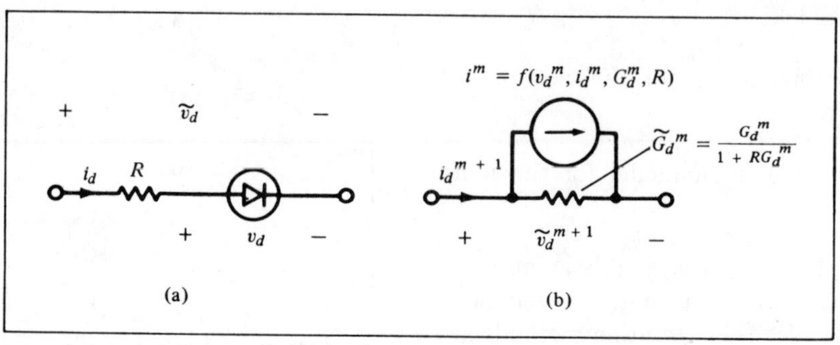

FIGURE P 5.7

(a) For the pinch-off region, we have

$$i_d = I_{dss}\left(1 - \frac{v_{gs}}{V_p}\right)^2$$

$$= g(v_{gs}) \qquad v_{gs} > V_p,\ v_{gd} < V_p$$

By expanding i_d^{m+1} about v_{gs}^m, find the companion model in terms of an independent current source between the d and s terminals and a linearized VCCS coupling the g-s and d-s terminals. (b) For the variable resistor region, we have

$$i_d = \frac{I_{dss}}{V_p^2}\,(2(v_{gs} - V_p)v_{ds} - v_{ds}^2)$$

$$v_{gs} > V_p,\ v_{gd} > V_p$$

Expand i_d^{m+1} about v_{gs}^m and v_{ds}^m and represent the companion model by a branch conductance, a VCCS, and an independent current source. (c) For the inverted region ($v_{gs} < V_p$, $v_{gd} > V_p$), the FET behaves as $i_d = I_{dss}(1 - v_{gd}/V_p)^2$; in the cutoff region ($v_{gs} < V_p$, $v_{gd} < V_p$), $i_d = 0$. Develop companion models for these cases.

5.15 A transistor can be made to behave as a nearly perfect constant current source. De-

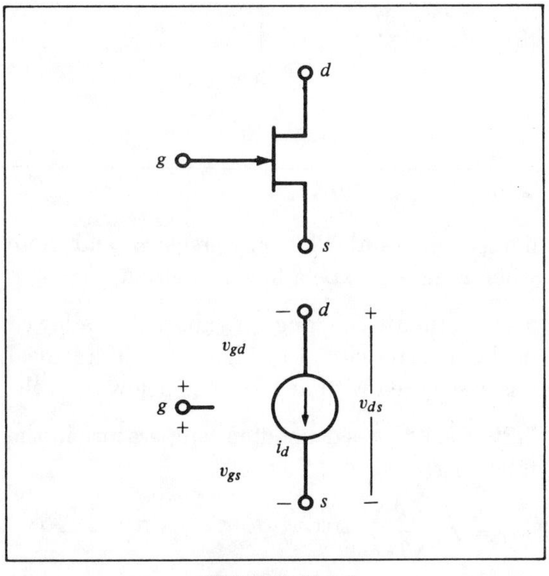

FIGURE P 5.8

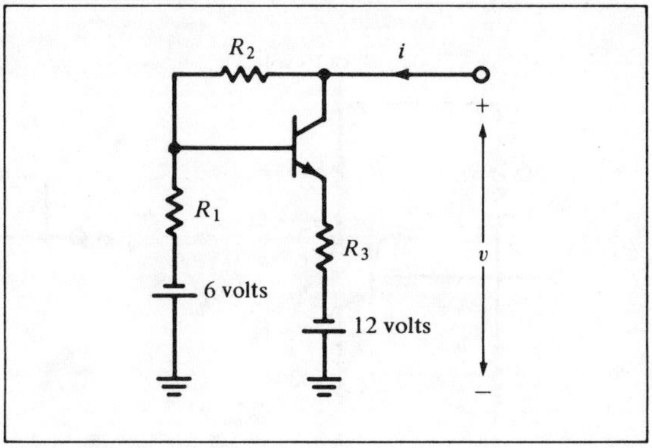

FIGURE P 5.9 $R_1 = 1000\ \Omega$, $R_2 = 100\ \mathrm{k}\Omega$, $R_3 = 300\ \Omega$, $\alpha = 0.98$, $I_s = 10^{-9}$ amps, $\lambda = 40\ \text{volts}^{-1}$.

termine the i-v characteristic of the circuit shown in Figure P 5.8 for $-9 \leqq v_s \leqq 1$. Denote the linear region of operation where the current source behavior is manifested. Use DNET or INVRT as a model program and use the transistor model of Figure 5.10(b).

5.16 A voltage regulator utilizes a transistor to reduce the slope of the i-v characteristic of a one-port. Determine the i-v characteristic of the regulated source in Figure P 5.10 for $0.1 \leqq i \leqq 1.1$ amps. Denote the linear region of operation. Use the transistor model of Figure 5.10(b). Use DNET or INVRT as a model program.

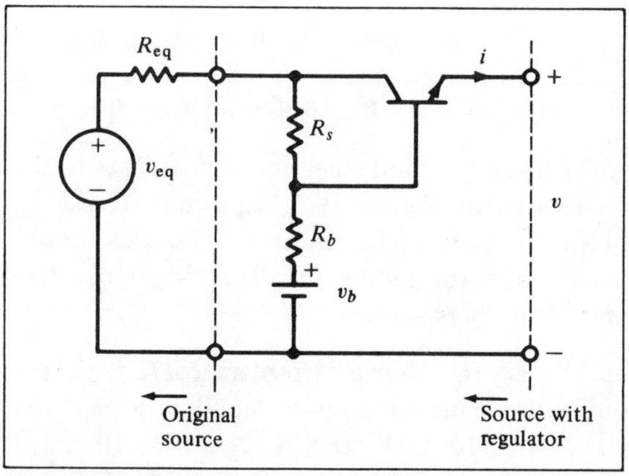

FIGURE P 5.10 $\alpha = 0.99$, $R_{eq} = R_b = 10\ \Omega$, $R_s = 700\ \Omega$, $v_{eq} = 30$ volts, $v_b = 20.35$ volts, $I_s = 10^{-9}$ amps, $\lambda = 40\ \text{volts}^{-1}$.

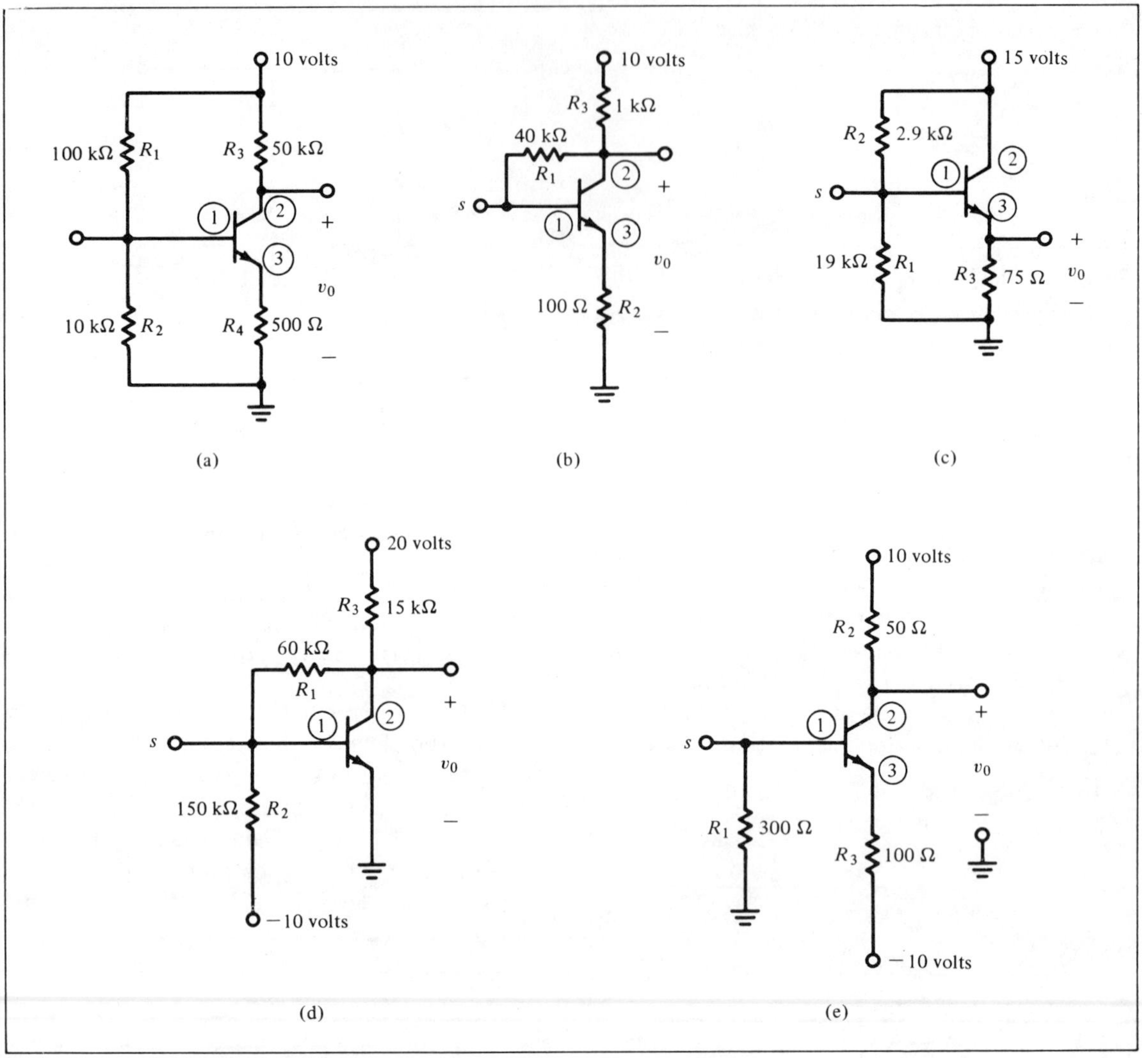

FIGURE P 5.11 Let $I_s = 10^{-9}$ amp, $\lambda = 40$ volts^{-1}, $\alpha_f = 0.98$, $\alpha_r = 0.5$.

5.17 (a)–(e) Find the dc value of v_0 in the networks in Figure P 5.11, using DNET or INVRT as a model program. Use the simplified transistor model of Figure 5.10(b). Use node numbers shown.

5.18 (a)–(e) Repeat Problem 5.17, with the following current sources applied at S: (a) $0 \leqq i_s \leqq 10^{-4}$, (b) $0 \leqq i_s \leqq 2.5 \times 10^{-4}$, (c) $0 \leqq i_s \leqq 5 \times 10^{-3}$, (d) $0 \leqq i_s \leqq 10^{-4}$, (e) $0 \leqq i_s \leqq 0.025$. Use DNET or INVRT as a model program. Plot v_0 versus i_s and note when v_0 ceases to be a linear function of i_s.

5.19 Estimate the range of values of r_{cc} allowed in the inverter circuit of Figure 5.17 if a logical **0** is associated with any voltage below 0.4 volt.

5.20 Real diodes depend on temperature in the following way:

$$i_d = I_{s0}\left(\frac{T}{T_0}\right)^3 \left(e^{14{,}000(T-T_0)/TT_0}\right)\left(e^{(\lambda T_0/T)v_d} - 1\right)$$

where T_0 is a reference temperature (usually 290°K); I_{s0} is the value of I_s at temperature T_0 (10^{-9} in our studies); T is the temperature in °K. (a) Show that when $T = T_0$, we obtain the familiar formula for i_d. (b) Study the effect of temperature on the transfer characteristic of the inverter circuit of Figure 5.17. Let the temperature be −55°C, 20°C, 55°C, and 100°C. Use the INVRT program of Figure 5.14.

5.21 A *bus* is a line or connection point over which information is sent between parts of a system. Usually, many parts of a computer can put information on the bus, suggesting the NOR circuit model of Figure P 5.12. Each transistor must be able to "pull down" the bus to a **0** logical state. Let $v_A = v_B = 0$ in Figure P 5.12.

Logical level **0** for $v \leqq 0.5$

Logical level **1** for $v \geqq 4.5$

Choose a value for R_c that permits a load current of 25 ma to flow when the bus is in a logical **1** state.

5.22 (See Appendix D.) The IGFET's of Figure P 5.13 have a pinch-off of 2 volts. Assume positive logic is used, with a logical **1** being 5 volts. (a) What logic or memory function could circuits of these configurations provide? Explain how each would function. (b) Add IGFET's to the gate of Figure P 5.13(b) to provide an "input" to the element. Explain how they would function.

5.23 The flip-flop can be set from a single input by adding a logical inverter as shown in Figure P 5.14. Construct a circuit configuration to implement this function. No component values are necessary, but you should explain your answer.

5.24 Show that the logical element of Figure P 5.15 functions as a flip-flop, that is, that it has two stable states and can be forced to a different stable state with either the $\overline{\text{SET}}$ or $\overline{\text{RESET}}$ inputs.

5.25 The flip-flop of Figure P 5.16 is augmented with two AND gates. With the inputs shown, plot Q for $t \geqq 0$. Why is the extra input called a clock?

5.26 (See Section 5.6.) The convergence of the Newton iteration can be studied for a single equation by expanding $F(\hat{x})$ in a Taylor's series with remainder about $F(x^m)$, where $F(\hat{x}) = 0$ (that is, $\hat{x}$ is the desired solution). We then have

$$F(\hat{x}) = F(x^m) + (\hat{x} - x^m)F'(x^m) + \tfrac{1}{2}(\hat{x} - x^m)^2F''(\xi)$$

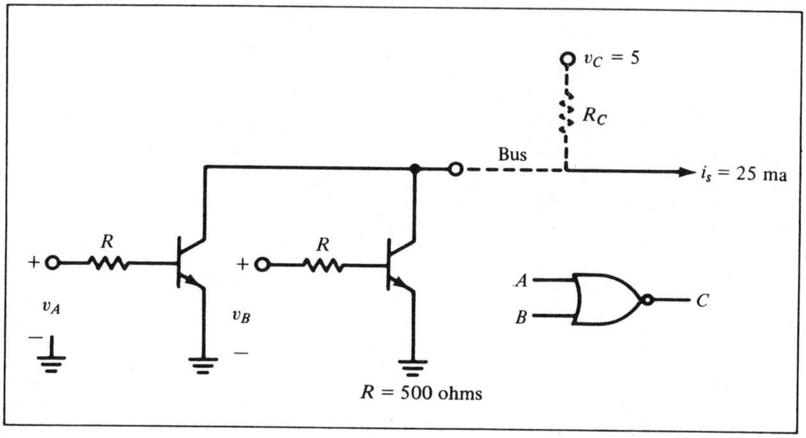

FIGURE P 5.12

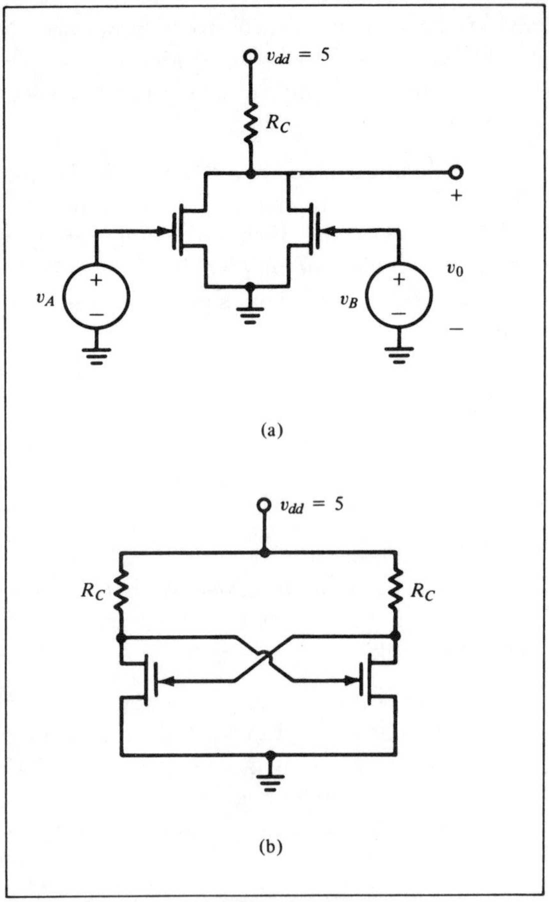

FIGURE P 5.13

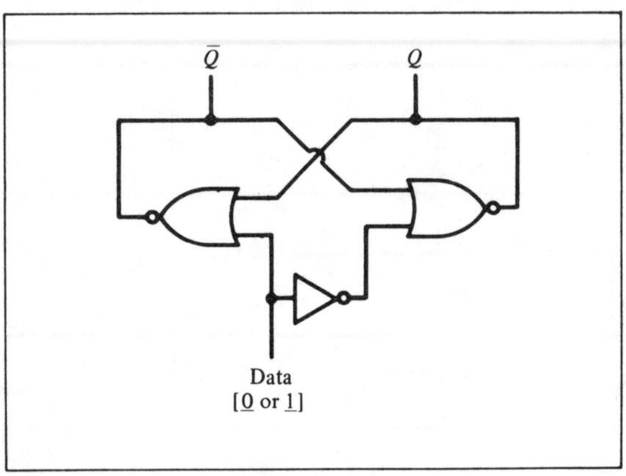

FIGURE P 5.14

where ξ lies in the interval $[x^m, \hat{x}]$. The Newton iteration formula is given as

$$x^{m+1} = x^m - \frac{F(x^m)}{F'(x^m)}$$

If $F''(x)$ can be considered a constant near $x = \hat{x}$ and if $F'(\hat{x}) \neq 0$, show that

$$\hat{x} - x^{m+1} = K(\hat{x} - x^m)^2$$

where K is a constant. This establishes the quadratic convergence of the method.

The following problems involve the use of the DCAP program. The intent is to verify the functional behavior of a network and to study the detailed performance of its components These could also be the basis for experiments by students in a related laboratory course. In the following problems, let $\alpha_f = 0.98$, $\alpha_r = 0.5$, $I_s = 10^{-9}$, $\lambda = 40$, $r_{cc} = 10$. Also, in 5.27 to 5.29, let a logical **0** correspond to the range $0 \leqq v \leqq 0.1$ volts, a logical **1**, $4.9 \leqq v \leqq 5$.

5.27 Calculate and plot the transfer characteristic of the NAND gate of Figure 5.21(a). Let

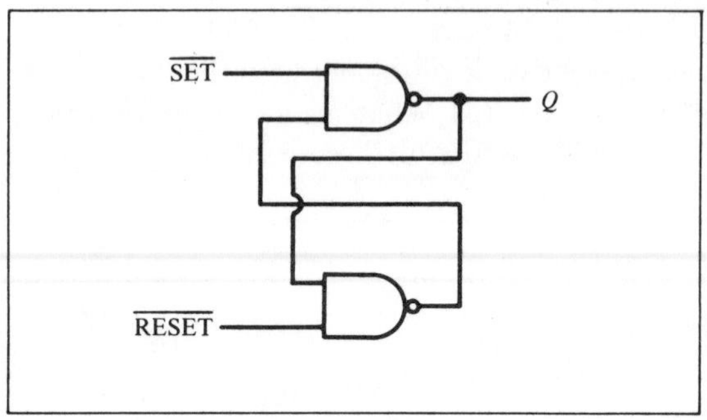

FIGURE P 5.15

$v_A = 5$ volts and step v_B between 0 and 0.5 volts using DCAP. Does the circuit realize the required logic function. If not, insert a diode in the base lead between the AND and NOT gates as shown in Figure P 5.17 and recalculate the transfer characteristic. Explain the function of the diode. Let $R = 500\ \Omega$, $v_c = 5$.

5.28 Repeat Problem 5.27 for the TTL (Transistor-Transistor-Logic) NAND gate of Figure 5.21(b).

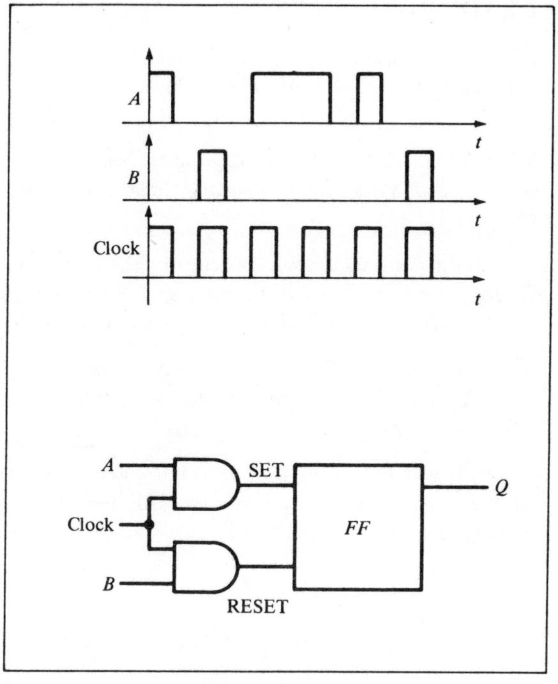

FIGURE P 5.16

5.29 The circuit of Figure P 5.18 is a NAND gate. Assume $r_{cc} = 0$. You are asked to determine the dc transfer characteristic between v_A and v_0 with $v_B = 5$. (a) Step v_A between 0 and 5 volts in 1-volt steps; then choose a step size sufficiently small to obtain at least four points on the transistion interval. (b) Remove the diode between the transistors and repeat the above calculation. From your results, explain the purpose of this diode. (c) Without using the computer, estimate the form of the transfer

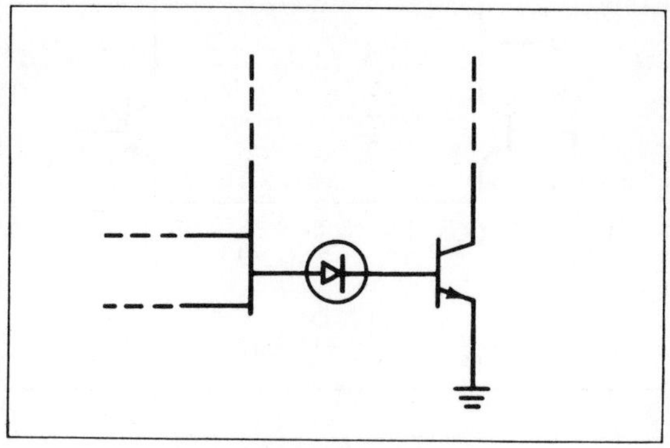

FIGURE P 5.17

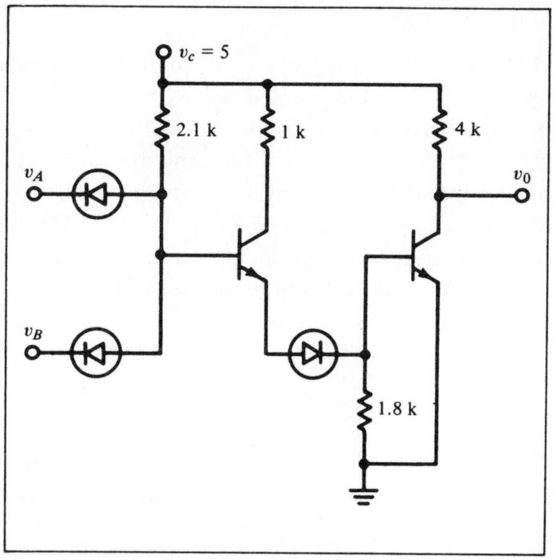

FIGURE P 5.18

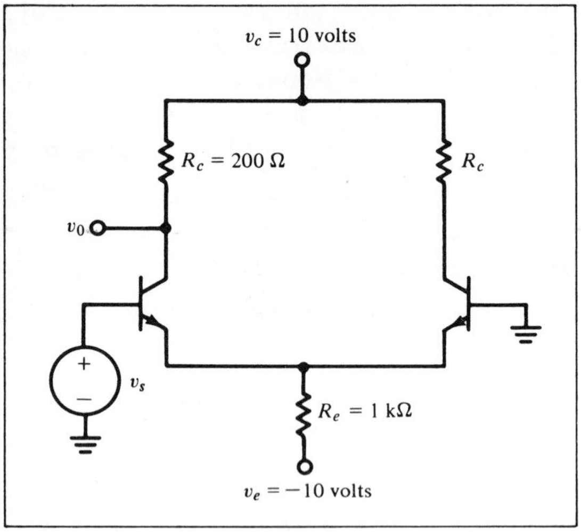

FIGURE P 5.19

characteristic with this diode replaced by *two* diodes in series.

5.30 An interesting class of digital circuits known as emitter-coupled logic is illustrated by the inverter of Figure P 5.19. Assume $I_s = 10^{-10}$, $\lambda = 35$. (a) Obtain the transfer characteristic by steeping v_s from −0.2 to 0.6 volts in 0.1 volt steps using DCAP. What are the output voltage levels corresponding to logical **0** and **1**, assuming positive logic? (b) Note on the transfer characteristic those regions where each transistor changes states (cutoff, linear, saturation, reverse). (c) A saturated transistor normally requires more time to switch between states than an unsaturated one. How would the above circuit rate in relative switching speed compared to the inverter of Figure 5.17? (d) The dc level of the output must be restored." Comment.

REFERENCES AND ADDITIONAL READING

5.1 J. N. HARRIS, *Digital Transistor Circuits*, (SEEC) Vol. 6. New York: John Wiley & Sons, Inc., 1966.

5.2 J. K. HAWKINS, *Circuit Design of Digital Computers*. New York: John Wiley & Sons, Inc., 1958. Chapter 5.

5.3 JACOB MILLMAN AND HERBERT TAUB, *Pulse, Digital, and Switching Waveforms*. New York: McGraw-Hill Book Company, Inc., 1969, Chapter 9.

5.4 D. A. CALAHAN, *Computer-Aided Network Design*. New York: McGraw-Hill Book Company, Inc., 1972, Chapter 3.

5.5 W. L. EVERITT, AND G. E. ANNER, *Communication Engineering*. New York: McGraw-Hill Book Company, Inc., 1956, Chapter 13.

6

Linearity and Its Significance; Small-Signal Analysis

6.1 Introduction

Numerical results can provide considerable insight into the behavior of a particular network by, for example, varying element values and observing the effect on the solution. However, they contribute negligibly to our knowledge about *classes* of networks. For example, those readers who worked any of the Problems 3.18 to 3.21 discovered from their computer solution that a transistor amplifier operates along a straight load line on the transistor output (i_c-v_{ce}) characteristic *even in those cases when we could not conveniently find the load-line equation* (Figure 6.1). Will this always occur? We suspect that the answer is yes if the transistor can be modeled by elements with linear *i-v* characteristics, but we have no sound basis for this conjecture. This chapter is concerned with such questions which involve the consequences of linearity.

Since our last discussion of linearity in Chapter 2, we have developed a number of general analytical and algorithmic techniques for network analysis. These in turn should allow us to express in more general terms the properties associated with networks of linear elements. For example, the concept of Thevenin and

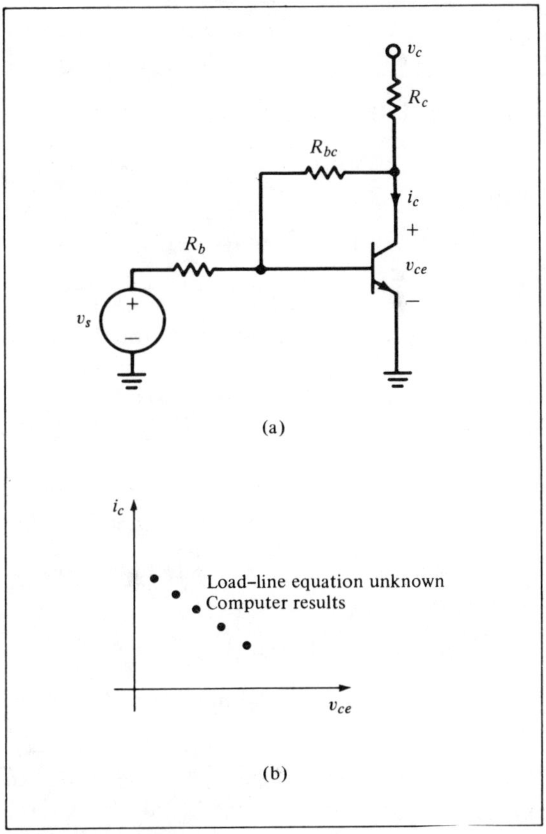

FIGURE 6.1 Problem 3.18 revisited. (a) Network; (b) numerical results.

Norton equivalents can be generalized beyond the simple networks of Chapter 2. Further, we will show that even nonlinear networks can share in these properties provided that we are willing to make certain reasonable and common "engineering approximations."

6.2 Linearity and Superposition

Let us consider the simple network shown in Figure 6.2. The mesh equations for this network are

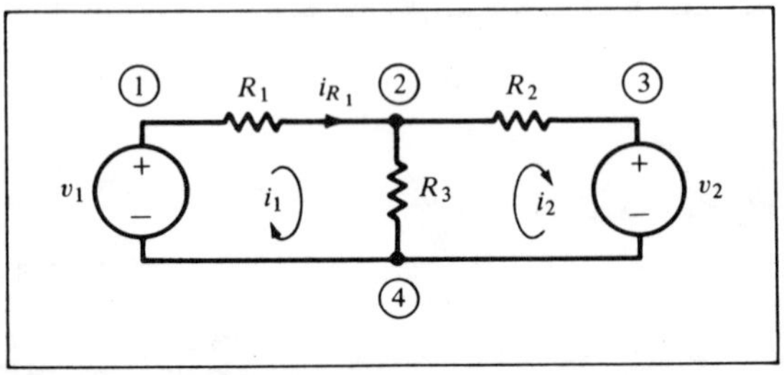

FIGURE 6.2 Superposition example.

$$
\begin{aligned}
(R_1 + R_3)i_1 - R_3 i_2 &= v_1 \\
-R_3 i_1 + (R_2 + R_3)i_2 &= -v_2
\end{aligned} \tag{6.1}
$$

These equations are easily solved for the mesh currents, i_1 and i_2, and from these we can find the currents in the three resistances. For example, we can obtain

$$i_1 = i_{R_1} = \frac{(R_2 + R_3)v_1 - R_3 v_2}{R_1R_2 + R_2R_3 + R_3R_1}$$

This expression can be viewed in terms of the sources v_1 and v_2 as follows:

$$
\begin{aligned}
i_{R_1} &= (\text{component due to } v_1) \\
&\qquad + (\text{component due to } v_2) \\
&= i_{R_1}{}^{(1)} + i_{R_1}{}^{(2)}
\end{aligned}
$$

Moreover, if v_2 is set to zero, then $i_R{}^{(2)}$ vanishes and $i_{R_1} = i_{R_1}{}^{(1)}$; conversely, if v_1 is set to zero, then $i_{R_1}{}^{(1)} = 0$. Thus, i_{R_1} is a *superposition* of responses due to each source.

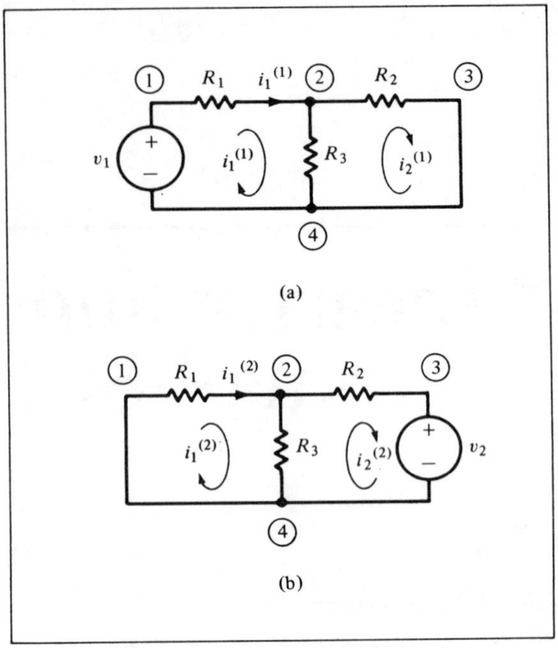

FIGURE 6.3 Separation of sources.

The process of setting sources to zero can be represented in the network as in Figure 6.3. A zero-valued voltage source is a short circuit, so that $i_{R_1}{}^{(1)}$ can be calculated from the network in Figure 6.3(a) with v_2 replaced by shorting nodes ③ and ④. The $i_{R_1}{}^{(2)}$ component is similarly determined in Figure 6.3(b).

This result is very easily generalized. The mesh equations for a network with n meshes have the general form

$$
\begin{aligned}
R_{11}i_1 + R_{12}i_2 + \cdots + R_{1n}i_n &= v_{s1} \\
R_{21}i_1 + R_{22}i_2 + \cdots + R_{2n}i_n &= v_{s2} \\
\vdots \qquad \vdots \qquad\quad \vdots \quad &\quad\ \vdots \\
R_{n1}i_1 + R_{n2}i_2 + \cdots + R_{nn}i_n &= v_{sn}
\end{aligned}
$$

In these equations, each of the v_{s_i} terms is a linear combination of the source voltages in a mesh, plus possibly terms of the form i_sR due to transformed current sources. The R_{ij} terms are sums of resistor values in the meshes, properly modified if controlled sources are present.

Any of the mesh currents above will be given by a linear combination of the form

Input source terms

$$i_j = \frac{\Delta_{1j}}{\Delta} v_{s_1} + \frac{\Delta_{2j}}{\Delta} v_{s_2} + \cdots + \frac{\Delta_{nj}}{\Delta} v_{s_n} \quad (6.2)$$

Resistors, controlled sources only

where, from Chapter 4, Δ is the determinant and Δ_{ij} is the ij cofactor of this determinant. Thus, each mesh current can be written as the sum of terms, each term being proportional to v_{s_i}, which in turn represents a summation of certain of the independent sources in the network. A similar result based on node analysis would be that each node voltage can be written as a linear combination of terms proportional to the independent sources in the network.

The generality of Equation (6.2) where each response is linearly related to each source value suggests the following result: In any network of (linear) resistors, (linear) controlled sources, and independent sources, each element current or voltage can be expressed as a sum of terms, each term being the current or voltage produced by a single independent source, with all the other independent sources in the network set to zero.

This is known as the *superposition principle* for resistive networks.

Although currents and voltages can be calculated using superposition, power cannot be; the power in an element is *not* the sum of the powers due to each source acting alone. To illustrate this, consider resistor R_1 of Figure 6.2. The current and voltage in this element are

$$i_{R_1} = i_{R_1}^{(1)} + i_{R_1}^{(2)}$$

$$v_{R_1} = v_{R_1}^{(1)} + v_{R_1}^{(2)}$$

each having a component produced by v_1 and a component produced by v_2. The power in this resistor is given by

$$\begin{aligned} p_{R_1} &= i_{R_1} v_{R_1} = (i_{R_1}^{(1)} + i_{R_1}^{(2)})(v_{R_1}^{(1)} + v_{R_1}^{(2)}) \\ &= i_{R_1}^{(1)} v_{R_1}^{(1)} + i_{R_1}^{(2)} v_{R_1}^{(1)} + i_{R_1}^{(1)} v_{R_1}^{(2)} \\ &\qquad + i_{R_1}^{(2)} v_{R_1}^{(2)} \\ &\neq i_{R_1}^{(1)} v_{R_1}^{(1)} + i_{R_1}^{(2)} v_{R_1}^{(2)} \end{aligned}$$

We can see from this equation that if we try to calculate power by superposition, we lose the "cross-product" terms, that is, the power resulting from the interaction of currents produced by one source with voltages produced by another source. In power calculations, we can use superposition to find currents and voltages, but the power must be determined using *total* element voltages and currents.

6.3 Generalized Thevenin and Norton Equivalents

In Chapter 2, we studied two methods for determining Thevenin-Norton equivalents: (1) by network reduction, and (2) by measurement of v_{oc} and i_{sc}. As we observed in Chapter 4, not all networks are reducible using the methods of Chapter 2 so that method (1) is not general. Further, we have yet to show that the *i-v* characteristic of a network of linear resistors and sources is a straight line, the assumption on which method (2) is based.

We now consider a method for computing one-port equivalents using mesh and node analysis. Because any network of linear resistors and sources can be analyzed with these methods, we will in effect be showing that the *i-v* characteristic is a straight line and that method (2) is indeed valid.

Consider the one-port model of Figure 6.4, where a current source i has been applied to the port. From the superposition principle, we know that v can be written as

$$\begin{aligned} v = {} & \text{(voltage due to } i) \\ & + \text{(voltage due to internal independent sources)} \end{aligned}$$

The second term is the port voltage when $i = 0$

(port open-circuited). This component is evidently $v_{oc} = v_{eq}$ studied in Chapter 2. The first term represents a measurement of the resistance R_{in} looking into the port when internal independent sources are set to zero. Therefore, we can write

$$v = R_{in}i + v_{eq}$$

But the equation is identical in form to the equation for a Thevenin equivalent

$$v = R_{eq}i + v_{eq}$$

Therefore, $R_{in} = R_{eq}$. For this reason, R_{eq} is frequently referred to as the "look-in" resistance.

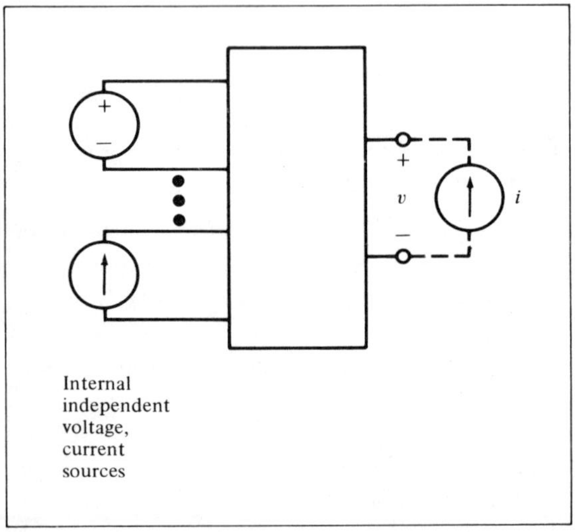

FIGURE 6.4 Model for derivation of Thevenin and Norton theorems.

We can obtain a similar result by applying a voltage source v at the port and solving for the Norton equivalent described by

$$i = \frac{v}{R_{eq}} - i_{eq}$$

Example 6.1

The Norton equivalent of the voltage divider of Figure 6.5 will be found by applying a voltage source v at the port and determining i.

First the mesh equations are written as

$$\begin{bmatrix} R_1 + R_2 & -R_2 \\ -R_2 & R_2 \end{bmatrix} \begin{bmatrix} i_1 \\ i_2 \end{bmatrix} = \begin{bmatrix} v_s \\ -v \end{bmatrix}$$

The mesh currents are given by

$$\begin{bmatrix} i_1 \\ i_2 \end{bmatrix} = \begin{bmatrix} \frac{\Delta_{11}}{\Delta} & \frac{\Delta_{21}}{\Delta} \\ \frac{\Delta_{12}}{\Delta} & \frac{\Delta_{22}}{\Delta} \end{bmatrix} \begin{bmatrix} v_s \\ -v \end{bmatrix}$$

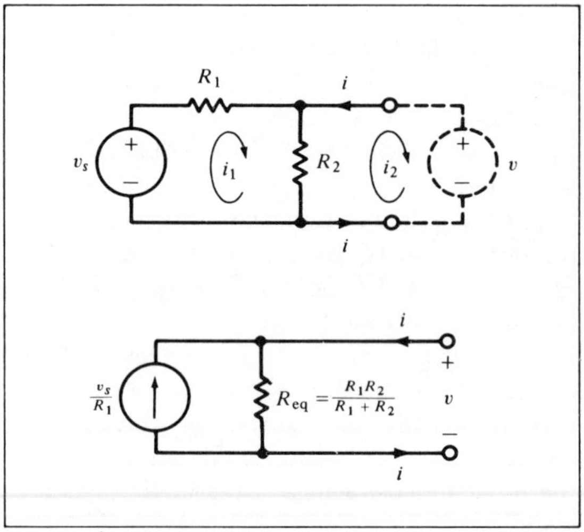

FIGURE 6.5 Voltage divider revisited.

Because $i = -i_2$, we have

$$i = \frac{-\Delta_{12}}{\Delta} v_s + \frac{\Delta_{22}}{\Delta} v$$

$$= \frac{-R_2}{R_1R_2} v_s + \frac{R_1 + R_2}{R_1R_2} v$$

$$= -i_{eq} + \frac{v}{R_{eq}}$$

6.4 Small-Signal Analysis

Introduction

The reader is surely familiar with the fact that a great many practical electrical circuits contain nonlinear elements such as diodes and transistors. It would be logical to conclude that the analysis of such circuits must always be performed with the assistance of the digital computer using the methods of Chapters 3 and 5. However, it is a fact that electronic circuits were designed long before widespread use of the digital computer as a design aid. We may wonder how these nonlinear elements were described in precomputer days.

The answer is that a *linear approximation* to these nonlinear elements is commonly used for design purposes. In this section, we wish to explore the nature of the linear approximation and the extent to which principles such as superposition and concepts such as the Thevenin equivalent can be exploited in the analysis of supposedly "nonlinear" networks.

Linear Approximation

The process of linearization is, of course, not new to us. In Chapters 3 and 5, the Newton iteration involved a set of equations which described a linear network with elements that were changed at each iteration. After convergence ($v_d^{m+1} = v_d^m$), the network model reverted to the original circuit model, as illustrated in Figure 6.6(a)–(c). We will term the model of Figure 6.6(c) the *dc model*, that is, the model with no signal applied.

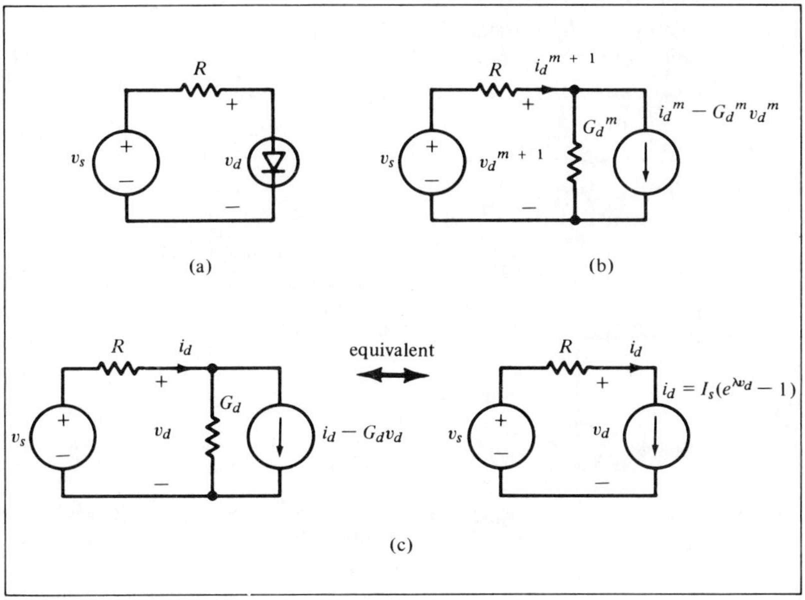

FIGURE 6.6 Illustration of linearized models. (a) Circuit; (b) companion model; (c) companion model after convergence (dc model).

Now suppose that a small voltage δv_s, representing a signal, is added to the source v_s. We define the resultant diode voltage as $v_d + \delta v_d$, so that δv_d is the *change* in the diode voltage due to the addition of the signal voltage δv_s. To calculate this change precisely, a new series of Newton iterations must seemingly be carried

TABLE 6.1

Evaluation of Error in Incremental Analysis (All Values in Volts)

$v_s + \delta v_s$	$v_d + \delta v_d^1$ (After 1 Iteration)	$v_d + \delta v_d$ (Precise Value)	Error (Absolute value)
10.0	0.4019268	0.4019268	0
10.2	0.4024464	0.4024411	5.3×10^{-6}
10.8	0.4040052	0.4039235	8.2×10^{-5}
13.2	0.4102402	0.4091061	1.1×10^{-3}

out using the digital computer. Consider, however, the possibility of *stopping after the first Newton iteration* and accepting the resultant error in δv_d. Because this requires the analysis of only one linear(ized) network, we can perform this analysis by hand and gain an insight into the network performance. We term such a calculation of the approximate response *incremental*, or more traditionally, *small-signal*, analysis.

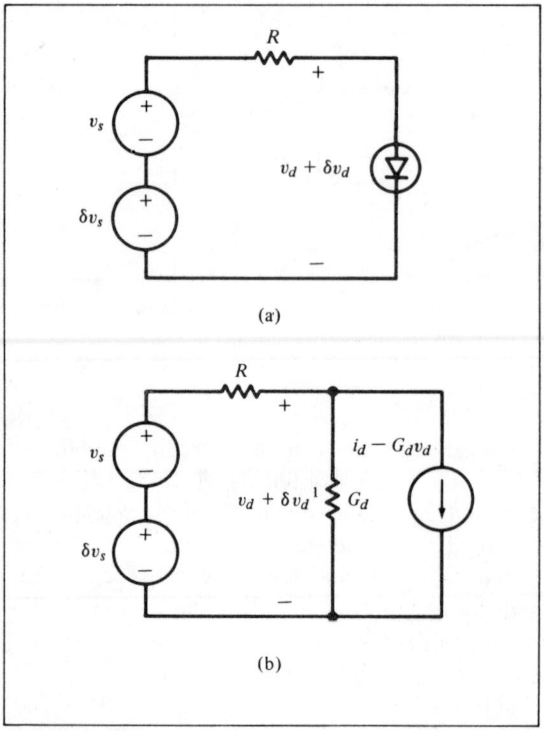

FIGURE 6.7 Network with signal applied. (a) Perturbed network; (b) model for first Newton iteration toward calculation of $v_d + \delta v_d$.

It is important to evaluate the extent of the approximation that we make in a typical incremental analysis. Figure 6.7(a) is the same network we analyzed in Chapter 3, Figure 3.6(b). With $v_s = 10$ volts, $R = 1000\ \Omega$, $I_s = 10^{-9}$, and $\lambda = 40$ in Figure 6.7(a), we can use the program of Figure 3.11 to find a precise dc value $v_d = 0.4019268$ volt. If we now add a voltage δv_s to v_s and begin a new series of Newton iterations, we obtain the results of Table 6.1. For a δv_s of value 0.2 and 0.8 volt, the error made in stopping after one iteration is surely negligible. Even with a value $\delta v_s = 3.2$ volts—32 percent of the value of v_s—the error of 0.11 percent is within most requirements for engineering accuracy. Our conclusion: A single linear network analysis is adequate to determine the effect of a small source voltage change on the response of this network.

Linearized Network Models

As we are about to discover, there is far more to be gained from this single linearization than a reduction in computation. Consider again the model of Figure 6.7(b), which we are now solving only once. Because this network is linear, the voltage $v_d + \delta v_d^1$ can be regarded as resulting from the sum of the responses due to sources v_s, $i_d - G_d v_d$, and δv_s *applied separately or in any combination.* In particular, let v_s and $i_d - G_d v_d$ be applied together and let the source voltage change δv_s be applied separately, as shown in Figure 6.8(b) and (c). Denote the

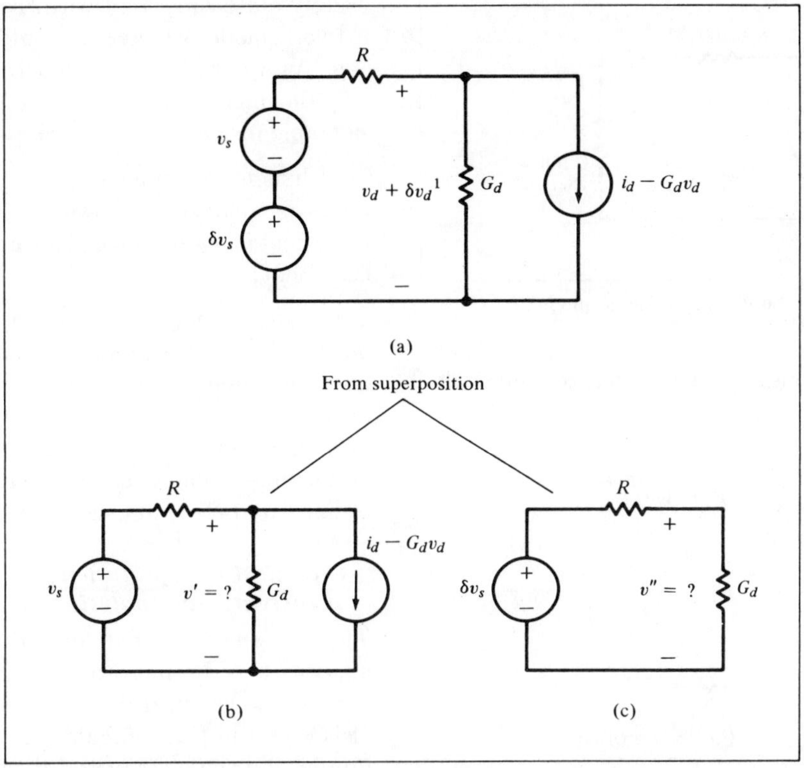

FIGURE 6.8 (a) Separation into dc and small-signal models. (b) dc model; (c) small-signal incremental model.

diode voltage responses to these sources as v' and v'', respectively. Note that the network of Figure 6.8(b) is merely the model with no voltage change (that is, the dc model) so that

$$v' = v_d$$

Now the superposition principle requires that the voltages v' and v'' sum to the diode voltage before separation of the network models; therefore,

$$v' + v'' = v_d + \delta v_d^1$$

so that

$$v'' = \delta v_d^1$$

Thus, the voltage across G_d in Figure 6.8(c) is precisely the *change* δv_d^1 from the dc value v_d. We can, therefore, separate the incremental from the dc component of v_d by analysis of the simple network of Figure 6.8(c). A little thought will reveal that *all* of the currents and voltages in Figure 6.8(c) are the incremental components of the corresponding currents and voltages in the original network. For this reason, the model of Figure 6.8(c) is known as the *incremental*, or *small-signal*, model.

The incremental model greatly facilitates the calculation of incremental changes in network currents and voltages. Continuing with the example at hand, we would find from the dc analysis of Table 6.1 that

$$G_d = \lambda I_s e^{\lambda v_d}$$

$$= 40 \times 10^{-9} \times e^{(40 \times 0.4019268)}$$

$$= 0.384 \ ℧$$

Any change in v_d in Table 6.1 can now be easily verified from Figure 6.9 by the voltage

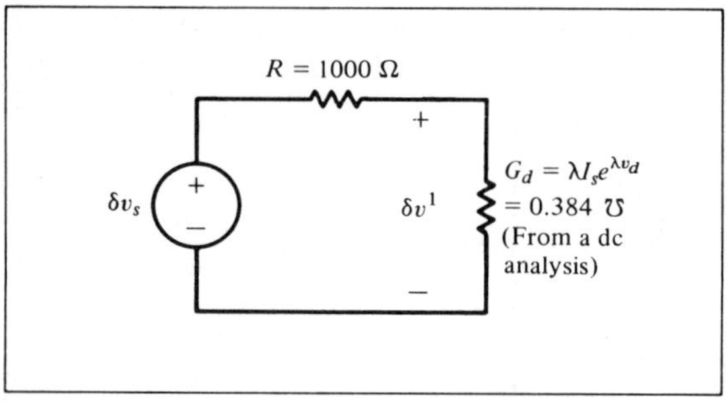

FIGURE 6.9 A small-signal model.

divider relationship. We have, to slide-rule accuracy,

$$\delta v_d^1 = \frac{1/G_d}{R + 1/G_d}\,\delta v_s$$

$$= \frac{\delta v_s}{RG_d + 1}$$

$$= \frac{3.2}{384 + 1}$$

$$= 0.00831 \text{ volt}$$

This compares with the computer-determined value of

$$(0.4102402 - 0.4019268) = 0.0083134 \text{ volt}$$

Among the features of the incremental model are:

1. Uniqueness, because there can be only *one* model representing the linearized behavior in the vicinity of the dc solution.
2. Ease of construction; any fixed sources such as v_s are associated with the dc model and have a zero value in the incremental model; diodes are represented by their linearized conductances; other elements remain intact.
3. Because it is linear, it lends itself to hand analysis using any of the methods of Chapters 2 and 4.
4. Slide-rule accuracy is adequate, because the result of calculation is only an approximation to the precise answer.
5. In practice, the quality of the answer is usually quite consistent with the quality of the network model which represents a real device.
6. As we will shortly illustrate in Section 6.5, the incremental model can provide considerable insight into network performance.

Although encouraged by the convenience it offers in analyzing a network response to an incremental source change, we should appreciate that:

1. Effects such as distortion which depend on nonlinearity cannot be determined; the linear incremental model will faithfully reproduce the shape of any $\delta v_s(t)$ or $\delta i_s(t)$.
2. A dc analysis must be performed before an incremental analysis to determine the value of linearized elements such as G_d.

Graphical Interpretation of Incremental Analysis

The dc analysis that precedes incremental analysis may be performed graphically, rather than numerically, from the dc network model. It is useful in these circumstances to represent graphically the incremental network solution.

From Chapter 3, we are familiar with the graphical constructions to find the exact solutions of both the perturbed and unperturbed networks. These are designated points C and A, respectively, in Figure 6.10(d), at the intersections of two load lines and the exponential diode characteristic. The linearized diode characteristic associated with the linearized network model of Figure 6.10(b) is represented by the tangent line to the nonlinear characteristic at the dc solution point. This tangent line is also used in the small-signal diode model, except that now the origin $(\delta v_d^1, \delta i_d^1) = (0, 0)$ *is taken at the solution of the dc model* (point A). By definition, because the tangent line must pass through this new origin, only the slope G_d of the tangent line appears in the small-signal model [see insert in Figure 6.10(d)]. Therefore, we can correctly write the linearized diode equation as

$$\delta i_d^1 = G_d \delta v_d^1$$

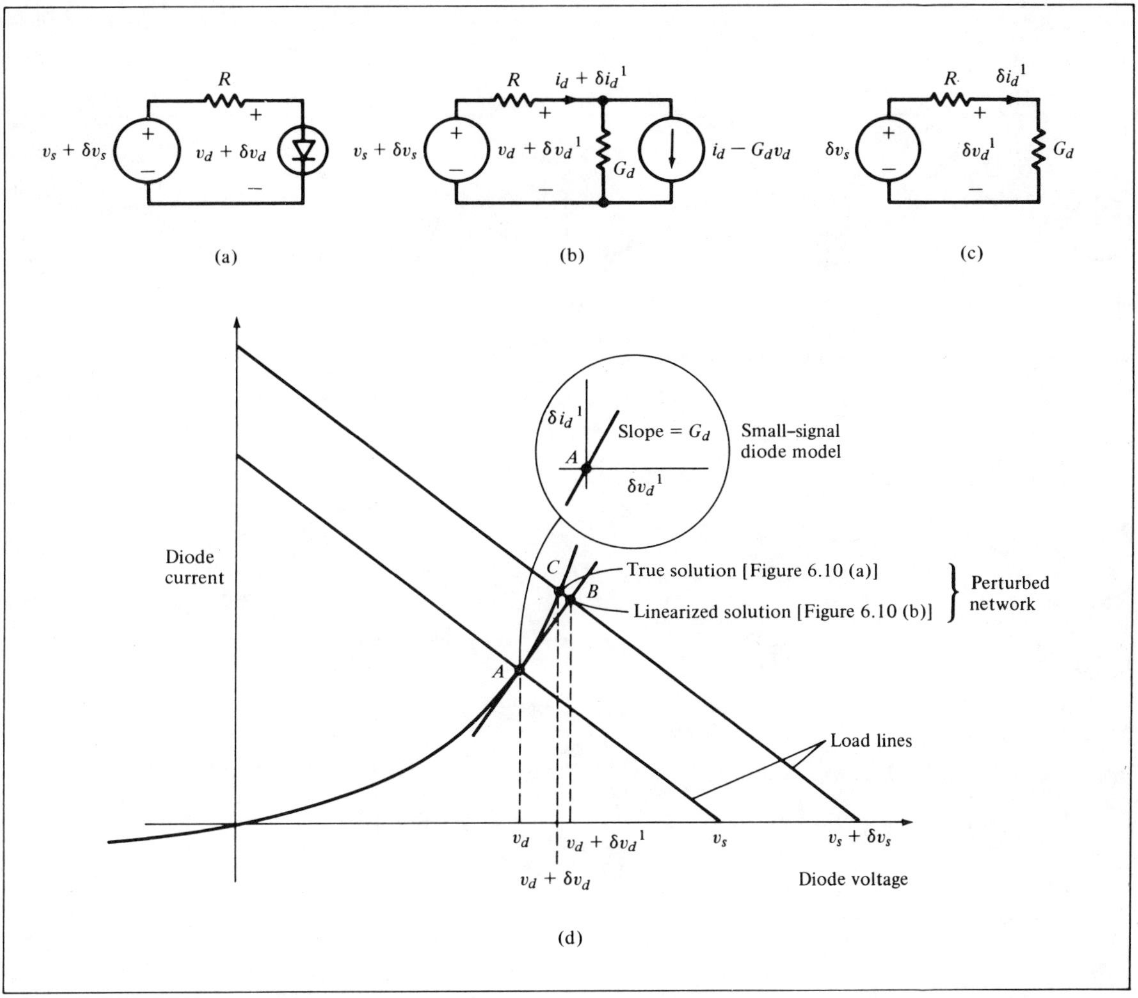

FIGURE 6.10 Graphical interpretations of solution models. (a) Nonlinear model; (b) linearized model; (c) small-signal model.

6.5 Applications of Incremental Analysis

Introduction

In Chapters 3 and 5, we performed a dc and gain analysis of simple amplifier circuits using elementary transistor models. Using the dc analysis methods of Chapter 5 and the incremental analysis techniques of this chapter, we are now in a position to make similar analyses on more complicated circuits with more complete models. In this section, we study a number of examples of such analysis, to make clear the power and generality of these combined methodologies.

In this section (and henceforth in the text), we will not normally distinguish between the approximate responses δv_0^1 or δi_0^1 and the precise values δv_0 and δi_0 (obtained after repeated Newton iteration). For notational compactness we will adopt the latter notation for the approximate response as well, except where the

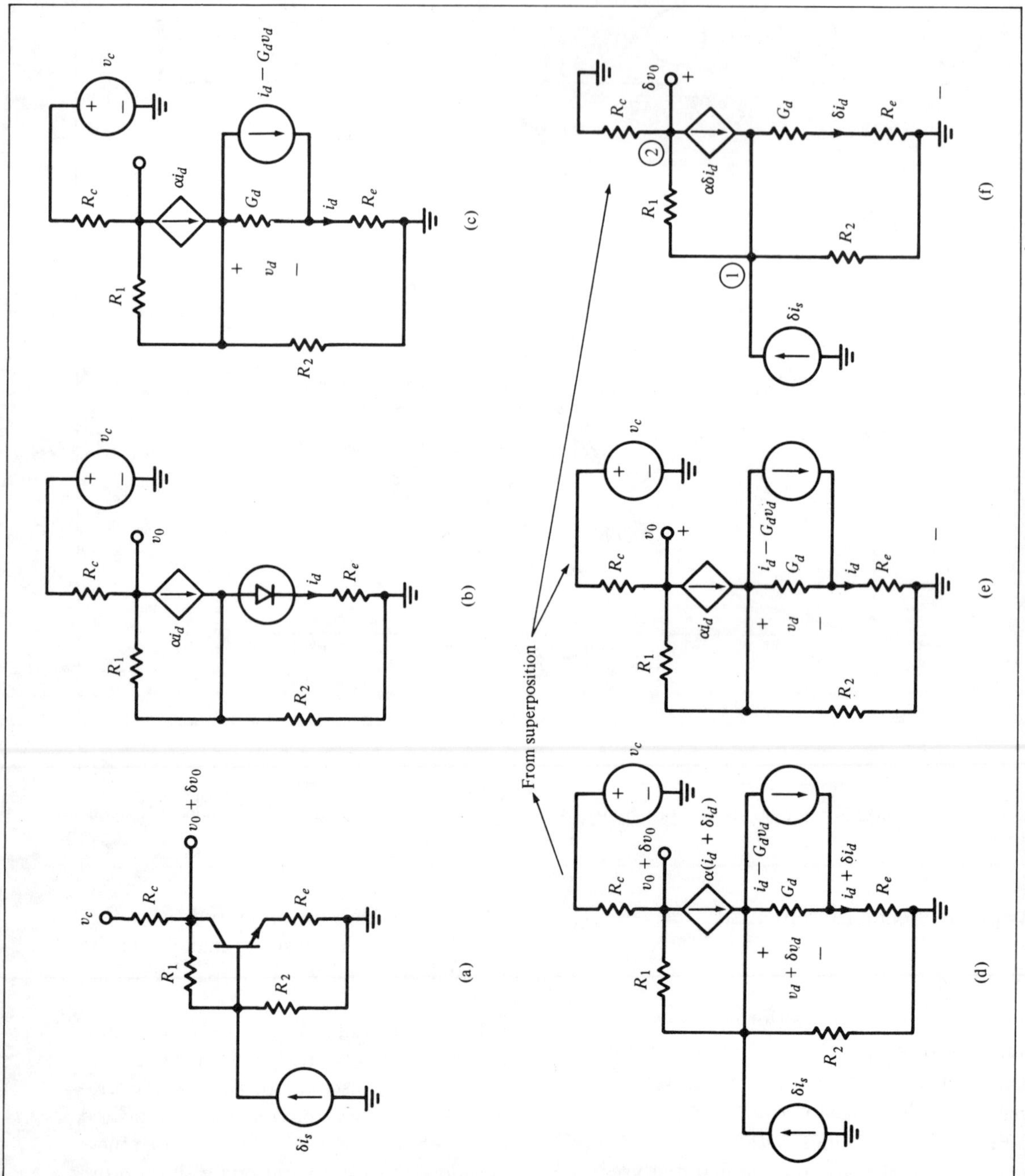

FIGURE 6.11 Large- and small-signal amplifier models. (a) Amplifier schematic with signal; (b) amplifier model without signal; (c) dc model after convergence; (d) complete linearized model with signal added; (e) dc model [same as (c)]; (f) incremental small-signal model.

accuracy of the incremental (small-signal) response is to be evaluated.

Small-Signal Analysis of Amplifiers

Our study of logic circuits in Chapter 3 revealed that these networks depended on the nonlinearity of the diode characteristic (the "on" and "off" behavior) to function. In contrast, normally amplifiers are designed to behave in an almost-linear manner so that signals are magnified but not distorted. Therefore, a small-signal model (exemplified by Figure 6.9) can be expected to be an excellent vehicle for predicting the response δv_0 of a well-designed amplifier to a signal δi_s or δv_s.

The example network we will study is depicted in Figure 6.11(a) and (b). First, the network is solved with the signal source δi_s set to zero, using Newton iteration as in Figure 5.8(a). The dc network model has the form of Figure 6.11(c) after convergence. Applying δi_s to the network results in the new linearized model of Figure 6.11(d). This model can be decomposed into two networks: (1) the first containing all independent sources responsible for the dc model of Figure 6.11(c) and (2) the second containing the signal source δi_s. From the superposition principle, the total (approximate) output voltage $v_0 + \delta v_0$ of Figure 6.11(d) can be written as the sum of the voltage v_0 of Figure 6.11(e) and δv_0 of Figure 6.11(f).

The incremental network model of Figure 6.11(f) can now be solved using any convenient method of linear network analysis. For example, the diode conductance G_d of Figure 6.11(f) can be combined with R_e to yield

$$G_{de} = \frac{1}{R_e + 1/G_d} = \frac{G_d}{1 + R_e G_d}$$

This reduces the number of nonreference nodes to two.

A node analysis using the node numbering shown proceeds in two steps, first regarding the controlled source $\alpha\delta i_d$ as independent, namely,

$$\begin{bmatrix} G_1 + G_2 + G_{de} & -G_1 \\ -G_1 & G_1 + G_c \end{bmatrix} \begin{bmatrix} \delta v_1 \\ \delta v_2 \end{bmatrix} = \begin{bmatrix} \delta i_s + \alpha\delta i_d \\ -\alpha\delta i_d \end{bmatrix} = \begin{bmatrix} \delta i_s + \alpha G_{de}\delta v_1 \\ -\alpha G_{de}\delta v_1 \end{bmatrix}$$

Transferring the terms containing unknown voltage δv_1 to the left-hand side of the equations gives

$$\begin{bmatrix} G_1 + G_2 + G_{de} - \alpha G_{de} & -G_1 \\ -G_1 + \alpha G_{de} & G_1 + G_c \end{bmatrix} \begin{bmatrix} \delta v_1 \\ \delta v_2 \end{bmatrix} = \begin{bmatrix} \delta i_s \\ 0 \end{bmatrix}$$

The solution for the incremental output voltage δv_0 routinely follows as

$$\begin{aligned} \delta v_0 &= \delta v_2 \\ &= \frac{\Delta_{12}}{\Delta}\delta i_s \\ &= \frac{(-\alpha G_{de} + G_1)\delta i_s}{(G_1 + G_2 + G_{de} - \alpha G_{de})(G_1 + G_c) - G_1(G_1 - \alpha G_{de})} \qquad (6.3) \\ &= \frac{(-\alpha G_{de} + G_1)\delta i_s}{G_c(G_1 + G_2 + G_{de} - \alpha G_{de}) + G_1(G_2 + G_{de})} \end{aligned}$$

TABLE 6.2

Comparison of Precise and Approximate Incremental Output

	Precise		Approximate	Error
δi_s (Amps)	$v_0 + \delta v_0^1$	δv_0^1	$\delta v_0 = 1.1359 \times 10^4 \delta i_s$	$\delta v_0^1 - \delta v_0$
-10^{-4}	5.9851	1.1271	1.1359	−0.0088
-5×10^{-5}	5.4241	0.5661	0.5679	−0.0018
0	4.8580	0.0000	0.0000	0.0000
5×10^{-5}	4.2884	−0.5696	−0.5679	−0.0017
10^{-4}	3.7162	−1.1418	−1.1359	−0.0059

If we use the element values of Figure 5.8(b), namely,

$$R_1 = 20{,}000\ \Omega \quad R_2 = 3000\ \Omega \quad R_c = 5000\ \Omega$$

$$R_e = 300\ \Omega \quad \alpha = 0.98 \quad v_c = 10 \text{ volts}$$

we find that, after Newton iteration, the dc model of Figure 6.11(e) contains a diode conductance $G_d = 0.0331$ ℧. Substitution of these values into Equation (6.3) gives

$$\delta v_0 = \frac{(-0.98(0.0331) + 5 \times 10^{-5})\delta i_s}{2 \times 10^{-4}(0.00044) + 5 \times 10^{-5}(0.00336)}$$

$$= -1.1359 \times 10^4 \delta i_s$$

We can therefore claim that the amplifier has an incremental, or small-signal, "gain" $\delta v_0/\delta i_s = -1.1359 \times 10^4$ units. (What are the units?)

The quality of the small-signal approximation can be checked by using the program of Figure 5.8(a) to perform the exact analysis of the original network of Figure 6.11(b) with δi_s applied. In other words, we could complete the series of Newton iterations which the small-signal model is viewed as initiating. For typical values of δi_s (in the range of ± 0.1 ma), the precise and approximate output voltage changes are shown in Table 6.2. The linear incremental model is seen to predict the output voltage to within 0.01 volt.

A further study of this example reveals that the quality of the approximation stems from the small effect on v_0 of changes in G_d (representing the diode). The diode resistance $R_d = 1/G_d = 30.2\ \Omega$ is in series with a 300Ω resistor. Thus, G_d can vary by a large percentage of 30.2Ω without appreciably affecting G_{de} and hence δv_0. In electronic circuit courses, the role of R_e in achieving this highly linear performance is studied in considerable detail.

Other Examples

To illustrate the insight obtained from a small-signal analysis, we consider several additional examples. It is assumed that the dc solution has been obtained from DCAP of Chapter 5, which prints out the small-signal diode resistance R_d and conductance G_d after the dc solution.

Example 6.2

The amplifier circuit of Figure 6.12(a) is to be analyzed to find its dc operating point and to determine the behavior in the vicinity of this operating point. In particular, the effect on the output/input ratio $\delta v_0/\delta i_s$ of adding a resistor R_0 across the output is to be found.

A dc analysis using the DCAP program of Chapter 5 yields the dc output voltage $v_0 = 2.61$ in Figure 6.12(b) and the small-signal diode resistance value $R_d = 3.38\Omega$ in Figure 6.12(c). From the small-signal model of Figure 6.12(d), the Thevenin equivalent is calculated. If the dc output voltage is added to the incre-

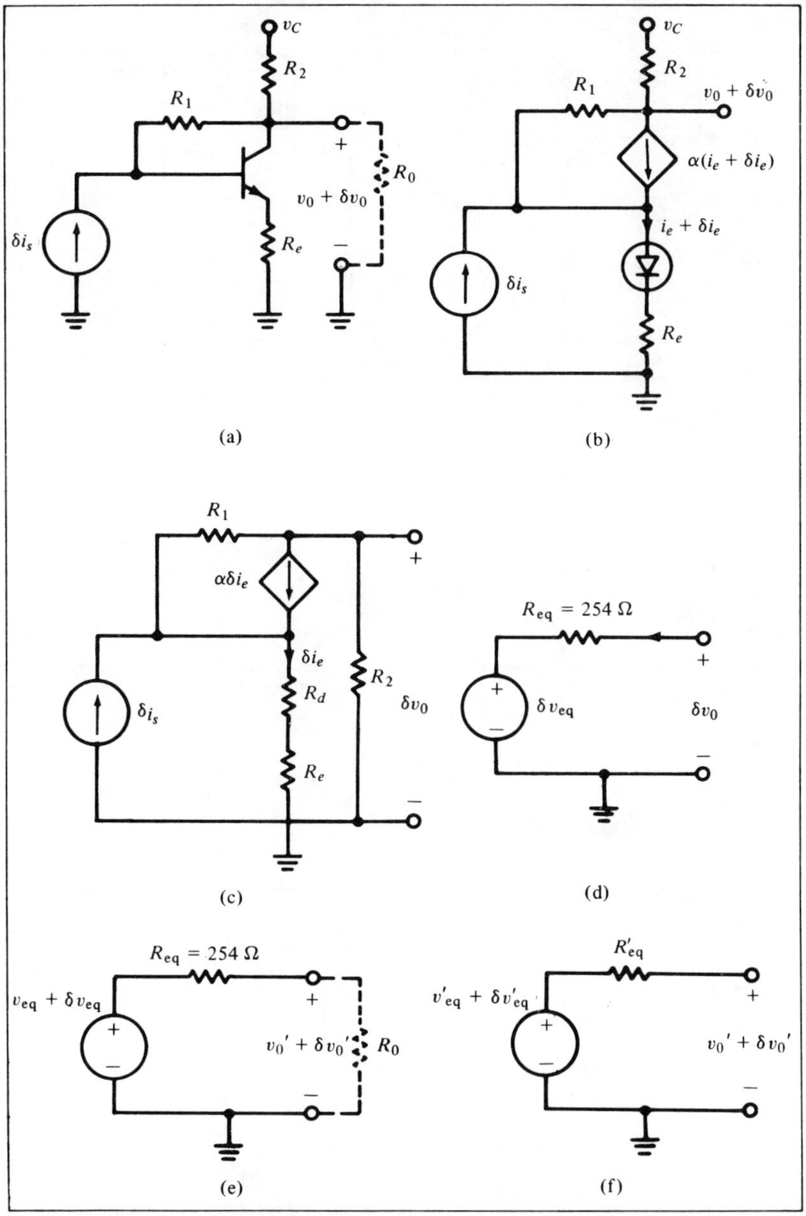

FIGURE 6.12 Amplifier analysis.

(a) Amplifier schematic:
$R_1 = 10\ \text{k}\Omega, \quad R_2 = 1\ \text{k}\Omega,$
$R_e = 100\ \Omega, \quad v_C = 10\ \text{volts};$

(b) nonlinear model and dc solution:
$\alpha = 0.98, \quad I_s = 10^{-9},$
$\lambda = 40,$
$$\left.\begin{aligned}\delta i_s &= 0\\ v_o &= 2.61\end{aligned}\right\}\text{dc solution from DCAP;}$$

(c) small-signal model:
$R_d = 3.38\ \Omega$ (from DCAP),
$R_{de} = R_d + R_e \cong 103\ \Omega;$

(d) small-signal Thevenin equivalent:
$$\delta v_0 = \frac{(G_1 - G_{de})\delta i_s}{G_1G_2 + G_1G_{de} + G_2G_{de}(1-\alpha)} = -8170\delta i_s = \delta v_{eq},$$
$$R_{eq} = \frac{G_1 + G_{de}(1-\alpha)}{G_1G_2 + G_1G_{de} + G_2G_{de}(1-\alpha)} = 254\ \Omega;$$
complete linearized Thevenin equivalent for network of (b):
$$v_{eq} + \delta v_{eq} = v_0 + \delta v_0 = 2.61 - 8170\delta i_s.$$

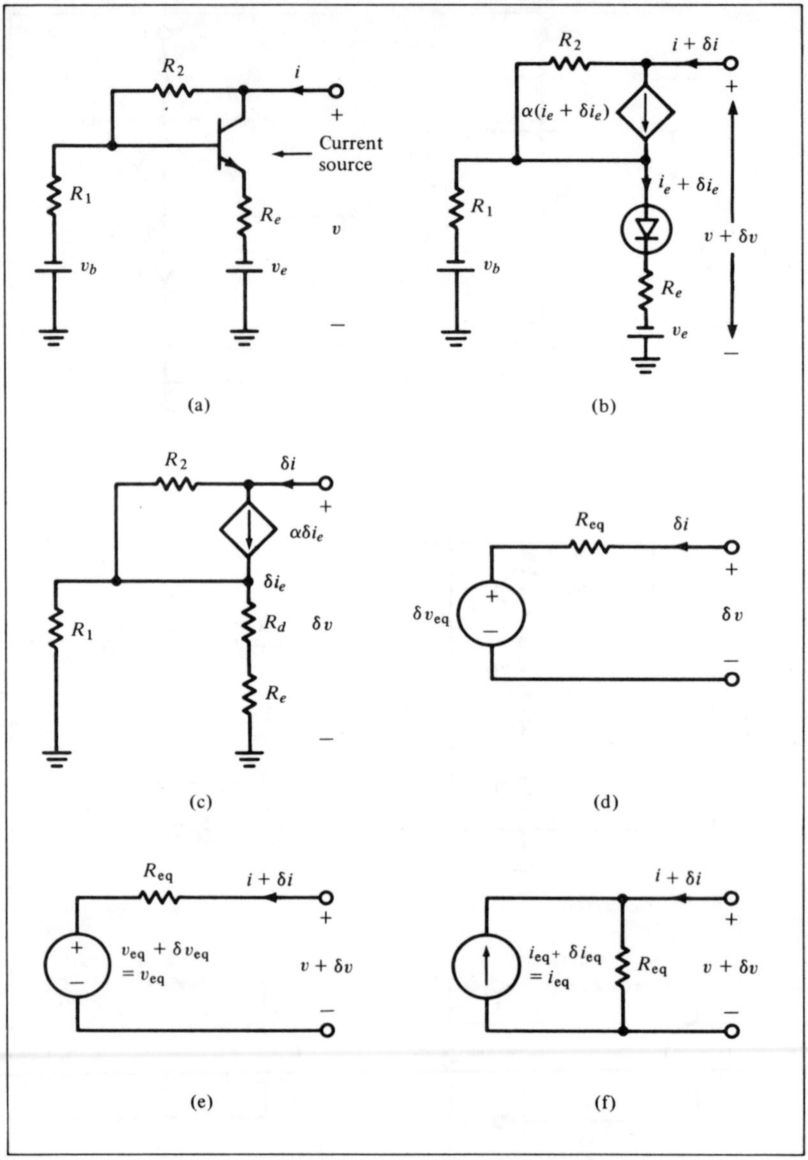

FIGURE 6.13 Current source equivalent.

(a) Current source schematic:
$R_1 = 1\ \text{k}\Omega, \quad R_2 = 10^5\ \Omega,$
$R_e = 300\ \Omega, \quad v_b = 6$ volts,
$v_e = 12$ volts;

(b) nonlinear model:
$\alpha = 0.98, \quad I_s = 10^{-9},$
$\lambda = 40,$
$i = 17.6$ ma, $v = 10$ volts } dc solution from DCAP;

(c) small-signal model:
$R_d = 1.39\ \Omega$ (from DCAP),
$R_{de} = R_d + R_e \cong 301\ \Omega$;

(d) small-signal Thevenin equivalent:

$$\delta v_{\text{eq}} = 0,$$

$$R_{\text{eq}} = \frac{G_1 + G_2 + G_{de}(1 - \alpha)}{G_1G_2 + G_2G_{de}} = 24{,}500\ \Omega = \delta v/\delta i;$$

(e) complete linearized Thevenin and Norton equivalents for network of (b) [v and i obtained from (b)]:

$$v_{\text{eq}} = v - iR_{\text{eq}} = 10 - 0.0176(24{,}500) = -420 \text{ volts};$$

(f)

$$i_{\text{eq}} = -\frac{420}{24{,}500} = -17.2 \text{ ma}.$$

mental Thevenin voltage generator, the resultant linearized equivalent of Figure 6.12(e) represents the complete (dc and small-signal) amplifier performance. As a check, when $i_s = 0$, the dc value of $v_0 = 2.61$ is observed at the output terminals of the equivalent network.

The effect of adding a resistor R_0 as shown in Figure 6.12(e) can be evaluated by a new Thevenin equivalent described by

$$R_{eq}' = \frac{R_0 R_{eq}}{R_0 + R_{eq}}$$

$$v_0' + \delta v_0' = v_{eq}' + \delta v_{eq}'$$

$$= (v_{eq} + \delta v_{eq}) \frac{R_0}{R_0 + R_{eq}} \quad (6.4)$$

as shown in Figure 6.12(f). The signal component of this expression

$$\delta v_0' = \frac{R_0}{R_{eq} + R_0} \delta v_{eq}$$

$$= \left(\frac{R_0}{R_0 + 254}\right) (-8170)\delta i_s$$

shows directly the effects of a load resistor on the $\delta v_0/\delta i_s$ ratio. For example, this ratio is nearly independent of the value R_0 if $R_0 > 1000\Omega$. Of course, the dc analysis may have to be checked to insure that the transistor is still operating in the linear region.

Example 6.3

The circuit of Figure 6.13(a) is claimed to be a close approximation to a current source. This implies that the current i is nearly independent of v.

To verify this claim, we apply a test dc voltage of 10 volts and using DCAP calculate the resultant dc current to be 17.6 ma as shown in Figure 6.13(b). This establishes one point on the i-v characteristic of the one-port. We examine the i-v behavior in the vicinity of this point with the help of the small-signal model of Figure 6.13(c). The complete linearized models of Figure 6.13(e) are determined to match the port behavior at $(v, i) = (10 \text{ volts}, 17.6 \text{ ma})$ and to have the slope of the small-signal model. It is interesting to note that:

1. When $v = 0$, the short-circuit current $i = -i_{eq} = 17.2$ ma of Figure 6.13(e) will flow; whereas when $v = 10$, the current $i = 17.6$ ma of Figure 6.13(b) flows; therefore, the port current will indeed remain nearly a constant, independent of v, so long as the incremental model is valid.

2. v_{eq} has a large negative value of -420 volts despite the fact that no voltage source in the original network is greater than 12 volts; would we measure this voltage if we leave the port open-circuited?

PROBLEMS

6.1 Which elements should be removed from the network of Figure P 6.1 so that the superposition principle applies at nodes ① and ②? This requires (for example) that doubling an applied voltage v_s will double the current i_s.

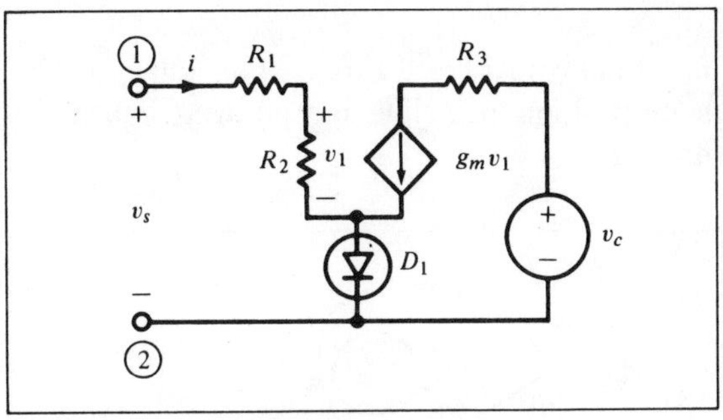

FIGURE P 6.1

6.2 Show that the voltage v_0 in Figure P 6.2 is 23/3 volts, using superposition. Use any convenient method of analysis.

6.3 Show that v_0 in Figure P 6.3 is -3.412 volts, using node analysis to compute v_0 when the current source is applied and mesh analysis when the voltage source is applied.

that is, R^2, R^3, . . . do not appear in the expression. This implies that any terms containing R^n must cancel if $n > 1$. [*Hint*: Initially replace R with a current source i_R; then write the node equations and determine

$$v = f_1(i_R, i)$$

$$v_R = f_2(i_R, i)$$

Then eliminate v_R and i_R.]

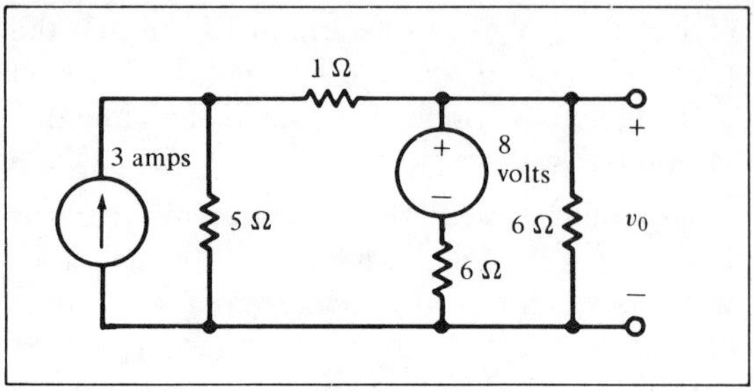

FIGURE P 6.2

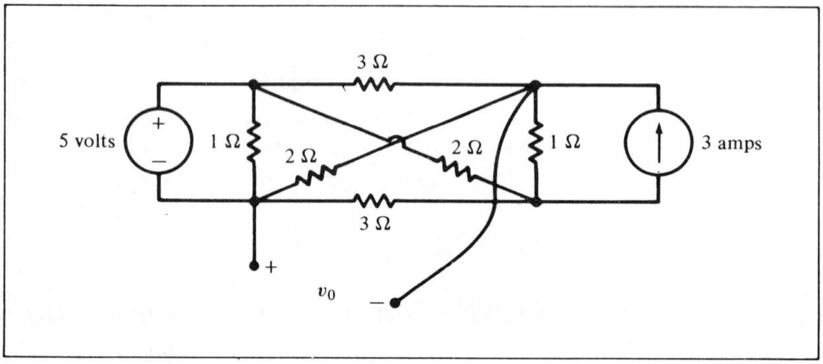

FIGURE P 6.3

6.4 Find v_0 in Figure P 6.4. Point out how the superposition principle is manifested in your answer.

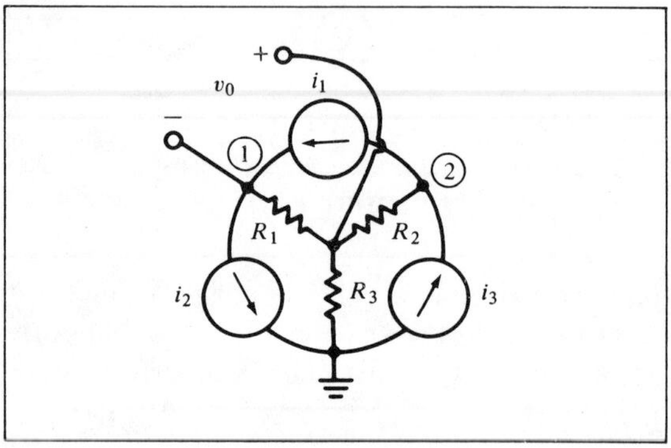

FIGURE P 6.4

6.5 Show that the input resistance of the network of Figure P 6.5(a) can be written as

$$R_{\text{eq}} = \frac{a + bR}{c + dR}$$

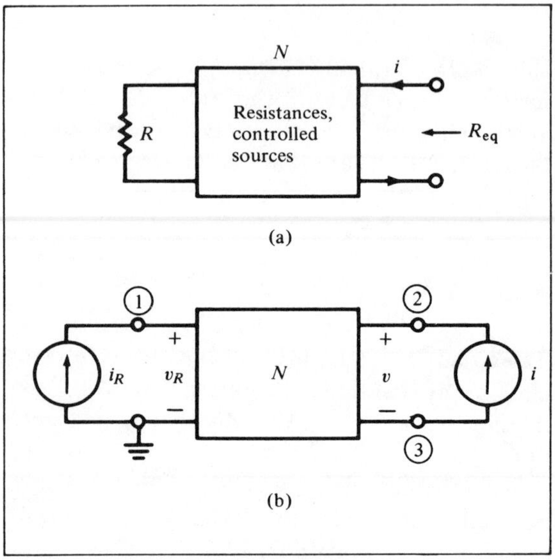

FIGURE P 6.5

6.6 Show that Thevenin equivalent of the one-port of Figure P 6.6 has an $R_{\text{eq}} = 0.4375\Omega$ and $v_{\text{eq}} = 0.0625$ volt. Use mesh or node analysis.

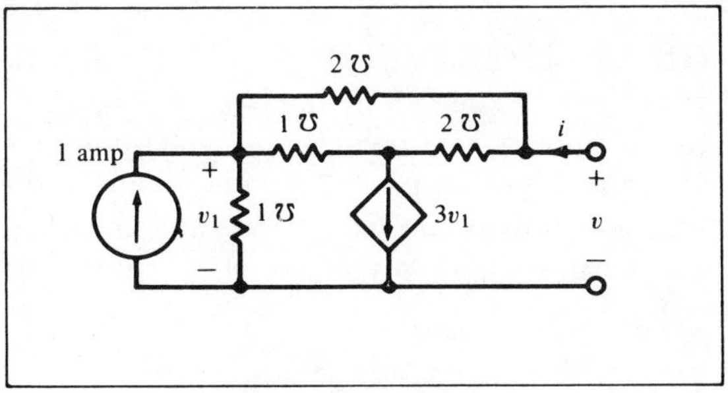

FIGURE P 6.6

6.7 Assume that mesh equations have been written for the network of Figure P 6.7. Find

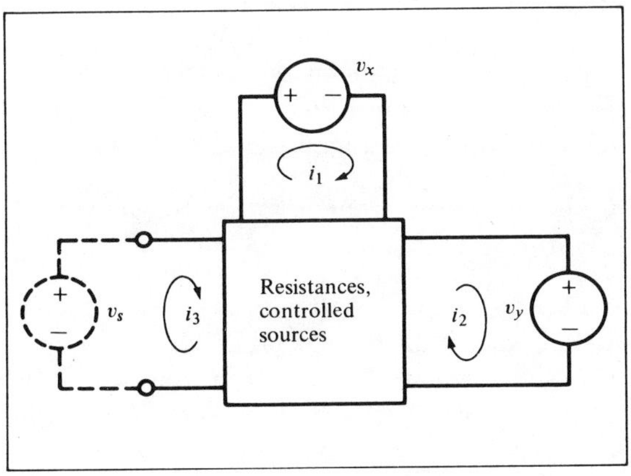

FIGURE P 6.7

the Thevenin and Norton equivalent across v_s as follows:

$$R_{\text{eq}} = f_1(\Delta, \Delta_{ij}\text{'s})$$

$$i_{\text{eq}} = f_2(\Delta, \Delta_{ij}\text{'s}, v_x, v_y)$$

$$v_{\text{eq}} = f_3(\Delta, \Delta_{ij}\text{'s}, v_x, v_y)$$

6.8 The formula for the Thevenin equivalent of the one-port of Figure P 6.8 was obtained in Chapter 3. Verify this result by node or mesh analysis. One source transformation is allowed.

6.9 Find the Thevenin or Norton equivalent for the one-port of Figure P 6.9. Apply a current or voltage source across the port shown and invoke the superposition principle to find R_{eq} and v_{eq} (or i_{eq}).

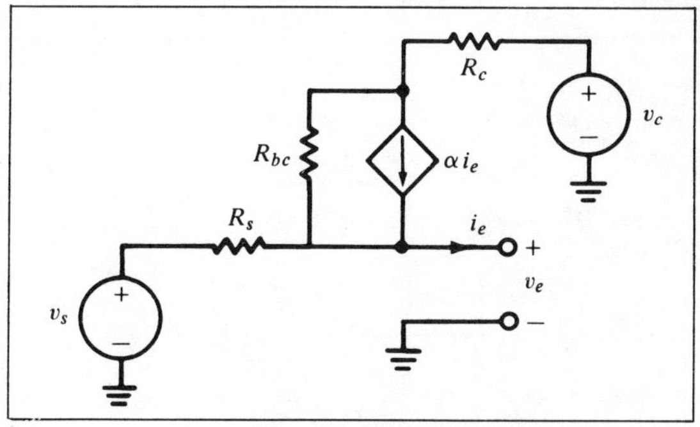

FIGURE P 6.8

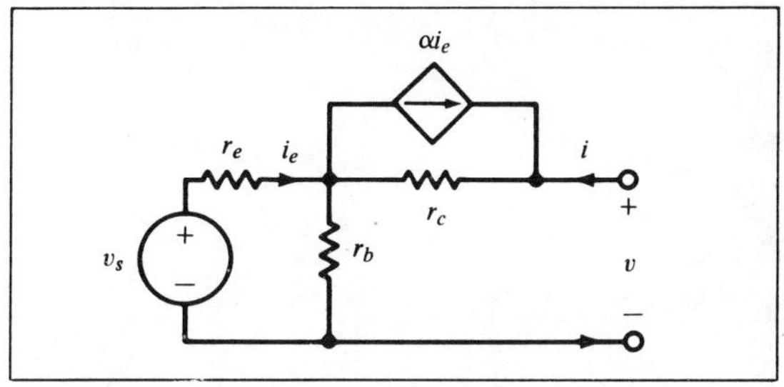

FIGURE P 6.9

6.10 A diode is described by

$$i_d = I_s(e^{\lambda v_d} - 1)$$

(a) Determine the small-signal conductance. (b) If the diode is conducting ($e^{\lambda v_d} \gg 1$), determine this conductance in ℧/ma.

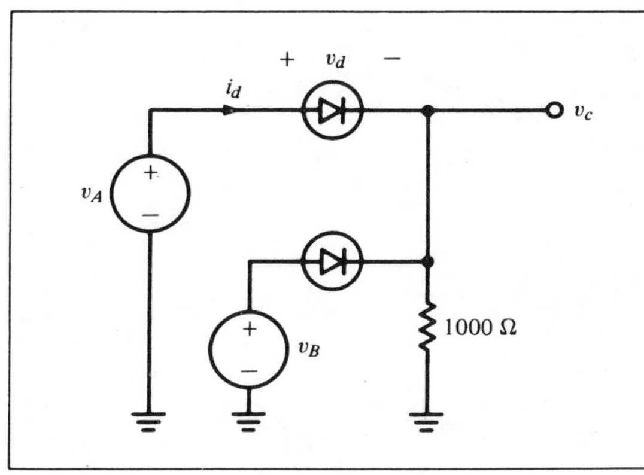

FIGURE P 6.10

Inputs

Logical	Voltage
0	0
1	3

$i_d = 10^{-9}(e^{40v_d} - 1)$

$v_d = 0.35$ when turned on.

6.11 Consider the OR circuit of Figure P 6.10 with inputs $v_A = 3$ volts and $v_B = 0$ volts. A small amount of "noise" (0.2 volt) appears at input A so that $v_A = 3.2$ volts. Find the resultant approximate change on the output v_C. Use small-signal analysis and show the small-signal model.

6.12 For the network of Figure P 6.11(a), use the transistor models of Figure 5.10 without the collector diode and without α_r. (a) Construct the dc model of the complete circuit. (b) Construct the companion network model of the complete circuit. (c) Write the node equations of the linearized model. (d) Construct the small-signal network model. (e) Write the node equations for the small-signal model. You should get the same node matrix as in (c). (f) Write $\delta v_O/\delta i_s$ in terms of the cofactors and determinant of the small-signal node matrix.

6.13 Repeat Problem 6.12 for the circuit of Figure P 6.11(b).

6.14 The amplifier circuit shown in Figure P 6.12 was studied in Chapter 5. (a) Construct the small-signal equivalent of the amplifier. Assume that (1) the emitter diode of the npn transistor has an incremental resistance of 59Ω, (2) the emitter diode of the pnp transistor has an incremental resistance of 11 Ω, and (3) the collector diodes of both transistors are re-

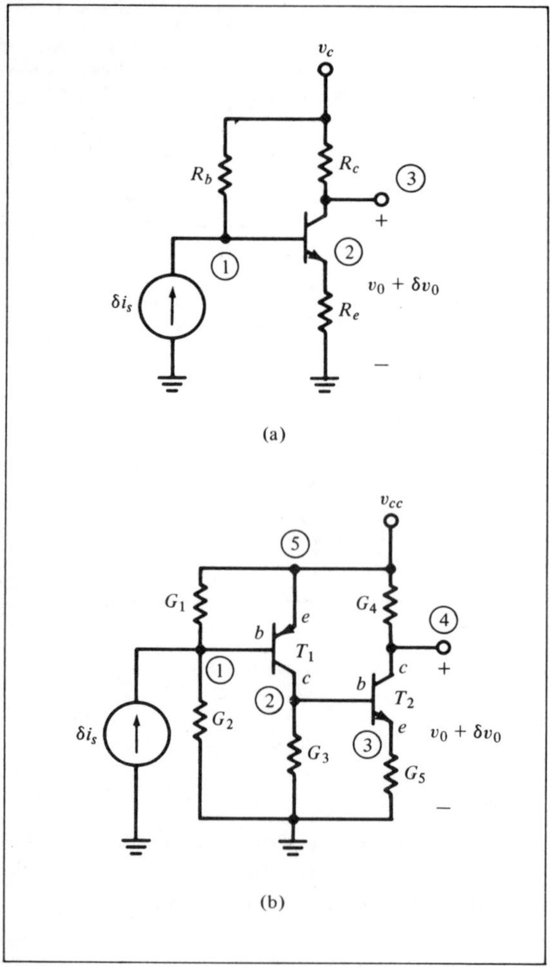

FIGURE P 6.11

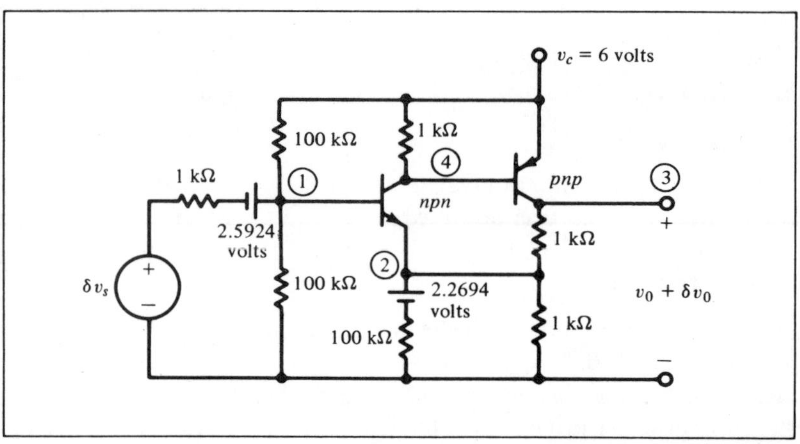

FIGURE P 6.12

verse-biased. (b) Using the node assignments shown, determine a numerical value for the small-signal gain $\delta v_0/\delta v_s$. Check this with the gain of 11.1 determined in Chapter 5.

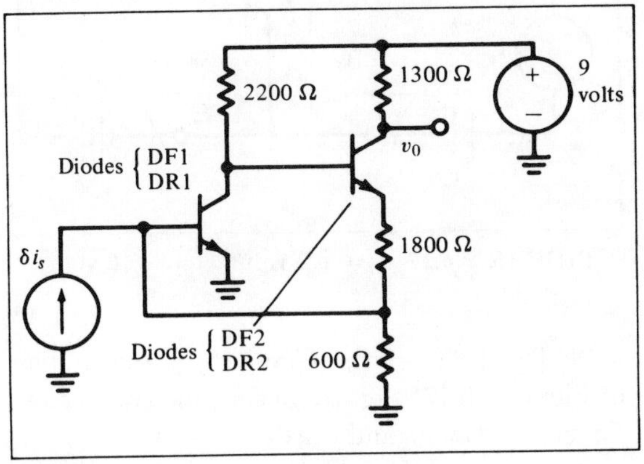

FIGURE P 6.13 $\alpha_{f_1} = \alpha_{f_2} = 0.98$; $\alpha_{r_1} = \alpha_{r_2} = 0.5$

6.15 Calculate the small-signal gain $\delta v_0/\delta i_s$ of the amplifier of Figure P 6.13 in two ways: (a) From the DCAP output of Figure P 6.14. (b) From the small-signal equivalent model, using the DCAP output to compute linearized component values.

6.16 The circuit of Figure P 6.15 will convert the $v_s - R_s$ nonideal source to a small-signal Thevenin equivalent with a reduced R_{eq}. Determine this small-signal Thevenin equivalent. A dc analysis yields a G_d of 0.0187℧ when $v_s = 1$ volt. Use $\alpha = 0.99$.

```
**********PRO 6.15
BRANCH NODE       ELEMENT       BRANCH CON-
 NAME    NOS.      VALUE        NAME    TROL
IS     0  2       0.
R2     2  0       6.000D 02
R3     1  3       2.200D 03
R4     5  2       1.800D 03
R5     1  4       1.300D 03
V1     1  0       9.000D 00
DF1    2  0       4.000D-01
IF1    3  2       9.800D-01     DF1      1
IR1    0  2       5.000D-01     DR1      1
DF2    3  5       4.000D-01
IF2    4  3       9.800D-01     DF2      1
IR2    5  3       5.000D-01     DR2      1
    DC ITERATION
 ITERATION      4 - 0 NODES
      0        8.1024D 00
      1        8.1183D 00
      2        8.1211D 00
      3        8.1211D 00
      4        8.1211D 00
      5        8.1211D 00
   DIODE INFORMATION AT DC SOLUTION
          DC INFO. (CHAPTER V)      INCREMENTAL INFO. (CHAPTER VI)
DIODE      VOLTAGE      CURRENT        CONDUCTANCE RESISTANCE
DF1     3.749D-01    3.255D-03       1.302D-01    7.682D 00
DR1    -1.578D 00   -1.000D-09       1.550D-35    6.450D 34
DF2     3.361D-01    6.899D-04       2.760D-02    3.624D 01
DR2    -6.168D 00   -1.000D-09       1.000D-60    1.000D 60
BRANCH NODE       ELEMENT       BRANCH CON-
 NAME    NOS.      VALUE        NAME    TROL
IS     0  2       1.000D-04
 ITERATION      4 - 0 NODES
      0        8.2449D 00
      1        8.2449D 00
   DIODE INFORMATION AT DC SOLUTION
          DC INFO. (CHAPTER V)      INCREMENTAL INFO. (CHAPTER VI)
DIODE      VOLTAGE      CURRENT        CONDUCTANCE RESISTANCE
DF1     3.755D-01    3.339D-03       1.336D-01    7.487D 00
DR1    -1.399D 00   -1.000D-09       1.981D-32    5.047D 31
DF2     3.323D-01    5.927D-04       2.371D-02    4.218D 01
DR2    -6.470D 00   -1.000D-09       1.000D-60    1.000D 60
```

FIGURE P 6.14 (a) $\delta i_s = 0$; (b) $\delta i_s = 10^{-4}$.

6.17 Determine the small-signal ratio $\delta v_0/\delta i_s$ for the amplifier of Figure P 6.16. When $\delta i_s = 0$, the emitter diode conductance is 0.098℧. Use $\alpha_f = 0.99$.

6.18. For the linearized transistor amplifier model of Figure 6.11(d), we can write

$$i_c = a_1 v_c + a_2 i_d + a_3 \delta i_s$$

$$v_{ce} = b_1 v_c + b_2 i_d + b_3 \delta i_s$$

Find the load-line equation relating i_c to v_{ce}.

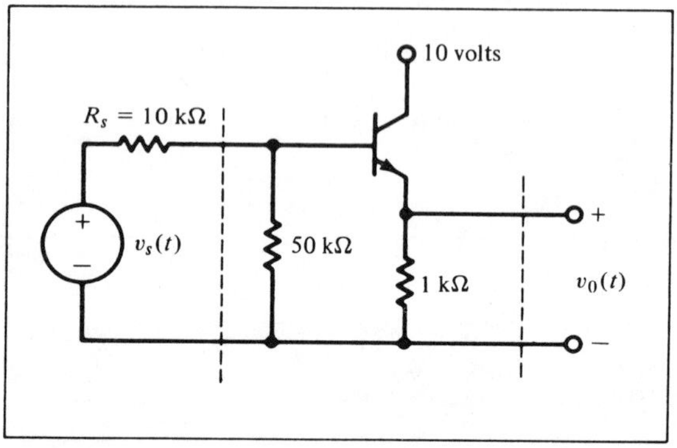

FIGURE P 6.15

Comment on the generality of this result (see Section 6.1).

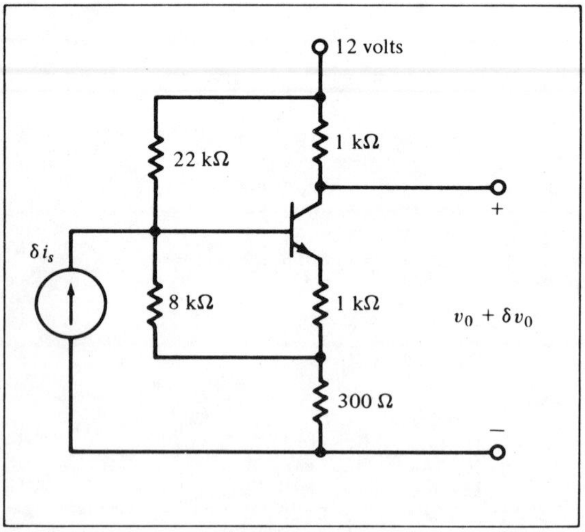

FIGURE P 6.16

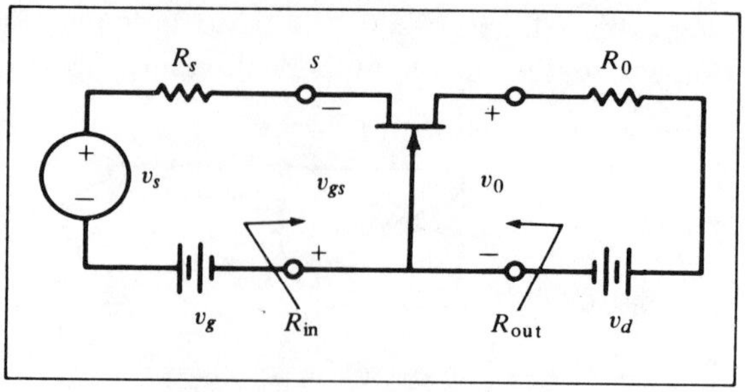

FIGURE P 6.17 $v_g = -2$ volts; $v_d = -10$ volts.

6.19 The FET of the common gate amplifier of Figure P 6.17 operates in the pinch off region. Given V_p, I_{dss}, r_d, and the dc value of $v_{gs}(=v_{gs0})$, find expressions for (a) the small signal equivalent circuit, (b) the small signal gain $\delta v_0/\delta v_s$, (c) the small signal input resistance R_{in}, (d) the

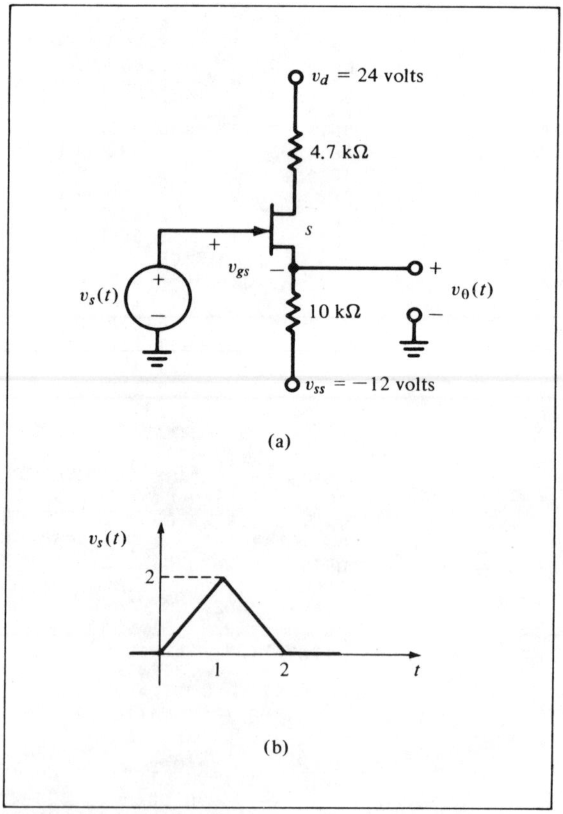

FIGURE P 6.18 (a) $V_p = -4$, $I_{dss} = 5$ ma, $r_d = 15$ k.

small signal output resistance R_{out}. (In the pinch-off region $\mu_d \triangleq \partial i_d/\partial v_{ds}$.)

6.20 The FET amplifier of Figure P 6.18 has a dc operating point given by $v_{gs} = -2.72$ volts. (a) Verify the operating point. (b) Construct the small signal equivalent network. (c) Evaluate the small signal voltage gain $\delta v_0(t)/v_s(t)$. (d) If $v_s(t)$ is given in Figure P 6.18(b), sketch $v_0(t)$, assuming small signal analysis is valid.

REFERENCES AND ADDITIONAL READING ON SUPERPOSITION

6.1 W. H. HAYT, JR., AND J. E. KEMMERLY, *Engineering Circuit Analysis.* New York: McGraw-Hill Book Company, Inc., 1963, Section 3.6.

6.2 R. E. SCOTT, *Elements of Linear Circuits.* Reading, Mass.: Addison-Wesley Publishing Company, Inc., 1960, Chapter 3.

7

Dynamic Circuit Elements

7.1 Introduction

Thus far we have been concerned with networks containing resistors, sources, and certain nonlinear resistive elements. In any interconnection of such elements, the network responses occur the instant the excitations are applied and so are constant if the excitations are constant. Mathematically, this means that all the network currents and voltages are related by algebraic linear and nonlinear equations.

In practice, few electrical circuits are adequately modeled with such resistive network elements. Inevitably, *dynamic* elements described by differential *i-v* relationships must be included in the network model, usually for one of the following reasons:

1. The *memory* function provided by dynamic elements is essential to the performance of circuits such as filters and oscillators. That is, dynamic elements are *intentionally* added to an otherwise resistive circuit and their values selected to obtain a prescribed design result.

2. When devices such as transistors are operated with rapidly varying signals, resistive device models may no longer be sufficient to describe certain internal physical phenomena that result in a storage (memory) effect within the device. In this case, *parasitic* dynamic elements must be added to the resistive device model, to properly account for the circuit performance.

In the remainder of this text, we will extend the analysis methods of previous chapters to include common dynamic network elements. This will involve the solution of differential, rather than algebraic, equations with a variety of applied signals. For the present, we content ourselves in this chapter with a study of these dynamic elements.

7.2 Capacitance

q-v Relationship

In 1745 P. van Musschenbroch at the University of Leyden filled a jar with water and inserted a metal rod through a stopper into the water. He observed that when he held the jar and touched the rod to an electrostatic generator he felt a "shock"; and after the jar was disconnected from the generator, if he touched the rod with his free hand, he felt another "shock." This demonstrated that somehow the "Leyden jar" was capable of storing electric energy! A year later John

Boers removed the water and simply coated the jar inside and outside with a thin metal film. Making an electrical connection to the inner film by a stopper-held rod and a chain, he observed the same Leyden Jar effect. This was the first capacitor.

The geometry of Figure 7.1, two concentric

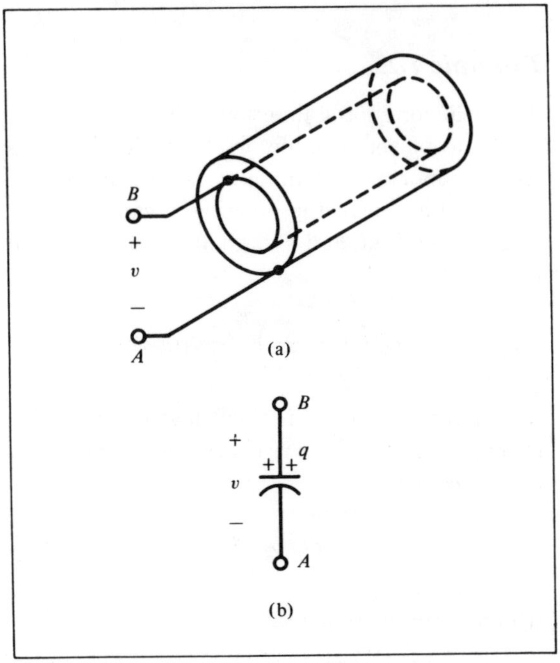

FIGURE 7.1 (a) Two metal cylinders separated by an insulator; (b) the capacitor symbol.

metal cylinders separated by an insulator, is that of the original Leyden jar. We know now that the particular geometry affects the magnitude, but not the nature, of the capacitance effect. This effect can be stated as follows: Let the conductors be designated A and B. An electric charge q can be transferred from A to B. When this is done, it is observed that *work must be done to transfer the charge,* and *the work required per unit charge transferred is a function of the total charge transferred.* Because the work required to transfer a charge between two points is just the voltage between the points, we may write

$$v = g(q)$$

or

$$q = f(v) \tag{7.1}$$

Any electrical device satisfying Equation (7.1) is a capacitor. When the two conductors are separated by a vacuum or many common solid or gaseous insulators, the relationship of Equation (7.1) is a linear one over the useful range of operation. Such a linear capacitor would have the q-v characteristic

$$q = Cv \tag{7.2}$$

for all v. C is the *capacitance* of the capacitor, and in the mks system, it is measured in farads.[1]

$$1 \text{ farad} = 1 \text{ coulomb/volt}$$

The farad happens to be a rather large capacitance; most elements that an electrical engineer encounters have capacitances in the range of 10^{-4} to 10^{-13} farad (f).

Figure 7.2 illustrates the q-v characteristic of a linear capacitor and a number of real capacitors. Figure 7.2(b) illustrates the characteristic up to some breakdown voltage. When the capacitor "breaks down," its electrical character changes radically. Usually it acts primarily as a nonlinear resistor in the breakdown region. For many insulating materials, real capacitors exhibit a nonlinear q-v relationship. The nonlinearity may be rather slight or very pronounced as shown in Figure 7.2(c).

For a nonlinear capacitor, there is no longer a single capacitance value that describes the q-v characteristic. However, at any point on the q-v characteristic, a small change in voltage produces a small change in charge.

$$\Delta q = \left.\frac{dq}{dv}\right|_{v=V_0} \Delta v$$

[1] The unit of capacitance honors Michael Faraday (1791–1867), British physicist and chemist who made fundamental discoveries concerning the functions of dielectrics (electrical insulators).

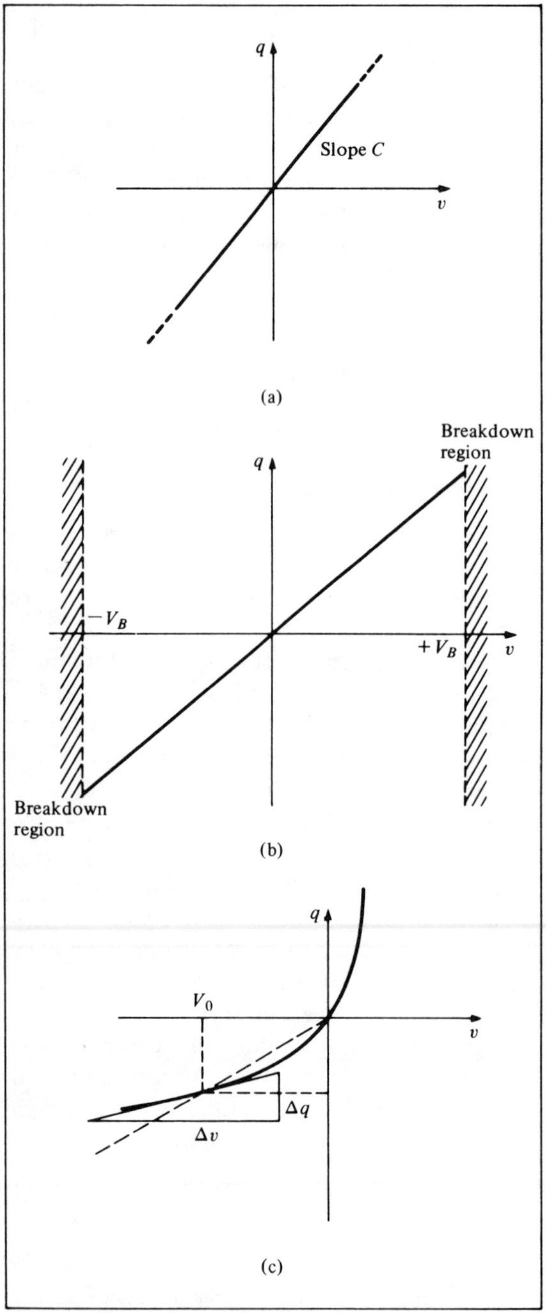

FIGURE 7.2 Typical charge-voltage characteristics. (a) Linear capacitor; (b) typical capacitor; (c) nonlinear capacitor.

Thus, the slope of the q-v characteristic relates Δq and Δv just as C relates q and v in an ideal linear capacitor. Therefore, we define

$$\left.\frac{dq}{dv}\right|_{v=V_0} = C(V_0) \tag{7.3}$$

that is, an *incremental capacitance* which is the slope of the q-v characteristic.

Example 7.1

The semiconductor junction, which is the cornerstone of semiconductor electronics, affords us an important example of nonlinear capacitance. At an ideal step junction, it can be shown that the voltage and charge are related by the equation

$$q = -Q_0\left(1 - \frac{v}{V_0}\right)^{1/2} \qquad (v \leqq 0) \tag{7.4}$$

Manufacturers of junction devices normally specify this nonlinear characteristic by the *barrier capacitance*

$$C_T \triangleq \frac{dq}{dv} \tag{7.5}$$

which is the incremental capacitance,

$$C_T = \frac{\frac{1}{2}(Q_0/V_0)}{(1 - (v/V_0))^{1/2}} = \frac{C_0}{(1 - (v/V_0))^{1/2}} \tag{7.6}$$

Current-Voltage Relationships for a Capacitor

Although capacitance is defined by a charge-voltage (q-v) relationship [Equation (7.1)], in a network analysis one is usually more interested in the flow of charge, that is, the current

$$i \triangleq \frac{dq}{dt} \tag{7.7}$$

If $q = f(v)$, Equation (7.7) becomes

$$i = \frac{d}{dt}f(v) = \frac{df}{dv}\frac{dv}{dt} \tag{7.8}$$

which can be written

$$i = C(v)\left(\frac{dv}{dt}\right) \tag{7.9}$$

For the linear case, C is a constant, and this reduces to

$$i = C\frac{dv}{dt} \tag{7.10}$$

From the above, the current depends on the *rate-of-change* of the voltage v, not on v itself or the past history of v. The standard reference directions and symbols for capacitance are shown in Figure 7.3.

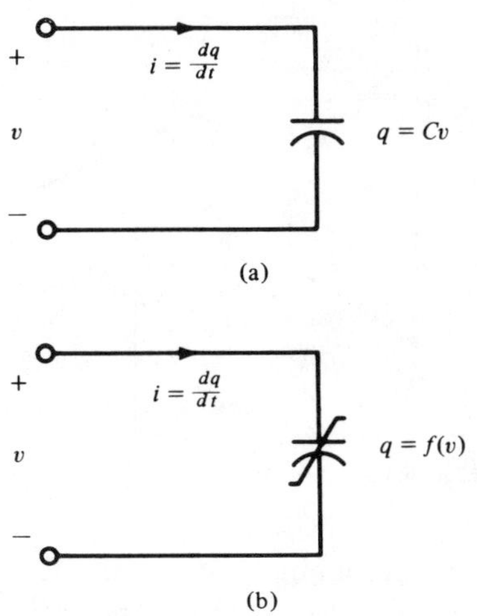

FIGURE 7.3 Network symbols and polarity conventions for: (a) linear capacitor; (b) nonlinear capacitor.

To express v as a function of i, Equation (7.10) is integrated to give

$$v(t) = \frac{1}{C}\int_{-\infty}^{t} i(\xi)\, d\xi \tag{7.11}$$

The integral in Equation (7.11) is just the charge on the capacitor at time t. If we are interested only in the condition of the capacitor after some initial time t_0, we can write

$$\begin{aligned} v(t) &= \frac{1}{C}\int_{-\infty}^{t_0} i(\xi)\, d\xi + \frac{1}{C}\int_{t_0}^{t} i(\xi)\, d\xi \\ &= v(t_0) + \frac{1}{C}\int_{t_0}^{t} i(\xi)\, d\xi \end{aligned} \tag{7.12}$$

The voltage across a capacitor depends only on the past history of $i(t)$; it does not depend on $i(t)$ or its derivative. In a very real sense, a capacitor "remembers" its current.

The two equations, (7.10) and (7.11), are for a linear capacitor. As we have already indicated, many real capacitors are linear or almost linear over wide ranges of charge and voltage. In addition, all nonlinear capacitors behave linearly for sufficiently small incremental currents and voltages.

Example 7.2

We are to find the capacitor current in the simple network of Figure 7.4, a linear capacitor connected to an ideal "triangular" voltage source.

The derivative of this voltage jumps from $+4 \times 10^5$ to -4×10^5 volts per second every 0.5 msec. Because the capacitance is 1 microfarad (μf), it follows from Equation (7.10) that the capacitor current jumps between

$$i_{\max} = 10^{-6} \times 4 \times 10^5 = 0.4 \text{ amp}$$

and

$$i_{\min} = -0.4 \text{ amp}$$

This current waveform is plotted directly below the voltage waveform in the same time scale.

Example 7.3

Let the voltage source of Figure 7.4 be replaced by a current source having the same waveform with a peak value of 1 amp. This network is given in Figure 7.5. The voltage response is to be found.

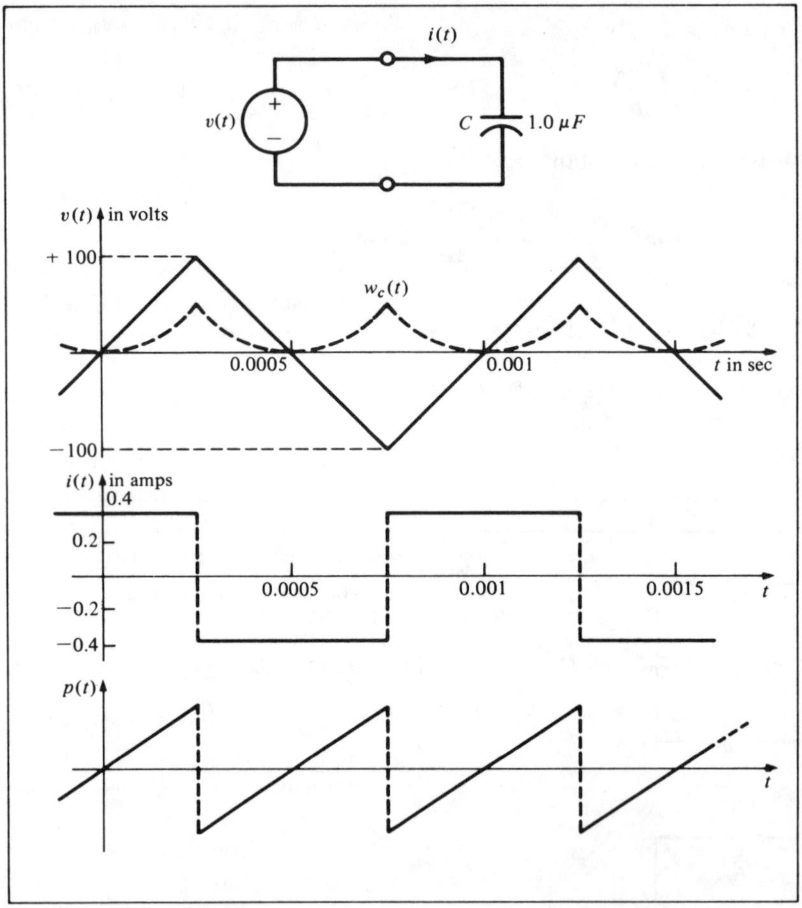

FIGURE 7.4 Response of linear capacitor to sawtooth voltage source.

The current waveform can be written

$$i(t) = 4000t \qquad 0 \leqq t \leqq \frac{1}{4000}$$

$$i(t) = 1 + 4000\left(\frac{1}{4000} - t\right)$$

$$\frac{1}{4000} \leqq t \leqq \frac{3}{4000}$$

$$i(t) = -1 + 4000\left(t - \frac{3}{4000}\right)$$

$$\frac{3}{4000} \leqq t \leqq \frac{5}{4000}$$

and so on. Assuming $v(0) = 0$, Equation (7.12) can be used to calculate $v(t)$. For $0 \leqq t \leqq 1/4000$,

$$v(t) = 10^6 \int_0^t 4000t\, dt = 2 \times 10^9 t^2$$

and for $1/4000 \leqq t \leqq 3/4000$

$$v(t) = 125 + 10^6 \int_{1/4000}^{t} (2 - 4000t)\, dt$$

$$= 125 + 10^6[2t - 2000t^2]_{1/4000}^{t}$$

$$= -250 + 2 \times 10^6 t - 2 \times 10^9 t^2$$

The voltage continues to rise up to 250 volts at $t = 1/2000$ sec where the current reverses. Then the voltage falls back to 125 volts at

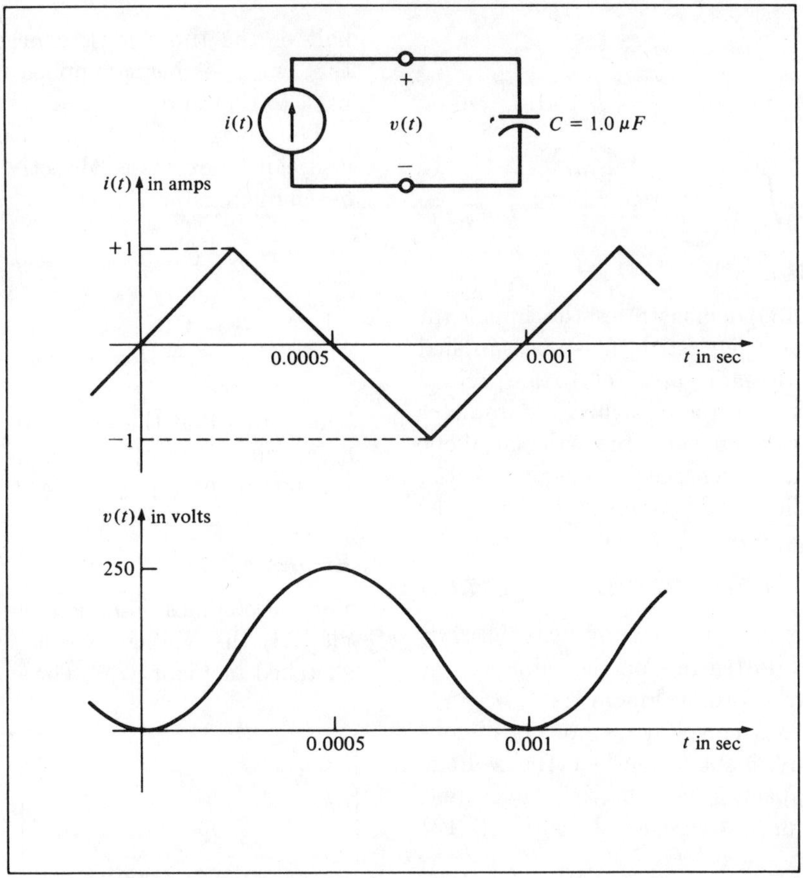

FIGURE 7.5 Response of linear capacitor sawtooth current source.

$t = 3/4000$ sec. The reader may find it instructive to calculate the final quarter cycle of the voltage waveform which carries the voltage back to 0 volts at $t = 1/1000$ sec. The complete waveform is sketched in Figure 7.5.

Power and Energy in Capacitors

For any element, the instantaneous power is given by

$$p(t) = v(t)i(t) \tag{7.13}$$

For the capacitor of Example 7.2, the power waveform is plotted below the voltage and current waveforms in Figure 7.4.

Of more significance for a capacitor is the energy it stores. We observe from Figure 7.4 that the power can be both positive- and negative-valued. This is in contrast to a linear resistor where the power always has the same sign and is always positive if the resistance R is positive. The energy $w_c(t)$ is the time integral of power; we shall see it is always positive for a capacitor. A capacitor can alternately store or release energy to the rest of the network over different periods of time.

A modest calculation verifies the above statements. The energy supplied to a capacitor between times t_0 and t is the integral of the power

$$w_c(t) \triangleq \int_{t_0}^{t} p(\xi)\, d\xi \tag{7.14}$$

Using Equations (7.10) and (7.13) this can be

written

$$
\begin{aligned}
w_c(t) &= \int_{t_0}^{t} v(\xi)i(\xi)\,d\xi = \int_{t_0}^{t} Cv(\xi)\frac{dv}{d\xi}\,d\xi \\
&= C\int_{v(t_0)}^{v(t)} v\,dv = \tfrac{1}{2}Cv^2\Big]_{v(t_0)}^{v(t)} \qquad (7.15) \\
&= \tfrac{1}{2}C[v^2(t) - v^2(t_0)]
\end{aligned}
$$

Equation (7.15) demonstrates the important result that the energy that must be supplied to charge a (linear) capacitor to a specified voltage depends on the capacitance parameter and the voltage magnitude. In particular, if one starts with a "discharged" capacitor [$v(t_0) = 0$], then the energy supplied to charge the capacitor to a potential $v(t)$ is

$$w_c(t) = \tfrac{1}{2}Cv^2(t) \qquad (7.16)$$

For Example 7.2, the energy waveform is sketched as a dotted line on the same axis as the voltage waveform in Figure 7.4.

Note that if the voltage source is disconnected at any instant t_0 when the voltage across the capacitor is nonzero, the current will instantly drop to zero, and Equation (7.12) becomes

$$v(t) = v(t_0) = \text{constant}$$

The voltage remains unchanged across the capacitor as long as the current is zero. Because the energy supplied to the capacitor does not change when the current is zero it is evident that a capacitor stores electric energy equal to

$$w_c = \tfrac{1}{2}Cv^2(t_0)$$

A charged capacitor stores electric energy just as a compressed spring stores mechanical energy. In subsequent chapters we will find this energy storage property a very useful one. If a charged capacitor is connected to a resistor, the stored electric energy can be converted into heat energy by the resistor. In the network of Figure 7.4, the stored electric energy is simply returned to the voltage source. The instantaneous power plot shows this clearly; half of the time, $p(t)$ is negative indicating that energy is being supplied by the capacitor back to the source.

The energy stored in a charged capacitor also can be expressed directly in terms of the q-v characteristic. The energy can be written as

$$
\begin{aligned}
w_c(t) &= \int_{t_0}^{t} vi\,d\xi = \int_{t_0}^{t} v\frac{dq}{d\xi}\,d\xi \\
&= \int_{q(t_0)}^{q(t)} v(q)\,dq
\end{aligned} \qquad (7.17)
$$

This shows that the energy required to charge a capacitor is directly expressable in terms of the area under a plot of $v(q)$ versus q.

Example 7.4

For the nonlinear junction capacitor of Example 7.1, the voltage versus charge curve is sketched in Figure 7.6. The energy that must be supplied to charge this capacitor from potential $-v_1$ to $-v_2$ is just the indicated shaded area.

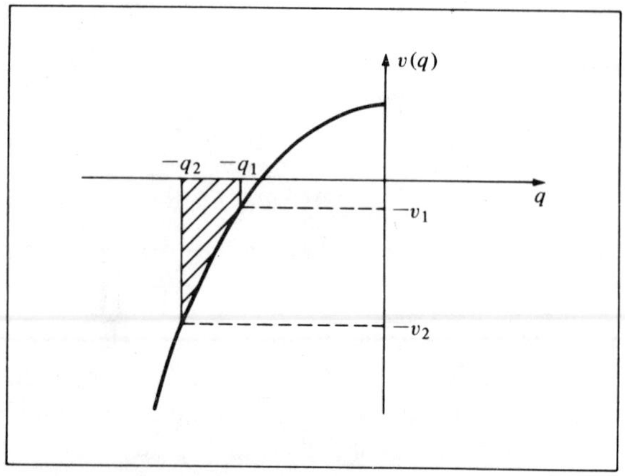

FIGURE 7.6 Illustration of stored energy determination.

7.3 Inductance

Whenever a current flows in a wire, a magnetic field is set up circling the wire as shown in

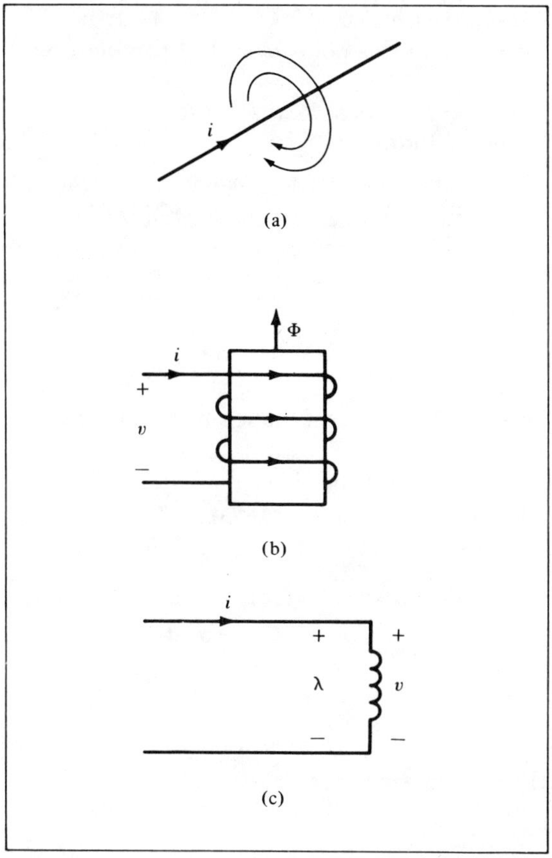

FIGURE 7.7 (a) Magnetic field around a straight wire; (b) an inductor; (c) the symbol for inductance.

Figure 7.7 (a). In general, the magnetic flux Φ is a function of the current generating it,

$$\Phi = f_1(i) \tag{7.18}$$

The function $f_1(i)$ depends on the magnetic properties of the medium in the vicinity of the current flow, and frequently it is just a straight line through the origin of the $\Phi - i$ plane. By winding a wire in a geometry such as the cylindrical coil of Figure 7.7 (b), it is possible to reinforce the flux inside the coil and to have the same flux link several turns of the coil. If a flux Φ links N turns, the *flux linkage* λ is

$$\lambda \triangleq N\Phi = f(i) \tag{7.19}$$

When the medium surrounding the coil or conductor is nonmagnetic, the flux linkage is proportional to i as shown in Figure 7.8 (a). In

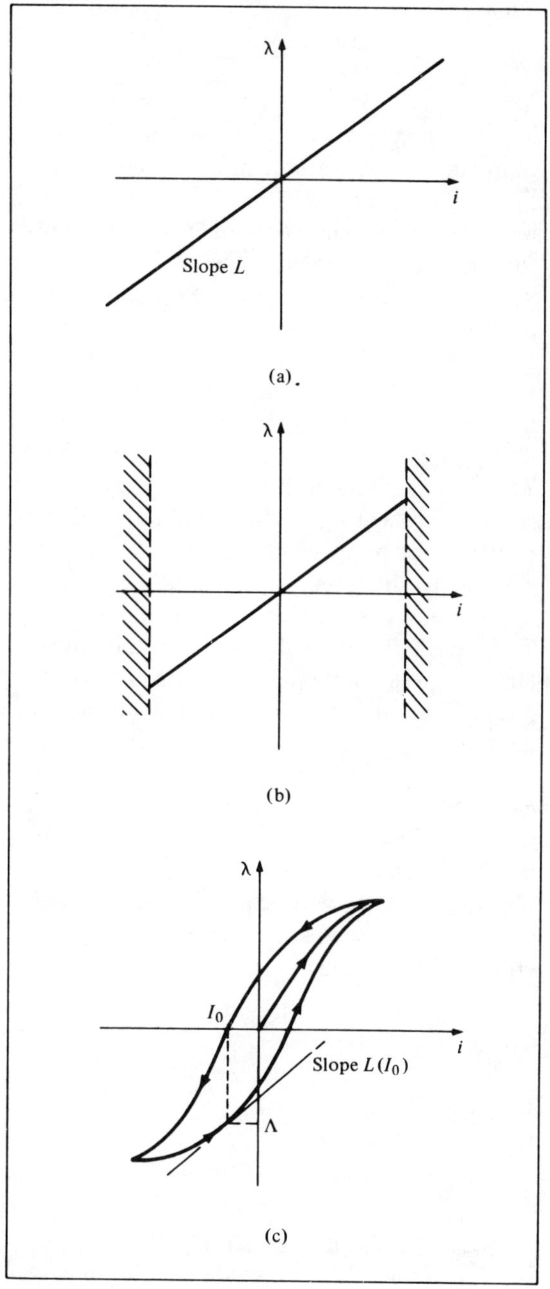

FIGURE 7.8 (a) Linear λ-i characteristic; (b) real inductor, nonmagnetic core; (c) real inductor, magnetic core.

this case we can write

$$\lambda = Li \tag{7.20}$$

where

$$L \triangleq \text{inductance in henrys}[2]$$

Equation (7.20) can be taken as defining a linear inductor, whereas Equation (7.19) is a more general definition which includes nonlinear characteristics encountered with certain core materials. The common network symbol for either case is given in Figure 7.7(c).

For a linear inductor, the flux linkage is proportional to the current through the inductor for all values of that current. For a real inductor, in the absence of magnetic materials, the flux linkages vary linearly with current over a finite range of currents as shown in Figure 7.8(b). The limits on the current range may be set by overheating of the conductors due to resistance or by mechanical deformation of the coil due to the forces which are set up because of the currents flowing in a magnetic field. To increase the magnetic field per unit current, ferromagnetic materials such as iron, nickel, or ferrite ceramics may be placed near the conductors. The use of such magnetic materials can increase the flux linkages for a specified current by factors of hundreds or thousands, but in addition, the λ-i plot usually becomes nonlinear and double-valued as sketched in Figure 7.8(c). As with capacitors, at any point on the λ-i characteristic an incremental parameter can be defined as

$$L(I_0) \triangleq \left.\frac{\partial \lambda}{\partial i}\right|_{i=I_0} \tag{7.21}$$

In general, this inductance is also a function of the segment of the double-valued curve on which we are operating; this, in turn, depends on the past history of the inductor current. We will explore this point more in Example 7.5.

Current-Voltage Relationships for an Inductor

The relationships of the magnetic flux linkage to terminal voltage was discovered by Michael Faraday. It is given by Faraday's law

$$v(t) = \frac{d\lambda(t)}{dt} \tag{7.22}$$

A voltage is "induced" in the inductor which is equal to the rate of change of the flux linkages in the inductor. Substituting Equation (7.19) into (7.22) and allowing a nonlinear inductance characteristic, the terminal voltage can be written

$$\begin{aligned} v(t) &= \frac{df(i)}{dt} = \frac{df}{di}\frac{di}{dt} \\ &= L(i)\frac{di}{dt} \end{aligned} \tag{7.23}$$

For the linear case this is simply

$$v(t) = L\frac{di(t)}{dt} \tag{7.24}$$

Alternately, if we wish to write the current as a function of the voltage, Equation (7.24) is integrated to give

$$i(t) = \frac{1}{L}\int_{-\infty}^{t} v(\xi)\,d\xi = \frac{\lambda(t)}{L} \tag{7.25}$$

This shows us that the time integral of the voltage at the terminals of a linear inductor equals the flux linkages for that inductor. For times after an arbitrary initial time t_0, Equation (7.25) can be written

$$\begin{aligned} i(t) &= \frac{1}{L}\int_{-\infty}^{t_0} v(\xi)\,d\xi + \frac{1}{L}\int_{t_0}^{t} v(\xi)\,d\xi \\ &= \frac{1}{L}\left[\lambda(t_0) + \int_{t_0}^{t} v(\xi)\,d\xi\right] \end{aligned} \tag{7.26}$$

[2] Joseph Henry (1797–1878), a U. S. engineer and physicist, invented the first telegraph, the first electromagnetic relay, and the first electromagnetic motor. He and Faraday independently discovered in 1832 the principle of electromagnetic self-induction.

Power and Energy in Inductors

Comparing inductance and capacitance, Equations (7.19), (7.24), and (7.26) with (7.1), (7.10), and (7.12), we see that there is a close analogy. Charge can be stored in a capacitor, and flux linkages can be stored in an inductor. The analogy also extends to energy and power. The power flow to the inductor of Figure 7.7(c) is

$$p(t) = v(t)i(t) \tag{7.27}$$

Therefore, the energy supplied between t_0 and t can be written

$$w_L(t) = \int_{t_0}^{t} v(\xi)i(\xi)\, d\xi \tag{7.28}$$

For a linear inductor, Equation (7.24) can be used to eliminate $v(t)$ giving

$$\begin{aligned} w_L(t) &= \int_{t_0}^{t} \left(L\frac{di(\xi)}{d\xi}\right) i(\xi)\, d\xi \\ &= L\int_{i(t_0)}^{i(t)} i\, di = L\left[\frac{i^2}{2}\right]_{i(t_0)}^{i(t)} \\ &= \tfrac{1}{2}L[i^2(t) - i^2(t_0)] \end{aligned} \tag{7.29}$$

Equation (7.29) demonstrates that the energy that must be supplied to establish a current in an inductor depends only on the inductance parameter and the current magnitude. At any instant the stored energy in a linear inductor is just

$$w_L = \tfrac{1}{2}Li^2 \tag{7.30}$$

which is analogous to the stored energy in a capacitor.

When a nonlinear inductor is encountered, Equation (7.24) is not applicable, but Equation (7.19) is. Recalling from Equation (7.22) that

$$d\lambda(t) = v(t)\, dt$$

the energy supplied to an inductor can be written

$$w_L(t) = \int_{0}^{t} v(\xi)i(\xi)\, d\xi = \int_{\lambda(t_0)}^{\lambda(t)} i(\lambda)\, d\lambda \tag{7.31}$$

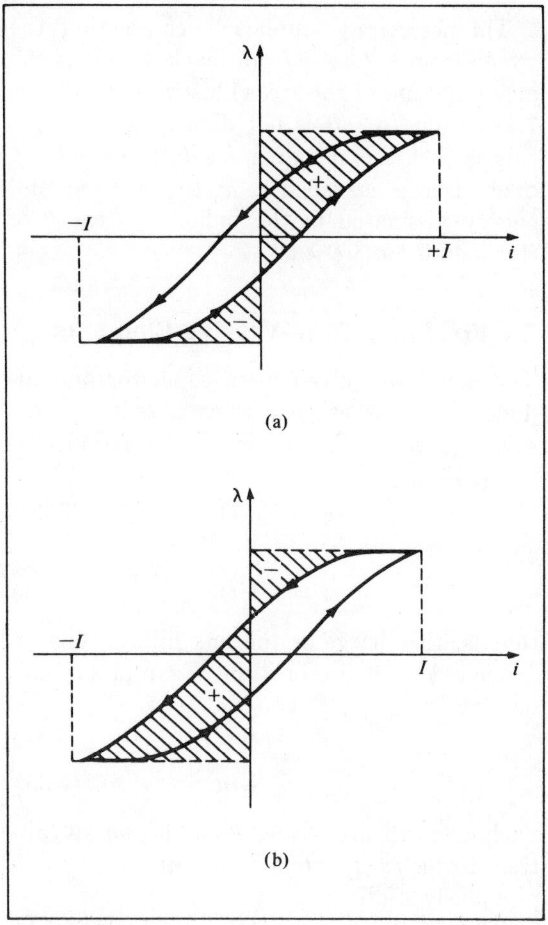

FIGURE 7.9 Energy supplied to iron core inductor by current varying from (a) $-I$ to $+I$; (b) $+I$ to $-I$.

Example 7.5

The λ-i characteristic of a typical iron core inductor excited by a current is sketched in Figure 7.9(a). We are to study the energy supplied to the inductor for a given $i(t)$.

As the current increases from $-I$ to $+I$, the energy supplied to the inductor is indicated by the cross-hatched area. [It should be noted that although in Figure 7.9, λ is plotted versus i, the energy is calculated by integrating $i(\lambda)$ with respect to λ.] If the current decreases from $+I$ to $-I$, the energy supplied to the inductor is indicated in Figure 7.9(b).

The net energy supplied in completing the sweep from $-I$ to $+I$ and back to $-I$ is the algebraic sum of the areas illustrated in Figure 7.9(a) and (b). It is not difficult to see that this is just the net area *inside* the λ-i loop. Such a loop is called a *hysteresis loop,* and energy dissipated in the inductor because of this loop is *hysteresis* loss.

7.4 Fixed and Time-Varying Elements

The definitions given for a capacitor and inductor are not as general as possible. Occasionally, we encounter elemental relationships of the form

$$q = f(v, t) \tag{7.32}$$

or

$$\lambda = f(i, t) \tag{7.33}$$

that is, the charge or the flux linkage has an explicit dependence on time. Examples would be

$$q_C = vq_0 \sin t \tag{7.34}$$

$$\lambda_L = i^2 + it \tag{7.35}$$

Such elements are termed *time-varying,* in contrast to the *fixed,* or *time-invariant,* elements we previously studied.

The voltage-current relationship for a capacitor is found by regarding voltage and time as independent variables in Equation (7.32); a total differentiation using the chain rule of partial differentiation then yields

$$dq = \frac{\partial q}{\partial v} dv + \frac{\partial q}{\partial t} dt$$

$$i = \frac{dq}{dt} = \frac{\partial q}{\partial v}\frac{dv}{dt} + \frac{\partial q}{\partial t}$$

for a capacitor and

$$v = \frac{\partial \lambda}{\partial i}\frac{\partial i}{dt} + \frac{\partial \lambda}{\partial t}$$

for an inductor. For the relationships of Equations (7.34) and (7.35), we would have

$$i(t) = q_0 \sin t \frac{dv}{dt} + vq_0 \cos t \tag{7.36}$$

$$v(t) = (2i(t) + t)\frac{di(t)}{dt} + i(t) \tag{7.37}$$

We observe that the partial derivatives $\partial q/\partial t$ or $\partial i/\partial t$ introduced terms that do not involve dv/dt or di/dt. These terms are therefore resistive in nature.

With few exceptions, we will in this text assume that elements are fixed. Many of the results we will obtain in later chapters will not apply to time-varying networks.

7.5 Mutual Inductance

Comparing the q-v relationships for capacitors and the λ-i relationships for inductors, we find remarkable correspondence in their analytic description. There is a very real *physical* difference, however, in that capacitance relates to a static field effects, whereas magnetic effects are due to moving charges (currents). Electrostatic fields always terminate on charges, while magnetic fields always form closed loops—the lines of magnetic flux—which terminate on themselves. This makes possible a "coupling" between magnetic circuits that has no analogy in electrostatic circuits.

Consider two coils located physically near to one another so that some of the flux generated by a current flow in one coil links the second coil. Figure 7.10 illustrates such a pair of coils wound on a common magnetic "core." Such a core is not essential to our discussion, but it does show a practical way of achieving large amounts of coupling. *We will assume that the magnetic material,* (if present) *does not introduce nonlinearity.* For many real coupled coils this is a reasonable assumption.

Connecting a current source to the first coil sets up a flux

$$\Phi_1 \sim i_1$$

TABLE 7.1

Summary of R, C, and L Elements

Element Value	Symbol and Reference Direction	Terminal Characteristics: Fixed, Linear	Fixed	General
Resistor	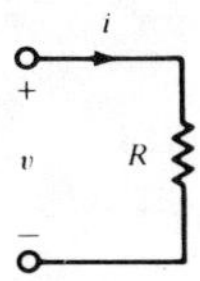	$v = Ri$	$v = f(i)$	$v = f(i, t)$
		$i = Gv$	$i = g(v)$	$i = g(v, t)$
Capacitor		$q = Cv$	$q = f(v)$	$q = f(v, t)$
	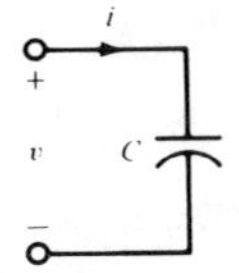	$v = \frac{1}{C} q$	$v = g(q)$	$v = g(q, t)$
		$i = \frac{dq}{dt}$	$i = \frac{\partial q}{\partial v}\frac{dv}{dt}$	$i = \frac{\partial q}{\partial v}\frac{dv}{dt} + \frac{\partial q}{\partial t}$
Inductor		$\lambda = Li$	$\lambda = f(i)$	$\lambda = f(i, t)$
	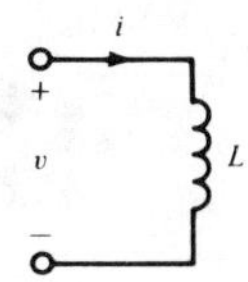	$i = \frac{1}{L}\lambda$	$i = g(\lambda)$	$i = g(\lambda, t)$
		$v = \frac{d\lambda}{dt}$	$v = \frac{\partial \lambda}{\partial i}\frac{di}{dt}$	$v = \frac{\partial \lambda}{\partial i}\frac{di}{dt} + \frac{d\lambda}{dt}$

If this flux links N_1 turns of the first coil, by Faraday's law [Equations (7.19) and (7.22)] we know the terminal voltage will be

$$v_1 \sim N_1 \frac{d\phi_1}{dt} \sim N_1^2 \frac{di_1}{dt} \tag{7.38}$$

The proportionality constant in Equation (7.38) is called the *self*-inductance of the first coil L_1

$$v_1 = L_1 \frac{di_1}{dt} \tag{7.39}$$

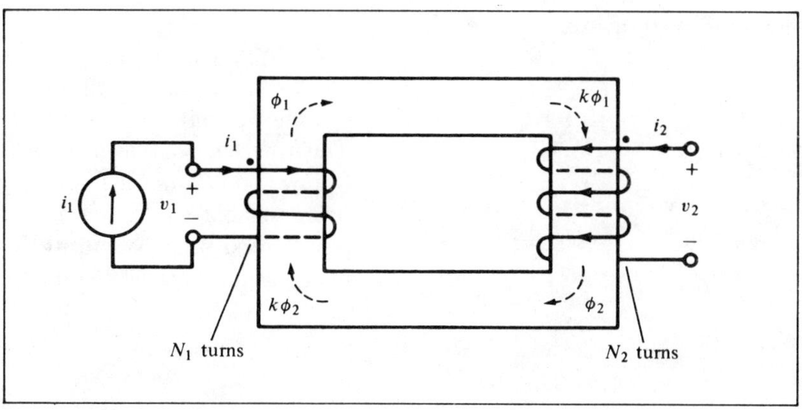

FIGURE 7.10 Two coils wound on a common magnetic core.

A fraction of the flux ϕ_1, $k\phi_1$, links the N_2 turns of the second coil, and changes of this flux produce a second voltage

$$v_2 \sim N_2 \frac{d(k\phi_1)}{dt} \sim kN_1N_2 \frac{di_1}{dt} \triangleq \pm M \frac{di_1}{dt} \quad (7.40)$$

The proportionality constant in this case is M, the *mutual* inductance between the two coils. Depending on how the second coil is wound, either sign is possible in Equation (7.40).

The sign preceding M is easily established once the positive current directions have been selected. We define the positive flux direction as the one produced by positive current. In the figure, i_1 and i_2 produce positive fluxes ϕ_1 and ϕ_2. The self-inductance voltage drop is defined to be positive. *If positive currents produce fluxes in the same direction*, the *mutual inductance has a positive sign*. Therefore, in Figure 7.10, for $i_2 = 0$,

$$v_1 = L_1 \frac{di_1}{dt}\bigg|_{i_2=0} \quad (7.41)$$

$$v_2 = M \frac{di_1}{dt}\bigg|_{i_2=0} \quad (7.42)$$

Whenever v_1 is positive, so is v_2; by convention *we indicate this by placing a dot on the positive terminals.*

If the current source is now shifted to the second coil, leaving $i_1 = 0$, a flux ϕ_2 is set up in the core

$$\Phi_2 \sim i_2$$

Then, exactly as before,

$$v_2 \sim N_2 \frac{d\phi_2}{dt} \sim \frac{di_2}{dt} \triangleq L_2 \frac{di_2}{dt}\bigg|_{i_1=0} \quad (7.43)$$

and

$$v_1 \sim N_1 \frac{d(k\phi_2)}{dt} \sim \frac{di_2}{dt} \triangleq \pm M \frac{di_2}{dt}\bigg|_{i_1=0} \quad (7.44)$$

For the present, we will take as an assumption that the fraction of ϕ_2 which links coil 1 is the same as the fraction of ϕ_1 which links coil 2 so that the mutual inductances are the same.[3]

Now, for linear magnetic materials the fluxes due to i_1 and i_2 and the voltages produced by their rates of change simply add to give the complete equations for a pair of coupled coils

$$v_1 = L_1 \frac{di_1}{dt} + M \frac{di_2}{dt} \quad (7.45)$$

$$v_2 = M \frac{di_1}{dt} + L_2 \frac{di_2}{dt} \quad (7.46)$$

The symbols for such a pair of coupled coils are shown in Figure 7.11(a).

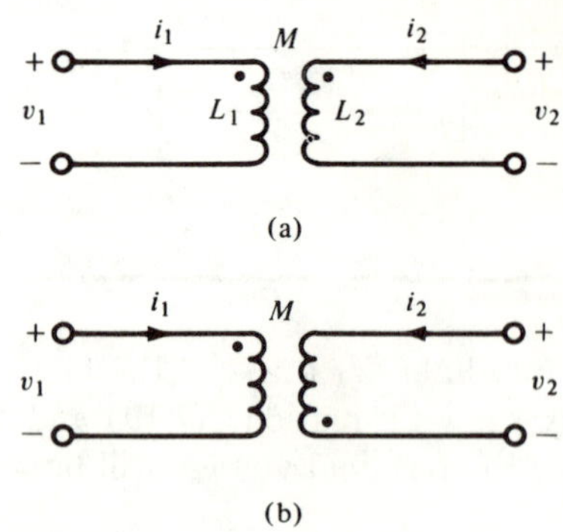

FIGURE 7.11 Pairs of coupled coils.

An appreciation for the significance of the dot notation follows from writing the equation for the mutual inductance in Figure 7.11(b). The difference between these two figures is that i_2 flows *into the dot* in Figure 7.11(a) but away from the dot in Figure 7.11(b). This can be represented by substituting $-v_2$ for v_2 and $-i_2$ for i_2 [Equations (7.45)–(7.46)]. The result is

$$v_1 = L_1 \frac{di_1}{dt} - M \frac{di_2}{dt} \quad (7.47)$$

[3] This is considered in greater detail in electromagnetic fields courses.

$$v_2 = -M\frac{di_1}{dt} + L_2\frac{di_2}{dt} \qquad (7.48)$$

In summary, if the currents i_1 and i_2 both flow *into* the dots or both flow away from the dots, the positive sign preceding M is used; if only one flows away from a dot, the negative sign of Equations (7.47)–(7.48) is appropriate. In any case, we will *always* regard M as a positive number. In engineering practice coupled coils are frequently called *transformers*.[4]

Example 7.6

By interconnecting a pair of coupled coils, we can achieve a variety of interesting networks. As an example, we are to analyze the network of Figure 7.12 to find the equivalent inductance L_{eq}.

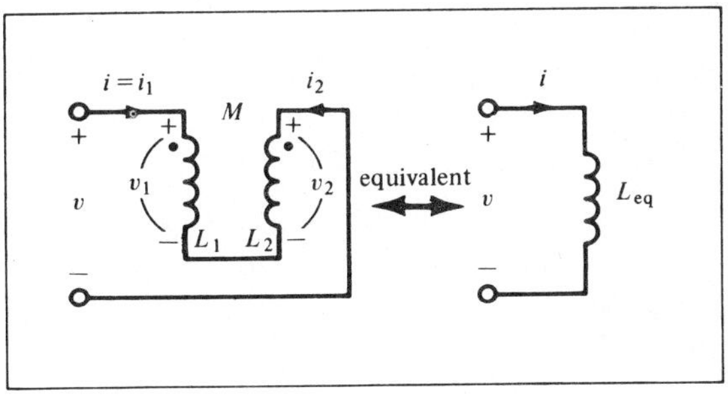

FIGURE 7.12 Series connections of two coupled coils.

The polarity dots and positive current directions in Figure 7.12 are the same as those in Figure 7.11(a), so Equations (7.45) to (7.46) apply. In addition,

$$i_1 = -i_2 = i \qquad (7.49)$$

$$v = -v_2 + v_1 \qquad (7.50)$$

Substituting from Equations (7.49) into Equations (7.45) and (7.46) gives

$$v_1 = (L_1 - M)\frac{di}{dt}$$

$$v_2 = (M - L_2)\frac{di}{dt}$$

which when subtracted yield

$$v = (L_1 + L_2 - 2M)\frac{di}{dt}$$

Thus, the coupled coils connected in series are equivalent to a single inductance

$$L_{eq} \triangleq L_1 + L_2 - 2M$$

By physically moving one coil relative to another, it is possible to vary M in magnitude. This is a common way of constructing a variable inductance for laboratory use.

[4] Transformers and more complex coupled coils are treated further in Chapter 16, Sections 16.1 and 16.2.

PROBLEMS

7.1 Inductors and capacitors in series and in parallel are shown in Figure P 7.1. Find a single-element equivalent in each case, assuming elements are linear.

7.2 (a) The two capacitors in Figure P 7.1(b) are now considered nonlinear. Their incremental capacitances are given by

$$C_1(v) = (v-1)^2 \qquad -0.8 \leqq v \leqq 0.8$$

$$C_2(v) = (v+1)^2 \qquad -0.8 \leqq v \leqq 0.8$$

Plot the equivalent single-element incremental capacitance and charge versus voltage for $-0.8 \leqq v \leqq 0.8$ volts. Assume that $q = 0$ if $v = 0$ for each capacitor. (b) If $C_1(v)$ is given as above, but $C_2(v) = C_2 =$ constant, what value of C_2 would be necessary to make the equivalent capacitance vary no more than 10 percent for the voltage range of 0 to .75 volt?

7.3 (a)–(b) In Figure P 7.2, find the indicated voltages, the charge on each capacitor, the total

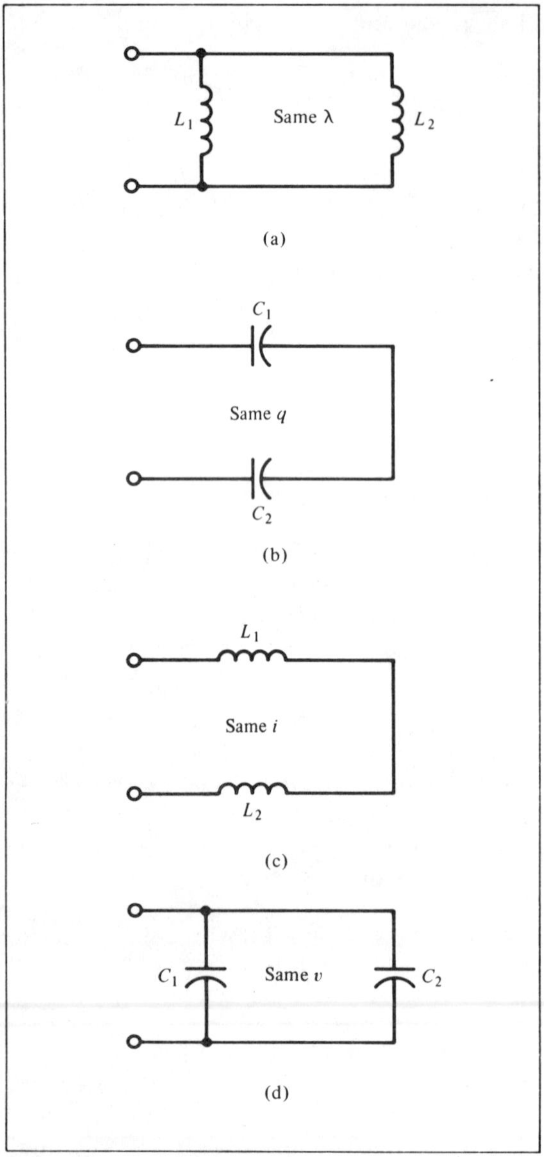

FIGURE P 7.1

capacitance across the voltage source, and the total stored energy.

7.4 The inductors of Figure P 7.3 have negligible resistance and are magnetically isolated from one another. Find each indicated current, the flux linkages in each inductor, the total

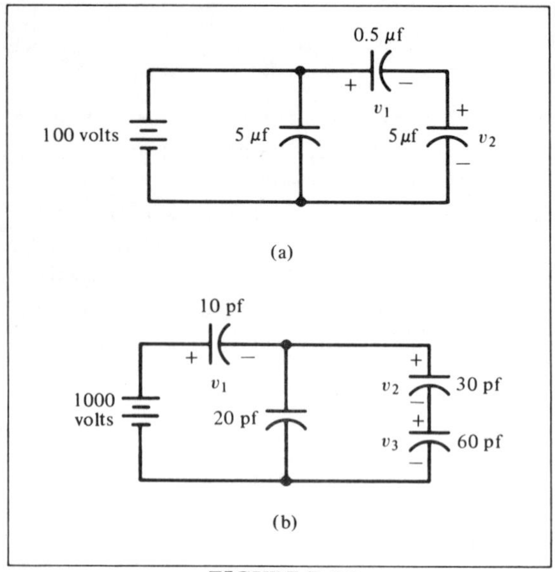

FIGURE P 7.2

inductance in series with the current source, and the total stored energy.

7.5 (a)–(b) The waveforms of Figure P 7.4 are inductor currents. Find the corresponding inductor voltages for $0 \leqq t \leqq 6$, if $L = 2\ h$.

7.6 (a)–(b) The waveforms of Figure P 7.5 are capacitor currents. Find the corresponding capacitor voltages. Assume that $C = 1\ \mu f$ and $v(0) = 0$.

7.7 The waveforms of Figure P 7.6 are associated with a constant capacitor of value $C = af$. Determine the capacitor charge (q), current (i), and voltage (v) in each case.

7.8 In the network of Figure P 7.7, the induc-

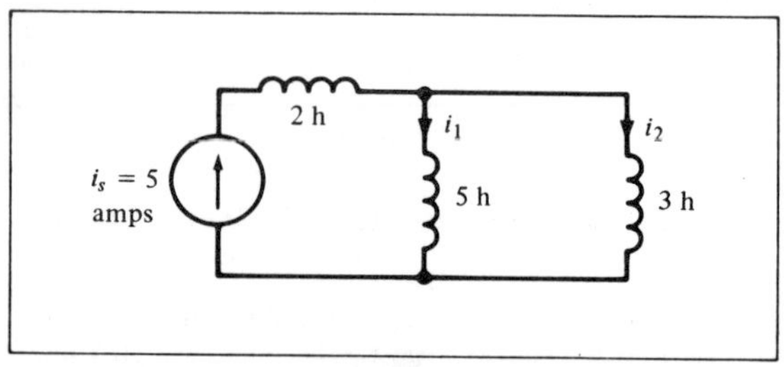

FIGURE P 7.3

tor voltage v is known to be e^{-t} for $t \geq 0$. At a certain time, the voltage v is 0.4 volt. At this time: (a) At what rate is i_L changing? (b) What is the flux linkage in the inductor? (c) What is the energy stored in the magnetic field of the inductor? (d) What is the rate of energy being taken from the magnetic field of the inductor? (e) At what rate is energy being dissipated in the resistor as heat?

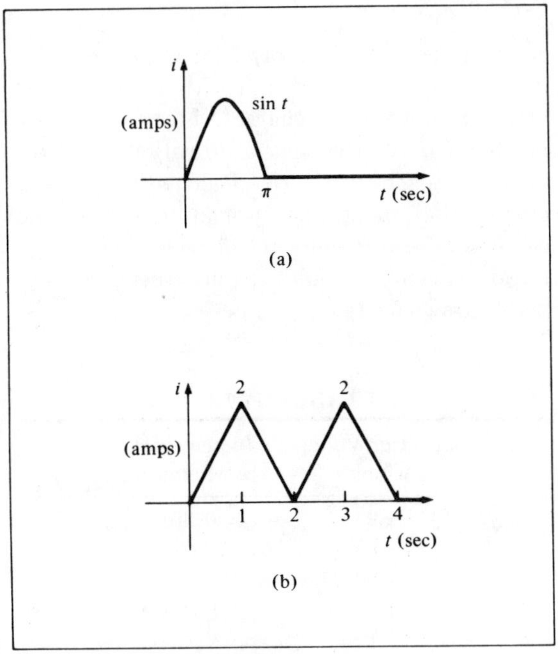

FIGURE P 7.4

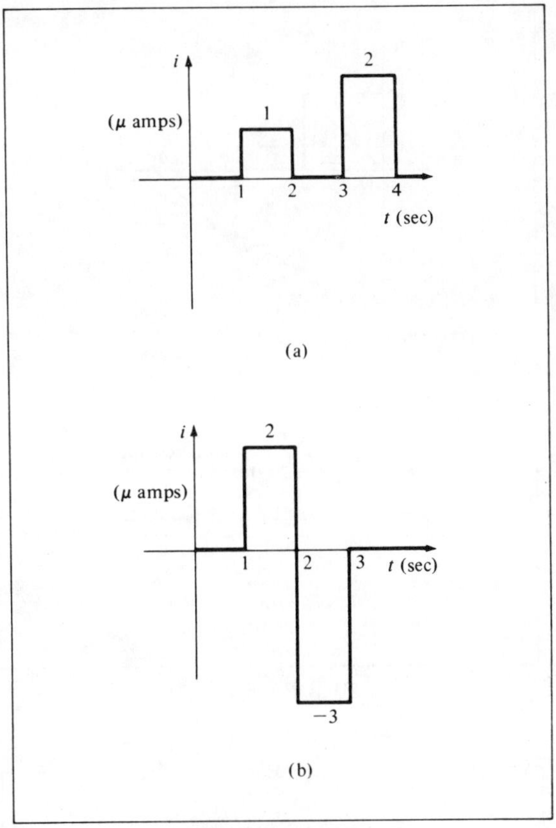

FIGURE P 7.5

7.9 The voltage across the capacitor in Figure P 7.8 is te^{-t}. Find $i(t)$ and $v_L(t)$. Determine $v_L(t)$ two different ways.

7.10 In Problem 7.9, determine the time when the energy stored in the capacitor is maximum.

7.11 For the q–v relationship of Figure P 7.9: (a) Find C. (b) Evaluate the energy stored in the capacitor when the voltage increases from $v = 0.2$ volt to 0.4 volt.

7.12 The power entering a linear capacitor is

$$p = \frac{d}{dt}\left(\tfrac{1}{2}Cv^2\right)$$

$$= Cv\frac{dv}{dt} \qquad (1)$$

If we define $C \triangleq \partial q/\partial v$, does (1) apply to a nonlinear capacitor? Justify your answer.

7.13 Using an instrument called a capacitance bridge, we can directly measure the incremental capacitance of capacitors in the laboratory. (Chapter 10 will provide the tools for analyzing the operation of such a bridge.) The results of a laboratory measurement of the incremental capacitance at a series of fixed voltages are

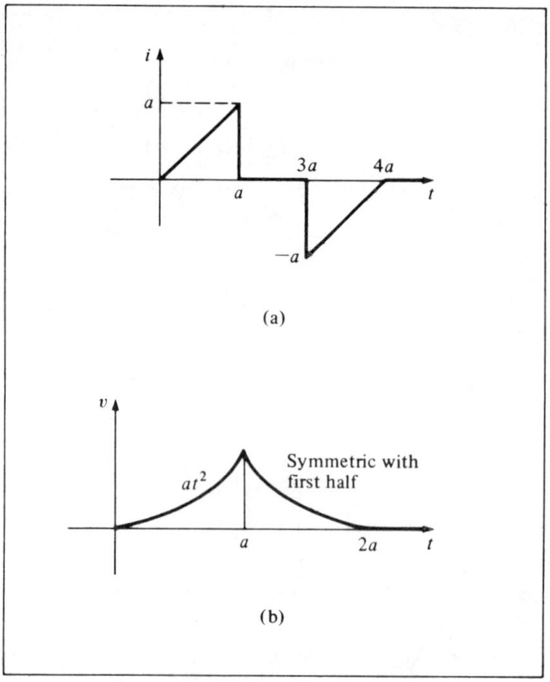

FIGURE P 7.6

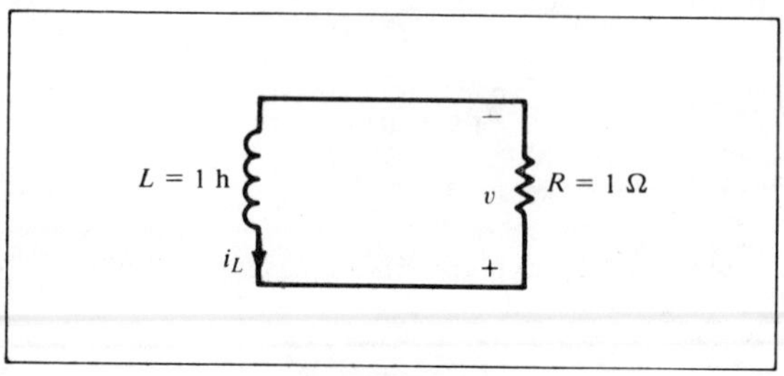

FIGURE P 7.7

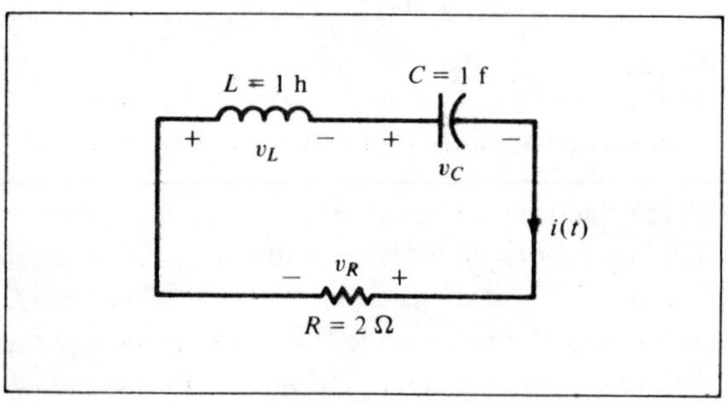

FIGURE P 7.8

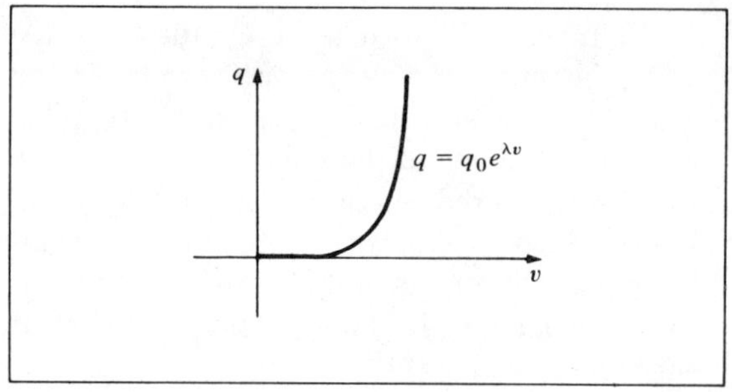

FIGURE P 7.9 $q_0 = 10^{-18}$ coulombs; $\lambda = 40$ volts^{-1}.

given in Table P 7.1. (a) Plot $C(v)$ from the data above and fit the data with a curve. Estimate and plot a q–v characteristic for this capacitor. [*Hint*: Recall the definition of incremental capacitance.] (b) Calculate the total energy that must be supplied to charge the capacitor from 0 volts to 25 volts. What size linear capacitor would require the same expenditure of energy?

TABLE P 7.1

Capacitor Voltage in Volts	Incremental C in Microfarads
+29	0.366
+25	0.414
+20	0.499
+15	0.611
+10	0.759
+ 5	0.912
+ 3	0.938
+ 2	0.942
+ 1	0.932
0	0.922
− 1	0.920
− 2	0.918
− 3	0.935
− 5	0.924
−10	0.780
−15	0.630
−20	0.503
−25	0.420
−29	0.351

7.14 The two inductors in Figure P 7.1(c) have incremental inductances given by

$$L_1(i) = L_0 - \frac{L_0 i^2}{2} \qquad (-1 \leqq i \leqq +1)$$

$$L_2(i) = \frac{2L_0}{1 + (i)^2} \qquad (-1 \leqq i \leqq +1)$$

with $\lambda(0) = 0$ for both inductors. (a) Find the incremental inductance of a single element equivalent inductor. (b) How much energy is stored in this equivalent inductor when $i = +1$ amp? -1 amp?

7.15 The double-valued $\lambda - i$ characteristic of a nonlinear inductor is shown in Figure

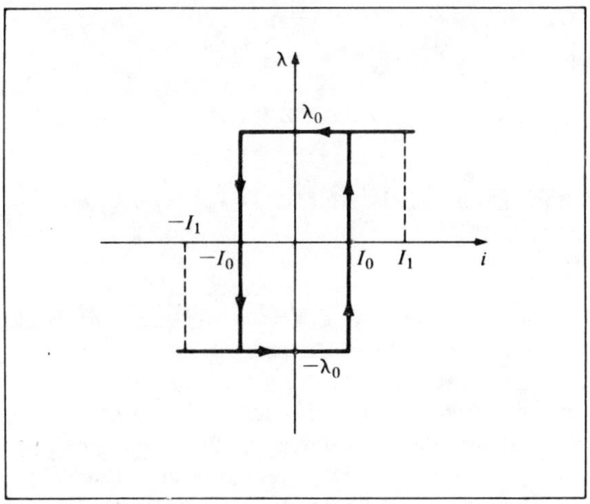

FIGURE P 7.10

P 7.10. The current begins at $i = -I_1$, reaches $+I_1$, and then returns to $-I_1$. Find the energy lost to hysteresis.

7.16 Certain ceramic dielectric capacitors exhibit a q–v characteristic which is multivalued as the applied voltage is cycled. A simplified approximation of such a characteristic is plotted in Figure P 7.11. (a) Calculate and

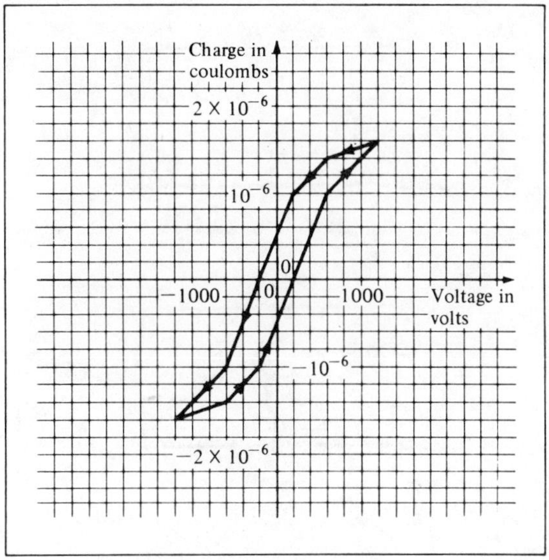

FIGURE P 7.11

plot the incremental capacitance as the voltage is varied from -1200 volts to $+1200$ volts and back. (b) Calculate the net energy that must be supplied by the source to cycle the voltage as indicated.

7.17 (a)–(c) Represent the coupled inductors of Figure P 7.12 by a single inductor.

7.18 (a) For the network of Figure P 7.13, write KVL around the two loops and obtain a single equation (a differential equation) relating i_2 and v_s. (b) If

$$v_s(t) = t \qquad (t \geqq 0)$$
$$= 0 \qquad (t < 0)$$

obtain a single differential equation for i_1. (The objective of this problem is to obtain differential equations; their solution will be treated in Chapter 8.)

7.19 Figure P 7.14 shows two cylindrical coils mounted coaxially so that one is free to slide inside the other. The mutual inductance between the coils is a maximum when the 1-2 coil

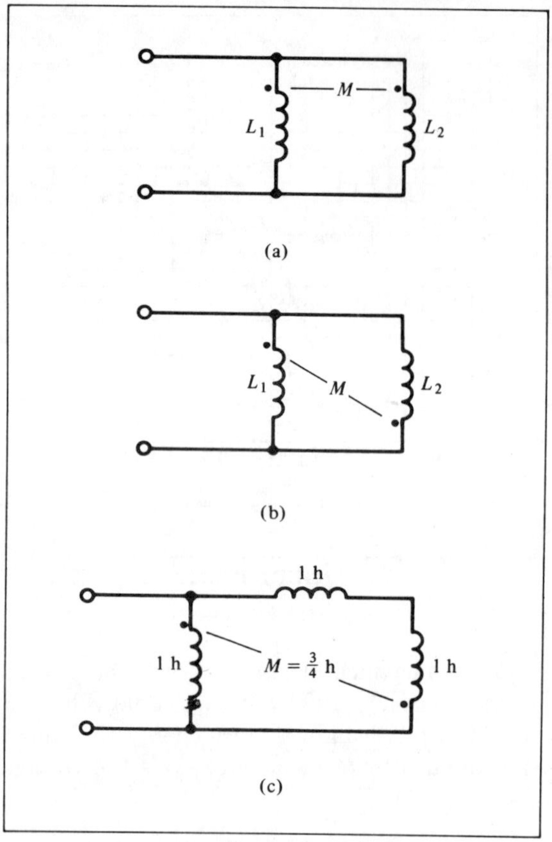

FIGURE P 7.12

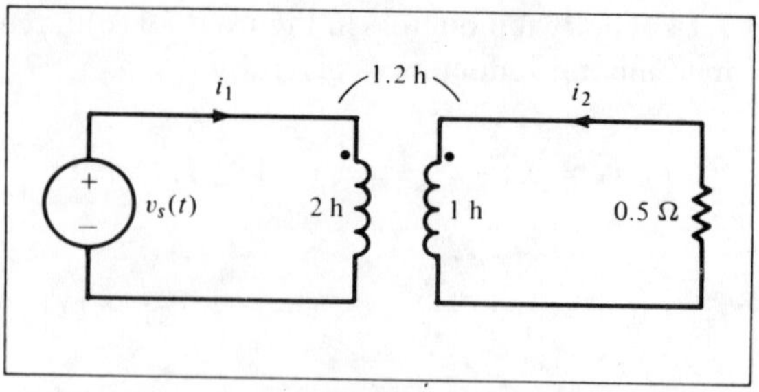

FIGURE P 7.13

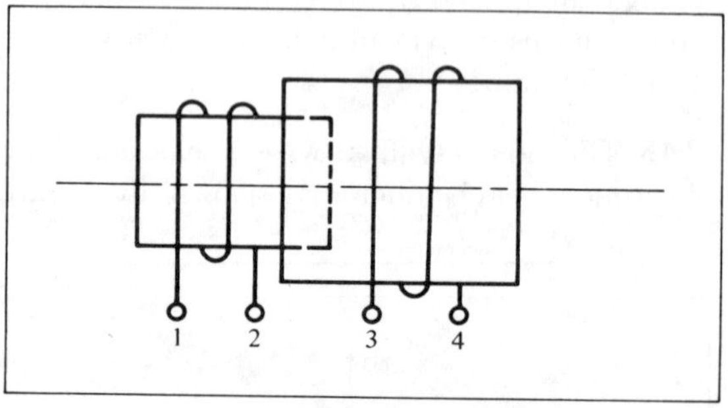

FIGURE P 7.14

is centered inside the 3-4 coil. The two coils are to be connected in series with one another and a current source. How should the series connection be made to produce an attractive force between the coils? [Hint: Physical systems try to minimize the stored energy.]

REFERENCES AND ADDITIONAL READING

7.1 W. H. HAYT, JR., AND J. E. KEMMERLY, *Engineering Circuit Analysis*. New York: McGraw-Hill Book Company, Inc., 1963, Chapter 4.

7.2 J. B. CRUZ AND M. E. VAN VALKENBURG, *Introductory Signals and Circuits*. Waltham, Mass.: Blaisdell Publishing Company, 1967, Chapter 4.

7.3 W. H. HAYT, JR., AND G. W. HUGHES, *Introduction to Electrical Engineering*. New York: McGraw-Hill Book Company, Inc., 1968, Chapter 6.

8

Solution of Simple Linear Dynamic Networks

8.1 Introduction

The *formulation* methods of Chapters 4 and 5 are quite general for resistive networks and can be easily extended to linear and nonlinear dynamic networks. However, the *solution* methods for dynamic systems can be quite different, especially when an analytic (versus a computer) solution is to be found. In this chapter, we examine analytic methods for solving simple dynamic networks. The principal intents are (1) to refresh the reader's mathematical acquaintance with the solution of differential equations, (2) to gain an initial insight into the forms of response that can be expected from dynamic networks, and (3) to introduce certain standard signal waveforms that will be used in the latter part of this text.

8.2 Single-Time-Constant Network

Introduction

The conceptually simplest dynamic network is a linear resistive network with a single capacitor or inductor. As illustrated in Figure 8.1, such a network can be reduced by means of a Thevenin or Norton equivalent to a network consisting of a single source, resistor, and capacitor.

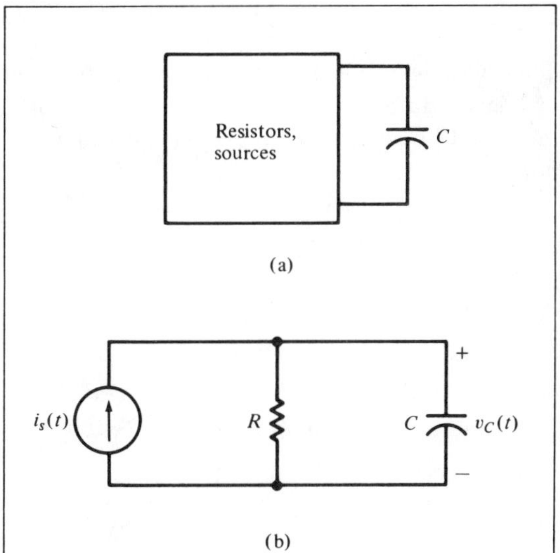

FIGURE 8.1 Reduction of single-capacitor network.

This network will exhibit a response $v_C(t)$ that is dependent on:

1. The values of R and C.
2. The form of the signal source $i_s(t)$.
3. The energy stored in the capacitor when the source $i_s(t)$ is first applied.

The latter is a new phenomenon for us to study, because the resistive networks we have studied until now responded only to input signals. Indeed, in the following analyses, we will distinguish between (1) the *zero-state* response, where no initial energy storage is present, and (2) the *zero-input* response, when no signal is present.

For the present, we assume that our problems are initiated at $t = 0$; that is, a signal $i_s(t)$ is applied that is given for $t \geq 0$, and a known amount of energy $w_C(t)$ is stored in the capacitor at $t = 0$. Because $w_C(0) = \frac{1}{2}Cv_C^2(0)$, this is equivalent to specifying $v_C(0)$.

Applying Kirchhoff's current law in Figure 8.1(b), we have

$$i_C(t) + i_R(t) = i_s(t)$$

or

$$\frac{C\,dv_C(t)}{dt} + \frac{v_C(t)}{R} = i_s(t) \tag{8.1}$$

This is a linear, constant coefficient, first-order ordinary differential equation. A network described by such an equation is called a *first-order network*.

Zero-Input Response

If $i_s(t) = 0$ in Equation (8.1), we are left with the *homogeneous* equation

$$\frac{C\,dv_C(t)}{dt} + \frac{v_C(t)}{R} = 0 \tag{8.2}$$

with $v_C(0)$ specified. By direct substitution, it is easy to verify that

$$v_C(t) = Ke^{-t/\tau} \tag{8.3}$$

is a solution, where $\tau = RC$. Although Equation (8.2) can be satisfied for any K, the initial condition on $v_C(t)$ requires

$$v_C(t) = v_C(0)e^{-t/RC} \tag{8.4}$$

This exponential decay is shown in Figure 8.2.

The quantity $\tau = RC$ (called the *time constant*) is especially important in this plot. First, dimensional analysis will verify that resistance (v/i) times capacitance $(i/dv/dt)$ does indeed have dimensions of time, namely,

$$\left(\frac{v}{i}\right) \times \left(\frac{i}{dv/dt}\right) \rightarrow \frac{v}{dv/dt} \rightarrow dt$$

If C is in farads and R in ohms, RC has the dimensions of seconds. Further, when $t = \tau$, the exponential curve reaches approximately 37 percent of the initial value; at four time constants $v_C(4\tau) = e^{-4} = 0.0184$, and the response can, for most practical purposes, be considered zero-valued.

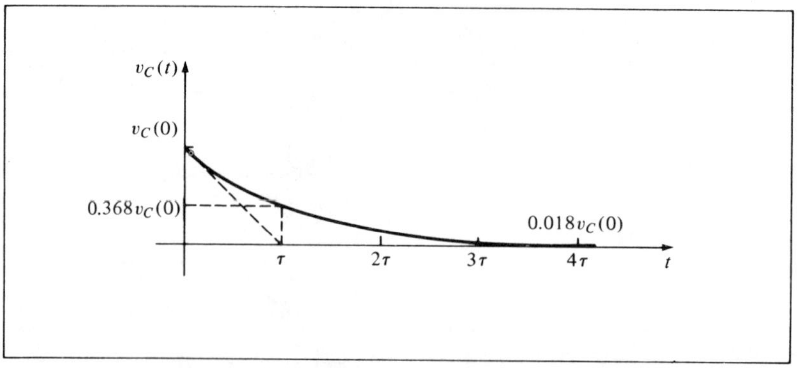

FIGURE 8.2 Exponential decay of RC network.

Example 8.1

The zero-input response $v_C(t)$ of Figure 8.3(a) is to be found in terms of the network elements and $v_C(0)$.

Because the zero-input response is defined as the response when sources are zero-valued, we set both $v_s(t)$ and $i_{eq}(t)$ in Figure 8.3 to zero for $t \geq 0$. We find, from Equation (8.4),

$$v_C(t) = v_C(0)e^{-t/R_{eq}C}$$

$$= v_C(0) \exp\left[-\left(\frac{R_1 + R_2}{R_1 R_2 C}\right)t\right] \qquad (8.5)$$

A completely analogous solution can be performed for the RL network of Figure 8.4. Beginning with the differential equation

$$L\frac{di_L(t)}{dt} + Ri_L(t) = v_s(t) \qquad (8.6)$$

obtained from writing KVL around the loop shown, we find the zero-input response $[v_s(t) = 0]$ to be

$$i_L(t) = i_L(0)e^{-t/\tau}$$

Here, the time constant is $\tau = L/R$ sec.

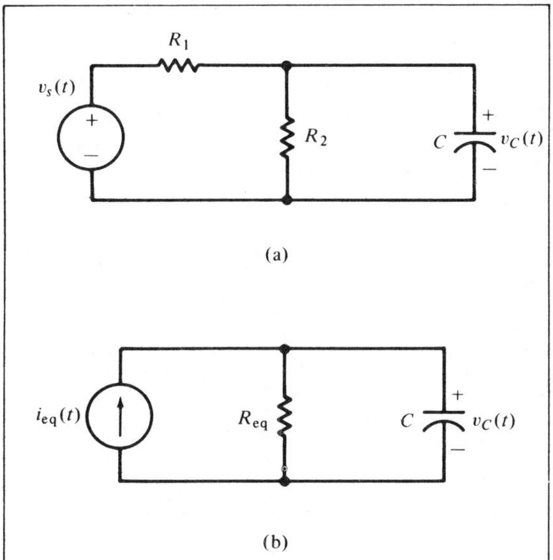

FIGURE 8.3 *RC* network example.

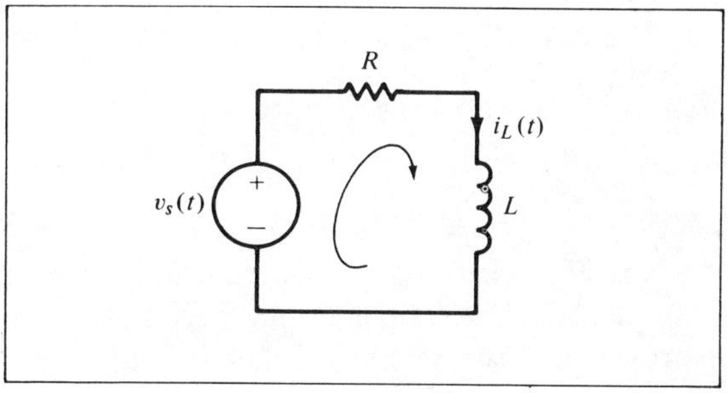

FIGURE 8.4 *RL* network.

Zero-State Response

We have purposely displayed the sources in the above examples to emphasize that only one component of the complete response has been found. To determine the response due to the source, we must specify the input waveform.

The simplest form of input is a signal that is constant for $t \geqq 0$. The RC network for $t \geqq 0$ is then illustrated in Figure 8.5 and is described by

$$C\frac{dv_C(t)}{dt} + \frac{v_C(t)}{R} = i_s(t) = I_s \qquad (t \geqq 0)$$

(8.7)

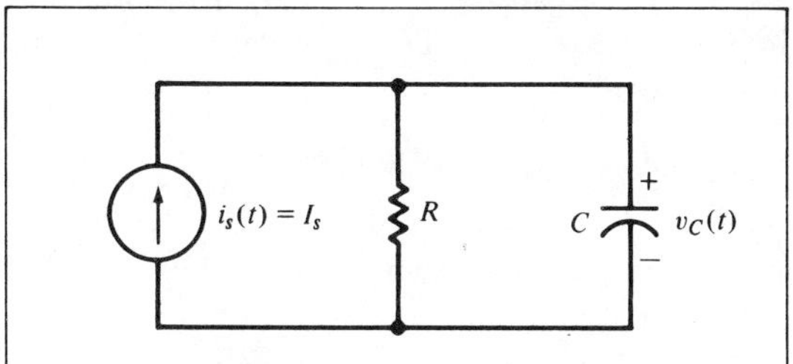

FIGURE 8.5 *RC* network ($t \geqq 0$).

with $v_C(0) = 0$ because of the zero-state assumption.

We have already observed that the solution of the *homogeneous equation* (8.2) has the form $v_{C_h}(t) = Ke^{-t/RC}$. Now let the solution voltage $v_C(t)$ with input applied be of the form

$$v_C(t) = v_{C_h}(t) + v_{C_p}(t) \qquad (8.8)$$

where $v_{C_p}(t)$ is the *particular* solution yet to be found. What we are assuming here is that the presence of the input will merely *add* a term to the zero-input solution $Ke^{-t/RC}$.

The first step in finding this particular solution is to substitute $v_{C_h}(t) + v_{C_p}(t)$ into the differential equation

$$C\frac{d}{dt}(v_{C_h}(t) + v_{C_p}(t)) + \frac{v_{C_h}(t) + v_{C_p}(t)}{R}$$

$$= C\frac{dv_{C_p}(t)}{dt} + \frac{v_{C_p}(t)}{R} = I_s$$

That is, the homogeneous component of the solution necessarily *vanishes* when we substitute it into the left-hand side.

To find the particular solution $v_{C_p}(t)$, we require a function that specifically satisfies the equation

$$C\frac{dv_{C_p}(t)}{dt} + \frac{v_{C_p}(t)}{R} = I_s \qquad (8.9)$$

Choosing $v_{C_p}(t) = RI_s$ obviously satisfies Equation (8.9). The complete solution is therefore

$$\begin{aligned} v_C(t) &= v_{C_p}(t) + v_{C_h}(t) \\ &= RI_s + Ke^{-t/RC} \qquad (t \geqq 0) \qquad (8.10) \end{aligned}$$

We have not used the zero-state initial condition. Substituting $v_C(0) = 0$ in Equation (8.10) gives $K = -RI_s$. The zero-state solution is then

$$v_C(t) = RI_s(1 - e^{-t/RC}) \qquad (t \geqq 0) \qquad (8.11)$$

This function is shown in Figure 8.6. Among its characteristics are:

1. The response is zero for $t = 0$, as required by the initial conditions.

2. The response approaches an asymptotic value (to be discussed shortly); the rise from zero to this value is exponential, with the same time constant as the zero-input response of Figure 8.2.

The latter observation is an important one. This exponential term will appear in the calculated response to *any* input. Its presence can be traced in each case to the homogeneous component of the response and is a property of the network itself. The constant K multiplying this term will, however, depend on the signal and the initial condition.

The RL network of Figure 8.4 has a similar solution. If $v_s(t) = V_s$, $t \geqq 0$, and $i_L(0) = 0$, the inductor current can be shown to have the form

$$i_L(t) = \frac{V_s}{R}(1 - e^{-Rt/L}) \qquad (t \geqq 0) \qquad (8.12)$$

Again, the zero-state response rises exponentially to a constant asymptotic value.

Complete Response

When *both* the initial conditions are nonzero *and* an input signal is applied, we obtain what is termed the *complete response*. We can demonstrate some interesting properties of this generalization of the zero-state and zero-input response by examining the RC network.

The analysis proceeds identically up through Equation (8.10), where we applied the initial condition. If we now require

$$v_C(t)\big|_{t=0} = v_C(0) = RI_s + K$$

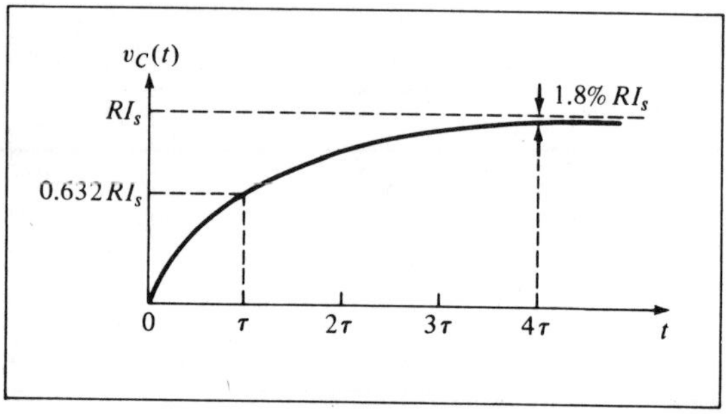

FIGURE 8.6 Zero-state response.

we obtain

$$K = v_C(0) - RI_s$$

The complete response follows as

$$v_C(t) = RI_s + (v_C(0) - RI_s)e^{-t/RC} \qquad (t \geqq 0) \tag{8.13}$$

It is easy to verify that, when $v_C(0) = 0$ and $I_s = 0$, this equation simplifies to the previously derived zero-state and zero-input responses, respectively. An even more interesting result follows from writing Equation (8.13) as

$$v_C(t) = \underbrace{v_C(0)e^{-t/RC}}_{\text{Zero-input}} + \underbrace{RI_s(1 - e^{-t/RC})}_{\text{Zero-state}} \tag{8.14}$$

Here, we observe that the complete response is the *sum* of zero-state and zero-input responses. *This is a general property of linear dynamic networks.* It follows from the fact that the initial condition can be viewed as a form of input signal; from the superposition principle of Chapter 6, the total response is then the sum of the responses due to both the true input signal and the initial condition pseudo-input. We will consider this aspect further in Chapter 11.

Another separation of the complete response is into *transient* and *steady-state* components. The response of Equation (8.13) is already in this form, namely,

$$v_C(t) = \underbrace{(v_C(0) - RI_s)e^{-t/RC}}_{\text{Transient}} + \underbrace{RI_s}_{\text{Steady-state}}$$

In general, the transient component includes those terms that decay with time; the steady-state component contains the remaining terms, which have the same form as the input signal (a constant in this case).

Example 8.2

A constant voltage source of $V_s = 12$ volts is applied to the network of Figure 8.7(a) at $t = 0$. With $v_C(0) = 4$ volts, $R = 1\Omega$, and

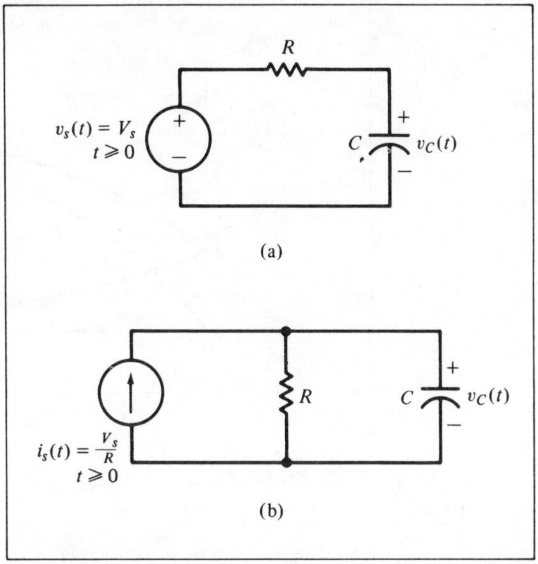

FIGURE 8.7 Example network for complete response.

$C = 5$ f, the response $v_C(t)$ is to be found for $t \geqq 0$.

We can convert this network to the equivalent one of Figure 8.7(b). The zero-state response of this network is the exponential rise to a constant

$$v_C^{(1)}(t) = V_s(1 - e^{-t/RC})$$

whereas the zero-input response is the decaying exponential

$$v_C^{(2)}(t) = v_C(0)e^{-t/RC}$$

The complete response is then

$$v_C(t) = v_C^{(1)}(t) + v_C^{(2)}(t)$$

or

$$\begin{aligned} v_C(t) &= v_C(0)e^{-t/RC} + V_s(1 - e^{-t/RC}) \\ &= 4e^{-0.2t} + 12(1 - e^{-0.2t}) \text{ volts,} \end{aligned}$$

which is plotted in Figure 8.8.

This example illustrates a useful procedure; it is often convenient to calculate separately the zero-input and zero-state responses and combine them, instead of calculating the entire solution in one step. Then, if either the initial conditions or the input change, only the cor-

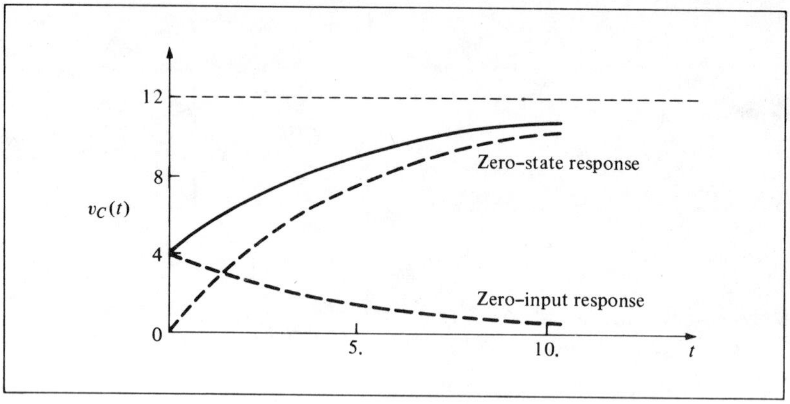

FIGURE 8.8 *RC* network solution.

responding part of the solution needs to be recalculated.

Determination of Initial Conditions

After the RC network reaches an asymptotic or equilibrium value (for practical purposes, after 4 to 5 time constants), none of its voltages or currents change. Therefore,

$$i_C(t) = C\frac{dv_C(t)}{dt} = 0 \tag{8.15}$$

and the capacitor behaves like an open circuit. By a similar argument, when the RL network reaches equilibrium, the inductor acts as a short circuit. This result is quite general: In any RC or RL network with constant excitation, when the network reaches equilibrium, all capacitors act as open circuits and all inductors as short circuits. This observation permits us to deduce the equilibrium values of a voltage or current *without determining the complete response.* An example is illustrated in Figure 8.9. The equilibrium value RI_s can be found by removing the capacitor from the RC network; the equilibrium voltage is then

$$v_C(\infty) = I_sR$$

which checks with Equation (8.13).

Another application of this principle is the determination of initial conditions. Frequently, the initial conditions are the result of the network remaining in some configuration long enough to reach equilibrium and then being switched to some new configuration at $t = 0$. It is only necessary, then, to treat all capacitors as open circuits, all inductors as short circuits, and analyze the resulting resistive network to determine initial capacitor voltages and inductor currents.

Figure 8.10 gives two simple examples; in both cases, it is assumed that the switch has been closed for a long time and is opened at $t = 0$. In Figure 8.10(a), with L acting as a short for $t < 0$, V_s appears across R_1, and the current V_s/R_1 all flows through L. In Figure 8.10(b), with C acting as an open circuit for $t < 0$, the voltage divider produces a voltage $V_sR_2/(R_1 + R_2)$ across C. When the switch is opened in either network at $t = 0$, the result is an RC or RL combination with initial energy storage.

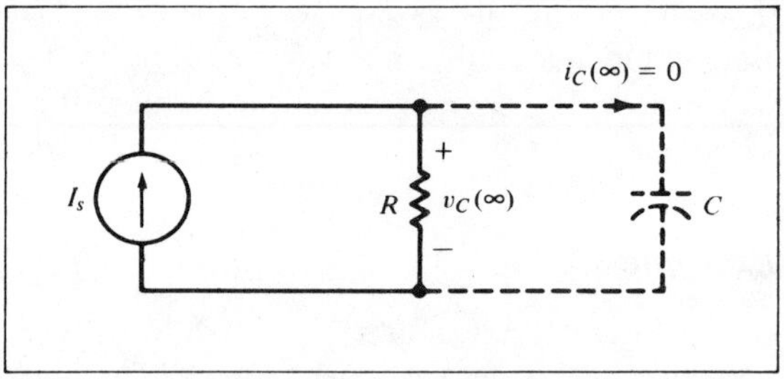

FIGURE 8.9 Equilibrium model.

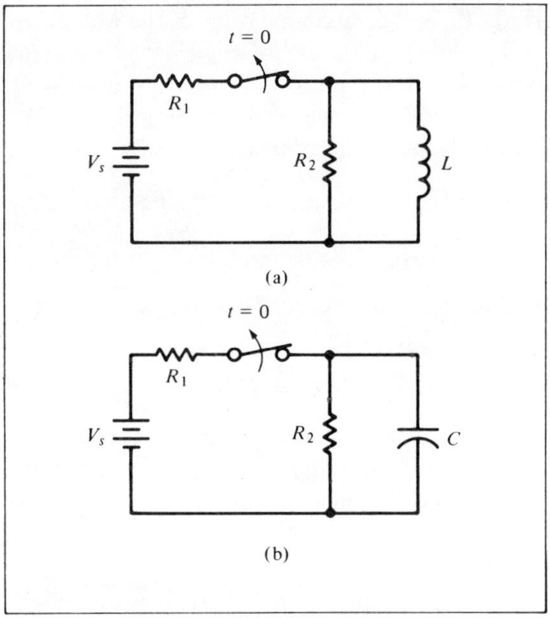

FIGURE 8.10 Networks with nonzero initial conditions.

Example 8.3

The networks of Figure 8.11 are allowed to reach equilibrium with the switch open; the switch is then closed at $t = 0$, as shown. We are to find the appropriate initial conditions, and $i_L(t)$ and $v_C(t)$ for $t \geqq 0$.

RC network. With C acting as an open circuit, the source current I_s flows through R_1 and R_2; the voltage across C is therefore

$$v_C(0) = I_s R_2 \tag{8.16}$$

The resistor connected to C with the switch closed is the parallel combination of R_1 and R_2; therefore,

$$\tau = C \frac{R_1 R_2}{R_1 + R_2} \tag{8.17}$$

The capacitor voltage for $t \geqq 0$ is therefore

$$v_C(t) = v_C(0)e^{-t/\tau}$$

$$= I_s R_2 e^{-t(R_1+R_2)/R_1R_2C}$$

RL network. With L acting as a short circuit, all the source current I_s will flow through it; therefore,

$$i_L(0) = I_s \tag{8.18}$$

The total resistance connected to L with the switch closed is determined from R_1 and R_2 in parallel; the time constant is therefore given by

$$\tau = L \frac{R_1 + R_2}{R_1 R_2} \tag{8.19}$$

The inductor current for $t \geqq 0$ is therefore

$$i_L(t) = I_s e^{-tR_1R_2/L_1(R_1+R_2)}$$

A crucial link in the above reasoning is the assumption that the inductor current or capacitor voltage has the same value before and after the switch is opened. (These times are often indicated as $t = 0^-$ and $t = 0^+$.) That this is ordinarily a good assumption is shown by the following argument.

Consider the equation for a capacitor

$$i(t) = C \frac{dv(t)}{dt} \tag{8.20}$$

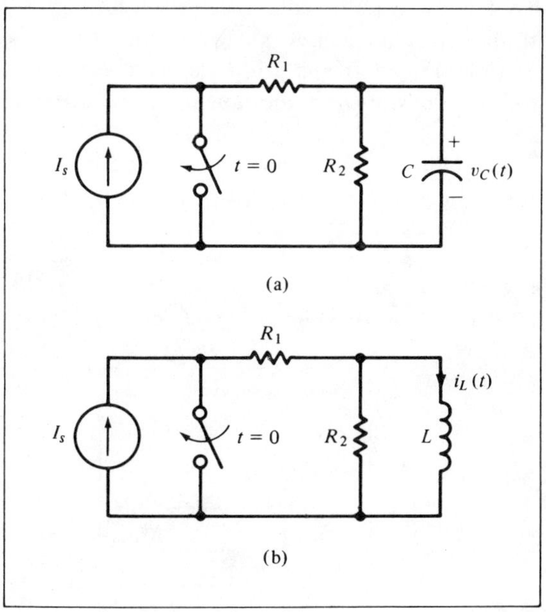

FIGURE 8.11 *RC* and *RL* networks with initial conditions. (a) *RC* circuit; (b) *RL* circuit.

It is clear from this relation that the faster the voltage across a capacitor changes, the larger the capacitor current must be. For the voltage to change by a nonzero amount in zero time would amount to changing the voltage at an infinite rate and would require an *infinite* current—a finite quantity of charge transferred to the capacitor in zero time. A similar argument shows that changing inductor current instantaneously requires an infinite voltage. As long as such currents and voltages remain finite, we know capacitor voltages and inductor currents must remain continuous. This is always the case in a physical circuit.

This is not to say that such discontinuities cannot occur in an idealized model, as for example by connecting a capacitor directly across a battery with no internal resistance or an inductor in series with a (ideal) current source. We will solve even these idealized problems shortly, after we have developed our analysis techniques.

Response to a Sinusoid

To gain additional experience in the solution of simple dynamic networks, we examine in this section the determination of the complete response to another common input signal: the sinusoid.

This signal is illustrated in detail in Figure 8.12, and is described by the equation

$$f(t) = \begin{cases} 0 & (t < 0) \\ F_m \cos(\omega t + \phi) & (t \geqq 0) \end{cases} \qquad (8.21)$$

Here, F_m is the *maximum value* of the waveform and ϕ is called the *phase angle*, or *phase shift*, indicating the amount that the cosine argument is shifted. We can represent by Equation (8.21) all functions of the general form

$$f(t) = \begin{cases} 0 & (t < 0) \\ A \cos \omega t + B \sin \omega t & (t \geqq 0) \end{cases} \qquad (8.22)$$

For example, if $A = 0$ in Equation (8.22), $f(t)$ can be written as

$$f(t) = B \sin \omega t = F_m \cos(\omega t + \phi)$$

where $F_m = B$ and $\phi = -\pi/2$. More generally, Equations (8.21) and (8.22) are related by trigonometric identities (to be shown in Chapter 10)

$$F_m = \sqrt{A^2 + B^2} \qquad \phi = -\tan^{-1}\frac{B}{A} \qquad (8.23)$$

$$A = F_m \cos\phi \qquad B = -F_m \sin\phi \qquad (8.24)$$

The calculation of the response to the above waveform will be demonstrated for the network in Figure 8.13. The differential equation for this network is

$$C\frac{dv_C(t)}{dt} + \frac{v_C(t)}{R} = I_{sm} \sin \omega t \qquad (t \geqq 0) \qquad (8.25)$$

with initial capacitor voltage $v_C(0)$.

The homogeneous solution is again given by $v_{C_h}(t) = Ke^{-t/RC}$, because the form of $v_{C_h}(t)$ does not depend on the excitation. The only change is in the particular solution. One way

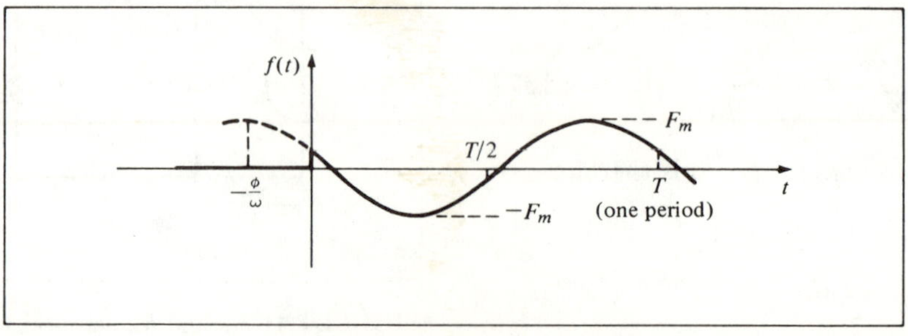

FIGURE 8.12 Sinusoidal signal.

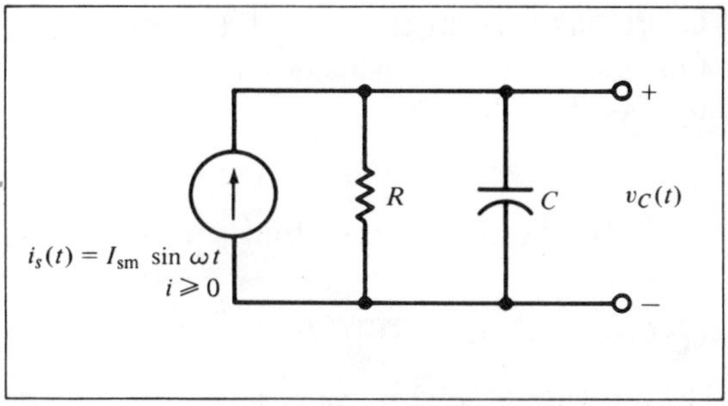

FIGURE 8.13 Dynamic network with sinusoidal input.

to find this response is to assume it to have the general sinusoidal form

$$v_{C_p}(t) = A \cos \omega t + B \sin \omega t \qquad (8.26)$$

Substituting this into Equation (8.25) and equating the sin ωt terms on the left to $I_{sm} \sin \omega t$ and the cos ωt terms to zero, we can solve for A and B. The result is

$$v_{C_p}(t) = I_{sm}R\left[\frac{-\omega RC \cos \omega t + \sin \omega t}{1 + (\omega RC)^2}\right] \qquad (8.27)$$

This particular solution must be added to the homogeneous solution

$$v_{C_h}(t) = Ke^{-t/RC} \qquad (8.28)$$

and K chosen to satisfy the initial condition requirement on $v_C(t)$. Adding Equations (8.27) and (8.28) and setting $t = 0$, we have

$$v_C(0) = K - \frac{\omega RC}{1 + (\omega RC)^2} I_{sm}R \qquad (8.29)$$

$$K = v_C(0) + \frac{\omega RC}{1 + (\omega RC)^2} I_{sm}R \qquad (8.30)$$

Therefore, the complete response is

$$v(t) = \underbrace{\left(v_C(0) + \frac{\omega RC}{1 + (\omega RC)^2} I_{sm}R\right) e^{-t/RC}}_{\text{Transient}}$$

$$+ \underbrace{I_{sm}R\left(\frac{-\omega RC \cos \omega t + \sin \omega t}{1 + (\omega RC)^2}\right)}_{\text{Steady state}} \qquad (8.31)$$

Written in this form, we again see that the first expression is a *transient* term that decays with increasing time, but the second is a *steady-state* term that follows the shape of the signal source $i_s(t)$ indefinitely. Converting the second term to the cosine form of Equation (8.21) using (8.23) gives the sinusoidal steady-state component as

$$v_{C_{ss}}(t) = \frac{I_{sm}R}{\sqrt{1 + (\omega RC)^2}}$$

$$\times \cos\left(\omega t - \tan^{-1}\left(\frac{1}{\omega RC}\right)\right) \qquad (8.32)$$

Although there is considerable insight to be gained from close study of Equation (8.32), we will defer this until Chapter 10, where an entire chapter is devoted to the topic of sinusoidal steady-state analysis.

8.3 Step Function and Step Response

Step Function

Just as we find it convenient to model circuits with familiar, idealized network elements, the idealization of waveforms provides an efficient language with which to describe the excitations and response of networks with energy-storage elements. Among the most important of these waveforms are the *singularity* functions. Each of these functions is discontinuous at one or more points, so that its derivative cannot be conveniently defined at that point.

One common singularity function is the *unit step function* of Figure 8.14, defined from

$$u(t) = \begin{cases} 0 & (t < 0) \\ 1 & (t > 0) \end{cases} \qquad (8.33)$$

For our purposes, the value at $t = 0$ is of no concern. This function has an effect similar to closing (or opening) a switch, as illustrated in Figure 8.14. This analogy is considered at length in the following examples.

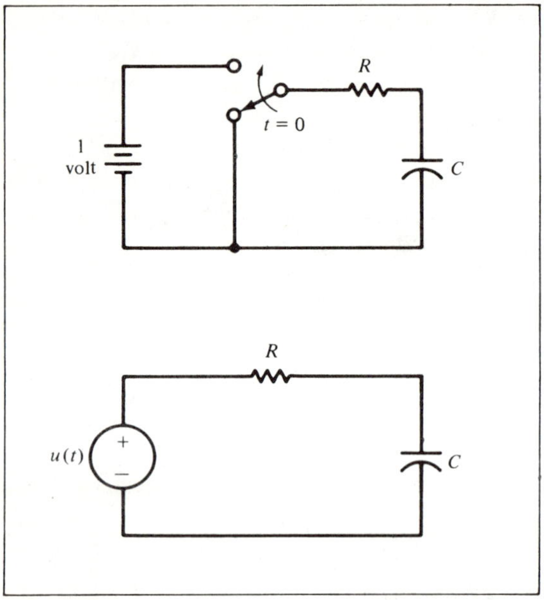

FIGURE 8.14 Step-response model of switching.

An Elementary Application

The value of the step function as a representational aid is illustrated in Figure 8.15 (a). Here, the RC network in the zero state with a constant input for $t > 0$ is modeled instead with a step input of the form

$$i_s(t) = I_s u(t) \qquad \text{(all } t\text{)} \qquad (8.34)$$

Similarly, the zero-state output can be written as

$$v_C(t) = I_s R(1 - e^{-t/RC})u(t) \qquad \text{(all } t\text{)} \qquad (8.35)$$

The presence of $u(t)$ in Equation (8.35) makes it unnecessary to indicate that the solution is valid only for $t > 0$.

The response of a network in the zero state to a unit step signal is known generally as the *(unit) step response*. We will often use the symbol $s(t)$ when we are discussing this response.

Shifting

It is quite important in composing mathematical models of physical problems to be able to represent signals and responses that are displaced or *shifted* in time relative to one another. The step function performs this task as follows.

Consider the function $u(t - t_0)$. To assist us

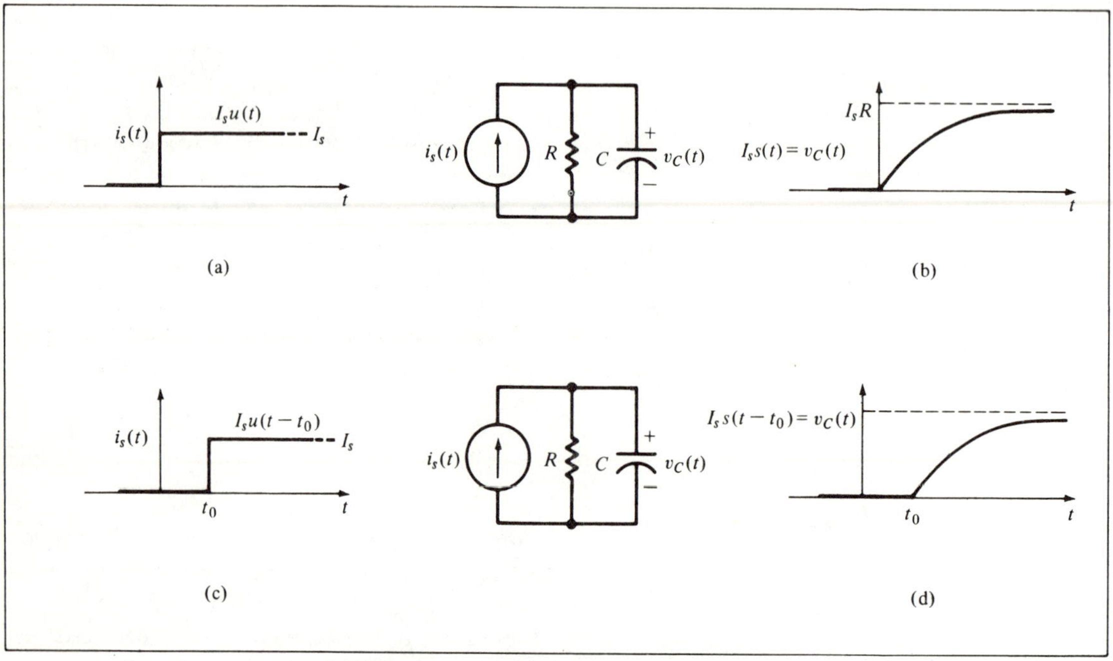

FIGURE 8.15 Unshifted and shifted step responses.

in visualizing the function, we define the new (shifted) time variable $t' = t - t_0$; then

$$u(t') = u(t - t_0) = \begin{cases} 0 & (t' < 0 \text{ or } t < t_0) \\ 1 & (t' > 0 \text{ or } t > t_0) \end{cases} \quad (8.36)$$

The function $u(t - t_0)$ thus has a discontinuity at $t = t_0$; that is, $u(t - t_0)$ is a shifted version of $u(t)$, as shown in Figure 8.15(c).

The response to such a shifted step will be simply $s(t - t_0)$, *a shifted version of the step response* as shown in Figure 8.15(b). (If the network is not in the zero state at $t = 0$, this result will not be valid, because the initial condition will begin to decay at $t = 0$.) We can write this response as

$$v_C(t) = I_s R(1 - e^{-(t-t_0)/RC})u(t - t_0) \quad (8.37)$$

In general, if a network has the step response $f(t)u(t)$, then the response to a step delayed by t_0 is $f(t - t_0)u(t - t_0)$. A common error made in this representation is to write the shifted response as $f(t)u(t - t_0)$; for example, we could *not* write in place of Equation (8.37)

$$v_C(t) = I_s R(1 - e^{-t/RC})u(t - t_0) \quad (8.38)$$

Pulse Train Analysis

The usefulness of the step function in signal representation can be further extended by

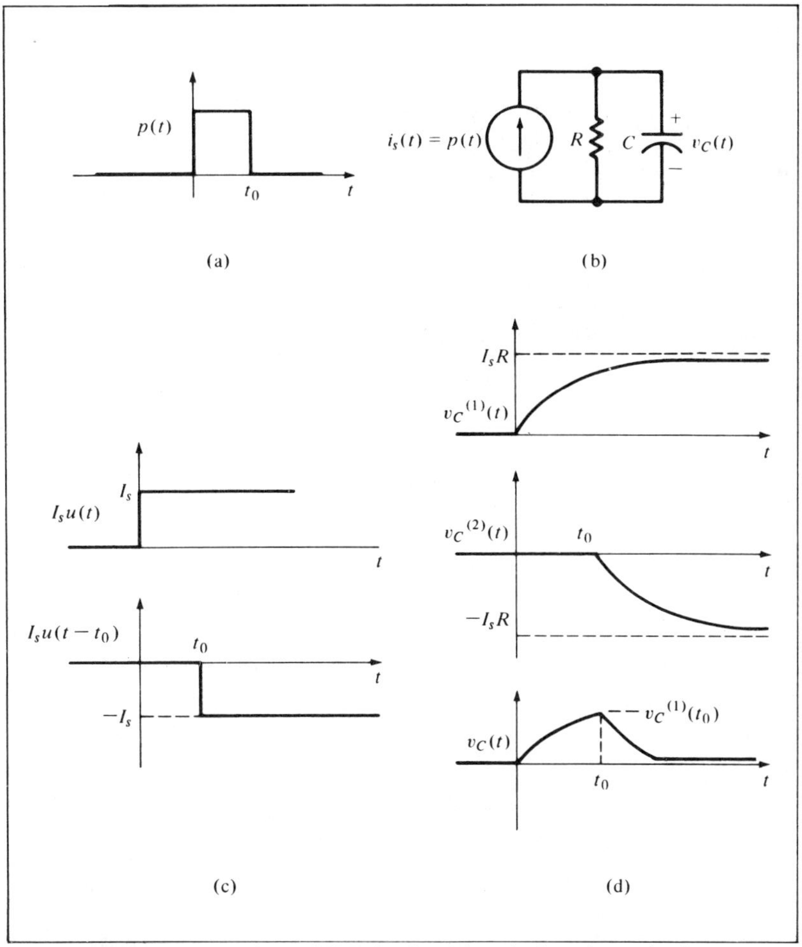

FIGURE 8.16 Analysis of *RC* circuit with a single pulse excitation.

exploiting the linearity of the differential equations that describe the simple dynamic networks that we have studied.

Consider, for example, the pulse function of Figure 8.16(a). As indicated in Figure 8.16(c), this function can be described in terms of a step function as

$$p(t) = I_s(u(t) - u(t - t_0)) \quad (8.39)$$

where t_0 is the pulse width. We may look upon Equation (8.39) as composed of two signals

$$p^{(1)}(t) = I_s u(t) \qquad p^{(2)}(t) = -I_s u(t - t_0) \quad (8.40)$$

with the responses shown in Figure 8.16(d)

$$v_C^{(1)}(t) = I_s R(1 - e^{-t/RC})u(t) \quad (8.41)$$

$$v_C^{(2)}(t) = -I_s R(1 - e^{-(t-t_0)/RC})u(t - t_0) \quad (8.42)$$

From the linearity of the network, we can express the total response as

$$v_C(t) = v_C^{(1)}(t) + v_C^{(2)}(t) \quad (8.43)$$

Equations (8.41) to (8.43) provide us with *both* a compact, explicit description of the response *and* insight into certain of its characteristics. For example, we can easily sketch $v_C(t)$ [Figure 8.16(d)]: for $0 \leqq t \leqq t_0$, $v_C(t) = v_C^{(1)}(t)$, and Equation (8.41) gives the response as the familiar rising exponential; for $t \geqq t_0$, the negative slope of $v_C^{(2)}(t)$ will always be larger in magnitude than the positive slope of $v_C^{(1)}(t)$, so $v_C(t)$ will have a negative slope as shown; the maximum value will then be

$$v_{C\max} = v_C(t_0) = v_C^{(1)}(t_0) = I_s R(1 - e^{-t_0/RC})$$

We can generalize this result for the case when a *sequence of pulses* are being applied to a network. The source $i_s(t)$ is represented by the summation of shifted steps

$$i_s(t) = \sum_{n=0,2,4,\ldots} I_s(u(t - t_n) - u(t - t_{n+1})) \quad (8.44)$$

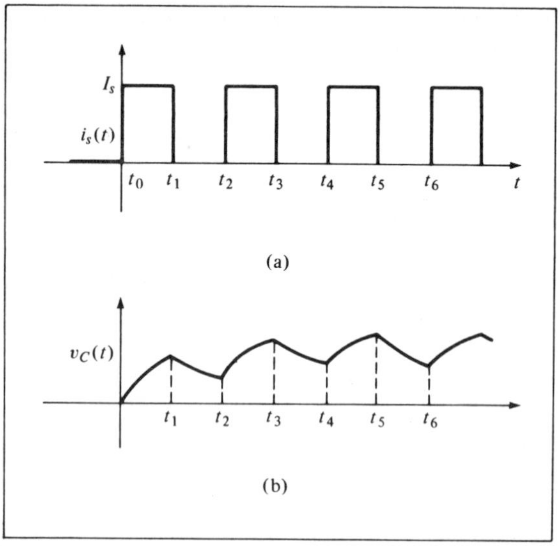

FIGURE 8.17 Pulse train response. (a) Signal; (b) response.

Each term in this summation produces a shifted output term; from the linearity of the network, these terms add to yield

$$v_C(t) = \sum_{n=0,2,4,\ldots} I_s R[(1 - e^{-(t-t_n)/RC})u(t - t_n) - (1 - e^{-(t-t_{n+1})/RC})u(t - t_{n+1})] \quad (8.45)$$

This result is sketched in Figure 8.17(b).

8.4 Impulse Function and Impulse Response

Definition of Impulse

A common dynamic analysis problem involves the determination of the response of a "slow" system to a narrow pulse as shown in Figure 8.18(a). An example would be the brief but intense forces of combustion that power an automobile. As we shall see, when the pulse width becomes small, the system response will depend on the *area* under the pulse rather than on the pulse width or height individually. Consequently, we will define an idealized signal known as an *impulse* with a value or *strength* equal to the area under the narrow pulse. The impulse function $\delta(t - t_0)$ is represented in Figure 8.18(c). Often such pulses and their impulse representations are used in pulse *trains* to model the transmission of signals which form the envelope of the pulses [Figure 8.18(b) and (d)].

Our mathematical description of the impulse is motivated by considering the approximation to a step function shown in Figure 8.19(a).

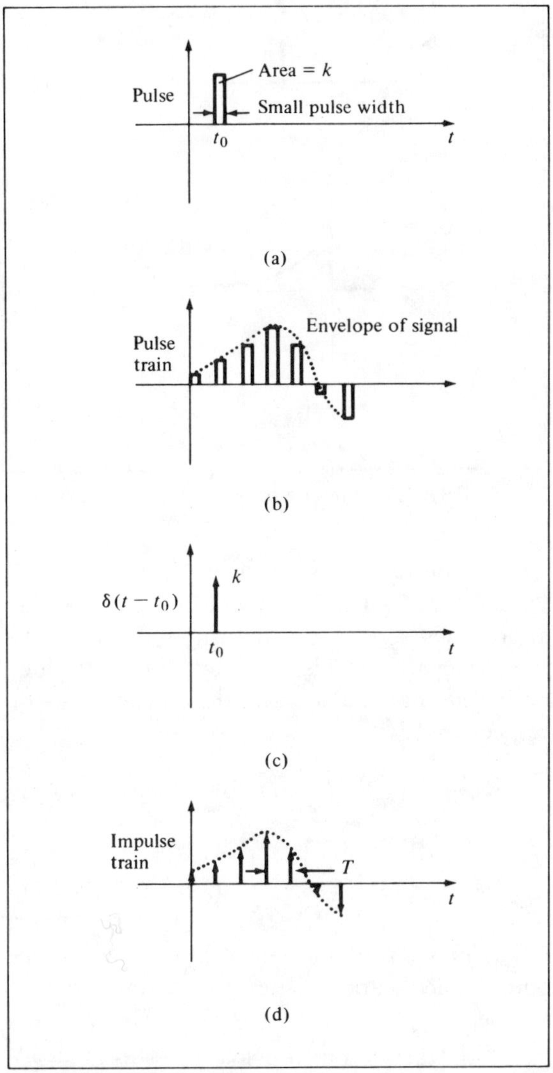

FIGURE 8.18 Pulse and impulse representations.

The function is defined by

$$u_\Delta(t) = \begin{cases} 0 & (t < 0) \\ t/\Delta & (0 < t < \Delta) \\ 1 & (t > \Delta) \end{cases} \qquad (8.46)$$

It has a well-defined derivative $p_\Delta(t)$ shown in Figure 8.19(b) and is given by

$$p_\Delta(t) = \begin{cases} 0 & (t < 0) \\ 1/\Delta & (0 < t < \Delta) \\ 0 & (t > \Delta) \end{cases} \qquad (8.47)$$

The area under the plot of $p_\Delta(t)$ is unity for all values of Δ. As we take Δ sufficiently small, the pulse approaches the form of the impulse function $\delta(t)$. However, we cannot mathematically take a limit, because $p_\Delta(0)$ does not exist for $\Delta = 0$.

An alternative mathematical representation is to define $\delta(t)$ from the requirement that

$$\begin{aligned} &(1) \quad \delta(t) = 0 \quad (t \neq 0) \\ &(2) \quad \int_{-\tau}^{\tau} \delta(t)\, dt = 1 \quad (\tau > 0) \end{aligned} \qquad (8.48)$$

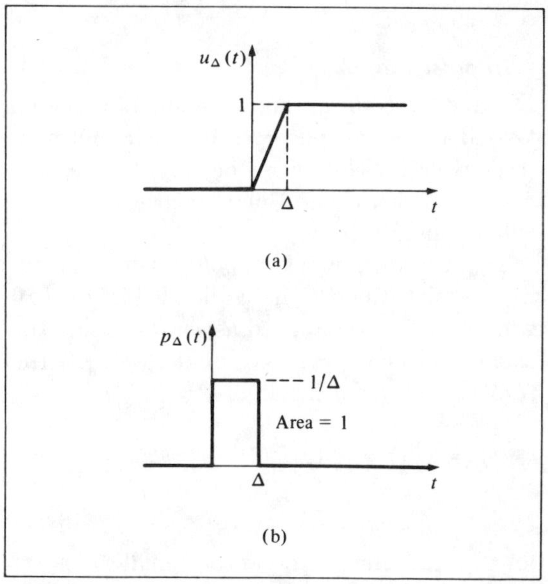

FIGURE 8.19 A step-function approximation and its derivative.

We can then claim that p_Δ approximates $\delta(t)$ in the sense that

$$(1) \quad \lim_{\Delta \to 0} p_\Delta(t) = 0 \qquad (t \neq 0)$$

$$(2) \quad \int_{-\tau}^{\tau} p_\Delta(t)\, dt = 1 \qquad (\tau > 0) \tag{8.49}$$

The unit step function $u(t)$ and the unit impulse $\delta(t)$ are related in the same manner that $u_\Delta(t)$ and $p_\Delta(t)$ are related. This is seen by observing

$$\int_{-\infty}^{t} \delta(\xi)\, d\xi = \begin{cases} 0 & (t < 0) \\ 1 & (t > 0) \end{cases}$$

$$= u(t) \tag{8.50}$$

Therefore, the step function appears to be the integral of the impulse function. We cannot, however, fully justify the corollary statement that

$$\delta(t) = \frac{du(t)}{dt} \qquad (\text{all } t) \tag{8.51}$$

because this would again involve taking the limit $\lim_{\Delta\to 0} p_\Delta(0)$. We will nonetheless accept its validity and show its usefulness in problem solving.

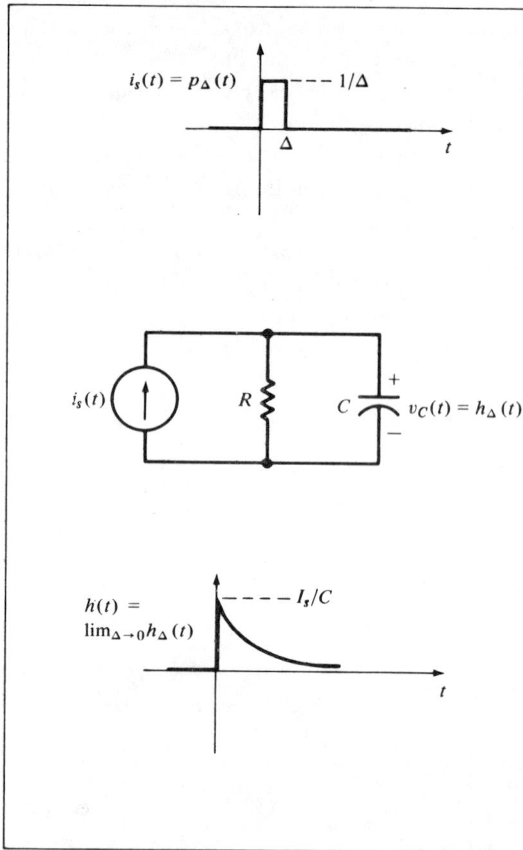

FIGURE 8.20 Impulse response determination.

Impulse Response

Dynamic circuits are most commonly characterized for design purposes by their unit step response $s(t)$. However, the *impulse response* $h(t)$ is far more useful mathematically, as we will see in Chapter 9.

To illustrate how this response is determined, we examine the RC network of Figure 8.20 with a pulse $p_\Delta(t)$ applied as shown. Using the example of Figure 8.16 as a guide, we write the response to such a signal as

$$h_\Delta(t) = v_C(t) = \frac{1}{\Delta}[I_s R(1 - e^{-t/RC}) - I_s R(1 - e^{-(t-\Delta)/RC})] \tag{8.52}$$

for $t \geq \Delta$. As Δ becomes smaller, $p_\Delta(t)$ approaches $\delta(t)$, so we can expect $h(t) = \lim_{\Delta\to 0} h_\Delta(t)$ if this limit exists. As the pulse width approaches zero, however, both the numerator and denominator of Equation (8.52) approaches zero. We can turn to L'Hôpital's rule to evaluate this limit by differentiating both the numerator and denominator with respect to Δ.

$$h(t) = \lim_{\Delta\to 0} \frac{d}{d\Delta}(I_s R e^{-(t-\Delta)/RC}) \tag{8.53}$$

$$= \frac{I_s}{C} e^{-t/RC} \qquad (t > 0) \tag{8.54}$$

The above procedure illustrates some rather remarkable characteristics of an impulse response.

1. Contrary to our previous assertion that a capacitor voltage or an inductor current could

not be discontinuous, we observe that the impulse, with what may be termed "infinite height" at $t = 0$, can indeed force such an instantaneous change. Therefore, we can use the impulse to affect an initial condition of capacitor voltage (or inductor current) by connecting impulse current sources in parallel with capacitors and impulse voltage sources in series with inductors. We will exploit this equivalence between impulse sources and initial conditions more in Chapter 11.

2. The impulse response calculation exemplified in Equation (8.52) can be generalized as

$$h(t) = \lim_{\Delta \to 0} \frac{[s(t) - s(t - \Delta)]}{\Delta} \quad (8.55)$$

where $s(t - \Delta)$ is the response to a unit step applied at Δ. Differentiating as we did in Equation (8.53) gives

$$h(t) = \lim_{\Delta \to 0} \frac{d}{d\Delta} [-s(t - \Delta)] \quad (8.56)$$

$$= \lim_{\Delta \to 0} \frac{d}{dt} s(t - \Delta) \quad (8.57)$$

$$= \frac{d}{dt} s(t) \quad (8.58)$$

That is, the impulse response is *the derivative of the step response*. This is illustrated by the following example.

Example 8.4

The zero-state response of the RL network of Figure 8.21 to a unit delayed impulse is to be found.

The response to a step delayed by an amount t_0 is routinely calculated as

$$s(t - t_0) = \frac{1}{R} (1 - e^{-(t-t_0)R/L}) u(t - t_0)$$

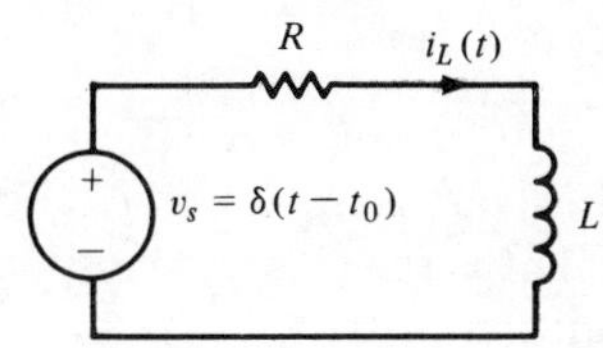

FIGURE 8.21 Impulse response example.

The derivative of $s(t - t_0)$ is

$$h(t - t_0) = \frac{d}{dt} s(t-t_0) = \frac{1}{L} e^{-(t-t_0)R/L} u(t - t_0) + \frac{1}{R} (1 - e^{-(t-t_0)R/L}) \delta(t - t_0) \quad (8.59)$$

because $du(t - t_0)/dt = \delta(t - t_0)$. The second term is zero-valued for all time. Only the first term remains, so that

$$h(t - t_0) = \frac{1}{L} e^{-(t-t_0)R/L} u(t - t_0) \quad (8.60)$$

The response decays exponentially from a value $1/L$ established at $t = t_0$ by the impulse.

PROBLEMS

8.1 Let the equation describing a first-order network be written

$$\frac{dr(t)}{dt} = \alpha r(t) + e(t)$$

where $e(t)$ denotes "excitation" and $r(t)$ denotes "response." The initial state $r(0)$ is given. (a) Find the zero-input response. (b) If $e(t) = E$ for $t > 0$, find the zero-state response. (c) Find the complete response from (a) and (b).

8.2 The zero-state response of the network of Figure P 8.1 to an unspecified $v_s(t)$ is $v_C(t) = V_1 e^{-t/RC}$. What is the complete response with an initial condition $v_C(0)$?

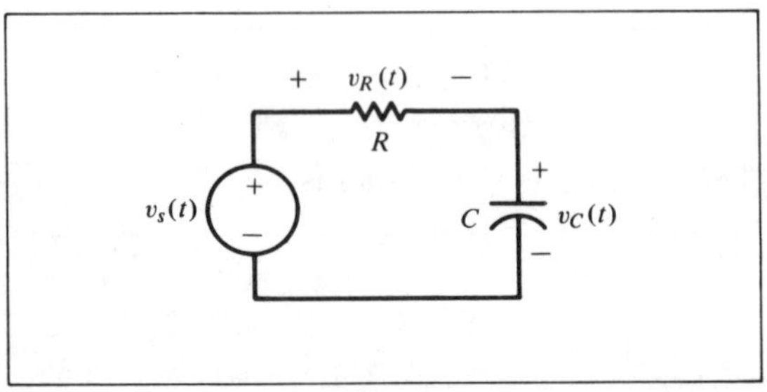

FIGURE P 8.1

8.3 Find $v_C(t)$ and $v_0(t)$ in Figure P 8.2 for: (*a*) $v_s(t) = 0$, $v_C(0) = 1$. (b) $v_s(t) = 1$, $t > 0$, $v_C(0) = 0$. (*c*) $v_s(t) = 1$, $t > 0$, $v_C(0) = 1$.

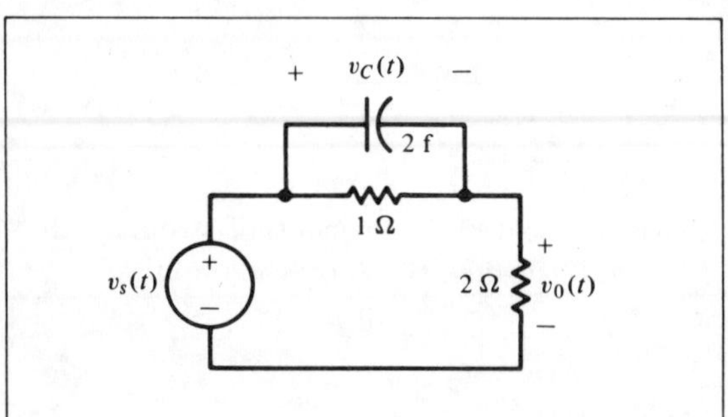

FIGURE P 8.2

8.4 In Figure P 8.3, $v_s(t) = 10$, $t > 0$, and $v_C(0) = 5$. Find $v_0(t)$ for $t > 0$. Identify the zero-state and zero-input portions of this response.

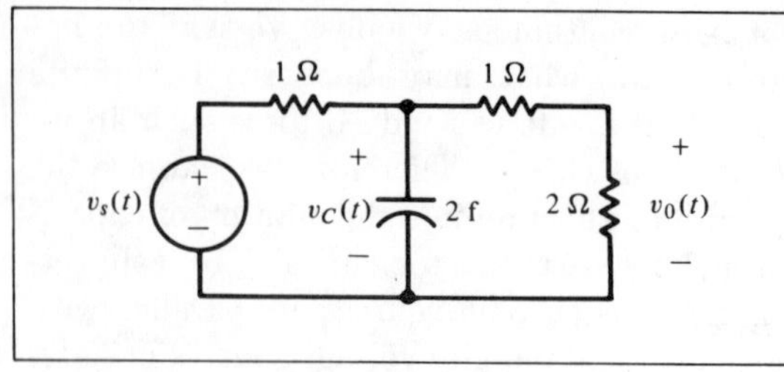

FIGURE P 8.3

8.5 For the network of Figure P 8.4, $i_s(t) = 1$, $t > 0$. (a) Develop an expression for the delay time t_0 required for $v_C(t)$ to reach a value V_0. (b) For what value of R is the delay minimized? Use intuition, not differentiation.

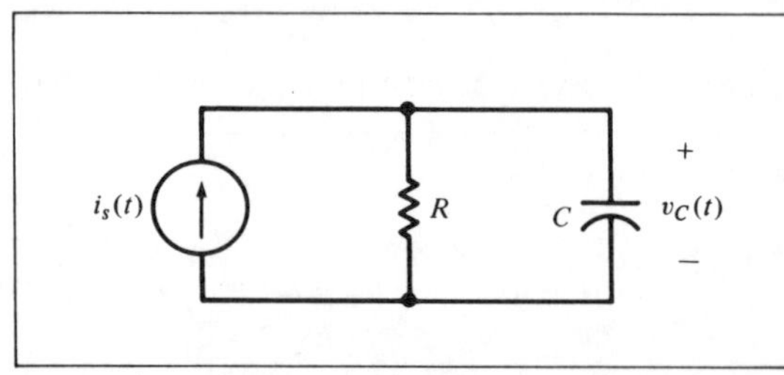

FIGURE P 8.4 $v_C(0) = 0$.

8.6 The network of Figure P 8.5 has reached equilibrium with the switch open. At $t = 0$ the switch is closed. (a) What are $i_1(0^-)$ and $v_1(0^-)$? (b) Determine $v_1(t)$ for $t > 0$.

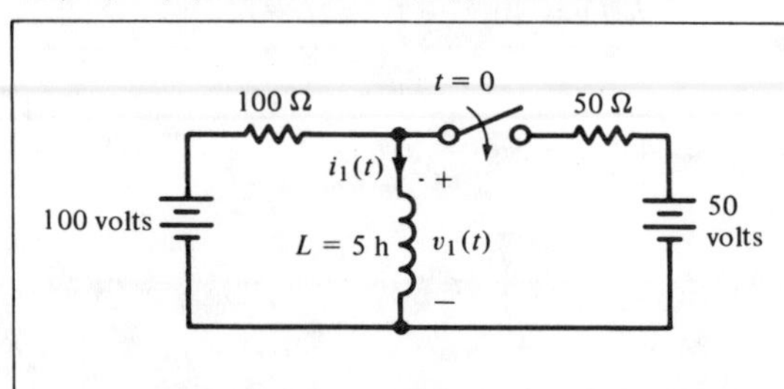

FIGURE P 8.5

8.7 The network of Figure P 8.6 is allowed to come to equilibrium with the switch open. The switch is closed at $t = 0$. (a) What is the value of the indicated current at $t = 0^-$? (b) Determine $i(t)$ for $t > 0$.

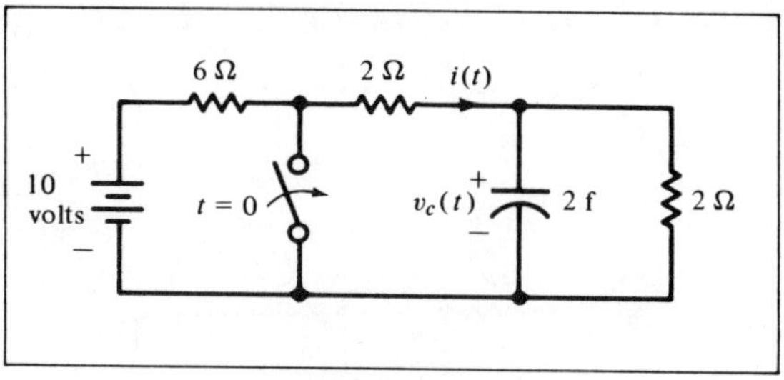

FIGURE P 8.6

8.8 The switch in Figure P 8.7 is thrown from A to B at $t = 0$. Plot $i_L(t)$ for $t > 0$ assuming that the network has reached a steady state when the switch is thrown. Clearly indicate on a plot the steady-state values and the time constant of the network.

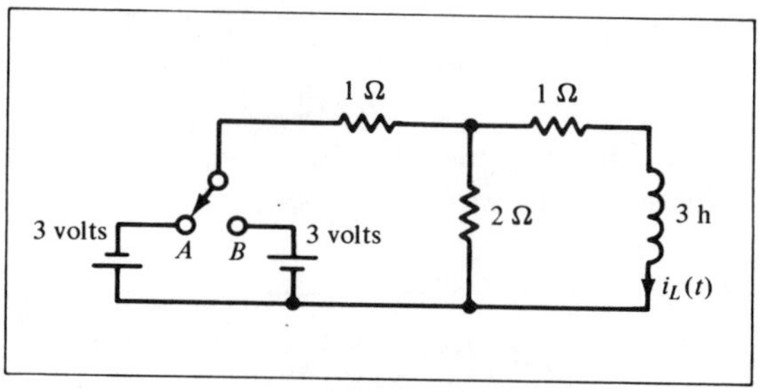

FIGURE P 8.7

8.9 The switch of Figure P 8.8 is closed at $t = 0$. Calculate $v_C(t)$ for $t > 0$. Is your answer unique?

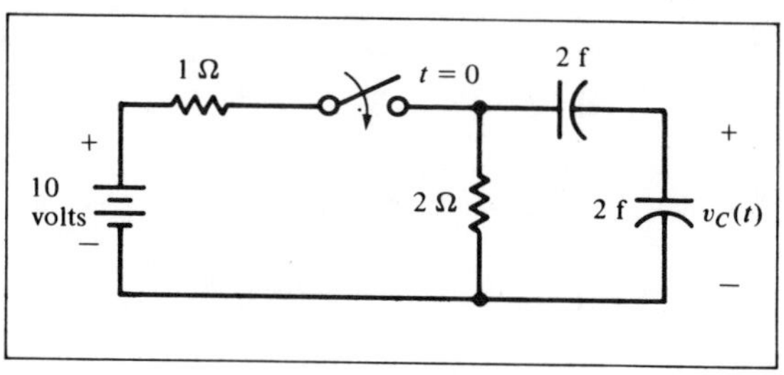

FIGURE P 8.8

8.10 The switch of Figure P 8.9 is closed at $t = 0$ after being open a long time, and at $t = 0.1$ sec the switch is opened again. (a) What is $v_L(0.1^-)$? (b) What is $v_L(0.1^+)$? (c) Calculate and plot $v_L(t)$ for $0 < t \leqq 0.2$ sec.

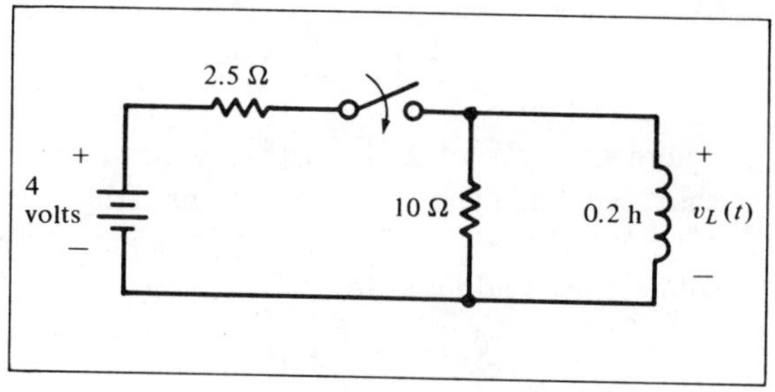

FIGURE P 8.9

8.11 The switch of Figure P 8.10 is closed after being open 1 min. Calculate how long it takes $i(t)$ to reach 90 percent of its equilibrium value.

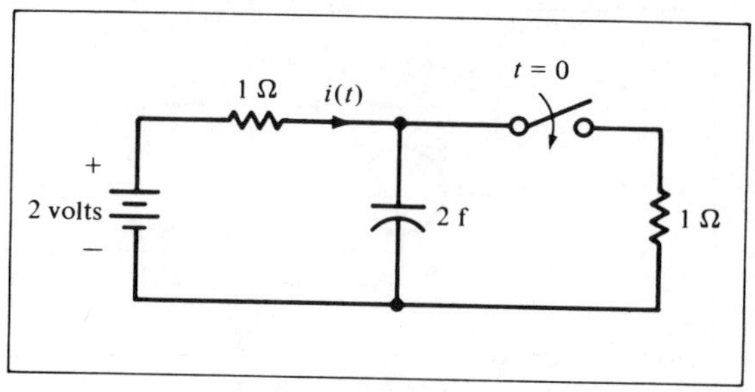

FIGURE P 8.10

8.12 The switch in Figure P 8.11 closes at $t = 0$ for 1 sec after having been open 10 sec. (a) Calculate $v(t)$ $(0 < t \leqq 1$ sec). (b) If the switch opens at $t = 1$ sec, what is $v(2)$?

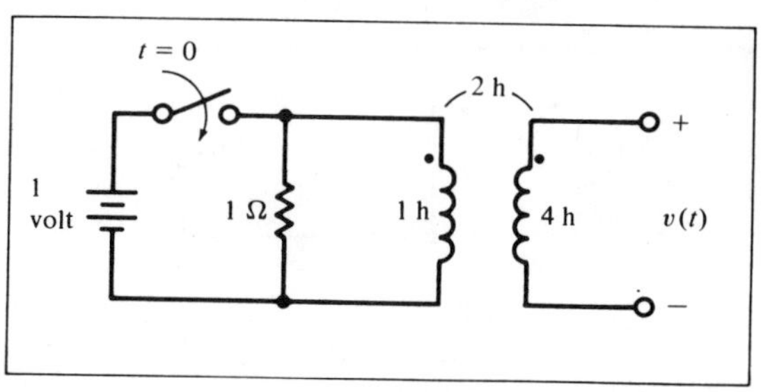

FIGURE P 8.11

8.13 Let an RC network be described by

$$\frac{C dv_C(t)}{dt} + \frac{v_C(t)}{R} = v_s(t)$$

where $v_s(t) = V_s e^{\gamma t}$ and $v_C(0)$ is given. (a) Show that the homogeneous solution has the form $Ke^{-t/RC}$. (b) Show that $K_\gamma e^{\gamma t}$ is a particular solution by finding a proper value for

$$K_\gamma (\gamma \neq -1/RC).$$

(c) Find the value of K in terms of the initial condition.

8.14 In Figure P 8.12, compute the current through the inductor with: (a) $v_s(t) = 3 \cos 2t$. (b) $v_s(t) = 3 \sin 2t$. (c) $v_s(t) = \cos 2t + \sin 2t$. (d) $v_s(t) = \cos(2t + \pi/6)$.

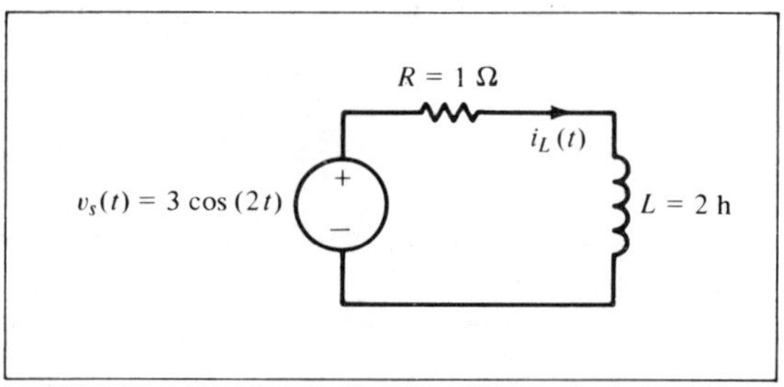

FIGURE P 8.12

8.15 In Figure P 8.13, find $v_L(t)$ and $i_0(t)$ for $t > 0$, for: (a) $i_s(t) = 2u(t)$, $i_0(0^-) = 1$. (b) $i_s(t) = 2\delta(t)$, $i_0(0^-) = 1$. (c) If $i_s(t) = 2\delta(t)$, find $i_0(0^-)$ so that $i_0(t) = 0$ for $t > 0$. Here, $i_0(0^-)$ represents the inductor current before the impulse occurs.

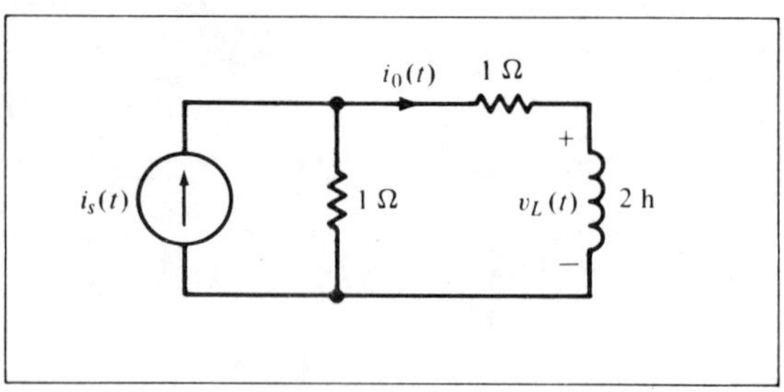

FIGURE P 8.13

8.16 Find $i_L(t)$ and $v_L(t)$ in Figure P 8.14 for $t > 0$ if $\alpha \neq 1$ and if: (a) $v_s(t) = 0$, $i_L(0) = 3$. (b) $v_s(t) = V_s$, $i_L(0) = 0$. (c) $v_s(t) = 2V_s$, $i_L(0) = -1$.

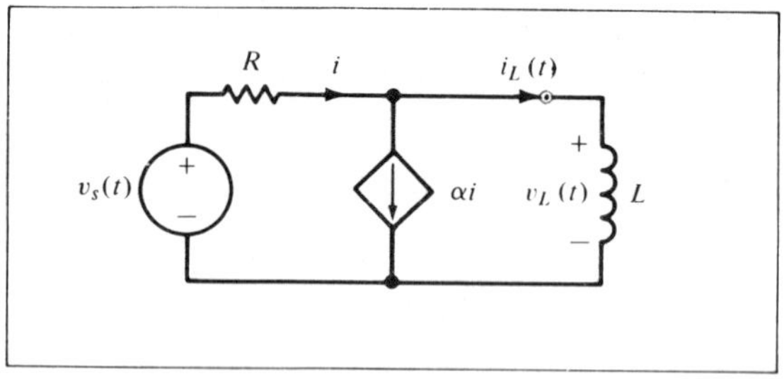

FIGURE P 8.14

8.17 Find the energy stored in the network of Figure P 8.15 when $t = 1$ sec.

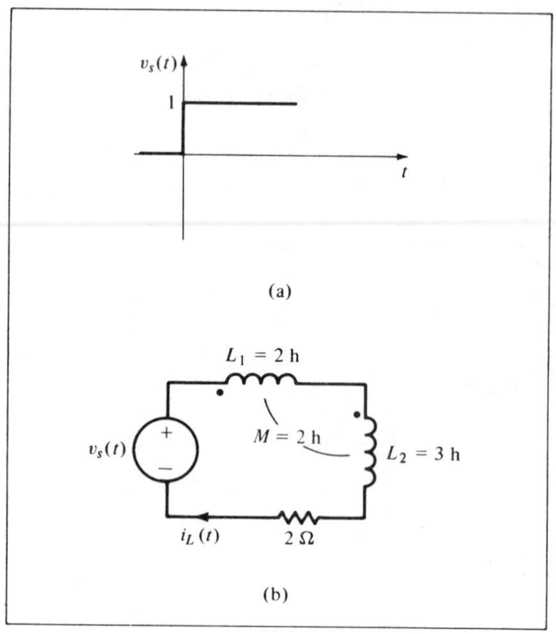

FIGURE P 8.15 (b) $i_L(0) = 0$.

8.18 (a–d) Find the unit step response for the networks of Figure P 8.16.

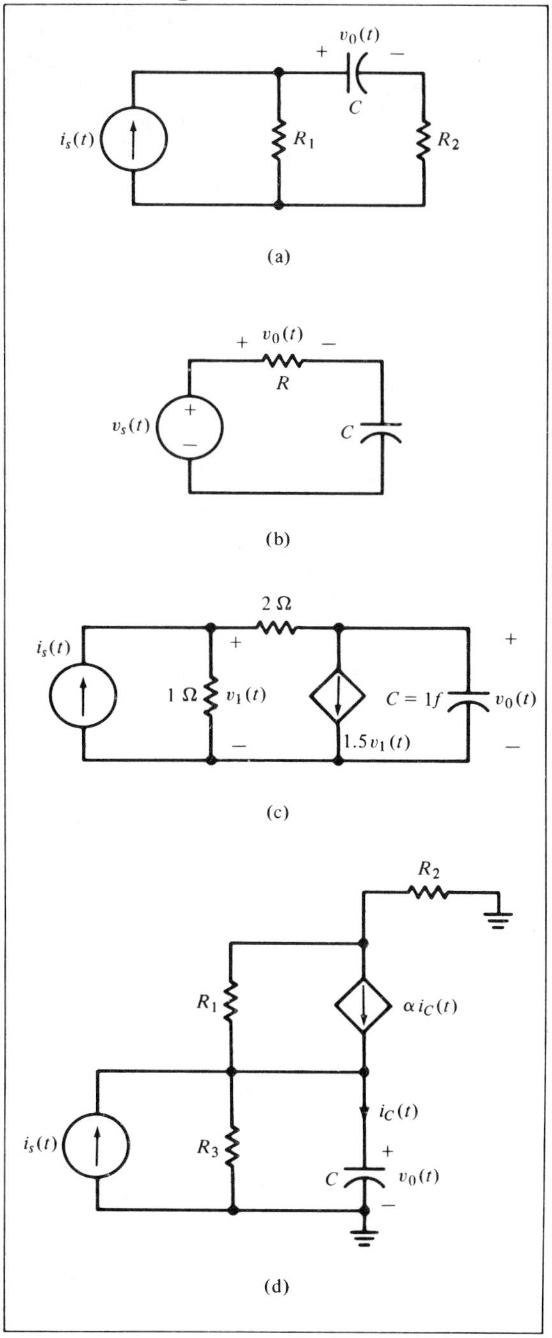

FIGURE P 8.16

8.19 In Figure P 8.17, consecutive positive and negative pulses of width $T = RC$ are applied to an RC network initially in the zero state. Find the negative pulse height V_s necessary to return $v_C(t)$ to the zero state at the end of the second pulse ($t = 2T$).

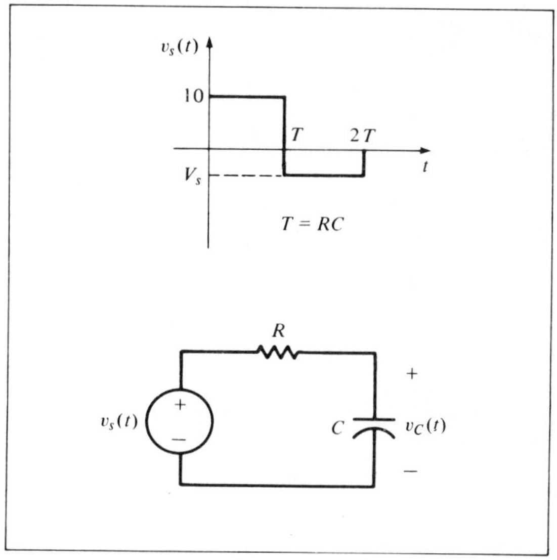

FIGURE P 8.17

8.20 The inductor current in Figure P 8.18 is decaying from an initial value $i_L(0) = 10$ amps.

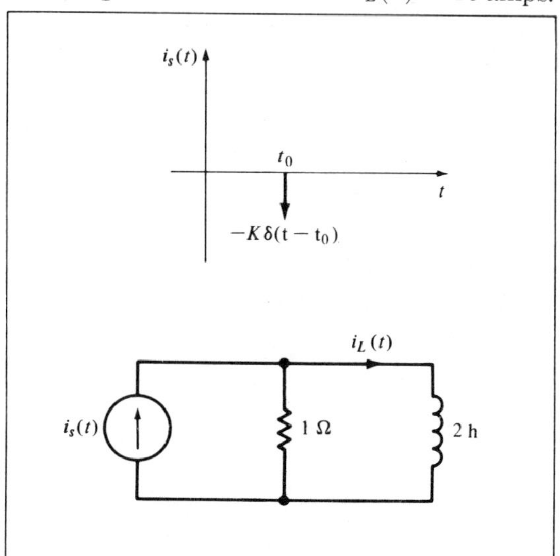

FIGURE P 8.18

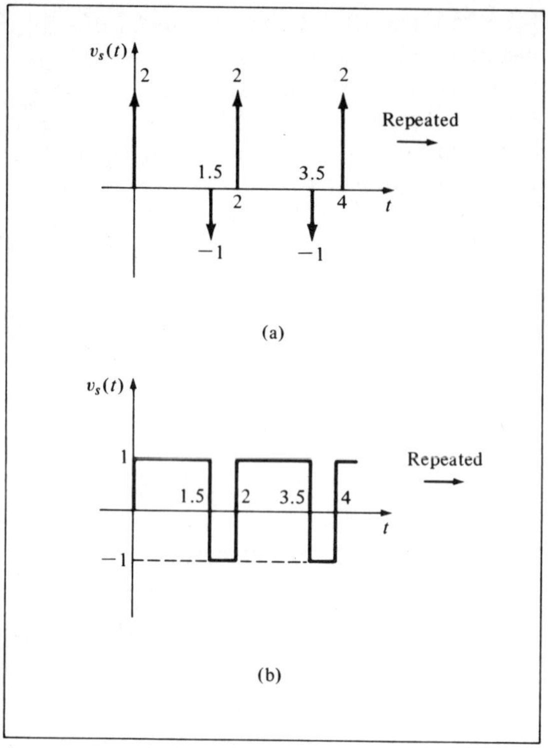

FIGURE P 8.19

We apply a negative-valued impulse at time t_0 so that $i_L(t) = 0$ for $t > t_0$. Find t_0.

8.21 (a–b) The repetitive signals of Figure P 8.19 are applied to the RC network of Figure P 8.17. Find the zero-state response for $v_C(t)$ for $t > 0$. Let $R = C = 1$.

REFERENCES AND ADDITIONAL READING

8.1 M. E. VAN VALKENBURG, *Network Analysis.* Englewood Cliffs, N.J.: Prentice-Hall, Inc., 1964, Chapter 4.

8.2 W. H. HAYT, JR. AND G. W. HUGHES, *Introduction to Electrical Engineering.* New York: McGraw-Hill Book Company, Inc., 1968, Chapter 7.

8.3 W. H. HAYT, JR. AND J. E. KEMMERLY, *Engineering Circuit Analysis.* New York: McGraw-Hill Book Company, Inc., 1963.

8.4 R. E. SCOTT, *Elements of Linear Circuits.* Reading, Mass.: Addison-Wesley Publishing Company, Inc., 1960, Chapter 8.

9

Second-Order Networks; Convolution

9.1 *RLC* Network Analysis

Our study has been restricted to networks containing a single energy-storage element, termed first-order networks. As we have seen, the response of such networks involves exponential functions of time. If we consider networks containing two energy-storage elements—for example, an inductor and a capacitor—the solution process becomes more complicated. Although solutions containing exponentially decaying terms do appear, other forms of response are possible. By acquaintance with all of these possible responses, we will be in an excellent position later to analyze networks with any number of energy-storage elements.

9.2 Equation Formulation

We will examine as a model problem the series *RLC* network of Figure 9.1. We can write the relationships for each element as

$$i_C = i = C\,\frac{dv_C}{dt} \qquad (9.1)$$

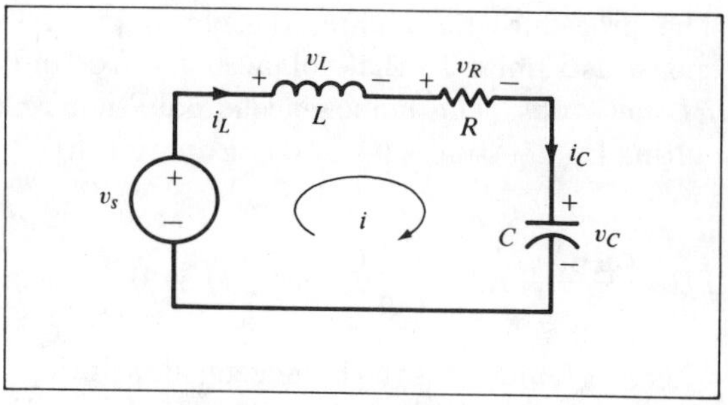

FIGURE 9.1 Series *RLC* network.

$$v_R = Ri = RC\,\frac{dv_C}{dt} \qquad (9.2)$$

$$v_L = L\,\frac{di}{dt} = LC\,\frac{d^2v_C}{dt^2} \qquad (9.3)$$

Writing KVL around the single loop yields

$$LC\,\frac{d^2v_C}{dt^2} + RC\,\frac{dv_C}{dt} + v_C = v_s(t) \qquad (9.4)$$

This is a second-order, constant-coefficient, linear ordinary differential equation in the unknown $v_C(t)$. From our knowledge of differential equations, we realize that *two* initial conditions, $v_C(0)$ and $(dv_C(t)/dt)_{t=0}$, must be known to obtain a solution. From our experience with RC networks, we can accept the requirement that $v_C(0)$ must be known. Further, from Equation (9.1), we can write

$$\left.\frac{dv_C(t)}{dt}\right|_{t=0} = \left.\frac{i_C(t)}{C}\right|_{t=0} = \frac{i_L(0)}{C} \tag{9.5}$$

Therefore, if we know $i_L(0)$ (again a reasonable requirement), we can easily determine the derivative $(dv_C(t)/dt)_{t=0}$. Note that $i_L(0)$ and $v_C(0)$ in turn determine the energy stored in the network at $t = 0$. We now show that from knowledge of (1) this initial energy storage and (2) the excitation $v_s(t)$ $(t \geqq 0)$ we can completely determine the solution variable $v_C(t)$ for $t \geqq 0$.

Zero-Input Response

The procedure for finding the zero-input response is similar to that followed for RC and RL networks. The homogeneous equation resulting from setting $v_s(t) = 0$ in Equation (9.4) is

$$LC\frac{d^2v_{C_h}(t)}{dt^2} + RC\frac{dv_{C_h}(t)}{dt} + v_{C_h}(t) = 0 \tag{9.6}$$

where $v_{C_h}(t)$ denotes the homogeneous solution. If we again let $v_{C_h}(t) = Ke^{st}$ where both K and s are to be determined, substitution into Equation (9.6) gives

$$(LCs^2 + RCs + 1)Ke^{st} = 0 \tag{9.7}$$

To satisfy Equation (9.7) for all t, we require

$$LCs^2 + RCs + 1 = 0 \tag{9.8}$$

This equation is termed the *characteristic equation* of the network; it has two solutions for s, namely,

$$s_{1,2} = -\frac{R}{2L} \pm \sqrt{\left(\frac{R}{2L}\right)^2 - \frac{1}{LC}} \tag{9.9}$$

Accordingly, the homogeneous solution must include terms of the form $K_1e^{s_1t}$ *and* $K_2e^{s_2t}$. Because the differential equation is linear, we write

$$v_{C_h}(t) = K_1e^{s_1t} + K_2e^{s_2t} \tag{9.10}$$

The constants s_1 and s_2 are known as the *characteristic*, or *natural*, *frequencies* of the network, because, being part of the homogeneous solution, they will appear in the solution for *any* excitation $v_s(t)$. The values for K_1 and K_2 are resolved by inserting the two initial conditions $v_C(0)$ and $dv_C(t)/dt)_{t=0}$.

$$v_C(0) = K_1 + K_2 \tag{9.11}$$

$$\left.\frac{dv_C(t)}{dt}\right|_{t=0} = s_1K_1 + s_2K_2 = \frac{i_L(0)}{C} \tag{9.12}$$

Solving these simultaneous equations for K_1 and K_2 gives

$$K_1 = \frac{1}{s_2 - s_1}\left(s_2v_C(0) - \frac{i_L(0)}{C}\right) \tag{9.13}$$

$$K_2 = \frac{1}{s_1 - s_2}\left(s_1v_C(0) - \frac{i_L(0)}{C}\right) \tag{9.14}$$

Equations (9.10), (9.13), and (9.14) represent the complete solution of the zero-input case for $v_C(t)$. The inductor current $i_{L_h}(t)$ is obtained from

$$i_{L_h}(t) = i(t) = \frac{C\,dv_{C_h}(t)}{dt}$$

$$= CK_1s_1e^{s_1t} + CK_2s_2e^{s_2t}$$

It is left as an exercise to show that

$$i_{L_h}(t)|_{t=0} = i_L(0)$$

using Equations (9.13) and (9.14).

Example 9.1

We are to determine $v_C(t)$ $(t \geqq 0)$ in Figure 9.2 for (a) $R = 3\Omega$, (b) $R = 2\Omega$, (c) $R = 1\Omega$.

(*a*) $R = 3\Omega$.

From Equation (9.8), the characteristic

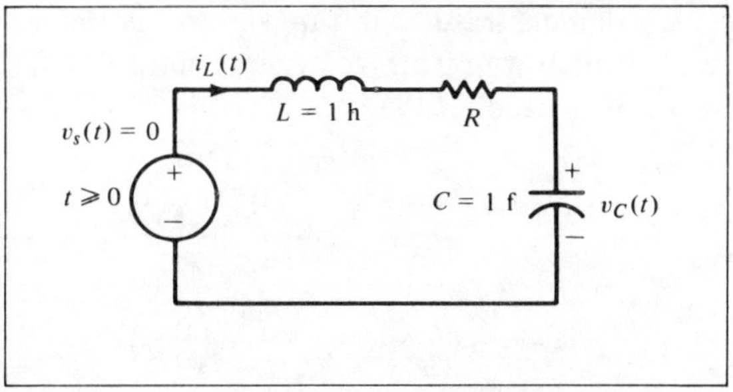

FIGURE 9.2 Zero-input *RLC* network example. $v_C(0) = 0$; $i_L(0) = 1$ amp.

equation is

$$s^2 + 3s + 1 = 0$$

with real solutions

$$\begin{aligned} s_{1,2} &= -1.5 \pm \sqrt{(1.5)^2 - 1} \\ s_1 &= -0.382 \qquad s_2 = -2.618 \end{aligned} \tag{9.15}$$

From Equations (9.13) and (9.14), the K's are determined from

$$K_1 = \frac{1}{-2.236}(-2.618(0) - 1) = 0.4474$$

$$K_2 = \frac{1}{2.236}(-0.382(0) - 1) = -0.4474$$

and the complete solution is

$$v_C(t) = 0.4474e^{-0.382t} - 0.4474e^{-2.618t} \qquad (t \geqq 0) \tag{9.16}$$

This response is plotted in Figure 9.3; we will discuss certain of its characteristics shortly.

(*b*) $R = 2\Omega$.

The natural frequencies are

$$s_{1,2} = -1$$

so that both roots are equal. In evaluating

$$v_C(t) = \frac{-i_L(0)/C}{s_2 - s_1}(e^{s_1 t} - e^{s_2 t})$$

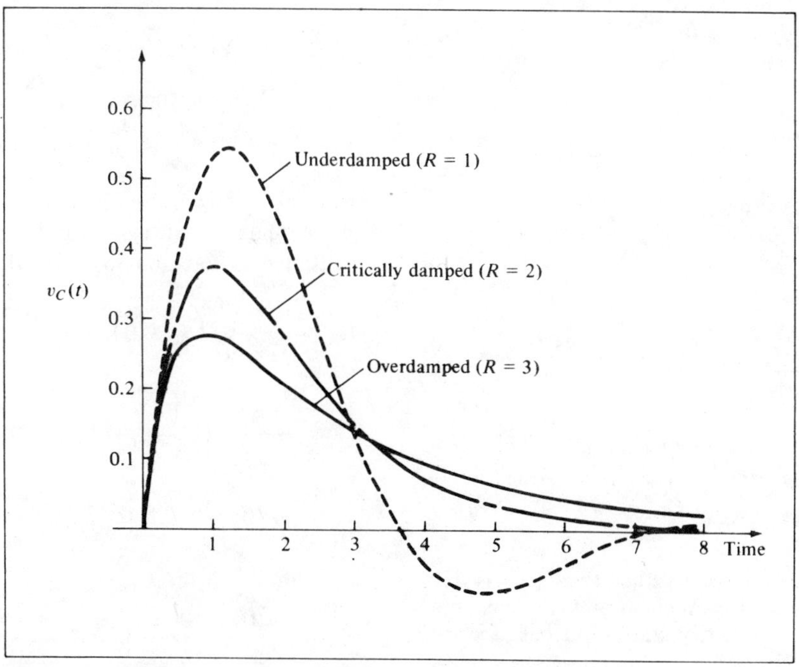

FIGURE 9.3 Zero-input response as a function of R.

for $s_1 = s_2$, we find that the expression becomes indeterminant. Applying L'Hôpital's rule, however,

$$v_C(t) = \left(\frac{-i_L(0)}{C}\right) \lim_{s_1 \to s_2} \frac{(d/ds_1)(e^{s_1 t} - e^{s_2 t})}{(d/ds_1)(s_2 - s_1)}$$

$$= \left(\frac{-i_L(0)}{C}\right) \lim_{s_1 \to s_2} (-te^{s_1 t})$$

$$= \frac{i_L(0)}{C} te^{s_2 t}$$

For the values of this problem, we have

$$v_C(t) = te^{-t} \qquad (t \geqq 0) \tag{9.17}$$

This response is plotted in Figure 9.3.

(c) $R = 1\Omega$.

The natural frequencies are[1]

$$s_{1,2} = -\tfrac{1}{2} \pm \sqrt{(0.5)^2 - 1}$$
$$= -\tfrac{1}{2} \pm j\frac{\sqrt{3}}{2} \tag{9.18}$$

and the K's are

$$K_1 = \frac{1}{j\sqrt{3}} \qquad K_2 = \frac{1}{-j\sqrt{3}}$$

The solution is now

$$v_C(t) = \frac{1}{j\sqrt{3}} e^{(-1/2 + j\sqrt{3}/2)t} + \frac{1}{-j\sqrt{3}} e^{(-1/2 - j\sqrt{3}/2)t}$$

This solution is not recognizable as belonging to a physical system where responses must be *real-valued* functions of time.

To demonstrate that the solution is indeed real-valued, we return to general notation. We write Equation (9.18) as

$$s_{1,2} = -\frac{R}{2L} \pm \sqrt{\left(\frac{R}{2L}\right)^2 - \frac{1}{LC}}$$

$$= -\frac{R}{2L} \pm j\sqrt{\frac{1}{LC} - \left(\frac{R}{2L}\right)^2}$$

$$= -\alpha \pm j\omega_d \tag{9.19}$$

so that $-\alpha$ and $\pm\omega_d$ are the real and imaginary parts of the roots. Then the solution becomes

$$v_C(t) = K_1 e^{s_1 t} + K_2 e^{s_2 t}$$
$$= K_1 e^{(-\alpha + j\omega_d)t} + K_2 e^{(-\alpha - j\omega_d)t} \tag{9.20}$$
$$= e^{-\alpha t}(K_1 e^{j\omega_d t} + K_2 e^{-j\omega_d t})$$

To proceed further, we use the Euler identity justified in Appendix B

$$e^{j\theta} = \cos\theta + j\sin\theta$$

Equation (9.20) then becomes

$$v_C(t) = e^{-\alpha t}(K_1(\cos\omega_d t + j\sin\omega_d t) + K_2(\cos\omega_d - j\sin\omega_d t)) \tag{9.21}$$
$$= e^{-\alpha t}(A\cos\omega_d t + B\sin\omega_d t)$$

where $A = K_1 + K_2$, $B = j(K_1 - K_2)$. To show that A and B are real constants, we evaluate Equations (9.13) and (9.14) as

$$K_1 = \frac{1}{s_2 - s_1}\left(s_2 v_C(0) - \frac{i_L(0)}{C}\right)$$

$$= \frac{1}{-j2\omega_d}\left[(-\alpha - j\omega_d)v_C(0) - \frac{i_L(0)}{\mathrm{C}}\right]$$

$$= \frac{v_C(0)}{2} - j\frac{1}{2\omega_d}\left(\alpha v_C(0) + \frac{i_L(0)}{C}\right)$$

$$K_2 = \frac{v_C(0)}{2} + j\frac{1}{2\omega_d}\left(\alpha v_C(0) + \frac{i_L(0)}{C}\right) = K_1^*$$

[1] In network analysis, j rather than i is used to denote the imaginary component of a complex number, because i is used to indicate current. See Appendix B.

Then

$$A = K_1 + K_2 = v_C(0) \qquad (9.22)$$

$$B = j(K_1 - K_2) = \frac{1}{\omega_d}\left(\alpha v_C(0) + \frac{i_L(0)}{C}\right) \qquad (9.23)$$

which are both real-valued. The solution is therefore of the form

$$v_C(t) = e^{-\alpha t}\left[v_C(0)\cos\omega_d t + \frac{1}{\omega_d}\left(\alpha v_C(0) + \frac{i_L(0)}{C}\right)\sin\omega_d t\right] \qquad (9.24)$$

Returning to the example, where from Equation (9.18)

$$\alpha = \frac{1}{2} \qquad \omega_d = \frac{\sqrt{3}}{2}$$

we have from Equations (9.22) and (9.23)

$$A = 0$$

$$B = \frac{2}{\sqrt{3}}$$

and the complete solution is

$$v_C(t) = \frac{2}{\sqrt{3}} e^{-\frac{1}{2}t}\sin\left(\tfrac{3}{2}t\right) \qquad (t \geqq 0) \qquad (9.25)$$

The three values of R in the above example result in three quite different forms of responses. These are shown in Figure 9.3. Based on the form of these responses, it is common to generalize the above results as follows:

1. *Overdamped case* (s_1 and s_2 are real and distinct). If $R^2 > 4L/C$, the quantity $(R/2L)^2 - 1/LC$ in Equation (9.9) is positive and s_1 and s_2 are both real and negative; that is,

$$s_1 = -\alpha_1 + j0 \qquad s_2 = -\alpha_2 + j0 \qquad (9.26)$$

where

$$\alpha_{1,2} = \frac{R}{2L} \pm \sqrt{\left(\frac{R}{2L}\right)^2 - \frac{1}{LC}} \qquad (9.27)$$

Figure 9.3 shows an example of the response; note that the solution rises rapidly from a zero initial value to a peak; this rise is due to the *faster* (smaller) time constant associated with the $e^{-2.618t}$ term of Equation (9.16). After this term has decayed, the term $e^{-0.382t}$ decays slowly as shown.

2. *Critically damped case* ($s_1 = s_2$, both real and negative). If $R^2 = 4L/C$, the radical is zero-valued and

$$s_{1,2} = -\frac{R}{2L} \qquad (9.28)$$

This response is more peaked than in the overdamped case.

3. *Underdamped case* (s_1 and s_2 are complex). If $R^2 < 4L/C$, we write the natural frequencies as

$$s_{1,2} = -\alpha \pm j\omega_d = -\frac{R}{2L} \pm j\sqrt{\frac{1}{LC} - \left(\frac{R}{2L}\right)^2} \qquad (9.29)$$

The resultant response is more peaked than the other responses and oscillates above and below the horizontal axis. This behavior is more evident in Figure 9.4(a), where the waveform is identified as a sinusoid oscillating within an envelope formed by the decaying exponential $e^{-\alpha t}$. Accordingly, we call this function a *decaying*, or *damped, sinusoid*. The frequency ω_d is called the *damped frequency*. If ω_d increases (by decreasing C, for example) the response will appear more oscillatory as in Figure 9.4(b), because the angle $\omega_d t$ will increase more rapidly with time. If α increases, the envelope will decay faster as shown in Figure 9.4(c).

Even when $i_L(0) \neq 0$, so that $B \neq 0$, we can recognize similar behavior by noting

$$\begin{aligned} e^{-\alpha t}(A\cos\omega_d t + B\sin\omega_d t) &= e^{-\alpha t}(C\cos\phi\cos\omega_d t - C\sin\phi\sin\omega_d t) \\ &= Ce^{-\alpha t}\cos(\omega_d t + \phi) \end{aligned} \qquad (9.30)$$

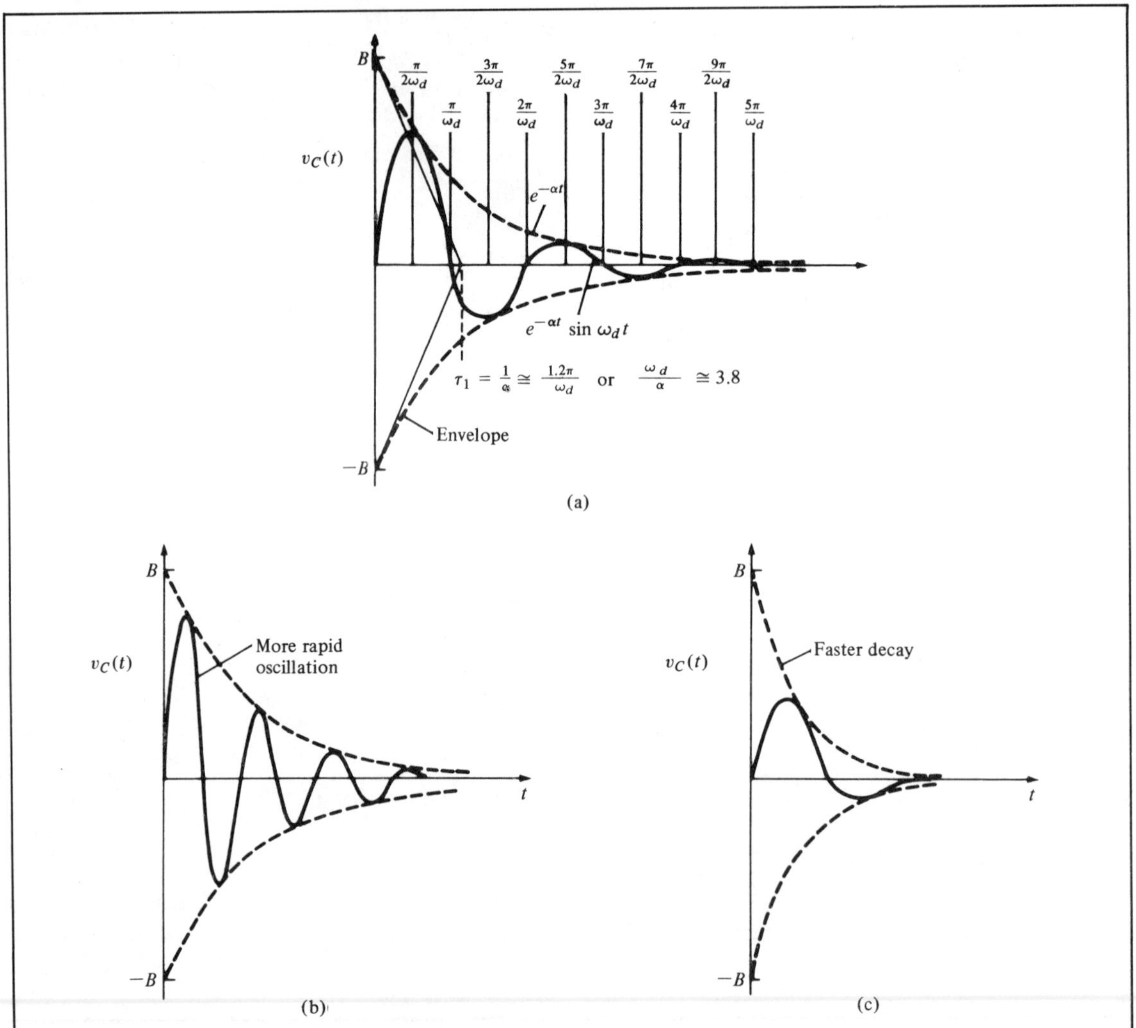

FIGURE 9.4 Damped sinusoid responses of the form $Be^{-\alpha t} \sin \omega_d t$. (a) Detailed plot; (b) ω_d doubled from (a); $\frac{\omega_d}{\alpha} = 7.6$; (c) α doubled from (a); $\omega_d/\alpha = 1.9$.

where

$$A = C \cos \phi \qquad B = -C \sin \phi$$

or

$$C = \sqrt{A^2 + B^2} \qquad \phi = -\tan^{-1}\left(\frac{B}{A}\right)$$

Again, we have a damped sinusoidal function, but shifted by ϕ rad or ϕ/ω_d sec.

Constant Input

If $v_s(t) = V_s$ $(t > 0)$, the differential equation becomes

$$LC \frac{d^2 v_C(t)}{dt^2} + RC \frac{dv_C(t)}{dt} + v_C(t) = V_s \quad (t > 0) \tag{9.31}$$

The particular solution satisfying this equation

then has the form $v_{C_p}(t) = V_s$ for $t > 0$, just as for the RC network of Figure 8.5. Equation (9.10) then becomes

$$v_C(t) = K_1 e^{s_1 t} + K_2^{s_2 t} + V_s \qquad (9.32)$$

Matching Equation (9.32) to the initial conditions gives

$$K_1 = \frac{1}{s_2 - s_1}\left[s_2(v_C(0) - V_s) - \frac{i_L(0)}{C}\right] \qquad (9.33)$$

$$K_2 = \frac{1}{s_1 - s_2}\left[s_1(v_C(0) - V_s) - \frac{i_L(0)}{C}\right] \qquad (9.34)$$

This response differs from the zero-input case only in that $v_C(0) - V_s$ of Equations (9.33) and (9.34) replaces $v_C(0)$ in Equations (9.13) and (9.14), and a constant V_s appears in Equation (9.32). Therefore, the solution proceeds identically to the zero-input case. For example, from Equation (9.24), we can write directly the solution of the underdamped case as

$$v_C(t) = e^{-\alpha t}\left[(v_C(0) - V_s)\cos\omega_d t + \frac{1}{\omega_d}\left(\alpha(v_C(0) - V_s) + \frac{i_L(0)}{L}\right)\sin\omega_d t\right] + V_s \ (t > 0) \qquad (9.35)$$

Unit Step and Impulse Responses

If $v_s(t) = u(t)$ and $v_C(0) = i_L(0) = 0$, the solution is the unit step response $v_C(t) = s(t)$. This is determined from Equation (9.32) to (9.34) as

$$s(t) = \left(\frac{s_2 e^{s_1 t}}{s_1 - s_2} + \frac{s_1 e^{s_2 t}}{s_2 - s_1} + 1\right)u(t) \qquad (9.36)$$

and yields for the overdamped case

$$s(t) = \left(\frac{\alpha_2}{\alpha_1 - \alpha_2}e^{-\alpha_1 t} + \frac{\alpha_1}{\alpha_2 - \alpha_1}e^{-\alpha_2 t} + 1\right)u(t) \qquad (9.37)$$

For the underdamped case we set $v_C(0) = i_L(0) = 0$ and $V_s = 1$ in Equation (9.35) to obtain

$$s(t) = \left(e^{-\alpha t}\left(-\cos\omega_d t - \frac{\alpha}{\omega_d}\sin\omega_d t\right) + 1\right)u(t) \qquad (9.38)$$

$$= \left(1 - \frac{\sqrt{\alpha^2 + \omega_d^2}}{\omega_d}e^{-\alpha t}\right)\cos\left(\omega_d t + \tan^{-1}\left(\frac{-\alpha}{\omega_d}\right)\right)u(t) \qquad (9.39)$$

If $v_s(t) = \delta(t)$, the impulse response is found from differentiating the step response, because $h(t) = ds(t)/dt$. This gives, for the overdamped case,

$$h(t) = \frac{\alpha_1\alpha_2}{\alpha_2 - \alpha_1}(e^{-\alpha_1 t} - e^{-\alpha_2 t})u(t) \qquad (9.40)$$

and for the underdamped case

$$h(t) = \left(\frac{\alpha^2 + \omega_d^2}{\omega_d}e^{-\alpha t}\sin\omega_d t\right)u(t) \qquad (9.41)$$

In the following example, we study the underdamped response in more detail.

Example 9.2

The unit step and impulse responses of the network of Figure 9.5 are to be found and displayed for $R = 2$, 1, 0.5, and 0.2.

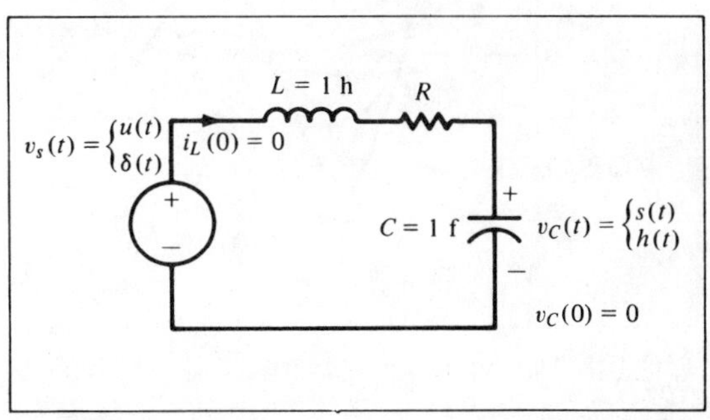

FIGURE 9.5 *RLC* series network with step and impulse excitation. $v_C(0) = 0$.

As an example, we choose $R = 1$. As we have shown, the natural frequencies are

$$s_{1,2} = -\frac{1}{2} \pm j\frac{\sqrt{3}}{2}$$

From Equations (9.39) and (9.41), the responses are

$$s(t) = \left(e^{-1/2t}\left(-\cos\frac{\sqrt{3}}{2}t - \frac{1}{\sqrt{3}}\sin\frac{\sqrt{3}}{2}t\right) + 1\right)u(t) = \left(1 - \frac{2}{\sqrt{3}}e^{-1/2t}\cos\left(\frac{\sqrt{3}}{2}t - 30°\right)\right)u(t)$$

$$h(t) = \frac{2}{\sqrt{3}}e^{-1/2t}\sin\left(\frac{\sqrt{3}}{2}t\right)u(t)$$

These solutions and solutions for the other values of R are displayed in Figure 9.6. Note that both the amplitude of the transient and the envelope of the decay are quite dependent on R, but the damped frequency ω_d remains relatively constant. The reader may wish to justify this observation by examining the solution when R is left as a variable.

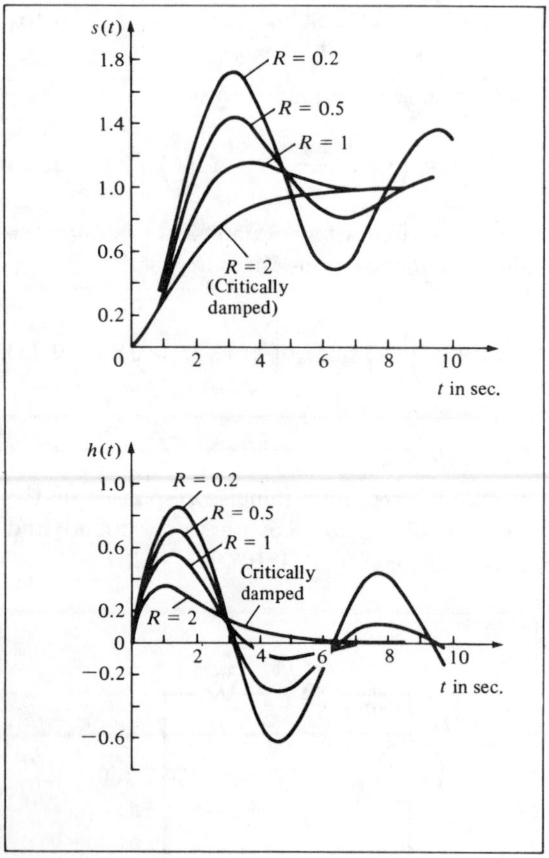

FIGURE 9.6 Step and impulse responses for Example 9.2.

9.3 Solution of Parallel *GCL* Network

From our observations in Chapter 8 about the similarity of *RC* and *RL* network equations, we can expect that the series *RLC* network is not the only dynamic network described by a second-order differential equation. For example, the parallel *GCL* network of Figure 9.7 is

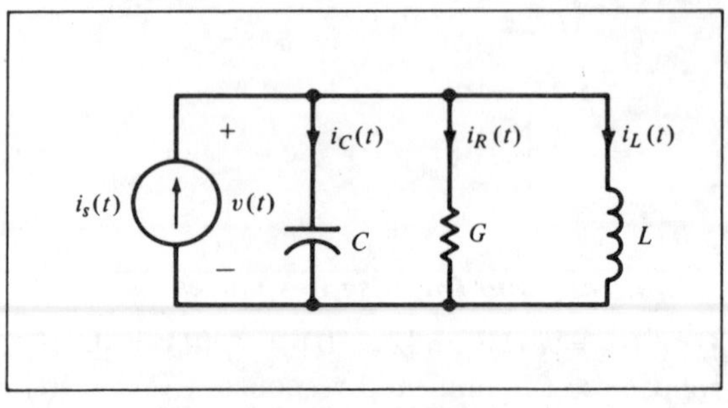

FIGURE 9.7 Parallel *GCL* network.

described by the KCL equation

$$i_C(t) + i_G(t) + i_L(t) = i_s(t)$$

Substituting $i_C = C\,dv/dt = LC\,d^2i_L/dt^2$ and $i_G = Gv = GL\,di_L/dt$ gives

$$LC\frac{d^2i_L(t)}{dt^2} + GL\frac{di_L(t)}{dt} + i_L(t) = i_s(t)$$

This equation is identical in form to the KVL

equation for a series RLC network, but the solution variable is now $i_L(t)$ rather than $v_C(t)$. All the methods used to analyze the series RLC network can be applied to this case.

9.4 Convolution

Introduction

In comparing the step and impulse responses in Chapter 8, it was mentioned that the latter is more commonly used as a mathematical description of the dynamic response of the system. In this section, we show, in particular, that *from the impulse response, we can find the response to any bounded excitation.* The mechanism for determining this response is the *convolution integral.*

Derivation of the Integral

Consider the waveforms of Figure 9.8. Here, $e(t)$ is an excitation and $e^*(t)$ is a "staircase" approximation to $e(t)$. We represent this approximation as

$$e^*(t) = \sum_{k=-\infty}^{\infty} e(k\Delta)(u(t-k\Delta) - u(t-k\Delta-\Delta)) \qquad (9.42)$$

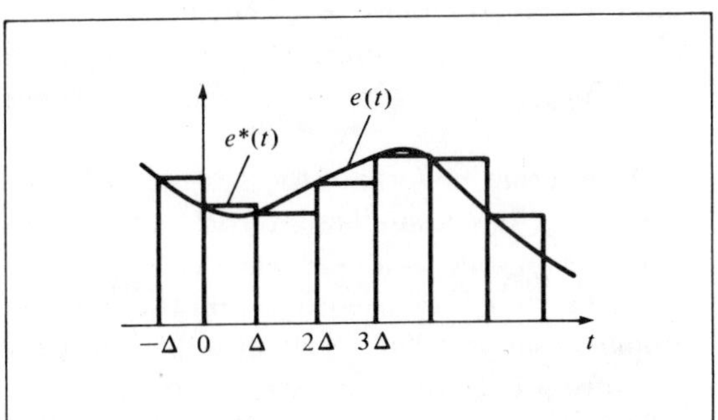

FIGURE 9.8 Excitation approximated by pulses.

Because the excitation $e^*(t)$ is represented in terms of step functions, the approximate response $r^*(t)$ can be represented in terms of the step response $s(t)$.

$$r^*(t) = \sum_{k=-\infty}^{\infty} e(k\Delta)(s(t-k\Delta) - s(t-k\Delta-\Delta))$$

$$= \sum_{k=-\infty}^{\infty} e(k\Delta)\left(\frac{s(t-k\Delta) - s(t-k\Delta-\Delta)}{\Delta}\right)\Delta \qquad (9.43)$$

If $\Delta \to 0$ and $k\Delta \to \tau$,

$$\frac{s(t-k\Delta) - s(t-k\Delta-\Delta)}{\Delta} \to \frac{ds(t-\tau)}{dt} \to h(t-\tau) \qquad (9.44)$$

provided $s(t)$ is continuous. As we pass to this limit, the quantities in Equation (9.43) become

$$k\Delta \to \tau \qquad \Delta \to d\tau$$

$$e(k\Delta) \to e(\tau) \qquad r^*(t) \to r(t)$$

Therefore the summation becomes an integral

$$r(t) = \int_{-\infty}^{\infty} e(\tau)h(t-\tau)\,d\tau \qquad (9.45)$$

By a simple change in variable $\xi = t - \tau$, we can also write

$$r(t) = \int_{-\infty}^{\infty} e(t-\xi)h(\xi)\,d\xi$$

$$= \int_{-\infty}^{\infty} e(t-\tau)h(\tau)\,d\tau \qquad (9.46)$$

because τ and ξ are dummy variables of integration.

Now consider the case where $e(t) = 0$ for $t < 0$. We assume that this implies $r(t) = h(t) = 0$ for $t < 0$. Such a system is termed a *causal* system, implying that a response cannot occur before an excitation is applied. Therefore, the lower limit of Equations (9.45) and (9.46) can be moved to $\tau = 0$ and the upper limit to $\tau = t$, giving

$$r(t) = \int_0^t e(\tau)h(t-\tau)\,d\tau \qquad (9.47)$$

$$= \int_0^t e(t-\tau)h(\tau)\,d\tau \qquad (9.48)$$

These integrals are both known as convolution integrals; they relate the response to the excitation through an integral involving the impulse response.

Example 9.3

We are to use convolution to find $v_C(t)$ for $t \geqq 0$ in Figure 9.9 when $v_s(t) = tu(t)$.

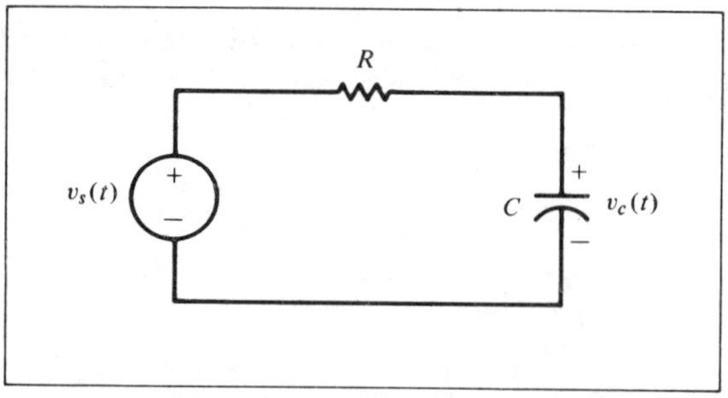

FIGURE 9.9 Network example to illustrate convolution.

We have already shown that the impulse response is [Equation (8.54) with $I_s = 1/R$]

$$h(t) = \frac{1}{RC}e^{-t/RC} \qquad (9.49)$$

For the excitation $e(t) = t$, the convolution integral of Equation (9.47) becomes

$$r(t) = \int_0^t e(\tau)h(t-\tau)\,d\tau$$

$$= \frac{1}{RC}\int_0^t \tau e^{-(t-\tau)/RC}\,d\tau$$

$$= \frac{e^{-t/RC}}{RC}\int_0^t \tau e^{\tau/RC}\,d\tau$$

$$= \frac{e^{-t/RC}}{RC}\left((RC)^2 e^{\tau/RC}\left(\frac{\tau}{RC}-1\right)\right)_0^t$$

$$= RCe^{-t/RC} + t - RC$$

$$= RC(e^{-t/RC} - 1) + t \qquad (t \geqq 0) \quad (9.50)$$

The convolution integral can be used to add to our understanding of dynamic response characteristics. As an example, in Figure 9.10, we construct a graphical representation of the integral involved in Example 9.3, namely,

$$r(t) = \int_0^t \underbrace{(\tau)}_{e(\tau)} \underbrace{(e^{-(t-\tau)/RC})}_{h(t-\tau)}\,d\tau \qquad (9.51)$$

To evaluate this integral, we must multiply $e(\tau)$ and $h(t-\tau)$ for each value of t and then find the area under the product for $0 \leqq \tau \leqq t$.

We first develop in three steps the plot of $h(t-\tau)$ in Figure 9.10(a)–(c). The resultant $h(t-\tau)$, when plotted versus the integration variable τ, is observed to be a shifted, mirrored image of $h(t)$.

The next step is to construct both $h(t-\tau)$ and $e(\tau)$ on the same plot, as shown in Figure 9.10(d). Here, the value of the integrand $e(\tau)h(t_i-\tau)$ is shown for three different values of t_i $(0, t_1, t_2)$. The area under the integrand as a function of τ is the value of the response $r(t_i)$. From the figure, we may view the impulse response as "moving through" the input as time t increases. We have stopped the "motion" for viewing at the three times shown.

The alternative form of the integral in Equation (9.48) has a similar interpretation; however, the product $e(t-\tau)h(\tau)$ is now interpreted as "moving" the excitation through the impulse response. The above example is shown in Figure 9.10(e) in this form.

The imagery associated with this moving process can be further developed with the

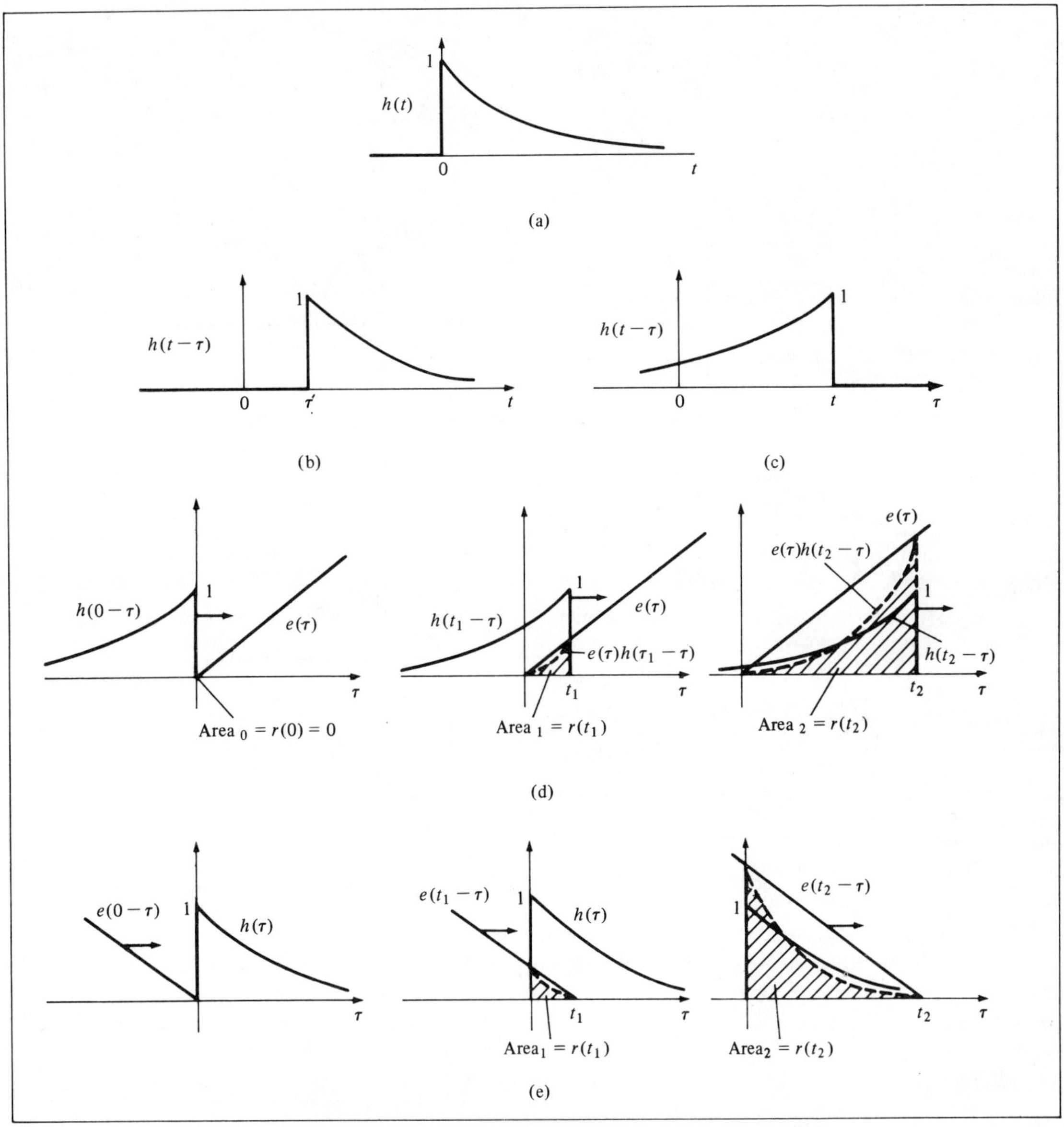

FIGURE 9.10 Representation of convolution integral. (a) Impulse response $h(t)$; (b) $h(t-\tau)$ as a function of t; (c) $h(t-\tau)$ as a function of τ; (d) convolution using Equation (9.47); (e) convolution using Equation (9.48).

examples shown in Figure 9.11. In Figure 9.11(a), an impulse delayed by t_0 is applied to a network with a damped sinusoidal impulse response. As the impulse $\delta(t - t_0 - \tau)$ moves along the τ-axis with increasing t, a zero-valued result is produced until $t = t_0$, since $h(\tau) = 0$ for $\tau < 0$. Then we have

$$r(t) = \int_0^t \delta(t - \tau - t_0)h(\tau)\, d\tau$$

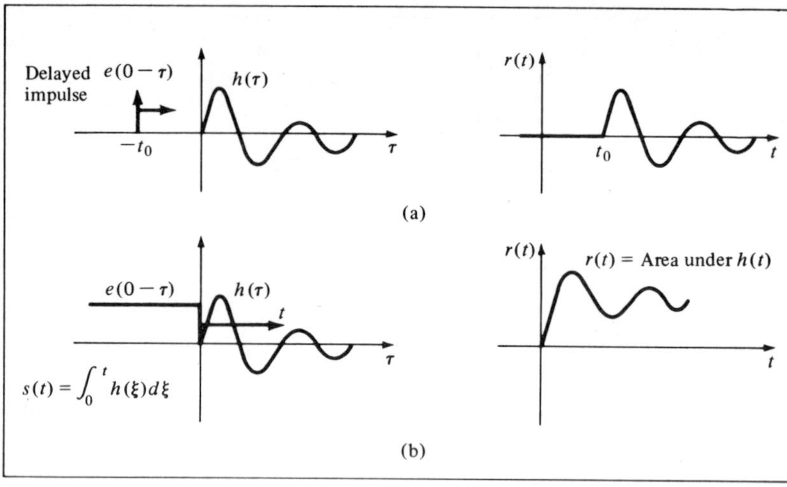

FIGURE 9.11 Example uses of convolution.

As a function of τ, the integrand is zero-valued except at $\tau = t - t_0$; at this value of τ, $h(\tau) = h(t-t_0)$ which is a constant. We can therefore remove this quantity from the integral to give

$$r(t) = h(t - t_0) \int_0^t \delta(t - \tau - t_0)\, d\tau \tag{9.52}$$
$$= h(t - t_0) \qquad (t > t_0)$$

because the integral of an impulse is unity. It follows from Equation (9.52) that the response is a shifted version of the impulse response. We witnessed a similar result in Example 8.3.

The step response is already known to be the integral of the impulse response. This is confirmed in Figure 9.11(b) by moving a unity-valued constant waveform through the impulse response as shown. The convolution integral then yields

$$s(t) = \int_0^t h(\tau)\, d\tau$$

as required.

Complete Response with Arbitrary Input

We have noted that the complete response is the sum of the zero-input and zero-state solutions. We can now evaluate from the convolution integral the zero-state response to any input; therefore, we can write the complete response with initial conditions as

$$r(t) = (\text{Zero-input response}) + \int_0^t e(\tau)h(t - \tau)\, d\tau \tag{9.53}$$

Let us apply this formula to the first-order networks of Chapter 8. These networks all are described by

$$\dot{r}(t) = \alpha r(t) + e(t)$$

with $r(0)$ given. The zero-input response is written as

$$r(t) = r(0)e^{\alpha t} \qquad (t > 0)$$

and the impulse response as

$$h(t) = e^{\alpha t} \qquad (t > 0)$$

Therefore, from Equation (9.53),

$$r(t) = r(0)e^{\alpha t} + \int_0^t e(\tau)e^{\alpha(t-\tau)}\, d\tau$$

$$= r(0)e^{\alpha t} + e^{\alpha t}\int_0^t e(\tau)e^{-\alpha\tau}\, d\tau \qquad (t > 0)$$

If the initial condition is given at $t = t_0$ rather

than $t = 0$, we can write

$$r(t) = r(t_0)e^{\alpha(t-t_0)} + e^{\alpha t}\int_{t_0}^{t} e(\tau)e^{-\alpha\tau}\,d\tau \qquad (t > t_0) \qquad (9.54)$$

Example 9.4

The RC network of Example 9.3 with an initial condition $v_C(0)$ has a negative ramp applied at t_0 sec. The response $v_C(t_0)$ is evaluated from Equation (9.50) as (see Figure 9.12)

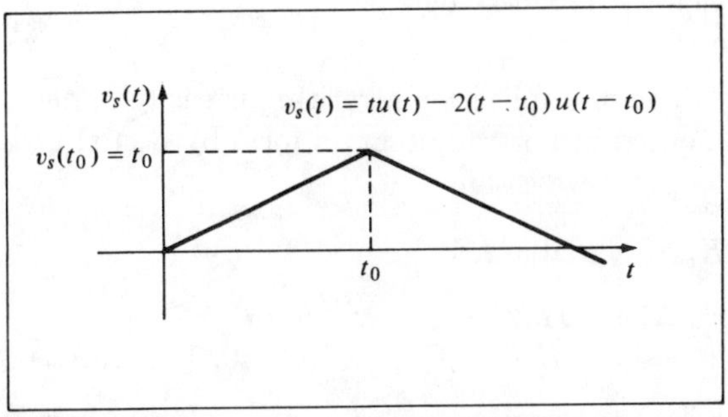

FIGURE 9.12 Excitation for RC network of Figure 9.9.

$$v_C(t_0) = RC(e^{-t_0/RC} - 1) + t_0$$

We are to find the response for $t > t_0$.

We find directly from Equation (9.54)

$$\begin{aligned} v_C(t) &= v_C(t_0)e^{-(t-t_0)/RC} \\ &\quad + e^{-t/RC}\int_{t_0}^{t} (2t_0 - \tau)e^{\tau/RC}\,d\tau \\ &= v_C(t_0)e^{-(t-t_0)/RC} \\ &\quad + e^{-t/RC}\left((2RCt_0 - RC\tau)e^{\tau/RC} + (RC)^2e^{\tau/RC}\right)_{t_0}^{t} \\ &= v_C(t_0)e^{-(t-t_0)/RC} + RC(2t_0 - t) + (RC)^2 \\ &\quad - (RCt_0 + (RC)^2)e^{-(t-t_0)/RC} \end{aligned}$$

Note that when $t = t_0$, the solution becomes simply $v_C(t_0)$.

APPLICATION

Filtering a Noisy Signal

Apart from its value as a conceptual aid in relating input-output quantities, the convolution integral is useful when a signal does not have a simple analytic representation. This occurs in separating a signal from noise. For example, the signal and noise might represent the communication from a space probe; *samples* of this waveform would be stored on a magnetic tape and post-processed by a computer after reception to eliminate as much as possible the noise inherent in such a transmission.

When the input is given in such tabular or sampled form, the convolution integral must be evaluated approximately. Representing the integrand as

$$e(\tau)h(t - \tau) = f(\tau) \qquad (9.55)$$

we must find the value of

$$r(t) = \int_0^t f(\tau)\,d\tau$$

that is, the area under $f(\tau)$, when f is given at discrete values $f(k\Delta)$, $k = 0, 1, 2, \ldots$.

A number of methods exist for evaluating the area under a curve specified at equally spaced values of the abscissa. One of the simplest is the trapezoidal rule, illustrated in Figure 9.13. Between successive values of time $t_k = k\Delta$ and $t_{k+1} = (k + 1)\Delta$, we represent the function by the average value $\frac{1}{2}[f(k\Delta) + f((k + 1)\Delta)]$. The area under this rectangular approximation is then

$$\begin{aligned} \int_{k\Delta}^{(k+1)\Delta} f(\tau)\,d\tau &\cong \{\text{Height}\}\{\text{Base}\} \\ &= \{\tfrac{1}{2}[f(k\Delta) + f((k + 1)\Delta)]\}\{\Delta\} \end{aligned}$$

This formula is illustrated in Figure 9.13 for $k = 0$. For integration over an interval $(0, t) = (0, K\Delta)$, the areas under all rectangular

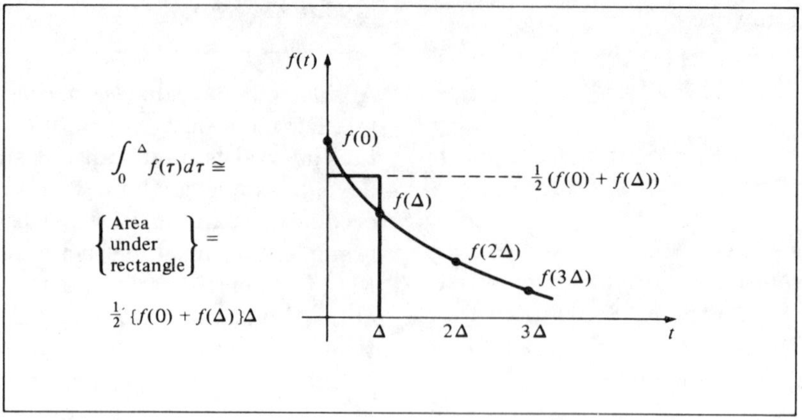

FIGURE 9.13 Example of integration using trapezoidal rule.

approximations add to yield the summation

$$\int_0^t f(\tau)\, d\tau \cong \sum_{k=0}^{(K-1)} \frac{1}{2}[f(k\Delta) + f((k+1)\Delta)]\Delta$$

$$= \frac{\Delta}{2} f(0) + \Delta \sum_{k=1}^{K-1} f(k\Delta) + \frac{\Delta}{2} f(K\Delta) \tag{9.56}$$

In terms of the convolution integral, we have, for $t = K\Delta$,

$$\int_0^t e(\tau)h(t-\tau)\, d\tau \cong \frac{\Delta}{2} e(0)h(K\Delta)$$

$$+ \Delta \sum_{k=1}^{K-1} e(k\Delta)h(K\Delta - k\Delta) + \frac{\Delta}{2} e(K\Delta)h(0) \tag{9.57}$$

This approximate response is designated $r^*(t)$.

The summation is easily organized into a computing algorithm as follows. Assume that we wish to determine $r^*(K\Delta)$ for $K = 0, 1, \ldots, K_m$. The values of $e(k\Delta)$ and $h(k\Delta)$ are first arranged as shown in Table 9.1. The summation of Equation (9.57) is formed from multiplication of the number pairs as if they represented digits in ordinary multiplication. The products at the endpoints are added to the result to give the approximate response function $r^*(K\Delta)$, as shown.

The multiplication algorithm is easily implemented in Fortran program form by nested DO loops. If we assign

K_m to variable KM

$e(k\Delta)$ to array position $E(k+1)$ $\quad (k = 0, 1, \ldots, K_m)$

$h(k\Delta)$ to array position $H(k+1)$ $\quad (k = 0, 1, \ldots, K_m)$

$r^*(k\Delta)$ to array position $R(k+1)$

then the multiplication algorithm of Table 9.1 is simply implemented with

```
      KMP1 = KM + 1
      DO 4 I = 2, KM
      DO 4 J = 2, KM
      JJ = I + J - 1
      IF(JJ. GT. KMP1) GO TO 4
      R(JJ) = R(JJ) + E(I)*H(J)
    4 CONTINUE
```

The response $r^*(k\Delta)$ cannot be calculated for $k > K_m$ because neither $e(k\Delta)$ nor $h(k\Delta)$ are assumed known for $k > K_m$. This is illustrated in Table 9.1 for $K_m = 4$, where $r^*(5\Delta)$ cannot

TABLE 9.1

Algorithm for Approximate Convolution

	$\cdots\ e(3\Delta)$	$e(2\Delta)$	$e(\Delta)$	
	$\cdots\ h(3\Delta)$	$h(2\Delta)$	$h(\Delta)$	
	$\cdots\ e(3\Delta)h(\Delta)$	$e(2\Delta)h(\Delta)$	$e(\Delta)h(\Delta)$	
	$\cdots\ e(2\Delta)h(2\Delta)$	$e(\Delta)h(2\Delta)$		
	$\cdots\ e(\Delta)h(3\Delta)$			
	$\cdots\ e(3\Delta)h(\Delta)$	$e(2\Delta)h(\Delta)$	$e(\Delta)h(\Delta)$	
$\sum_{k=1}^{K-1} e(k\Delta)h(K\Delta - k\Delta)$	$\cdots + e(2\Delta)h(2\Delta)$	$+e(\Delta)h(2\Delta)$		
	$\cdots + e(\Delta)h(3\Delta)$			
$+\frac{1}{2}e(0)h(K\Delta)$	$\cdots + \frac{1}{2}e(0)h(4\Delta)$	$+\frac{1}{2}e(0)h(3\Delta)$	$+\frac{1}{2}e(0)h(2\Delta)$	$+\frac{1}{2}e(0)h(\Delta)$
$+\frac{1}{2}e(K\Delta)h(0)$	$\cdots + \frac{1}{2}e(4\Delta)h(0)$	$+\frac{1}{2}e(3\Delta)h(0)$	$+\frac{1}{2}e(2\Delta)h(0)$	$+\frac{1}{2}e(\Delta)h(0)$
Response $r^*(K\Delta)$	$\cdots\ r^*(4\Delta)$	$r^*(3\Delta)$	$r^*(2\Delta)$	$r^*(\Delta)$
	$K = 4$	$K = 3$	$K = 2$	$K = 1$

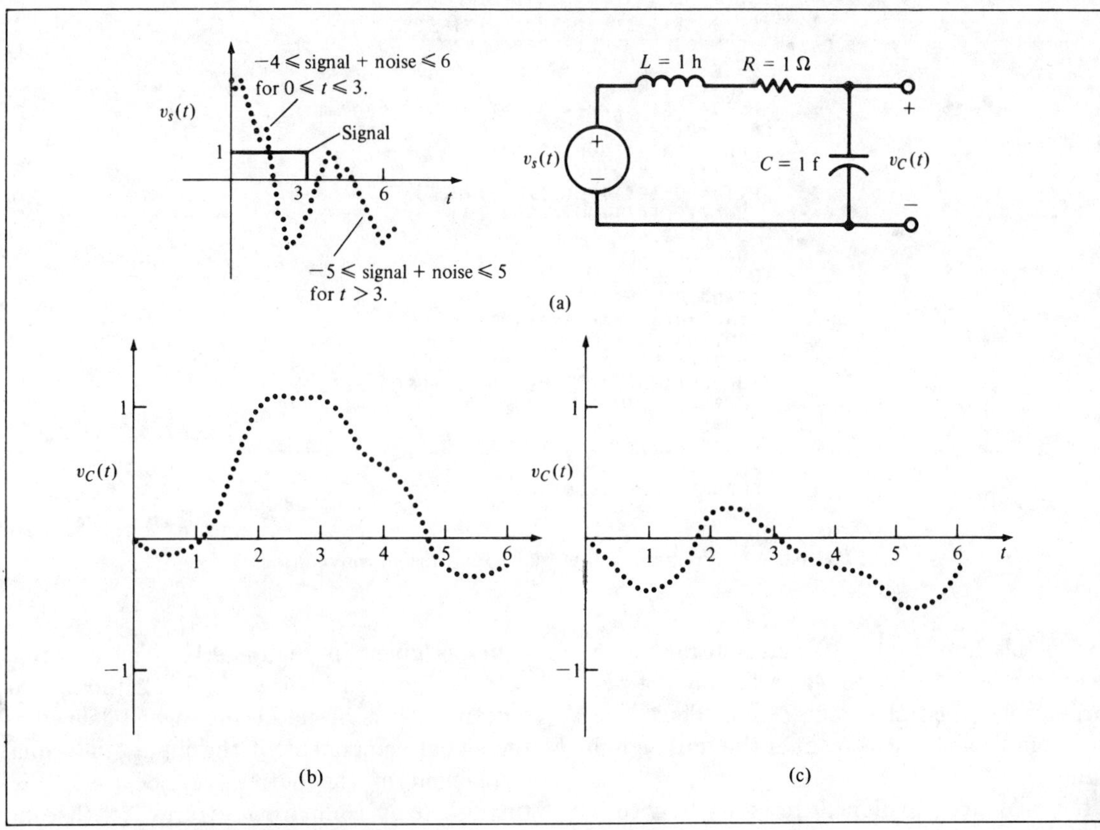

FIGURE 9.14 Filter and a noisy signal. (a) Example of signal and filter; (b) output with pulse; (c) output without pulse.

be found without knowledge of $e(5\Delta)$ and $h(5\Delta)$.

An interesting application of the algorithm is illustrated in Figure 9.14. Here a unit pulse is imbedded in a noisy background. The effect of noise is obtained by choosing numbers from a uniformly-distributed random number generator, available at most computing centers. These numbers have equal probability of having any value between zero and one. Multiplying this number by 10 and subtracting 5 yields a number with equal probability of occurrence between the values ± 5. Thus, the noise can be as large as 5 times the unit signal value.

It would be quite difficult to verify by visual inspection the presence of such a small signal amid the large noisy background. However, if we insert a *filter* such as the series *RLC* network of Figure 9.14(a) with impulse response

$$h(t) = \sqrt{2}e^{-0.707t}\sin 0.707t \qquad (9.58)$$

we find that the pulse can be made far more obvious. (This choice can be justified on theoretical grounds, but the statement of the theory is beyond the scope of this book.)

A computer program to perform the convolution of the $e(t)$ and $h(t)$ described above is given in Figure 9.15. The flow chart for this figure is given in Figure 9.16. The resultant $r^*(t)$ is shown in Figure 9.14(b). Although the output pulse is by no means as well-defined as the signal component of the input, the signal component of the output can be easily distinguished by comparing Figures 9.14(b) and 9.14(c). It should be pointed out that the form of the output signal will vary considerably, depending on the random numbers selected.

```
      DIMENSION E(70),H(70),R(70)
      INIT=1
      TMAX=6.
      T=.1
      KM=TMAX/T
      KMP1=KM+1
C****  COMPOSE INPUT
      DO 1 I=1,KMP1
      E(I)=0.
      IF(I.LE.30)E(I)=1.
    1 E(I)=E(I)+10.*URAND(INIT)-5.
C****  COMPOSE IMPULSE RESPONSE
      DO 2 I=1,KMP1
      TIM=(I-1)*T
    2 H(I)=1.414*EXP(-.707*TIM)*SIN(.707*TIM)
      R(1)=0.
C****  ENDPOINT ALGORITHM
      DO 3 I=2,KMP1
    3 R(I)=.5*(E(1)*H(I)+E(I)*H(1))
C****  MULTIPLICATION ALGORITHM
      DO 4 I=2,KM
      DO 4 J=2,KM
      JJ=I+J-1
      IF(JJ.GT.KMP1)GO TO 4
      R(JJ)=R(JJ)+E(I)*H(J)
    4 CONTINUE
      WRITE(6,5)
    5 FORMAT(1H0,7X,'TIME',10X,'RESPONSE',/)
      DO 6  I=1,KMP1
      TIM=(I-1)*T
    6 WRITE(6,7)TIM,R(I)
      RETURN
    7 FORMAT(2X,2E15.4)
      END
```

FIGURE 9.15 Program to filter noisy input using convolution.

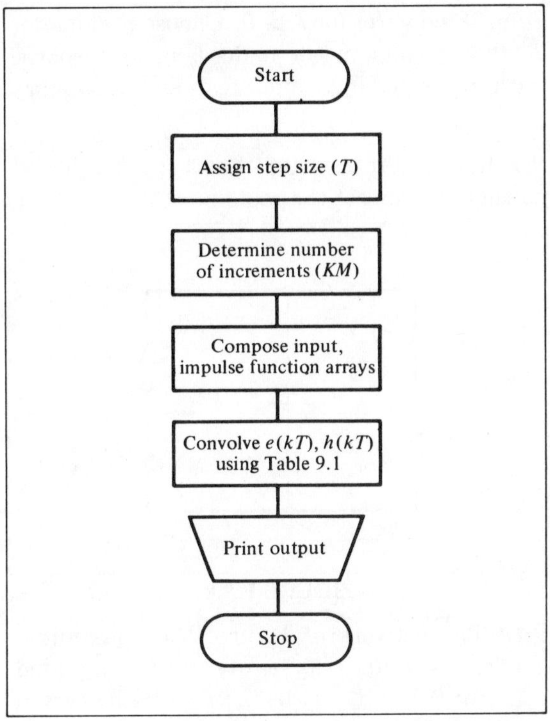

FIGURE 9.16 Flow chart for program of Figure 9.15.

The presence of the signal can, however, be sensed by visual observation of outputs with the same set of random numbers, with and without the signal present such as Figures 9.14(b) and 8.14(c).

PROBLEMS

9.1 The switch in Figure P 9.1 is closed at $t = 0$. Find $v_C(t)$ for $t > 0$ if (*a*) $R = 5000\ \Omega$ and (*b*) $R = 2000\ \Omega$. In each case $v_C(0^-) = 0$.

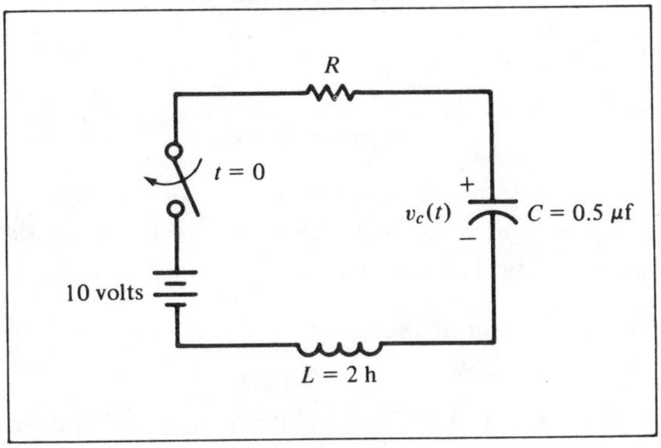

FIGURE P 9.1

9.2 Find $i_L(t)$ in Figure P 9.2 if $i_s(t) = 10$ ma, $t > 0$, and $i_L(0) = v_C(0) = 0$.

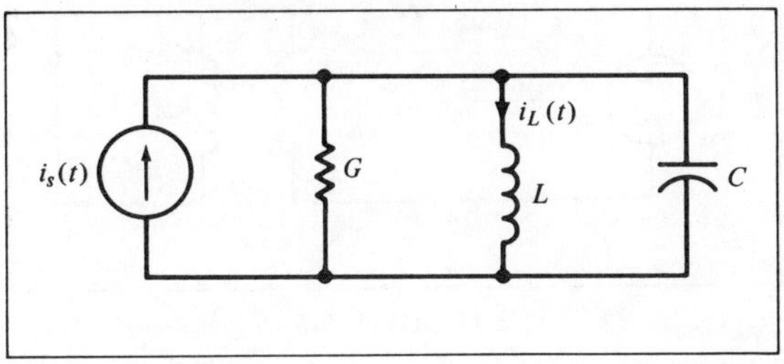

FIGURE P 9.2 $G = 0.1$ ℧; $C = 0.05\ \mu$f; $L = 5\ \mu h$.

9.3 A capacitor C is added to the series network of Figure P 9.3. Will this added capacitance cause the natural frequencies to become overdamped or underdamped? Justify your answer.

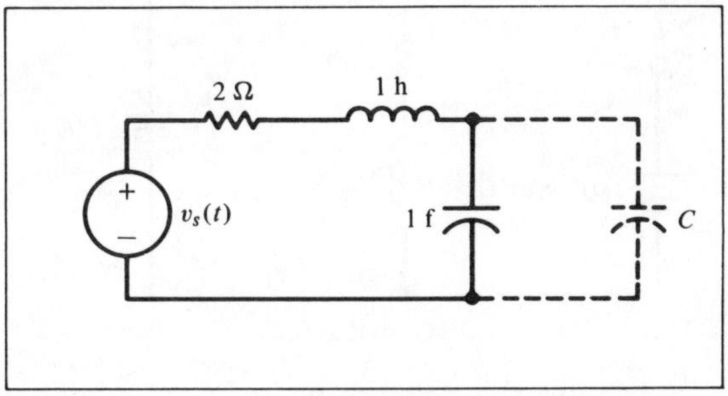

FIGURE P 9.3

9.4 A controlled source is added to the series network of Figure P 9.4. Will this controlled source cause the natural frequencies to become overdamped or underdamped? Justify your answer.

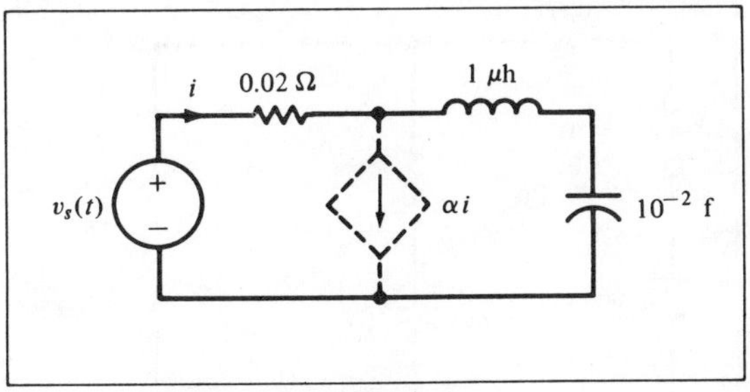

FIGURE P 9.4 $0 < \alpha < 1$.

9.5 In Figure P 9.5, solve for the value of M that results in critical damping.

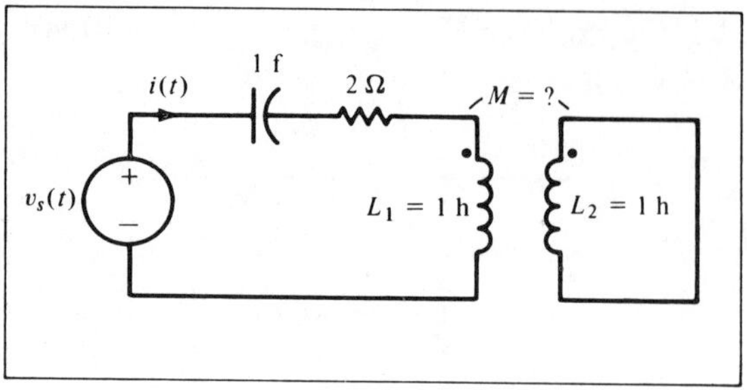

FIGURE P 9.5

9.6 The switch in Figure P 9.6 is opened at

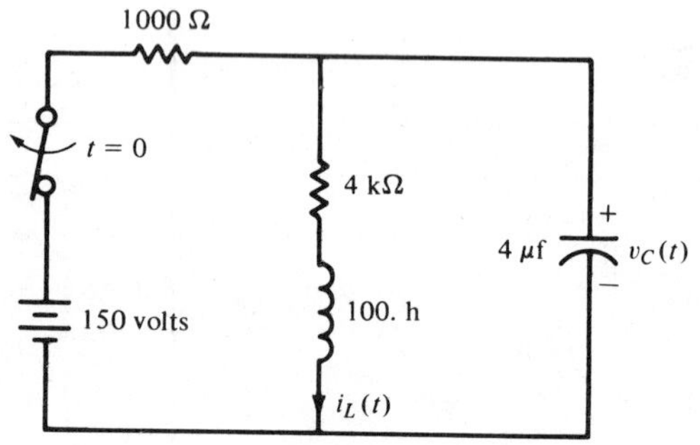

FIGURE P 9.6

$t = 0$, after the network has reached equilibrium. Find $v_C(t)$ $(t > 0)$.

9.7 In Problem 9.6, find $i_L(t)$.

9.8 The switch in Figure P 9.7 is opened at $t = 0$ after the network has reached equilibrium. Find $v_C(t)$ for $t > 0$. Choose a value for R that results in an underdamped response with roots having equal real and imaginary parts.

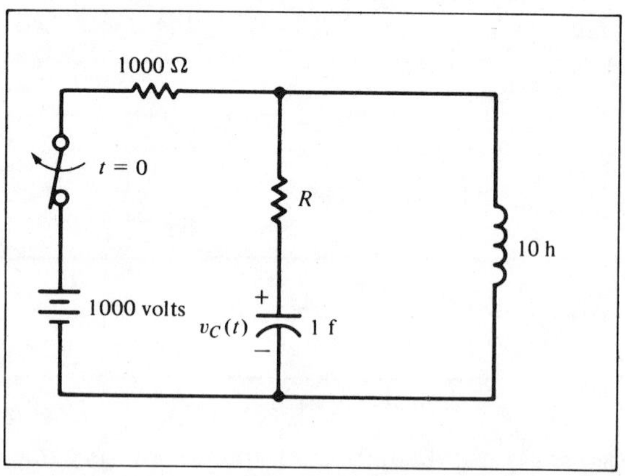

FIGURE P 9.7

9.9 Repeat Problem 9.8, but with R selected so that the sum of the roots is -5.

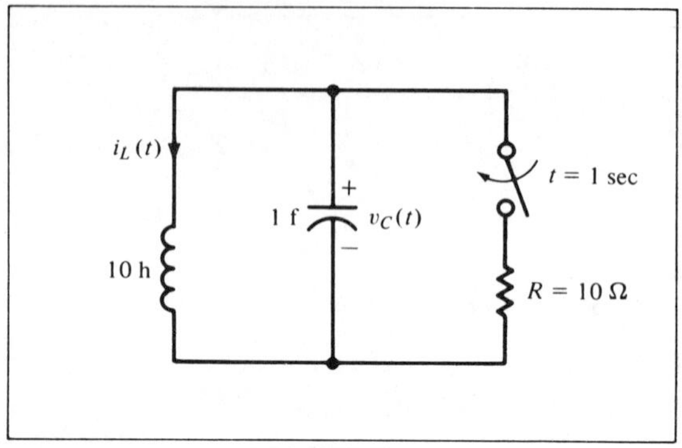

FIGURE P 9.8

9.10 The network of Figure P 9.8 has initial conditions $i_L(0) = 0$, $v_C(0) = 5$. (a) Find $i_L(t)$ for $0 < t \leqq 1$. (b) At $t = 1$, the switch is closed; find $i_L(t)$ for $t > 1$.

9.11 In the network of Figure P 9.9, the capacitor voltages are both zero at $t = 0$ when the switch is suddenly closed. Calculate $v_2(t)$ for

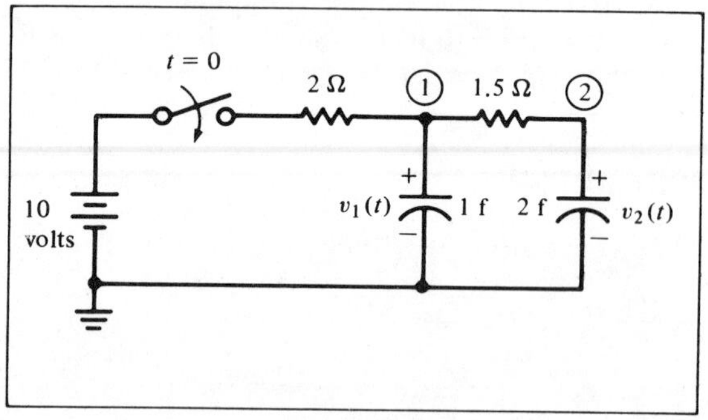

FIGURE P 9.9

$t > 0$. [*Hint*: Write KCL at nodes ① and ② and combine the equations to obtain a single differential equation for $v_2(t)$.]

9.12 For the network in Figure P 9.9, calculate $v_1(t)$ for $t > 0$.

9.13 Prior to $t = 0$, the 10f capacitor of Figure

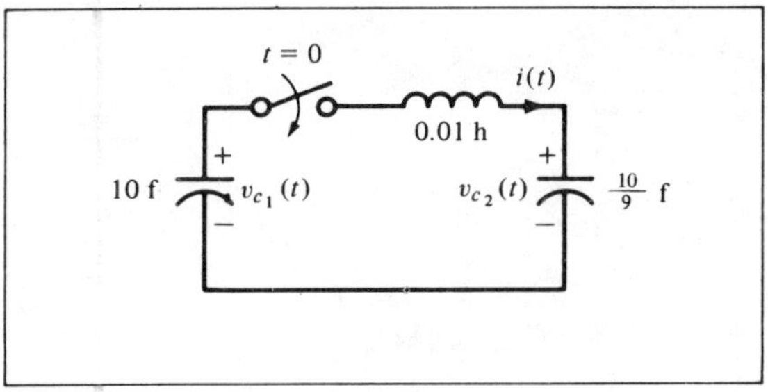

FIGURE P 9.10

P 9.10 is charged to 10 volts and there is no energy stored in the rest of the network. The switch is closed at $t = 0$. (a) Calculate $i(t)$ and the time required to reach its maximum value. (b) Calculate the maximum value of $i(t)$ and the energy stored in the coil when $i(t)$ is a maximum. How does this energy compare with the initial energy stored in the 10-f capacitor? What is the energy stored in the coil when the current reaches its minimum value? (c) What is the maximum voltage developed across the 10/9-f capacitor?

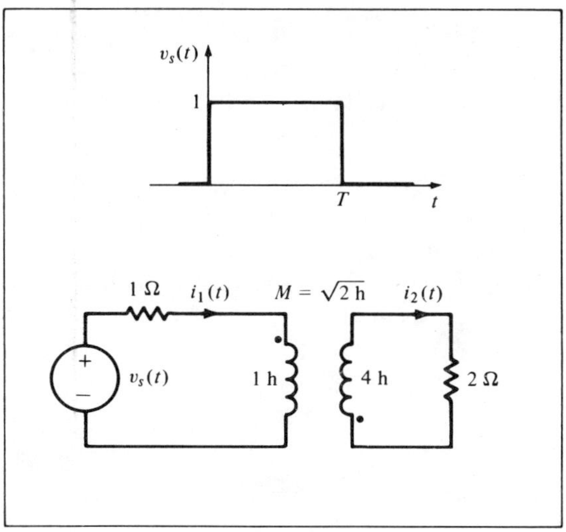

FIGURE P 9.11

9.14 For the network of Figure P 9.11: (a) Using KVL obtain a pair of equations relating $i_1(t)$, $i_2(t)$, and the voltage source $v_s(t)$; reduce these equations to a single differential equation relating $i_2(t)$ and $v_s(t)$. (b) If $v_s(t)$ has the waveform sketched at the top in Figure P 9.11, determine the minimum value for T that will assure that $i_2(t)$ reaches its maximum value; assume zero-valued initial condition. (c) Assuming that T has the value calculated in (b), calculate $i_2(t)$ for $t \geqq T$.

9.15 Compute the step and impulse responses of the network of Figure P 9.12 for $R = 2\ \Omega$.

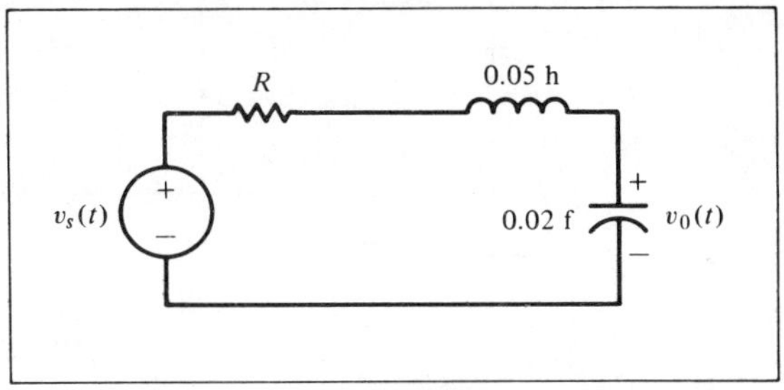

FIGURE P 9.12

Compare this response with those of Figure 9.6. Find the values of R which compare similarly with the other responses of Figure 9.6.

9.16 Sketch all the responses developed in Problem 9.15.

9.17 The unit step response of the network of Figure P 9.13 has an "overshoot" as illustrated

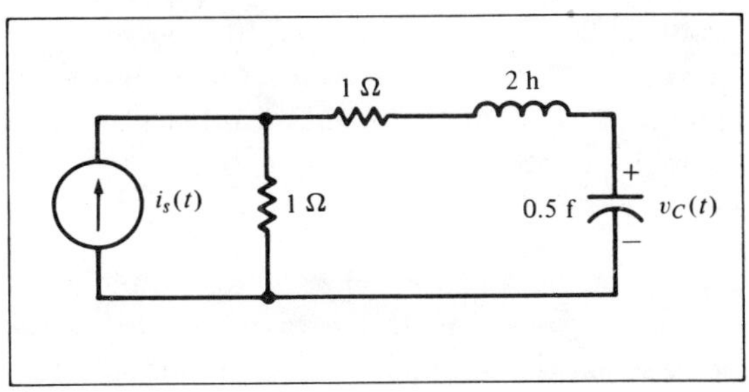

FIGURE P 9.13

in Figure 9.6 for $R = 1\ \Omega$. Verify that a maximum of $v_C(t)$ occurs at

$$t = (1/\omega_d)(\tan^{-1}(-\alpha/\beta) - \phi)$$

where $\omega_d = \sqrt{3}/2$, $\alpha = 0.5$, $\phi = 5\pi/6$.

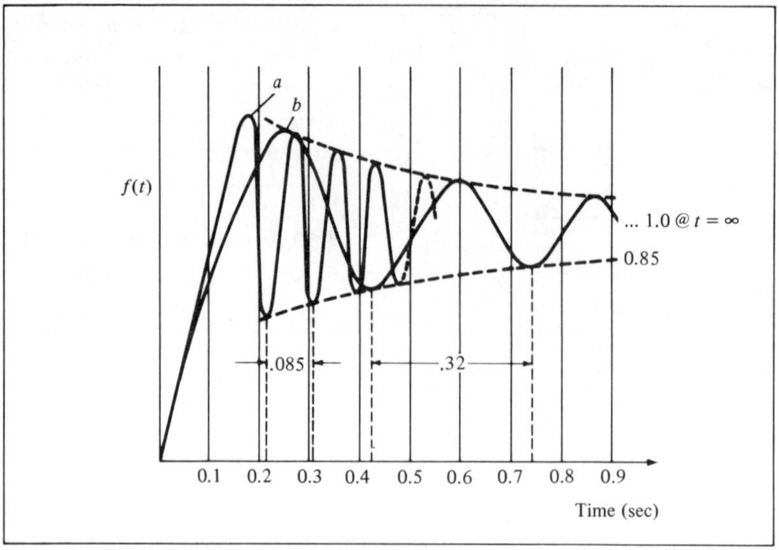

FIGURE P 9.14

9.18 The step responses of two second-order circuits are shown in Figure P 9.14. For each case, determine approximate values for α and ω_d.

9.19 (a)–(c) The signals of Figure P 9.15 are applied to the parallel GCL network of Figure 9.7 with $G = 0.00667$ ℧, $C = 10$ pf, $L = 1$ μh. Find $i_L(t)$, assuming zero-valued initial conditions.

9.20 The circuit of Figure P 9.16 is a simplified linear, incremental model of a two-stage FET amplifier. Use the convolution integral to calculate the amplifier's impulse response $[v_0(t)$ when $v_s(t) = \delta(t)]$.

9.21 The impulse response of an amplifier is $\cos tu(t)$. Use convolution to calculate the response of the amplifier to an input $e^{-t}u(t)$. (The amplifier can be assumed to behave as a linear network.)

9.22 (a)–(c) Determine by convolution the voltage $v_C(t)$ for $t > 0$ for the signals of Figure P 9.17(a)–(c).

9.23 Compare the result you obtain in Problem 9.22(a) with response of a hypothetical network with an impulse response

$$h(t) = u(t) - u(t - T)$$

to an input of the form

$$v_s(t) = \frac{e^{-t/T}}{T} u(t).$$

9.24 As shown in Figure P 9.18, an arbitrary excitation $e(t)$ can be approximated by a sum of step functions rather than the series of pulses shown in Figure 9.8. If we assume that $e(t) = 0$ for $t = 0$,

$$e^*(t) = e(0)u(t) + \left(\frac{e(\Delta) - e(0)}{\Delta}\right) u(t - \Delta)\Delta$$

$$+ \left(\frac{e(2\Delta) - e(\Delta)}{\Delta}\right) u(t - 2\Delta)\Delta + \cdots$$

(a) Write this sum in the form of Equation (9.42) and write a sum for $r^*(t)$ analogous to Equation (9.43). By letting $\Delta \to d\tau$, derive an alternative integral representation of $r(t)$ in terms of a network's step response $s(t)$ and the derivative of the excitation $e(t)$. (b) Apply

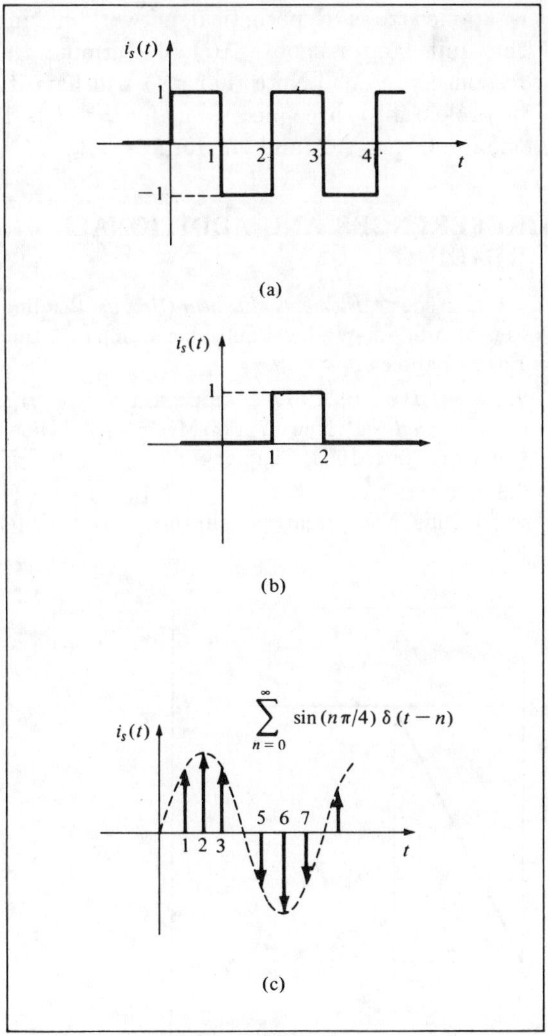

FIGURE P 9.15

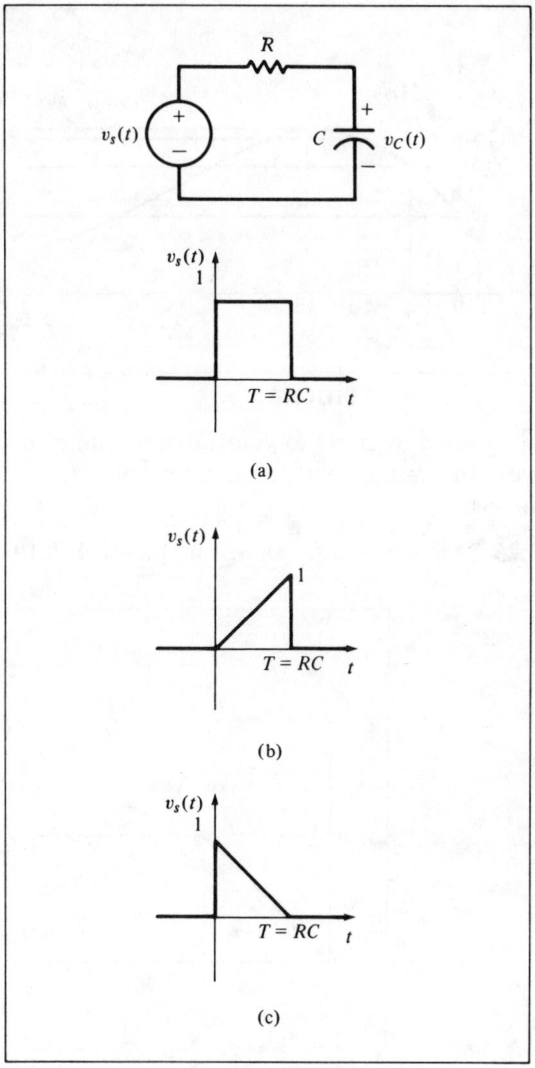

FIGURE P 9.17

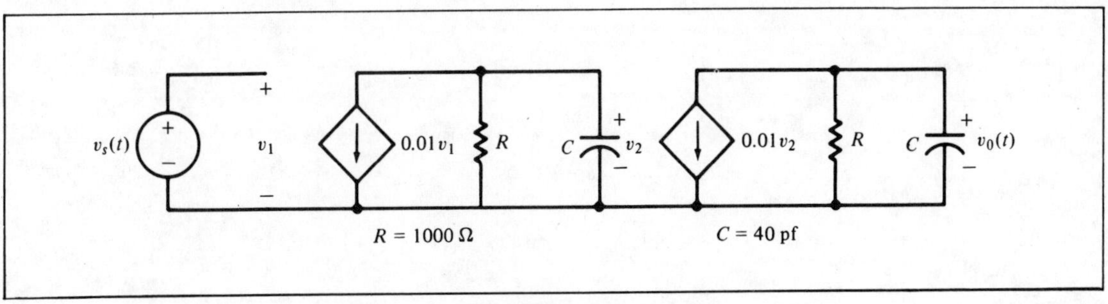

FIGURE P 9.16

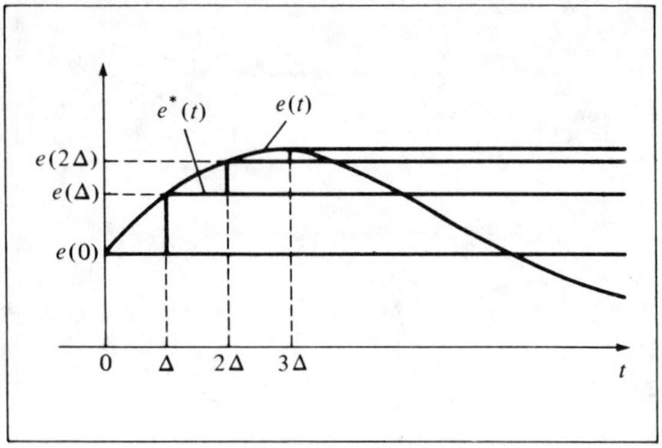

FIGURE P 9.18

integration by parts to your integral and compare the results with Equations (9.47) and (9.48).

9.25 The voltage $v(t)$ shown in Figure P 9.19 is applied to a hypothetical network having the unit step response $s(t)$. Determine the response to $v(t)$. [Note that $r(t)$ will have to be broken into three pieces, one for $0 < t \leqq 1$, one for $1 < t \leqq 2$, and one for $t > 2$.]

REFERENCES AND ADDITIONAL READING

9.1 R. E. SCOTT, *Elements of Linear Circuits.* Reading, Mass.: Addison-Wesley Publishing Company, Inc., 1960, Chapter 9.

9.2 W. H. HAYT, JR. AND J. E. KEMMERLY, *Engineering Circuit Analysis.* New York: McGraw-Hill Book Company, Inc., 1963, Chapter 8.

9.3 M. E. VAN VALKENBURG, *Network Analysis,* Englewood Cliffs, N.J.: Prentice-Hall, Inc., 1964, Chapter 9.

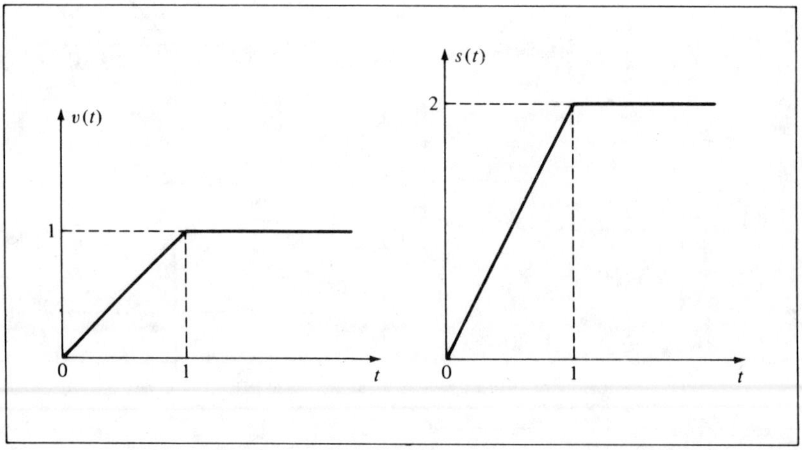

FIGURE P 9.19

10

Sinusoidal Steady-State Analysis

10.1 Introduction

The signals that we used in Chapters 8 and 9 were of a very simple type, usually either a step or an impulse. Another important class of signals are *periodic* signals defined by

$$f(t) = f(t + T) \qquad (\text{all } t) \qquad (10.1)$$

That is, $f(t)$ is repetitive every T sec; the quantity T is accordingly called the *period* of the signal. Any signal that is not periodic is *aperiodic*. A function satisfying $f(t) = 0$, $t < 0$ would be aperiodic, because Equation (10.1) must be satisfied for all t. Examples are shown in Figure 10.1.

An especially important form of periodic signal (and the subject of this chapter) is the sinusoid of Figure 10.1(c). The applications of this waveform are many.

1. The response of many electrical and mechanical systems are sinusoidal in nature; examples are the vibrating string, the pendulum, and networks that contain only inductors and capacitors.
2. A sinusoid is the waveform commonly used in the transmission of information and energy; for example, electrical power is supplied to households by a sinusoidal waveform.
3. As discussed in Appendix C, other periodic signals can be represented in terms of sinusoids by means of a *Fourier series*.

Besides these pragmatic uses, several mathematical tools and certain network concepts related to sinusoidal analysis will be useful in our later network analysis studies.

10.2 Characteristics of Periodic Signals

Average Value

We recall from elementary calculus that the average value of any function $f(t)$ over an interval $[t_1, t_2]$ is given by

$$F_{\text{av}} \triangleq \frac{1}{t_2 - t_1} \int_{t_1}^{t_2} f(t)\, dt$$

In the case of a periodic signal, this integral has the same value so long as it is taken over

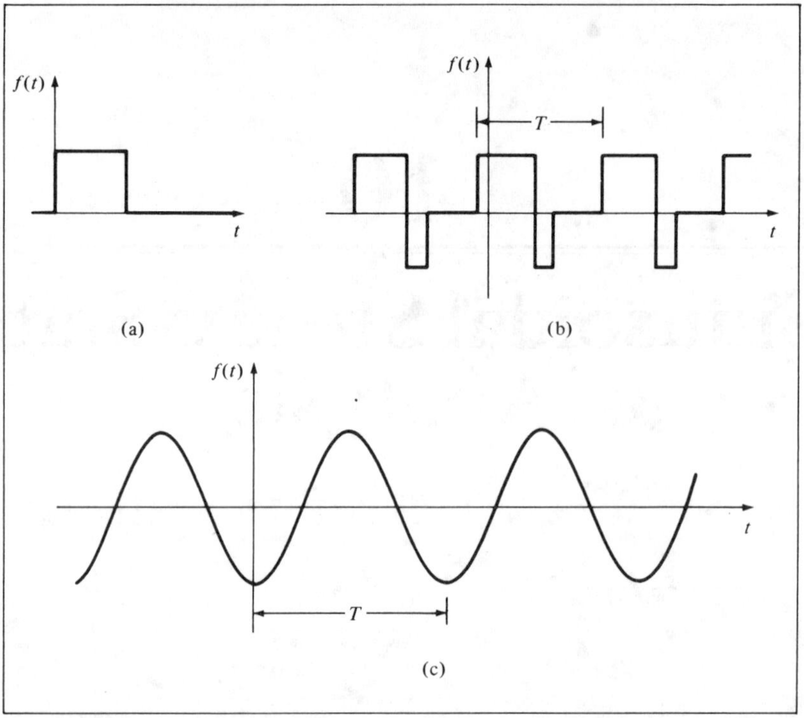

FIGURE 10.1 Examples of signal classification. (a) Aperoidic signal; (b) periodic signal; (c) periodic sinusoidal signal.

any complete period. We indicate this by writing

$$F_{av} \triangleq \frac{1}{T}\int_{t_1}^{t_1+T} f(t)\,dt \qquad (\text{all } t_1) \qquad (10.2)$$

Effective (rms) Value

If we apply the definition (10.2) to calculate the average *power* absorbed in a network, we arrive at the concept of *effective value*. In particular, consider a periodic current $i(t)$ flowing through a resistor of value R. The instantaneous power absorbed in the resistor is therefore

$$p(t) = i^2(t)R$$

This is the rate at which electric energy is converted to heat energy in the resistor. The average rate is calculated from Equation (10.2) as

$$P_{av} = \frac{1}{T}\int_{t_1}^{t_1+T} i^2(t)R\,dt$$

The same average power would be dissipated in the resistor if a constant current I_{dc} were applied, provided

$$I_{dc}^2R = \frac{1}{T}\int_{t_1}^{t_1+T} i^2(t)R\,dt \qquad (10.3)$$

This constant current *which would give the same average heating effect* in the resistor R as the periodic current $i(t)$ is defined as the *effective value* of $i(t)$. Solving Equation (10.3), this effective value is seen to be

$$I_{dc} = I_{eff} \triangleq \sqrt{\frac{1}{T}\int_{t_1}^{t_1+T} i^2(t)\,dt} \qquad (10.4)$$

Similar definitions apply to the effective values of periodic voltage, charge, or flux linkage waveforms. This value is also called the *root mean square*, or *rms*, value, from the squaring and averaging operations observed in Equation (10.4).

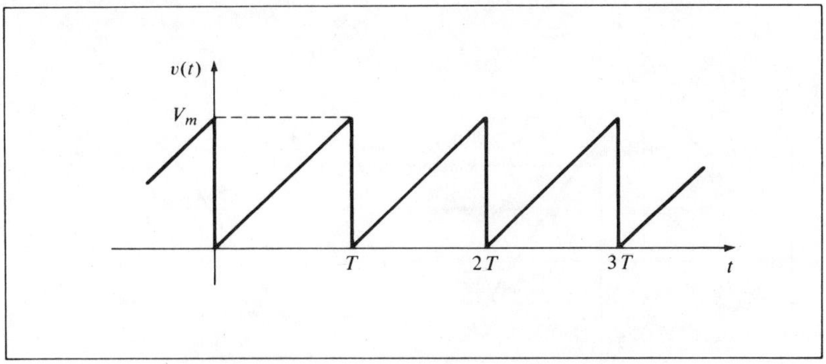

FIGURE 10.2 Periodic signal waveform.

Example 10.1

We are to determine the average and effective values of the signal voltage of Figure 10.2.

This sawtooth voltage waveform can be written

$$v(t) = \left(\frac{V_m}{T}\right)t \qquad (0 \leqq t < T)$$

The average value is determined from Equation (10.2) by

$$V_{av} = \frac{1}{T}\int_0^T v(t)\,dt$$

$$= \frac{1}{T}\int_0^T \frac{V_m}{T}\,t\,dt$$

$$= \frac{V_m}{T^2}\left[\frac{t^2}{2}\right]_0^T$$

$$= \frac{V_m}{2} \text{ volts}$$

which agrees with an observation of the area under $v(t)$ over one period. The effective value of $v(t)$ is

$$V_{eff} = \sqrt{\frac{1}{T}\int_0^T v^2(t)\,dt}$$

$$= \sqrt{\frac{1}{T}\int_0^T \left(\frac{V_m}{T}\right)^2 t^2\,dt}$$

$$= \sqrt{\frac{V_m^{\ 2}}{T^3}\int_0^T t^2\,dt}$$

$$= \sqrt{\frac{V_m^{\ 2}}{T^3}\left(\frac{t^3}{3}\right)\Bigg|_0^T}$$

$$= \frac{V_m}{\sqrt{3}} \text{ volts}$$

which is somewhat larger than the average value.

10.3 Introduction to Sinusoidal Steady-State Analysis

The Sinusoid

A sinusoidally varying quantity has the general form

$$f(t) = F_m \cos(\omega t + \phi) \qquad (10.5)$$

$$= F_m \sin\left(\omega t + \phi + \frac{\pi}{2}\right) \qquad (10.6)$$

where F_m = amplitude or maximum value; $\omega = 2\pi f$ = the angular frequency in radians per second, and ϕ = phase angle in radians or degrees (the radian measure is mathematically preferable; the degree is used more commonly by engineers). Expanding Equation (10.5) by simple trigonometry gives

$$f(t) = F_m \cos\phi \cos\omega t - F_m \sin\phi \sin\omega t \qquad (10.7)$$

$$= A \cos\omega t + B \sin\omega t \qquad (10.8)$$

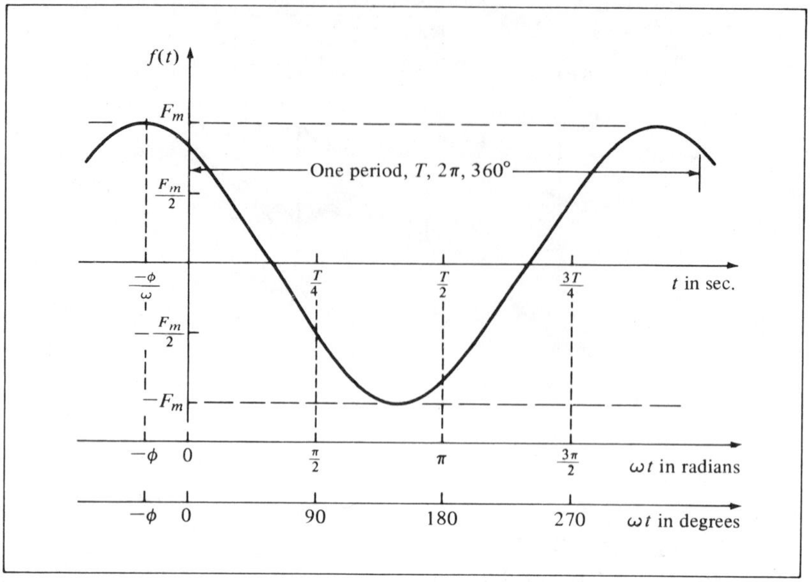

FIGURE 10.3 A general sinuoisdal waveform.

where $B = -F_m \sin \phi$; $A = F_m \cos \phi$. If the sinusoid is given in the form of Equation (10.8), we can obtain the compact representation of Equation (10.5) by calculating

$$F_m = \sqrt{A^2 + B^2} \tag{10.9}$$

and

$$\phi = -\tan^{-1}\left(\frac{B}{A}\right) \tag{10.10}$$

This waveform is plotted versus t and ωt in Figure 10.3. Because the sinusoid assumes identical positive and negative values over a period, its average value is zero. It is left as an exercise to show that its effective value is $F_{\text{eff}} = F_m/\sqrt{2}$.

In handling sinusoidal waveforms it is convenient to have representations in addition to the trigonometric forms of Equations (10.5) and (10.8). These are based on the Euler[1] identity

$$e^{j\theta} = \cos \theta + j \sin \theta \tag{10.11}$$

(See Appendix B for further discussion on complex arithmetic.) This identity gives us two alternative ways of expressing sinusoids. For the cosine

$$\cos \theta = \frac{e^{j\theta} + e^{-j\theta}}{2} \tag{10.12}$$

or

$$\cos \theta = \text{Re}[e^{j\theta}] \tag{10.13}$$

where Re[ ] means "the real part of." For the sine

$$\sin \theta = \frac{e^{j\theta} - e^{-j\theta}}{2j} \tag{10.14}$$

or

$$\sin \theta = \text{Im}[e^{j\theta}] \tag{10.15}$$

where Im[ ] means "the imaginary part of."

Consider a sinusoidal current

$$i(t) = I_m \cos(\omega t + \phi_i) \tag{10.16}$$

Substituting Equation (10.13) into (10.16) we

[1] Leonhard Euler (1707–1783) was a brilliant Swiss mathematician. His work introduced the symbols e, π, and i (for $\sqrt{-1}$), and he is responsible for the analytic treatment of the trigonometric functions as numerical ratios rather than as geometric lines.

write

$$i(t) = I_m \operatorname{Re}[e^{j(\omega t + \phi_i)}]$$
$$= I_m \operatorname{Re}[e^{j\phi_i} e^{j\omega t}]$$

Because I_m is a real constant, it can be taken inside the brackets to yield

$$i(t) = \operatorname{Re}[I_m e^{j\phi_i} e^{j\omega t}]$$
$$= \operatorname{Re}[I e^{j\omega t}] \quad (10.17)$$

Here $I = I_m e^{j\phi_i}$ is a complex constant. This constant, with the frequency ω, completely represents the sinusoid Equation (10.16) because it contains both constants (I_m and ϕ_i) of the original function of time.

Example 10.2

The sinusoid

$$i(t) = 5\cos(377t + 60°)$$
$$= \operatorname{Re}[5e^{j60°} e^{j377t}]$$
$$= \operatorname{Re}[I e^{j377t}]$$

is represented by the single constant

$$I = 5e^{j60°}$$
$$= 5(\cos 60° + j \sin 60°)$$
$$= 5\left(\frac{1}{2} + j\frac{\sqrt{3}}{2}\right)$$
$$= \frac{5}{2} + j5\frac{\sqrt{3}}{2}$$
$$= 2.50 + j4.34$$

A complex constant such as I that represents a sinusoidal function of time is called a *phasor*. This complex constant is often displayed graphically as shown in Figure 10.4(c). Here, the imaginary parts and real parts of I are plotted on vertical and horizontal axes, respectively, and the resultant point indicated by an arrow from the origin.

It will be observed that in using phasor representations *either* the peak *or* the rms values of the variables can be employed. In general, we will use peak values in our analysis.

(a) *Time:*

$$i(t) = I_m \cos(\omega t + \phi_i) = \operatorname{Re}[I e^{j\omega t}]$$

(b) *Phasor:*

$$I = I_m e^{j\phi_i}$$
$$= I_m \cos\phi_i + jI_m \sin\phi_i$$

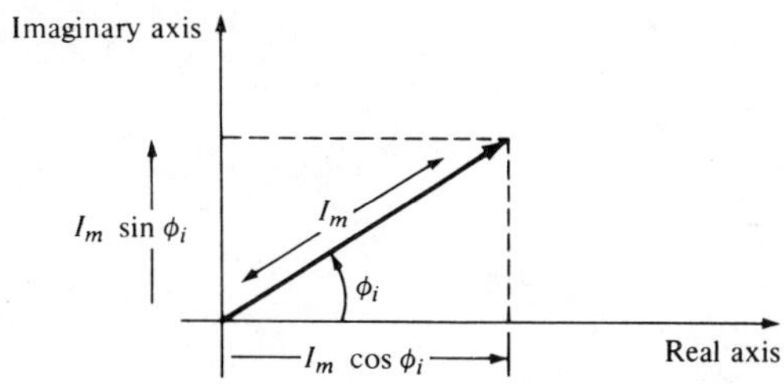

(c)

FIGURE 10.4 Phasor representation of a sinusoid.

Rotating Phasor

The complex time function

$$\hat{i}(t) = I e^{j\omega t} \quad (10.18)$$

where

$$I = I_m e^{j\phi_i} \quad (10.19)$$

is an example of a *rotating phasor*. This is illustrated in Figure 10.5; at any instant t_1, $\hat{i}(t_1)$ is the phasor I rotated by an additional angle ωt_1. The projection of $\hat{i}(t)$ on the real axis is just $i(t)$ as written in Equation (10.17). Thus, a sinusoid can be viewed as the projection on the real axis of a complex phasor rotating with a constant angular velocity

$$\frac{d}{dt}(\omega t + \phi_i) = \omega \quad (10.20)$$

As long as we are dealing with sinusoids at a single, fixed frequency, the fixed phasor of Figure 10.4(c) is just as useful as the rotating phasor of Figure 10.5. When more than a single frequency is present or when the frequency is

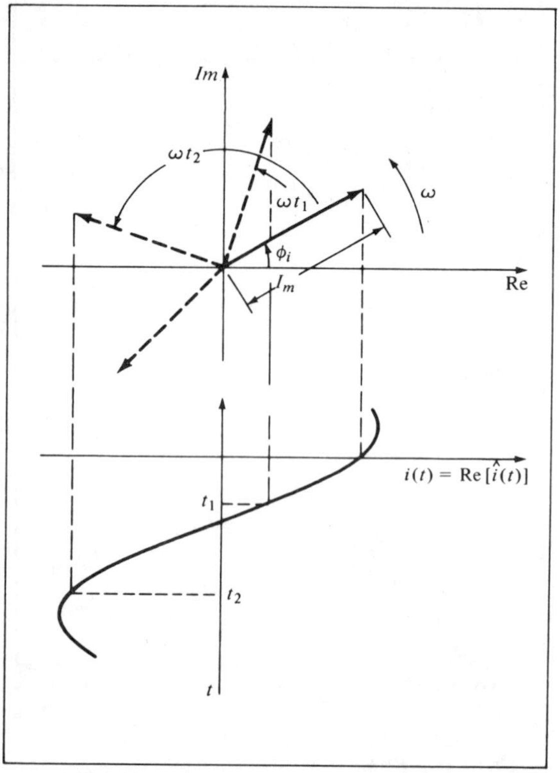

FIGURE 10.5 Illustration of rotating phasor and its sinusoidal projection on the real axis.

allowed to vary, as is often the case in communications systems, the rotating phasor point of view can give valuable insight into the nature of the resulting waveforms. We will use only the fixed phasor representations in this text.

Sinusoidal Steady State

As explained in Section 8.2, the steady-state component of a function of time is that part which does not decay to zero as $t \to \infty$. For example, a response of the form

$$r(t) = \underbrace{A \cos \omega t + B \sin \omega t}_{\text{sinusoidal steady state}} + \underbrace{Ce^{-\alpha t}}_{\text{transient}} + \underbrace{D}_{\text{dc steady state}} \tag{10.21}$$

has two steady-state terms [see Equation (8.31)]. Once we have obtained the dc term by shorting inductors and removing capacitors, we next might wish to determine the sinusoidal steady-state term. This is known as sinusoidal steady-state analysis and is the subject of the remainder of this chapter.

We recall from dc analysis that a network must contain a constant (dc) source to produce a dc response. The same is usually true for a sinusoidal steady-state response. Indeed, we will assume in our later analysis that, as $t \to \infty$, only sinusoidal currents and voltages are present in the network. If dc sources are present, then, from the superposition principle of Chapter 6, these can be separately applied and the dc response term separately evaluated.

10.4 Representation of Elements in the Sinusoidal Steady State

Introduction

As we have seen in our computer analysis, it is often possible to avoid writing nonlinear algebraic equations by making use of network models. In succeeding sections we will show that by introducing appropriate models we can simplify the solution for a network's sinusoidal steady-state response. Specifically, we can reduce calculation of the *particular response* [see Equation (8.27)] to the solution of a set of algebraic equations with complex-valued coefficients. Further, this set of equations can be obtained *without writing the differential equations of the network*. Finally, all of the systematic methods of solving sets of simultaneous algebraic equations in resistive circuits can be applied directly to the solution of networks in the sinusoidal steady state.

We begin at the element level, and develop models for resistors, capacitors, and inductors in the sinusoidal steady state. In every case we will find that the sinusoidal character is preserved (a sinusoidal current requires a sinusoidal voltage, and vice versa), and the relationship between the current and voltage waveforms can be represented by simple phasor equations.

Resistors

In the sinusoidal steady state, we can write the resistor current and voltage as

$$i_R(t) = I_m \cos(\omega t + \phi_i) = \text{Re}[I_R e^{j\omega t}] \tag{10.22}$$

$$v_R(t) = V_m \cos(\omega t + \phi_v) = \text{Re}[V_R e^{j\omega t}]$$

where $I_R = I_m e^{\phi_i}$, $V_R = V_m e^{j\phi_v}$. Then Ohm's law gives

$$V_m \cos(\omega t + \phi_v) = RI_m \cos(\omega t + \phi_i) \tag{10.23}$$

or

$$\text{Re}[\underbrace{V_m e^{\phi_v}}_{V_R} e^{j\omega t}] = \text{Re}[R \underbrace{I_m e^{\phi_i}}_{I_R} e^{j\omega t}]$$

Equation (10.23) can be satisfied for all values of time if and only if

$$V_R = RI_R \tag{10.24}$$

This equation has the same form as Ohm's law; however, it now relates the phasors V_R and I_R rather than the real-valued functions of time from which it was derived [Equations (10.22)]. These phasor and time-domain relationships for a fixed, linear resistor are illustrated in Figure 10.6(a).

Inductors

We define the inductor current and voltage as

$$i_L(t) = I_m \cos(\omega t + \phi_i) = \text{Re}[I_L e^{j\omega t}] \tag{10.25}$$

$$v_L(t) = V_m \cos(\omega t + \phi_v) = \text{Re}[V_L e^{j\omega t}] \tag{10.26}$$

where $I_L = I_m e^{j\phi_i}$, $V_L = V_m e^{j\phi_v}$. For a fixed linear inductor, we know

$$v_L(t) = L \frac{di_L(t)}{dt} \tag{10.27}$$

Substituting from Equations (10.25) and (10.26) gives

$$\text{Re}[V_L e^{j\omega t}] = L \frac{d}{dt} \{\text{Re}[I_L e^{j\omega t}]\}$$

$$= \text{Re}\left[LI_L \frac{d}{dt} e^{j\omega t}\right] \tag{10.28}$$

$$= \text{Re}[j\omega L I_L e^{j\omega t}] \tag{10.29}$$

Equation (10.29) requires that

$$V_L = (j\omega L) I_L$$

or equivalently that

$$V_m e^{j\phi_v} = (j\omega L) I_m e^{j\phi_i}$$

$$= \omega L I_m e^{j(\phi_i + 90^\circ)}$$

Therefore, the amplitudes of current and voltage are related by

$$V_m = \omega L I_m \tag{10.30}$$

and the phases by

$$\phi_v = \phi_i + 90^\circ \tag{10.31}$$

We can now substitute Equations (10.30) and (10.31) into (10.26) to yield

$$v_L(t) = \omega L I_m \cos(\omega t + \phi_i + 90^\circ) \tag{10.32}$$

We see that an inductor alters the amplitude of $i_L(t)$ by ωL and shifts its phase by 90°. This is indicated in Figure 10.6(b) both in the complex phasor plane and as a function of time. The phase shift causes the inductor voltage to reach its maximum value one-quarter cycle, or 90°, before the current maximum occurs. We say that in an inductor the sinusoidal voltage *leads* the current by 90°.

Capacitors

The current-voltage equation for a fixed, linear capacitor is

$$i_C(t) = C \frac{dv_C(t)}{dt} \tag{10.33}$$

This form is similar to the equation for an inductor (10.27). We can conclude without a

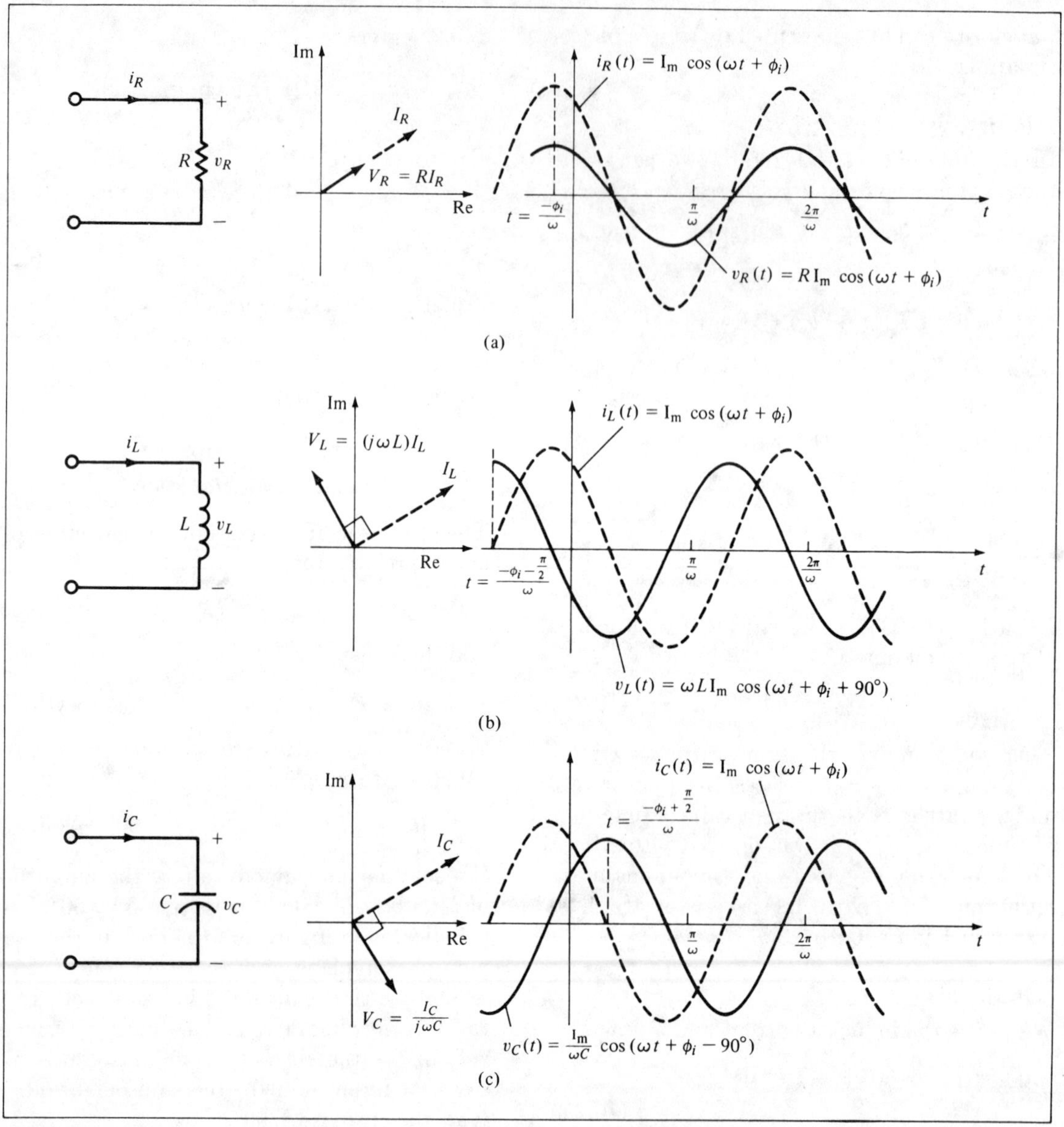

FIGURE 10.6 Phasor and time-domain relationships between current and voltage in single elements. (a) Resistor; (b) inductor; (c) capacitor.

formal derivation, therefore, that the phasor equation for a capacitor is

$$I_C = j\omega C V_C$$

or

$$V_C = \frac{1}{j\omega C} I_C$$

$$= -\frac{j}{\omega C} I_C$$

$$= \frac{1}{\omega C} I_m e^{j(\phi_i - 90^\circ)}$$

A capacitor voltage is shifted by $-90°$ from the current, rather than $+90°$ as occurring with an inductor; the capacitor voltage *lags* the current by 90°. The phasor and time-domain representations of this relation are shown in Figure 10.6(c).

10.5 Sinusoidal Steady-State Network Analysis

Consider the network of Figure 10.7(a), a series *RLC* network excited by a sinusoidal voltage source. After sufficient time, the transient terms in the response $i(t)$ will effectively vanish, and all currents and voltages will be sinusoidal. The validity of this assumption will be verified if both the element relations studied above *and* the network connection laws—KVL in this case—can be satisfied with only sinusoidal currents and voltages.

If all currents and voltages are sinusoidal, the steady-state relationship between the phasors I and V of each element type is valid. In particular, we write

$$V_R = RI \qquad (10.34)$$

$$V_L = j\omega LI \qquad (10.35)$$

$$V_C = \frac{-j}{\omega C} I \qquad (10.36)$$

where $I = I_m e^{j\phi_i}$ is the phasor representation of the loop current

$$i(t) = I_m \cos(\omega t + \phi_i) \qquad (10.37)$$

For the Figure 10.7(a), routine application of KVL yields

$$v_L(t) + v_C(t) + v_R(t) - v_s(t) = 0 \qquad (10.38)$$

In terms of general sinusoidal voltages, this equation can be written

$$V_{Lm}\cos(\omega t + \phi_L) + V_{Cm}\cos(\omega t + \phi_C) + V_{Rm}\cos(\omega t + \phi_R) - V_s\cos(\omega t + \phi_s) = 0 \qquad (10.39)$$

or alternatively

$$\mathrm{Re}[\underbrace{V_{Lm}e^{j\phi_L}}_{V_L}e^{j\omega t}] + \mathrm{Re}[\underbrace{V_{Cm}e^{j\phi_C}}_{V_C}e^{j\omega t}] + \mathrm{Re}[\underbrace{V_{Rm}e^{j\phi_R}}_{V_R}e^{j\omega t}] - \mathrm{Re}[\underbrace{V_{sm}e^{j\phi_s}}_{V_s}e^{j\omega t}] = 0 \qquad (10.40)$$

Collecting all bracketed terms together (a procedure the reader may wish to justify), Equation (10.40) reduces to

$$\mathrm{Re}[V_L e^{j\omega t} + V_C e^{j\omega t} + V_R e^{j\omega t} - V_s e^{j\omega t}] = 0$$

or

$$\mathrm{Re}[(V_L + V_C + V_R - V_s)e^{j\omega t}] = 0 \qquad (10.41)$$

This equation can be satisfied for all time t if and only if

$$V_L + V_C + V_R - V_s = 0 \qquad (10.42)$$

This phasor equation is evidently the phasor equivalent of the original KVL equation (10.38). This suggests that we can construct a phasor network model as shown in Figure 10.7(b). The KVL equations of this network then yield Equation (10.42) directly without the above involved derivation.

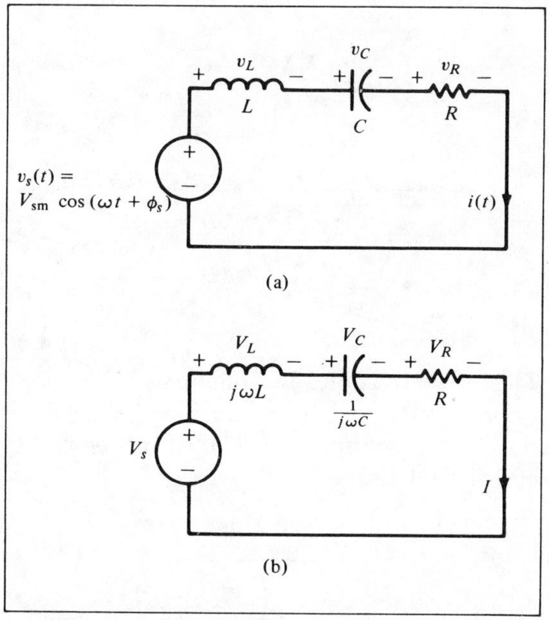

FIGURE 10.7 Series network analysis. (a) Original network; (b) phasor model.

The unknown current phasor I in Figure 10.7(b) now can be found without recourse to the original network. Combining Equations (10.34) to (10.36) with (10.42) gives

$$V_s = j\omega LI - \frac{j}{\omega C}I + RI$$

$$\tag{10.43}$$

$$V_s = \left[j\omega L - \frac{j}{\omega C} + R\right]I$$

Solving for I gives

$$I = \frac{V_s}{j\omega L - \dfrac{j}{\omega C} + R}$$

$$= \frac{V_{sm}e^{j\phi_s}}{R + j\left(\omega L - \dfrac{1}{\omega C}\right)}$$

$$= \frac{V_{sm}}{\sqrt{R^2 + \left(\omega L - \dfrac{1}{\omega C}\right)^2}} e^{j(\phi_s - \tan^{-1}[(\omega L - 1/\omega C)/R])}$$

$$= I_m e^{j\phi_i}$$

This verifies that KVL, Equation (10.38), can indeed be satisfied by the assumed sinusoidal steady-state solution, (10.37). It further shows how the amplitude

$$I_m = \frac{V_{sm}}{\sqrt{R^2 + \left(\omega L - \dfrac{1}{\omega C}\right)^2}} \tag{10.44}$$

and the phase

$$\phi_i = \phi_s - \tan^{-1}\left[\frac{\omega L - \dfrac{1}{\omega C}}{R}\right] \tag{10.45}$$

of that solution are related to the network element values and the amplitude, phase, and frequency of the voltage excitation.

Example 10.3

Let the signal of Figure 10.7 be described by

$$V_{sm} = 2 \text{ volts} \qquad \omega = 3 \text{ rad/sec} \qquad \phi_s = 0°$$

and let the element values be given by

$$R = 1\,\Omega \qquad L = \tfrac{1}{2}\text{ h} \qquad C = \tfrac{2}{3}\text{ f}$$

We are to find $i(t)$.

The phasor I is given by

$$I = \frac{V_s}{j(\omega L) - j(1/\omega C) + 1}$$

$$= \frac{2}{j(\frac{3}{2}) - j(\frac{1}{2}) + 1}$$

$$= \frac{2}{1 + j1} \times \frac{1 - j1}{1 - j1}$$

$$= \frac{2 - j2}{2}$$

$$= 1 - j1$$

$$= \sqrt{2}\, e^{-j45°}$$

This result is the phasor representation of the sinusoidal steady-state time-domain solution

$$i(t) = \sqrt{2}\cos(3t - 45°)$$

Impedance and Admittance

The method of sinusoidal steady-state analysis illustrated in Example 10.3 is highly suggestive of a more general procedure, using the resistive network analysis methods of Chapters 2 and 4. In this section, we formalize this procedure through the concepts of *impedance* and *admittance* and application of KCL and KVL to phasor network models.

Examining Equations (10.34) to (10.36), we find that in each case the element phasor voltage (V) and current (I) are related by an expression of the form

$$V = ZI \tag{10.46}$$

The Z term is called the *impedance* of the element. For the resistor, inductor, and capacitor the impedances are

$$Z_R = R$$
$$Z_L = j\omega L \quad (10.47)$$
$$Z_C = \frac{1}{j\omega C}$$

The reciprocal of impedance is given the symbol $Y = 1/Z$ and is called *admittance.* We observe that resistance and conductance are each a special form of impedance and admittance. In the general case we can write

$$Z = \frac{V}{I} = \frac{1}{Y}$$

where V and I are phasor representations of any voltage-current pair. The impedance Z is a complex-valued quantity which is often written

$$Z = R + jX \quad (10.48)$$

where $R \triangleq \text{Re}[Z]$ is the resistance and $X \triangleq \text{Im}[Z]$ is called the *reactance.* From Equation (10.47) it is evident that the impedance of a single inductor or capacitor is a reactance. The real and imaginary parts of a complex admittance also have conventional names and symbols

$$Y = G + jB \quad (10.49)$$

where $G \triangleq \text{Re}[Y]$ is the conductance and $B \triangleq \text{Im}[Y]$ is called the *susceptance.*

Generalization of Phasor Network Analysis

The concepts of impedance and admittance are useful in constructing network models to assist in the solution process. A *phasor network model* is constructed by replacing each element by its impedance given by Equation (10.47). Each current and voltage in the network is the phasor equivalent of the sinusoidal currents and voltages in the original network. This is illustrated in Figure 10.7. KVL and KCL place constraints on an interconnected network, and we have indicated that *the same KCL and KVL must be satisfied by the currents and voltages in the original network and in the phasor network model.* That is, if

$$\sum_{\text{Closed loop}} v_k(t) = 0$$

around a loop in the original network, then substituting $v_k = V_{mk}\cos(\omega t + \phi_k)$ results in the phasor network equation

$$\sum_{\text{Closed loop}} V_{mk}e^{j\phi_k} = \sum_{\text{Closed loop}} V_k = 0$$

The same relationship applies to the KCL equations in both networks.

These results allow us to analyze phasor network models with impedances and admittances using the same formal methods we adopted in resistive network analysis, independently of the type of impedances (Z_R, Z_L, or Z_C) involved. The only difference will be the use of complex instead of real numbers in the numerical calculations.

The complete solution of a sinusoidal steady-state problem can be conveniently divided into three steps, as the following example shows.

Example 10.4

The parallel *GCL* network of Figure 10.8 is a common form of network. A sinusoidal cur-

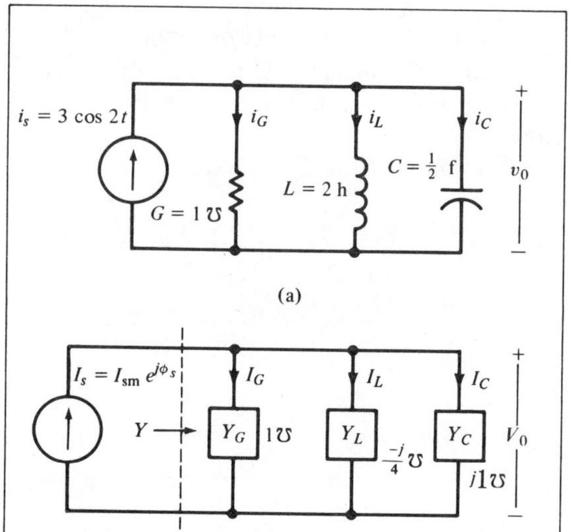

FIGURE 10.8 Parallel *GCL* network and its phasor model. (a) Time domain model; (b) phasor model.

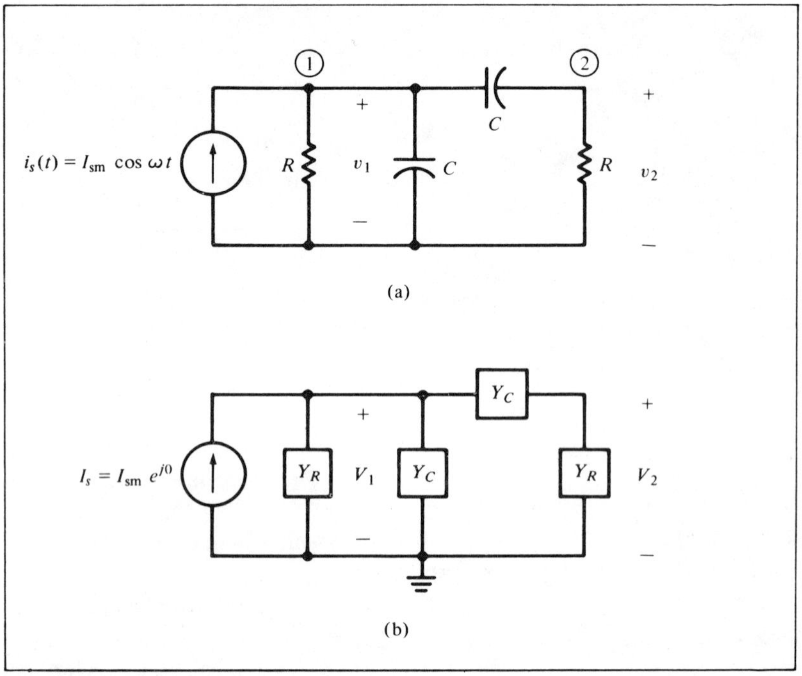

FIGURE 10.9 Two-node sinusoidal steady-state problem. (a) Time-domain model; (b) phasor model.

rent source is applied, and it is desired to determine the steady-state voltage $v_0(t)$.

Step 1. The phasor network model is drawn and analyzed using any method of analysis. We can treat the admittances of Figure 10.8(b) like conductances; from the simple analysis methods of Chapter 3, we have

$$Y = Y_G + Y_L + Y_C$$

This allows us to write

$$V_0 = \frac{I_s}{Y}$$

$$= \frac{I_s}{Y_G + Y_L + Y_C}$$

Step 2. The phasor representation of each impedance and admittance is inserted. This can be performed either symbolically or numerically, such as

$$V_0 = \frac{I_{sm}e^{j\phi_s}}{G + \dfrac{1}{j\omega L} + j\omega C}$$

or more directly

$$V_0 = \frac{3e^{j0°}}{1 + \dfrac{1}{j4} + j1}$$

$$= \frac{3e^{j0°}}{1 + j0.75}$$

$$= \frac{3e^{j0°}}{1.25e^{j36.8°}}$$

$$= 2.4e^{-j36.8°}$$

Step 3. The phasor representation of the

response is converted back to the time domain. In this case

$$v_0(t) = 2.4 \cos(2t - 36.8°)$$

This is a real-valued function of time and is the answer to the original network analysis problem.

In the following example, we use the analysis methods of Chapter 4 to solve a more complicated phasor network model.

Example 10.5

The sinusoidal steady-state response $v_1(t)$ is to be found for the network of Figure 10.9(a). In this example, we use the node analysis procedure of Chapter 4.

Step 1. The node equations are of the form

$$\begin{bmatrix} Y_R + 2Y_C & -Y_C \\ -Y_C & Y_C + Y_R \end{bmatrix} \begin{bmatrix} V_1 \\ V_2 \end{bmatrix} = \begin{bmatrix} I_s \\ 0 \end{bmatrix}$$

We find V_1, the phasor representation of $v_1(t)$, in symbolic form by inverting the node matrix

$$\begin{bmatrix} V_1 \\ V_2 \end{bmatrix} = \frac{1}{\Delta} \begin{bmatrix} Y_C + Y_R & Y_C \\ Y_C & Y_R + 2Y_C \end{bmatrix} \begin{bmatrix} I_s \\ 0 \end{bmatrix}$$

where $\Delta = Y_R^2 + 3Y_CY_R + Y_C^2$. The solution for V_1 is

$$V_1 = \frac{(Y_C + Y_R)I_s}{Y_R^2 + 3Y_CY_R + Y_C^2}$$

$$= \frac{(j\omega C + 1/R)I_s}{(1/R)^2 + 3j\omega C/R + (j\omega C)^2}$$

$$= \frac{R(j\omega RC + 1)I_s}{1 - \omega^2R^2C^2 + j3\omega RC}$$

Step 2. This is as far as the solution can be carried without knowledge of the element values (R and C) and the frequency of $i_s(t)$. If $R = C = 1$ and

$$i_s(t) = \sqrt{2} \cos t$$

then $\omega = 1$ and

$$V_1 = \frac{(j + 1)\sqrt{2}}{j3} = \frac{\sqrt{2}\, e^{j45°} \sqrt{2}}{3e^{j90°}}$$

$$= \tfrac{2}{3}e^{-j45°}$$

Step 3. The corresponding time-domain function is

$$v_1(t) = \tfrac{2}{3} \cos(t - 45°)$$

Application of Linear Network Principles

We can infer from the linearity of the I–V phasor relationships of each element that all

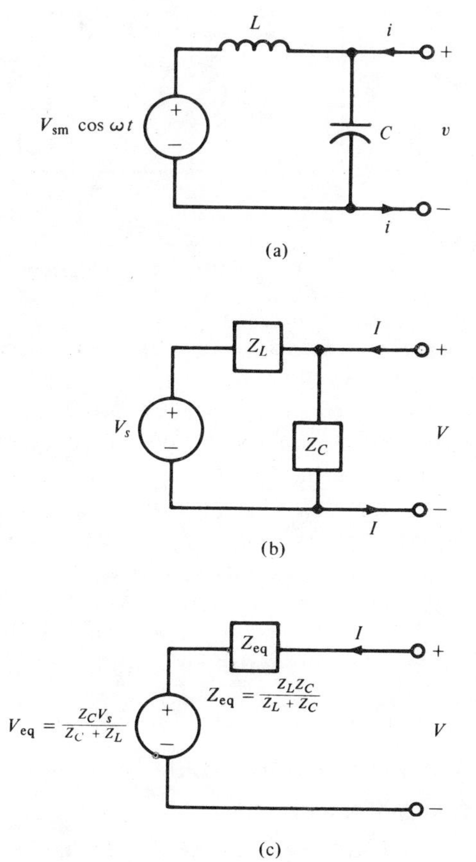

FIGURE 10.10 A sinusoidal steady-state Thevenin equivalent. (a) Time-domain model; (b) phasor model; (c) Thevenin equivalent.

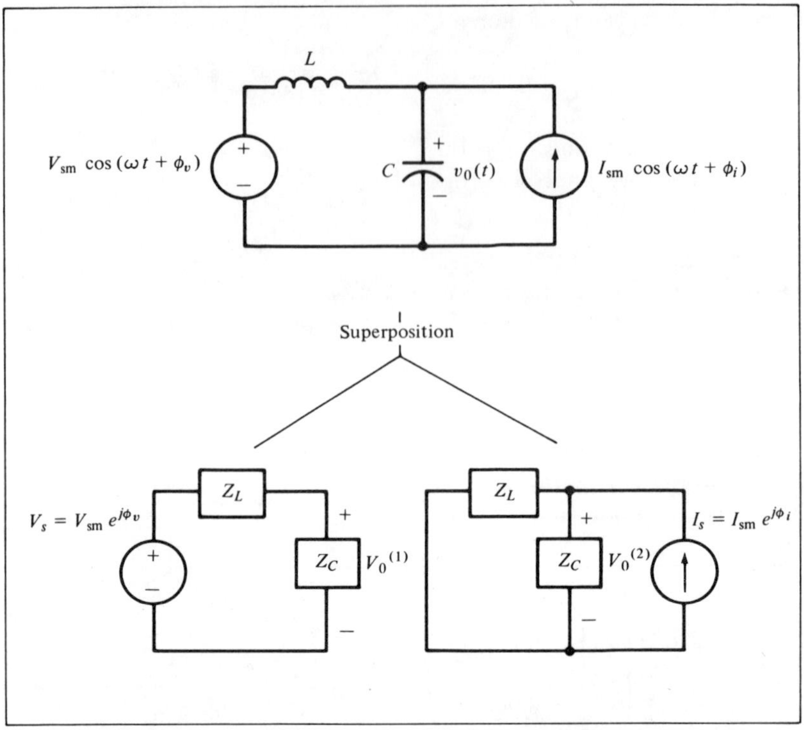

FIGURE 10.11 Application of the superposition principle.

the concepts of linear resistive network analysis apply. For example, we can define and determine a Thevenin equivalent as shown in Figure 10.10. Another result of linearity is the superposition principle of Chapter 6. As illustrated in Figure 10.11, we can apply sources *of the same radian frequency* (ω) either simultaneously or one at a time. The resultant phasor voltage V_0 is the sum of components $V_0^{(1)}$ and $V_0^{(2)}$. How would the superposition principle be used when sources have different frequencies?

10.6 Power in the Sinusoidal Steady State

Average Power

As we have indicated, one of the principal uses of the sinusoid waveform is in the transmission of energy. Accordingly, the calculation of power absorbed or generated is an important concept in sinusoidal steady-state analysis.

Consider a one-port network with the current and voltage definitions shown in Figure 10.12. We can represent $v(t)$ and $i(t)$ in the

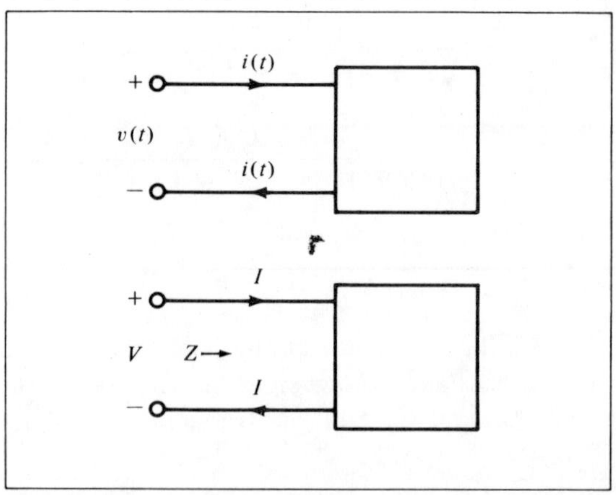

FIGURE 10.12 References for determination of power.

forms

$$v(t) = V_m \cos(\omega t + \phi_v) = \text{Re}[Ve^{j\omega t}]$$

$$i(t) = I_m \cos(\omega t + \phi_i) = \text{Re}[Ie^{j\omega t}]$$

where V and I are phasors. It is convenient to use the alternative forms

$$v(t) = \tfrac{1}{2}(V_m e^{j(\omega t+\phi_v)} + V_m e^{-j(\omega t+\phi_v)})$$
$$= \tfrac{1}{2}(V_m e^{j\phi_v}e^{j\omega t} + V_m e^{-j\phi_v}e^{-j\omega t})$$
$$= \tfrac{1}{2}(Ve^{j\omega t} + V^* e^{-j\omega t})$$
$$i(t) = \tfrac{1}{2}(Ie^{j\omega t} + I^* e^{-j\omega t})$$

where the asterisk denotes the complex conjugate.

Now consider the instantaneous power absorbed in the network:

$$p(t) = v(t)i(t)$$
$$= \tfrac{1}{4}(Ve^{j\omega t} + V^*e^{-j\omega t})(Ie^{j\omega t} + I^*e^{-j\omega t})$$
$$= \tfrac{1}{4}(VI^* + V^*I + VIe^{j2\omega t} + V^*I^*e^{-j2\omega t}) \tag{10.50}$$

$$= \tfrac{1}{4}(V_m I_m e^{j(\phi_v-\phi_i)} + V_m I_m e^{j(\phi_i-\phi_v)}) + \tfrac{1}{4}(V_m I_m e^{j(2\omega t+\phi_v+\phi_i)} + V_m I_m e^{-j(2\omega t+\phi_v+\phi_i)}) \tag{10.51}$$

$$= \tfrac{1}{2}V_m I_m \cos(\phi_v - \phi_i) + \tfrac{1}{2}V_m I_m \cos(2\omega t + \phi_v + \phi_i)$$
$$= V_{\text{eff}}I_{\text{eff}} \cos(\phi_v - \phi_i) + V_{\text{eff}}I_{\text{eff}} \cos(2\omega t + \phi_v + \phi_i) \tag{10.52}$$

The instantaneous power consists of two terms: (1) a constant, and (2) a double-frequency sinusoid alternating at twice the frequency of $v(t)$ and $i(t)$. Expanding the second term in Equation (10.52) by the double-angle formula we can write

$$p(t) = V_{\text{eff}}I_{\text{eff}} \cos\theta(1 + \cos(2\omega t + 2\phi_i)) - V_{\text{eff}}I_{\text{eff}} \sin\theta \sin(2\omega t + 2\phi_i) \tag{10.53}$$

where

$$\theta = \phi_v - \phi_i \tag{10.54}$$

is the phase angle *between* the voltage and the current. The two components of this equation are plotted in Figure 10.13. The first term is always positive, oscillating between 0 and $2V_{\text{eff}}I_{\text{eff}} \cos\theta$ at a frequency twice that of the current and voltage. The average value of this power component represents the average energy flow from left to right in Figure 10.12.

The second component of Equation (10.53) is plotted in Figure 10.13(b). This is also a fluctuating term with a frequency $2f$, but it has zero average value. When this term is positive, energy is flowlng into the network; when it is negative, energy is flowing out of the network! In fact, *the second power term in Equation (10.53) is just the rate of change of the energy stored in the network elements.* Because $p(t)$ is periodic, the associated average value is

$$P_{\text{av}} = \frac{1}{T}\int_0^T p(t)\,dt$$

which, using Equation (10.52), becomes

$$P_{\text{av}} = \frac{1}{T}\int_0^T [V_{\text{eff}}I_{\text{eff}} \cos(\phi_v - \phi_i) + V_{\text{eff}}I_{\text{eff}}\cos(2\omega t + \phi_v + \phi_i)]dt \tag{10.55}$$
$$= V_{\text{eff}}I_{\text{eff}} \cos(\phi_v - \phi_i)$$

Here the second term in the integral vanishes because it has a zero average value over the period $T = 2\pi/\omega$. This is directly evident from the form of Equation (10.53) and the plots of Figure 10.13.

Complex Power

We may also represent P_{av} in terms of the phasors I and V. First we write

$$V = V_m e^{j\phi_v} = \sqrt{2}\, V_{\text{eff}} e^{j\phi_v}$$

$$I^* = I_m e^{-j\phi_i} = \sqrt{2}\, I_{\text{eff}} e^{-j\phi_i}$$

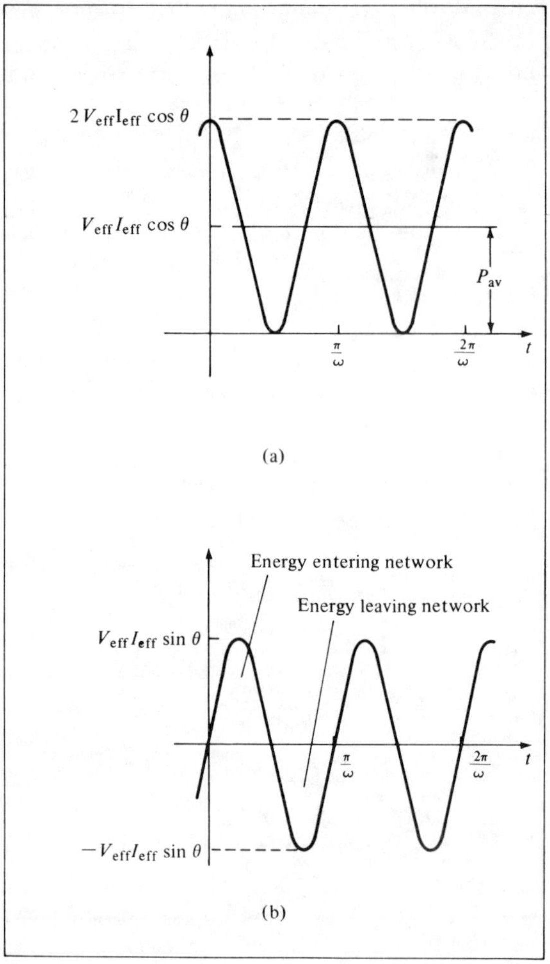

FIGURE 10.13 Plots of the two components of power flow in the sinusoidal steady state ($\phi_i = 0$).

Then we define

$$S \triangleq \tfrac{1}{2}VI^* \tag{10.56}$$

where S is known as the *complex* (or *phasor*) power. This can be written

$$S = \tfrac{1}{2}V_m e^{j\phi_v} I_m e^{-j\phi_i}$$

$$= \frac{V_m I_m}{2}\cos(\phi_v - \phi_i) + j\frac{V_m I_m}{2}\sin(\phi_v - \phi_i) \tag{10.57}$$

Comparing Equation (10.57) with (10.54) and (10.55) we see that the complex power can be written

$$S = P + jQ \tag{10.58}$$

where

$$Q \triangleq \tfrac{1}{2}Im[VI^*] = \frac{V_m I_m}{2}\sin(\phi_v - \phi_i)$$

$$= V_{eff} I_{eff} \sin\theta \tag{10.59}$$

$$P \triangleq \tfrac{1}{2}\,\mathrm{Re}[VI^*] = V_{eff} I_{eff}\cos\theta \tag{10.60}$$

Here, $P = P_{av}$ is the average or *active* power with which we are familiar, and Q is termed the *reactive* power. The average power, therefore, depends on the effective values of the voltage and currents and the *cosine of the angle between them*. This quantity

$$\cos(\phi_v - \phi_i) = \cos\theta = \mathrm{pf} \tag{10.61}$$

is defined as the *power factor* of the circuit with the convention that it is said to be *lagging* if the current lags the voltage, as it does in an inductive circuit. Thus, a power factor of 0.707 *leading* means that the current leads the voltage by $\cos^{-1}(0.707) = 45°$.

The quantity

$$Q = V_{eff} I_{eff} \sin\theta$$

measures the fluctuating component of power and is expressed in *volt-amperes reactive*, or *vars*. The sign convention chosen here allows one to distinguish between inductive and capacitive networks by the sign of Q. A positive value for Q implies an inductive network.

Example 10.6

The current and voltage flowing into a one-port network are given by

$$i(t) = 3\cos(2t + 45°)$$

$$v(t) = 2\cos(2t + 30°)$$

From the phasors

$$I = 3e^{j45°}$$

$$V = 2e^{j30°}$$

we determine

$$S = \tfrac{1}{2}VI^*$$
$$= \tfrac{1}{2}(2e^{j30°})(3e^{-j45°})$$
$$= 3\cos(-15°) + j3\sin(-15°)$$

We now solve

$$P = 3\cos(-15°)$$
$$= 2.898 \text{ watts}$$
$$Q = 3\sin(-15°)$$
$$= -0.75 \text{ vars}$$

The power factor is $\cos(15°) = 0.966$ leading.

Maximum Power Transfer Theorem

A common power calculation encountered in sinusoidal steady-state analysis of networks is the determination of the load adjustment which extracts the maximum average power from a real source, that is, an ideal source and its Thevenin equivalent impedance.

Referring to the Thevenin equivalent model of Figure 10.14, the expression for the power absorbed in Z_L is found as follows. The phasor current I is given as

$$I = \frac{V_{eq}}{Z_{eq} + Z_L}$$

$$= \frac{V_{eq}}{(R_{eq} + R_L) + j(X_{eq} + X_L)}$$

The average power absorbed in Z_L is then

$$P_{av} = \tfrac{1}{2}\,\text{Re}[V_L I^*]$$
$$= \tfrac{1}{2}\,\text{Re}[Z_L I I^*]$$
$$= \tfrac{1}{2}\,|I|^2\,\text{Re}[Z_L]$$

because II^* is the squared magnitude of the phasor I and a real number. We can therefore write

$$P_{av} = \frac{1}{2}\frac{|V_{eq}|^2 R_L}{(R_{eq} + R_L)^2 + (X_{eq} + X_L)^2} \tag{10.62}$$

In seeking the values of R_L and X_L that maximize this expression, we observe that X_L appears only in the denominator. The minimum value of the denominator occurs when $X_L = -X_{eq}$, for any value of R_L. This determines the desired value of X_L. The value of R_L can now be chosen to minimize the simplified expression

$$P_{av} = \frac{1}{2}\frac{|V_{eq}|^2 R_L}{(R_{eq} + R_L)^2} \tag{10.63}$$

When $R_L = 0\Omega$ and $R_L = \infty\,\Omega$, this expression is zero; for every other positive value of R_L, the expression is positive. This establishes the existence of a maximum in the interval $[0, \infty]$. Differentiation of Equation (10.63) to obtain the maximum gives

$$\frac{d}{dR_L}(P_{av}) = \tfrac{1}{2}|V_{eq}|^2 \left\{\frac{(R_{eq} + R_L)^2 - 2(R_{eq} + R_L)R_L}{(R_{eq} + R_L)^4}\right\} = 0$$

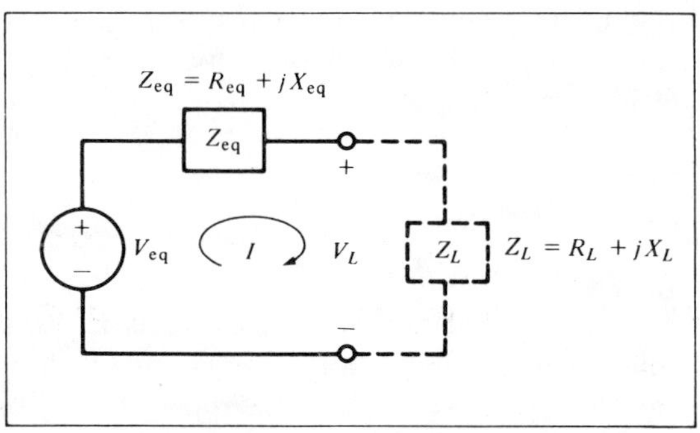

FIGURE 10.14 A Thevenin equivalent with load.

yielding

$$R_{eq}^2 - R_L^2 = 0$$

The only positive-valued solution for R_L is

$$R_L = R_{eq} \tag{10.64}$$

Therefore, regardless of the voltage source value V_{eq}, the maximum power transfer to the load Z_L occurs when

$$R_L = R_{eq}$$
$$X_L = -X_{eq} \tag{10.65}$$

This is the maximum power transfer theorem. The value of this maximum power is

$$(P_{av})_{max} = \frac{1}{2}\left(\frac{|V_{eq}|^2}{4R_{eq}}\right) \quad \text{(if } |V_{eq}| \text{ is peak value)} \quad (10.66)$$

$$= \frac{|V_{eq}|^2}{4R_{eq}} \quad \text{(if } |V_{eq}| \text{ is rms value)} \quad (10.67)$$

It is called the *available power* of the equivalent source. No matter what connection is made to the source, this is the maximum power we can possibly extract from it.

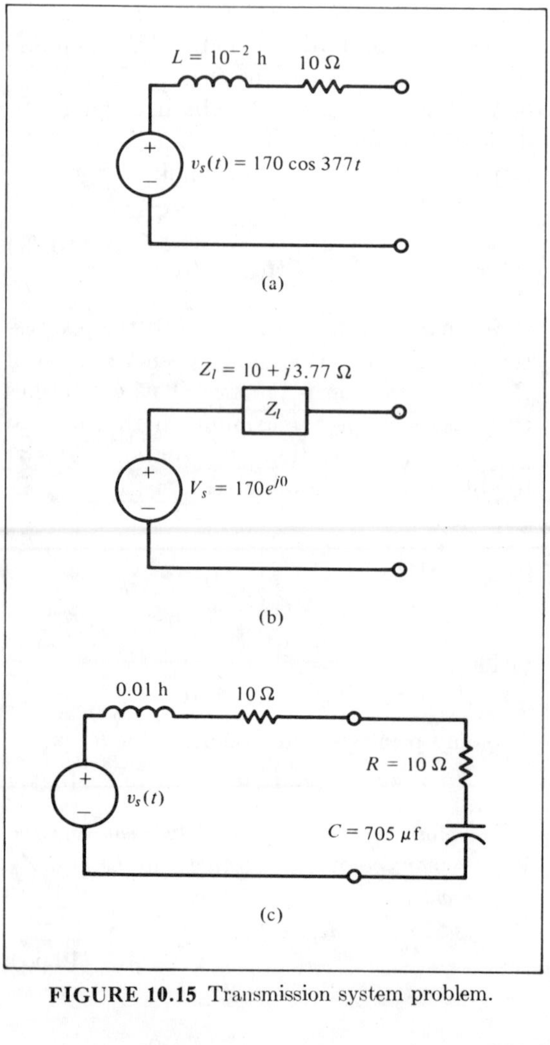

FIGURE 10.15 Transmission system problem.

Example 10.7

The model of a signal transmission system is shown in Figure 10.15; the source may represent an electric generator, and the series resistor and inductor could model the characteristics of a signal line. We are to find the maximum power transfer.

From the phasor model, we conclude that the maximum power is obtained from the system when the load impedance has the value

$$R_L = R_{eq} = 10$$

$$X_L = -X_{eq} = -j3.77$$

If we are asked to realize this value of X_L, we could set

$$-j3.77 = \frac{-j}{\omega C} = \frac{-j}{377C}$$

The capacitance value is then found to be 7.05×10^{-4} f. The power absorbed under these conditions is then

$$P_{av} = \frac{(170)^2}{80} = 361 \text{ watts}$$

It is important to realize that this choice of C maximizes power at only one frequency; if ω changes, then the impedance of L changes, and the value of C necessary for cancellation of X_{eq} also changes.

10.7 Scaling

Impedance Scaling

As was pointed out in Chapter 5, it is often convenient to change all currents and/or voltages by a constant *scale factor* in performing an analysis. A similar procedure applies to networks being analyzed in the sinusoidal steady state; it is now termed *impedance scaling*.

The idea of impedance scaling is a simple one. If all of the impedances in a network are multiplied by a fixed factor k, the new network will behave exactly the same as the original except that the ratios of voltage to current will all be multiplied by k. The necessary changes in the element values of an RLC network are

$$R \to kR$$
$$L \to kL \qquad (10.68)$$
$$C \to C/k$$

For example, the impedance of a series RLC network is given by

$$\frac{V_s}{I} = Z = R + j\left(\omega L - \frac{1}{\omega C}\right) \qquad (10.69)$$

If all three elements are scaled as indicated in (10.68), we get the network of Figure 10.16. The impedance of this scaled network is

$$Z' = kR + j\left(\omega kL - \frac{k}{\omega C}\right) = kZ \qquad (10.70)$$

If the excitation voltage is the same in both the original and scaled networks, the current that flows in the scaled network will be reduced by the factor $1/k$.

Frequency Scaling

The impedances of inductors and capacitors are a function of the sinusoidal steady-state

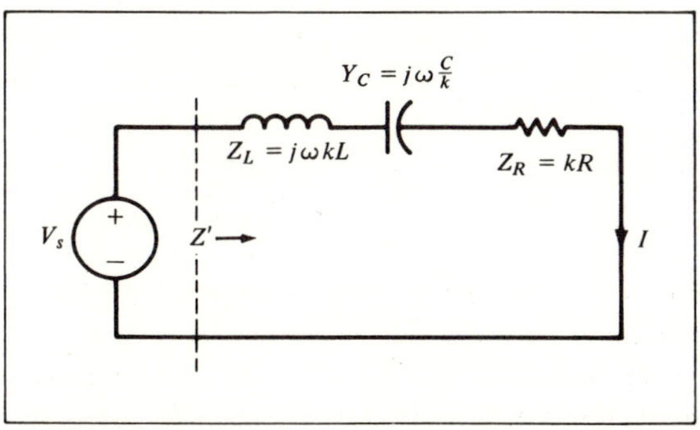

FIGURE 10.16 Series network after impedance scaling.

frequency. If we reduce the value of an inductor by k', the impedance of that inductor at a particular frequency ω_1 is

$$Z(j\omega_1) = j\omega_1\left(\frac{L}{k'}\right)$$

Clearly, if we increase the frequency to $k'\omega_1$, the impedance is

$$Z(jk'\omega_1) = jk'\omega_1\left(\frac{L}{k'}\right) = j\omega_1 L$$

which is the impedance of the original inductor at ω_1! Similarly, for a capacitor, reducing its value by k' does not change the impedance if the frequency is increased by k'. Therefore, if for every inductor and capacitor in a network we scale the values

$$L \to \frac{L}{k'}$$
$$\qquad (10.71)$$
$$C \to \frac{C}{k'}$$

the resulting network will behave exactly as the original one at the scaled frequency $\omega \to k'\omega$.

Frequency scaling also scales time. This can be seen from the time-domain element relations. For an inductor scaled by (10.71),

$$v_L' = \frac{L}{k'}\frac{di_L'}{dt}$$

which can be written

$$v_L' = L\frac{di_L'}{d(t/k')}$$

Therefore, *scaling the frequency by k' is the same as scaling the time by $1/k'$.*

Example 10.8

In Example 10.4, phasor analysis was illustrated for the parallel GCL circuit of Figure 10.17(a). The excitation frequency is very low, 0.319 hertz (hz), and one of the element values, $C = \frac{1}{2}$ f, is impractically large. Suppose we use

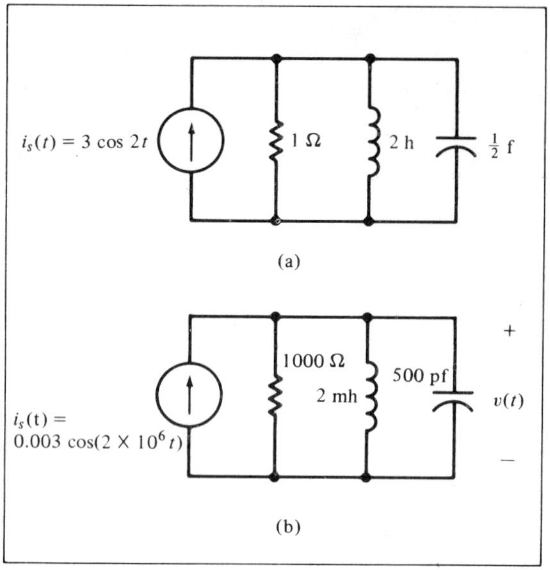

FIGURE 10.17 Scaling example. (a) Before scaling; (b) after scaling ($k = 10^3$, $k' = 10^6$).

frequency scaling to raise the frequency to 319 khz and impedance scaling to reduce the required excitation current, for the same voltage response, to 3 ma. The required frequency scaling is

$$\frac{319 \times 10^3}{0.319} = 10^6 = k'$$

and the required impedance scaling is

$$\frac{3}{3 \times 10^{-3}} = 10^3 = k$$

Applying both scalings gives

$$G \rightarrow G' = \frac{1}{10^3} = 10^{-3} \text{ ℧}$$

$$C \rightarrow C' = \frac{1}{2} \times \frac{1}{10^6} \times \frac{1}{10^3} = 500 \text{ pf}$$

$$L \rightarrow L' = 2 \times \frac{1}{10^6} \times 10^3 = 2 \times 10^{-3} \text{ h}$$

The complete scaled network is given in Figure 10.17(b).

Directly from the analysis of Example 10.4 we know that

$$v(t) = 2.40 \cos(2 \times 10^6 t - 36.8°)$$

APPLICATION

Load-Flow Problem

The economical, reliable transmission of electrical power is an important area of study. The demands for electrical services vary regularly over a 24-hr period and increase over longer periods of time; in addition, unexpected disturbances can occur due to component failure and environmental factors. The predictable changes can be accommodated by design of the sort we will discuss; the unexpected interruptions must often be met with short term, relatively inefficient use of emergency resources.

The problem of allocation of generators to meet power demands is an example of the *load-flow problem*. In this problem, the transmission of power is viewed as occurring between nodes of an electrical network representing the complete power system (see Figure 10.18). These nodes are connected by impedances representing the characteristics of the transmission lines. Usually, the impedances are inductive in nature and so have a positive imaginary part. Each node is connected to a

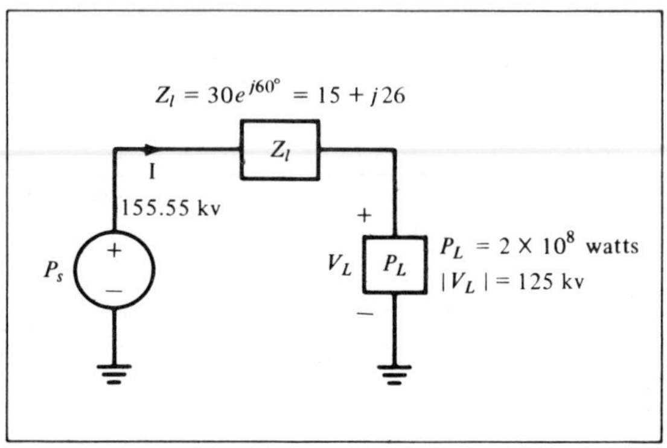

FIGURE 10.18 Load flow problem.

```
C***   PROGRAM TO SOLVE LOADFLOW PROBLEM OF FIGURE 10.18
C      USING NEWTON'S METHOD (LOADFLO)
      IMPLICIT REAL*8(A-H,O-Z)
      REAL *8 IR,II,IM,IP
      DIMENSION A(4,4),B(4)
      M=0
C***   ASSIGN GENERATOR VOLTAGE
      VS=1.5555D5
C***   ASSIGN INITIAL VALUES TO REAL AND IMAGINARY PARTS
C      OF LOAD VOLTAGE AND CURRENT
      VLR=125D3
      VLI=0.D0
      IR=2000.
      II=0.D0
      WRITE(6,6)
    6 FORMAT(4X,'ITERATION    LOAD VOLTAGE',7X,'LOAD CURRENT',4X,'GENERATOR'/
     1 12X,' MAGNITUDE   PHASE  MAGNITUDE   PHASE    POWER')
4     DO 2 J=1,4
      B(J)=0.D0
      DO 2 K=1,4
2     A(J,K)=0.D0
C***   CONSTRUCT PARTIAL DERIVATIVE MATRIX
      A(1,1)=-1.D0
      A(1,3)=-15.D0
      A(1,4)=26.D0
      A(2,2)=-1.D0
      A(2,3)=-26.D0
      A(2,4)=-15.D0
      A(3,1)=IR
      A(3,2)=II
      A(3,3)=VLR
      A(3,4)=VLI
      A(4,1)=-2.D0*VLR
      A(4,2)=-2.D0*VLI
C***   CONSTRUCT RIGHT HAND SIDE
      B(1)=VLR-VS+15.D0*IR-26.D0*II
      B(2)=VLI+26.D0*IR+15.D0*II
      B(3)=-IR*VLR-II*VLI+2.D8
      B(4)=VLR**2+VLI**2-25.D10/16.D0
C***   SOLVE LINEARIZED EQUATIONS
      CALL REDUCE(A,4,B,4)
      ERR=0.D0
      DO 5 J=1,4
5     ERR=ERR+B(J)**2
C***   UPDATE ITERANTS
      VLR=VLR+B(1)
      VLI=VLI+B(2)
      IR=IR+B(3)
      II=II+B(4)
      VM=DSQRT(VLR*VLR+VLI*VLI)
      VP=180.D0/3.14159D0*DATAN(VLI/VLR)
      IM=DSQRT(IR*IR+II*II)
      IP=180.D0/3.14159D0*DATAN(II/IR)
      POWER=VS*IR
      WRITE(6,3)M,VM,VP,IM,IP,POWER
    3 FORMAT(5X,I3,4X,F10.3,2X,F5.1,2X,F10.5,2X,F5.1,2X,F10.0)
C***   CONVERGENCE TEST
      IF(ERR.LT.1.D0)STOP
      M=M+1
      GO TO 4
      END
```

FIGURE 10.19 Program to solve load flow problem.

reference or ground node by an admittance that represents the load on the power system at that point.

A simple two-node power system is shown in Figure 10.18, where phasors are effective values. A 155.55-kv generator provides power

ITERATION	LOAD VOLTAGE MAGNITUDE	PHASE	LOAD CURRENT MAGNITUDE	PHASE	GENERATOR POWER
0	130596.477	-16.8	1619.71147	-8.9	248880000.
1	125170.738	-15.2	1592.65301	-16.7	237337051.
2	125000.234	-15.3	1600.49983	-16.7	238423020.
3	125000.000	-15.3	1600.51063	-16.7	238424514.

FIGURE 10.20 Newton iteration results.

to a load of 200 Mw (a typical value for a city of 1 million population) at 125 kv. The problem is to determine the current I and the power P_s which the generator produces.

The specification of the average power at the load rather than a value for the admittance to ground makes the problem nonlinear. To see this, we write

$$F_1 = -V_L + V_s - IZ_l = 0 \tag{10.72}$$

$$F_2 = (-200 \times 10^6) + \text{Re}[I^*V_L] = 0 \tag{10.73}$$

$$F_3 = -|V_L|^2 + (1.25 \times 10^5)^2 = 0 \tag{10.74}$$

Both Equations (10.73) and (10.74) are nonlinear in I and/or V_L.

If we examine all the components of these complex quantities, we observe that there are only four unknowns:

1. Real and imaginary parts of V_L.
2. Real and imaginary parts of I.

Because Equation (10.72) represents two real-valued equations and Equations (10.73) and (10.74) each represent one equation, there are four nonlinear equations in four unknowns.

Such systems of equations ($F_i = 0$) may be solved by linearization of the equations and iteration of the linearized equations using Newton's method (see Section 5.6). The reader is invited to write the real-and imaginary-part equations and perform this linearization. A program to perform the iteration is given in Figure 10.19, with the variables ordered as Re[V_L], Im[V_L], Re[I], Im[I]. The results are shown in Figure 10.20 at each iteration. The final load and generator current is 1600 amps. The flow chart for Figure 10.19 is shown in Figure 10.21.

In practice, the generator voltage would be adjusted to reduce generator power as much as possible because the fuel consumed is a function of P_s. Therefore, rather than conducting a single analysis as shown in Figure 10.20, we

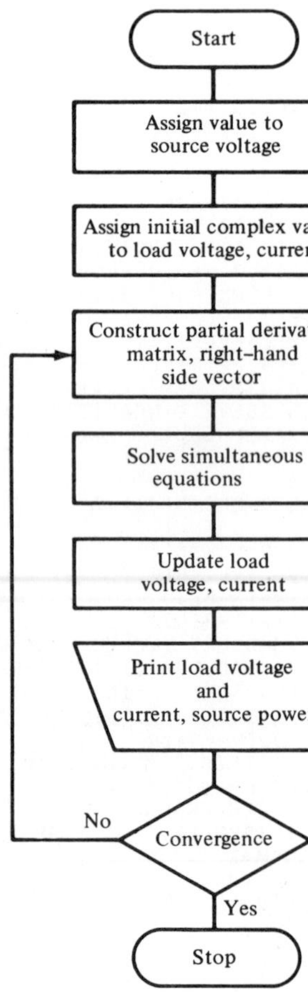

FIGURE 10.21 Flow chart of program of Figure 10.19.

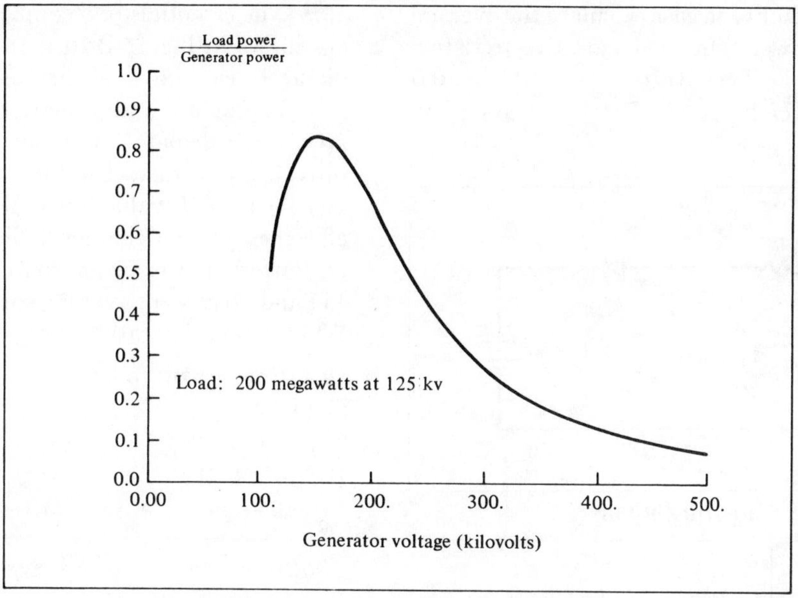

FIGURE 10.22 Power efficiency analysis.

could vary V_s and determine the resultant generator power. For the problem at hand, the resultant plot is shown in Figure 10.22, where the efficiency

$$\eta = \frac{\text{load power}}{\text{generator power}}$$

$$= 2 \times \frac{10^8}{P_s}$$

is shown. Our original choice of $V_s = 155.55$ kv is found to be optimum, with $\eta = 0.86$. Note that this does not occur at maximum power transfer, because the load current and load voltage are nearly in phase, whereas Z_l has a large phase angle. (Why does the curve of Figure 10.22 end at a $V_s \cong 110$ kv?)

PROBLEMS

10.1 Each of the voltage waveforms specified below are periodic outside of the specified time interval. Sketch each waveform *carefully* for at least two periods and calculate the average and effective values.

(a) $v(t) = V_m \sin(\omega t) \quad \left(0 \leqq t \leqq \frac{\pi}{\omega}\right)$

$v(t) = 0 \quad \left(\frac{\pi}{\omega} \leqq t \leqq \frac{2\pi}{\omega}\right)$

(b) $v(t) = V_m \cos(\omega t) \quad \left(-\frac{\pi}{2\omega} \leqq t \leqq \frac{\pi}{2\omega}\right)$

$v(t) = 0 \quad \left(\frac{\pi}{2\omega} \leqq t \leqq \frac{3\pi}{2\omega}\right)$

(c) $v(t) = V_m \sin(\omega t) \quad \left(0 \leqq t \leqq \frac{2\pi}{\omega}\right)$

$v(t) = 0 \quad \left(\frac{2\pi}{\omega} \leqq t \leqq \frac{4\pi}{\omega}\right)$

(d) $v(t) = V_m \,|\cos(\omega t)|$ (for all t).

10.2 The currents flowing in a 10-Ω resistor are

given below. For each case calculate the average electrical power supplied to the resistor. (a) $i(t) = 1 + \cos(377t)$. (b) $i(t) = 0.5 + \sqrt{2}\cos(377t) + \cos(754t)$. (c) $i(t) = |10\cos(377t)|$.

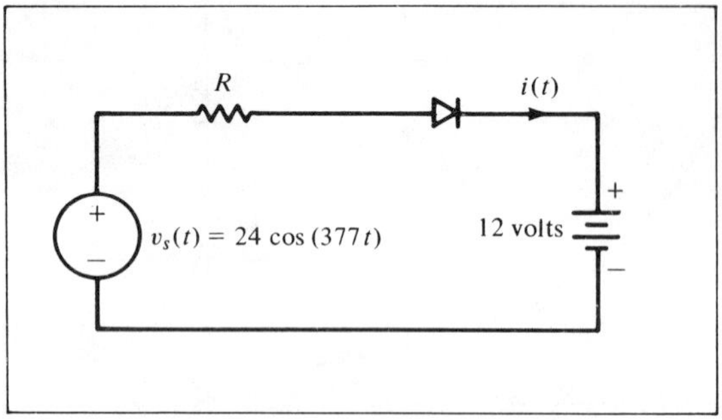

FIGURE P 10.1

10.3 The essentials of a simple battery charger are shown in Figure P 10.1. (a) Assuming that diode D can be modelled as an *ideal diode* [see Figure 3.1 and Equation (3.2)], sketch $i(t)$ and calculate the value of R needed to provide an average charging current of 5 amps. (b) With the R value found above, what is the efficiency of this charger (efficiency = power to battery/power from source)? (c) Repeat (a) and (b) for an average current of 10 amps. What dissipation rating must the resistor have?

10.4 Convert the following complex numbers to polar form.

(a) $3 + j4$.
(b) $-5 + j12$.
(c) $10 - j5$.
(d) $-2 - j10$.
(e) $j10$.
(f) $-j2$.

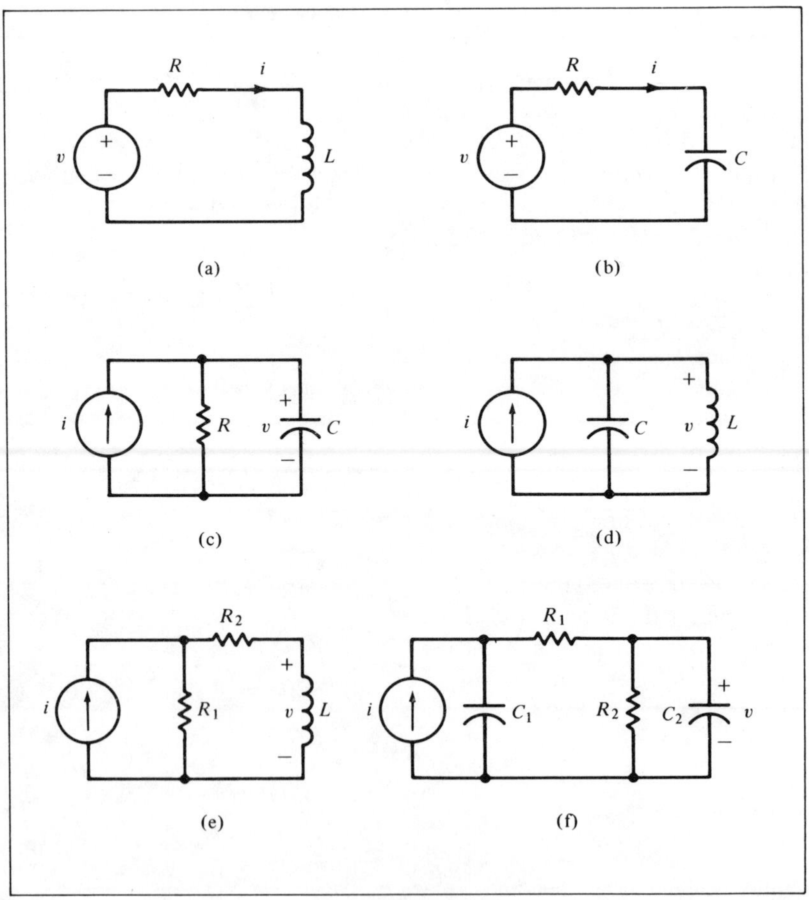

FIGURE P 10.2

10.5 Convert the following complex numbers to rectangular form.

(a) $3e^{j30°}$ (d) $-1e^{j90°}$
(b) $7e^{-j45°}$ (e) $-10e^{j215°}$
(c) $8e^{-j27°}$ (f) $0.2e^{-j90°}$

10.6 Represent the following sinusoidal functions of time in phasor form.

(a) $2 \cos 3t$. (e) $12 \cos(10t + 315°)$.
(b) $5 \cos(2t + 30°)$. (f) $-2 \cos(0.1t)$.
(c) $\sin 10t$. (g) $2 \sin(3t - 45°)$.
(d) $10 \cos(100t - 45°)$.

10.7 (a)–(f) Let V and I be the phasor representations of v and i in the following networks. Determine whenever possible whether V leads or lags I. If you cannot determine this, explain why. Your answer must be independent of the element values, which are assumed positive (Figure P 10.2), but may depend on ω.

10.8 (a)–(c) Three networks that are sometimes used for the compensation of feedback amplifiers are shown in Figure P 10.3. For each network, determine the values of ω for which the output voltage $v_0(t)$ lags the input voltage $v_s(t)$ in the sinusoidal steady state.

10.9 (a)–(i) Verify the phase calculations used to determine the sinusoidal steady-state outputs shown in each part of Figure P 10.4. Also find the magnitude in each case and the corresponding output function of time.

10.10 Find $v_0(t)$ in the steady state for the network of Figure P 10.5.

10.11 If $v_s(t) = \sqrt{2} \cos(2t - 45°)$, what value of C_0 will maximize the sinusoidal steady-state current through R_0 (Figure P 10.6)? [*Hint:* Apply Thevenin's theorem to the network to the left of the dotted lines.]

10.12 For the network in Figure P 10.7, $i_1(t) = 2 \cos 100t$, $i_2(t) = 2\sqrt{2} \cos(100t - 45°)$, and $i_3(t) = 5 \sin 100t$. Find an expression for $i(t)$ that has the form $A \cos(100t + \theta)$.

10.13 The network in Figure P 10.8 has been operating in the steady state with the switch S as shown. At $t = 0$ the switch is moved to ②. Find $v_0(t)$ for $t > 0$.

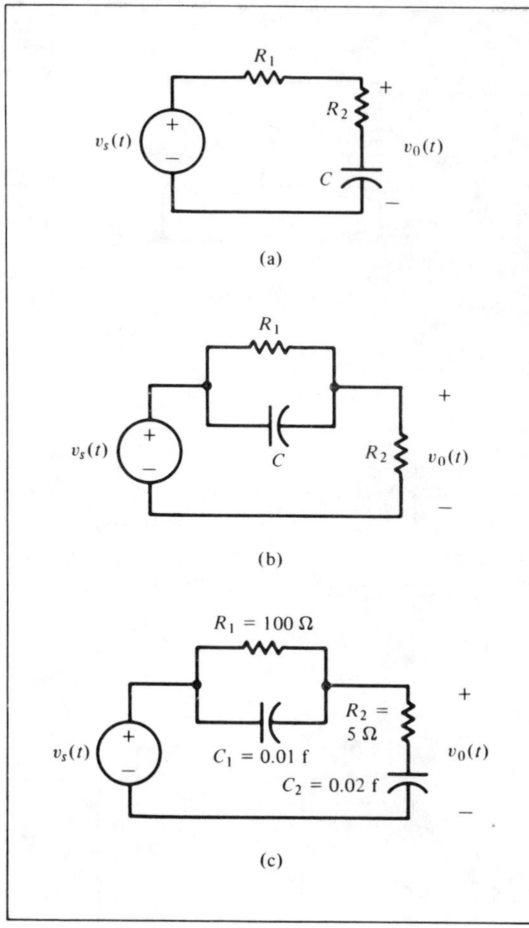

FIGURE P 10.3

10.14 (a)–(c) Construct the complete sinusoidal steady-state Thevenin equivalents of the networks of Figure P 10.9. The phase of Z_{eq} is given as a check.

10.15 (a)–(c) Find the maximum power that can be extracted from the port terminals of the networks of Figure P 10.9.

10.16 The equivalent circuit of a one-port in the sinusoidal steady state is shown in Figure P 10.10 at $\omega = 1$ rad/sec. What can you de-

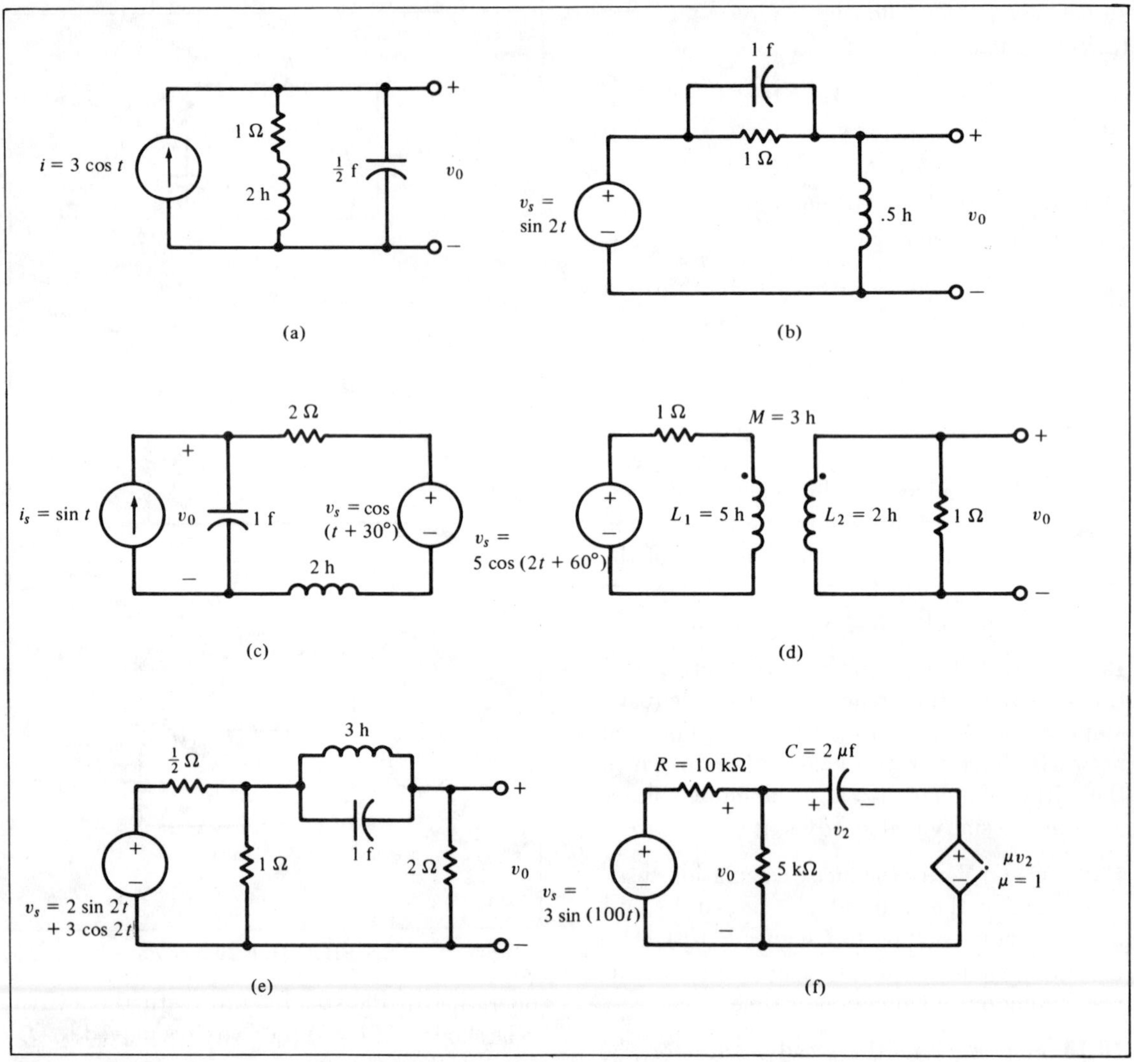

FIGURE P 10.4 (a) $V_0/I_s = Ae^{-j26.5°}$; (b) $V_0/V_s = Ae^{j18.4°}$; (c) $V_0/V_s = A_1e^{j243.4°}$, $V_0/I_s = A_2e^{-j71.6°}$; (d) $V_0/V_s = Ae^{-j12.1°}$; (e) $V_0/V_s = Ae^{j13.2°}$; (f) $V_0/V_s = Ae^{-18.4°}$.

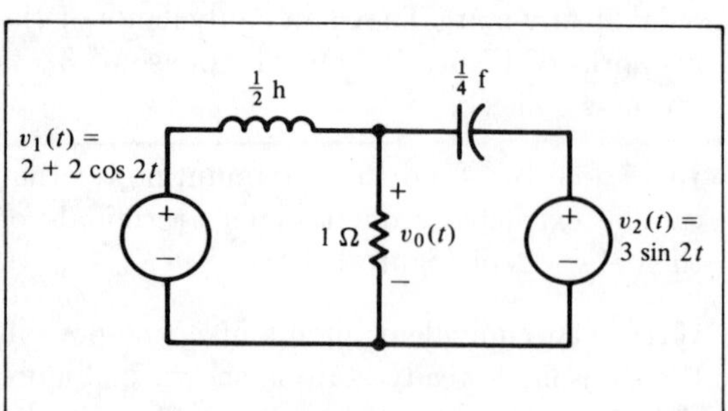

FIGURE P 10.5

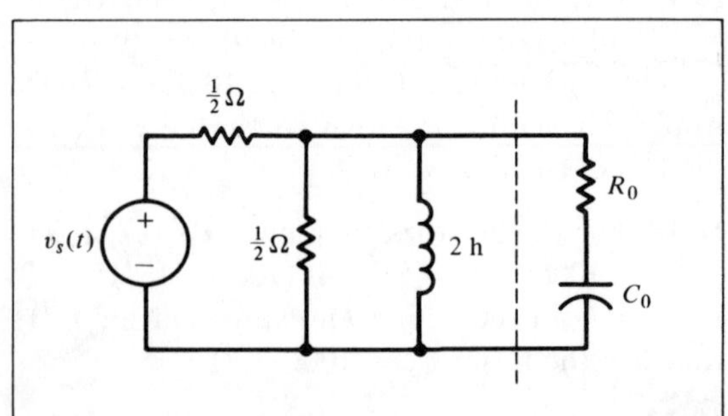

FIGURE P 10.6

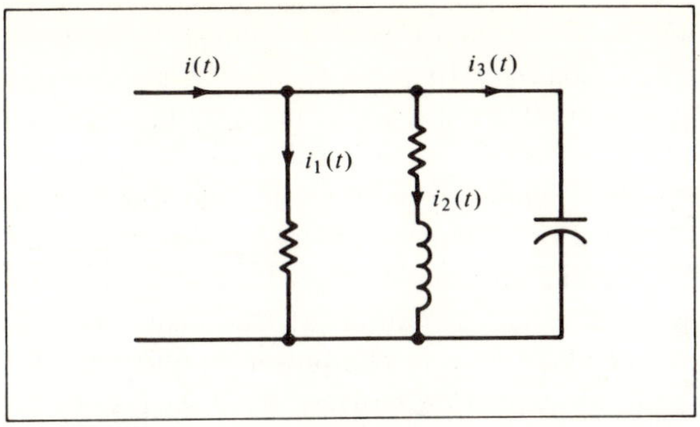

FIGURE P 10.7

duce about the equivalent circuit when $\omega = 2$ rad/sec?

10.17 At a *single* sinusoidal frequency, the two networks in Figure P 10.11 have the same impedance Z. Develop equations expressing $C_p(R_s, C_s, \omega)$ and $R_p(R_s, C_s, \omega)$ and their inverses $C_s(C_p, R_p, \omega)$ and $R_s(C_p, R_p, \omega)$ under this assumption of equivalence. Under what conditions is $C_s \cong C_p$, and when this is true how are R_s and R_p related? (Note that similar relations can be developed for the equivalence between series and parallel RL circuits at a single frequency.)

10.18 (a) Show for the network of Figure P 10.12 that $V_0 = 0$ if $Z_1 Z_2 = Z_4 Z_3$. (b) If R_1 and R_2 are fixed, R_3 and C are adjustable, and if R_4 and L are unknown, then R_3 and C can be adjusted until $V_0 = 0$. For these conditions, find R_4 and L in terms of the other element values and the radian frequency ω. This is termed a "bridge" circuit and is quite useful in measuring component values such as R_4 and L.

10.19 Repeat Problem 10.18 when $Z_3 = R + 1/j\omega C$.

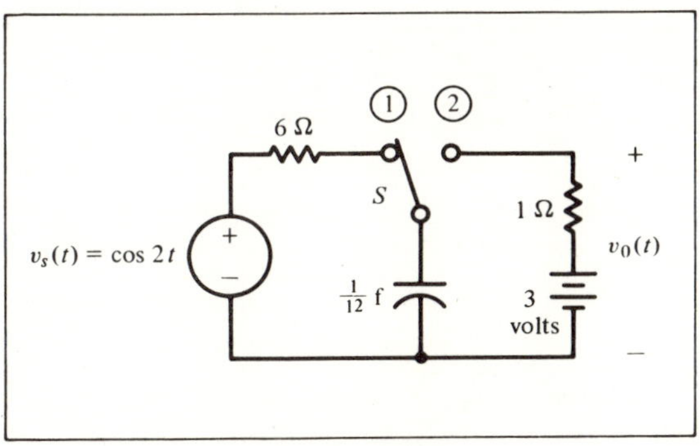

FIGURE P 10.8

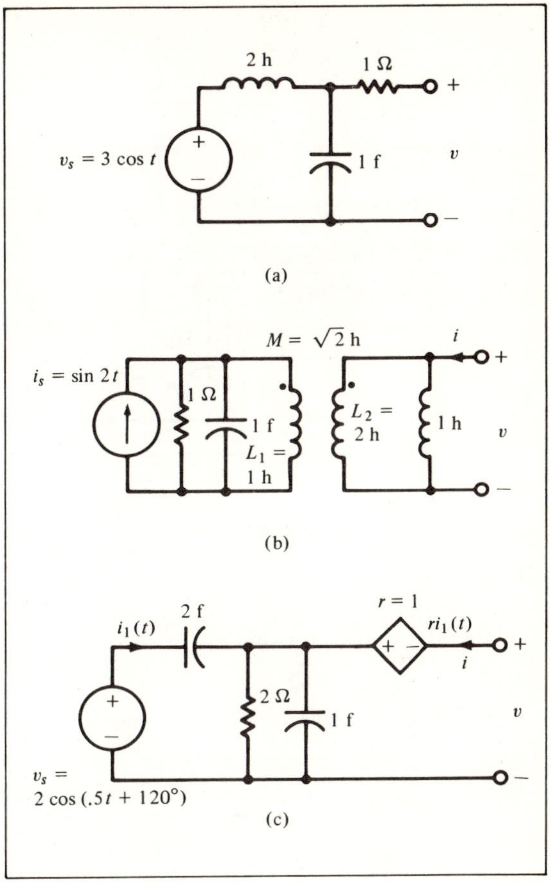

FIGURE P 10.9 (a) $Z_{eq} = Ae^{-j63.4°}$; (b) $Z_{eq} = Ae^{-j26.6°}$; (c) $Z_{eq} = Ae^{-j28.4°}$.

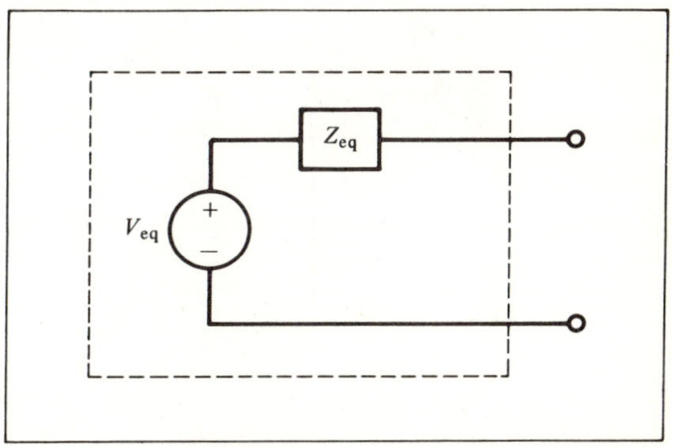

FIGURE P10.10 $Z_{eq} = 1 - j1$; $V_{eq} = e^{j0}$.

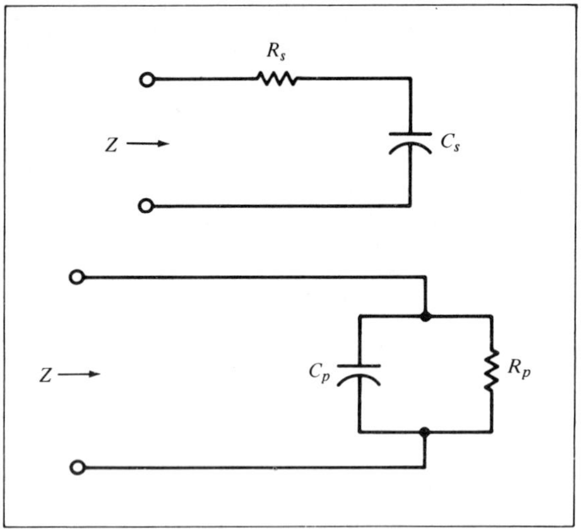

FIGURE P 10.11

10.20 The inductor in Figure P 10.13 is adjusted until the average power dissipated is 3600 watts. Find the value of L and the power factor.

10.21 A 60-hz power system is modelled in Figure P 10.14. When the effective voltage at the load is 440 volts, it absorbs an average power of 50 kw at a power factor of 0.8 logging. (a) If switch S is open, calculate the required *effective* generator voltage V_1 and the power supplied by that generator. (b) Determine the capacitor that must be placed across the load to raise the power factor at the load to 1.0. Recalculate V_1 and the power supplied by V_1 with this capacitor in place. (c) If the generator V_2 is synchronized with V_1, determine the effective value of V_2 and its phase angle relative to V_1 so that no current will flow when S is closed.

10.22 (a) What capacitor would you have to connect in series with a 115-volt soldering iron (which acts like a resistor) to reduce the heat supplied to the iron from 100 to 50 watts? Assume that the house wall plug acts as a 60-hz sinusoidal voltage source. (b) Repeat (a), replacing the capacitor by an inductor. Consider what would happen if you used each

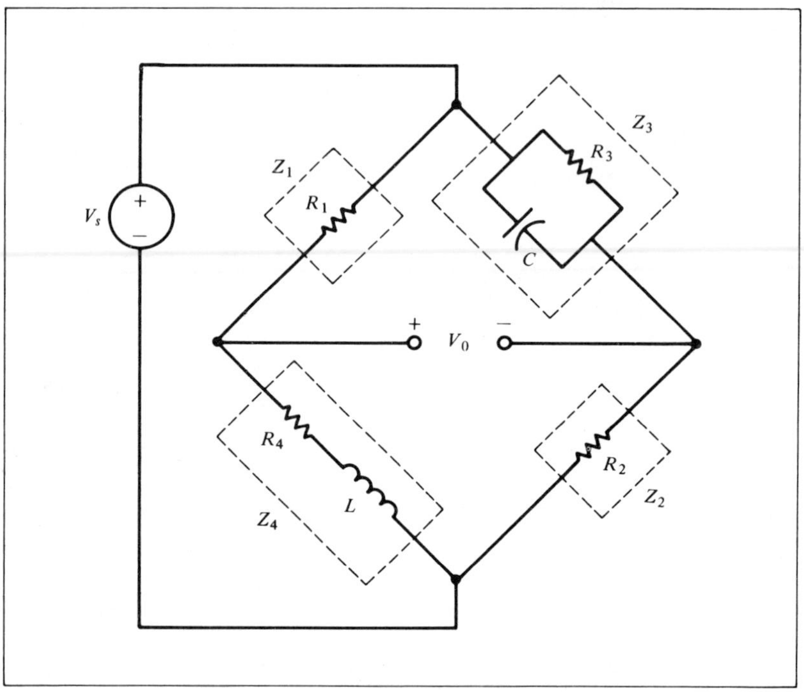

FIGURE P 10.12

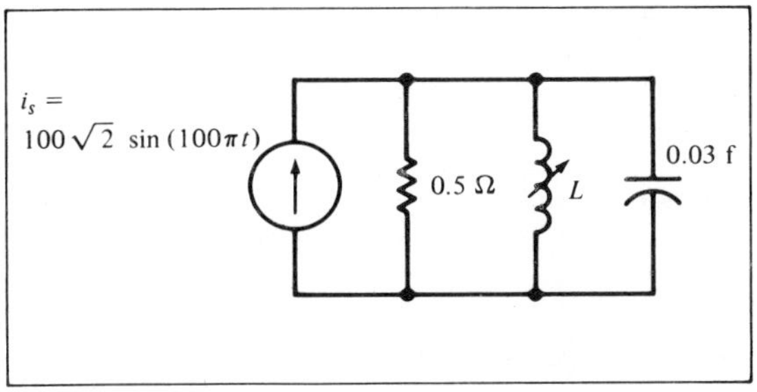

FIGURE P 10.13

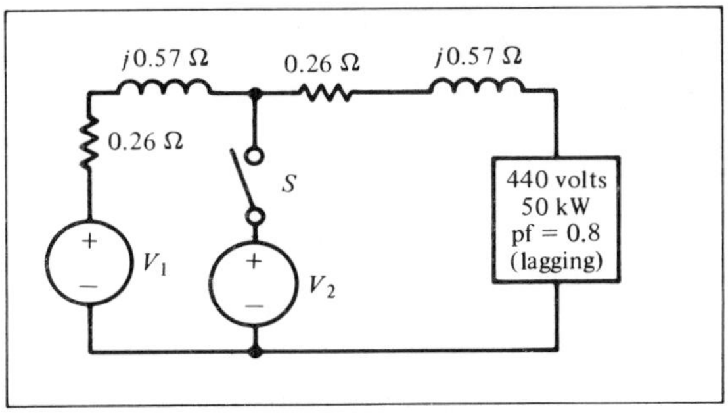

FIGURE P 10.14

circuit in an aircraft which has a 400-hz 115-volt ac system.

10.23 We know that maximum power is delivered to a load when the source and load resistors are equal. When they are unequal, an LC network can be used to make unequal source and load resistors "appear" equal to each other. Show that in Figure P 10.15, the 5-Ω load

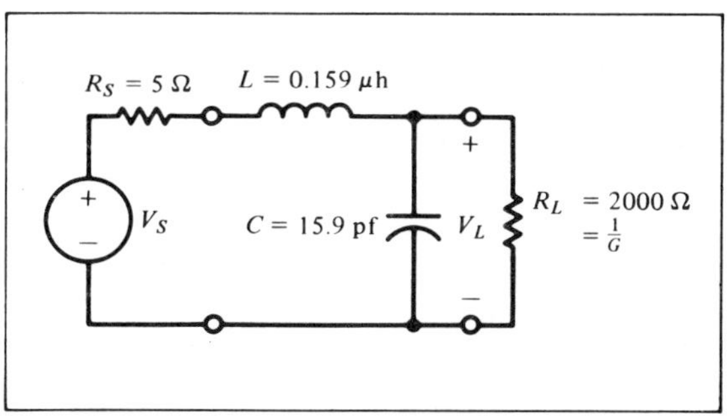

FIGURE P 10.15

appears as a 2000-Ω resistance to R_L, at a frequency of 10^8 hz. Is this match still preserved at 10^7 hz? Determine the power consumed in R_L at each frequency if $v_s(t) = \sqrt{2}\cos(\omega t)$.

10.24 (a) The voltage source in Figure P 10.16 is sinusoidal with an effective value of 0.1 volt and a frequency of 100 Mhz. Determine the values needed for C_L and R_L to maximize the

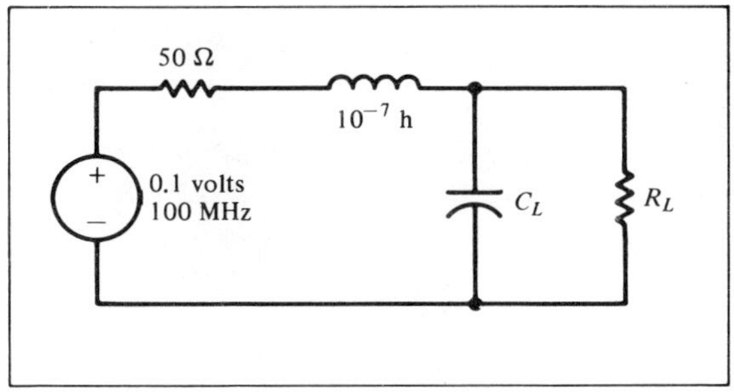

FIGURE P 10.16

power dissipated in R_L. What is this maximum power? (b) If $C_L = 0$, what value of R_L will maximize the power supplied to R_L? What is the maximum power in this case?

10.25 (a)–(c) In Figure P 10.17, a current $i(t) = 3\cos(t)$ results in the sinusoidal steady-state voltage $v(t) = 12\cos(t - 45°)$. Find scaled values of R, L, and C so that: (a) $v(t)$

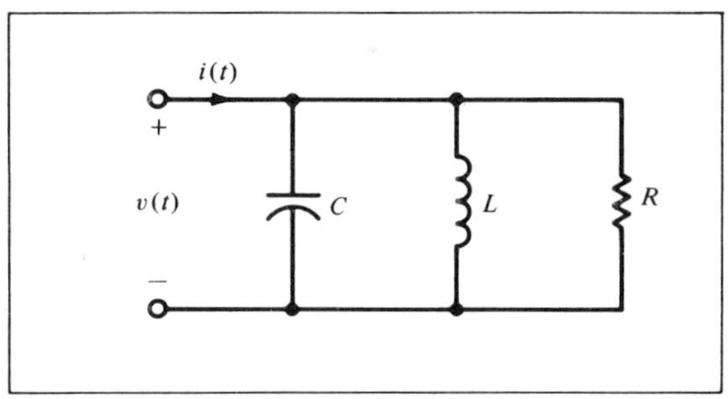

FIGURE P 10.17

$= 6\cos(t - 45°)$ for the same $i(t)$ as above. (b) $v(t) = 0.12\cos(10^4 t - 45°)$ for $i(t) = 3\cos(10^4 t)$. (c) $v(t) = V\cos(\omega t - 45°)$ for $i(t) = 12\cos(\omega t)$.

10.26 The network of Figure P 10.4(f) is scaled so that $C = 2\,f$ and $R = 1\ \Omega$. Show the scaled values of all other elements in the network, including the independent voltage source (which maintains its peak value of 3 volts after scaling).

10.27 Use frequency and impedance scaling to alter the LC matching network of Figure P 10.15 so that a 100-Ω source resistor appears as 40,000 Ω to R_L at a frequency of 10^7 hz.

REFERENCES AND ADDITIONAL READING

10.1 W. H. HAYT, JR., AND G. W. HUGHES, *Introduction to Electrical Engineering.* New York: McGraw-Hill Book Company, Inc., 1968, Chapter 8.

10.2 M. E. VAN VALKENBURG, *Network Analysis,* Englewood Cliffs, N.J.: Prentice-Hall, Inc., 1964. Chapter 12.

10.3 W. H. HAYT, JR., AND J. E. KEMMERLY, *Engineering Circuit Analysis.* New York: McGraw-Hill Book Company, Inc., 1963, Chapter 9.

10.4 R. E. SCOTT, *Elements of Linear Circuits.* Reading, Mass.: Addison-Wesley Publishing Company, Inc., 1960, Chapter 10.

11

The Laplace Transform

11.1 Introduction

From our experience with elementary first- and second-order networks, we might guess that more complex networks can result in many simultaneous linear differential equations. To solve such systems of equations, we introduce in this chapter the Laplace transform. This analysis tool enables us to convert differential equations to algebraic equations, as we did in Chapter 10. Unlike the methods of Chapter 10, use of the Laplace transform does not require sinusoidal excitation; more general waveforms are allowed. Indeed, we will be able to consider sinusoidal steady-state analysis as a special case of our new methodology.

The *Laplace transform* of a function $f(t)$ defined for $t \geqq 0$, is given by

$$\mathcal{L}\{f(t)\} = F(s) \triangleq \int_0^\infty f(t)\epsilon^{-st}\,dt \quad (11.1)$$

where s is a complex variable $\sigma + j\omega$. $F(s)$ and $f(t)$ are known as a *transform* pair.

The following set of conditions are *sufficient* to insure the existence of the integral:

1. In any finite interval, $f(t)$ has at most a finite number of finite (step-type) discontinuities and is continuous elsewhere.
2. There exist constants M and c such that $|f(t)| < M\epsilon^{ct}$ for all t greater than some t_0.

A function satisfying condition 1 is said to be *continuous almost everywhere*, and one satisfying condition 2 is said to be of *exponential order*. A function not of exponential order is e^{e^t}; however, virtually all functions encountered in the analysis of lumped networks satisfy these two conditions. If they are satisfied, the Laplace transform exists for $\sigma > c$.

11.2 Derivation of Transform Pairs

Elementary Functions

Fortunately, the number of different functions of time that are encountered in the transient solution of lumped circuits is small. In this section, we derive the transforms of most of these common functions; the remaining are summarized in Table 11.1.[1]

[1] A much more extensive tabulation of useful transform pairs is given in Appendix A of reference 11.6.

Unit step. If $f(t) = u(t)$, then

$$F(s) = \int_0^\infty u(t)e^{-st}\,dt = \int_0^\infty e^{-st}\,dt$$

$$= \frac{e^{-st}}{-s}\bigg|_0^\infty = \frac{1}{s} \tag{11.2}$$

Note that $\sigma = \text{Re}[s] > 0$ is necessary so that the evaluation at the upper limit

$$\lim_{t\to\infty} \{e^{-st}\} = \lim_{t\to\infty} \{e^{(-\sigma-j\omega)t}\}$$

$$= \lim_{t\to\infty} \{e^{-\sigma t}e^{-j\omega t}\}$$

will be zero because $e^{-\sigma t}$ becomes zero.

Exponential decay. If $f(t) = e^{-\alpha t}$, where α is real, then

$$F(s) = \int_0^\infty e^{-\alpha t}e^{-st}\,dt$$

$$= \left[\frac{e^{-(s+\alpha)t}}{-(s+\alpha)}\right]_0^\infty \tag{11.3}$$

$$= \frac{1}{s+\alpha}$$

for $\sigma + \alpha > 0$.

If α is replaced by a complex number $\alpha + j\beta$, then the same form of result

$$F(s) = \frac{1}{s+\alpha+j\beta} \tag{11.4}$$

is obtained for $\sigma + \alpha > 0$. We will find a use for this result shortly.

Unit impulse. If $f(t) = \delta(t)$, then

$$F(s) = \int_0^\infty \delta(t)e^{-st}\,dt$$

We must be concerned here with the limits of integration, because the lower limit occurs simultaneously with the impulse. Because the impulse has zero value for $t > 0$, we will understand the lower limit to be chosen to *include*

TABLE 11.1

(a) Summary of Laplace Transform Relationships

Elementary Functions	
Function	**Transform**
$\delta(t)$	1
$u(t)$	$\frac{1}{s}$
$e^{-\alpha t}$	$\frac{1}{s+\alpha}$
$\sin \beta t$	$\frac{\beta}{s^2+\beta^2}$
$\cos \beta t$	$\frac{s}{s^2+\beta^2}$
t^n	$\frac{n!}{s^{n+1}}$
$t^n e^{-\alpha t}$	$\frac{n!}{(s+\alpha)^{n+1}}$
$\frac{1}{\beta}e^{-\alpha t}\sin \beta t$	$\frac{1}{(s+\alpha)^2+\beta^2}$

Properties ($\mathcal{L}[f(t)] = F(s)$)	
Function	**Transform**
$f_1(t) + f_2(t)$	$F_1(s) + F_2(s)$
$kf(t)$	$kF(s)$
$e^{-\alpha t}f(t)$	$F(s+\alpha)$
$f(t-t_0)u(t-t_0)$	$e^{-t_0 s}F(s)$
$\frac{df(t)}{dt}$	$sF(s) - f(0)$
$\int_{-\infty}^{t} f(\xi)\,d\xi$	$\frac{F(s)}{s} + \frac{f^{-1}(0)}{s}$

Initial Value: $\lim_{t\to 0} f(t) = \lim_{s\to\infty} sF(s)$

Final Value: $\lim_{t\to\infty} f(t) = \lim_{s\to 0} sF(s)$

TABLE 11.1 (continued)

(b) Table of Inverse Transforms involving a Complex Pair of Poles (for a single pole at $s = 0$, let $\gamma = 0$).

$F(s)$	$f(t)$
$\dfrac{1}{(s+\alpha)^2+\beta^2}$	$\dfrac{1}{\beta} e^{-\alpha t} \sin \beta t$
$\dfrac{s+a_0}{(s+\alpha)^2+\beta^2}$	$\dfrac{1}{\beta}[(a_0-\alpha)^2+\beta^2]^{1/2} e^{-\alpha t} \sin(\beta t+\psi)$ $\psi = \tan^{-1}\dfrac{\beta}{a_0-\alpha}$
$\dfrac{1}{(s+\gamma)[(s+\alpha)^2+\beta^2]}$	$\dfrac{e^{-\gamma t}}{(\gamma-\alpha)^2+\beta^2} + \dfrac{1}{\beta[(\gamma-\alpha)^2+\beta^2]^{1/2}} e^{-\alpha t} \sin(\beta t-\psi)$ $\psi = \tan^{-1}\dfrac{\beta}{\gamma-\alpha}$
$\dfrac{s+a_0}{(s+\gamma)[(s+\alpha)^2+\beta^2]}$	$\dfrac{a_0-\gamma}{(\alpha-\gamma)^2+\beta^2} e^{-\gamma t} + \dfrac{1}{\beta}\left[\dfrac{(a_0-\alpha)^2+\beta^2}{(\gamma-\alpha)^2+\beta^2}\right]^{1/2} e^{-\alpha t} \sin(\beta t+\psi)$ $\psi = \tan^{-1}\dfrac{\beta}{a_0-\alpha} - \tan^{-1}\dfrac{\beta}{\gamma-\alpha}$
$\dfrac{s^2+a_1 s+a_0}{(s+\gamma)[(s+\alpha)^2+\beta^2]}$	$\dfrac{\gamma^2-a_1\gamma+a_0}{(\alpha-\gamma)^2+\beta^2} e^{-\gamma t} + \dfrac{1}{\beta}\left[\dfrac{(\alpha^2-\beta^2-a_1\alpha+a_0)^2+\beta^2(a_1-2\alpha)^2}{(\gamma-\alpha)^2+\beta^2}\right]^{1/2} e^{-\alpha t} \sin(\beta t+\psi)$ $\psi = \tan^{-1}\dfrac{\beta(a_1-2\alpha)}{\alpha^2-\beta^2-a_1\alpha+a_0} - \tan^{-1}\dfrac{\beta}{\gamma-\alpha}$
$\dfrac{1}{(s^2+\lambda^2)[(s+\alpha)^2+\beta^2]}$	$\dfrac{1}{[(\beta_0{}^2-\lambda^2)^2+4\alpha^2\lambda^2]^{1/2}}\left[\dfrac{1}{\lambda}\sin(\lambda t-\psi_1) + \dfrac{1}{\beta} e^{-\alpha t} \sin(\beta t-\psi_2)\right]$ $\psi_1 = \tan^{-1}\dfrac{2\alpha\lambda}{\beta_0{}^2-\lambda^2}$; $\psi_2 = \tan^{-1}\dfrac{-2\alpha\beta}{\alpha^2-\beta^2+\lambda^2}$; $\beta_0{}^2 = \alpha^2+\beta^2$
$\dfrac{s+a_0}{(s^2+\lambda^2)[(s+\alpha)^2+\beta^2]}$	$\dfrac{1}{\lambda}\left[\dfrac{a_0{}^2+\lambda^2}{(\beta_0{}^2-\lambda^2)^2+4\alpha^2\lambda^2}\right]^{1/2} \sin(\lambda t+\psi_1) + \dfrac{1}{\beta}\left[\dfrac{(a_0-\alpha)^2+\beta^2}{(\beta_0{}^2-\lambda^2)^2+4\alpha^2\lambda^2}\right]^{1/2} e^{-\alpha t} \sin(\beta t+\psi_2)$ $\psi_1 = \tan^{-1}\dfrac{\lambda}{a_0} - \tan^{-1}\dfrac{2\alpha\lambda}{\beta_0{}^2-\lambda^2}$; $\psi_2 = \tan^{-1}\dfrac{\beta}{a_0-\alpha} - \tan^{-1}\dfrac{-2\alpha\beta}{\alpha^2-\beta^2+\lambda^2}$ $\beta_0{}^2 = \alpha^2+\beta^2$

TABLE 11.1 (continued)

(b) Table of Inverse Transforms involving a Complex Pair of Poles (for a single pole at $s = 0$, let $\gamma = 0$).

$F(s)$	$f(t)$
$\dfrac{s^2 + a_1 s + a_0}{(s^2 + \lambda^2)[(s+\alpha)^2 + \beta^2]}$	$\dfrac{1}{\lambda}\left[\dfrac{(a_0 - \lambda^2)^2 + a_1^2\lambda^2}{(\beta_0^2 - \lambda^2)^2 + 4\alpha^2\lambda^2}\right]^{1/2} \sin(\lambda t + \psi_1) + \dfrac{1}{\beta}\left[\dfrac{(\alpha^2 - \beta^2 - a_1\alpha + a_0)^2 + \beta^2(a_1 - 2\alpha)^2}{(\beta_0^2 - \lambda^2)^2 + 4\alpha^2\lambda^2}\right]^{1/2} e^{-\alpha t}\sin(\beta t + \psi_2)$ $\psi_1 = \tan^{-1}\dfrac{a_1\lambda}{a_0 - \lambda^2} - \tan^{-1}\dfrac{2\alpha\lambda}{\beta_0^2 - \lambda^2}$ $\psi_2 = \tan^{-1}\dfrac{\beta(a_1 - 2\alpha)}{\alpha^2 - \beta^2 - a_1\alpha + a_0} - \tan^{-1}\dfrac{-2\alpha\beta}{\alpha^2 - \beta^2 + \lambda^2}$ $\beta_0^2 = \alpha^2 + \beta^2$

the impulse in the integral. Thus, the integral might better be written

$$\int_{0^-}^{\infty} \delta(t)e^{-st}\,dt$$

Then

$$F(s) = \int_{0^-}^{0^+} \delta(t)e^{-st}\,dt = e^{-s(0)} = 1 \qquad (11.5)$$

To proceed further, we note that the Laplace transformation is a linear operator; that is,

$$\mathcal{L}[f_1(t) + f_2(t)] = \int_0^\infty (f_1(t) + f_2(t))e^{-st}\,dt$$

$$= \int_0^\infty f_1(t)e^{-st}\,dt + \int_0^\infty f_2(t)e^{-st}\,dt$$

$$= \mathcal{L}[f_1(t)] + \mathcal{L}[f_2(t)]$$

That is, the Laplace transform of a sum of two functions is the sum of their Laplace transforms. Similarly, it may be shown that

$$\mathcal{L}[kf(t)] = k\mathcal{L}[f(t)]$$

We apply this result in the following derivation.

Sinusoid. If

$$f(t) = \sin \beta t = \frac{e^{j\beta t} - e^{-j\beta t}}{2j}$$

then

$$F(s) = \frac{1}{2j}\{\mathcal{L}[e^{j\beta t}] - \mathcal{L}[e^{-j\beta t}]\}$$

From Equation (11.4), we have

$$F(s) = \frac{1}{2j}\left\{\frac{1}{s - j\beta} - \frac{1}{s + j\beta}\right\}$$

$$= \frac{\beta}{s^2 + \beta^2}$$

for $\sigma > 0$. Similarly, it can be shown that $\mathcal{L}[\cos \beta t] = s/(s^2 + \beta^2)$.

Elementary Operators

The above transforms illustrate the Laplace transform's feature of converting certain signal waveforms into algebraic functions of s. A complementary aspect is its ability to represent common *operations* such as shifting, differentiation, and integration in similar algebraic form.

s-Plane shift. If $F(s) = \mathcal{L}[f(t)]$, then

$$\mathcal{L}[e^{-\alpha t}f(t)] = \int_0^\infty f(t)e^{-\alpha t}e^{-st}\,dt$$

$$= \int_0^\infty f(t)e^{-(s+\alpha)t}\,dt$$

$$= F(s+\alpha) \qquad (11.6)$$

Therefore, multiplication of any signal by $e^{-\alpha t}$ is analogous in the s-plane to shifting by an amount α. For example, because $\mathcal{L}[\sin\beta t] = \beta/(s^2+\beta^2)$

$$\mathcal{L}[e^{-\alpha t}\sin\beta t] = \frac{\beta}{(s+\alpha)^2+\beta^2}$$

$$= \frac{\beta}{s^2+2\alpha s+\alpha^2+\beta^2} \qquad (11.7)$$

This will be a highly useful transform-pair in our later analysis.

Time shift. The function $f(t-t_0)u(t-t_0)$ is a shifted version of $f(t)u(t)$ and is necessarily zero for $t < t_0$. To find

$$\mathcal{L}[f(t-t_0)u(t-t_0)]$$

$$= \int_0^\infty f(t-t_0)u(t-t_0)e^{-st}\,dt$$

we make the change in variable $t' = t - t_0$, leaving

$$\mathcal{L}[f(t-t_0)u(t-t_0)]$$

$$= \int_{-t_0}^\infty f(t')u(t')e^{-s(t'+t_0)}\,dt' \qquad (11.8)$$

Because $f(t')u(t') = 0$ for $t' < 0$, the lower limit of integration can be made zero with no effect on the value of the integral. Equation (11.8) then becomes

$$\mathcal{L}[f(t-t_0)u(t-t_0)] = e^{-t_0 s}\int_0^\infty f(t')e^{-st'}\,dt'$$

$$= e^{-t_0 s}F(s)$$

We conclude that if a waveform is delayed by t_0 sec, its Laplace transform is multiplied by $e^{-t_0 s}$. For example, a delayed unit step $u(t-t_0)$ has a transform $F(s) = e^{-t_0 s}/s$.

Example 11.1

The Laplace transform of the signal of Figure 11.1 is found directly from its representation in

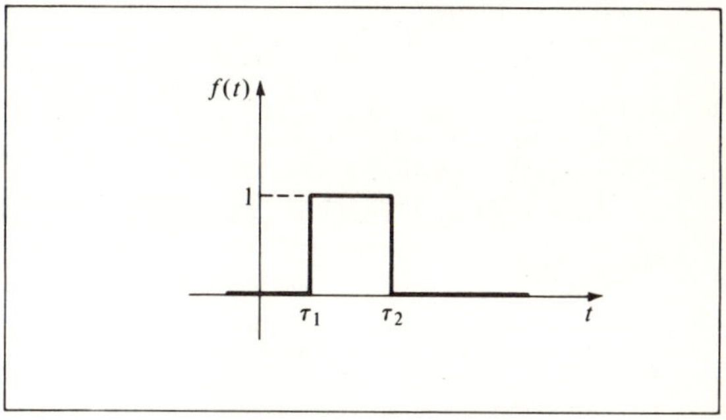

FIGURE 11.1 Delayed pulse.

terms of component elementary waveforms. From Chapter 8, we write

$$f(t) = u(t-\tau_1) - u(t-\tau_2)$$

The transform of the unit step being $1/s$, we can immediately write

$$F(s) = \frac{e^{-\tau_1 s}}{s} - \frac{e^{-\tau_2 s}}{s}$$

Differentiation. The derivative operator is transformed as follows:

$$\mathcal{L}\left[\frac{d}{dt}f(t)\right] = \int_0^\infty \left[\frac{d}{dt}f(t)\right]e^{-st}\,dt$$

Integrating by parts, we define

$$u = e^{-st} \qquad dv = \frac{d}{dt}f(t)\,dt = df(t)$$

$$du = -se^{-st}\,dt \qquad v = f(t)$$

so that the integral becomes

$$\mathcal{L}\left[\frac{d}{dt}f(t)\right] = e^{-st}f(t)\Big|_0^\infty + s\int_0^\infty f(t)e^{-st}\,dt \qquad (11.9)$$

For σ sufficiently large, the first term vanishes at $t = \infty$; this leaves

$$\mathcal{L}\left[\frac{d}{dt}f(t)\right] = sF(s) - f(0)$$

There is an ambiguity associated with this equation when $f(t)$ is discontinuous at $t = 0$. Usually, we will understand our problem to begin at $t = 0^+$, so that

$$\mathcal{L}\left[\frac{d}{dt} f(t)\right] = sF(s) - f(0^+) \qquad (11.10)$$

Occasionally, where an impulse $\delta(t)$ is included in the problem description, we extend the lower limit of Equation (11.9) to $t = 0^-$ to capture the impulse; then

$$\mathcal{L}\left[\frac{d}{dt} f(t)\right] = sF(s) - f(0^-)$$

As Problem 11.12 shows, either choice is acceptable so long as we are consistent.

Higher derivatives are routinely evaluated as follows. Substituting df/dt for f in Equation (11.9), we can write

$$\mathcal{L}\left[\frac{d^2f}{dt^2}\right] = e^{-st}\frac{df}{dt}\bigg|_0^\infty + s\int_0^\infty \frac{df}{dt} e^{-st}\,dt$$

$$= -\dot{f}(0) + s(sF(s) - f(0))$$

$$= s^2F(s) - sf(0) - \dot{f}(0)$$

This process can be continued to give the general formula

$$\mathcal{L}\left[\frac{d^nf}{dt^n}\right] = s^nF(s) - \sum_{r=0}^{n-1} s^{n-r-1} f^r(0)$$

where $f^r(0)$ is the rth derivative of $f(t)$ at $t = 0$.

Integration. The integral operation is transformed as follows:

$$\mathcal{L}\left[\int_{-\infty}^{t} f(\xi)\,d\xi\right]$$

$$= \mathcal{L}\left[\int_{-\infty}^{0} f(\xi)\,d\xi + \int_{0}^{t} f(\xi)\,d\xi\right]$$

The first term is a constant because neither limit is a function of time. Therefore, from Equation (11.2),

$$\mathcal{L}\left[\int_{-\infty}^{0} f(\xi)\,d\xi\right] = \frac{f^{-1}(0)}{s}$$

where

$$f^{-1}(0) \triangleq \int_{-\infty}^{0} f(\xi)\,d\xi$$

The transform of the second term is obtained by integration by parts, letting

$$u = \int f(t)\,dt \qquad dv = e^{-st}\,dt$$

$$du = f(t)\,dt \qquad v = -\frac{1}{s}e^{-st}$$

Then we have

$$\mathcal{L}\left[\int_0^t f(\xi)\,d\xi\right] = \left[-\frac{1}{s}e^{-st}\int_0^t f(\xi)\,d\xi\right]_0^\infty$$

$$+ \frac{1}{s}\int_0^\infty f(t)e^{-st}\,dt$$

$$= \frac{1}{s}F(s)$$

The complete transform therefore has the form

$$\mathcal{L}\left[\int_{-\infty}^{t} f(\xi)\,d\xi\right] = \frac{F(s)}{s} + \frac{f^{-1}(0)}{s} \qquad (11.11)$$

Example 11.2

We are to find $\mathcal{L}[t]$ by integrating the unit step function.

For $f(t) = u(t)$, $F(s) = 1/s$. Further, the value of the integral

$$f^{-1}(0) = \int_{-\infty}^{0} u(\xi)\,d\xi = 0$$

because $u(t) = 0$ for $t < 0$. Therefore, from Equation (11.11)

$$\mathcal{L}[t] = \mathcal{L}\left[\int_{-\infty}^{t} u(\xi)\,d\xi\right]$$

$$= \frac{1}{s}\left(\frac{1}{s}\right) + \frac{1}{s}(0) = \frac{1}{s^2}$$

Repeated application of the integration formula allows us to write

$$\mathcal{L}[t^n] = \frac{n!}{s^{n+1}}$$

Further, if we now apply the shifting rule of Equation (11.6), we obtain what for us will be the important result

$$\mathcal{L}[e^{-\alpha t}t^n] = \frac{n!}{(s+\alpha)^{n+1}}$$

Two other operations useful in the interpretation of Laplace transforms involve relating the limiting values of $f(t)$ as $t \to 0$ and as $t \to \infty$ to the transform $F(s)$. This allows us to determine the initial and final values of $f(t)$ without determining the entire inverse transform.

Initial value. To find $\lim_{t\to 0}\{f(t)\}$, consider taking $\lim_{s\to\infty}$ of the derivative transform of Equation (11.10)

$$\lim_{s\to\infty}\left\{\int_{0^+}^{\infty}\frac{df}{dt}e^{-st}\,dt\right\} = \lim_{s\to\infty}\{sF(s) - f(0^+)\}$$

Assuming that limiting and integration operators can be interchanged, we have

$$\lim_{s\to\infty} e^{-st} = 0$$

and the left-hand side vanishes. The right-hand side then yields

$$\lim_{s\to\infty}\{sF(s)\} = f(0^+)$$

Final value. Using the differentiation formula and assuming $(df/dt)_{t=0}$ exists

$$\lim_{s\to 0}\left\{\int_{0}^{\infty}\frac{df}{dt}e^{-st}\,dt\right\} = \lim_{s\to 0}\{sF(s) - f(0)\} \tag{11.12}$$

Exchanging the integration and limiting operations, we have $\lim_{s\to 0}\{e^{-st}\} = 1$ and

$$\int_0^\infty \left(\frac{df}{dt}\right)dt = \lim_{t\to\infty}\left\{\int_0^t\left(\frac{df}{dt}\right)dt\right\}$$

$$= \lim_{t\to\infty}\left\{\int_{f(0)}^{f(t)} df\right\} \tag{11.13}$$

$$= \lim_{t\to\infty}\{f(t) - f(0)\}$$

Combining Equations (11.12) and (11.13) therefore yields

$$\lim_{t\to\infty}[f(t)] = \lim_{s\to 0}[sF(s)]$$

It is well to point out that the interchange of integration and limiting operations is valid only if $sF(s)$ is analytic on and to the right of the imaginary axis in the complex s-plane. For example, it does not apply to $\mathcal{L}[\sin\beta t] = s/(s^2+\beta^2)$, because the roots of the denominator polynomial are at $\pm j\beta$ on the imaginary axis. This is a reasonable restriction because $\sin\beta t$ oscillates indefinitely and so does not approach a limiting value.

Example 11.3

If $f(t) = e^{-\alpha t}$, then $F(s) = 1/(s+\alpha)$ and

$$\lim_{t\to 0}\{e^{-\alpha t}\} = \lim_{s\to\infty}\left\{s\left(\frac{1}{s+\alpha}\right)\right\}$$

$$= 1$$

and

$$\lim_{t\to\infty}\{e^{-\alpha t}\} = \lim_{s\to 0}\left\{s\left(\frac{1}{s+\alpha}\right)\right\}$$

$$= 0$$

Both limits check with the limiting values of $e^{-\alpha t}$.

11.3 Determination of the Inverse Transform

Introduction

Before continuing with the development of mathematical preliminaries, let us summarize how we are progressing toward our goal of a general linear dynamic network solution procedure. Table 11.2 compares the solution of a sinusoidal steady-state network problem with the solution using Laplace transforms. Beginning with the time-domain function, we

TABLE 11.2

Comparison of Solution Procedures

		Solution Process		
	Time Domain	Transform	Network Solution	Inverse Transform
Sinusoidal Steady State	$A\cos(\omega t+\phi)$	$Ae^{j\phi}$	$Be^{j\theta}$	$B\cos(\omega t+\theta)$
Laplace	$Ae^{-\alpha t}$	$\dfrac{A}{s+\alpha}$	$F(s)$	$f(t)$

first determine a transformed representation of all signals. The network output is then found in a transformed representation (for example, as a phasor), which is then converted to the time domain. This latter conversion to the time domain is trivial in the case of sinusoidal steady-state analysis; however, the analogous *inverse transform* determination we are about to study can be computationally quite involved.

There is no particular difficulty in computing the inverse transform so long as $F(s)$ appears in Table 11.1 (a). However, transforms such as

$$F(s) = \frac{s^2 + a_1 s + a_0}{(s+\gamma)[(s+\alpha)^2+\beta^2]} \tag{11.14}$$

will commonly arise as the result of a network analysis. We must be able to compute $f(t)$ from this function before we can claim to have a general solution method.

The Inverse Laplace Transform

The basis for evaluating the inverse transform from $F(s)$ is by the inversion integral that defines, together with Equation (11.1), $F(s)$ and $f(t)$ as a transform pair.

$$f(t) = \mathcal{L}^{-1}[F(s)] \triangleq \frac{1}{2\pi j}\int_{c-j\infty}^{c+j\infty} F(s)e^{st}\,ds \tag{11.15}$$

It can be shown that if $F(s)$ is obtained from $f(t)$ using Equation (11.1), then $f(t)$ can be recovered from $F(s)$ using Equation (11.15).

Because Equation (11.15) involves integration in the complex variable s, the integration must be performed as a *contour integral.* Fortunately, for the functions of interest to us, we can carry out the algebra to evaluate this integral without becoming involved in more general contour integral evaluation. The reader who wishes a formal mathematical treatment of the topic should examine references on complex variable theory, such as Reference 11.4.

Partial Fraction Expansion

It should be noted from Table 11.1(a) that the transforms of the differential and integral operators involve either s or s^{-1}. An integro-differential equation, with higher-order derivatives, will transform into a polynomial or ratio of polynomials in s. Such ratios are known as *rational* functions of s. The transform of a linear, constant-coefficient, differential equation always involves rational functions. Therefore, the transform of any solution variable [such as $F(s)$ in Equation (11.14)] will always be a rational function. In the following section, we study methods of decomposing such functions into the simple functions that appear in Table 11.1(a). These functions can then be individually transformed back into the time domain to give the solution.

Let us begin by considering a Laplace-transformed function of the form

$$F(s) = \frac{K(s-s_3)}{s(s-s_1)(s-s_2)} \tag{11.16}$$

where s_1, s_2, and s_3 are distinct (nonequal) constants. Now suppose $F(s)$ could be written in the form

$$F(s) = \frac{k_0}{s} + \frac{k_1}{s - s_1} + \frac{k_2}{s - s_2} \qquad (11.17)$$

Then, from the linearity of the Laplace transform, we would have

$$\begin{aligned} f(t) &= f_0(t) + f_1(t) + f_2(t) \\ &= \mathcal{L}^{-1}[F_0(s) + F_1(s) + F_2(s)] \end{aligned} \qquad (11.18)$$

Comparing Equations (11.17) and (11.18) and consulting Table 11.1, we would find

$$\begin{aligned} f(t) &= \mathcal{L}^{-1}[F(s)] \\ &= k_0 + k_1 e^{s_1 t} + k_2 e^{s_2 t} \qquad (t \geqq 0) \end{aligned}$$

To find the values of the k_i, we multiply Equations (11.16) and (11.17) by $(s - s_i)$ and take the limit as $s \to s_i$. The limit of every term of the form

$$\lim_{s \to s_i} \left\{ (s - s_i) \frac{k_i}{s - s_j} \right\} = 0 \qquad (i \neq j)$$

so that

$$k_1 = \lim_{s \to s_i} \{ (s - s_i) F(s) \}$$

provided that all s_i are distinct.

A summation of the form of Equation (11.17) is called a *partial fraction expansion* of $F(s)$.

Example 11.4

We are to find the k_i in the partial fraction expansion

$$\frac{10(s+2)(s+5)}{s(s+1)(s+3)} = \frac{k_0}{s} + \frac{k_1}{s+1} + \frac{k_2}{s+3}$$

To find k_0, we multiply both sides by s to give

$$\left.\frac{10(s+2)(s+5)}{(s+1)(s+3)}\right|_{s=0} = \left(k_0 + \frac{k_1 s}{s+1} + \frac{k_2 s}{s+3}\right)_{s=0}$$

or

$$\frac{10(2)(5)}{(1)(3)} = k_0 = \frac{100}{3}$$

Similarly,

$$\begin{aligned} k_1 &= (s+1)F(s)\,|_{s=-1} \\ &= \left.\frac{10(s+2)(s+5)}{s(s+3)}\right|_{s=-1} \\ &= -20 \\ k_2 &= (s+3)F(s)\,|_{s=-3} \\ &= \left.\frac{10(s+2)(s+5)}{s(s+1)}\right|_{s=-3} \\ &= -10/3 \end{aligned}$$

The denominator roots (0, -1, and -3 above) are known as *poles* of $F(s)$; the numerator roots are known as *zeros*. We will designate these roots as p_i and z_i, respectively.

This expansion procedure applies as well to rational functions with complex-valued poles and zeros, as the following shows.

Example 11.5

We deduce from the above that the inverse transform of $F_1(s)/(s + \alpha)$ has a term of the form $F_1(-\alpha)e^{-\alpha t}$. We are to find the term corresponding to a complex pair of poles of the form

$$F(s) = \frac{F_1(s)}{(s+\alpha)^2 + \beta^2}$$

Expanding in partial fractions we have

$$F(s) = \frac{k_1}{s + \alpha - j\beta} + \frac{k_2}{s + \alpha + j\beta} + \cdots$$

where

$$k_1 = (s + \alpha - j\beta)F(s)\,|_{s=-\alpha+j\beta}$$

$$= \left.\frac{F_1(s)}{s + \alpha + j\beta}\right|_{s=-\alpha+j\beta}$$

$$= \frac{F_1(-\alpha + j\beta)}{2j\beta} = \frac{|F_1(-\alpha + j\beta)|}{2j\beta} e^{j\phi}$$

where $\phi = \arg F_1(-\alpha + j\beta)$, and

$$k_2 = (s + \alpha + j\beta)F(s)\,|_{s=-\alpha-j\beta}$$

$$= \frac{F_1(-\alpha - j\beta)}{-2j\beta} = \frac{|F_1(-\alpha - j\beta)|}{-2j\beta} e^{-j\phi}$$

$$= k_1^*$$

because $|F_1(-\alpha - j\beta)| = |F_1(-\alpha + j\beta)|$. Then

$$f(t) = k_1 e^{(-\alpha+j\beta)t} + k_2 e^{(-\alpha-j\beta)t}$$

$$= \frac{|F_1(-\alpha + j\beta)|}{2j\beta} e^{j\phi} e^{(-\alpha+j\beta)t}$$

$$- \frac{|F_1(-\alpha - j\beta)|}{2j\beta} e^{-j\phi} e^{(-\alpha-j\beta)t}$$

$$= \frac{|F_1(-\alpha + j\beta)|}{\beta} e^{-\alpha t} \sin(\beta t + \phi) \tag{11.19}$$

using the Euler identity.

Summarizing the above procedure, the magnitude $|F_1(-\alpha + j\beta)|$ and the argument ϕ of the transform $F(s)$ are calculated *after removal of the complex pole pair*. From these, the corresponding term in the response is then calculated using Equation (11.19). As a result, it is easy to write by inspection the form of the inverse transform of a quite complicated transform $F(s)$. For example, the inverse transform of

$$F(s) = \frac{s^2 + a_1 s + a_0}{(s^2 + \lambda^2)((s + \alpha)^2 + \beta^2)}$$

with two pairs of complex poles is

$$f(t) = \frac{1}{\lambda} \overbrace{\left| \left(\frac{s^2 + a_1 s + a_0}{(s + \alpha)^2 + \beta^2} \right)_{s=j\lambda} \right|}^{k_\lambda} \sin(\lambda t + \phi_\lambda)$$

$$+ \frac{1}{\beta} \overbrace{\left| \left(\frac{s^2 + a_1 s + a_0}{s^2 + \lambda^2} \right)_{s=-\alpha+j\beta} \right|}^{k_\beta} e^{-\alpha t} \sin(\beta t + \phi_\beta)$$

where $\phi_\lambda = \arg k_\lambda$, $\phi_\beta = \arg k_\beta$.

As an exercise, the reader may find it useful to justify one or more of the inverse transforms in Table 11.1(b). This table will in turn be useful in solving certain of the problems at the end of the chapter.

Multiple Poles

Consider a Laplace transform of the form

$$F(s) = \frac{K(s - z_1)}{s^2(s - p_1)^2(s - p_2)} \tag{11.20}$$

The poles at $s = 0$ and $s = p_1$ are of the form $(s - p_j)^n$ where $n = 2$; these are known as *second-order* multiple poles. The partial fraction expansion is of the form

$$F(s) = \frac{k_{0,1}}{s} + \frac{k_{0,2}}{s^2} + \frac{k_{1,1}}{s - p_1}$$

$$+ \frac{k_{1,2}}{(s - p_1)^2} + \frac{k_2}{s - p_2} \tag{11.21}$$

where a double subscript notation is used for k's associated with multiple poles. All of the terms in this expansion have inverse transforms appearing in Table 11.1; in this case,

$$f(t) = (k_{0,1} + k_{0,2}t + k_{1,1}e^{p_1 t}$$

$$+ k_{1,2}te^{p_2 t} + k_2 e^{p_2 t})u(t)$$

The determination of $k_{0,2}$, $k_{1,2}$, and k_2 follow procedures similar to those discussed above.

$$k_2 = \lim_{s \to p_2} (s - p_2)F(s) = \frac{K(p_2 - z_1)}{p_2^2(p_2 - p_1)^2}$$

$$k_{1,2} = \lim_{s \to p_1} (s - p_1)^2 F(s) = \frac{K(p_1 - z_1)}{p_1^2(p_1 - p_2)}$$

$$k_{0,2} = \lim_{s \to 0} s^2 F(s) = \frac{K(-z_1)}{(-p_1)^2(-p_2)}$$

The constants $k_{0,1}$ and $k_{2,1}$ cannot be found in this way. For example, if we multiply both sides of (11.21) by s and set $s = 0$, we obtain

$$sF(s)\Big|_{s=0} = k_{0,1} + \frac{k_{0,2}}{s}\Big|_{s=0}$$

which does not exist. However, suppose we multiply both sides of Equation (11.21) by s^2, differentiate the result with respect to s, and then set $s = 0$.

$$\left\{\frac{d}{ds}\left[(s^2)\left(\frac{k_{0,1}}{s} + \frac{k_{0,2}}{s^2} + \frac{k_{1,1}}{s - p_1} + \frac{k_{1,2}}{(s - p_1)^2} + \frac{k_2}{s - p_2}\right)\right]\right\}_{s=0} = k_{0,1}$$

Similarly,

$$k_{1,1} = \left\{\frac{d}{ds}[(s - p_1)^2 F(s)]\right\}_{s=p_1}$$

The generalization of this procedure proceeds as follows. If $F(s)$ is a rational function of s and has a pole of order n, then the terms of the partial fraction expansion associated with that pole are of the form

$$F(s) = \frac{k_{i,1}}{s - p_i} + \frac{k_{i,2}}{(s - p_i)^2} + \cdots + \frac{k_{i,n}}{(s - p_i)^n} + F_i(s)$$

where $F_i(s)$ contains all the terms not involving p_i. As we have seen

$$k_{i,n} = (s - p_i)^n F(s)\big|_{s=p_i}$$

To find $k_{i,n-1}$, we imitate the example by differentiating $(s - p_i)^n F(s)$ as

$$\left\{\frac{d}{ds}[(s - p_i)^n F(s)]\right\}_{s=p_i} = \Big\{(n - 1)(s - p_i)^{n-2}k_{i,1} + (n - 2)(s - p_i)^{n-3}k_{i,2} + \cdots + k_{i,n-1}$$

$$+ (s - p_i)^{n-1}\left[nF(s) + (s - p_i)\frac{d}{ds}F(s)\right]\Big\}_{s=p_i}$$

All terms except $k_{i,n-1}$ vanish to leave

$$k_{i,n-1} = \left[\frac{d}{ds}(s - p_i)^n F(s)\right]_{s=p_i}$$

Continued differentiation with respect to s yields the general formula

$$k_{i,r} = \frac{1}{(n - r)!}\left\{\frac{d^{(n-r)}}{ds^{(n-r)}}[(s - p_i)^n F(s)]\right\}_{s=p_i} \tag{11.22}$$

Example 11.6

We are to find the partial fraction expansion of

$$F(s) = \frac{3s^2 + 11s + 11}{(s + 1)^2(s + 2)}$$

$$= \frac{k_{1,1}}{s + 1} + \frac{k_{1,2}}{(s + 1)^2} + \frac{k_2}{s + 2}$$

The coefficients $k_{1,2}$ and k_2 are routinely evaluated as 3 and 1, respectively. Then

$$k_{1,1} = \lim_{s \to -1} \frac{d}{ds}(s + 1)^2 F(s)$$

$$= \lim_{s \to -1} \frac{d}{ds}\left(\frac{3s^2 + 11s + 11}{s + 2}\right)$$

$$= \lim_{s \to -1} \frac{3s^2 + 12s + 11}{(s + 2)^2} = 2$$

so the complete solution is

$$F(s) = \frac{2}{s + 1} + \frac{3}{(s + 1)^2} + \frac{1}{s + 2}$$

The inverse transform is then

$$f(t) = 2e^{-t} + 3te^{-t} + e^{-2t}$$

11.4 Laplace Solution of Linear Dynamic Networks

Introduction

In using the Laplace transform to analyze dynamic networks, we again have a choice between formal methods that are quite broad in application and methods convenient for electrical networks. The distinction, similar to that observed in computer-aided and steady-state analysis, is based on whether (1) the network equations are written as differential equations and then solved using the Laplace transformation, or whether (2) the transformation is carried out at the branch level and then KVL and KCL are used to relate the transformed currents and voltage.

We will consider briefly the formal method [(1) above] for simple networks and then examine network-oriented procedures in considerable detail. The formal approach is taken up again in Chapter 13.

Transformation of Network Equations

The process of applying Laplace transforms to network equations is straightforward. We merely write the differential (or integro-differential) equations, transform them term-by-term, solve for the transform of the desired quantity, and then take the inverse transform using a partial fraction expansion.

Example 11.7

The network of Figure 11.2 is to be solved for $v_C(t)$, $t > 0$.

This network can be described by the differential equation

$$RC\frac{dv_C(t)}{dt} + v_C(t) = V_s u(t) \qquad (11.23)$$

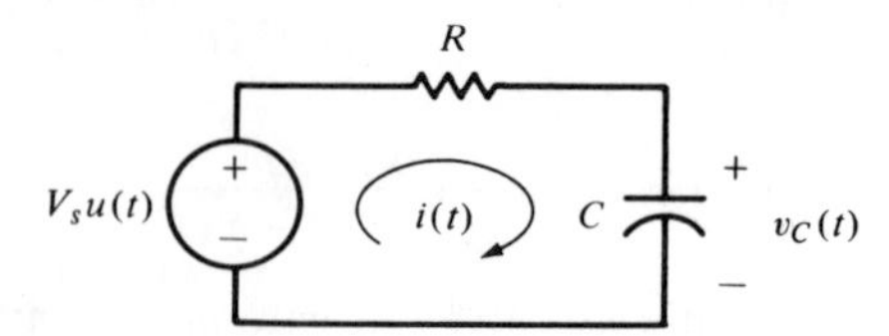

FIGURE 11.2 *RC* network with step input.

or by the integrodifferential equation

$$Ri(t) + \frac{1}{C}\int_0^t i(\xi)d\xi + v_C(0) = V_s u(t) \qquad (11.24)$$

Including the initial condition in the integral, we could also write

$$Ri(t) + \frac{1}{C}\int_{-\infty}^t i(\xi)\,d\xi = V_s u(t) \qquad (11.25)$$

1. *Differential equation solution.* Transforming Equation (11.23), we obtain

$$RC[sV_C(s) - v_C(0)] + V_C(s) = \frac{V_s}{s}$$

This can be solved for $V_C(s)$ to give

$$V_C(s) = \frac{V_s}{s(RCs+1)} + \frac{RCv_C(0)}{RCs+1}$$

When this is expanded in partial fractions,

$$V_C(s) = \frac{V_s}{s} - \frac{V_s}{s+1/RC} + \frac{v_C(0)}{s+1/RC} \qquad (11.26)$$

the inverse-transform can be written term-by-term in the form

$$v_C(t) = V_s(1 - e^{-t/RC}) + v_C(0)e^{-t/RC} \qquad (t > 0) \qquad (11.27)$$

This is a familiar result from Chapter 8.

2. *Integrodifferential equation solution.* Trans-

forming Equations (11.24) or (11.25), we obtain

$$RI(s) + \frac{I(s)}{sC} + \frac{v_C(0)}{s} = \frac{V_s}{s} \qquad (11.28)$$

or

$$I(s) = \frac{C(V_s - v_C(0))}{RCs + 1}$$

$$LC(s^2V_C(s) - sv_C(0) - \dot{v}_C(0)) + RC(sV_C(s) - v_C(0)) + V_C(s) = 1 \qquad (11.32)$$

Because

$$v_C(t) = \frac{1}{C}\int_0^t i(\xi)\, d\xi + v_C(0) \qquad (11.29)$$

we can write

$$\begin{aligned} V_C(s) &= \frac{I(s)}{sC} + \frac{v_C(0)}{s} \\ &= \frac{V_s - v_C(0)}{s(RCs+1)} + \frac{v_C(0)}{s} \qquad (11.30) \\ &= \frac{V_s}{s} - \frac{V_s}{s + 1/RC} + \frac{v_C(0)}{s + 1/RC} \end{aligned}$$

which agrees with Equation (11.26).

Example 11.8

The impulse response of the RLC network of Figure 11.3 is to be found assuming that the element values result in an underdamped response.

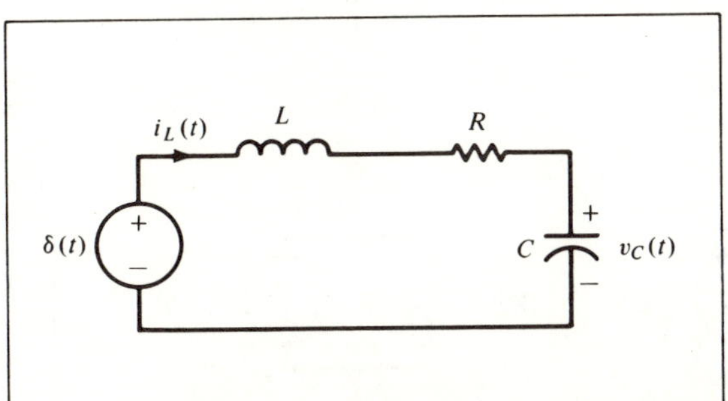

FIGURE 11.3 Series RLC network.

The differential equation is given by

$$LC\frac{d^2v_C(t)}{dt^2} + RC\frac{dv_C(t)}{dt} + v_C(t) = \delta(t) \qquad (11.31)$$

The transform of both sides is taken, using $t = 0^-$ for the initial state due to the presence of the impulse.

Solving for $V_C(s)$ and using $i_L(0) = C\dot{v}_C(0)$, we obtain

$$V_C(s) = \frac{(RC + LCs)v_C(0) + Li_L(0) + 1}{LCs^2 + RCs + 1} \qquad (11.33)$$

Because $v_C(0) = i_L(0) = 0$ for the impulse response calculation, the transform becomes

$$\begin{aligned} V_C(s) &= \frac{1/LC}{s^2 + \frac{R}{L}s + \frac{1}{LC}} \\ &= \frac{1/LC}{\left(s + \frac{R}{2L}\right)^2 + \frac{1}{LC} - \frac{R^2}{4L^2}} \end{aligned}$$

The inverse transform is obtained directly from Table 11.1 by identifying

$$(s+\alpha)^2 + \beta^2 = \left(s + \frac{R}{2L}\right)^2 + \frac{1}{LC} - \frac{R^2}{4L^2}$$

so that

$$s_{1,2} = -\alpha \pm j\beta = -\frac{R}{2L} \pm j\sqrt{\frac{1}{LC} - \left(\frac{R}{2L}\right)^2} \qquad (11.34)$$

provided

$$\frac{1}{LC} > \left(\frac{R}{2L}\right)^2$$

Then we have

$$V_C(s) = \frac{K\beta}{(s+\alpha)^2+\beta^2} \qquad (11.35)$$

where $K = 1/(LC\beta)$, so the inverse transform is

$$v_C(t) = Ke^{-Rt/2L}\sin\left(\sqrt{\frac{1}{LC}-\left(\frac{R}{2L}\right)^2}\,t\right) \qquad (t>0) \qquad (11.36)$$

If, for example, $R = L = C = 1$ as in Example 9.1(c), $\alpha = \frac{1}{2}$, $\beta = \sqrt{3}/2$, and the solution is

$$v_C(t) = \frac{2}{\sqrt{3}}e^{-t/2}\sin\frac{\sqrt{3}}{2}t \qquad (t>0) \qquad (11.37)$$

Element Models in the s-Domain

Although the above is a workable procedure, the writing of the differential equation is unnecessarily cumbersome when the network is more complex. We can simplify the process of obtaining the transformed network equation relationships as we did in Chapter 10 and develop a transformed model of the entire network.

The time-domain relationships for the R, L, and C elements are

$$v_R(t) = Ri_R(t) \qquad (11.38)$$

$$v_L(t) = L\frac{di_L(t)}{dt} \qquad (11.39)$$

$$v_C(t) = \frac{1}{C}\int_0^t i_C(\xi)\,d\xi + v_C(0) \qquad (11.40)$$

Taking the Laplace transform of each of these equations, we obtain

$$V_R(s) = RI_R(s) \qquad (11.41)$$

$$V_L(s) = sLI_L(s) - Li_L(0) \qquad (11.42)$$

$$V_C(s) = \frac{1}{sC}I_C(s) + \frac{1}{s}v_C(0) \qquad (11.43)$$

These last three equations can be treated as element relationships in the s-domain, that is, as directly relating $V(s)$ and $I(s)$. We can, in fact, construct network models for each relationship, as shown in Figure 11.4(a). The

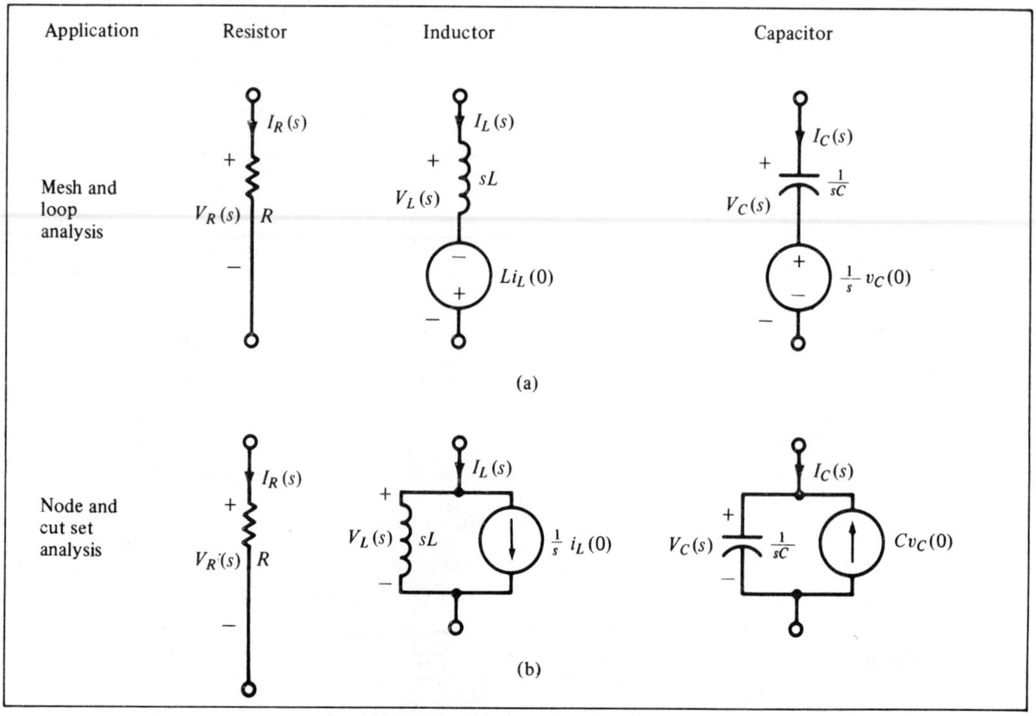

FIGURE 11.4 s-plane models of elements.

network variables are now $V(s)$ and $I(s)$, and the element symbols are symbols for the s-domain generalizations of Ohm's law. That is, the resistor symbol now stands for the relationship

$$V_R(s) = RI_R(s) \tag{11.44}$$

the inductor symbol for

$$V_L(s) = sLI_L(s) \tag{11.45}$$

and the capacitor symbol for

$$V_C(s) = \frac{1}{sC} I_C(s) \tag{11.46}$$

The initial conditions, which give rise to additive terms in Equations (11.42) and (11.43), are accounted for by the voltage sources in Figures 11.4(a). In doing this, we have made use of the fact that the Kirchhoff current and voltage laws hold in the s-domain just as they do in the time domain; voltages in series therefore add in the conventional manner.

If we replace each element in a network by its s-domain model, we obtain a network in which the variables $V(s)$ and $I(s)$ are related by Kirchhoff's laws and *algebraic* relations of the form (11.44) to (11.46). The analysis of such a network is only slightly more complicated than the analysis of a resistive network, the principal difference being that polynomials, rather than constants, must be manipulated.

Example 11.9

The RC network of Figure 11.2 is to be solved using branch models of the elements.

Figure 11.5 shows the s-domain analog of Figure 11.2. From this network, we can immediately write the mesh equation

$$\left(R + \frac{1}{sC}\right) I(s) = \frac{V_s}{s} - \frac{v_C(0)}{s} \tag{11.47}$$

which yields

$$I(s) = \frac{V_s - v_C(0)}{R\left(s + \dfrac{1}{RC}\right)} \tag{11.48}$$

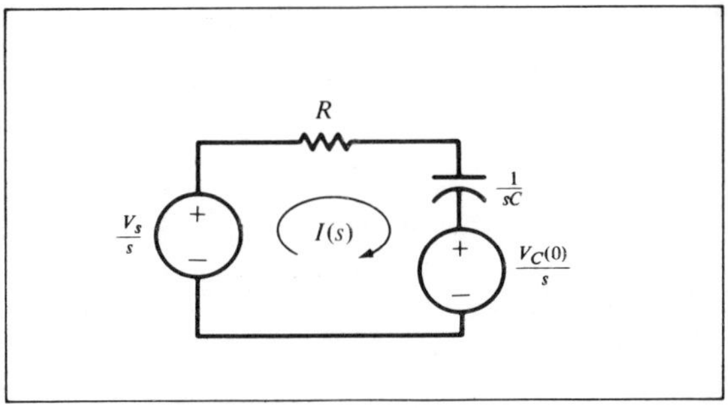

FIGURE 11.5 s-domain model of RC network.

This agrees with Example 11.7.

The models shown in Figure 11.4(a) are not the only ones possible. We can solve Equations (11.41) to (11.43) for current to obtain

$$I_R(s) = \frac{1}{R} V_R(s) \tag{11.49}$$

$$I_L(s) = \frac{1}{sL} V_L(s) + \frac{1}{s} i_L(0) \tag{11.50}$$

$$I_C(s) = sCV_C(s) - Cv_C(0) \tag{11.51}$$

The s-domain network models for these relationships are shown in Figure 11.4(b). The models of Figure 11.4(b) will be found to be more convenient than those of Figure 11.4(a) when node equations are to be written.

Example 11.10

The network of Figure 11.6 is to be solved for $v(t)$.

We form the s-domain model in Figure 11.6(b); from this network, we write

$$\left(\frac{1}{sL} + \frac{1}{R}\right) V(s) = \frac{I_s}{s} - \frac{i_L(0)}{s}$$

from which we obtain the transform of the inductor voltage

$$V(s) = \frac{I_s - i_L(0)}{s + \dfrac{R}{L}} R$$

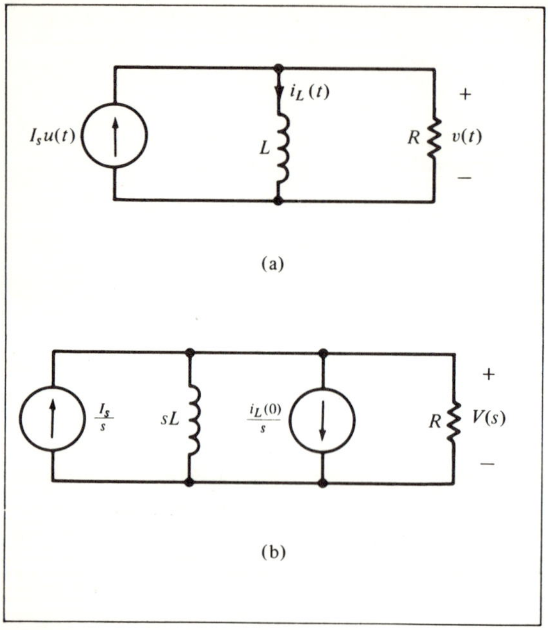

FIGURE 11.6 *RL* network models. (a) Time domain; (b) *s*-domain.

Finally, taking the inverse transform gives

$$v(t) = [I_s - i_L(0)]Re^{-Rt/L} \qquad (t > 0) \tag{11.52}$$

The significant advantage of the *s*-domain network models appears when we formulate equations for multiloop or multinode networks. Formulating the equations directly in the *s*-domain is generally much more straightforward than writing equations in the time domain and then transforming.

Example 11.11

The network of Figure 11.7 is to be analyzed to find $i_s(t)$.

Writing mesh equations, we have

$$(sL_1 + R_1)I_1 - R_1 I_2 = \frac{V_s}{s} + L_1 i_{L_1}(0) \tag{11.53}$$

$$-R_1 I_1 + (sL_2 + R_1 + R_2)I_2 = L_2 i_{L_2}(0) \tag{11.54}$$

These equations must be solved simultaneously. It must be admitted that the algebra becomes a bit tedious even in this relatively simple example. To simplify matters, we will assume that $i_{L_1}(0) = i_{L_2}(0) = 0$ and solve for $I_2(s)$; after some manipulation we obtain

$$I_2(s) = \frac{V_s/R_2}{s\left[s^2 \dfrac{L_1 L_2}{R_1 R_2} + s\left(\dfrac{L_1}{R_1} + \dfrac{L_1}{R_2} + \dfrac{L_2}{R_2}\right) + 1\right]} \tag{11.55}$$

If $v_s(t) = V_s u(t)$, $L_1 = 1$ mh, $L_2 = 9$ mh, $R_1 = 3$ kΩ, and $R_2 = 24$ kΩ, $I_2(s)$ becomes

$$I_2(s) = \frac{12/24 \times 10^3}{s\left[s^2 \dfrac{9 \times 10^{-6}}{72 \times 10^6} + s\left(\dfrac{10^3}{3 \times 10^3} + \dfrac{10^{-3}}{24 \times 10^3} + \dfrac{9 \times 10^{-3}}{24 \times 10^3}\right) + 1\right]}$$

$$= \frac{4 \times 10^9}{s(s^2 + 6 \times 10^6 s + 8 \times 10^{12})} = \frac{4 \times 10^9}{s(s + 2 \times 10^6)(s + 4 \times 10^6)}$$

This is expanded in partial fractions to yield

$$I_2(s) = \frac{0.5 \times 10^{-3}}{s} - \frac{10^{-3}}{s + 2 \times 10^6} + \frac{0.5 \times 10^{-3}}{s + 4 \times 10^6}$$

The inverse transform is

$$i_2(t) = 0.5 \times 10^{-3}(1 - 2e^{-2 \times 10^6 t} + e^{-4 \times 10^6 t})u(t) \tag{11.56}$$

The above example provides an opportunity to check the final value rule. From the network, the equilibrium value of $i_2(t)$ is V_s/R_2, obtained by replacing inductors by short circuits

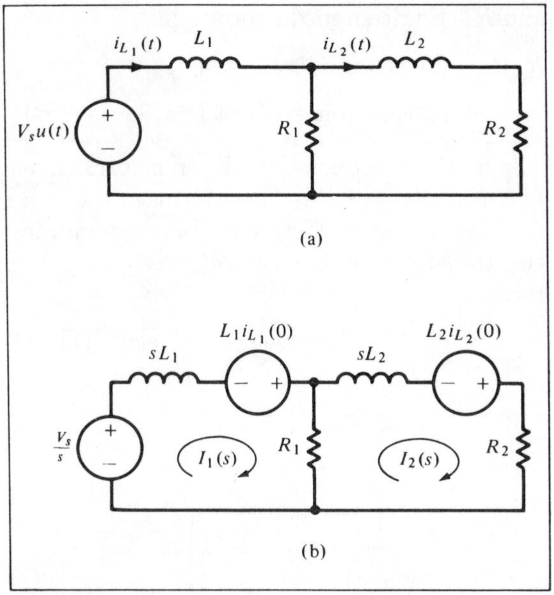

FIGURE 11.7 Two-mesh RL network models. (a) Time domain; (b) s-domain.

and capacitors by open circuits. We obtain the same result from Equation (11.55) as

$$\lim_{s\to 0}\left[sI_2(s)\right] = \frac{V_s}{R_2}$$

11.5 Zero-State Analysis

Introduction

Determination of the zero-state response, illustrated in Example 11.11, is by far the most common problem in dynamic network analysis.

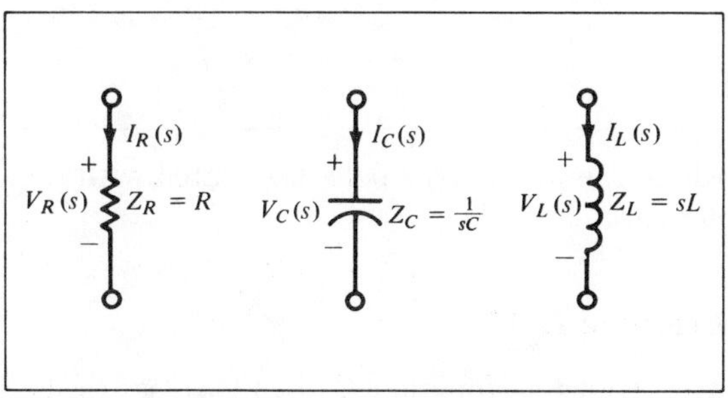

FIGURE 11.8 Zero-state models.

With zero-valued initial conditions, a number of simplifications are possible that facilitate analysis and provide insight into network properties.

Elemental Models

With initial conditions set to zero, the branch models of Figure 11.4 simplify to those shown in Figure 11.8. The absence of sources means we can write for each element equations of the form

$$V(s) = Z(s)I(s) \qquad I(s) = Y(s)V(s) \tag{11.57}$$

where $Z(s)$ is termed an impedance and $Y(s)$ an admittance.

Inspection of the impedances shown in Figure 11.8 reveals that, in each case, we can replace s by $j\omega$ and obtain the impedances used in sinusoidal steady-state analysis. We will exploit this relationship shortly.

Network Functions

The ratio of the Laplace transform of a network response to the transform of a network excitation is termed a *network function.* An example is

$$\frac{I_2(s)}{V_s(s)} = H(s)$$

$$= \frac{1/R_2}{s^2\,\dfrac{L_1L_2}{R_1R_2} + \left(\dfrac{L_1}{R_1} + \dfrac{L_1}{R_2} + \dfrac{L_2}{R_2}\right)s + 1} \tag{11.58}$$

from Equation (11.55). Generally, a network function can relate either a current or voltage response to a current or voltage excitation. Thus, the ratio can be an impedance (voltage/current), an admittance (current/voltage), or a simple ratio (current/current or voltage/voltage). If the current and voltage are associated with a single port, then the ratio is a *driving point* function; otherwise, the ratio $H(s)$ is a *transfer* function.

Because they represent zero-state responses, network functions have a number of useful properties.

1. The replacement of s by $j\omega$ in a network function yields a complex-valued function of ω that relates sinusoidal steady-state phasor representations of the excitation and response. This follows from the similar observations about elemental impedance models. For example, if $v_s(t) = V_m \cos(\omega t + \phi_v)$ in Figure 11.7, the sinusoidal steady-state component of the output is $i_{2ss}(t) = I_m \cos(\omega t + \phi_i)$ where

$$I_m = |H(j\omega)|\, V_m$$

$$\phi_i = \phi_v + \arg H(j\omega)$$

and

$$H(j\omega) = \frac{1/R_2}{-\omega^2 \dfrac{L_1 L_2}{R_1 R_2} + \left(\dfrac{L_1}{R_1} + \dfrac{L_1}{R_2} + \dfrac{L_2}{R_2}\right) j\omega + 1}$$

2. Let the excitation be denoted $e(t) = \mathcal{L}^{-1}[E(s)]$ and the response $r(t) = \mathcal{L}^{-1}[R(s)]$. Either can be a current or a voltage. Then we can write

$$R(s) = E(s)H(s)$$

$$= \frac{P(s)}{Q(s)} \frac{N(s)}{D(s)}$$

where P, Q, N, and D are all polynomials in s. For example, the right-hand side of Equation (11.55) can be written

$$I_2(s) = \left(\frac{V_s}{s}\right)\left(\frac{1/R_2}{s^2 \dfrac{L_1 L_2}{R_1 R_2} + s\left(\dfrac{L_1}{R_1} + \dfrac{L_1}{R_2} + \dfrac{L_2}{R_2}\right) + 1}\right) \quad (11.59)$$

The partial fraction expansion of $R(s)$ will contain a term for each root of $Q(s)$ and a term for each root of $D(s)$. These denominator roots, which are the *poles* of $R(s)$, therefore determine the kinds of time variation occurring in $r(t)$. The terms in the partial fraction expansion of $R(s)$ may be grouped according to their origin and $R(s)$ written in the form

$$R(s) = [\text{terms from roots of } Q(s)] + [\text{terms from roots of } D(s)] \quad (11.60)$$

The first bracket contains the transform of the *forced response*, that is, those terms of the same form as the input. The second bracket contains the transform of the *natural response* of the form

$$k_1 e^{p_1 t} + k_2 e^{p_2 t} + \cdots + k_n e^{p_n t} \quad (11.61)$$

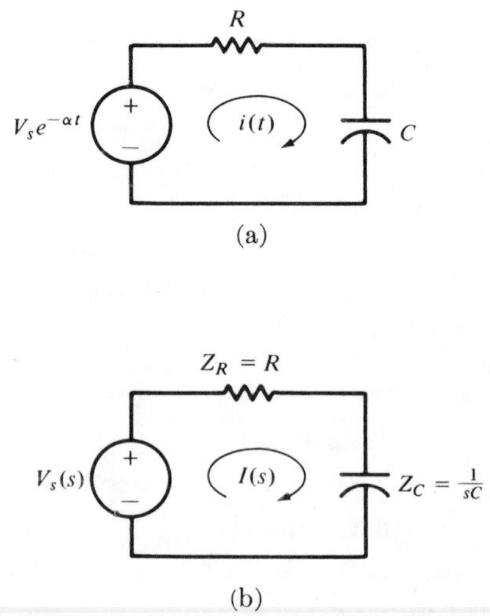

FIGURE 11.9 *RC* network models. (a) Time domain; (b) *s*-domain.

where the p_i's are the poles associated with the network.

Example 11.12

The transfer function $I(s)/V_s(s)$ is to be found for the network of Figure 11.9(b); the response to $v_s(t) = V_s e^{-\alpha t}$ is also desired for $t > 0$.

The transfer function is routinely calculated from Figure 11.9(b) as

$$H(s) = \frac{I(s)}{V_s(s)} = \frac{1}{Z_R + Z_C}$$

$$= \frac{1}{R + \dfrac{1}{sC}}$$

$$= \frac{s/R}{s + \dfrac{1}{RC}}$$

The response to $v_s(t)$ is then obtained from

$$I(s) = V_s(s)H(s)$$

$$= \left(\frac{V_s}{s+\alpha}\right)\left(\frac{s/R}{s + 1/RC}\right)$$

If $\alpha \neq 1/RC$ so that there is no multiple pole at $-\alpha$, the response will be of the form

$$i(t) = k_1 e^{-\alpha t} + k_2 e^{-t/RC}$$

where

$$k_1 = \alpha V_s C/(RC\alpha - 1),$$

$$k_2 = V_s/R(-RC\alpha + 1)$$

The first term is due to the excitation; the latter arises from the natural frequency of the network at $s = -1/RC$.

A special case for which the forced and natural responses cannot be separated occurs when $E(s)$ and $H(s)$ have coincident poles. For example, if $\alpha = 1/RC$ above, then we have

$$i(t) = \frac{V_s}{R}(1 - \alpha t)\epsilon^{-\alpha t} = \frac{V_s}{R}\left(1 - \frac{t}{RC}\right)\epsilon^{-t/RC}$$

where neither term can be identified with either the forced or the natural response.

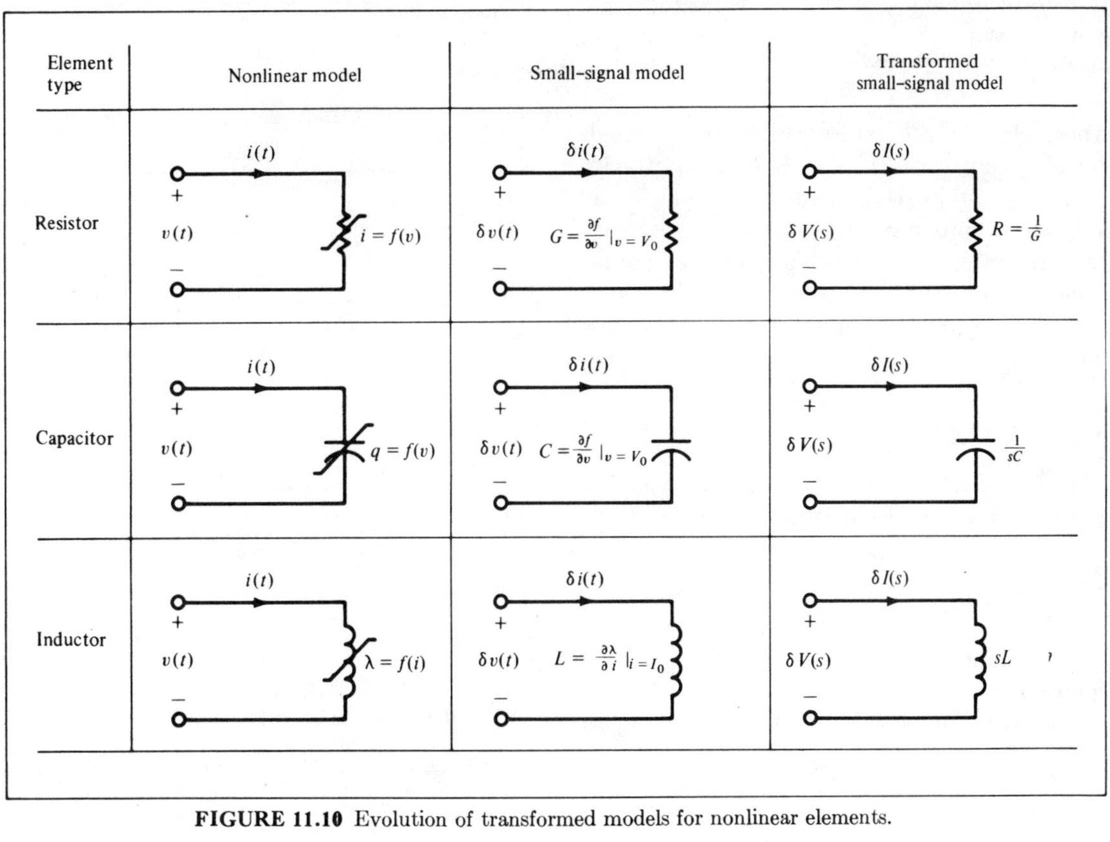

FIGURE 11.10 Evolution of transformed models for nonlinear elements.

11.6 Small-Signal Dynamic Analysis

Because all physical circuits are nonlinear if the elements are operated over a sufficiently wide range of voltage or current, all linear network analysis necessarily involves the use of incremental models for the physical components. As we have seen, *small-signal analysis* is a special case of this incremental analysis when the network is analyzed around a dc operating point.

Because any small-signal *response* is necessarily zero-valued when the *signal* is zero-valued, such a response is a zero-state response. We can come to the same conclusion by examining the small-signal models for resistors, inductors, and capacitors in Chapter 7. As illustrated in Figure 11.10, the small-signal relationship for each element does not explicitly involve initial (dc) conditions. Therefore, the transformed model of the element is of the general form

$$\delta V(s) = Z(s)\delta I(s)$$

where $\delta V(s)$ and $\delta I(s)$ are transforms of small signal quantities $\delta v(t)$ and $\delta i(t)$, respectively.

Having established small-signal analysis as a form of zero-state calculation, we can use the properties of network functions to obtain insight into the transient and sinusoidal steady-state solution of small-signal network models.

Example 11.13

The response $i(t)$ of the nonlinear network of Figure 11.11(a) to inputs of the form

(1) $v_s(t) = 10 + u(t)$

(2) $v_s(t) = 10 + \sin \omega t u(t)$

is to be found. In (2), only the steady-state response is to be found.

The equations of the circuit are of the form

$$Ri(t) + L(i)\frac{di(t)}{dt} = v_s(t)$$

where $L(i) = \partial\lambda/\partial i = 6i$. The equation, being nonlinear, has no obvious analytic solution. Noting that the input change is small at $t = 0$ (from 10 to 11 volts), we decide to linearize the circuit about the dc operating point and obtain an approximate solution by small-signal analysis.

The equilibrium (dc) solution for $t < 0$ is obtained by shorting the inductor as in Figure 11.11(b); the dc current is readily found as $I_0 = 10 \text{ volts}/1000\Omega = 0.01$ amp. The lin-

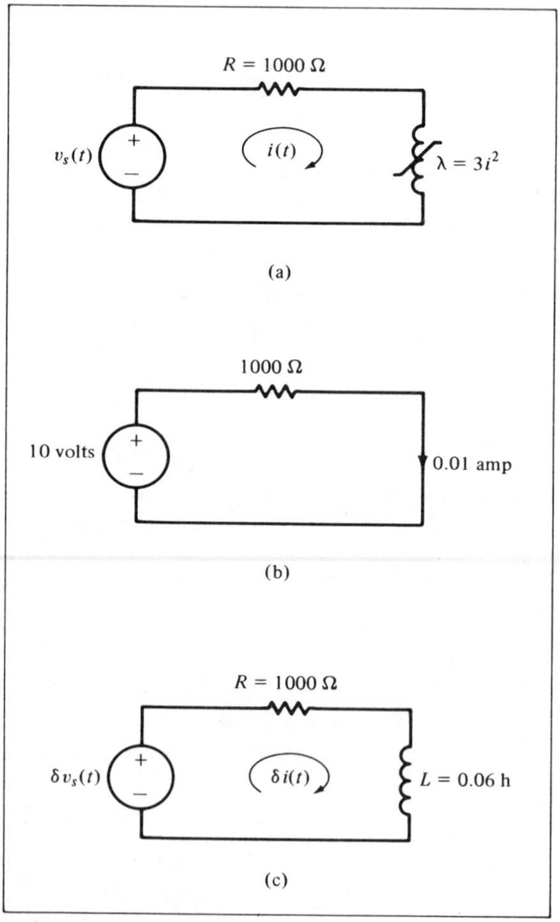

FIGURE 11.11 Small-signal analysis of a nonlinear network. (a) Nonlinear network; (b) dc model; (c) small-signal model; (d) step signal; (e) step response; (f) sinusoidal input; (g) transient and steady-state response.

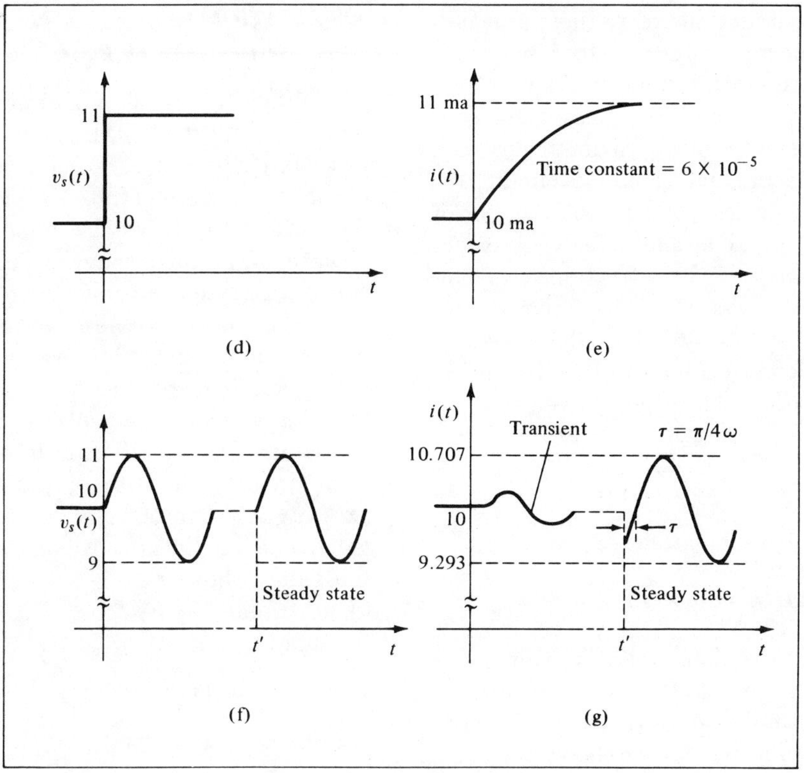

FIGURE 11.11 (continued).

earized value of $L(I_0)$ is

$$(\partial\lambda/\partial i)_{i=I_0} = 6I_0 = 0.06 \text{ h}$$

The small-signal model then has the form of Figure 11.11(c). The small-signal analysis routinely proceeds by evaluating the admittance

$$Y(s) = \frac{\delta I(s)}{\delta V_s(s)} = \frac{1}{Z_L + Z_R} = \frac{1}{sL + R} \tag{11.62}$$

Then the transformed response is written as $\delta I(s) = Y(s)\delta V_s(s)$ or

$$\delta I(s) = \frac{1}{L}\frac{\delta V_s(s)}{s + R/L} \tag{11.63}$$

1. *Step response*. The small-signal component of the waveform of Figure 11.11(d) is

$$(11 - 10)u(t) = u(t)$$

Equation (11.63) then becomes

$$\delta I(s) = \frac{1}{0.06}\frac{1}{s(s + 1/(6 \times 10^{-5}))}$$
$$= \frac{10^{-3}}{s} + \frac{-10^{-3}}{(s+1/(6\times10^{-5}))}$$

The inverse transform is

$$\delta i(t) = 10^{-3}(1 - e^{-t/(6\times10^{-5})}) \qquad (t > 0)$$

and the approximate total response is

$$i(t) = I_0 + \delta i(t)$$
$$= 10^{-2} + 10^{-3}(1 - e^{-t/(6\times10^{-5})}) \qquad (t > 0) \tag{11.64}$$

The original circuit therefore has a small-signal time constant of 6×10^{-5} sec; this, together with the equilibrium values of 10 ma and 11 ma for $i(0)$ and $i(\infty)$, provides considerable insight into the approximate dynamic behavior of the original nonlinear circuit. Indeed, this is a typical example of the use of "engineering approximation" to solve for practical purposes an otherwise intractable problem.

2. *Steady-state response.* The approximate steady-state response to a small sinusoidal signal is obtained in a similar way. Substituting $s = j\omega$ in Equation (11.62), we have

$$Y(j\omega) = \frac{1/R}{j\omega(L/R) + 1} \tag{11.65}$$

It is clear from this equation that if

$$\omega L/R = 6 \times 10^{-5}\omega$$

is significantly less than unity in value, the inductor exerts no influence on the steady-state response, and it can be replaced by a short ($L = 0$). If, for illustration, we choose $\omega = R/L$ so that the inductor *is* significant, then

$$Y(j\omega) = \frac{1/R}{1 + j1} = 0.707\frac{1}{R}e^{-j45°}$$

The total steady-state response to an input of the form ($t>0$)

$$v_s(t) = 10 + \sin\left(\frac{R}{L}t\right) \tag{11.66}$$

would then be

$$i(t) = \underbrace{0.01}_{\text{dc steady state}} + \underbrace{0.707 \times 10^{-3}\sin\left(\frac{R}{L}t - 45°\right)}_{\text{Sinusoidal steady state}} \tag{11.67}$$

Recall that because we are determining the *steady-state* response, transient terms of the form $e^{-Rt/L}$ do not appear in Equation (11.67). This is illustrated in Figure 11.11(g).

APPLICATION

Dynamic Response of Two-Stage Amplifier

The simple network of Example 11.12 is intended to illustrate the small-signal analysis procedure in detail. A far more typical example of this method is illustrated in Figure 11.12(a)–(b), where the two-stage amplifier studied in Chapters 5 and 6 is converted to a small-signal equivalent. We are comfortable in performing this linearization, because the amplifier should perform in an almost linear manner.

In practice, the batteries introduced in Chapter 6 would be replaced by capacitors. A dc analysis (with capacitors replaced by open circuits) yields the initial capacitor voltages $(v_{C_1})_0 = 2.5924$ volts and $(v_{C_2})_0 = 2.2694$ volts, and all linearized diode resistances are the values shown in Figure 11.12(b). We are to determine the response of the amplifier to the inputs

$$v_s(t) = 0.05u(t)$$

$$v_s(t) = 0.02\sin(\omega t)u(t)$$

Because the network is relatively large, we resort to a computer analysis to obtain the small-signal transfer function

$$\frac{\delta V_0(s)}{\delta V_s(s)} = \frac{11.1266s^2 + 18{,}793.1s}{s^2 + 9786.72s + 197{,}239} \tag{11.68}$$

$$= \frac{11.1266s(s + 1689)}{(s + 9766)(s + 20.19)} = H(s)$$

1. *Step input.* The transient response to a step of 0.05 volt is obtained from

$$\delta V_0(s) = \frac{0.05}{s}H(s)$$

giving

$$\delta v_0(t) = 0.46105e^{-9766t} + 0.09525e^{-20.19t}$$

At $t = 0^+$ we obtain a signal of

$$\delta v_0(0^+) = 0.46105 + 0.09525$$

$$= 0.55630$$

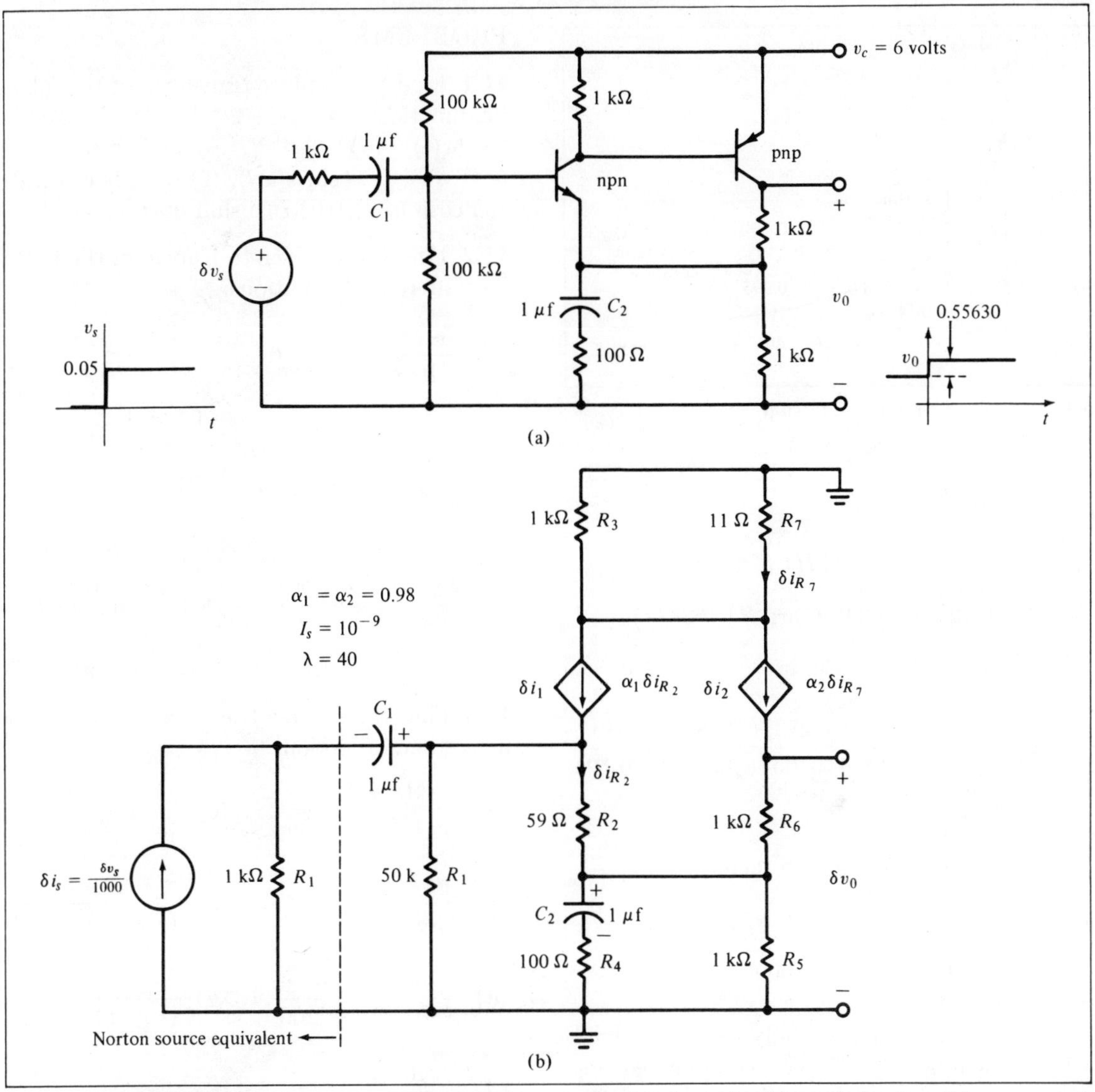

FIGURE 11.12 Dynamic amplifier analysis. (a) Circuit schematic; (b) small-signal model.

This value yields a gain of 11.127 found in Equation (5.10). (Why?) The transient response discloses two time constants of value 1/9766 sec and 1/20.19 sec. Thus, in 1/9766 sec, the first term will have been decayed to $1/e$ of its initial value, giving us a notion of the rate of decay of the response. A display of the small-signal response is given in Figure 11.13. Eventually, the small-signal response becomes essentially zero-valued, as the capacitors charge to a different voltage level resulting from the constant input signal.

2. *Sinusoidal input.* We choose

$$\omega = 2\pi \times 1000$$

so that

$$\delta v_s(t) = 0.02 \sin(2\pi \times 1000t)$$

We can then calculate the total steady-state

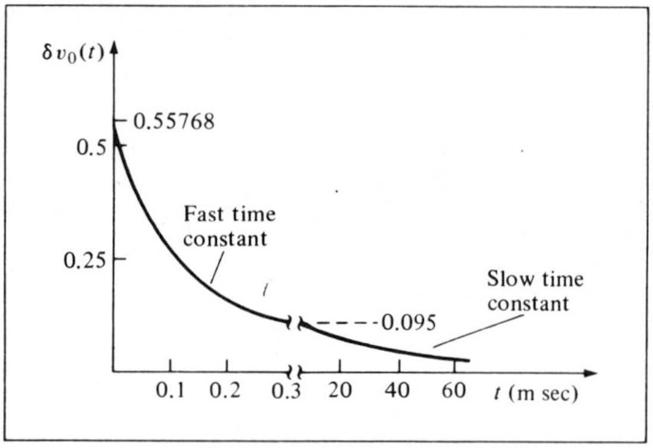

FIGURE 11.13 Small signal step response to $\delta v_s = 0.05\,u(t)$.

response to be

$$v_0(t) = 4.57596 + 0.02\,|\,H(j6283)\,| \times \sin(2\pi \times 1000t + \arg H(j6283))$$

where

$$H(j1000) = \frac{11.1266(j6283)(j6283 + 1689)}{(j6283 + 9766)(j6283 + 20.19)} = 6.233e^{j42.38°}$$

The steady-state response is then

$$v_s(t) = 4.57596 + 0.12466 \sin(2\pi \times 1000t + 42.38°)$$

Again, observe that not only have we solved what was originally a very complicated set of nonlinear differential equations, but we have extracted only those parts of the solution (the steady state) that we required. The linearizing approximations that we made can be shown to be quite reasonable (by understanding the design procedures that produced this network).

The computer program used to generate the transfer function of Equation (11.68) is the CORNAP program described in Reference 11.5. This program will also compute the step response and the sinusoidal steady-state response displayed above.

PROBLEMS

11.1 Find the Laplace transform of the following functions. (a) $(5 - 2e^{-2t})u(t)$. (b) $2\delta(t) - 3e^{-7t}u(t)$. (c) $(8e^{-1000t} - 5e^{-2000t})u(t)$. (d) $te^{-500t}u(t)$. (e) $t \sin \omega t u(t)$. (f) $(t-1) \cos 2tu(t)$. (g) $\cosh tu(t)$. (h) $e^{-\alpha t} \sinh \beta tu(t)$.

11.2 Find the Laplace transform of the following differential equations.

(a) $\dfrac{dv(t)}{dt} + v(t) = 0$ $\quad (v(0) = 1)$.

(b) $\dfrac{2\,dv(t)}{dt} + 3v(t) = e^{-3t}u(t)$ $\quad (v(0) = 2)$.

(c) $\dfrac{5\,d^2v(t)}{dt^2} + \dfrac{3\,dv(t)}{dt} - 2v(t) = \sin 6tu(t)$ $\quad (v(0) = 0,\ \dot{v}(0) = 3)$.

11.3 Find the inverse transform of the following rational functions.

(a) $\dfrac{(s+6)}{(s+2)(s+5)}$.

(b) $\dfrac{10(s+2)}{(s)(s+5)}$.

(c) $\dfrac{100(s+50)}{s^2 + 201s + 200}$.

(d) $\dfrac{1}{(s+3)^2}$.

(e) $\dfrac{1}{(s^2+3)^2}$.

(f) $\dfrac{(s+3)}{(s+1)^2(s+2)}$.

(g) $\dfrac{(s+3)}{(s+1)^3(s+2)}$.

(h) $\dfrac{(s+5)}{s^2(s^2+3)}$.

11.4 Find the Laplace transform of (a) $u(t) - u(t-1)$. (b) $e^{-(t-2)}u(t-2)$. (c) $e^{-(t-2)}u(t)$. (d) $t[u(t) - u(t-2)]$. (e) $\sin 3tu(t-1)$.

11.5 Find the initial values of the inverse transforms of Problem 11.3 directly from the transform.

11.6 Find the final values of the inverse transforms of Problem 11.3 directly from the transform.

11.7 Consider the function

$$f(t) = \sum_{n=0,2,4,\ldots}^{\infty} f_1(t - nT)(u(t - nT) - u(t - nT - T))$$

Show that the transform of $f(t)$ is

$$F(s) = \frac{\mathcal{L}[f_1(t)]}{1 + e^{-Ts}}$$

11.8 Find $v_L(t)$, $t > 0$, in Figure P 11.1, given $v_C(0) = i_L(0) = 0$. Use:

(a) $R = 5000\ \Omega$.
(b) $R = 4000\ \Omega$.
(c) $R = 2000\ \Omega$.

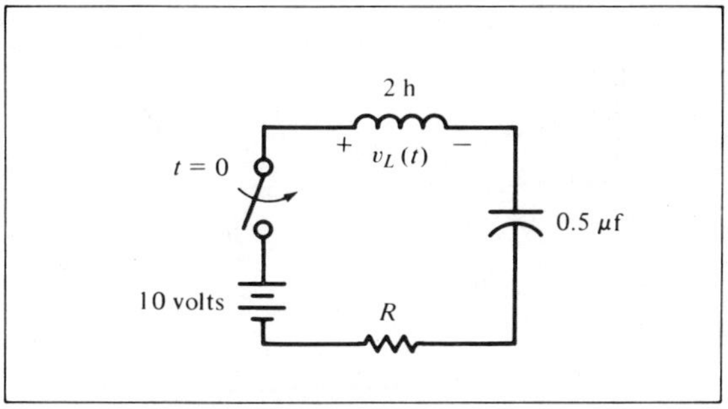

FIGURE P 11.1

11.9 Find $v_0(t)$ in Figure P 11.2.

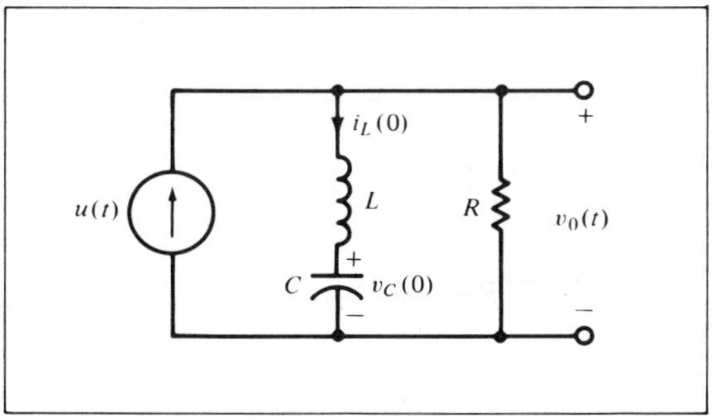

FIGURE P 11.2 $R = 1\ \Omega$; $L = 1$ h; $C = 1$ f; $v_C(0) = 1$; $i_L(0) = 0$.

11.10 Find $v_{C_2}(t)$ in Figure P 11.3. Assume that the switch was closed for $t < 0$ so that the network had reached equilibrium.

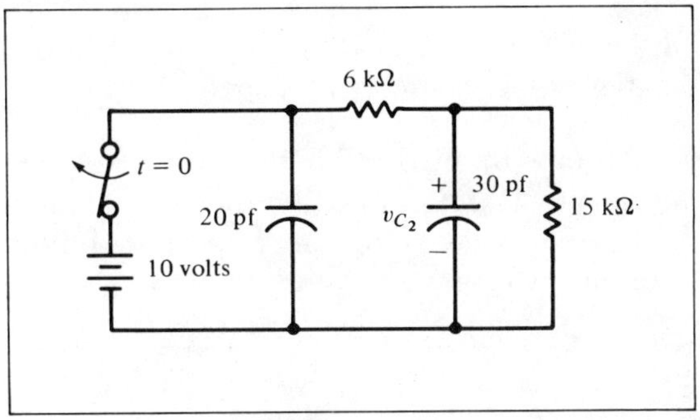

FIGURE P 11.3

11.11 Find $v_C(t)$ for $t > 0$ in Figure P 11.4. Assume $v_C(0^-) = 0$.

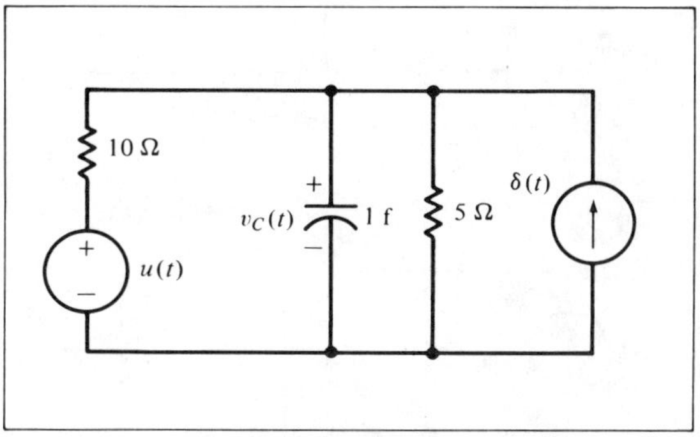

FIGURE P 11.4

11.12 Calculate the voltage on the capacitor in Figure P 11.4 for $t = 0^+$, after the impulse has passed. Then apply the Laplace transform with a lower integration limit of $t = 0^+$ and compute $v_C(t)$, $t > 0$. Is this the same result as in Problem 11.11?

11.13 In Figure P 11.5, find a set of initial con-

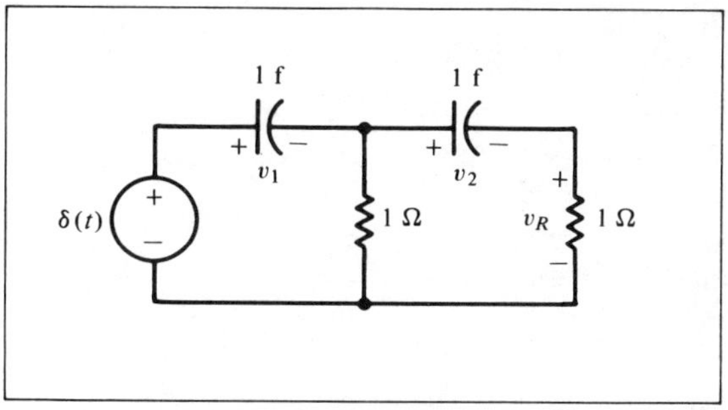

FIGURE P 11.5

ditions [$v_1(0^-)$ and $v_2(0^-)$] that result in the voltage v_R being simply an impulse $K\delta(t)$

11.14 (a)–(b) Find $v_0(t)$ in the networks shown in Figure P 11.6. Identify the natural and forced components of the responses. Initial capacitor voltages are zero.

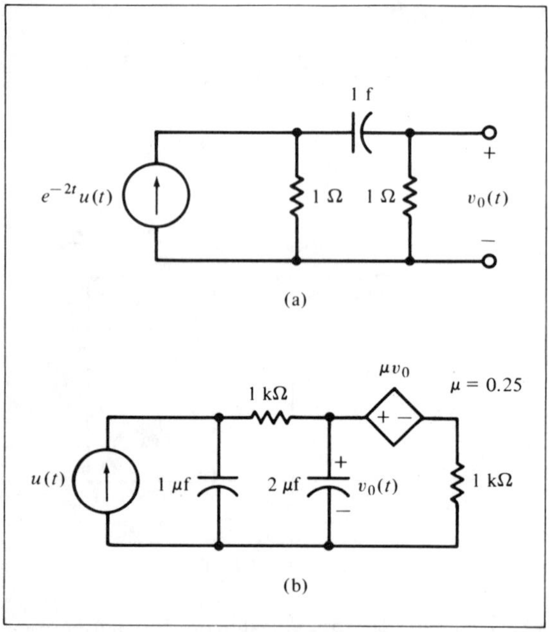

FIGURE P 11.6

11.15 The network functions below are to be used to evaluate the sinusoidal steady-state response $r(t)$ to an excitation of the form $\cos \omega t$, where ω is given.

(a) $H(s) = \dfrac{s}{s+1}$ $(\omega = 1)$.

(b) $H(s) = \dfrac{1}{s^2+10}$ $(\omega = 2)$.

(c) $H(s) = \dfrac{10s}{(s+1)^2+1}$ $(\omega = 2)$.

(d) $H(s) = \dfrac{e^{-s}}{s+1}$ $(\omega = 1)$.

11.16 You are asked to show that the sinusoidal steady-state response of $v_0(t)$ of Figure *P* 11.7 is zero when $\omega = 1/\sqrt{(C_1R_1R_2 - L)/LC_1C_2R_1R_2}$. This is to be shown in the following steps. (a) Write the mesh equations for the network model of Figure P 11.7(b). (b) Solve for the network function $V_0(s)/V_s(s)$ from the mesh equations, using any technique you wish. Obtain the null frequency from the numerator of this function.

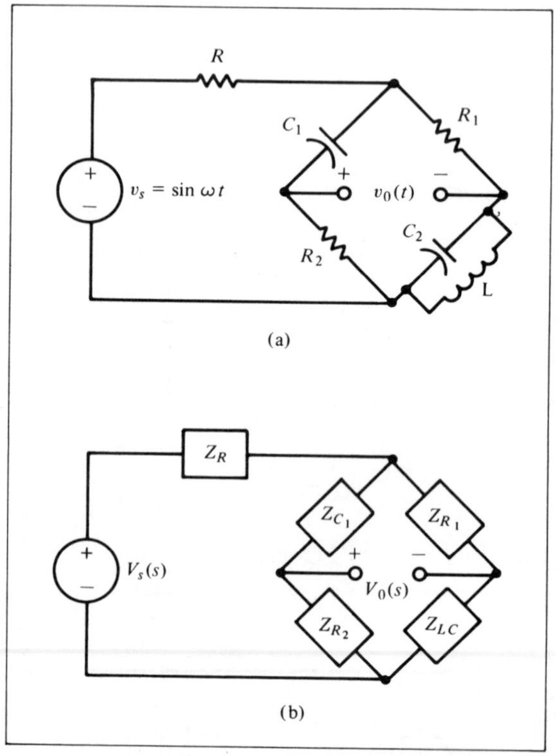

FIGURE P 11.7 Bridge circuit.

11.17 The reader will notice that by combining the relationship for the Laplace transform of a derivative (11.10) and the initial value theorem, we can always obtain the first term in the power series expansion of the zero-state response of a network. (Recall the coefficients of a power series are related to the function's derivatives.) (a) Demonstrate this by calculating the first term in the power series for $v_0(t)$ in the two networks shown in Figure

P 11.8. (b) Verify your results in (a) for the network (a) by calculating $v_0(t)$ for all $t > 0$.

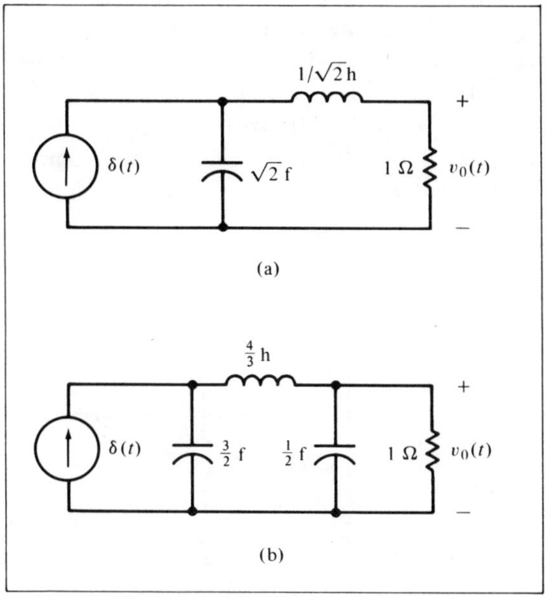

FIGURE P 11.8

11.18 A real diode as shown in Figure P 11.9 has, in addition to the familiar exponential i_c–v_d characteristic, a q_d–v_d characteristic to account for internal charge storage effects. Assume that the network shown operates at a dc diode voltage of $v_d = v_{d0}$. A small change δv_s occurs in v_s. (a) Construct the small-signal equivalent network. (b) Show that the small-signal natural frequency of the diode is at $s = -1/\tau$, for any v_{d0}. (For this evaluation, assume that R_s is very large, so that the diode is connected to a small-signal open circuit.) (c) Find the voltage transfer function $\delta V_d(s)/\delta V_s(s)$ in terms of the element values of the small-signal model. What is the small-signal natural frequency of the network? Do not assume a large R_s. (d) If $\delta v_s = Au(t)$ where A is small, find $\delta v_d(t)$ and $\delta i_d(t)$.

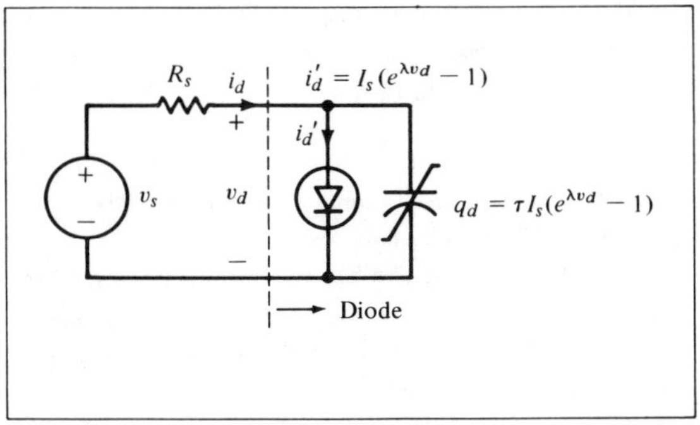

FIGURE P 11.9

11.19 The circuit of Figure P 11.10 has an operating point defined by the two 5-volt sources shown. The DCAP program is used to compute the small-signal emitter diode resistance at this operating point as 16.1 Ω. The q–v equation for the emitter diode (see Problem 11.18) is given by $q_d = 10^{-17}(e^{40v_d} - 1) = \tau I_s(e^{\lambda v_d} - 1)$. (a) Construct the small-signal equivalent circuit, including a numerical value for the small-signal capacitance $\partial q_d/\partial v_d$ at the operating point. (b) Evaluate the small-signal voltage transfer function $\delta V_0(s)/\delta V_s(s)$. (c) If $\delta v_s(t) = 0.2u(t)$, find $\delta v_0(t)$. (d) If $\delta v_s(t) = 0.2 \cos 10^8 t$, find the sinusoidal steady-state response of $\delta v_0(t)$ and the complete steady-state response of $v_0(t)$.

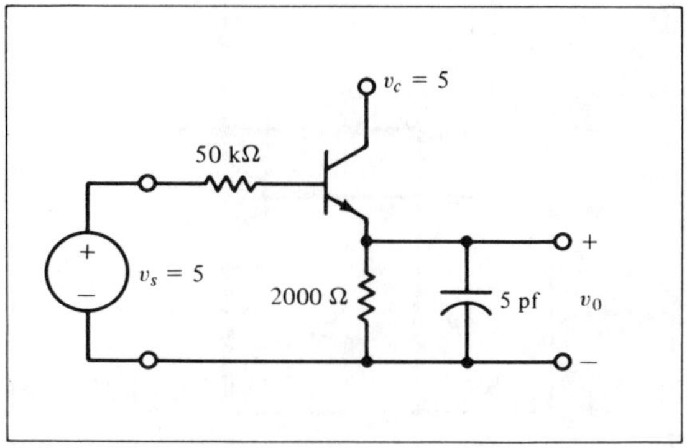

FIGURE P 11.10 $\alpha = 0.98$, $I_s = 10^{-9}$, $v_0 = 3.08$ volts at operating point.

11.20 (a–c) In the following problems, you are asked to determine the small-signal response of a nonlinear dynamic circuit. Use the transistor and diode small-signal models of Chapter 6. (a) Find $\delta V_0(s)/\delta V_s(s)$ from Figure P 11.11. (b) Determine $\delta V_0(s)/\delta I_s(s)$ in the network

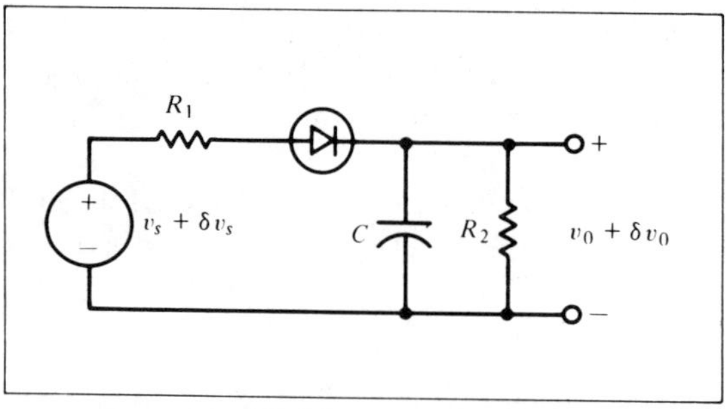

FIGURE P 11.11

shown in Figure P 11.12. Assume that the transistor operates in the normal (forward)

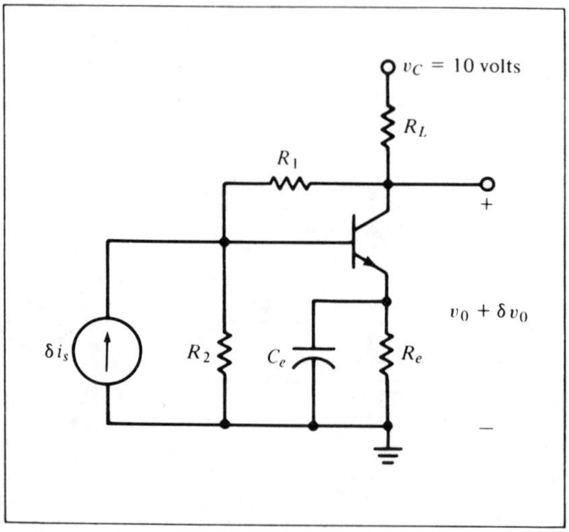

FIGURE P 11.12

region. (c) Determine $\delta V_0(s)/\delta V_s(s)$ in the network shown in Figure P 11.13.

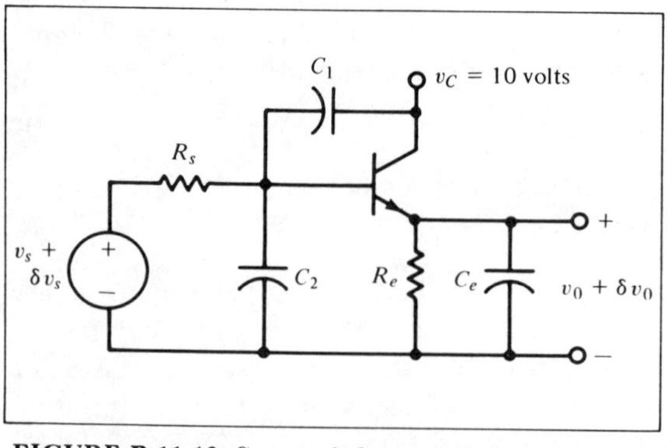

FIGURE P 11.13 $C_1 = 1$ pf, $C_2 = 10$ pf, $C_e = 0.005\ \mu f$, $R_s = 200\ \Omega$, $R_e = 1\ k\Omega$, $G_d = 0.03$ ℧, $\alpha = 0.98$.

11.21 The dynamic behavior of the amplifier of Figure P 11.14 is to be evaluated using small-signal analysis at the dc operating point, $v_{gs} = -1.15$, $v_0 = 14.5$ $v_{ds} = 4.98$. (a) Construct the small-signal equivalent network. (b) Evaluate the voltage transfer function $H(s) = \delta V_0(s)/V_s(s)$. (c) Determine $\delta v_0(t)$ and $v_0(t)$ if $v_s(t) = 0.1u(t)$. Assume small-signal analysis is valid.

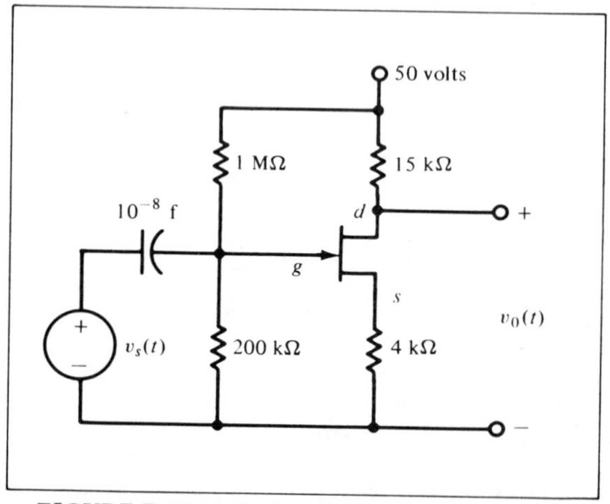

FIGURE P 11.14 $I_{dss} = 10^{-3}$, $V_p = -4$, $r_d = \infty$.

REFERENCES AND ADDITIONAL READING

11.1. M. E. VAN VALKENBURG, *Network Analysis.* Englewood Cliffs, N.J.: Prentice-Hall, Inc., 1964, Chapter 9.

11.2. R. E. SCOTT, *Elements of Linear Circuits.* Reading, Mass.: Addison-Wesley Publishing Company, Inc., 1960, Chapter 14.

11.3. B. J. LEON AND P. A. WINTZ, *Basic Linear Networks For Electrical and Electronics Engineers.* New York: Holt, Rinehart and Winston, Inc., 1970, Chapter 12.

11.4. R. V. CHURCHILL, *Introduction to Complex Variables and Applications.* New York: McGraw-Hill Book Company, Inc., 1948.

11.5. C. POTTLE, "State-Space Techniques for General Active Network Analysis," in *System Analysis by Digital Computer* (edited by Kuo and Kaiser). New York: John Wiley & Sons, Inc., 1965.

11.6 M. E. GARDNER AND J. L. BARNES, *Transients in Linear Systems,* John Wiley & Sons, Inc., 1942.

12

s-Domain Analysis

12.1 Introduction

Engineering specifications on dynamic networks are seldom given in terms of the poles and zeros of the network function. Usually, either the sinusoidal steady-state or the transient characteristics are prescribed. We may wonder, then, why we should devote an entire chapter to s-domain (pole-zero) analysis. The justification lies in the central role that pole-zero analysis plays in predicting *both* the transient and sinusoidal steady-state response. Thus, by knowing certain characteristics of either of the two responses, we are able to estimate the poles and zeros and from these obtain related properties of the other response.

In this chapter, we will consider these s-plane relationships to the time and steady-state response, with emphasis on the latter. Although we will consider only network applications, the fact that the development will be based on the network function

$$H(s) = \frac{K \prod_{j=1}^{m} (s - z_j)}{\prod_{i=1}^{n} (s - p_i)}$$

implies that any linear dynamic system described by a rational transfer function in s can be similarly analyzed.

12.2 Time-Domain Relationships

Simple Components of the Transient Response

The impulse response associated with a network function $H(s)$ with simple poles at $p_1, p_2, \ldots, p_n$ has the form

$$\mathcal{L}^{-1}[H(s)] = h(t) = \sum_{i=1}^{n} k_i e^{p_i t}$$

where complex-valued p_i are accompanied by their conjugates. Each term in the summation contributes to the transient response one of the functions shown in Figure 12.1: (1) a growing or decaying exponential, (2) a growing or decaying sinusoid, or (3) an oscillating sinusoid. Of course, the relative magnitudes of these terms are determined by the k_i, which in turn are governed by the zero positions. In the extreme ease, if a zero z_j is equal to a pole p_i, that pole will have a zero-valued k_i and so will not appear in the above response.

Stability

A consequence of the explicit appearance of each pole in the impulse response is that, if

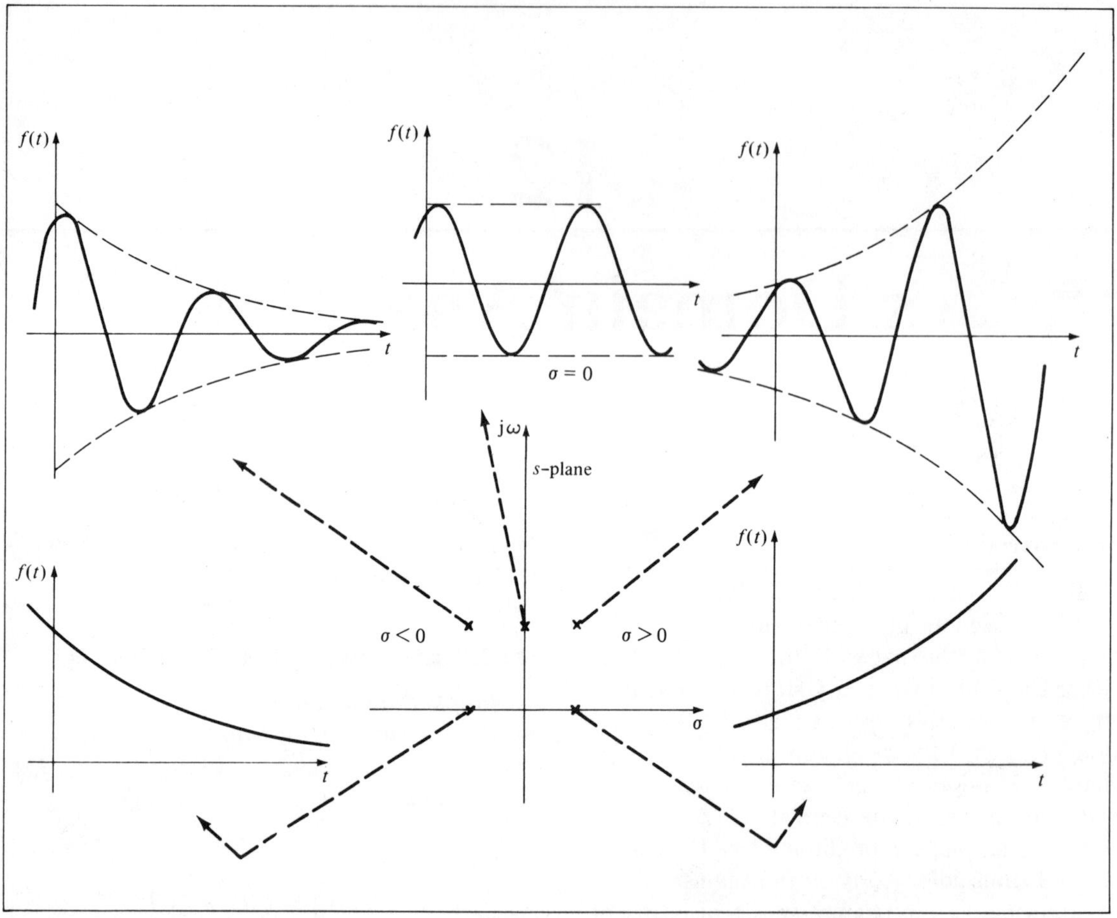

FIGURE 12.1 Real time functions associated with possible values of the complex frequency variable s.

$\text{Re}[p_i] = -\alpha_i > 0$ for *any* i, the response will have a term of the form $k_i e^{-\alpha_i t} e^{j\beta_i t}$ that *grows with increasing time.* Such a network response is termed *unstable.* If all $-\alpha_i < 0$, then

$$\lim_{t \to \infty} \{h(t)\} = 0$$

and the network is *stable.* However, if one or more of the poles have a zero-valued real part, then either of two possibilities exist.

1. If all the poles are *simple,* then they will contribute terms of the form

$$k_i \cos(\beta_i t + \phi_i)$$

to the transient response. These terms oscillate indefinitely. Such a network response is termed *conditionally* stable.

2. If any of these imaginary-axis poles are *double poles,* as for example,

$$H(s) = \frac{P(s)}{Q(s)(s^2 + \beta_i^2)^2}$$

then a term will exist in the transient response of the form

$$(k_{i,1} + k_{i,2}t) \cos(\beta_i t + \phi_i)$$

and the response will *grow* with time. A similar result applies for j-axis poles of higher multi-

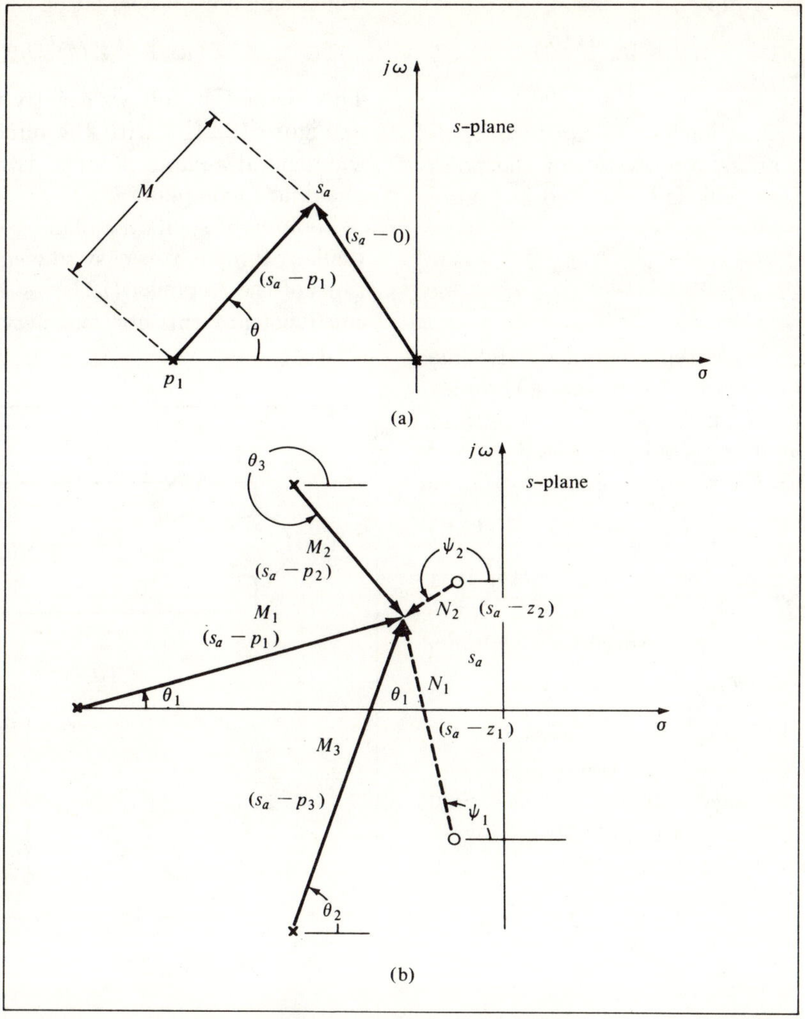

FIGURE 12.2 Graphical representation of numerator and denominator factors of $H(s_a)$. (a) A single real-axis pole; (b) a three-pole example.

plicity. All related network responses will be *unstable*.

We see from the above that knowledge of the poles of a network function gives us considerable insight into the qualitative transient behavior. Quantitative evaluation inevitably necessitates that the partial fraction expansion of Chapter 11 be carried out; frequently a digital computer will be used.

12.3 Graphical s-Plane Representation of $H(s)$

Evaluation in the s-Plane

Perhaps the most attractive feature of the pole-zero description of a network function is the ease with which we can visualize and estimate the value of $H(s)$ at a particular complex frequency s_a. An evaluation of $H(s)$

at $s = s_a$ has the form

$$H(s_a) = K\frac{(s_a - z_1)\dots(s_a - z_m)}{(s_a - p_1)\dots(s_a - p_n)} \quad (12.1)$$

Each factor in Equation (12.1) can be graphically shown in an s-plane sketch of the poles and zeros of $H(s)$. This is illustrated in Figure 12.2(a) for the function

$$H(s) = \frac{K}{s(s - p_1)}$$

The term $(s_a - p_1)$ is represented by the line from $s = p_1$ to $s = s_a$ as shown. It is a complex number having the magnitude of the length of that line, M, and an argument equal to the angle made with the positive σ-axis, θ.

$$(s_a - p_1) = Me^{j\theta} \quad (12.2)$$

In exactly the same fashion, each factor in a more complex $H(s_a)$ can be determined. Figure 12.2(b) illustrates a typical three-pole example,

$$H(s) = \frac{K(s - z_1)(s - z_2)}{(s - p_1)(s - p_2)(s - p_3)}$$

Solid lines are associated with the real pole and the complex conjugate pair of poles, and dashed lines are used for the complex conjugate zero pair. From the figure we see that

$$|H(s_a)| = \frac{N_1 \times N_2}{M_1 \times M_2 \times M_3} \quad (12.3)$$

and the angle is

$$\arg(H(s_a)) = \psi_1 + \psi_2 - \theta_1 - \theta_2 - \theta_3 \quad (12.4)$$

Frequency Response

The above calculation has use in evaluation of the sinusoidal steady-state response and in the evaluation of the partial fraction expansions of Chapter 11. If a one-port with an impedance $Z(s)$ is excited by a sinusoidal current source

$$i(t) = \text{Re}[Ie^{j\omega_a t}] \quad (12.5)$$

then from Chapter 10, the sinusoidal steady-state port voltage is

$$v_{ss}(t) = \text{Re}[Z(j\omega_a)Ie^{j\omega_a t}] \quad (12.6)$$

which calls for an evaluation of

$$Z(j\omega_a) = Z(s)|_{s=j\omega_a} \quad (12.7)$$

This evaluation follows exactly the illustration in Figure 12.2(b) with the minor change that the general s-plane point s_a is replaced by an imaginary-axis point $j\omega_a$.

Looking at such an s-plane sketch, for a particular example, it is easy to visualize the variations of the response (12.6) as ω_a is varied. We can illustrate this for the network of Figure 12.3(a).

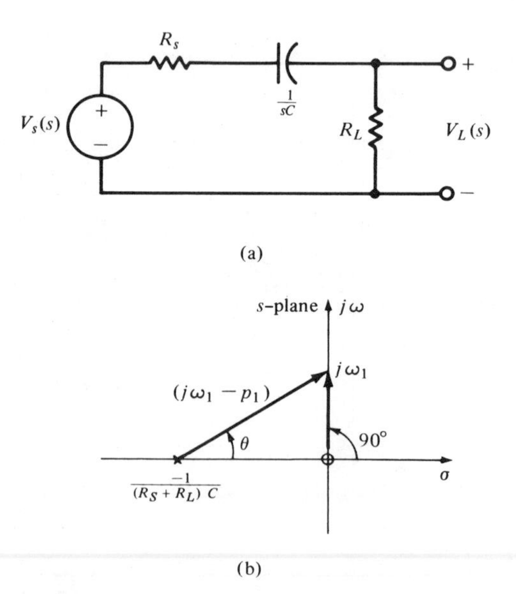

FIGURE 12.3 (a) An RC coupling network; (b) the pole-zero pattern for V_L/V_s evaluated at $s = j\omega_1$.

By the voltage divider rule

$$V_L = \frac{R_L}{R_s + R_L + \dfrac{1}{sC}} V_s$$

so the transfer function is

$$\frac{V_L}{V_s}(s) = \frac{R_L}{R_s + R_L}\frac{s}{s + \dfrac{1}{(R_s + R_L)C}} \quad (12.8)$$

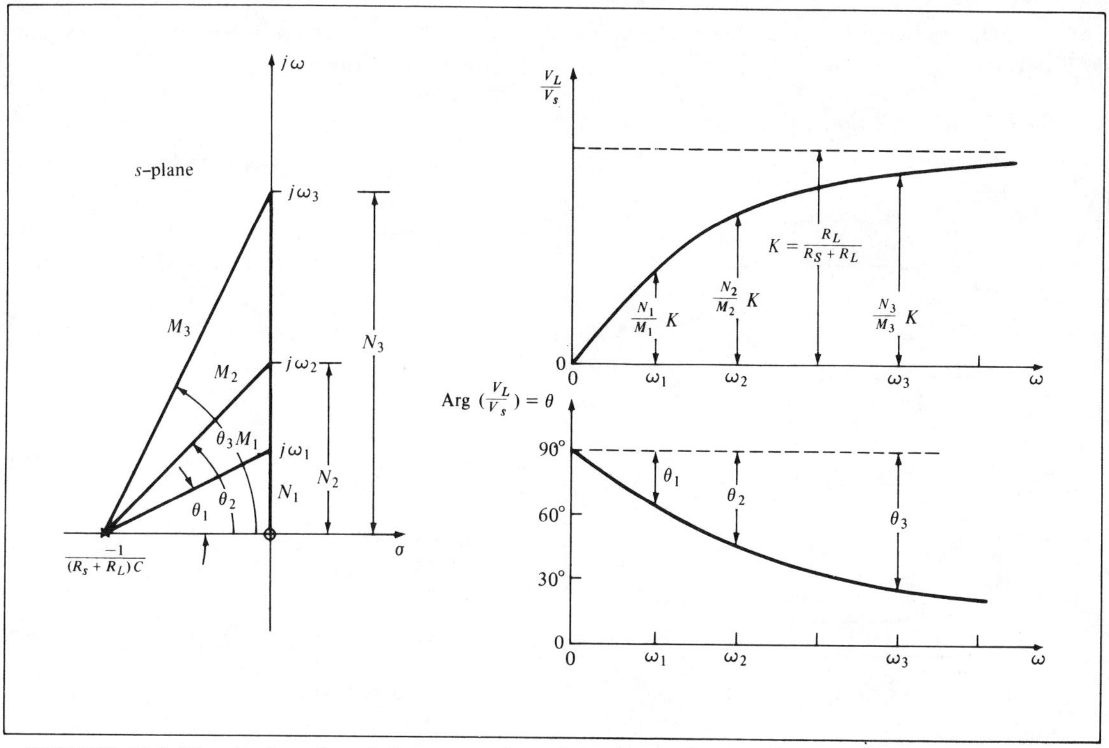

FIGURE 12.4 Magnitude and angle frequency dependence from pole-zero diagram of an *RC* coupling circuit.

This function has a single zero at the origin and a simple pole at

$$s = p_1 = \frac{-1}{(R_s + R_L)C}$$

The sinusoidal steady-state response of $v_L(t)$ is obtained by evaluating $V_L(j\omega)/V_s(j\omega)$. The magnitude and angle of $V_L(j\omega)/V_s(j\omega)$ can be visualized from Figure 12.3(b) *as a function of* ω. For example, as ω approaches zero, the magnitude of the numerator approaches zero while the denominator approaches the horizontal line $1/(R_1 + R_2)C$; therefore,

$$\lim_{\omega \to 0} |V_L(j\omega)/V_s(j\omega)| = 0$$

and

$$\lim_{\omega \to 0} \theta(j\omega) = \pi/2.$$

Similarly, as ω becomes very large, the numerator and denominator factors in Figure 12.3(b) approach a pair of equal parallel lines at right angles to the σ-axis. Therefore, at high frequencies, the magnitude $|V_L/V_s|$ becomes a constant and its angle becomes zero. At intermediate frequencies, the magnitude and angle of the transfer ratio is inferred from Figure 12.3(b) as illustrated in Figure 12.4. The two plots of Figure 12.4 are known as the *frequency response* of the network.

12.4 Resonance

We have seen in Section 12.3 how s-plane sketches of the poles and zeros of a network function can be used to visualize the magnitude and angle of the network's sinusoidal steady-state frequency response. This is particularly useful in understanding the properties of *resonant* networks.

Two networks that commonly demonstrate

this resonance phenomenon are the series RLC and parallel GCL networks of Figure 12.5.

The network function that relates the response to the excitation of these networks is

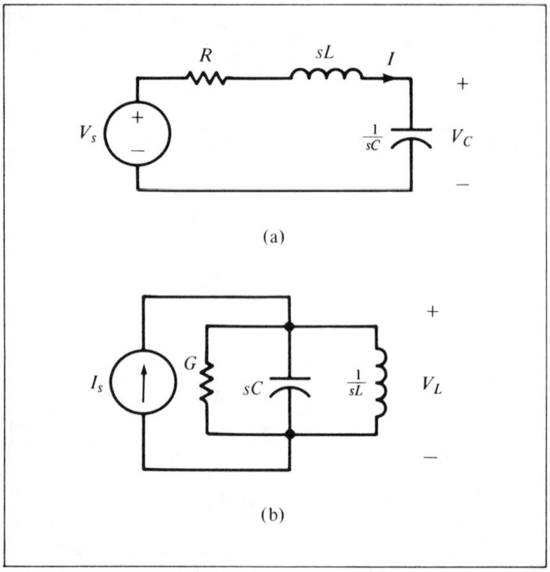

FIGURE 12.5 s-domain models of series RLC and parallel GCL networks.

the admittance of the series network

$$\frac{I}{V_s} = Y(s) = \frac{1}{R + sL + \dfrac{1}{sC}} \tag{12.9}$$

and the impedance of the parallel network

$$\frac{V}{I_s} = \frac{1}{G + sC + \dfrac{1}{sL}} \tag{12.10}$$

As functions of s, Equations (12.9) and (12.10) both can be put in the form

$$H(s) = K\frac{s}{(s - p_1)(s - p_2)} \tag{12.11}$$

This $H(s)$ has zeros at $s = 0$ and $s = \infty$ [since $\lim_{s\to\infty} H(s) = 0$] and a pair of poles that can be negative real or complex conjugates. For example, for the parallel network the poles are

$$p_{1,2} = -\frac{G}{2C} \pm \sqrt{\left(\frac{G}{2C}\right)^2 - \frac{1}{LC}} \tag{12.12}$$

The poles of the series network that were found in Chapter 9 have an identical form. Further, we can write

$$p_{1,2} = -\alpha \pm j\omega_d \tag{12.13}$$

provided

$$\omega_d \triangleq \sqrt{\omega_0^2 - \alpha^2} \tag{12.14}$$

$$\omega_0 = \frac{1}{\sqrt{LC}} \tag{12.15}$$

The magnitude of the real parts are given for the parallel network as

$$\alpha_{GCL} = \frac{G}{2C} \tag{12.16}$$

and for the series as

$$\alpha_{RLC} = \frac{R}{2L} \tag{12.17}$$

For both networks the value of α can be varied by changing the value of the resistive element. When this is done, the poles move along the s-plane locus shown in Figure 12.6. For example, in the series resonant case, when $R = 0$ the poles lie on the $j\omega$-axis at $\pm j\omega_0$. Then as R increases, $-\alpha$, the real part, becomes negative, and the poles move along a semicircular path until they meet on the negative real axis when $\alpha = \omega_0$. Any further increase of R causes the poles to split and move along the real axis approaching zero and infinity as R approaches infinity.

The above study of the locus of the poles shows that for α sufficiently small, the poles are near the imaginary axis point $j\omega_0$. In the following sections, we study the steady-state response under this condition.

Suppose the response in Figure 12.5(a) is the voltage across the capacitor instead of the cur-

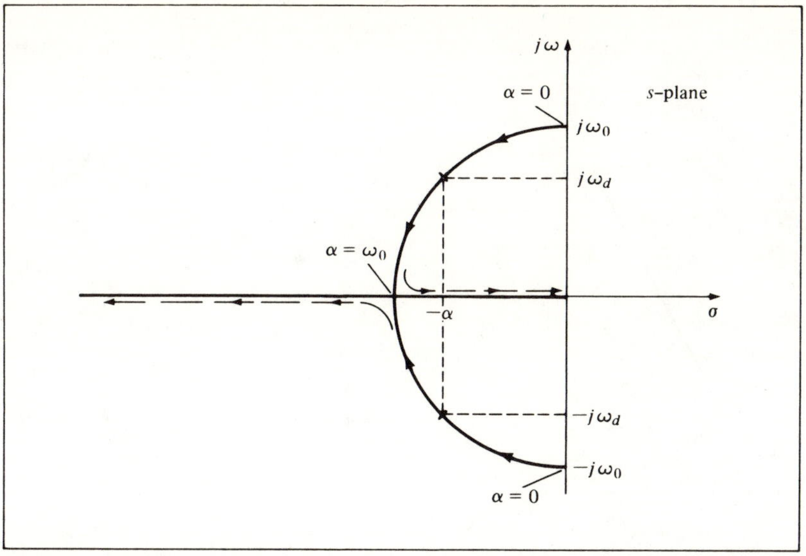

FIGURE 12.6 Locus of poles for series (parallel) resonant networks as $R(G)$ is increased from zero to infinity.

rent. Then the transfer function V_C/V_s is

$$\frac{V_C}{V_s}(s) = Y(s)\frac{1}{sC} = \frac{1}{s^2LC + sRC + 1}$$

$$= \frac{1}{LC}\frac{1}{(s - p_1)(s - p_2)} = H(s) \tag{12.18}$$

Here, the poles are unchanged from Equation (12.13), but now both zeros are at $s = \infty$. Two possible pole patterns and vector representations of the denominator factors of Equation (12.18) at a particular sinusoidal steady-state frequency ω_a are given in Figure 12.7(a), (b). From these sketches, we can visualize how the magnitude of the voltage ratio, (12.18), varies between the two cases illustrated. In Figure 12.7(a), we can see that neither pole dominates the frequency response. As ω_a is varied, both denominator factors vary nearly equally. The resulting frequency response is sketched in Figure 12.7(c). In Figure 12.7(b), however, this is not the case. For values of $j\omega_a$ near the pole p_1, the line between $j\omega_a$ and p_1 varies rapidly in both length and angle, while the other factor $(j\omega_a - p_2)$ changes relatively little. Because $(j\omega_a - p_1)$ is in the denominator of Equation (12.18), the magnitude of V_C/V_s will peak near $\omega_a = \omega_0$, as sketched in Figure 12.7(d).

While both response curves in Figure 12.7(c) and (d) represent the voltage ratio of a series *RLC* circuit, we term the circuit with the peaked response of Figure 12.7(d) a *resonant* circuit. It is evident that such a response results whenever a network pole is "close enough" to the $j\omega$-axis of the s-plane to dominate $H(j\omega)$. The parallel *GCL* network can produce a similar resonant or peaked effect of the voltage $V(j\omega)$ if G is chosen sufficiently small.

Detailed Analysis

Quantitatively, we are generally concerned with two properties of resonant networks: (1) the maximum value at the peak and (2) the sharpness or selectivity of the resonant peak. In this section, we delve into the resonance phenomenon in greater detail to develop a set

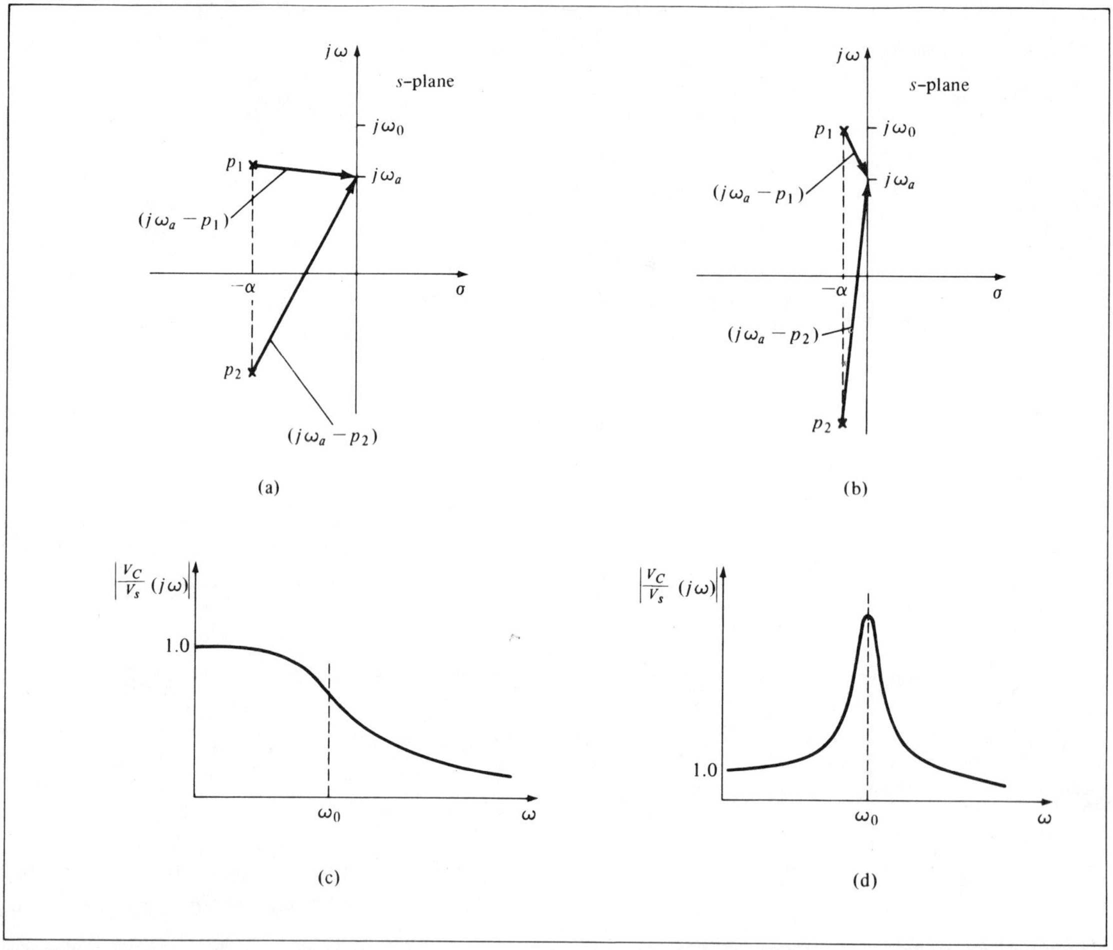

FIGURE 12.7 Pole patterns of a series resonant circuit for (*a*) $\alpha = \omega_0/\sqrt{2}$; (b) $\alpha \ll \omega_0$; (c) and (d) corresponding frequency response curves.

of parameters that describe these effects; these descriptions, like the resonance effect itself, are encountered throughout most engineering systems.

Historically, resonance was studied in terms of the average power supplied by the source [see Equation (10.60)]

$$\begin{aligned} P_{\text{av}} &= \text{Re}[V_s(j\omega)I^*(j\omega)] \\ &= \text{Re}[Z(j\omega)I(j\omega)I^*(j\omega)] \\ &= \text{Re}[Z(j\omega)]|I(j\omega)|^2 \end{aligned}$$

where effective values of current and voltage are used.

For the series *RLC* circuit of Figure 12.5(a) $\text{Re}[Z(j\omega)] = R$ and $I(j\omega) = V_s(j\omega)Y(j\omega)$, so

$$P_{\text{av}} = R|V_s(j\omega)|^2|Y(j\omega)|^2$$

and from Equation (12.9)

$$P_{\text{av}} = \frac{R|V_s(j\omega)|^2}{R^2 + \left(\omega L - \dfrac{1}{\omega C}\right)^2}$$

The denominator has a minimum value when $\omega L - 1/\omega C = 0$ or

$$\omega = \omega_0 = \frac{1}{\sqrt{LC}}$$

The maximum average power is therefore

$$P_{\mathrm{av}_m} = \frac{|V_s|^2}{R}$$

Thus, the maximum value occurs when the impedance of the inductor and the capacitor precisely cancel one another, leaving only the resistor of value R effective in the network. Accordingly, ω_0 is called the resonant frequency of the network.

The most common description of the selectivity of the network is the *bandwidth* or, equivalently, the *half-power frequencies*. These frequencies $\omega_{1,2} = \omega_{\mathrm{hp}}$ are defined by setting P_{av} to one-half of its maximum value. Solving for the resulting ω, we get

$$\tfrac{1}{2}P_{\mathrm{av}_m} = \frac{|V_s|^2}{2R} = \frac{R|V_s|^2}{R^2 + (\omega_{\mathrm{hp}}L - 1/\omega_{\mathrm{hp}}C)^2}$$

This equation can be solved to yield

$$\omega_{\mathrm{hp}}L - \frac{1}{\omega_{\mathrm{hp}}C} = \pm R \qquad (12.19)$$

From the definition of ω_0 we have

$$\frac{1}{C} = \omega_0^2 L$$

so that C may be eliminated in Equation (12.19) to yield

$$\omega_{\mathrm{hp}}L \mp R - \frac{\omega_0^2 L}{\omega_{\mathrm{hp}}} = 0$$

Multiplication by $\omega_{\mathrm{hp}}/\omega_0^2$ and division by L then produces

$$\left(\frac{\omega_{\mathrm{hp}}}{\omega_0}\right)^2 \mp \frac{1}{Q}\left(\frac{\omega_{\mathrm{hp}}}{\omega_0}\right) - 1 = 0 \qquad (12.20)$$

where

$$Q = \frac{\omega_0 L}{R} \qquad (12.21)$$

Solution of Equation (12.20) yields the four solutions

$$\frac{\omega_{\mathrm{hp}}}{\omega_0} = \pm\frac{1}{2Q} \pm \sqrt{\frac{1}{(2Q)^2} + 1} \qquad (12.22)$$

Two of the four solutions are simply the negatives of the two positive-valued solutions

$$\frac{\omega_1}{\omega_0} = \sqrt{\left(\frac{1}{2Q}\right)^2 + 1} - \frac{1}{2Q} \qquad (12.23)$$

and

$$\frac{\omega_2}{\omega_0} = \sqrt{\left(\frac{1}{2Q}\right)^2 + 1} + \frac{1}{2Q} \qquad (12.24)$$

Because they were derived from the condition of delivering $\frac{1}{2}$ the maximum power to R, ω_1 and ω_2 are known as the *lower and upper half-power frequencies,* respectively.

The bandwidth $\mathrm{BW} = \omega_2 - \omega_1$ is the range of frequency between ω_1 and ω_2 where the peak power occurs. We can then write from Equations (12.23) and (12.24)

$$Q = \frac{\omega_0}{\mathrm{BW}} \qquad (12.25)$$

(We will shortly discover the generality of this expression for Q.)

This equation shows that for a fixed $\omega_0 = 1/\sqrt{LC}$, the higher the value of Q, the smaller will be the bandwidth and the more selective the network will be.

Example 12.1

We are to determine the frequency selective characteristics of the network of Figure 12.8.

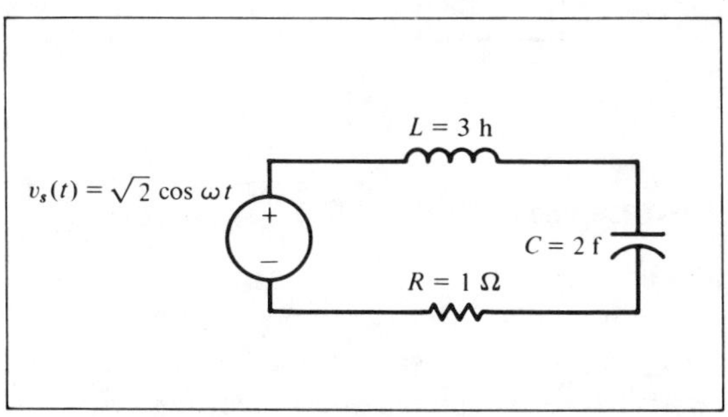

FIGURE 12.8 Resonant network example.

We determine in order

$$\omega_0 = \frac{1}{\sqrt{3 \times 2}} = \frac{1}{\sqrt{6}}$$
$$= 0.408 \text{ rad/sec}$$
$$Q = \frac{\omega_0 L}{R} = \frac{3}{\sqrt{6}} = 1.23$$
$$\frac{\omega_{hp}}{\omega_0} = \sqrt{\frac{1}{4Q^2} + 1} \pm \frac{1}{2Q}$$
$$= \sqrt{\frac{1}{6} + 1} \pm \frac{1}{\sqrt{6}}$$
$$= 1.488,\ 0.672$$

giving

$$\omega_1,\ \omega_2 = 0.608,\ 0.274 \text{ rad/sec}$$

Because $P_{av} = R|V_s|^2|Y(j\omega)|^2 = R|Y(j\omega)|^2$,

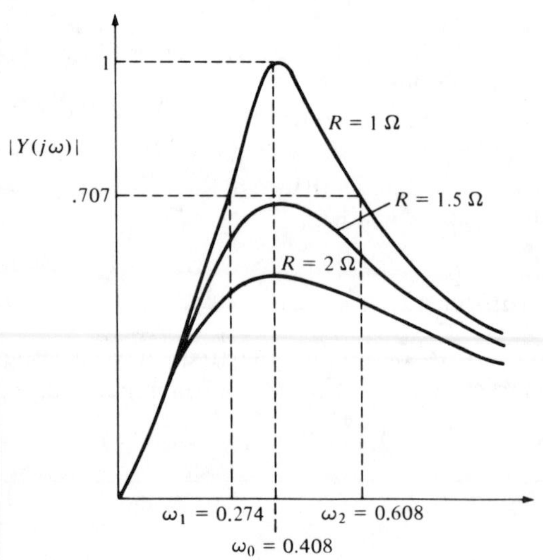

FIGURE 12.9 Response of a typical resonant circuit.

we can display the resonance with a plot of $|Y(j\omega)|$. This is shown in Figure 12.9 for $R = 1\Omega$. For comparison purposes, the responses for two other values of R are displayed as well. We observe for the series circuit that the larger the resistance, the smaller the Q and the less peaked the response.

More Complex Resonant Circuits

Although our discussion in this section has concentrated on networks with a single pair of conjugate complex poles near the j-axis, it should also be recognized that it is entirely possible to have more than one pair of high-Q poles in more complex networks. A few examples of other common resonant networks and typical pole-zero patterns suitable for qualitative analysis are shown in Figure 12.10. The networks shown in (a), (b), and (d) are typical of those found in the radio-frequency and intermediate-frequency portions of a superhetrodyne receiver. They are usually designed for two objectives: maximum power transfer and frequency selection, or filtering. The networks having a *single pair of dominant complex conjugate poles* are called *single-tuned*, whereas the network shown in Figure 12.10(d) is referred to as a *double-tuned* network. Figure 12.10(c) shows a network that can be adjusted for both a series and a parallel resonance. The high-Q series resonance produces a sharp dip in the response at ω_1, and the parallel resonance produces the peak at ω_2. Such a circuit is useful in situations where one wishes to minimize a strong interfering radio signal in relation to a weak signal at an adjacent frequency (Reference 12.5).

Double-tuned circuits are common in radio and television systems. They are conveniently analyzed and designed with the help of their pole-zero patterns.

12.5 Dominant Pole Analysis

Frequency Domain Analysis

Having established the existence of the resonance phenomenon in simple series RLC and parallel GCL networks, we will develop this important concept further for more general forms of networks. Our assumption will be that a *single high-Q dominant pole* exists, and this pole is solely responsible for the resonance

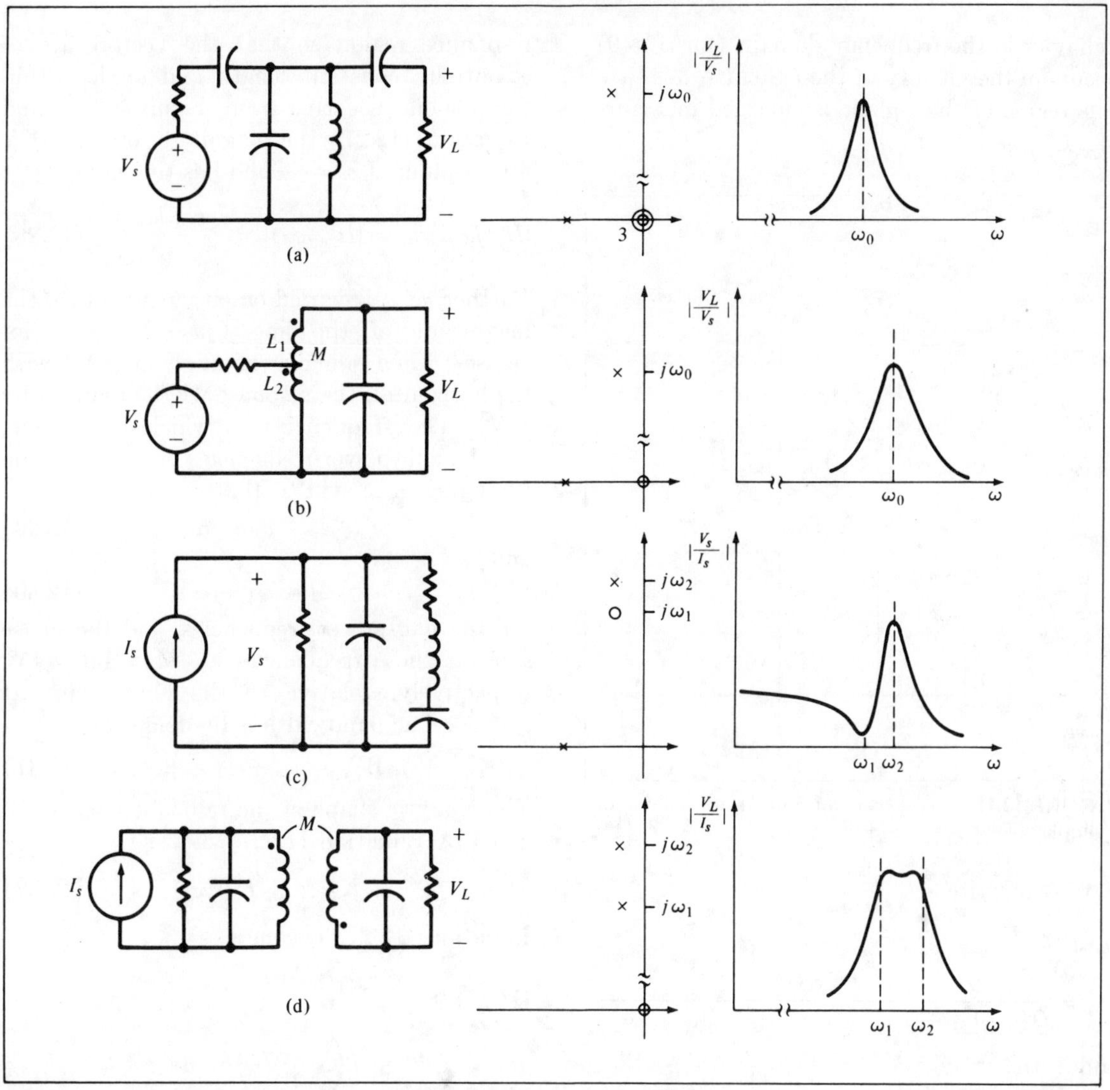

FIGURE 12.10 Examples of practical resonant circuits, their pole-zero patterns, and typical magnitude versus frequency response curves. (a) and (b) Single tuned matching circuits; (c) a simple filter; (d) a double tuned coupling circuit.

effect. This is exactly the case in the series RLC and parallel GCL networks already considered, and it is often approximately true for much more complex networks. We wish to gain as much quantitative information as we can concerning this case, because in very large networks (hundreds of nodes) it may not be economically feasible to compute more than this single pole, even with a digital computer. Also, this pole may be associated with a particular device in the network about which we wish quantitative information.

The network function must have the form

$$H(s) = \frac{P(s)}{Q(s)(s + \alpha - j\omega_d)(s + \alpha + j\omega_d)} \tag{12.26}$$

where the complex pair has been factored from the denominator for study. The resonant behavior in the frequency domain (for $\omega > 0$) occurs in the vicinity of the pole at $-\alpha + j\omega_d$; this region of the s-plane is enlarged in Figure 12.11.

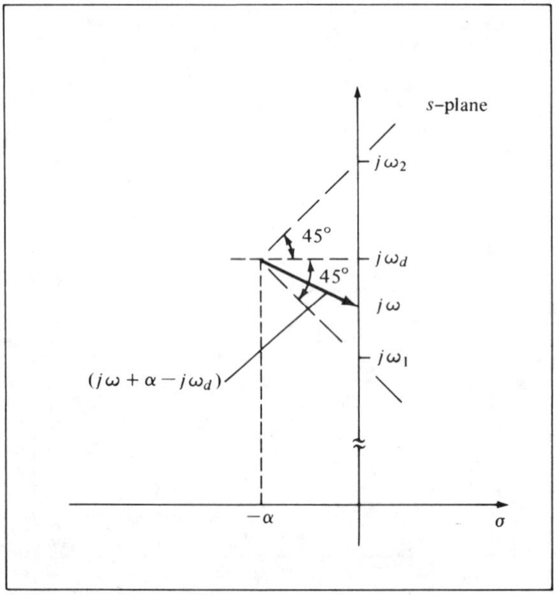

FIGURE 12.11 Enlargement of s-plane in vicinity of a simple pole at $s = -\alpha + j\omega_d$.

We can write

$$H(j\omega) = \frac{P(j\omega)}{Q(j\omega)[\alpha + j(\omega + \omega_d)][\alpha + j(\omega - \omega_d)]}$$

and in this vicinity

$$H(j\omega) \cong \frac{P(j\omega_d)}{Q(j\omega_d)2j\omega_d} \frac{1}{[\alpha + j(\omega - \omega_d)]}$$

$$= \frac{K(j\omega_d)}{\alpha + j(\omega - \omega_d)} \tag{12.27}$$

In Equation (12.27), all of the numerator and denominator factors of $H(j\omega)$ except the one associated with the pole at $s = -\alpha + j\omega_d$ have been replaced by their value at $s = j\omega_d$ and written $K(j\omega_d)$. This is valid if all the other poles and zeros are sufficiently far from the resonance region so that the vectors are of essentially constant length and angle within the region. Clearly, from Figure 12.11 and Equation (12.27) the magnitude of $H(j\omega)$ is a maximum at $\omega = \omega_d$ and has the value

$$|H(j\omega)|_{\max} = |H(j\omega_d)| = \left|\frac{K(j\omega_d)}{\alpha}\right| \tag{12.28}$$

Further, as ω is varied on either side of ω_d, the magnitude of the line $(j\omega + \alpha - j\omega_d)$ increases; when it makes an angle of $\pm 45°$ with the horizontal, the response will be reduced by $1/\sqrt{2}$. The frequencies at which this occurs are the half-power frequencies ω_1 and ω_2. From the figure, it is evident that

$$\omega_1 = \omega_d - \alpha \tag{12.29}$$

and

$$\omega_2 = \omega_d + \alpha \tag{12.30}$$

are the half-power frequencies, and the phase shift at these frequencies is $-45°$ and $+45°$, respectively, relative to the phase shift at $\omega = \omega_d$. The bandwidth is by definition

$$\text{BW} = \omega_2 - \omega_1 = 2\alpha \tag{12.31}$$

The analytic shape of the resonant response is given by Equation (12.27). Letting

$$K_d e^{j\theta_d} \triangleq K(j\omega_d) \tag{12.32}$$

Equation (12.27) becomes

$$\text{H}(j\omega) \cong \frac{(K_d/\alpha)e^{j\theta_d}}{1 + j\left(\dfrac{\omega - \omega_d}{\alpha}\right)}$$

$$= \frac{K_d/\alpha}{\sqrt{1 + \left(\dfrac{\omega - \omega_d}{\alpha}\right)^2}} \tag{12.33}$$

$$\times \left[\exp j\left(\theta_d - \tan^{-1}\left(\frac{\omega - \omega_d}{\alpha}\right)\right)\right]$$

The magnitude and angle of this function are sketched on a normalized frequency scale $(\omega - \omega_d)/\omega_d$ in Figure 12.12. This scale has

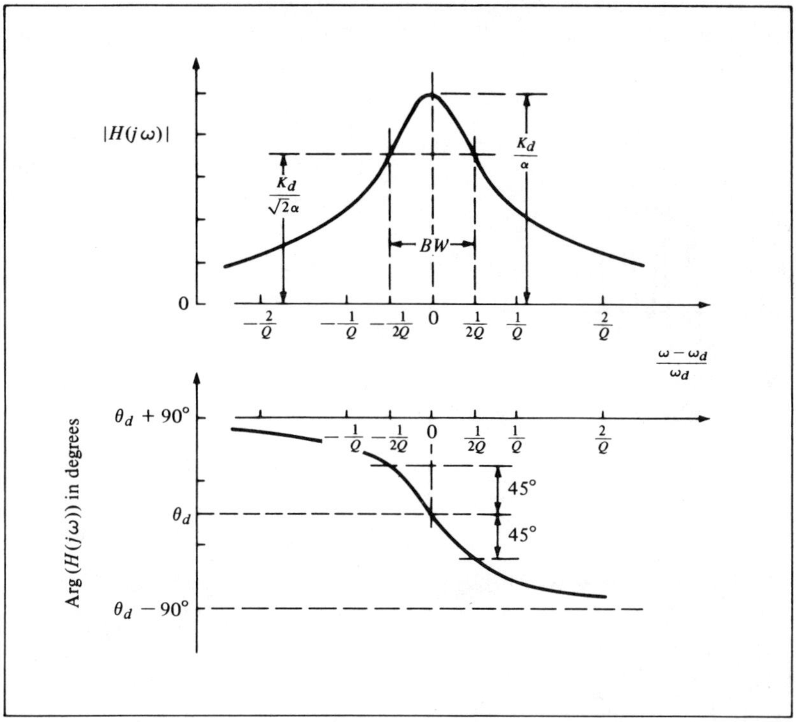

FIGURE 12.12 Magnitude and angle versus frequency for a single-pole resonance.

the effect of making the half-power frequencies displaced $\pm 1/2Q$ around a normalized frequency of zero, where

$$Q = \frac{\omega_0}{\mathrm{BW}} \cong \frac{\omega_d}{\mathrm{BW}} = \frac{\omega_d}{2\alpha} \tag{12.34}$$

As an example, consider the function

$$H(s) = \frac{(s+2)}{(s+1)((s+0.01)^2+1)} \tag{12.35}$$

evaluated at $\omega = \omega_d + \Delta\omega = 1 + \Delta\omega$. We have

$$H(j(1+\Delta\omega)) = \frac{j(1+\Delta\omega)+2}{(j(1+\Delta\omega)+1)(j(1+\Delta\omega)+0.01+j1)(j(1+\Delta\omega)+0.01-j1)}$$

$$\cong \frac{j1+2}{(j1+1)(j2)} \times \frac{1}{0.01+j\Delta\omega}$$

$$= \frac{0.7906e^{-j108.05^\circ}}{0.01+j\Delta\omega}$$

$$= \frac{79.06}{\sqrt{1+(100\Delta\omega)^2}}\, e^{j(-108.05^\circ - \tan^{-1}100\Delta\omega)}$$

which is analogous to Equation (12.33). The peak magnitude occurs when $\Delta\omega = 0$. What is Q?

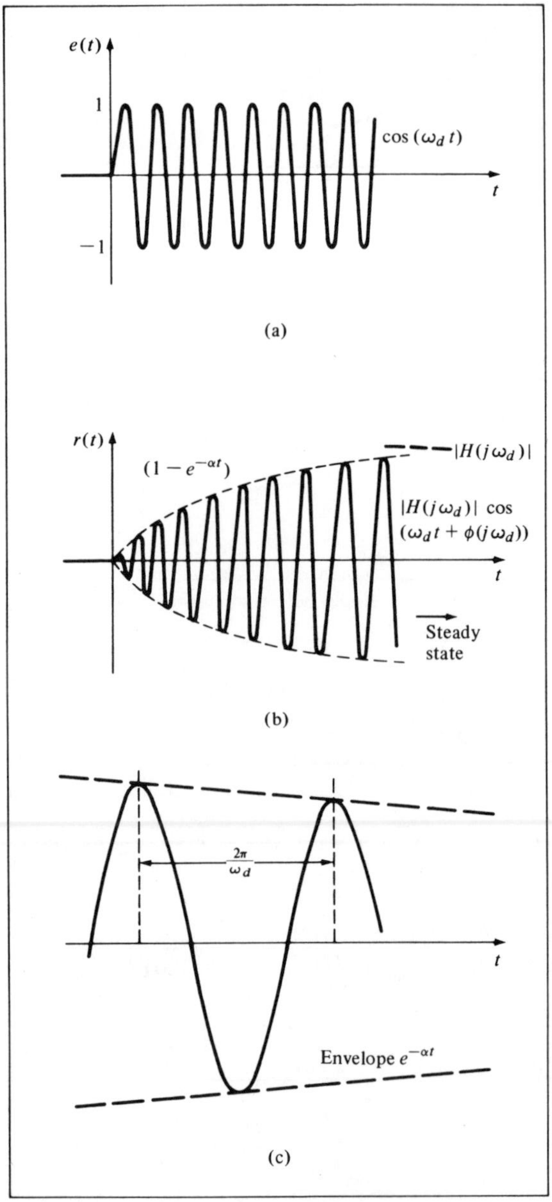

FIGURE 12.13 The time response associated with a pair of poles at $s = -\alpha \pm j\omega_d$. (a) Input; (b) complete response [Equation (12.37)]; (c) natural response.

Time Relationships

The most common excitation for a resonant network is a sinusoidal step, or pulse, as shown in Figure 12.13(a). For example, suppose a sinusoid of frequency $\omega = \omega_d = 1$ is applied to the function $H(s)$ above. The response function is then

$$R(s) = \frac{s}{s^2 + \omega_d^2} H(s)$$

$$= \frac{s}{s^2 + 1} \times \frac{(s + 2)}{(s + 1)((s + 0.01)^2 + 1)}$$

$$= \frac{-0.252}{(s + 1)} + \frac{1}{2}\frac{79.06e^{j108.05°}}{(s + j1)}$$

$$+ \frac{1}{2}\frac{79.06e^{-j108.05°}}{(s - j1)}$$

$$- \frac{1}{2}\frac{79.14e^{j108.32°}}{s + 0.01 + j1} - \frac{1}{2}\frac{79.14e^{-j108.32°}}{s + 0.01 - j1}$$

The inverse transform has the form

$$r(t) = -0.252e^{-t}$$
$$+ 79.06 \cos(t - 108.05°)$$
Steady state
$$- 79.14e^{-0.01t} \cos(t - 108.32°) \quad (12.36)$$
Dominant transient

A study of the response yields the following easily generalized information:

1. The sinusoidal steady-state term has the magnitude and angle predicted from the frequency response.
2. The transient term is dominated by the slowly decaying sinusoid resulting from the natural frequency at $s = -\alpha \pm j\omega_d$. We have termed this the natural response. This term is nearly identical in size to the steady-state term at $t = 0$, but it has the opposite sign.
3. The steady-state and transient response, having nearly the same amplitude and phase, can be grouped together to form the approxi-

mate response

$$r(t) \cong 79.06(1 - e^{-0.01t}) \cos(t - 108.05°) \qquad (12.37)$$

when the other transient terms are ignored because of their small size. This response is sketched in Figure 12.13(b) for the general α and ω. The rather remarkable result is that the $(1 - e^{-\alpha t})$ term is seen to form an *envelope* around the rapidly varying $\cos(\omega_d t + \phi)$ response. The transient response, then, is determined by (1) the amplitude of the frequency response characteristic $|H(j\omega_d)|$, which gives the asymptotic amplitude of the response, and (2) the real part of the dominant pole, which determines how fast the response achieves the steady state value. The phase $\phi(j\omega_d)$ contributes negligibly to the macroscopic properties of the response.

Computationally, the calculation is a trivial one once the dominant pole is found, since the frequency response $H(j\omega_d)$ requires only one solution of the network equations with $s = j\omega_d$. Indeed, if we are given a plot of the frequency response so that the resonant frequency and bandwidth can be estimated, the dominant pole position can be found from Equation (12.31) as

$$\alpha \pm j\omega_d \cong \frac{\text{BW}}{2} \pm j\omega_0$$

In summary, then, the approximate transient response can be determined completely from the frequency response, which of course can be found from solving the network equations repeatedly as ω is stepped with sufficient range and detail to insure capture of the resonance phenomena.

The results have application to cases where more than a single complex pole pair are involved. In such a case, the poles clustered around a point $j\omega_d$ on the $j\omega$-axis result in an envelope response to an input $\cos \omega_d t u(t)$ much like a cluster around $\omega = 0$ would react to a step input $\cos 0t\, u(t) = u(t)$. This is investigated in Problem 12.21.

Energy Relationships

The natural response is often quite informative in determining properties of circuit components and devices which account for the resonance. In one cycle of the natural response its amplitude drops from

$$e^{-\alpha t_1} \quad \text{to} \quad e^{-\alpha(t_1+2\pi/\omega_d)} = e^{-2\pi\alpha/\omega_d}e^{-\alpha t_1}$$

The amplitude drops by the factor

$$e^{-2\pi\alpha/\omega_d} \cong 1 - \frac{2\pi\alpha}{\omega_d} = 1 - \frac{\pi}{Q} \qquad (12.38)$$

with each cycle of the natural response, or *the amplitude drops* $100\pi/Q$ *percent per cycle*. Thus, the higher the Q the longer it takes for the transient to die out after an applied signal is reduced to zero or for the response to build up to its steady-state value when a signal is suddenly applied.

The concept of Q has an even more general physical significance which is related to the *energy* stored in a resonant network. To develop insight into this property, we return to the series RLC network of Figure 12.5(a). The current through the inductor is

$$I = \frac{(1/L)s}{s^2 + \dfrac{R}{L}s + \dfrac{1}{LC}} V_s \qquad (12.39)$$

and the voltage across the capacitor is

$$V_C = \frac{1/LC}{s^2 + \dfrac{R}{L}s + \dfrac{1}{LC}} V_s \qquad (12.40)$$

Therefore, at the resonant frequency $s = j\omega_d \cong j\omega_0 = j1/\sqrt{LC}$

$$I(j\omega_0) = \frac{V_s}{R}$$

and

$$V_C = -j\frac{\omega_0 L}{R}V_s = -jQV_s \qquad (12.41)$$

which shows that there is a 90° phase shift between the inductor current and the capacitor voltage. Recalling that the energy stored in these elements can be written

$$\text{energy in inductor} = w_L(t) = \frac{1}{2}Li_L^2(t)$$

and

$$\text{energy in capacitor} = w_C(t) = \frac{1}{2}Cv_C^2(t)$$

we see that the 90° phase shift between I and V_C means that the stored energy in the inductor is a maxmum when the stored energy in the capacitance is zero, and vice versa. This is characteristic of resonant systems excited at the resonant frequency; in the steady state there is a stored energy that oscillates back and forth between the inductors and the capacitors. In high Q circuits this stored energy is large compared to the average energy supplied per cycle of the input, and it is the necessity of supplying this large energy that explains the gradual buildup to the steady state illustrated in Fig. 12.13(b). Since the series network has only a single capacitor, the peak energy stored in it is also the total stored energy in the network.

Returning to the natural response, then, if at one voltage peak the stored energy is

$$w_C(t_1) = w_{\text{total}} = \frac{1}{2}CA^2e^{-2\alpha t_1} \qquad (12.42)$$

at the next peak, $2\pi/\omega_d$ sec later, the energy will be reduced to

$$w_C\left(t_1 + \frac{2\pi}{\omega_d}\right) = \frac{1}{2}CA^2\epsilon^{-2\alpha t_1}e^{-4\pi\alpha/\omega_d} \qquad (12.43)$$

The energy loss per cycle is

$$w_C(t_1) - w_C\left(t_1 + \frac{2\pi}{\omega_d}\right)$$

$$= \frac{1}{2}CA^2e^{-2\alpha t_1}(1 - e^{-4\pi\alpha/\omega_d}) \qquad (12.44)$$

which can be written

$$(1 - e^{-2\pi/Q}) = \frac{\text{energy loss per cycle}}{\text{total stored energy}} \qquad (12.45)$$

For $Q > 20$,

$$1 - e^{-2\pi/Q} \cong \frac{2\pi}{Q}$$

Therefore Equation (12.45) can be written

$$Q_{\text{circuit}} \triangleq 2\pi\left(\frac{\text{total stored energy}}{\text{energy loss per cycle}}\right) \qquad (12.46)$$

This definition is useful because *it can be applied to circumstances in which we cannot even define current and voltage.* For example, it is useful in dealing with distributed electromagnetic systems, such as waveguides and antennas, as well as a variety of nonelectrical systems.

12.6 Bode Diagrams

Logarithmic Gain

In previous sections of this chapter, a number of plots of the magnitude and angle of a network function were given as a function of the sinusoidal frequency. Such plots are particularly useful in the analysis and design of control and information-handling circuits. When a logarithmic scale is used for the magnitude and the frequency axes, certain useful symmetries and simplifications arise, particularly when the poles and zeros lie on the real axis of the s plane. H. W. Bode made extensive use of such logarithmic scales in his classic study of feedback amplifiers, and such plots are known as *Bode plots,* or *Bode diagrams* (Reference 12.4).

The natural logarithm (to the base e) of a complex function

$$H(j\omega) = M(\omega)e^{j\phi(\omega)}$$

where $M(\omega)$ and $\phi(\omega)$ are real functions of ω, is itself the complex function

$$\ln(H(j\omega)) = \ln M(\omega) + j\phi(\omega)$$

The imaginary part of $\ln(H(j\omega)$ is just the angle of $H(j\omega)$ in radians, and the real part is called the *gain*, or the *logarithmic gain*. When $H(j\omega)$ is a current or a voltage ratio, its units are *nepers*. Engineers usually prefer to measure angles in degrees and gains in *decibels* (db). The decibel gain is defined by

$$\text{gain in db} = 20 \log M(j\omega) \qquad (12.47)$$

where "log" implies to the base 10. Conversion to these engineering units is obtained by

$$\text{number of db} = 8.68 \times \text{number of nepers} \qquad (12.48)$$

$$\text{number of degrees} = 57.3 \times \text{number of radians} \qquad (12.49)$$

The convenience of the logarithmic measure relates to the fact that the logarithm reduces products and quotients to sums and differences. Consider, for example, the factored form of a network function

$$H(s) = K\frac{(s - z_1)\ldots(s - z_m)}{(s - p_1)\ldots(s - p_n)} \qquad (12.50)$$

The logarithm of this function is

$$\begin{aligned}\ln[H(s)] &= \ln K + \ln(s - z_1) \\ &\quad + \cdots + \ln(s - z_m) \\ &\quad - \ln(s - p_1) - \cdots - \ln(s - p_n)\end{aligned} \qquad (12.51)$$

For $s = j\omega$, each term in parentheses will be a complex number of the polar form $re^{j\theta}$. We then have for each term $\ln(re^{j\theta}) = \ln r + \ln(e^{j\theta}) = \ln r + j\theta$. The expansion of Equation (12.51) then becomes

$$\begin{aligned}\ln[H(j\omega)] &= \ln K + \ln|j\omega - z_1| + \cdots \\ &\quad - \ln|j\omega - p_1| - \cdots \\ &\quad + j\{\arg(j\omega - z_1) + \cdots \\ &\quad - \arg(j\omega - p_1) - \cdots\}\end{aligned} \qquad (12.52)$$

The overall logarithmic gain is the sum and difference of the gains of the individual factors of $H(j\omega)$ in exactly the same way that the overall angle is the sum and difference of the individual factor angles! This is a worthwhile simplification.

Each factor in Equation (12.50) is associated with a single pole or zero of $H(s)$. A study of the logarithmic gain and angle of typical factors allows us to develop a set of component curves which can subsequently be added and subtracted to build up complex gain and angle curves. Because the polynomials that are the numerator and denominator of $H(s)$ have real coefficients, the poles and zeros must be real, complex conjugate pairs or imaginary pairs. Real poles and zeros are particularly easy to handle, and fortunately many useful network functions have only real poles and zeros.

A single real zero at $-\sigma_1$ gives rise to the term $\ln(s + \sigma_1)$ in the $\ln[H(s)]$. For $s = j\omega$ this becomes

$$\ln(j\omega + \sigma_1) = \ln \sigma_1 + \ln\left|1 + \frac{j\omega}{\sigma_1}\right| + j\tan^{-1}\left(\frac{\omega}{\sigma_1}\right) \qquad (12.53)$$

and with decibel gain and degrees this would be written

$$20 \log \sigma_1 + 20 \log\left|1 + \frac{j\omega}{\sigma_1}\right| + j57.3 \tan^{-1}\left(\frac{\omega}{\sigma_1}\right) \qquad (12.54)$$

The first two terms contribute to the gain and the last term contributes to the phase shift of $H(j\omega)$. If $H(s)$ has zeros at $-\sigma_1, -\sigma_2, = \ldots, -\sigma_m$ and poles at $-\sigma_a, -\sigma_b, = \ldots, -\sigma_n$, the complete decibel gain can be written

$$\begin{aligned}20 \log|H(j\omega)| &= 20 \log K \cdot \frac{\sigma_1\sigma_2\ldots\sigma_m}{\sigma_a\sigma_b\ldots\sigma_n} \\ &\quad + 20 \log\left|1 + \frac{j\omega}{\sigma_1}\right| + \cdots \\ &\quad + 20 \log\left|1 + \frac{j\omega}{\sigma_m}\right| \\ &\quad - 20 \log\left|1 + \frac{j\omega}{\sigma_a}\right| - \cdots \\ &\quad - 20 \log\left|1 + \frac{j\omega}{\sigma_n}\right|\end{aligned} \qquad (12.55)$$

For this same function the angle in degrees is

$$\arg(H(j\omega)) = 57.3\tan^{-1}\left(\frac{\omega}{\sigma_1}\right) + 57.3\tan^{-1}\left(\frac{\omega}{\sigma_2}\right) + \cdots + 57.3\tan^{-1}\left(\frac{\omega}{\sigma_m}\right) - 57.3\tan^{-1}\left(\frac{\omega}{\sigma_a}\right) - \cdots - 57.3\tan^{-1}\left(\frac{\omega}{\sigma_n}\right) \quad (12.56)$$

Magnitude for Real Poles and Zeros

Consider a typical term in the gain equation (12.55)

$$G_1 = 20\log\left|1 + \frac{j\omega}{\sigma_1}\right| = 20\log\left(1 + \frac{\omega^2}{\sigma_1^2}\right)^{1/2}$$

$$= 10\log\left(1 + \frac{\omega^2}{\sigma_1^2}\right) \quad (12.57)$$

As long as $\omega \ll \sigma_1$

$$G_1 \approx 10\log(1) = 0 \text{ db} \quad (12.58)$$

which on a plot of gain versus log frequency is a horizontal straight line. On the other hand, when $\omega \gg \sigma_1$, Equation (12.57) reduces to

$$G_1 \approx 20\log\left(\frac{\omega}{\sigma_1}\right) = 20\log\omega - 20\log\sigma_1 \quad (12.59)$$

which is also a *straight line* on a *log-frequency scale*. The slope of this line can be seen from the first term in Equation (12.59). At $\omega = 2\omega_1$ this term becomes

$$20\log(2\omega_1) = 20\log(\omega_1) + 20\log(2) = 20\log(\omega_1) + 6 \text{ db}$$

showing a 6 db increase in the gain between ω_1 and $2\omega_1$. The slope of $G_1(\omega)$ is therefore +6 *db per octave*. Since a 10:1 increase in frequency adds 20 db to G_1 of Equation (12.59), this slope is also 20 *db per decade*. Equations (12.58) and (12.59) show that a plot of G_1 on a logarithmic-frequency scale is asymptotic to a pair of straight lines, when ω is far above or far below σ_1. Equating (12.58) and (12.59) we find the lines intersect at $\omega = \sigma_1$, as shown in Figure 12.14. The frequency σ_1 is called the *break frequency*, or *corner frequency*.

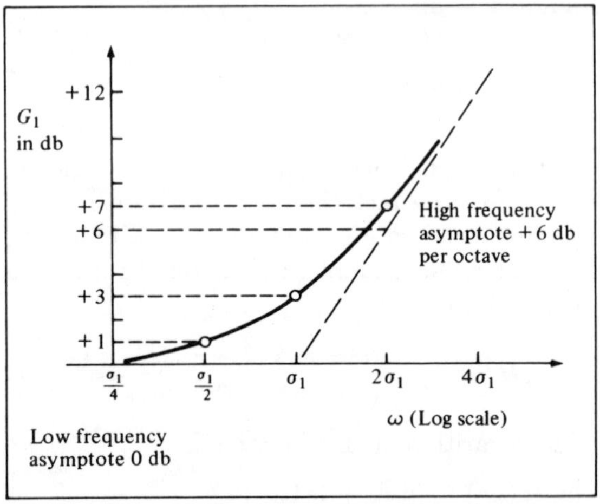

FIGURE 12.14 Plot of Equation (12.57) and its high and low-frequency asymptotes, Equations (12.59) and (12.58).

Figure 12.14 also shows the exact value of G_1, given by Equation (12.57). It is apparent that the asymptotes are very close to the exact curve as long as the frequency is more than an octave above or below σ_1. The exact gain, the

TABLE 12.1

Gain of a Simple Real Zero Factor

Frequency	Gain G_1 (db)	Asymptotic Gain (db)	Difference (db)
$\sigma_1/4$	0.26	0	0.26
$\sigma_1/2$	0.96	0	0.96
σ_1	3.01	0	3.01
$2\sigma_1$	6.99	6	0.99
$4\sigma_1$	12.3	12	0.3

asymptotic gain, and the difference are tabulated in Table 12.1. For most engineering purposes the "correction" that must be applied to the asymptotic straight lines is just 3 db at the *break frequency*, 1 db one octave above and below the break, and zero elsewhere. The ease with which we can sketch a gain curve is illustrated by Example 12.2.

Example 12.2

The *RC* network of Figure 12.15 is frequently used to boost the high frequencies in audio sys-

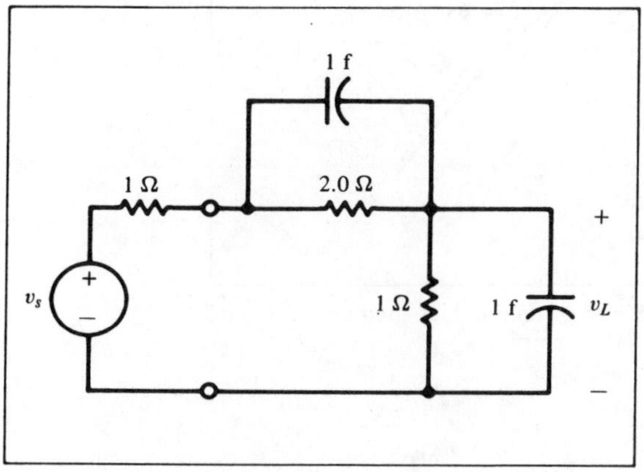

FIGURE 12.15 High-frequency filter.

tems. We are to verify this by sketching the Bode plot of

$$G = 20 \log \left| \frac{V_L}{V_s} \right|$$

The voltage ratio can be written down directly because the network is a simple voltage divider.

$$\frac{V_L}{V_s}(s) = \frac{\dfrac{1}{1+s}}{1 + \dfrac{2}{1+2s} + \dfrac{1}{1+s}}$$

$$= \frac{1+2s}{4+7s+2s^2}$$

For a Bode plot the convenient factored form is

$$\frac{V_L}{V_s}(s) = \frac{1}{4}\frac{(1+s/0.5)}{\left(1+\dfrac{4s}{7-\sqrt{17}}\right)\left(1+\dfrac{4s}{7+\sqrt{17}}\right)}$$

This voltage ratio has a zero at $s = -0.5$ and poles at $s = -0.719$ and -2.78 rad/sec. Furthermore, the low-frequency constant value is $20 \log(0.25) = -12$ db. The asymptotic gain is therefore the sum of the four asymptotic gains shown in Figure 12.16. The sum is drawn as a dotted line. Applying the corrections of Table 12.1, we can easily obtain the true curve which is the solid line. The asymptotes predict a 3-db gain increase, and the true curve is just shy of a 2-db "boost." The reader may find it interesting to try to modify the element values of the network to give larger increases in the high-frequency gain relative to the low-frequency value.

Phase of Real Poles and Zeros

The phase associated with a typical real axis factor is

$$\phi_1 = \arg\left(1 + \frac{j\omega}{\sigma_1}\right) = \tan^{-1}\left(\frac{\omega}{\sigma_1}\right) \qquad (12.60)$$

The value of this angle in degrees is tabulated in the second column of Table 12.2 at one octave intervals for three octaves on either side of $\omega = \sigma_1$. This angle is also plotted in Figure 12.17. The concept of simple asymptotes is still useful for the angle curve. As $\omega \to 0$, $\phi_1 \to 0$, and as $\omega \to \infty$, $\phi_1 \to 90°$. The asymptotic phase is therefore a constant 0° below $\omega = \sigma_1$ and a constant 90° above $\omega = \sigma_1$. The true curve is an odd function about 45° at $\omega = \sigma_1$. This is indicated in Table 12.2. If the phase shift is $\phi_1 = \theta_k$ at $\omega = k\sigma_1$ $(k < 1)$, then at $\omega = \sigma_1/k$ $\phi_1 = 90 - \theta_k$.

ϕ_1 is plotted on a log-frequency scale in Figure 12.17 and the asymptotic phase is indicated by *dotted* lines. Also drawn in this figure, with *dashed* lines, is the approximation of a constant phase slope of 45° per decade from

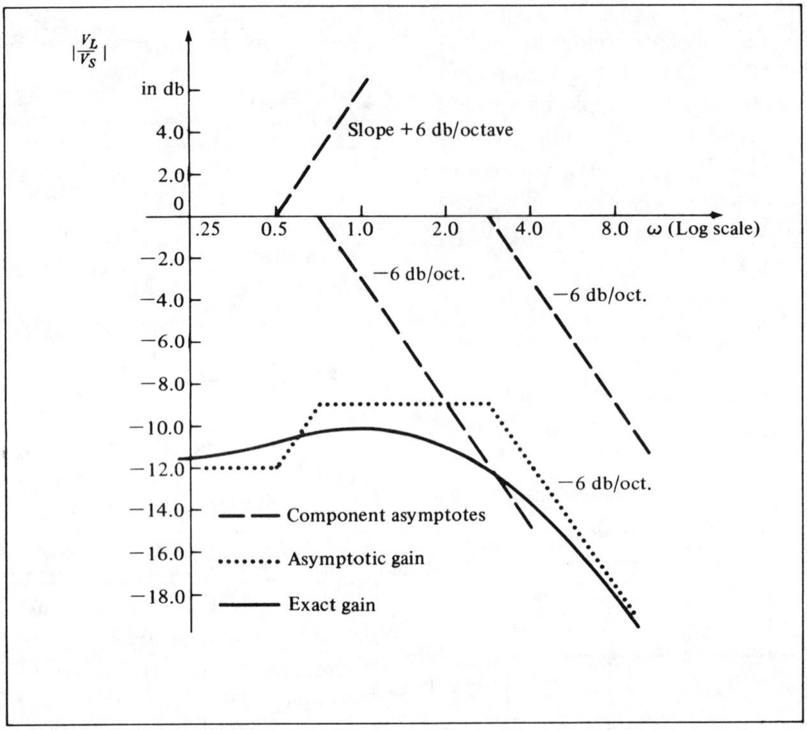

FIGURE 12.16 Asymptotic and exact Bode plot of an *RC* high-frequency "boost" circuit.

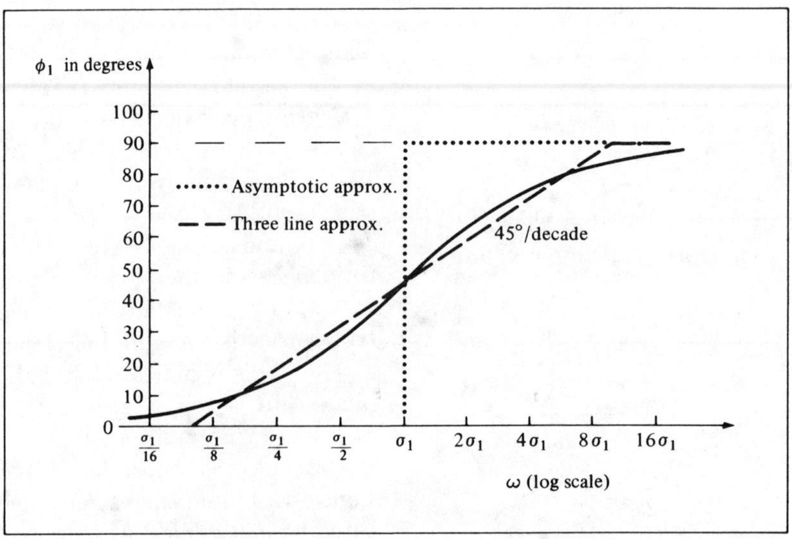

FIGURE 12.17 Phase shift produced by a single zero at $s = -\sigma_1$.

TABLE 12.2

Phase of a Simple Real Zero Factor

Frequency ω	Phase Shift ϕ_1	Approximate Phase 45°/Decade	Difference
$\sigma_1/8$	7.1°	4.5°	+2.6°
$\sigma_1/4$	14.0°	18°	−4.0°
$\sigma_1/2$	26.6°	31.5°	−4.9°
σ_1	45°	45°	0
$2\sigma_1$	90° − 26.6 = 63.4°	58.5°	+4.9°
$4\sigma_1$	90 − 14 = 76°	72°	+4.0°
$8\sigma_1$	90 − 7.1 = 82.9°	85.5°	−2.6°

one decade below to one decade above the *break* frequency. As this plot and the last column in Table 12.2 indicates, this approximation, which has three straight line segments, is considerably better than the asymptotic approximation and is good enough for many engineering applications. It is important to observe that the frequency range over which the phase deviates substantially from its asymptotic values is much wider than the range for which the gain deviates from its asymptotes. The gain transition is essentially complete in two octaves, whereas the phase transition is spread over six octaves.

Example 12.3

We will calculate the phase shift of the RC network analyzed in Example 12.2.

The voltage transfer ratio of that network is

$$\frac{V_L}{V_s}(j\omega) = \frac{1}{4}\,\frac{(1 + j2\omega)}{(1 + j1.390\omega)(1 + j0.36\omega)} \tag{12.61}$$

so that the phase shift is given by

$$\phi_L(j\omega) = \tan^{-1}(2\omega) - \tan^{-1}(1.390\omega) - \tan^{-1}(0.36\omega) \tag{12.62}$$

The asymptotic components of this phase shift, the three segment approximations, and the sum of these approximations are plotted in Figure 12.18(a). The three components of the exact phase (12.62) and their algebraic sum is plotted in Figure 12.18(b).

Complex Conjugate Poles and Zeros

A pair of complex conjugate zeros at $-\alpha \pm j\omega_d$ gives a numerator factor

$$s^2 + 2\alpha s + \alpha^2 + \omega_d^2 = s^2 + 2\alpha s + \omega_0^2$$
$$= s^2 + 2\xi\omega_0 s + \omega_0^2$$

where $\xi = \alpha/\omega_0$ is termed the *damping ratio.* This is usually normalized by dividing through by the zero-frequency value ω_0^2

$$F(s) = 1 + 2\xi\left(\frac{s}{\omega_0}\right) + \left(\frac{s}{\omega_0}\right)^2 \tag{12.63}$$

For $s = j\omega$ the magnitude and phase of Equation (12.63) are

$$|F(j\omega)| = 10 \log\left[\left(1 - \left(\frac{\omega}{\omega_0}\right)^2\right)^2 + \left(2\xi\frac{\omega}{\omega_C}\right)^2\right] \text{db} \tag{12.64}$$

and

$$\phi(\omega) = \tan^{-1}\left(\frac{2\xi\dfrac{\omega}{\omega_0}}{1 - \left(\dfrac{\omega}{\omega_0}\right)^2}\right) \tag{12.65}$$

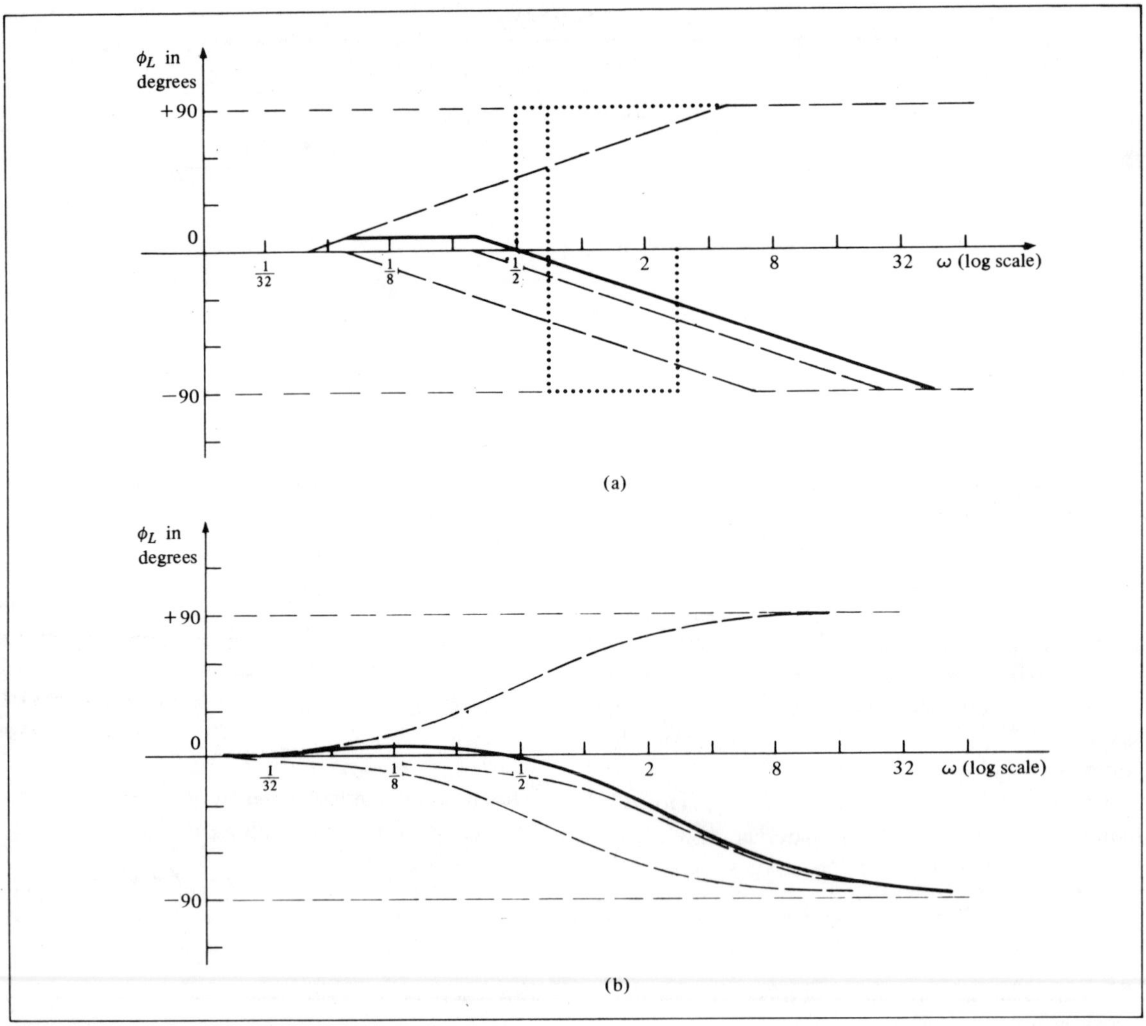

FIGURE 12.18 (a) Asymptotic and three line segment phase approximations; (b) exact phase shift of *RC* high-frequency "boost" circuit.

As in the case of real poles or zeros these terms can be plotted and added on a logarithmic-frequency scale, but now the damping ratio ξ is a parameter which complicates things somewhat. A typical set of plots is shown in Figure 12.19.

Despite the increased complexity, several useful observations can be made.

1. As in the case of real roots, the low-frequency asymptotic gain is 0 db, and the phase shift is 0°.

2. For $\omega/\omega_0 \gg 1$ the gain approaches

$$|F(j\omega)| \approx 40 \log\left(\frac{\omega}{\omega_0}\right) \text{db} \qquad (12.66)$$

and the phase approaches

$$\phi_2(\omega) \approx 180° \qquad (12.67)$$

The high-frequency asymptotic gain is a straight line with a slope of +12 db per octave or +40 db per decade that intersects 0 db at $\omega = \omega_0$.

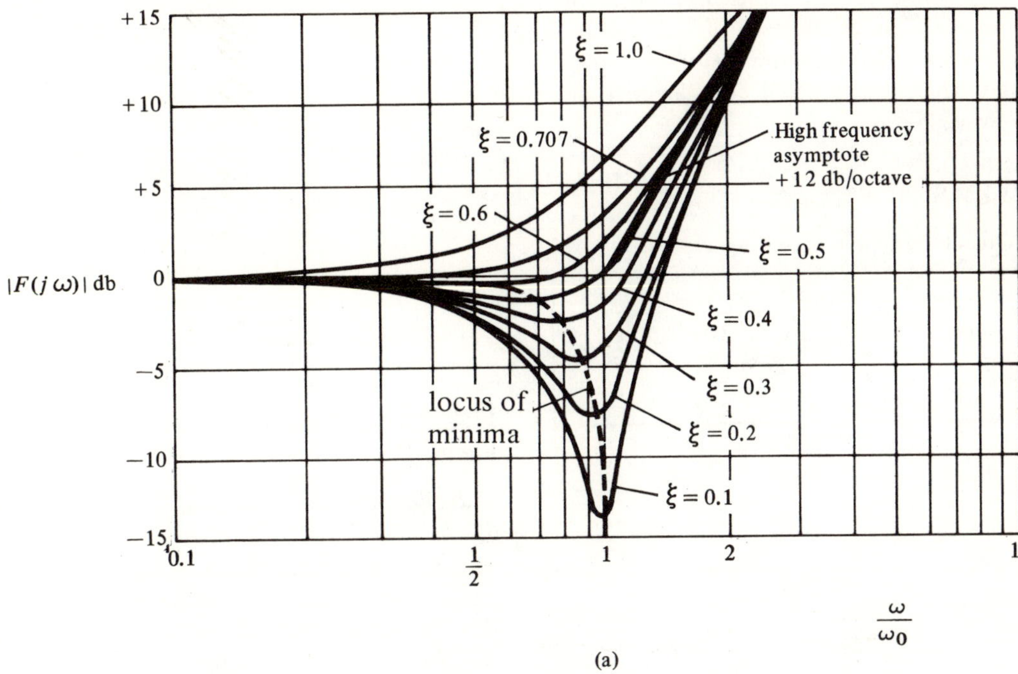

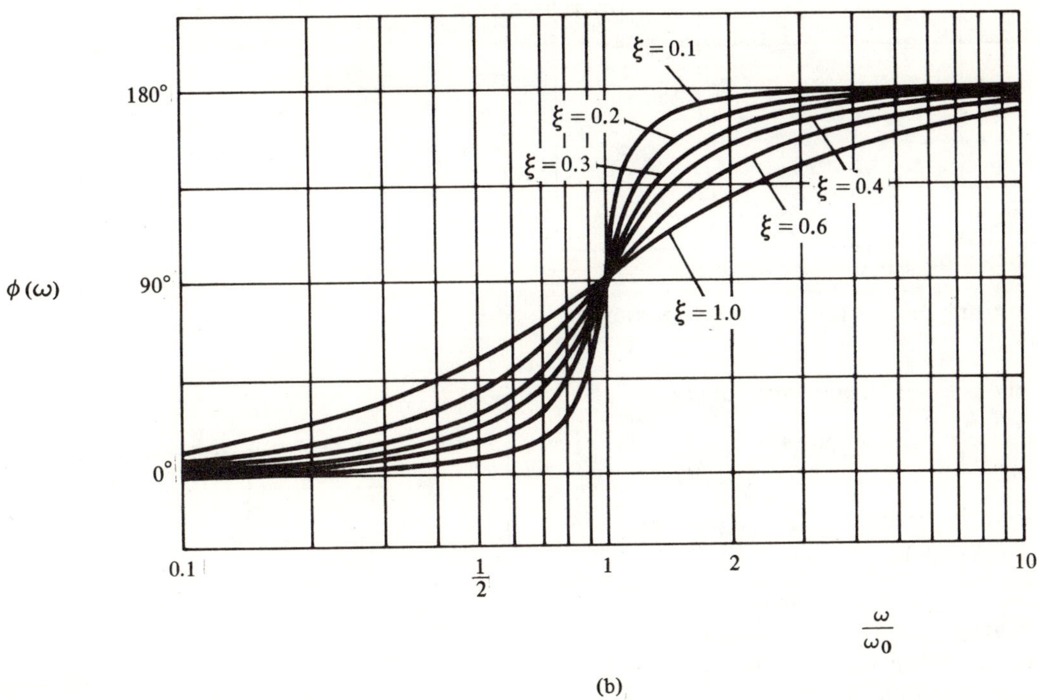

FIGURE 12.19 Magnitude and phase responses given by Equations (12.64) and (12.65) for several values of ζ.

3. The phase shift curve has transitions from 0 to 180° passing through 90° at the break frequency $\omega = \omega_0$. As in the case of real factors, the phase has odd symmetry about the break frequency when it is plotted on a logarithmic frequency scale.

PROBLEMS

12.1 Which of the pole-zero diagrams of Figure P 12.1 represent unstable network response? Explain.

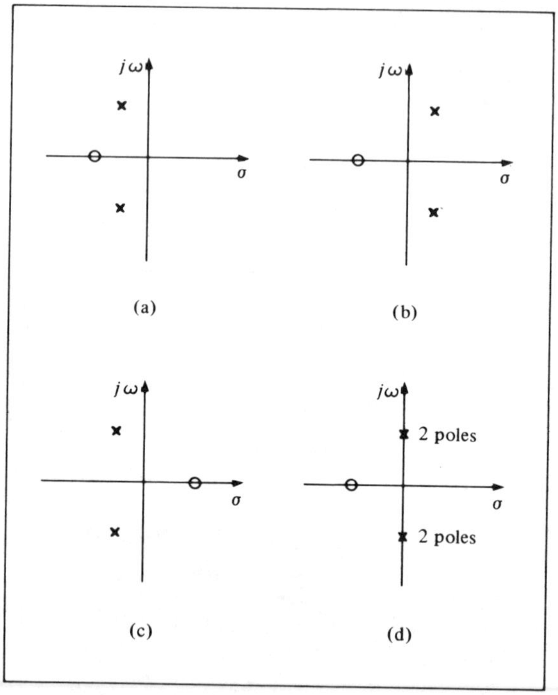

FIGURE P 12.1

12.2 The poles of the network function

$$H(s) = \frac{1}{(s^2 + 1)(s^2 + 100)}$$

are shown in Figure P 12.2. Which pair will dominate the impulse response?

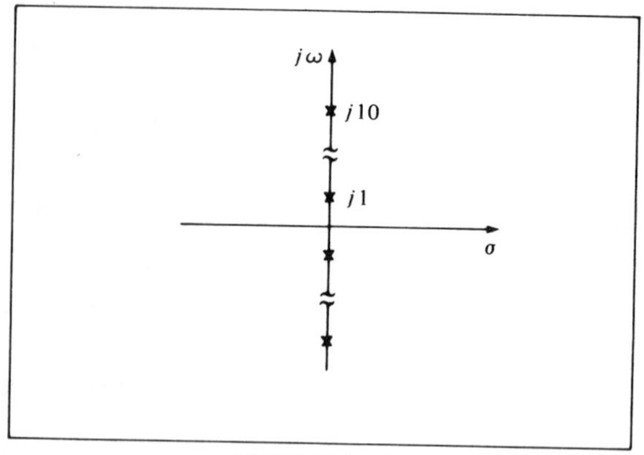

FIGURE P 12.2

12.3 In Figure P 12.3(a) determine the value of resistance to obtain: (a) A Q of 10. (b) A bandwidth of 0.1 rad/sec. (c) A half-power frequency of 0.3 rad/sec.

12.4 Repeat Problem 12.3 for the network of Figure P 12.3(b).

12.5 The network of Figure P 12.4 is an incremental (small-signal) model of an amplifier with resistive feedback. (a) Calculate the poles and zeros of the transfer impedance $V_0(s)/I_s(s)$. (b) Sketch the path of the poles in the s-plane as G_f increases from zero. (c) Determine the value of G_f needed to place a pair of complex conjugate poles on lines which make a ± 30°

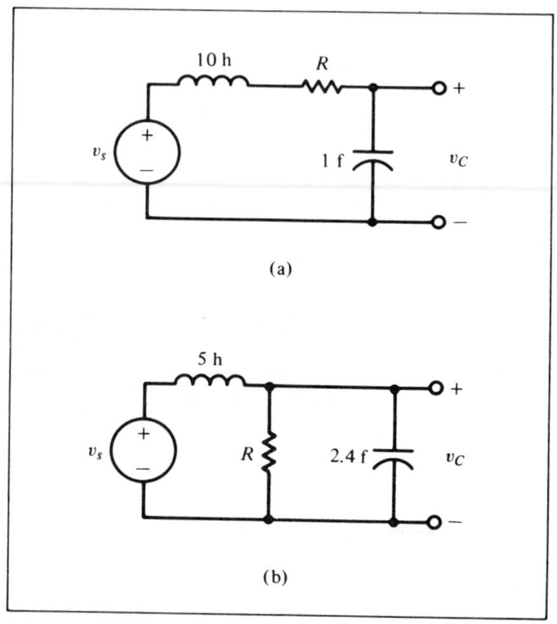

FIGURE P 12.3

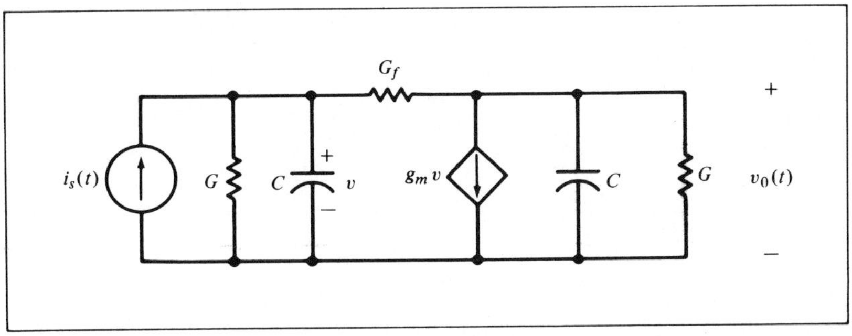

FIGURE P 12.4

angle with the $j\omega$-axis. What is the value of $V_0(0)/I_s(0)$ in this case?

12.6 Find $I_0(s)/I_s(s)$ in Figure P 12.5. Determine the half-power frequencies, the band-

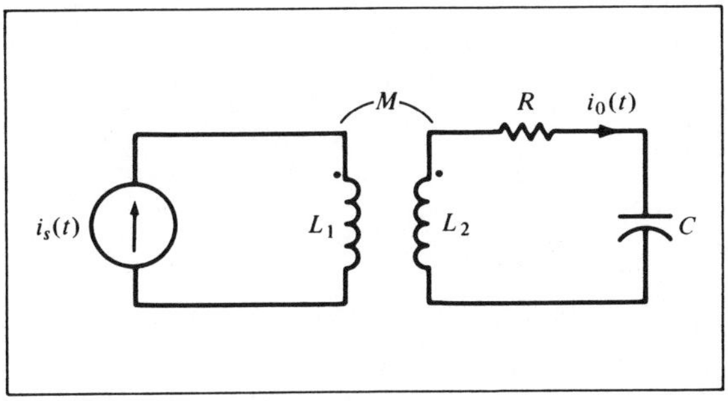

FIGURE P 12.5

width, and the Q in terms of the element values. Assume that $Q \gg 1$.

12.7 In Figure P 12.6, define $L_1C_1 = L_2C_2 = 1/\omega_0^2$, $Q_1 = \omega_0 L_1/R_1$, $Q_2 = \omega_0 L_2/R_2$. Find the

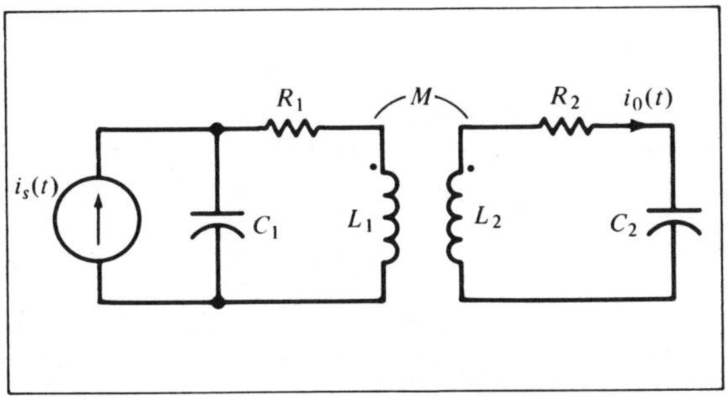

FIGURE P 12.6

value for M that maximizes the current gain $|I_0(j\omega_0)/I_s(j\omega_0)|$.

12.8 In Figure P 12.7, choose a value for α that gives a Q of ∞.

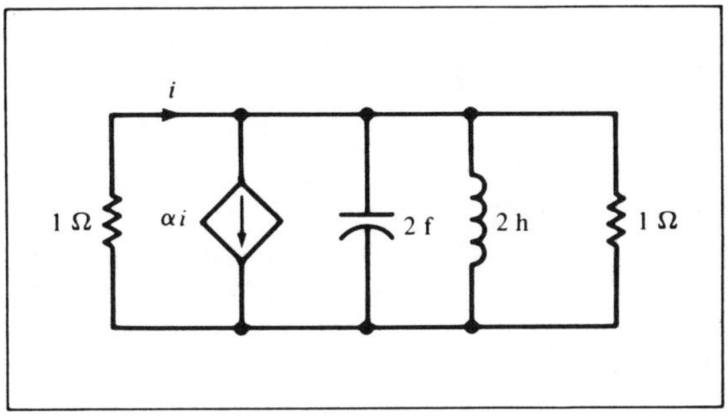

FIGURE P 12.7

12.9 The impedance $Z(s)$ of the network of Figure P 12.8 has the zero and poles shown in the accompanying s-plane sketch. (a) If $Z(0) = 1\ \Omega$, what are the parameter values? If $i_1(t) = 1 + \cos 100t$, what is $v_0(t)$? (b) What is the BW of the resonance curve between its half-power points? What is the value of Q? Check this value by computing it from the stored and energy loss functions [see Equation (12.46)].

12.10 (a) If the two parallel resonant circuits in Figure 12.10(d) are identical high-Q circuits ($Q > 25$), calculate the approximate pole locations for $M > 0$, but small. (b) Estimate the coefficient of coupling needed to produce a double-peaked response as shown in the figure, if for $k = M/L = 0$ the poles lie at $-0.025 \pm j1.0$. (c) If both resistors are 1 Ω, determine the circuit parameters needed in (b).

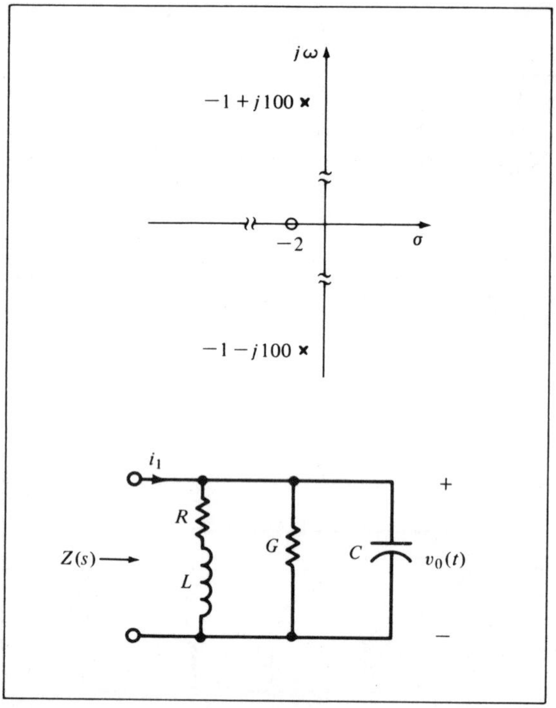

FIGURE P 12.8

12.11 Sketch the magnitude responses (db versus log ω) associated with the following rational functions:

(a) $\dfrac{3(s+10)}{(s+6)(s+5)}$.

(b) $\dfrac{10s}{(s+2)}$.

(c) $\dfrac{s^2}{(s+2)^2}$.

(d) $\dfrac{(s+5)}{(s+100)(s+1000)}$.

(e) $\dfrac{15s}{(2s+3)(6s+9)}$.

(f) $\dfrac{s(s^2-s+1)}{(s+3)(s^2+s+1)}$.

12.12 A number of pole-zero patterns are shown in Figure P 12.9. Sketch the magnitude response (db versus log ω) in each case. K is defined in Equation (12.50).

12.13 Compare the magnitude and phase responses associated with the complex pole pairs of Figure P 12.10.

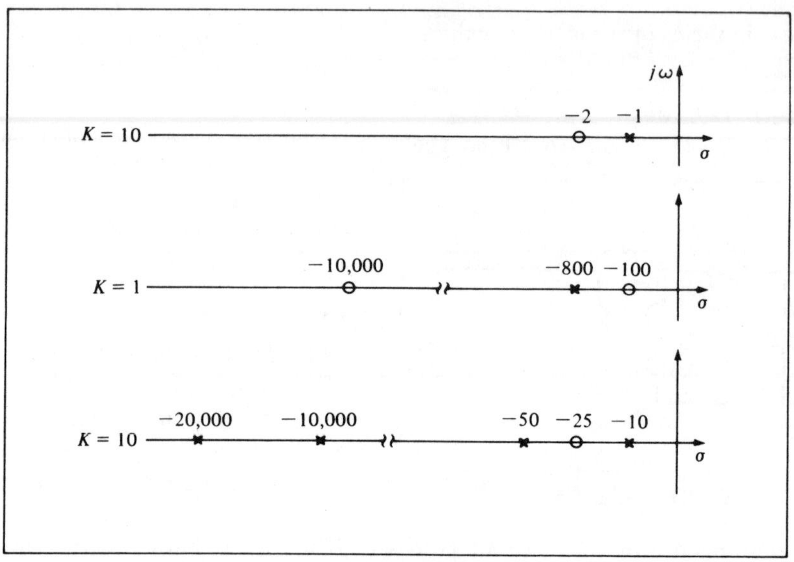

FIGURE P 12.9

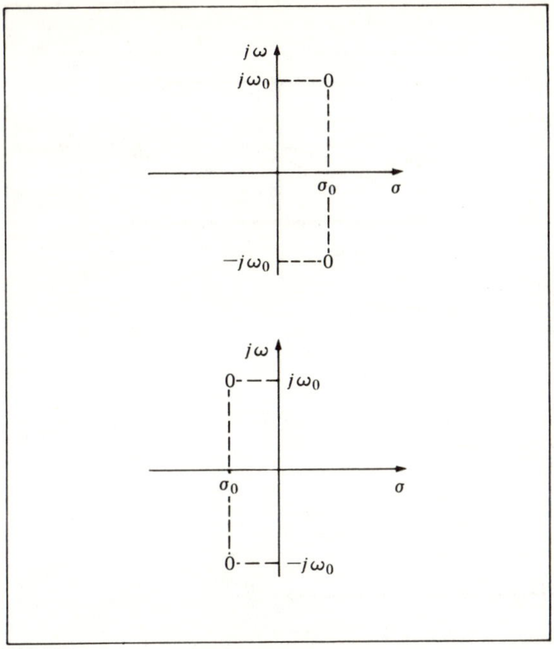

FIGURE P 12.10

12.14 (a)–(b) Find the transfer function $H(s)$ from the Bode plots of Figure P 12.11.

12.15 (a)–(b) Sketch the phase responses associated with the transfer functions found from Figure P 12.11. Assume that no poles or zeros have positive real parts.

12.16 (a)–(d) Sketch the Bode plots of the magnitude responses associated with the networks of Figure P 12.12.

12.17 (a)–(c) For each of the networks of Figure P 12.13, plot the pole-zero patterns for the transfer function, and draw the Bode diagrams. In networks (a) and (b), assume that the break points are at least a decade apart. These three networks are sometimes used for the compensation of feedback systems; for each network determine (estimate from Bode diagrams) the values of ω for which the steady-state output voltage $v_0(t)$ lags the sinusoidal input voltage $v_s(t)$.

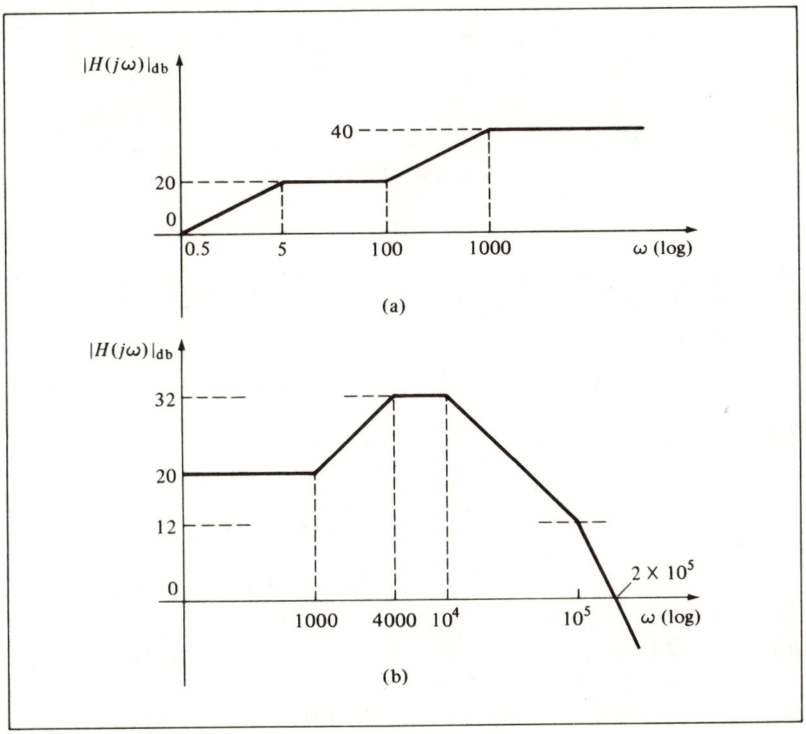

FIGURE P 12.11

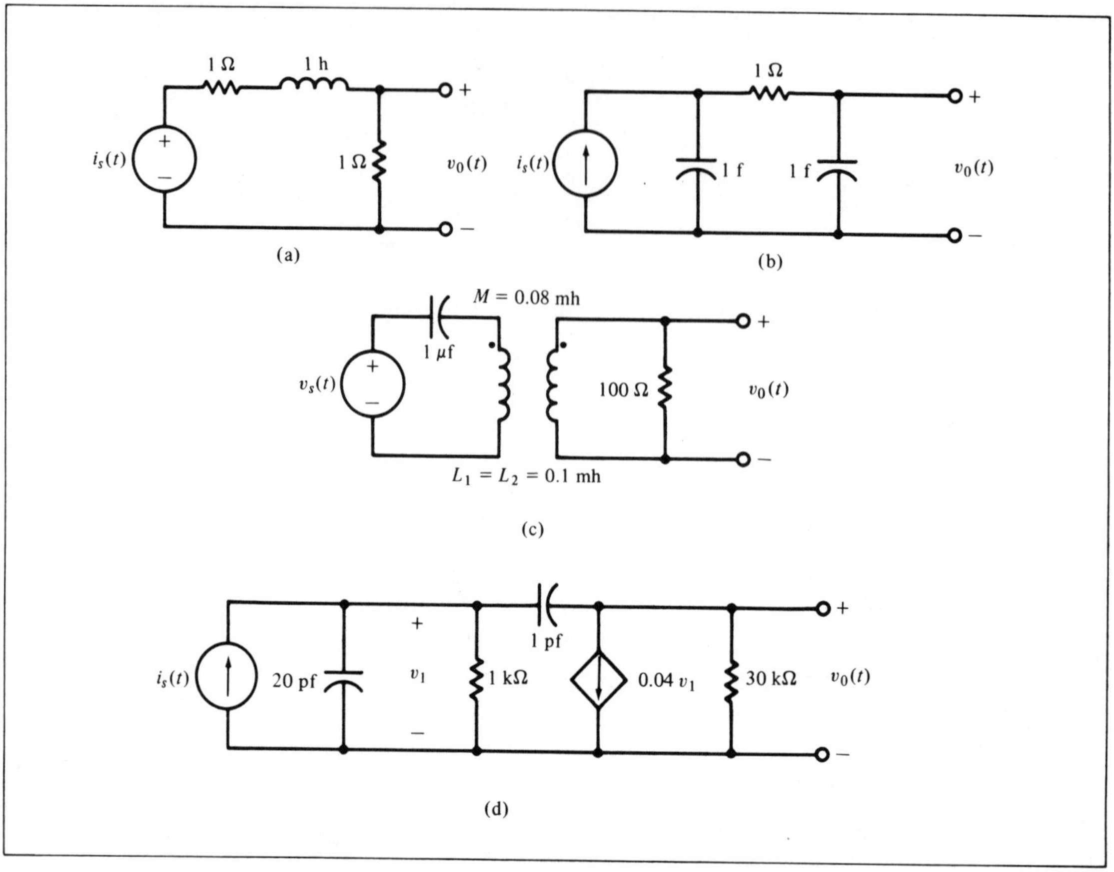

FIGURE P 12.12

12.18 The Bode diagram associated with the network of Figure P 12.14(a) is shown in Figure P 12.14(b). (a) Verify that

$$\omega_1 = \frac{1}{(C_1 + C_2)R}.$$

(b) Determine ω_2; find A in db. (c) Find the unit step response of the network.

12.19 The Bode diagram associated with the network of Figure P 12.15(a) is shown in Figure P 12.15(b). (a) Verify that one of the poles is located at $p_i = -1/(R_2 + R_L)C_C$. (b) Determine the position of the zero; find A in db.

12.20 A network with the transfer function

$$H(s) = \frac{1.04}{[(s + 0.01)^2 + 1.01^2][(s + 0.01)^2 + 0.99^2]}$$

is excited with an input

$$e(t) = \cos t$$

The resultant time function as determined by

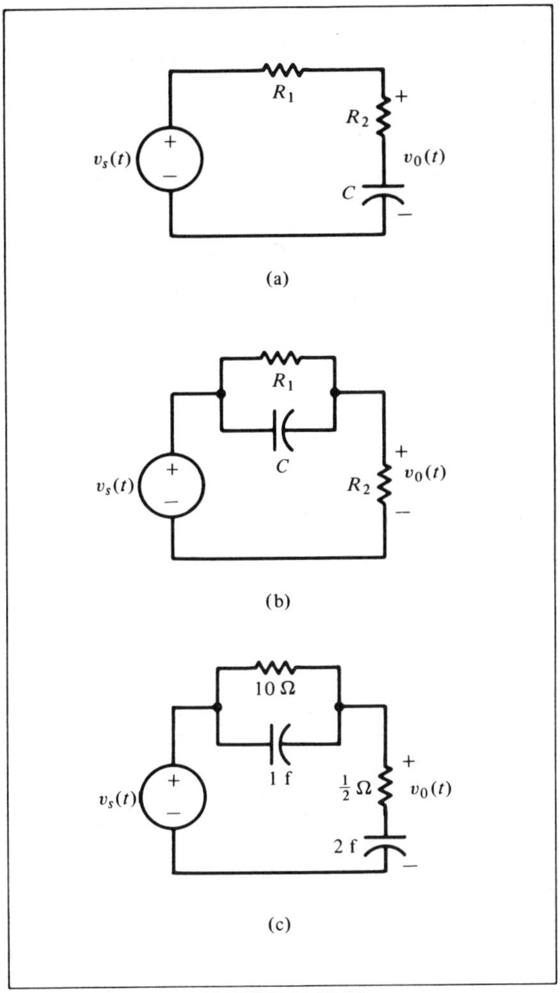

FIGURE P 12.13

computer is

$$r(t) = [907e^{-0.01t} \cos (1.01t - 44.3°) + 930e^{-0.01t} \cos (0.99t + 45.7°) + 1300 \cos (t + 181.4°)]u(t)$$

(a) Compute $|H(j1)|$, the sinusoidal steady state response for $\omega = 1$ and check with the transient response given. (b) Sketch the cluster of poles near $s = j1$ and use that sketch to estimate and sketch $|H(j\omega)|$ near $\omega = 1$.

12.21 The band-pass transfer function of Prob-

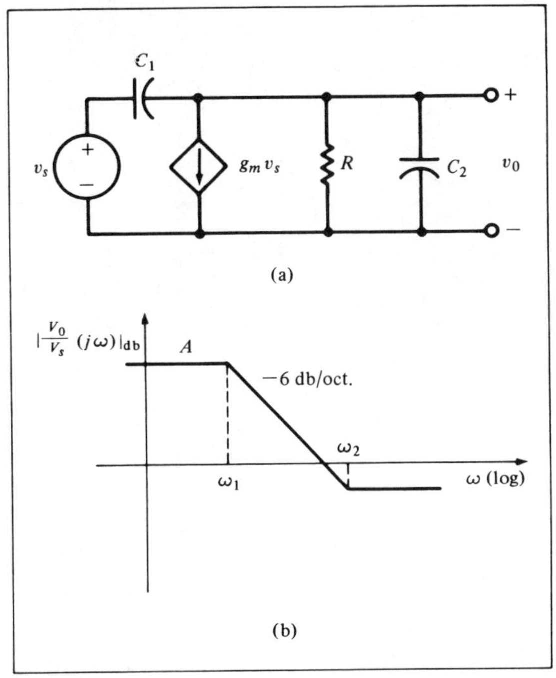

FIGURE P 12.14

lem 12.20 has four poles, two clustered near $s = j1$ and two near $s = -j1$. The objective of this problem is to compare the time and frequency responses of that function with the corresponding responses of the low-pass transfer function

$$H_{lp}(s) = \frac{0.25}{(s + 0.01 + j0.01)(s + 0.01 - j0.01)}$$

(a) Sketch the magnitude of H_{lp} near $\omega = 0$. How does this frequency response compare with the band-pass response near $\omega = 1$ in Problem 12.20? (b) Calculate the response of the low-pass network with the above transfer function to

$$e(t) = u(t).$$

(c) Express $r(t)$ in Problem 12.20 as an envelope function times the input; that is, $r(t) = R(t) \cos t$. (*Hint:* Write $\cos (0.99t + \Theta) = \cos (t - 0.01t + \Theta)$ and use trigonometric

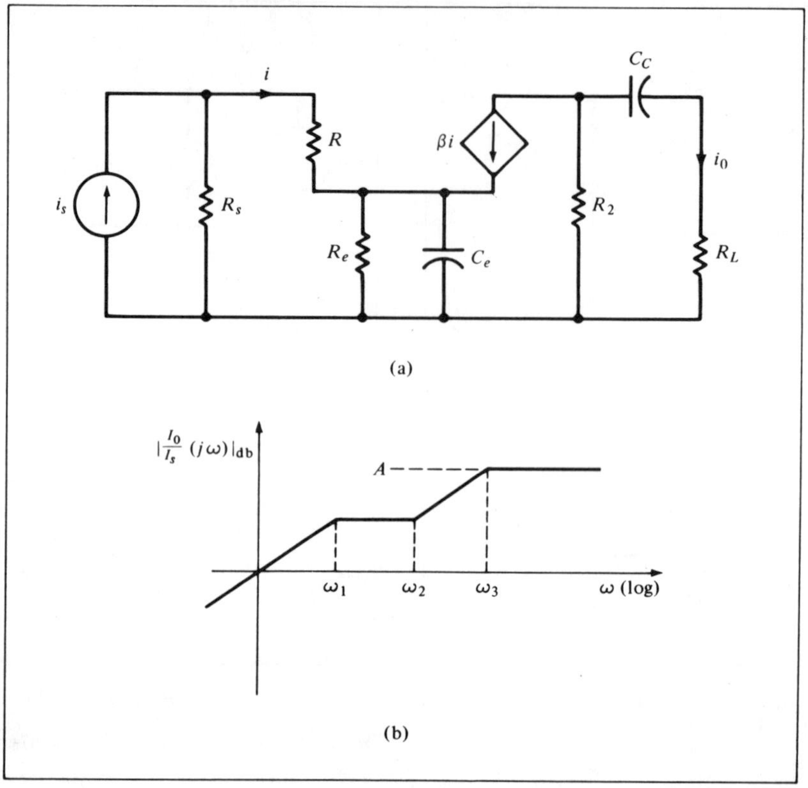

FIGURE P 12.15

identities for cos $(A + B)$); compare $R(t)$ with the step response found in (b). Sketch $R(t)$.

REFERENCES AND ADDITIONAL READING

12.1 M. E. VAN VALKENBURG, *Network Analysis.* Englewood, N.J.: Prentice-Hall, Inc., 1964, Chapter 10.

12.2 J. B. CRUZ AND M. E. VAN VALKENBURG, *Introductory Signals and Circuits.* Waltham, Mass.: Blaisdell Publishing Company, 1967, Chapter 11.

12.3 C. M. CLOSE, *The Analysis of Linear Circuits.* New York: Harcourt, Brace & World, Inc., 1966, Chapter 6.

12.4 H. W. BODE, *Network Analysis and Feedback Amplifier Design.* Princeton, N.J.: D. Van Nostrand Company, Inc., 1945.

12.5 F. F. KUO, *Network Analysis and Synthesis.* New York: John Wiley and Sons, Inc., 1964 (Second Edition), Chapter 7.

13

State Variable Analysis

13.1 Introduction

In this chapter, we take one of our frequent diversions from procedure-oriented topics. Indeed, were we interested only in the solution of electrical networks, we could justifiably ignore the subject of this section. As we have seen in Chapter 11 and will see in Chapter 14, networks of *any* complexity can be conveniently analyzed either by hand or by computer using methods akin to the node analysis procedures of Chapter 4. In contrast, the network characterization of this chapter can be quite awkward to formulate and not particularly easy to solve.

However, just as in Chapter 4 we studied those topological properties that electrical networks share with other network-like systems, in this section we examine a characteristic that dynamic networks have in common with nearly every form of natural- and human-made system: the property of *state*.

When we describe the state of a system, we try to characterize all of those system properties that permit us to predict how the system will respond to external excitations. To appreciate how broad this notion is, consider:

1. In Chapter 3, we described the state of AND or OR circuits in terms of the voltages present at the input terminals; if one of the voltages changed, we could then predict the effect on the output.
2. From a characterization of the "state of the economy" or a person's "state of mind," we could hope to gauge the effect of an external economic or behavioral stimulus.

For a dynamic system, this description of state takes the form of a set of variables whose knowledge at any time t_0 allows us to determine the response to any signal input for $t > t_0$. Such variables are termed *state variables*.

We can guess the identity of a set of state variables for electrical networks from our acquaintance with the initial conditions required for the solution of a series RLC network. From Figure 13.1, we recall that knowledge of

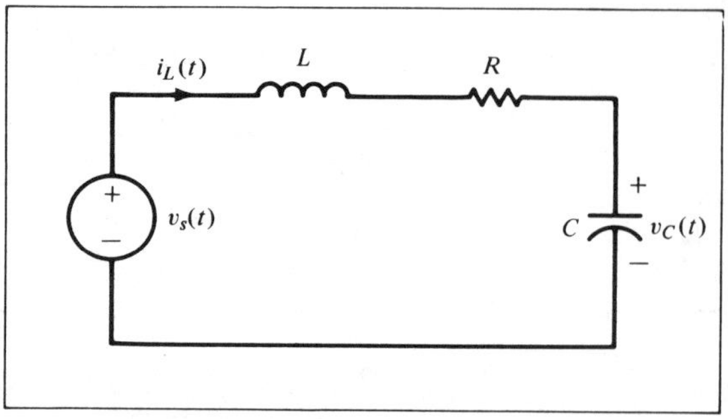

FIGURE 13.1 Series *RLC* network revisited.

(1) $i_L(0)$ and $v_C(0)$, and (2) the input voltage permitted solution of the differential equations of the network. A little thought will reveal that the initial time $t = 0$ was arbitrary, and that knowledge of $v_L(t_0)$ and $v_C(t_0)$ and $v_s(t)$ for $t \geqq t_0$ would similarly allow a solution for $t > t_0$.

Accordingly, we can usually denote the set of *all* $v_{C_i}(t)$ and $i_{L_j}(t)$ as a state variable set. Occasionally, if the network contains a loop of capacitors as in Figure 13.2(a), capacitor voltages are not independent (for example, $v_{c_1} = v_{c_2} + v_{c_3}$) and not all capacitor voltages need be included in the state variable set. Similarly, a node with only inductors connected to it [Figure 13.2(b)] results in dependent inductor currents.[1] We will not study either of these special cases in this text; the reader may examine Reference 13.1 for such discussion.

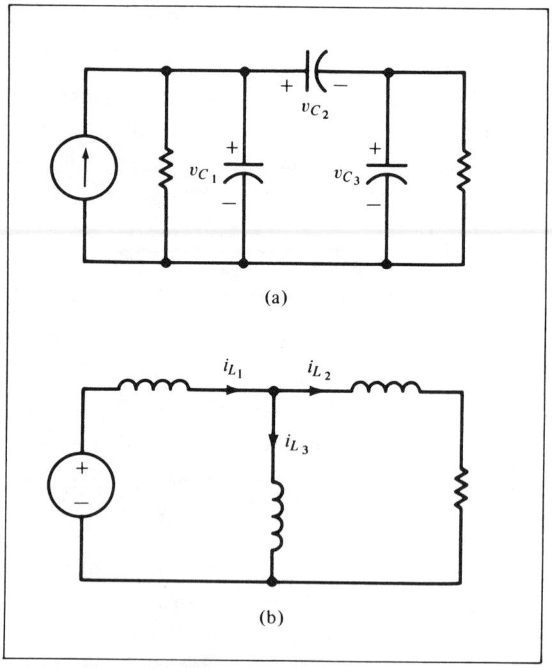

FIGURE 13.2 Examples of "capacitor-loop" and "inductor-node" degeneracy.

13.2 Equation Formulation

State Equations for Linear Networks

The *state equations* of a network are a set of simultaneous first-order differential equations relating the state variables. For example, the state equations for the series *RLC* network are of the form

$$\frac{dv_C(t)}{dt} = \frac{1}{C} i_L(t) \tag{13.1}$$

$$\frac{di_L(t)}{dt} = -\frac{1}{L} v_C(t) - \frac{R}{L} i_L(t) + \frac{1}{L} v_s(t) \tag{13.2}$$

These first-order equations involve only $v_C(t)$, $i_L(t)$, and their first derivatives as unknowns.

One way of writing the state equations for a given network is to treat each capacitor as a *known* voltage source $v_{C_i}(t)$ and each inductor as a *known* current source $i_{L_j}(t)$. The resulting network, which contains only resistors and sources, is then solved for the capacitor currents and inductor voltages; these quantities will, in general, depend on the capacitor voltages, the inductor currents, and the inputs. Finally, making the substitutions

$$i_{C_i}(t) = C_i \frac{dv_{C_i}(t)}{dt} \tag{13.3}$$

$$v_{L_j}(t) = L_j \frac{di_{L_j}(t)}{dt} \tag{13.4}$$

yields equations in the form (13.1) to (13.2).

[1] More generally, a cut-set of inductors will produce a similar degeneracy.

Example 13.1

As a simple example of this procedure, we write the state equations for the network shown in Figure 13.3(a).

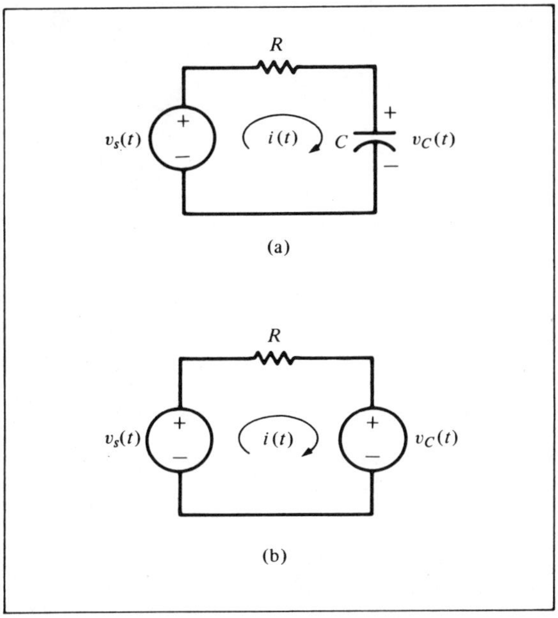

FIGURE 13.3 Network for state-variable analysis. (a) Original network; (b) modified network.

There is only one state variable $v_C(t)$ for this network, because there is only one energy-storage element. When the capacitor is replaced by a voltage source $v_C(t)$, the modified network of Figure 13.3(b) results. We now find by inspection that

$$i(t) = i_C(t) = \frac{v_s(t) - v_C(t)}{R} \qquad (13.5)$$

Substituting for $i_C(t)$ and dividing by C, we obtain

$$\frac{dv_C(t)}{dt} = -\frac{1}{RC}v_C(t) + \frac{1}{RC}v_s(t) \qquad (13.6)$$

which has the desired form.

The state variables are not a topologically complete set of variables, because we cannot always directly obtain any other network variables from them without further analysis. We often must write an *output equation.* For example, in Figure 13.3(a), the current $i(t)$ is given as

$$i(t) = -\frac{1}{R}v_C(t) + \frac{1}{R}v_s(t) \qquad (13.7)$$

In this case, the equation is a simple one. In some instances, finding it may require considerable additional analysis; it is, however, in general necessary for a complete network description.

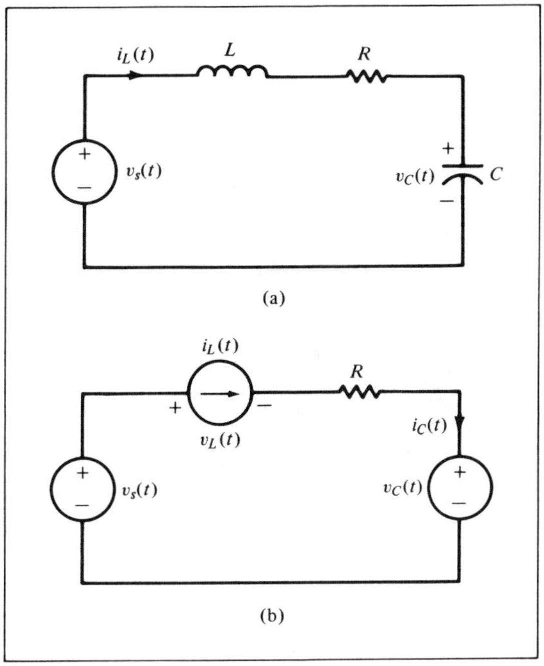

FIGURE 13.4 Second-order network for state-variable aualysis. (a) Original network; (b) modified network.

Example 13.2

We are to derive Equations (13.1) and (13.2) for the series *RLC* network shown in Figure 13.4.

From the modified network, it is easily seen that

$$i_C(t) = i_L(t) \tag{13.8}$$

$$v_L(t) = v_s(t) - Ri_L(t) - v_C(t) \tag{13.9}$$

Substituting and rearranging, we obtain

$$\frac{dv_C(t)}{dt} = \frac{1}{C} i_L(t) \tag{13.10}$$

$$\frac{di_L(t)}{dt} = -\frac{1}{L} v_C(t) - \frac{R}{L} i_L(t) + \frac{1}{L} v_s(t) \tag{13.11}$$

or in matrix form

$$\frac{d}{dt}\begin{bmatrix} v_C(t) \\ i_L(t) \end{bmatrix} = \begin{bmatrix} 0 & \frac{1}{C} \\ -\frac{1}{L} & -\frac{R}{L} \end{bmatrix}\begin{bmatrix} v_C(t) \\ i_L(t) \end{bmatrix} + \begin{bmatrix} 0 \\ \frac{1}{L} \end{bmatrix} v_s(t) \tag{13.12}$$

Choosing $v_C(t)$ as an output quantity, we can write formally the (trivial) output equation

$$v_C(t) = [1 \quad 0]\begin{bmatrix} v_C(t) \\ i_L(t) \end{bmatrix} + \begin{bmatrix} 0 \\ 0 \end{bmatrix} v_s(t) \tag{13.13}$$

In summary of the above two examples, fixed linear dynamic networks can usually be described by equations of the matrix form

State equation:

$$\dot{\mathbf{x}} = \mathbf{Ax} + \mathbf{Bu} \tag{13.14}$$

Output equation:

$$\mathbf{y} = \mathbf{Cx} + \mathbf{Du} \tag{13.15}$$

where $\mathbf{x}$ is a vector of capacitor voltages and inductor currents, with entries x_k; $\mathbf{u}$ is a vector of sources, with entries u_j; $\mathbf{y}$ is a vector of outputs, with entries y_i; and $\mathbf{A}$, $\mathbf{B}$, $\mathbf{C}$, and $\mathbf{D}$ are constant matrices.

By substitution of variables, the *first-order* matrix equation (13.14) can be reduced to a single *high-order* differential equation of the form

$$\frac{d^n y_i(t)}{dt^n} + a_1 \frac{d^{n-1} y_i(t)}{dt^{n-1}} + \cdots + a_{n-1} y_i(t) = b_1 \frac{d^{n-1} u_j(t)}{dt^{n-1}} + b_2 \frac{d^{n-2} u_j(t)}{dt^{n-2}} + \cdots + b_{n-1} u_j(t) \tag{13.16}$$

where $y_i(t)$ and $u_j(t)$ are any two outputs and inputs, respectively, provided all other inputs are zero-valued. For example, the familiar second-order differential equation describing $v_C(t)$ can be obtained from Equation (13.12) by substituting from Equation (13.10)

$$i_L(t) = i_C(t) = \frac{C dv_C(t)}{dt} \tag{13.17}$$

yielding

$$C \frac{d^2 v_C(t)}{dt^2} = -\frac{1}{L} v_C(t) - \frac{R}{L} C \frac{dv_C(t)}{dt} + \frac{1}{L} v_s(t) \tag{13.18}$$

Multiplying by L and rearranging we obtain

$$LC \frac{d^2 v_C(t)}{dt^2} + RC \frac{dv_C(t)}{dt} + v_C(t) = v_s(t) \tag{13.19}$$

which is simply the KVL equation for the network.

In the following example, we formulate the equations of a network with three energy-storage elements. We will later solve the equations by a digital computer program.

Example 13.3

The state equations for the network of Figure 13.5(a) are to be found.

For the modified network, we have from KCL,

$$i_{C_1} = i_s - i_L$$

$$i_{C_2} = i_L - i_R = i_L - \frac{v_{C_2}}{R}$$

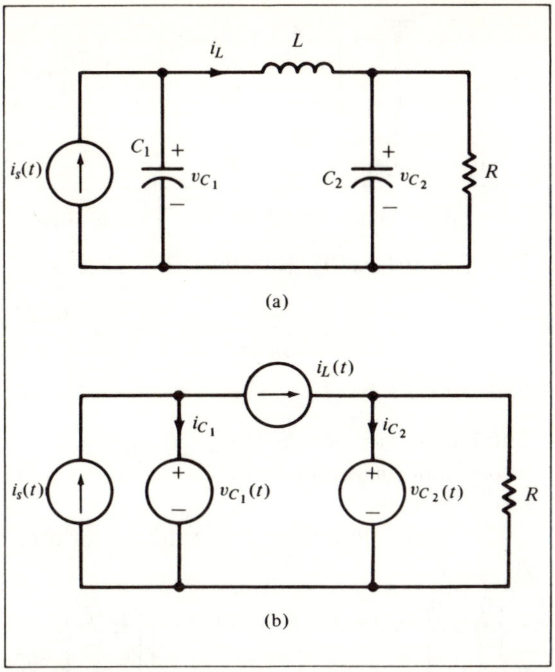

FIGURE 13.5 A dynamic network and its resistive model.

and from KVL,

$$v_L = v_{C_1} - v_{C_2}$$

Substituting for i_{C_1}, i_{C_2}, and v_L, and arranging in matrix form, we have

$$\frac{d}{dt}\begin{bmatrix} v_{C_1} \\ v_{C_2} \\ i_L \end{bmatrix} = \begin{bmatrix} 0 & 0 & -1/C_1 \\ 0 & -1/RC_2 & 1/C_2 \\ 1/L & -1/L & 0 \end{bmatrix}\begin{bmatrix} v_{C_1} \\ v_{C_2} \\ i_L \end{bmatrix} + \begin{bmatrix} 1/C_1 \\ 0 \\ 0 \end{bmatrix} i_s \quad (13.20)$$

which is of the general form $\dot{\mathbf{x}} = \mathbf{Ax} + \mathbf{Bu}$.

In summary, we observe that the state variable analysis of a network is more concerned with the form of the network equations, as first-order differential equations, rather than with specifying a simple, organized procedure to obtain these equations. In Example 13.3 the state equations were relatively easy to formulate because the modified network contains mainly sources. The complexity of the *resistive* part of the network determines the difficulty of writing state equations more than the number of energy-storage elements. The reader interested in general state equation formulation procedures should see Reference 13.1.

State Equations for Nonlinear Networks

We recall from Chapter 7 that nonlinear capacitors and inductors can be described either by

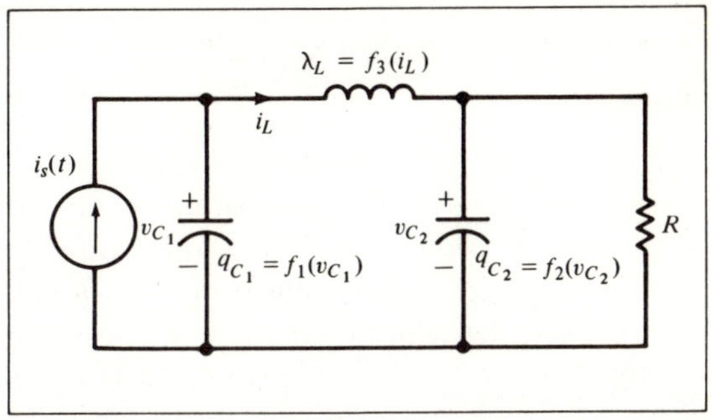

FIGURE 13.6 Network with nonlinear L and C.

(1) their q-v and λ-i equations (respectively) or (2) their i-v equations as

$$i_C = \frac{df(v_C)}{dt} = \frac{\partial f(v_C)}{\partial v_C}\frac{dv_C}{dt} = C(v_C)\frac{dv_C}{dt} \quad (13.21)$$

$$v_L = \frac{df(i_L)}{dt} = \frac{\partial f(i_L)}{\partial i_L}\frac{di_L}{dt} = L(i_L)\frac{di_L}{dt} \quad (13.22)$$

These equations, except for the dependence of C and L on v and i, are of the same form as their linear counterparts. Therefore, state equations can be constructed for networks with nonlinear energy-storage elements by using the procedure applied to linear networks. For example, comparing Figures 13.5 and 13.6, we can write similarly to Equation (13.16),

$$\frac{d}{dt}\begin{bmatrix} v_{C_1} \\ v_{C_2} \\ i_L \end{bmatrix} = \begin{bmatrix} 0 & 0 & -1/C_1(v_{C_1}) \\ 0 & -1/RC_2(v_{C_2}) & 1/C_2(v_{C_2}) \\ 1/L(i_L) & -1/L(i_L) & 0 \end{bmatrix}\begin{bmatrix} v_{C_1} \\ v_{C_2} \\ i_L \end{bmatrix} + \begin{bmatrix} 1/C_1(v_{C_1}) \\ 0 \\ 0 \end{bmatrix} i_s \qquad (13.23)$$

The solution of such nonlinear equations can be performed only by computer, which is discussed in Section 13.5.

13.3 Laplace Solution of the State Equations

For fixed linear dynamic networks, the state and output equations are linear, constant coefficient equations; consequently, they readily admit to a solution using Laplace transform. If we apply the Laplace transformation to the state and output equations of (13.14) to (13.15), we have

$$s\mathbf{X}(s) - \mathbf{x}(0) = \mathbf{AX}(s) + \mathbf{BU}(s) \qquad (13.24)$$

$$\mathbf{Y}(s) = \mathbf{CX}(s) + \mathbf{DU}(s) \qquad (13.25)$$

where $\mathbf{x}(0)$ is a vector specifying the initial inductor currents and capacitor voltages.

Rearranging Equation (13.24) yields

$$[s\mathbf{I} - \mathbf{A}]\mathbf{X}(s) = \mathbf{BU}(s) + \mathbf{x}(0)$$

or

$$\begin{aligned}\mathbf{X}(s) &= [s\mathbf{I} - \mathbf{A}]^{-1}[\mathbf{BU}(s) + \mathbf{x}(0)] \\ &= \mathbf{\Phi}(s)[\mathbf{BU}(s) + \mathbf{x}(0)] \qquad (13.26)\end{aligned}$$

where $\mathbf{I}$ is the unit matrix and where $\mathbf{\Phi}(s) \triangleq [s\mathbf{I} - \mathbf{A}]^{-1}$. The output vector $\mathbf{Y}(s)$ is then

$$\mathbf{Y}(s) = \mathbf{C\Phi}(s)\mathbf{x}(0) + [\mathbf{C\Phi}(s)\mathbf{B} + \mathbf{D}]\mathbf{U}(s) \qquad (13.27)$$

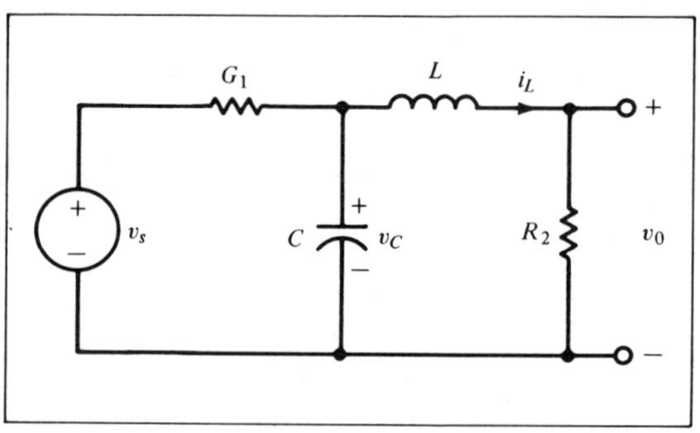

FIGURE 13.7 Example for Laplace solution.

Example 13.4

The Laplace transform of $v(t)$ in Figure 13.7 is to be found, assuming all initial conditions are zero-valued.

The state equations for the network are given by

$$\frac{d}{dt}\begin{bmatrix} i_L(t) \\ v_C(t) \end{bmatrix} = \underbrace{\begin{bmatrix} -R_2/L_1 & 1/L_1 \\ -1/C_1 & -G_1/C_1 \end{bmatrix}}_{\mathbf{A}}\begin{bmatrix} i_L(t) \\ v_C(t) \end{bmatrix} + \underbrace{\begin{bmatrix} 0 \\ G_1/C_1 \end{bmatrix}}_{\mathbf{B}} v_s(t)$$

and the output equation is

$$v_0(t) = \underbrace{[R_2 \quad 0]}_{\mathbf{C}}\begin{bmatrix} i_L(t) \\ v_C(t) \end{bmatrix} + \underbrace{[0]}_{\mathbf{D}} v_s$$

The transformed equations then have the form

$$s\begin{bmatrix} I_L \\ V_C \end{bmatrix} - \begin{bmatrix} i_L(0) \\ v_C(0) \end{bmatrix} = \begin{bmatrix} -R_2/L_1 & 1/L_1 \\ -1/C_1 & -G_1/C_1 \end{bmatrix}\begin{bmatrix} I_L \\ V_C \end{bmatrix} + \begin{bmatrix} 0 \\ G_1/C_1 \end{bmatrix} V_s(s) \qquad (13.28)$$

The matrix $[\mathbf{\Phi}(s)]$ is given by

$$\mathbf{\Phi}(s) = [s\mathbf{I} - \mathbf{A}]^{-1}$$

$$= \begin{bmatrix} s + R_2/L_1 & -1/L_1 \\ 1/C_1 & s + G_1/C_1 \end{bmatrix}^{-1}$$

$$= \frac{1}{\Delta(s)} \begin{bmatrix} s + G_1/C_1 & 1/L_1 \\ -1/C_1 & s + R_2/L_1 \end{bmatrix}$$

where

$$\Delta(s) = s^2 + \left(\frac{R_2}{L_1} + \frac{G_1}{C_1}\right)s + \frac{(R_2G_1 + 1)}{L_1C_1}$$

The zero-state output is then, from Equation (13.27),

$$V_0(s) = \left([R_2 \;\; 0]\, \mathbf{\Phi} \begin{bmatrix} 0 \\ G_1/C_1 \end{bmatrix} + [0] \right) V_s(s)$$

$$= \frac{(R_2G_1)/(L_1C_1)V_s(s)}{s^2 + (R_2/L_1 + G_1/C_1)s + (R_2G_1 + 1)/(L_1C_1)} \qquad (13.29)$$

The solution for $\mathbf{Y}(s)$, as given in Equation (13.27) and exemplified in Equation (13.29), has several interesting features.

1. The poles of *all* output function transforms are identical, and are determined from setting the determinant of $\mathbf{\Phi}(s)$ to zero. These poles are just the *natural frequencies of the network* identified in Chapters 8 and 9 for simple networks. Note, however, that a pole may be cancelled by an identical zero, so that the associated exponential or sinusoidal term in the transient response may indeed be missing.

2. The *number* of natural frequencies is equal to the number of state variables. In contrast, the number of node, mesh, or other topologically based equations has no relationship to the number of natural frequencies.

3. Equation (13.26) shows the output is the sum of responses due to the initial conditions $\mathbf{x}(0)$ and the transform of the input $\mathbf{U}(s)$. This agrees with the view of an initial condition as a form of independent source, as discussed in Chapter 11.

4. Because the state equations have a standard form of first-order, simultaneous, linear differential equations, the above solution procedure can be applied to *any* dynamic system that can be similarly described. This includes social and economic systems that have no network analogy.

13.4 Direct Solution of the State Equations

Given a first-order matrix differential equation of the form of the state equations

$$\dot{\mathbf{x}} = \mathbf{A}\mathbf{x} + \mathbf{B}\mathbf{u} \qquad (13.30)$$

it should not be surprising that a solution exists generalizing the single variable case

$$x(t) = e^{\alpha(t-t_0)}x(t_0) + e^{\alpha t}\int_{t_0}^{t} e^{-\alpha\tau}f(\tau)\,d\tau$$

of Section 9.5. This generalization is written

$$\mathbf{x}(t) = e^{\mathbf{A}(t-t_0)}\mathbf{x}(t_0) + e^{\mathbf{A}t}\int_{t_0}^{t} e^{-\mathbf{A}\tau}\,\mathbf{BU}(\tau)\,d\tau \qquad (13.31)$$

We explain this solution as follows:

1. Just as the quantity $e^{\alpha t}$ is defined from the power series expansion

$$e^{\alpha t} = 1 + \alpha t + \frac{(\alpha t)^2}{2!} + \frac{(\alpha t)^3}{3!} + \cdots + \frac{(\alpha t)^r}{r!} + \cdots$$

the *matrix* exponential $e^{\mathbf{A}t}$ is defined for a

square $n \times n$ matrix $\mathbf{A}$ by

$$e^{\mathbf{A}t} = \mathbf{I} + \mathbf{A}t + \frac{\mathbf{A}^2t^2}{2!} + \frac{\mathbf{A}^3t^3}{3!} + \cdots \frac{(\mathbf{A}t)^r}{r!} + \cdots \tag{13.32}$$

Thus, $e^{\mathbf{A}t}$ is itself an $n \times n$ matrix.

2. With zero input ($\mathbf{u} = \mathbf{0}$), Equation (13.30) becomes

$$\dot{\mathbf{x}} = \mathbf{A}\mathbf{x}$$

If $\dot{\mathbf{x}}(t) = e^{\mathbf{A}(t-t_0)}\mathbf{x}(t_0)$ is to be the solution as Equation (13.31) indicates, then we must have

$$\frac{d}{dt}\left(e^{\mathbf{A}(t-t_0)}\mathbf{x}(t_0)\right) = \mathbf{A}e^{\mathbf{A}(t-t_0)}\mathbf{x}(t_0)$$

Expanding both sides in a power series expansion, we have

$$\frac{d}{dt}\left(\mathbf{I} + \mathbf{A}(t - t_0) + \frac{\mathbf{A}^2(t-t_0)^2}{2!} + \cdots + \frac{\mathbf{A}^r(t-t_0)^r}{r!} + \cdots\right)\mathbf{x}(t_0)$$

$$= \mathbf{A}\left(\mathbf{I} + \mathbf{A}(t - t_0) + \frac{\mathbf{A}^2(t-t_0)^2}{2!} + \cdots + \frac{\mathbf{A}^r(t-t_0)^r}{r!} + \cdots\right)\mathbf{x}(t_0) \tag{13.33}$$

It is easily seen by differentiation that this equation is indeed satisfied.

3. Having checked the zero-input term in Equation (13.33), we can validate the zero-state response by differentiation of Equation (13.31) to see if $\dot{\mathbf{x}} = \mathbf{A}\mathbf{x} + \mathbf{B}\mathbf{u}$ is satisfied with $\mathbf{x}(t_0) = \mathbf{0}$. Differentiating Equation (13.31) by parts, we have

$$\frac{d}{dt}\left[e^{\mathbf{A}t}\int_{t_0}^{t} e^{-\mathbf{A}\tau}\mathbf{B}\mathbf{u}(\tau)\,d\tau\right] = \underbrace{(\mathbf{A}e^{\mathbf{A}t})\left(\int_{t_0}^{t} e^{-\mathbf{A}\tau}\mathbf{B}\mathbf{u}(\tau)\,d\tau\right)}_{\mathbf{x}} + \underbrace{(e^{\mathbf{A}t})(e^{-\mathbf{A}t}}_{\mathbf{I}}\mathbf{B}\mathbf{u}(t)) = \mathbf{A}\mathbf{x} + \mathbf{B}\mathbf{u}$$

as required.

The above is not proposed as a practical means of solving the state equations, anymore than is the convolution integral, on which the above zero-state solution is based. The principal problem that arises is the determination of the matrix exponential $e^{\mathbf{A}t}$. The power series of Equation (13.32) does not converge rapidly for large t and is never used for calculation purposes. Perhaps it is enough to observe that $e^{\mathbf{A}t}$ can be written in terms of familiar functions. We show this by comparing the Laplace solution of Equation (13.26) with the direct solution of the zero-input case. We obtain the result, with $t_0 = 0$,

$$e^{\mathbf{A}t} = \mathcal{L}^{-1}[\mathbf{\Phi}(s)] = \mathcal{L}^{-1}[(s\mathbf{I} - \mathbf{A})^{-1}] \tag{13.34}$$

If, for example,

$$\mathbf{A} = \begin{bmatrix} -2 & \sqrt{2} \\ \sqrt{2} & -1 \end{bmatrix}$$

then

$$(s\mathbf{I} - \mathbf{A})^{-1} = \begin{pmatrix} s+2 & -\sqrt{2} \\ -\sqrt{2} & s+1 \end{pmatrix}^{-1} = \begin{pmatrix} \dfrac{s+1}{\Delta(s)} & \dfrac{\sqrt{2}}{\Delta(s)} \\ \dfrac{\sqrt{2}}{\Delta(s)} & \dfrac{s+2}{\Delta(s)} \end{pmatrix} \tag{13.35}$$

where $\Delta(s) = s^2 + 3s$. The inverse transform of Equation (13.35) is taken term-by-term to yield

$$e^{\mathbf{A}t} = \begin{pmatrix} \mathcal{L}^{-1}\left[\dfrac{s+1}{s(s+3)}\right] & \mathcal{L}^{-1}\left[\dfrac{\sqrt{2}}{s(s+3)}\right] \\ \mathcal{L}^{-1}\left[\dfrac{\sqrt{2}}{s(s+3)}\right] & \mathcal{L}^{-1}\left[\dfrac{s+2}{s(s+3)}\right] \end{pmatrix}$$

$$= \begin{pmatrix} \frac{1}{3} + \frac{2}{3}e^{-3t} & \frac{\sqrt{2}}{3} - \frac{\sqrt{2}}{3}e^{-3t} \\ \frac{\sqrt{2}}{3} - \frac{\sqrt{2}}{3}e^{-3t} & \frac{2}{3} + \frac{1}{3}e^{-3t} \end{pmatrix} \quad (t \geqq 0)$$

We now observe the elegance hidden in the direct solution of Equation (13.31). The entire matrix expression is functionally no more than an organizational scheme for manipulation of the exponential and sinusoidal functions that constitute the basic response functions of linear networks and systems.

13.5 Numerical Solution

Numerical Integration

A great variety of numerical solution methods exist for solution of first-order differential equations. The most sophisticated of these appear as subroutine packages in scientific program libraries available through most computing centers. We will consider an elementary algorithm known as the *forward Euler* method. Its simplicity will allow us to understand the details of a computer solution program.

Suppose we are given the single nonlinear differential equation

$$\dot{x} = f(x, t) \tag{13.36}$$

with initial condition $x(t^0)$; this equation is to be solved for x in the interval (t^0, t_{max}).

In obtaining a computer solution, we will discretize the independent variable, time, and identify a number of specific times t^n in the solution interval (t^0, t_{max}) where we wish to calculate the solution $x(t^n)$. We then can write at these points

$$\dot{x}(t^n) = f(x(t^n), t^n) \tag{13.37}$$

Besides the differential equation, we need an *integration rule* that relates successive values of $x(t^n)$ and $\dot{x}(t^n)$. One of the simplest rules is the *forward Euler* formula

$$x(t^{n+1}) = x(t^n) + T\dot{x}(t^n) - \epsilon^{n+1} \tag{13.38}$$

where $T = t^{n+1} - t^n$ is the *step size* and ϵ is the error made in the approximation. This calculation is illustrated in Figure 13.8 for $n = 0$.

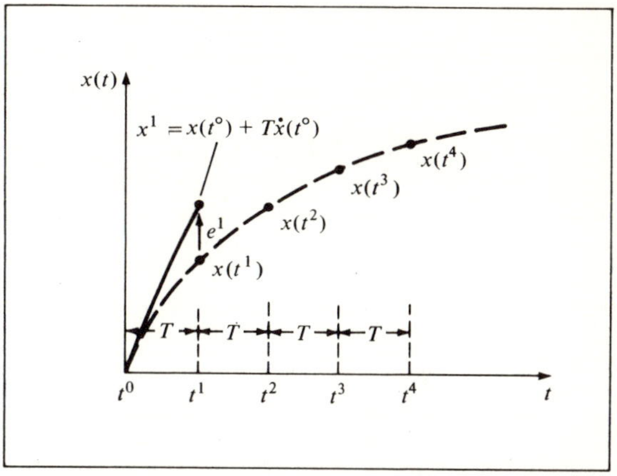

FIGURE 13.8 Illustration of discretization (forward Euler).

The figure shows how the value of x at time t^{n+1} is estimated from values of x and the derivative $\dot{x}$ at time t^n by a straight-line representation of the solution curve from t^n to t^{n+1}.

Unfortunately, the error ϵ^{n+1} cannot be evaluated, so we accept the approximate equation given by

$$x^{n+1} = x^n + T\dot{x}^n \tag{13.39}$$

Here the x^n and $\dot{x}^{n+1}$ symbols are used to indicate the approximate values we will calculate in our numerical solution, in contrast to the *exact* solution symbols $x(t^n)$ and $\dot{x}(t^{n+1})$ which cannot, in general, be found. Equation (13.39) is known as the *forward Euler* integration rule. Similarly, the discretized form of the state equation is

$$\dot{x}^n = f(x^n, t^n) \tag{13.40}$$

Combining the state equation (13.40) and the integration rule (13.39), we obtain the solution algorithm

$$x^{n+1} = x^n + Tf(x^n, t^n) \tag{13.41}$$

which constitutes the basis for our program.

Given the initial condition $x^0 = x(t^0)$, x^1 is

found from Equation (13.41) with $n = 0$; then n is incremented by 1, and the process is repeated. For example, the linear differential equation

$$\dot{x}(t) = \alpha x(t) + u(t) = f(x(t), t)$$

with $x(0) = x^0$, results in the solution formula

$$x^{n+1} = x^n + T(\alpha x^n + 1)$$
$$= (1 + T\alpha)x^n + T \qquad (n = 0, 1, \cdots)$$

Note that the solution formula of Equation (13.41) applies to linear *and* nonlinear equations. For a set of simultaneous equations,

$$\dot{\mathbf{x}} = f(\mathbf{x}, t) \qquad (13.42)$$

the solution formula (13.41) becomes

$$\mathbf{x}^{n+1} = \mathbf{x}^n + T[f(\mathbf{x}^n, t^n)] \qquad (13.43)$$

Example 13.5

The network state equations of Equation (13.20) are to be solved numerically using the forward Euler rule. The following element values are to be used: $R = 1000\Omega$, $L = 10^{-6}$ h, $C_1 = 10^{-12}$ f, and $C_2 = 2 \times 10^{-12}$ f. The input $i_s(t)$ is shown in Figure 13.9; $v_{C_2}{}^n$ is the output quantity. An appropriate T and $t_{\max}$ are to be found.

The solution formula (13.43) becomes

$$\begin{bmatrix} v_{C_1}{}^{n+1} \\ v_{C_2}{}^{n+1} \\ i_L{}^{n+1} \end{bmatrix} = \begin{bmatrix} v_{C_1}{}^{n} \\ v_{C_2}{}^{n} \\ i_L{}^{n} \end{bmatrix} + T\left\{\begin{bmatrix} 0 & 0 & -1/C_1 \\ 0 & -1/RC_2 & 1/C_2 \\ 1/L & -1/L & 0 \end{bmatrix}\begin{bmatrix} v_{C_1}{}^{n} \\ v_{C_2}{}^{n} \\ i_L{}^{n} \end{bmatrix} + \begin{bmatrix} 1/C_1 \\ 0 \\ 0 \end{bmatrix} i_s(t^n)\right\} \qquad (13.44)$$

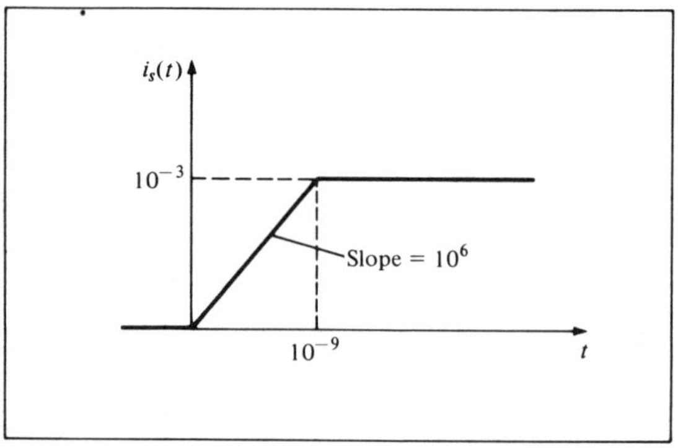

FIGURE 13.9 Signal input. $i_s(t) = 10^6 t u(t) - 10^6 \times (t - 10^{-9})u(t - 10^{-9})$.

A Fortran program to perform the step-by-step solution is shown in Figure 13.10; the flow chart is displayed in Figure 13.11. An appropriate value for T can be guessed by observing the time constants of pairs of RC, RL, and LC elements in the network. For example, $RC_1 = 10^{-9}$ sec, $L/R = 10^{-9}$ sec, and $\sqrt{LC_1} = 10^{-9}$ sec. We decide to take 100 time steps within 10^{-9} sec, resulting in a step size of $10^{-9}/100 = 10^{-11}$ sec. Rather than printing $v_{C_2}{}^n$ for each time step, we print only every fiftieth value, that is, every 0.5×10^{-9} sec. The value for $t_{\max}$ can be obtained by experimentation as follows. From Chapter 8, we calculate the equilibrium value of $v_{C_2}(t)$ to be $(i_s(\infty))R = 1$ volt. Therefore, when $v_{C_2}{}^n$ approaches this value, we can terminate execution. Alternately, we can choose a value related to the approximate network time constants determined above, in this case 10 time constants or 10^{-8} sec.

The results are shown in Figure 13.12. Note that the response is quite different from that obtained from first-order and second-order networks; it is not identifiable either as an exponential or damped sinusoidal. This occurs, of course, because the network has *three* energy-storage elements.

Error

The error made in adopting the forward Euler rule ϵ^{n+1} requires some discussion. If $x(t^n)$ is continuous, then from Figure 13.8, ϵ^{n+1} can be made as small as desired by choosing

```
C***    THIS PROGRAM SOLVES THE STATE EQUATIONS OF EXAMPLE 13.5
C       USING FORWARD EULER INTEGRATION
      REAL L
      DIMENSION X(3),DERX(3),C(2)
      COMMON R,L,C
C***    ASSIGN ELEMENT VALUES
      R=1000.
      L=1.E-6
      C(1)=1.E-12
      C(2)=2.E-12
C***    ASSIGN INTEGRATION PARAMETERS
      TSTEP=.01E-9
      TMAX=10.E-9
      NPRINT=50
      NX=3
C***    DEFINE INITIAL CONDITIONS
      DO 1 J=1,3
1     X(J)=0.
      TIME=0.
      WRITE(6,2)
2     FORMAT(/'   SOLUTION OF EXAMPLE 13.5'//'    TIME        VC2')
      CALL FEULER(X,DERX,NX,TIME,TSTEP,TMAX,NPRINT)
      STOP
      END
C***    SUBROUTINE FEULER INTEGRATES A SET OF SUMULTANEOUS
C       FIRST-ORDER DIFFERENTIAL EQUATIONS FROM AN INITIAL
C       TIME TO TMAX, USING FORWARD EULER RULE
      SUBROUTINE FEULER(X,DERX,NX,TIME,TSTEP,TMAX,NPRINT)
      DIMENSION X(NX),DERX(NX)
C***    EVALUATE AND PRINT X, DX/DT AT INITIAL TIME
      CALL FCT(X,DERX,TIME)
      CALL OUTP(X,DERX,TIME)
C***    ENTER INTEGRATION LOOP; PERFORM NPRINT STEPS W/OUT PRINTING
1     DO 3 K=1,NPRINT
      CALL FCT(X,DERX,TIME)
      DO 2 J=1,NX
2     X(J)=X(J)+DERX(J)*TSTEP
3     TIME=TIME+TSTEP
      CALL OUTP(X,DERX,TIME)
      IF(TIME.GT.TMAX)RETURN
      GO TO 1
      END
C***    SUBROUTINE FCT IS SUPPLIED BY USER TO DEFINE STATE EQNS.
      SUBROUTINE FCT(X,DERX,TIME)
      REAL L
      DIMENSION X(3),DERX(3),C(2)
      COMMON R,L,C
      CUR=1.E6*TIME
      IF(TIME.GE.1.E-9)CUR=1.E-3
      DERX(1)=(-1./C(1))*X(3)+(1./C(1))*CUR
      DERX(2)=(-1./(R*C(2)))*X(2)+(1./C(2))*X(3)
      DERX(3)=(1./L)*X(1)+(-1./L)*X(2)
      RETURN
      END
C***    SUBROUTINE OUTP IS SUPPLIED BY USER TO PRINT OUTPUT VARIABLES
      SUBROUTINE OUTP(X,DERX,TIME)
      DIMENSION X(3),DERX(3)
      WRITE(6,1)TIME,X(2)
1     FORMAT(2E12.4)
      RETURN
      END
```

FIGURE 13.10 Fortran program to solve Example 13.5.

a sufficiently small step size T. Usually, the accumulated error made in integrating from t^0 to t_{max} decreases *linearly* with the step size; for example, to decrease this error by a factor of 10 requires one-tenth the step size and hence 10 times the number of applications of the

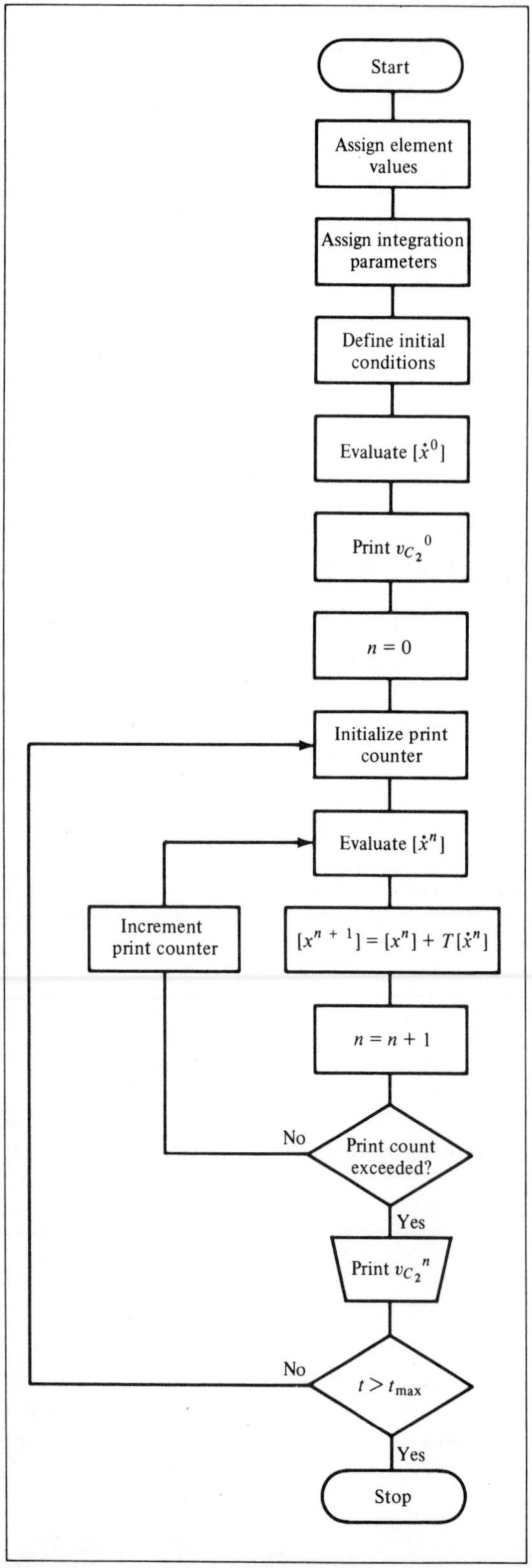

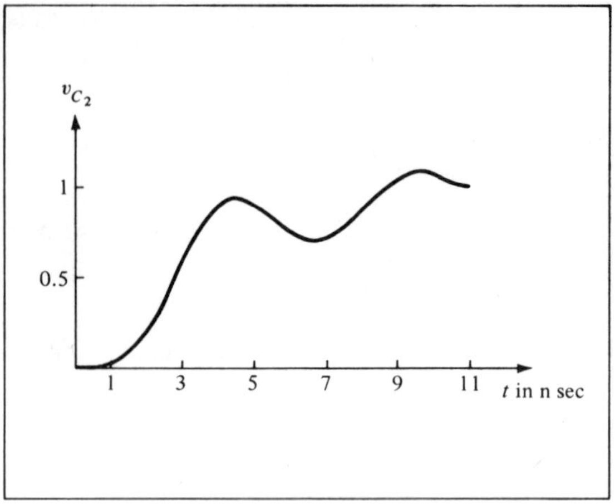

FIGURE 13.12 Solution of Example 13.5.

solution formula (13.44). This is an undesirable result; the accumulated error of a "good" integration rule should vary as T^r where $r \geqq 2$.

Just as important, as users we would like to be able to specify the error and have the step size selected *automatically* by the integration subroutine. A "good" integration subroutine should also be able to estimate the error ϵ^{n+1} at each time t^n and adjust the step size accordingly. Thus, the step size will change as the solution proceeds from t^0 to t_{max}. This is especially important in analysis of networks with *both* fast and slow time constants, because we would like to choose different time steps to capture both the fast and slow response terms with equal accuracy.

The program of Example 13.5 is purposefully written so that the inaccurate forward Euler rule is contained in a subroutine that could be replaced by a more sophisticated algorithm, with automatic error control. Thus, the FEULER subroutine need not be composed as in Figure 13.10; the user may simply regard it as a "black box" that requires the user supply subroutines FCT to evaluate derivatives of the solution variables and OUTP for printing. An example of this is given in the Solutions Manual.

FIGURE 13.11 Flow chart for program to solve Example 13.5.

PROBLEMS

13.1 (a)–(h) Find the state and output equation of the networks of Figure P 13.1.

13.2 The switch of Figure P 13.2 is moved from a to b at $t = 0$, after the network has achieved a steady state. (a) From the state equations, obtain the differential equation of

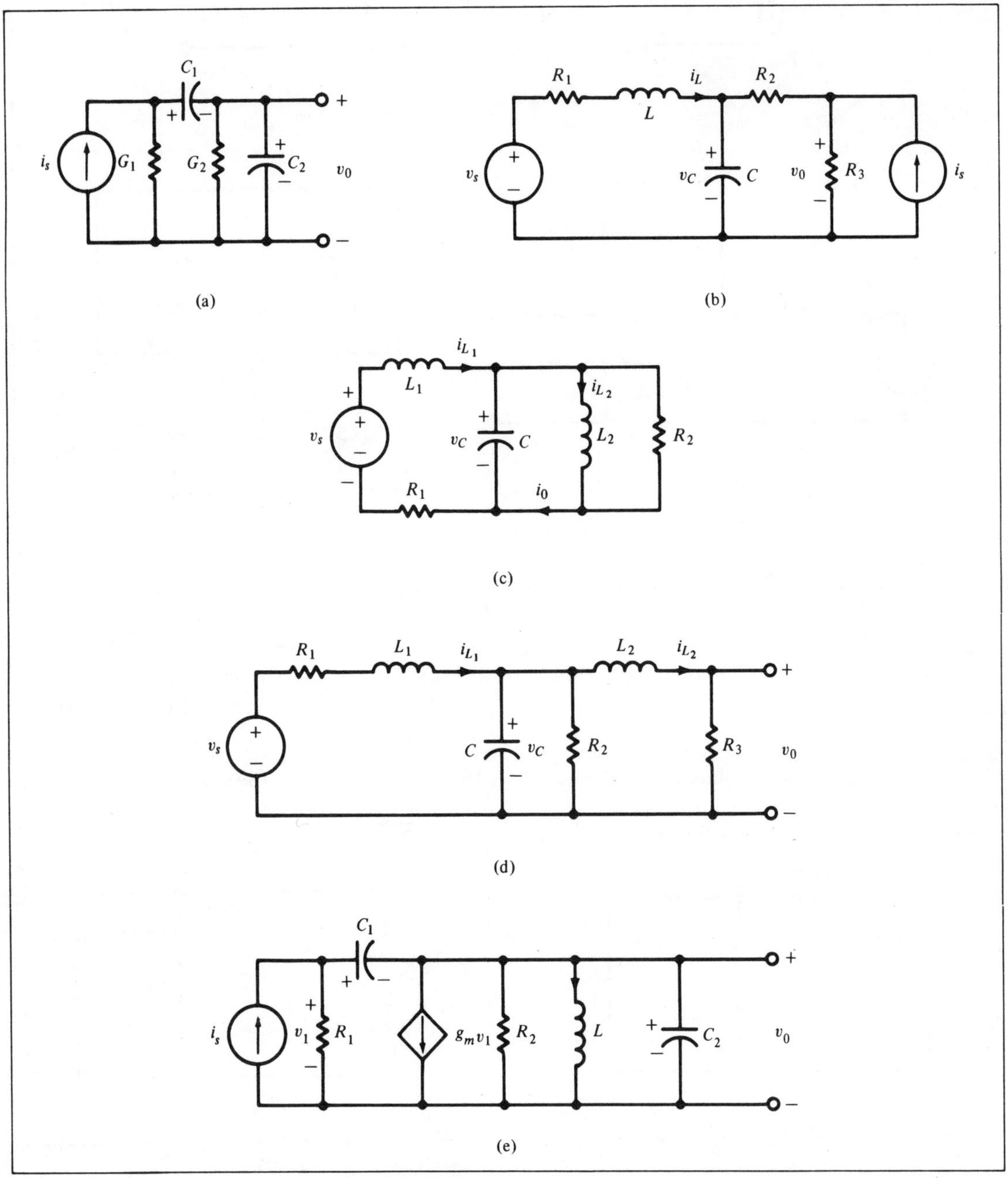

FIGURE P 13.1 State equation only.

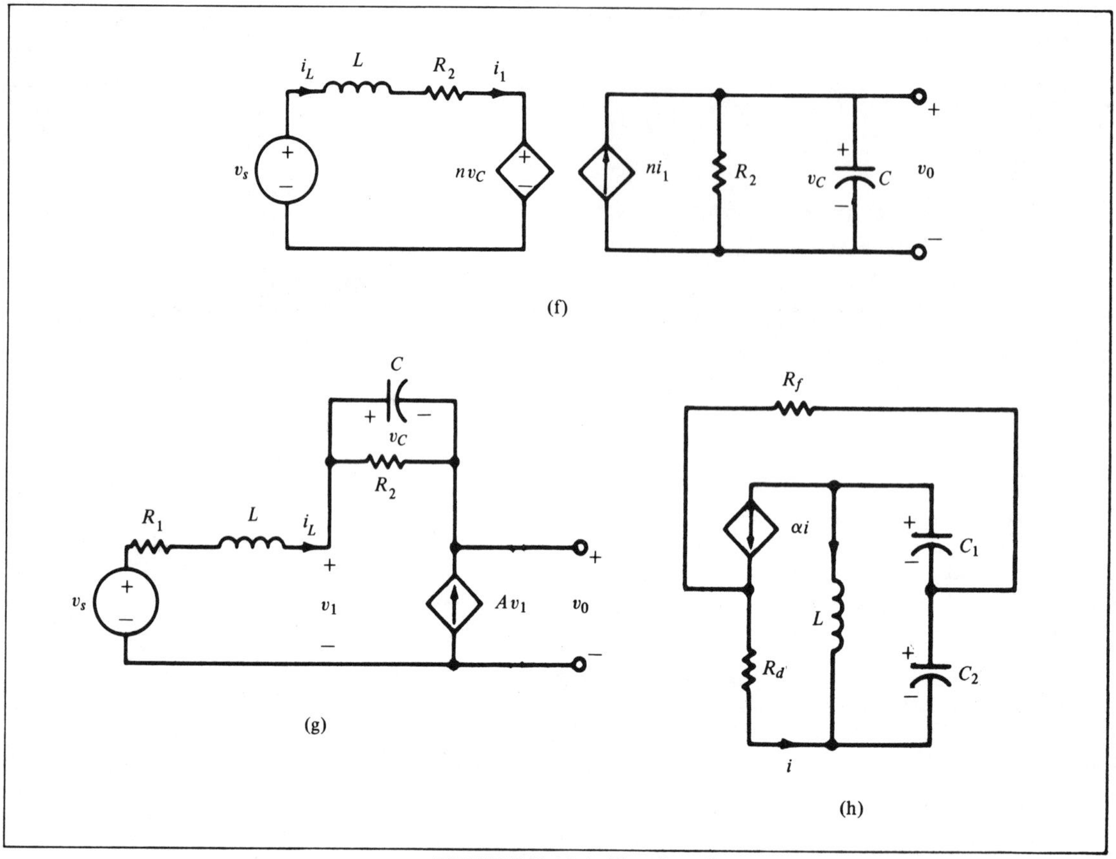

FIGURE P 13.1 (Continued)

the circuit in terms of $i_L(t)$. (b) Find $i_L(t)$, $t > 0$.

13.3 The switch in Figure P 13.3 is closed at $t = 0$ after the network has reached equilibrium. (a) From the state equations, obtain the differential equation of the circuit in terms of $v_c(t)$. (b) Find $v_c(t)$, $t > 0$.

13.4 (a)–(b) Write the state and output equa-

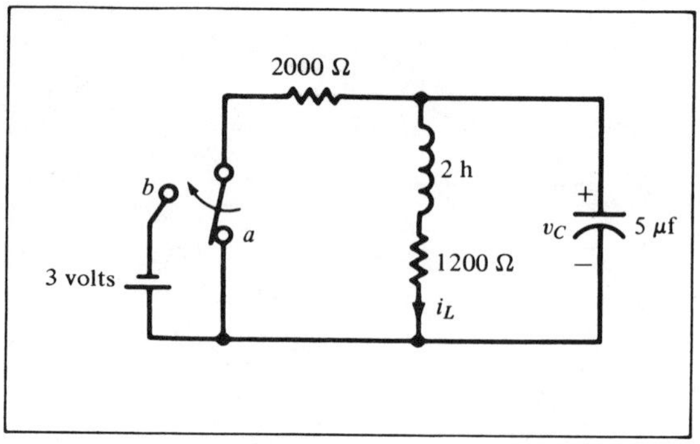

FIGURE P 13.2

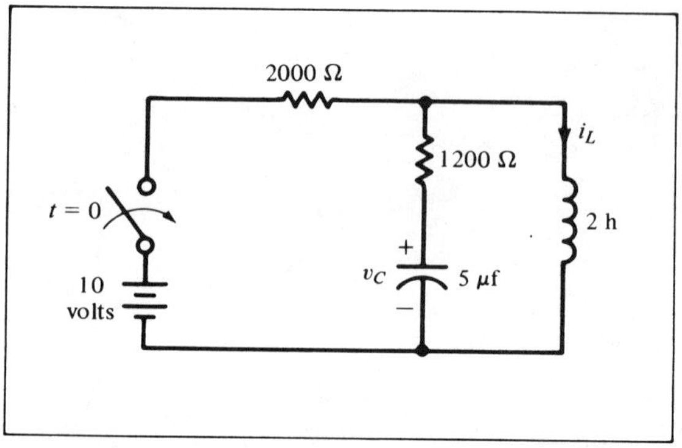

FIGURE P 13.3

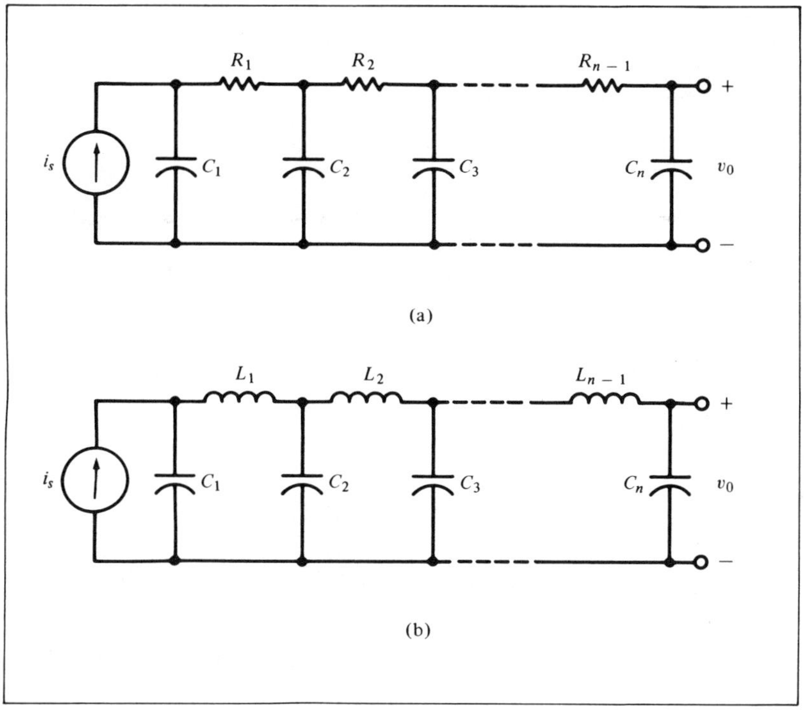

FIGURE P 13.4

tions for the ladder networks of Figure P 13.4.

13.5 Both capacitors of Figure P 13.5 have a charge voltage characteristic given by

$$q = k^3 v^3$$

Write state and output equations for this network using voltages as state variables. Repeat using charges as state variables.

13.6 A current source is applied to the network of Figure P 13.6 as shown; the voltage output is also indicated. (a) Show two other locations for the current source and the voltage output which preserve the natural frequencies. (b) Indicate where a voltage source could be located to preserve the natural frequencies.

13.7 The state equations of a network are given in (a) below. Find k in the parts (b) and (c) that result in the same natural frequencies.

$$\dot{x}_1 = -100x_1 + x_2$$

$$\dot{x}_2 = 0x_1 - x_2$$

(a)

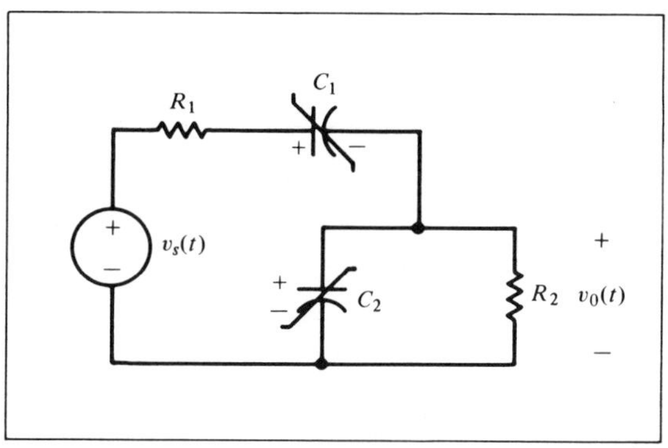

FIGURE P 13.5

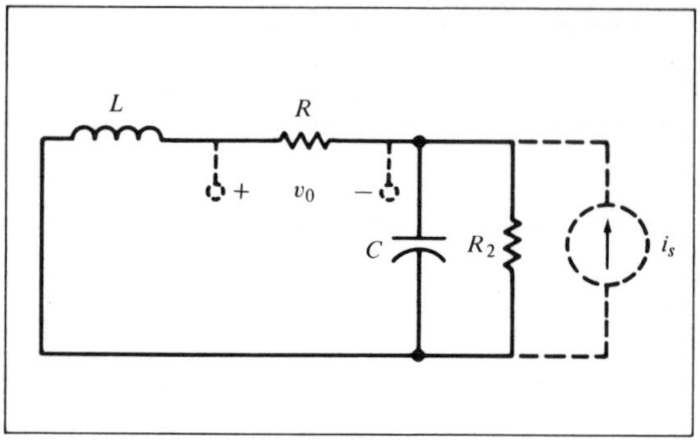

FIGURE P 13.6

$$\dot{x}_1 = -10x_1 + 10x_2$$
$$\dot{x}_2 = kx_1 - 91x_2$$

(b)

$$\dot{x}_1 = -50.5x_1 + kx_2$$
$$\dot{x}_2 = 100x_1 - 50.5x_2$$

(c)

13.8 Can a network of resistors and capacitors contain more state variables (and hence more natural frequencies) than node voltage variables? If inductors are allowed, does this change your answer? A counterexample may help.

13.9 Write the state equations of the network of Figure P 13.7 before and after L_2 is added. Has the inductor resulted in an additional natural frequency? If so, where is this frequency? What connection of capacitors would yield a similar result?

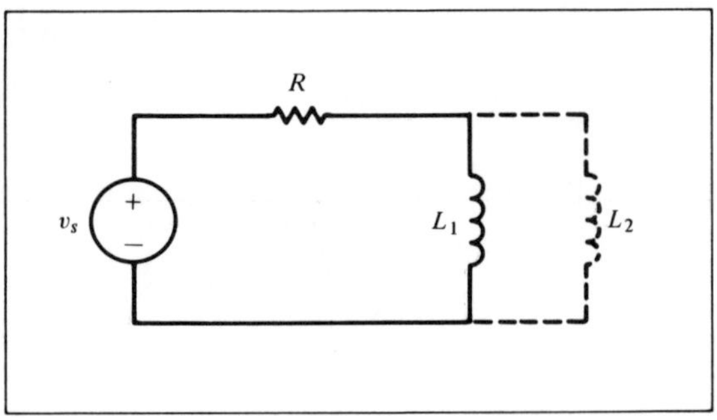

FIGURE P 13.7

13.10 If there are n state variables, m inputs, and p outputs, what are the dimensions of **A**, **B**, **C**, and **D**?

13.11 Find $e^{\mathbf{A}t}$, where **A** is given by the following matrices. Determine the natural frequencies in each case.

(a) $[1]$.

(b) $\begin{bmatrix} 1 & 2 \\ 3 & 0 \end{bmatrix}$.

(c) $\begin{bmatrix} -1 & 2 \\ 2 & -1 \end{bmatrix}$.

13.12 If **A**, **B**, **C**, and **D** are given by the following matrices, find $\mathbf{x}(t)$ and $\mathbf{y}(t)$ in terms of $\mathbf{x}(0)$. All inputs are the unit step $u(t)$. Use either Laplace or direct solution methods.

(a) $\mathbf{A} = \begin{bmatrix} -2 & 1 \\ 1 & -3 \end{bmatrix}$ $\quad \mathbf{B} = \begin{bmatrix} 1 \\ 0 \end{bmatrix}$

$\mathbf{C} = [1 \quad 2]$ $\quad \mathbf{D} = [0]$

(b) $\mathbf{A} = \begin{bmatrix} 1 & 2 \\ -1 & -1 \end{bmatrix}$ $\quad \mathbf{B} = \begin{bmatrix} 1 \\ 3 \end{bmatrix}$

$\mathbf{C} = [1 \quad 6]$ $\quad \mathbf{D} = [1]$

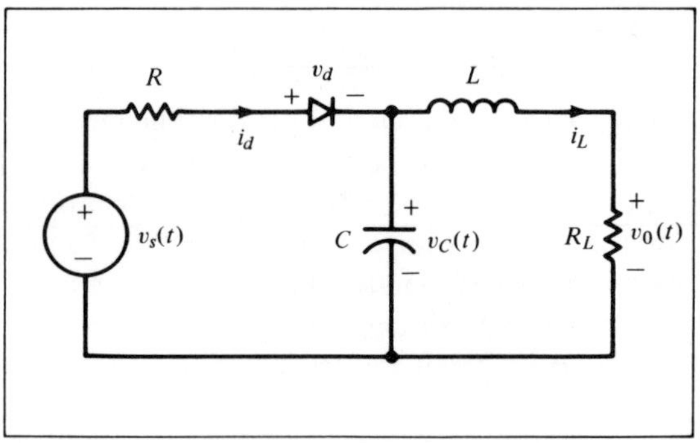

FIGURE P 13.8

13.13 In the network of Figure P 13.8 above the diode is described by the two equations

$$i_d = \dot{q} + I_s(e^{\lambda v_d} - 1)$$

$$v_d = \frac{1}{\lambda} \ln\left(1 + \frac{q}{\tau I_s}\right)$$

where q represents the charged stored in the diode when it is conducting. Write the state and output equations if $v_s(t)$ is a voltage source and $v_0(t)$ and $v_L(t)$ are the desired responses.

Problem 13.13 and related problems are solved in the Instructor's Manual using the IBM SSP integration package.

REFERENCES AND ADDITIONAL READING

13.1 C. A. DESOER AND E. S. KUH, *Basic Circuit Theory*. New York: McGraw-Hill Book Company, Inc., 1969, Chapter 10.

14

Transient Analysis by Computer

14.1 Introduction

Traditionally, the analysis of a nonlinear dynamic network is performed by linearizing the network equations and then using one of the methods of this text applicable to linear network analysis. Example 11.13 is an illustration of this procedure. This approach:

1. Usually permits a hand solution with associated insight into the dependence of time constants (α) and oscillation frequencies (ω_d) on element values (see Example 11.13).

2. Is relatively inexpensive to perform by computer, compared with a direct solution of the nonlinear differential network equations.

However, there is an increasing tendency in engineering problem solving to obtain the most accurate solution possible, thus removing the uncertainty about the quality of linearizing approximations. The speed of current and next-generation computers and the sophistication of modern computing algorithms suggest the feasibility of such analysis. Another compelling argument for transient analysis using nonlinear models is that certain types of circuits depend on nonlinear effects for their operation (for example, the digital circuits of Chapter 3). For these, linearized models are far less reliable and are occasionally misleading.

In this chapter, we study computer-oriented methods that can be applied both to linear and nonlinear dynamic network analysis. As we will see, these methods still involve approximations, but the error in this approximation *can be made as small as desired* at a cost of additional computing effort.

14.2 Analysis of Linear Dynamic Networks

Discretization

An analytic solution involving exponentials, sinusoids, and so on, is not possible when the equations are nonlinear. Instead, we will obtain from a computer solution a *table of values* of the solution variable, which we represent graphically in Figure 14.1. This process of representing a continuous variable by a sample of its values is termed *discretization*. That is, the variable $v(t^i)$, $i = 0, 1, \ldots,$ is a discretized version of $v(t)$.

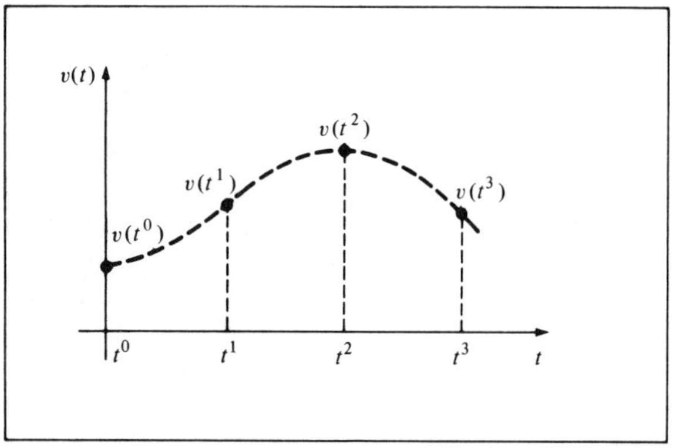

FIGURE 14.1 Example of discretized variable.

Derivative Approximation

The numerical solution of a differential equation $\dot{v} = f(v)$ is developed as follows. Given an initial condition $v(t^0)$, we solve successively for

$$v(t^1) = g_1(v(t^0))$$
$$v(t^2) = g_2(v(t^0), v(t^1))$$
$$v(t^3) = g_3(v(t^0), v(t^1), v(t^2)) \qquad (14.1)$$
$$\vdots$$

That is, the solution is developed with increasing time, each new solution point $v(t^{n+1})$ being a function g_i of one or more of the solution points already found. Because the differential equation at any time t^{n+1} relates only variables at that time point [that is, $\dot{v}(t^{n+1}) = f(v(t^{n+1}))$], some "glue" is needed relating successive solution points to obtain the relationships promised in Equation (14.1). This relationship is provided by replacing the derivative operator $\dot{v}(t^{n+1})$ with an approximate numerical formula known as an *integration rule.*

Assume that we are at time t^n in the solution process, and that we know the solution value $v(t^n)$. We wish to find $v(t^{n+1})$. Let us approximate the derivative at t^{n+1} by

$$\dot{v}(t^{n+1}) \cong \frac{v^{n+1} - v^n}{t^{n+1} - t^n} \triangleq \dot{v}^{n+1} \qquad (14.2)$$

The notations v^{n+1} and v^n represent *approximations* to the true values $v(t^{n+1})$, $v(t^n)$ which we wish to find. If the *time step* $T = t^{n+1} - t^n$ is constant for all n, then we can write Equation (14.2) as

$$\dot{v}^{n+1} = \frac{v^{n+1} - v^n}{T} \qquad (14.3)$$

This formula is known as the *backward Euler* approximation because the derivative approximation uses past values of v. [Contrast this with the *forward Euler* approximation of Equation (13.39).]

Let us apply this approximation to the simple problem of finding $v_C(t)$ in Figure 14.2. The differential equation is

$$\frac{dv_C(t)}{dt} = \frac{1}{RC}\,(-v_C(t) + I_sR) \qquad (t \geqq 0) \qquad (14.4)$$

Discretization requires that we solve

$$\dot{v}_C(t)\big|_{t=t^{n+1}} = \frac{1}{RC}\,(-v_C(t) + I_sR)\big|_{t=t^{n+1}} \qquad (n = 0, 1, \ldots)$$

Replacing $\dot{v}_C(t^{n+1})$ by its approximate form in Equation (14.3) gives

$$\frac{v_C{}^{n+1} - v_C{}^n}{T} = \frac{1}{RC}\,(-v_C{}^{n+1} + I_sR)$$

or

$$v_C{}^{n+1} = \frac{v_C{}^nRC + I_sRT}{T + RC} \qquad (n = 0, 1, \ldots) \qquad (14.5)$$

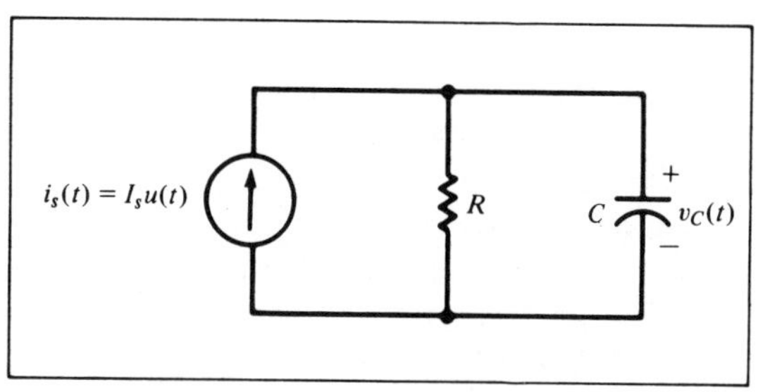

FIGURE 14.2 Simple *RC* network example.

Substituting $n = 0$ initially, we solve successively

$$v_C^1 = \frac{v_C^0 RC + I_s RT}{T + RC} = g_1(v_C^0)$$

$$v_C^2 = \frac{v_C^1 RC + I_s RT}{T + RC} = g_2(v_C^1)$$

$$v_C^3 = \frac{v_C^2 RC + I_s RT}{T + RC} = g_3(v_C^2)$$

$$\vdots$$

as indicated in Equation (14.1).

The analytic solution of Equation (14.4) with $v_c^0 = 0$, $I_s = 1$, and $RC = 1$ is given by

$$v_C(t) = 1 - e^{-t} \qquad (t \geqq 0) \qquad (14.6)$$

A comparison of this analytic solution with the numerical solution using Equation (14.5) is shown in Figure 14.3 and Table 14.1. In Figure

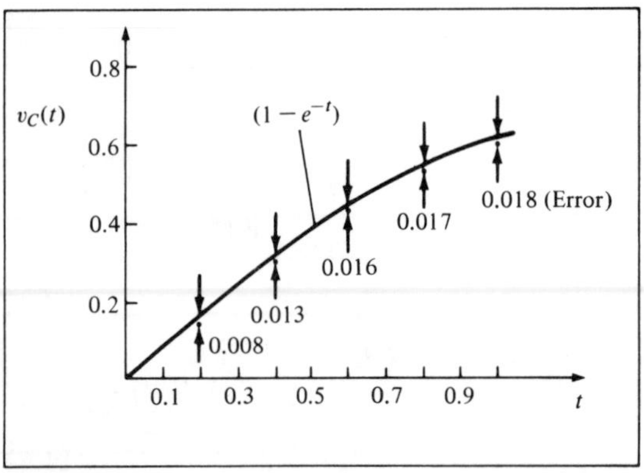

FIGURE 14.3 Comparison of exact and approximate solution ($T = 0.1$).

14.3, the error in the numerical solution is displayed versus time; in Table 14.1, the error at $t = 1$ is displayed versus the step size T. These results show a number of interesting properties of this derivative approximation.

1. The error in the solution is seen in Table 14.1 to decrease *linearly* with T. Thus, to increase the accuracy by a factor of 10 requires 10 times the number of steps and 10 times the computation effort.

TABLE 14.1
Comparison of Analytic and Numerical Solutions at $t = 1$. $(1 - e^{-1}) = 0.632121$

Step Size (T)	Calculated Value v_C^n	Error $(1 - e^{-1}) - v_C^n$
0.1	0.614456	0.017665
0.05	0.623111	0.009010
0.02	0.628472	0.003649
0.01	0.630289	0.001832
0.005	0.631203	0.000918
0.002	0.631753	0.000368
0.001	0.631937	0.000184

2. The error is always positive in Figure 14.3. This point is amplified in Figure 14.4 where the

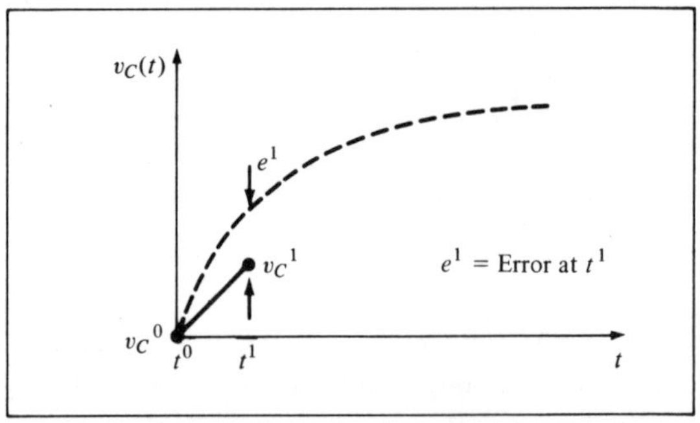

FIGURE 14.4 Error with backward Euler formula.

first solution value v_C^1 is shown to lie *under* the rising exponential characteristic. This results in the approximation being *stable* in a numerical sense, because if the successive values were to "overshoot" the exponential solution, the accumulated errors could become prohibitively large.[1]

Companion Models

We have often used element models to represent a solution procedure and then have

[1] See the forward Euler approximation in Figure 13.8. A mathematical justification of this observation about the stability of the backward Euler approximation is beyond the scope of this text. See reference 14.1.

formulated network equations directly from the resultant network models. The companion model of Chapter 5 and the impedance models of Chapters 10 and 12 are examples. This procedure is valid because any set of network equations are formulated from KVL and KCL equations (which do not change form in a solution procedure) and the element relationships; therefore, transforming only the latter suffices in representing the solution process.

Similarly, the KCL and KVL equations do not change form in the discretization process used to solve Equation (14.4). For example, a KCL equation such as $i_j(t) + i_k(t) = 0$ becomes simply $i_j^{n+1} + i_k^{n+1} = 0$. The element relationships are routinely derived from the derivative approximation of Equation (14.3) as follows:

$$\text{Resistor:} \quad v^{n+1} = Ri^{n+1} \tag{14.7}$$

$$\text{Capacitor:} \quad i^{n+1} = \frac{C}{T}(v^{n+1} - v^n) \tag{14.8}$$

$$\text{Inductor:} \quad v^{n+1} = \frac{L}{T}(i^{n+1} - i^n) \tag{14.9}$$

We can obtain network models for these elements by observing that all currents and voltages denoted $(\ )^{n+1}$ are unknowns at the nth time step; all variables denoted $(\ )^n$ are known. Thus, we can regard Equation (14.8) as describing a resistive model relating v^{n+1} and i^{n+1} by writing it in the form

$$i^{n+1} = \underbrace{\frac{C}{T} v^{n+1}}_{\substack{\text{Current}\\ \text{through resistor}}} + \underbrace{\frac{C}{T} v^n}_{\substack{\text{Current}\\ \text{through source}}} \tag{14.10}$$

As an example of the use of companion models, consider again the simple RC network of Figure 14.2, which is repeated in Figure 14.5(a). We construct a companion network model with the same structure as the original network, but with the companion model substituted for all elements. To facilitate analysis on a node basis, the source is transformed, giving the network of Figure 14.5(c). Analysis of the companion network of Figure 14.5(c)

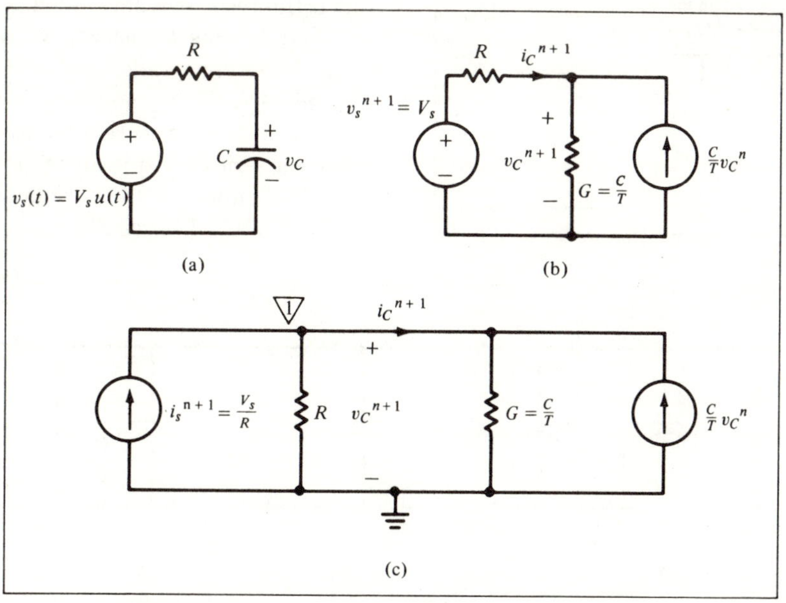

FIGURE 14.5 Steps in evolution of companion model.

yields

$$v_C^{n+1} = \frac{\dfrac{V_s}{R} + \dfrac{C}{T} v_C^n}{\dfrac{1}{R} + \dfrac{C}{T}} \qquad (n = 0, 1, \ldots) \tag{14.11}$$

which is the same as Equation (14.5) if $V_s = I_s R$.

APPLICATION

RC Transmission Line

A common method of producing a resistor in an integrated circuit is shown in Figure 14.6.

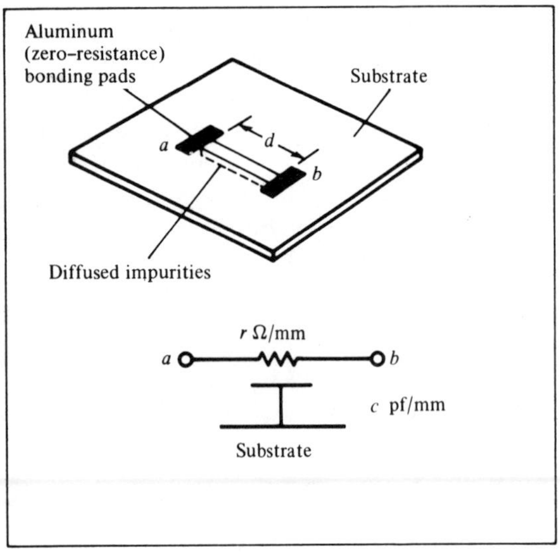

FIGURE 14.6 Diffused resistor and symbol.

Impurities such as boron or phosphorus are *diffused* into a semiconductor material called a *substrate*. This diffusion is performed through a mask, producing a long, thin line (length d) of impurities that will act as a resistor under the proper circuit design conditions. This resistor is characterized by a

$$\text{Resistance per unit length} = r\ \Omega/\text{mm}$$

The proximity of this distributed resistor to the substrate produces a capacitive effect between them, described by a

$$\text{Capacitance per unit length} = c\ \text{pf/mm}$$

Therefore, if we apply a step voltage at one end of the resistor, the step does not immediately appear at the other end. Instead, this signal is delayed until the distributed capacitance along the line is charged to the value of the step voltage. The signal which then appears at the other end of the line will not have the appearance of a step, but will have an exponential-like rise to an asymptotic value. This transient may or may not be significant, depending on the other transient effects in the circuit in which the resistor is imbedded.

To study this transient, we model the *distributed* resistor by a *lumped* model as shown in Figure 14.7. This model is composed of n "π-sections"; each section models a part of the line by a series resistor and two shunt capacitors. The values of these elements are chosen so that the total series resistance of the line is

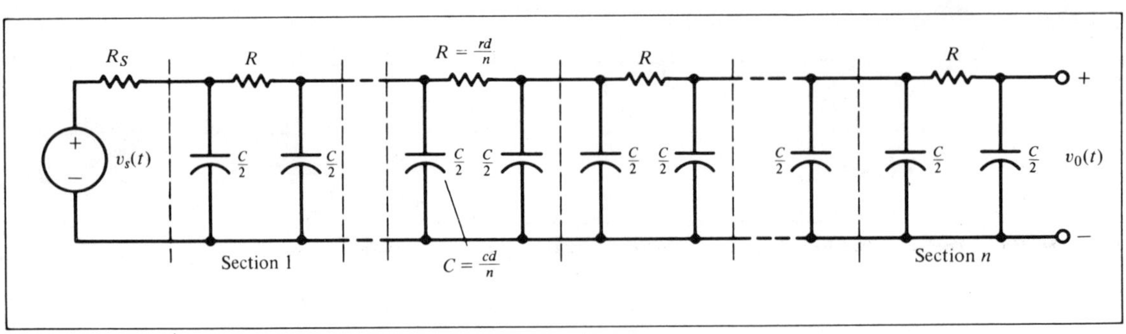

FIGURE 14.7 Lumped approximation to *RC* line.

$$R_T = \sum_{i=1}^{n} R = \sum_{i=1}^{n} \frac{rd}{n}$$

$$= rd = \left(\frac{\text{resistance}}{\text{unit length}}\right)(\text{length}) \qquad (14.12)$$

and the total capacitance is

$$C_T = \sum_{i=1}^{n} \left(\frac{C}{2} + \frac{C}{2}\right) = \sum_{i=1}^{n} C = \sum_{i=1}^{n} \frac{cd}{n}$$

$$= cd = \left(\frac{\text{capacitance}}{\text{unit length}}\right)(\text{length}) \qquad (14.13)$$

This total capacitance can be viewed as the capacitance that would exist between the line and the substrate if *both* were perfect conductors (that is, $R = 0$).

The companion model of a typical (rth) section is shown in Figure 14.8. The node equations of this section are

$$\begin{bmatrix} \frac{C}{2T} + \frac{1}{R} & -\frac{1}{R} \\ -\frac{1}{R} & \frac{C}{2T} + \frac{1}{R} \end{bmatrix} \begin{bmatrix} v_r^{n+1} \\ v_{r+1}^{n+1} \end{bmatrix} = \begin{bmatrix} \frac{C}{2T} v_r^n \\ \frac{C}{2T} v_{r+1}^n \end{bmatrix} \qquad (14.14)$$

When these sectional equations are put together and the effects of R_s and v_s are included, the node equations of the entire companion model have the form exemplified in Equation (14.15) for $r = 4$.

$$\begin{bmatrix} \frac{1}{R_s} + \frac{1}{R} + \frac{C}{2T} & -\frac{1}{R} & 0 & 0 & 0 \\ -\frac{1}{R} & \frac{2}{R} + \frac{C}{T} & -\frac{1}{R} & 0 & 0 \\ 0 & -\frac{1}{R} & \frac{2}{R} + \frac{C}{T} & -\frac{1}{R} & 0 \\ 0 & 0 & -\frac{1}{R} & \frac{2}{R} + \frac{C}{T} & -\frac{1}{R} \\ 0 & 0 & 0 & -\frac{1}{R} & \frac{C}{2T} \end{bmatrix} \begin{bmatrix} v_1^{n+1} \\ v_2^{n+1} \\ v_3^{n+1} \\ v_4^{n+1} \\ v_5^{n+1} \end{bmatrix} = \begin{bmatrix} v_s^{n+1}/R_s + \frac{C}{2T} v_1^n \\ \frac{C}{T} v_2^n \\ \frac{C}{T} v_3^n \\ \frac{C}{T} v_4^n \\ \frac{C}{2T} v_5^n \end{bmatrix} \qquad (14.15)$$

The flow chart of a computer program (TNET1) to solve Equation (14.15) at each time step is given in Figure 14.9. The Fortran program is given in Figure 14.10(a). This program is similar in structure to the DNET program, in that subroutines are provided to:

1 Read the input data (subroutine READ)
2. Set up the node equations (subroutine NODEQ).
3. Solve the node equations (subroutine REDUCE).
4. Print the voltage across the designated output nodes (subroutine OUTPUT), at each time step.

The node matrix is constructed from the companion network model by choosing values for r, c, d, and the step size T. In particular, choosing

$$r = 1000\ \Omega/\text{mm}$$

$$c = 1.2\ \text{pf/mm}$$

$$d = 1.5\ \text{mm}$$

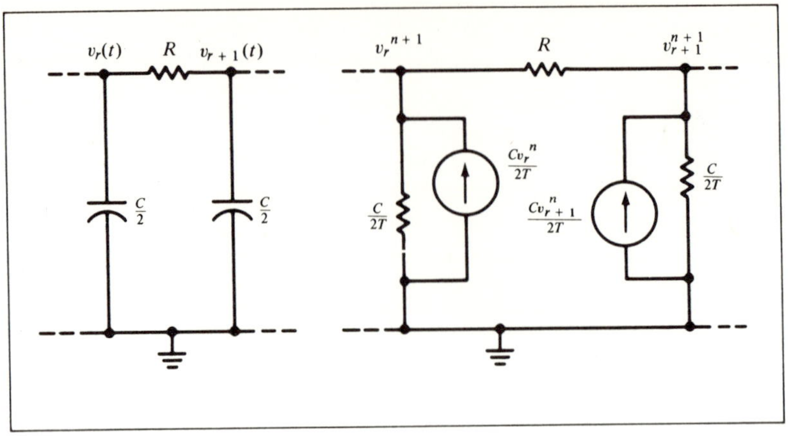

FIGURE 14.8 rth section and companion model.

$$v_s = 1 \text{ volt}$$

$$R_s = 500\ \Omega$$

for input data results in the calculated response $v_0(t)$ of Figure 14.10(c) for $0 \leqq t \leqq 10^{-8}$ sec. The output is observed to achieve 90 percent of its final value in 5×10^{-9} sec or 5 nsec. This could be a significant delay in a high-speed switching circuit.

14.3 Analysis of Nonlinear Dynamic Networks

Introduction

As we have seen in Chapters 11 and 12, powerful analytical methods exist for the analysis of linear dynamic networks. These methods do not suffer from the inaccuracies evident in Table 14.1. We cannot, then, justify our present approach if we restrict our attention to linear networks. Our ambition is to find the transient behavior of nonlinear networks.

As in Section 14.2, we will seek to find companion network models that reduce the problem to the analysis of a linear resistive network. A summary of our accomplishments so far in developing companion network models is shown in Figure 14.11. Unfortunately, the simple network of Figure 14.12(a) is beyond our present capability, because it contains both nonlinear resistive elements (a diode) and linear energy storage elements (a capacitor). However, the techniques for iterative analysis of resistive networks may be combined with those for step-by-step transient analysis of linear dynamic networks to yield an algorithm for solving such a network.

The development of our computing algorithm proceeds as follows. Ignoring the presence of nonlinear resistive elements, we replace energy storage elements by approximate resistive models as we did in Section 14.2. Although the resulting network is nonlinear and so cannot be solved directly, it still offers an approximation to the original network. That is, as $T \to 0$, the derivative approximation for capacitors

$$C \left.\frac{dv}{dt}\right|_{t=t^{n+1}} \cong C\,\frac{(v^{n+1} - v^n)}{T}$$

becomes better, and in the limit it becomes exact. Because this is the *only* approximation involved in the resistive model, the model of Figure 14.12(c) may be made to represent the original network as closely as desired in the vicinity of t^n by choosing T sufficiently small.

The solution of the nonlinear resistive network now proceeds iteratively using Newton iteration. The solution becomes notationally cumbersome, because an iteration counter

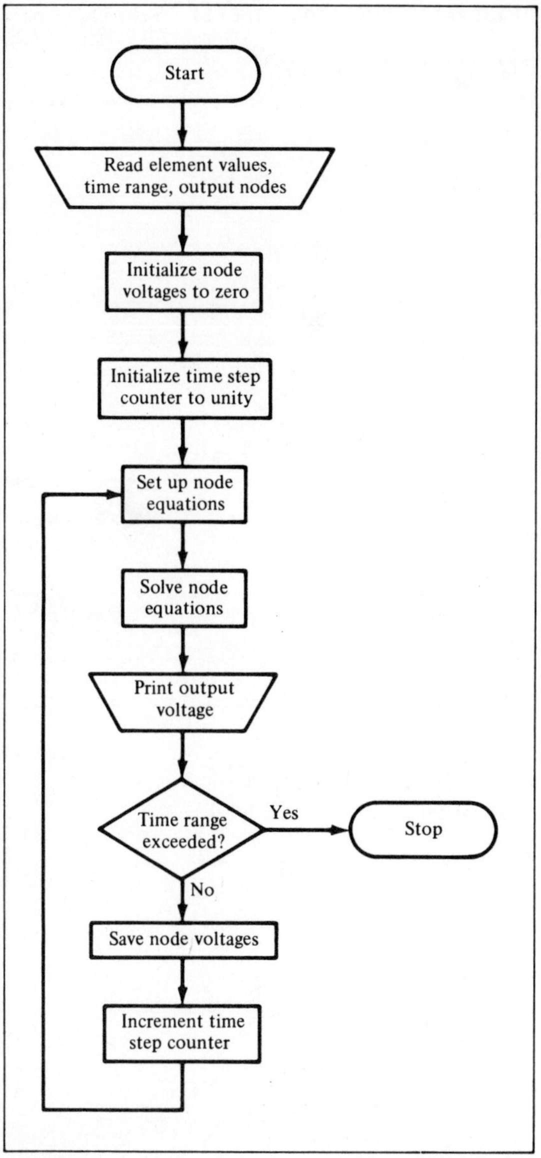

FIGURE 14.9 Flow chart of program to solve Equation (14.15).

(m = 0, 1, . . .) must now be used to indicate which parts of the companion model are updated at the $(m + 1)$st Newton iteration and which are changed less frequently at each time step (indicated by n). We handle this notational problem by using a double superscript for variables which are dependent both on m and n. Thus, the companion model representing both the discretization and linearization of the original network is shown in Figure 14.12(c). The Newton iteration formula for this network is then given by

$$v_d^{n+1,m+1} = \frac{C\dfrac{v_C^n}{T} + \dfrac{v_s^{n+1}}{R} - i_d^{n+1,m} + G_d^{n+1,m} v_d^{n+1,m}}{\dfrac{1}{R} + G_d^{n+1,m} + \dfrac{C}{T}} \qquad (14.16)$$

where $v_d = v_C$.

The solution proceeds in the following manner. We assume that the capacitor voltage v_C^n is known at $t = 0$ $(n = 0)$. A new time step is initiated by evaluating $v_s^{n+1} = v_s^1$. The diode voltage is now calculated by iteration: because the capacitor voltage and the diode voltage are equal, we take as our first guess $(m = 0)$ in the iteration $v_d^{1,0} = v_C^0$. We can then calculate

$$G_d^{1,0} = I_s \lambda e^{\lambda v_d^{1,0}} \qquad (14.17)$$

$$i_d^{1,0} - G_d^{1,0} v_d^{1,0} = I_s(e^{\lambda v_d^{1,0}} - 1) - G_d^{1,0} v_d^{1,0} \qquad (14.18)$$

The right side of Equation (14.16) is now known, and we can calculate $v_d^{1,1}$. The iteration is continued in this manner until the diode voltage does not change appreciably between iterations, that is, until

$$(v_d^{1,M} - v_d^{1,M-1})^2 \leqq \epsilon$$

where M is the number of the final iteration. The first time step has now been taken. To proceed, we update sources v_s^{n+1} and Cv_C^n/T. The final value of the diode voltage from the previous time step is used as the initial guess for the new iteration; that is,

$$v_d^{n+1,0} = v_d^{n,M}$$

for each n.

```
C****  TRANSIENT ANALYSIS OF INTEGRATED RESISTOR (TNET1)
      IMPLICIT REAL*8(A-H,O-Z)
      REAL*8 G(10,10),I(10),V(10),VSAV(10)
      DIMENSION NODOUT(2,5)
      NDIM=10
    1 N=0
C****  READ INPUT DATA
      CALL READ(NN,NODOUT,NOUT,T,TMAX)
      TIME=0.D0
      DO 6 J=1,NN
    6 V(J)=0.D0
      GO TO 2
C****  SET UP NODE EQUATIONS
    3 CALL NODEQ(G,V,I,VSAV,TIME,T,NN,NDIM)
C****  SOLVE NODE EQUATIONS
      CALL REDUCE(G,NN,I,NDIM)
C****  V(J)-I(J) IS THE INCREMENT IN THE JTH NODE VOLTAGE
      DO 10 J=1,NN
   10 V(J)=I(J)
C****  CALCULATE AND PRINT OUTPUT VOLTAGES
      IF(TIME.GE.TMAX)STOP
C****  STORE NODE VOLTAGES FOR UPDATING CAPACITORS
    2 DO 5 J=1,NN
    5 VSAV(J)=V(J)
      CALL OUTPUT(NODOUT,NOUT,V,TIME)
      N=N+1
      TIME=TIME+T
      GO TO 3
    8 WRITE(6,9)
    9 FORMAT(' TOO MANY ITERATIONS')
      STOP
      END
      SUBROUTINE READ(NN,NODOUT,NOUT,T,TMAX)
      IMPLICIT REAL*8(A-H,O-Z)
      DIMENSION NODOUT(2,1)
      COMMON R,C,D,VS,RS
      READ(5,1)R,C,D,VS,RS,T,TMAX
      READ(5,3)NN,NOUT,(NODOUT(1,J),NODOUT(2,J),J=1,NOUT)
      WRITE(6,2)R,C,D,VS,RS
      RETURN
    1 FORMAT(6F10.0)
    2 FORMAT(' R =',D11.4,' C =',D11.4,' D =',D11.4/
     &' VS =',D11.4,' RS =',D11.4)
    3 FORMAT(20I2)
      END
      SUBROUTINE NODEQ(G,V,I,VSAV,TIME,T,NN,NDIM)
      IMPLICIT REAL*8 (A-H,O-Z)
      REAL*8 V(1),G(NDIM,1),I(1),VSAV(1)
      COMMON R,C,D,VS,RS
      DATA INIT/0/
C****  ZERO NODE MATRIX AND CURRENT VECTOR
      DO 1 K=1,NN
      I(K)=0.D0
      DO 1 J=1,NN
    1 G(J,K)=0.D0
      IF(INIT.NE.0)GO TO 2
```

FIGURE 14.10 (a) Program to solve lumped *RC* transmission line model of integrated resistor; (b) input data for program of (a); (c) plot of step response of integrated resistor.

Complete Solution of Nonlinear Dynamic Networks

Most dynamic network analysis is carried out beginning from an equilibrium state. An example is shown in Figure 14.13 where, for $t < 0$, the network is assumed to have attained an equilibrium state in which no voltages or currents are changing appreciably. The response to the step in $v_s(t)$ then begins from this value of $v_C(0)$. We intend to show that both the

```
      NSECT=NN-1
      CL=C*D/(NSECT*2.D0*T)
      GL=NSECT/(R*D)
C****  LOAD NODE MATRIX
    2 G(1,1)=1.D0/RS
      I(1)=VS/RS
      DO 3 J=1,NSECT
      G(J,J)=G(J,J)+CL+GL
      G(J,J+1)=-GL
      G(J+1,J)=-GL
      G(J+1,J+1)=G(J+1,J+1)+CL+GL
      I(J)=I(J)+CL*VSAV(J)
    3 I(J+1)=I(J+1)+CL*VSAV(J+1)
      INIT=1
      RETURN
      END
      SUBROUTINE OUTPUT(NODOUT,NOUT,V,TIME)
      IMPLICIT REAL*8 (A-H,O-Z)
      DATA INIT/0/
      DIMENSION NODOUT(2,1),V(1),VOUT(5)
      IF(INIT.NE.0)GO TO 3
      WRITE(6,4)(NODOUT(1,J),NODOUT(2,J),J=1,NOUT)
    3 DO 2 I=1,NOUT
      VOUT(I)=0.D0
      N1=NODOUT(1,I)
      N2=NODOUT(2,I)
      IF(N1.NE.0)VOUT(I)=VOUT(I)+V(N1)
      IF(N2.NE.0)VOUT(I)=VOUT(I)-V(N2)
    2 CONTINUE
      WRITE(6,1)TIME,(VOUT(I),I=1,NOUT)
      INIT=1
      RETURN
    1 FORMAT(6D12.3)
    4 FORMAT('    TIME      ',5(I2,' -',I2,' NODES'))
      END
      SUBROUTINE REDUCE(A,N,B,NDIM)
      IMPLICIT REAL*8(A-H,O-Z)
      DIMENSION A(NDIM,1),B(1)
C****  THIS SUBROUTINE SOLVES A*X=B BY GAUSS ELIMINATION
C****  N IS THE MATRIX SIZE; X IS RETURNED IN B; A IS DESTROYED
C****  CONVERT A TO UPPER TRIANGULAR FORM
      NM1=N-1
      IF(N.EQ.1)GO TO 8
      DO 1 J=1,NM1
      IJ=J+1
      DO 4 K=IJ,N
    4 A(J,K)=A(J,K)/A(J,J)
      B(J)=B(J)/A(J,J)
      DO 1 I=IJ,N
      ALPHA=A(I,J)
      DO 2 K=IJ,N
    2 A(I,K)=A(I,K)-ALPHA*A(J,K)
```

FIGURE 14.10 (Continued)

equilibrium state for $t < 0$ and the transient response for $t \geqq 0$ can be calculated from the same equation.

From Chapter 8, we recall that the dc analysis of a network of inductors and capacitors begins by replacing these elements with short and open circuits, respectively. With this in mind, return to the iteration equation for the transient analysis, given by Equation (14.16). Setting $C = 0$ and $v_s^{n+1} = V_1$, we have for $t < 0$

$$v_d^{n+1,m+1} = \frac{\dfrac{V_1}{R} - i_d^{n+1,m} + G_d^{n+1,m} v_d^{n+1,m}}{\dfrac{1}{R} + G_d^{n+1,m}} \qquad (14.19)$$

```
    1 B(I)=B(I)-ALPHA*B(J)
    8 B(N)=B(N)/A(N,N)
C****  PERFORM BACK SUBSTITUTION
      IF(N.LT.2)GO TO 7
      DO 3 J=2,N
      LAST=N-J+1
      LASTP1=LAST+1
      DO 3 I=LASTP1,N
    3 B(LAST)=B(LAST)-A(LAST,I)*B(I)
    7 RETURN
      END
```

(a)

```
1000.     1.2D-12,  1.5       1.        500.       2.D-10,
1.D-8,
 5 1 5 0
```

(b)

FIGURE 14.10 (Continued)

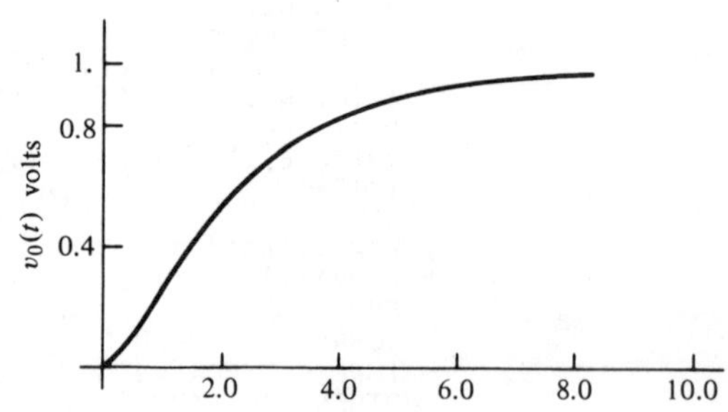

(c)

FIGURE 14.10 (Continued)

This equation defines the Newton iteration associated with the dc analysis of a simple resistor-diode network. Note that the superscript $n + 1$ is superfluous and can be dropped, because Equation (14.19) no longer contains a term to be updated with time. When this iteration in m converges, then Equation (14.16) is used for transient response calculations ($t \geqq 0$), beginning with the $v_C{}^0$ determined from the dc iteration.

APPLICATION

Rectifier-Filter Example

The circuit of Figure 14.14(a) is a rectifier-filter that converts a signal (usually a sinusoid) with a zero average value to a nearly constant (dc) value. Its performance can be understood by removing the capacitor. When the input voltage is positive, the diode conducts and nearly all of $v_s(t)$ appears across the output as $v_O(t)$. The small distortion produced by the diode nonlinearity often can be ignored, resulting in the half-wave rectified sinusoid of Figure 14.14(c). When the capacitor is added, it charges when the diode is conducting and discharges through the 10000-Ω resistor when the diode is nonconducting. The time constants for charging (R_sC) and discharging (R_LC) usually differ by several orders of magnitude. When $R_LC \gg T_0$, the slow rate of discharge causes the output to "ripple" around a constant value as shown in Figure 14.14(d).

The companion model for this circuit is shown in Figure 14.15. The node equations are easily written as

$$\begin{bmatrix} \dfrac{1}{R_s} + G_d{}^{n+1,m} & -G_d{}^{n+1,m} \\ -G_d{}^{n+1,m} & G_d{}^{n+1,m} + \dfrac{1}{R_L} + \dfrac{C}{T} \end{bmatrix} \begin{bmatrix} v_1{}^{n+1,m+1} \\ v_2{}^{n+1,m+1} \end{bmatrix} = \begin{bmatrix} \dfrac{v_s{}^{n+1}}{R_s} - i_d{}^{n+1,m} + G_d{}^{n+1,m}v_d{}^{n+1,m} \\ \dfrac{C}{T}v_2{}^n + i_d{}^{n+1,m} - G_d{}^{n+1,m}v_d{}^{n+1,m} \end{bmatrix} \tag{14.20}$$

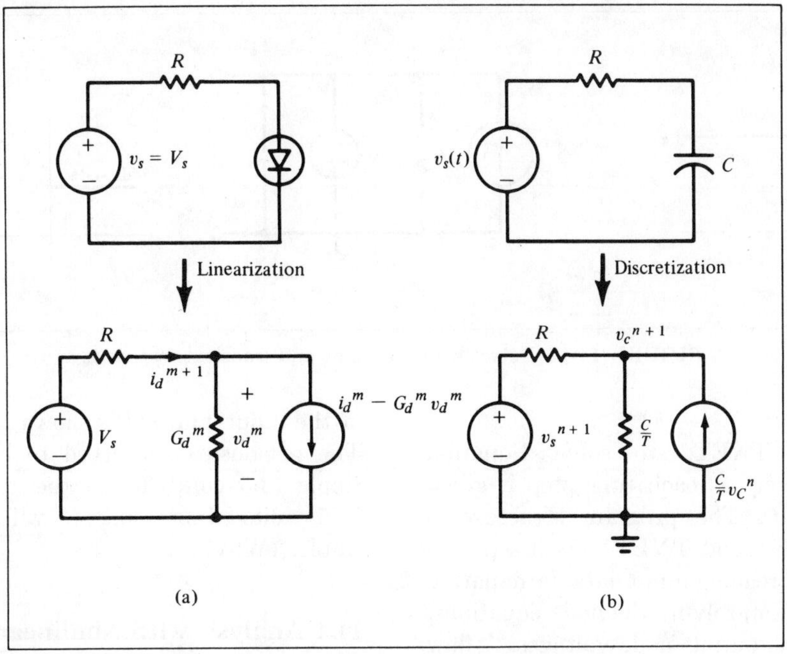

FIGURE 14.11 Companion models we have studied. (a) Nonlinear, dc; (b) linear, dynamic.

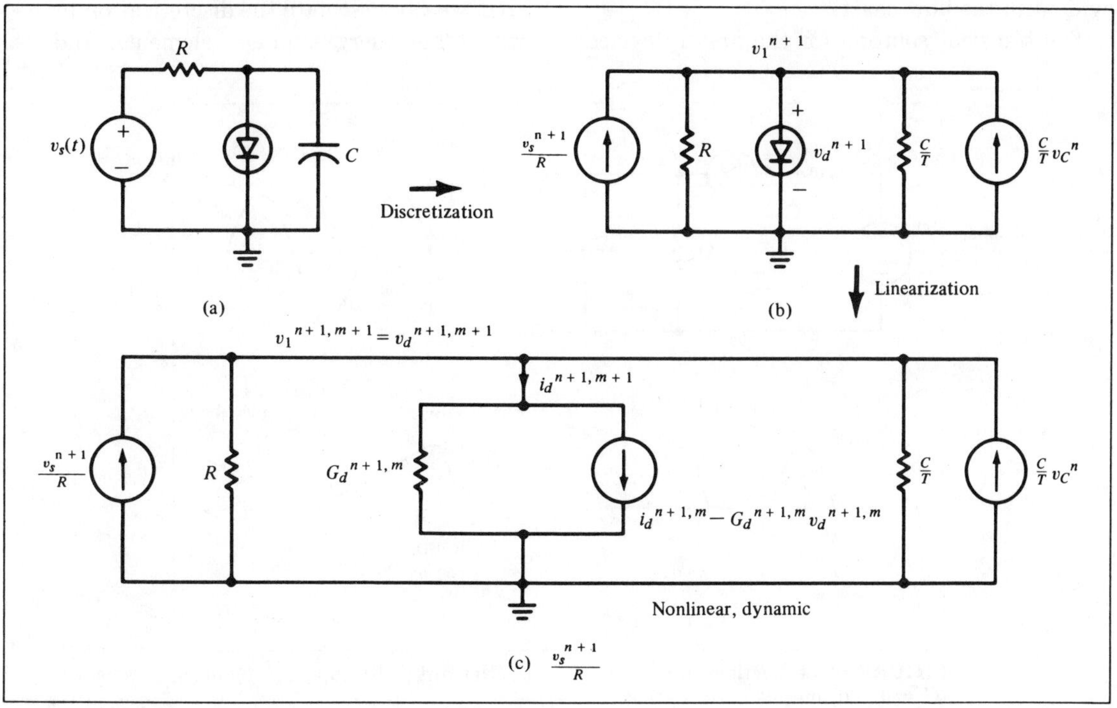

FIGURE 14.12 Combination of nonlinear and dynamic companion models.

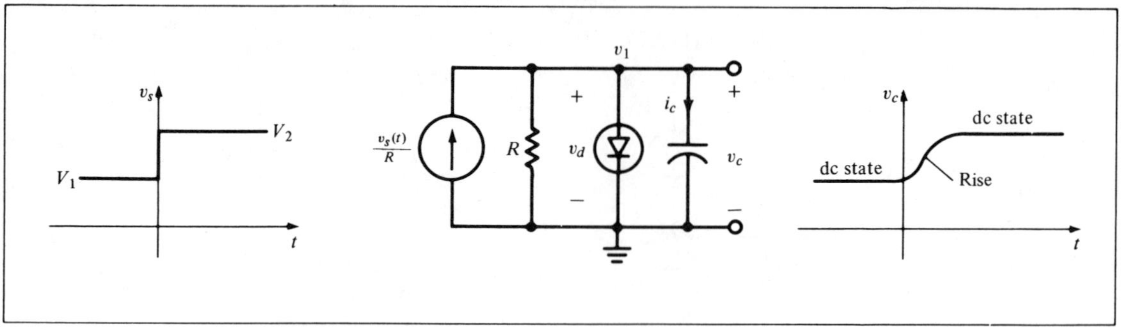

FIGURE 14.13 Network with both dc and transient responses.

where $v_d^{n+1,m} = v_1^{n+1,m} - v_2^{n+1,m}$.

A program TNET2 to solve Equation (14.20) iteratively at each time step is given in Figure 14.16. The program structure is similar to DNET and TNET1, with separate subroutines for reading input data, formulating the node equation, solving the node equations, and printing the output node voltages. A flow chart is shown in Figure 14.17. The presence of the Newton iteration loop within the loop formed by the time-domain discretization is evident in the flow chart.

The transient solution for the first $1\frac{1}{2}$ cycles of the input sinusoid is shown in Figure 14.18. The response is observed to have a sizeable ripple (3.5 volts) about the average value of 7.75 volts. Increasing C will decrease this ripple. (Why?)

14.4 Analysis with Nonlinear Dynamic Elements

Introduction

So long as the nonlinear elements are resistive, we can associate the discretization process with the energy-storage elements and the

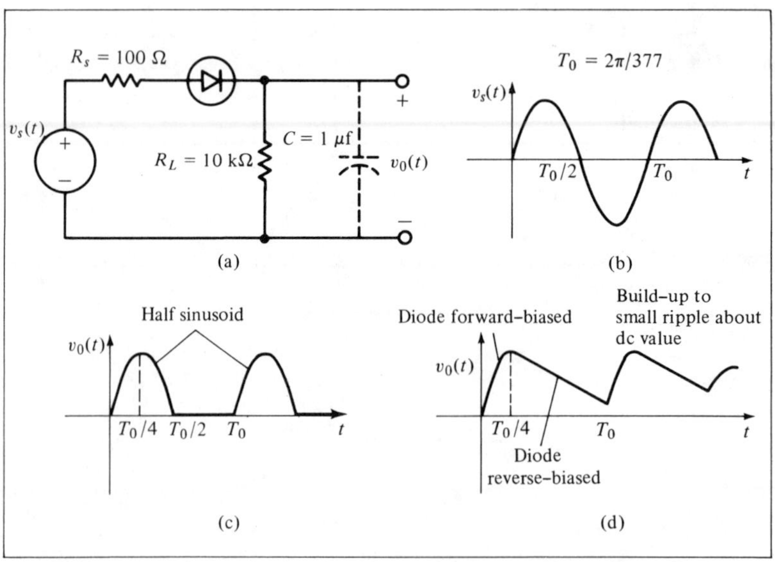

FIGURE 14.14 Rectifier-filter example. (a) Network; (b) input; (c) output ($C = 0$); (d) output ($C = 1\ \mu$f).

linearization process with the resistive elements. However, when capacitors (inductors) are nonlinear functions of charge (flux-linkage), both discretization and linearization must be performed on the same element. The procedure of the previous section still applies, however, as the following illustrates.

Consider a nonlinear capacitor described by

$$q = f(v) \qquad i = \frac{dq}{dt} \tag{14.21}$$

As before, we discretize time by using the backward Euler derivative approximation to dq/dt.

$$i^{n+1} = \frac{q^{n+1} - q^n}{T} \tag{14.22}$$

Here, q^n is known from the last time step. However, q^{n+1} must be formed through iteration by successive linearization around $q^{n+1,m}$, analogously to resistive nonlinearities.

$$q^{n+1,m+1} = q^{n+1,m} + \left(\frac{\partial q}{\partial v}\right)^{n+1,m} (v^{n+1,m+1} - v^{n+1,m})$$

Substituting $q^{n+1,m+1}$ for q^{n+1} in Equation (14.22), we have

$$\begin{aligned} i^{n+1,m+1} &= \left(\frac{1}{T}\frac{\partial q}{\partial v}\right)^{n+1,m} v^{n+1,m+1} \\ &\quad - \frac{1}{T}\left(q^n - q^{n+1,m} + \left(\frac{\partial q}{\partial v}\right)^{n+1,m} v^{n+1,m}\right) \\ &= \frac{1}{T} C(v^{n+1,m}) v^{n+1,m+1} - \frac{1}{T}(f(v^n) \\ &\quad - f(v^{n+1,m}) + C(v^{n+1,m}) v^{n+1,m}) \end{aligned} \tag{14.23}$$

where $C = \partial f/\partial v$.

Equation (14.23) describes the companion model of Figure 14.19. This model can therefore be inserted in place of the nonlinear capacitor in the network model from which the Newton iteration equations are developed. It is left as an exercise to show that when $q = Cv$, Equation (14.23) becomes identical to that for a linear capacitor.

Example 14.1

A simplified dynamic model for a real diode is shown in Figure 14.20. The companion model for the network and the iteration equation for

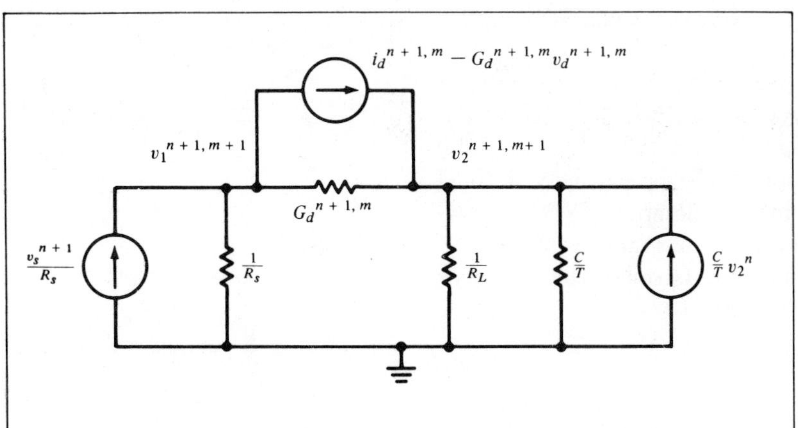

FIGURE 14.15 Companion model of rectifier-filter.

```
C****  TRANSIENT ANALYSIS OF RECT.-FILTER (TNET2)
      IMPLICIT REAL*8(A-H,O-Z)
      REAL*8 G(10,10),I(10),V(10),VSAV(10)
      DIMENSION NODOUT(2,5)
      NDIM=10
    1 M=0
      N=0
C****  READ INPUT DATA
      CALL READ(NN,NODOUT,NOUT,T,TMAX)
      TIME=0.D0
      DO 6 J=1,NN
    6 V(J)=0.D0
C****  SET UP NODE EQUATIONS
    3 CALL NODEQ(G,V,I,VSAV,TIME,T,NN,NDIM)
C****  SOLVE NODE EQUATIONS
      CALL REDUCE(G,NN,I,NDIM)
C****  V(J)-I(J) IS THE INCREMENT IN THE JTH NODE VOLTAGE
      M=M+1
      ERR=0.D0
      DO 2 J=1,NN
      ERR=ERR+(V(J)-I(J))**2
    2 V(J)=I(J)
C****  TEST CONVERGENCE CRITERION
      IF(ERR.GT.1.D-6.AND.M.LT.50)GO TO 3
      IF(M.EQ.50)GO TO 8
C****  CALCULATE AND PRINT OUTPUT VOLTAGES
      CALL OUTPUT(NODOUT,NOUT,V,TIME)
      IF(TIME.GE.TMAX)STOP
C****  STORE NODE VOLTAGES FOR UPDATING CAPACITORS
      DO 5 J=1,NN
    5 VSAV(J)=V(J)
      M=0
      N=N+1
      TIME=TIME+T
      GO TO 3
    8 WRITE(6,9)
    9 FORMAT(' TOO MANY ITERATIONS')
      STOP
      END
      SUBROUTINE READ(NN,NODOUT,NOUT,T,TMAX)
      IMPLICIT REAL*8(A-H,O-Z)
      DIMENSION NODOUT(2,1)
      COMMON RS,RL,C,VSM,W,VDI
      READ(5,1)RS,RL,C,VSM,W,VDI,T,TMAX
      READ(5,3)NN,NOUT,(NODOUT(1,J),NODOUT(2,J),J=1,NOUT)
      WRITE(6,2)RS,RL,C,VSM,W
      RETURN
```

FIGURE 14.16 (a) Program to solve rectifier-filter; (b) input data for program of (a).

the voltage v_d are to be developed. (Note that τ is in units of time, because it relates charge and current.)

The companion model is routinely derived up to the point of determining $G_C{}^{n+1,m} = C(v^{n+1,m})/T$ and $i_C{}^{n+1,m}$ of Figure 14.21.

From the companion model of the nonlinear capacitor, we can write

$$G_C{}^{n+1,m} = \frac{1}{T}\left(\frac{\partial f}{\partial v_d}\right)^{n+1,m}$$

$$= \frac{1}{T}\, C\,(v_d{}^{n+1,m})$$

$$= \frac{\tau}{T}\,\lambda I_s e^{\lambda v_d{}^{n+1,m}} + \frac{C_T}{T}\, v_d{}^{n+1,m}$$

$$i_C{}^{n+1,m} = \frac{1}{T}\,\left(f(v_d{}^n) - f(v_d{}^{n+1,m})\right)$$

$$+ G_C{}^{n+1,m} v_d{}^{n+1,m}$$

$$= \frac{1}{T}\,[\tau I_s(e^{\lambda v_d{}^n} - 1) + C_T v_d{}^n$$

$$- \tau I_s(e^{\lambda v_d{}^{n+1,m}} - 1)$$

$$- C_T v_d{}^{n+1,m}] + G_C{}^{n+1,m} v_d{}^{n+1,m}$$

```
    1 FORMAT(6F10.0)
    2 FORMAT(' RS =',D11.4,' RL =',D11.4,' C =',D11.4/
     1' VSM =',D11.4,' W =',D11.4)
    3 FORMAT(20I2)
      END
      SUBROUTINE NODEQ(G,V,I,VSAV,TIME,T,NN,NDIM)
      IMPLICIT REAL*8 (A-H,O-Z)
      REAL*8 G(NDIM,1),I(1),V(1),LAMDA,VSAV(1),INPUT
      COMMON RS,RL,C,VSM,W,VDI
      DATA INIT/0/
      DATA CSAT/1.D-9/,LAMDA/40.D0/
C****  ZERO NODE MATRIX AND CURRENT VECTOR
      DO 1 K=1,NN
      I(K)=0.D0
      DO 1 J=1,NN
    1 G(J,K)=0.D0
C****  CALCULATE DIODE PARAMETERS
      VDIODE=DMIN1(V(1)-V(2),.5D0)
      VS=VSM*DSIN(W*TIME)
C****  APPLY INITIAL DIODE VOLTAGE IF ON FIRST ITERATION
      IF(INIT.EQ.0)VDIODE=VDI
      DEX=DEXP(LAMDA*VDIODE)
      GD=LAMDA*CSAT*DEX
      CUR=CSAT*(DEX-1.D0)-GD*VDIODE
C****  LOAD NODE MATRIX
      G(1,1)=GD+1.D0/RS
      G(1,2)=-GD
      G(2,1)=-GD
      G(2,2)=GD+1.D0/RL
      IF(TIME.NE.0.D0)G(2,2)=G(2,2)+C/T
      I(1)=-CUR
      I(2)=CUR
      IF(TIME.EQ.0.D0)GO TO 2
      I(1)=I(1)+VS/RS
      I(2)=I(2)+C*VSAV(2)/T
    2 INIT=1
      RETURN
      END
      SUBROUTINE OUTPUT(NODOUT,NOUT,V,TIME)
      IMPLICIT REAL*8 (A-H,O-Z)
      DATA INIT/0/
      DIMENSION NODOUT(2,1),V(1),VOUT(5)
      IF(INIT.NE.0)GO TO 3
      WRITE(6,4)(NODOUT(1,J),NODOUT(2,J),J=1,NOUT)
    3 DO 2 I=1,NOUT
      VOUT(I)=0.D0
      N1=NODOUT(1,I)
      N2=NODOUT(2,I)
      IF(N1.NE.0)VOUT(I)=VOUT(I)+V(N1)
      IF(N2.NE.0)VOUT(I)=VOUT(I)-V(N2)
    2 CONTINUE
      WRITE(6,1)TIME,(VOUT(I),I=1,NOUT)
      INIT=1
      RETURN
    1 FORMAT(6D12.3)
    4 FORMAT('    TIME      ',5(I2,' -',I2,' NODES'))
```

FIGURE 14.16 (Continued)

The iteration equation is then given by

$$v_d^{n+1,m+1} = \frac{v_s^{n+1}G_s + i_C^{n+1,m} - i_d^{n+1,m} + G_d^{n+1,m}v_d^{n+1,m}}{G_s + G_d^{n+1,m} + G_C^{n+1,m}} \tag{14.24}$$

It is interesting to observe the response of such a diode to a step input. Typical values for a real diode are

$$\tau = 200 \text{ psec} \qquad I_s = 10^{-9} \text{ amps}$$
$$\lambda = 40 \text{ volts}^{-1}$$

With $R_s = 10^4\ \Omega$ and $v_s(t) = 10u(t)$, the volt-

```
      END
      SUBROUTINE REDUCE(A,N,B,NDIM)
      IMPLICIT REAL*8(A-H,O-Z)
      DIMENSION A(NDIM,1),B(1)
C****  THIS SUBROUTINE SOLVES A*X=B BY GAUSS ELIMINATION
C****  N IS THE MATRIX SIZE; X IS RETURNED IN B; A IS DESTROYED
C****  CONVERT A TO UPPER TRIANGULAR FORM
      NM1=N-1
      IF(N.EQ.1)GO TO 8
      DO 1 J=1,NM1
      IJ=J+1
      DO 4 K=IJ,N
    4 A(J,K)=A(J,K)/A(J,J)
      B(J)=B(J)/A(J,J)
      DO 1 I=IJ,N
      ALPHA=A(I,J)
      DO 2 K=IJ,N
    2 A(I,K)=A(I,K)-ALPHA*A(J,K)
    1 B(I)=B(I)-ALPHA*B(J)
    8 B(N)=B(N)/A(N,N)
C****  PERFORM BACK SUBSTITUTION
      IF(N.LT.2)GO TO 7
      DO 3 J=2,N
      LAST=N-J+1
      LASTP1=LAST+1
      DO 3 I=LASTP1,N
    3 B(LAST)=B(LAST)-A(LAST,I)*B(I)
    7 RETURN
      END
```

(a)

```
100.       30000.     .000001    10.        377.       0.
.0004      .08
 2 2 2 0 1 2
```

(b)

FIGURE 14.16 (Continued)

age $v_d(t)$ can be calculated using Equation (14.24). The results are shown in Figure 14.22 for a number of values of C_T ($0 \leqq C_T \leqq 0.02$ pf). For $C_T = 0$, the response is observed to be unlike the exponential characteristics that we have come to expect of single-time-constant networks. We observe $v_d(t)$ to have a very sharp rise, because the diode behaves essentially as an open circuit if $v_d < 0.2$. As C_T is increased, the familiar exponential characteristic returns as the R_sC_T time constant dominates the initial part of the response. These results would have been very difficult to obtain using the small-signal approximation. A direct solution of the circuit equations is necessary. The remainder of the response ($t >$ 10 psec) can be deduced by approximate methods as discussed in Problem 14.15.

Transistor Circuit Analysis

We have, after fourteen chapters, arrived at the point where we can conduct by hand or by the digital computer the analysis of nearly any form of electrical network. This methodology has been carefully developed, from our early experiences with Thevenin and Norton transformations, through the use of formal procedures for resistive and dynamic network analysis, and the notions of linearization and discretization as the bases for obtaining an approximate solution of nonlinear, dynamic problems. We have applied these procedures to the analysis of two principal classes of circuits—amplifiers and digital circuits. In particular, the simple transistor circuits of Figure 14.23 have been studied rather extensively.

It is fitting, as we terminate the branch-oriented analysis of networks, to expose the power and elegance of the methods available to us by solving several common practical problems:

1. The transient response of the inverter circuit of Figure 14.23 (a).

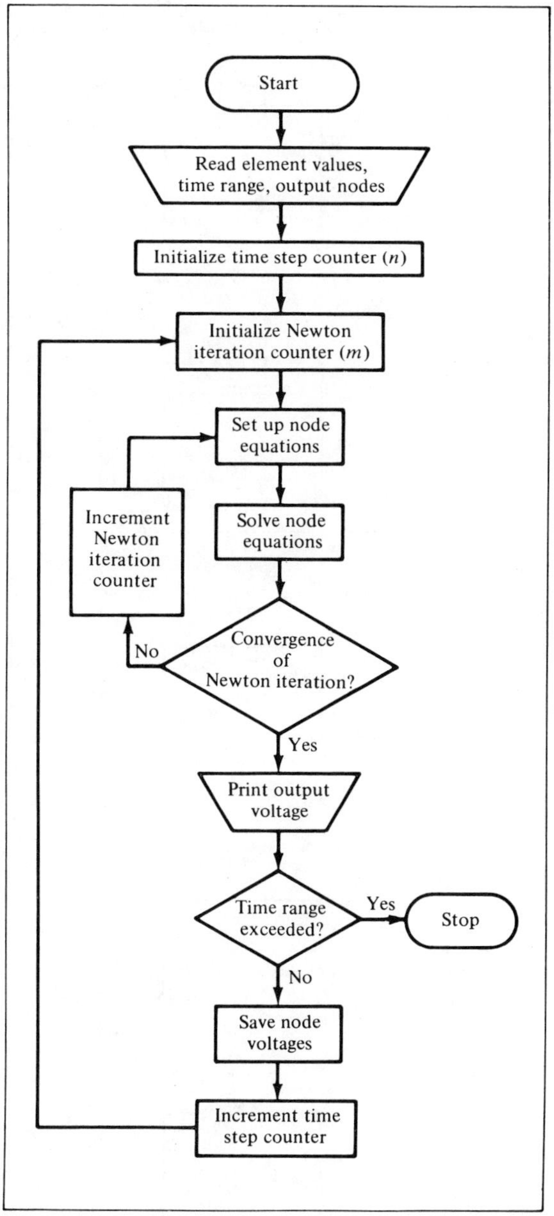

FIGURE 14.17 Flow chart for program of Figure 14.16.

2. The small-signal response of the amplifier circuit of Figure 14.23 (b).

As we will see, the methods of this text make the solution of these networks remarkably similar.

Before beginning, it is worthwhile examining Table 14.2, a summary of the *RLC* network elements models. As we have seen, these are the primitive representations from which we can build a set of network equations to perform any analysis.

The Transistor Inverter

From the model of the diode in Figure 14.20, we can deduce the nonlinear dynamic model of the transistor. Corresponding to the familiar emitter and collector diodes in the resistive model of the transistor, the transistor dynamic model has two nonlinear capacitors, as shown in Figure 14.23 (c), of the form

$$q = \tau I_s(e^{\lambda v} - 1) + C_T v$$

The capacitor C_f is associated with the emitter diode and C_r with the collector diode.

The linearized form of this equation yields two capacitors given by

$$C_f^* = \tau\lambda I_s e^{\lambda v_f *} + C_T$$

$$C_r^* = \tau\lambda I_s e^{\lambda v_r *} + C_T$$

where the "*" indicates that C and v in each case are functions of counters n and m. For simplicity it is assumed the diodes are identical. From these quantities and the linearized diode conductances of the form

$$G_f^* = \lambda I_s e^{\lambda v_f *}$$

$$G_r^* = \lambda I_s e^{\lambda v_r *}$$

we can construct almost by inspection the node matrix of the entire companion model. The result is shown in Figure 14.23 (e). The determination of the right-hand side is, of course, a more involved process which is left as an exercise.

A modified form of the INVRT program of

TABLE 14.2 Glossary of element models.

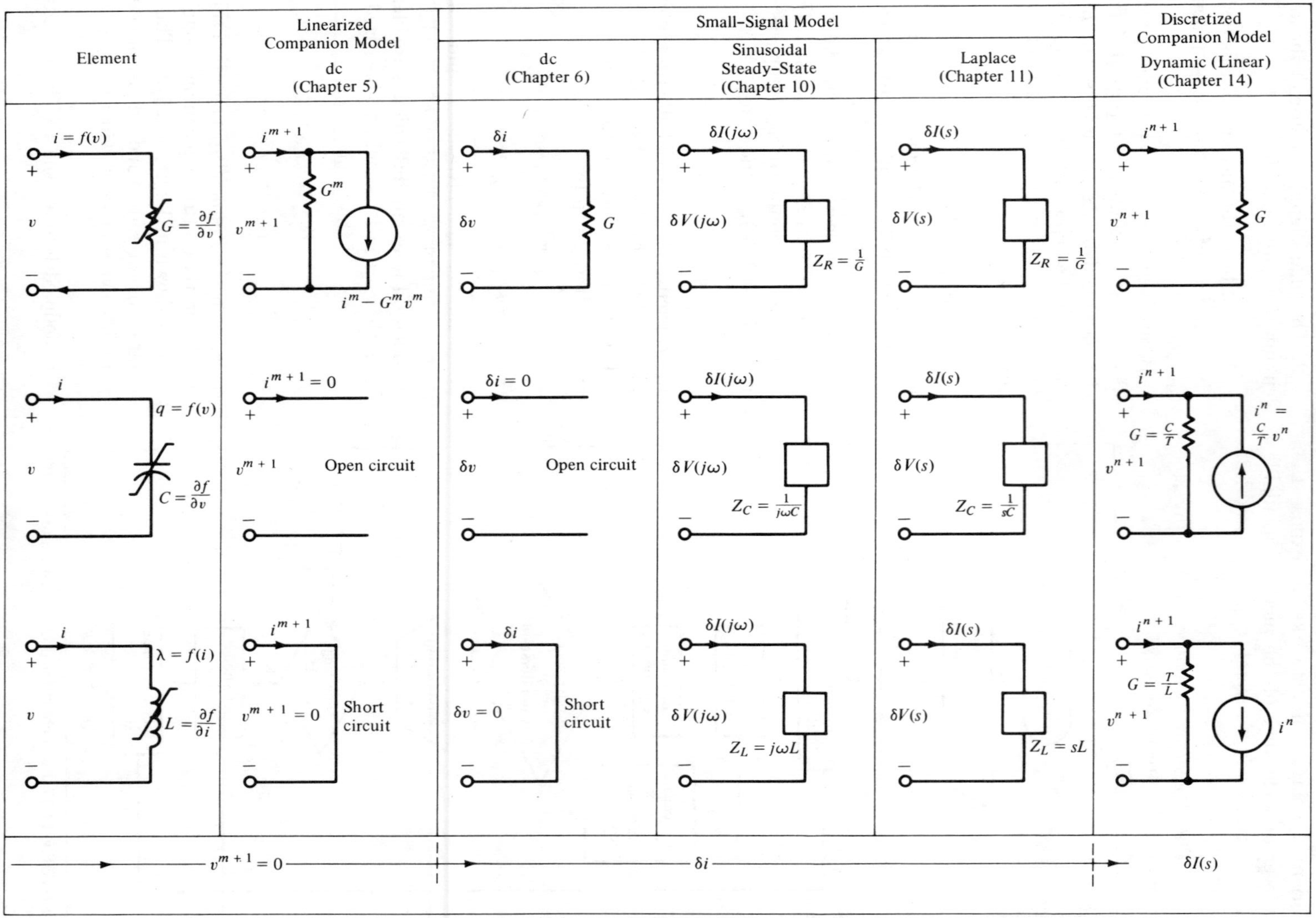

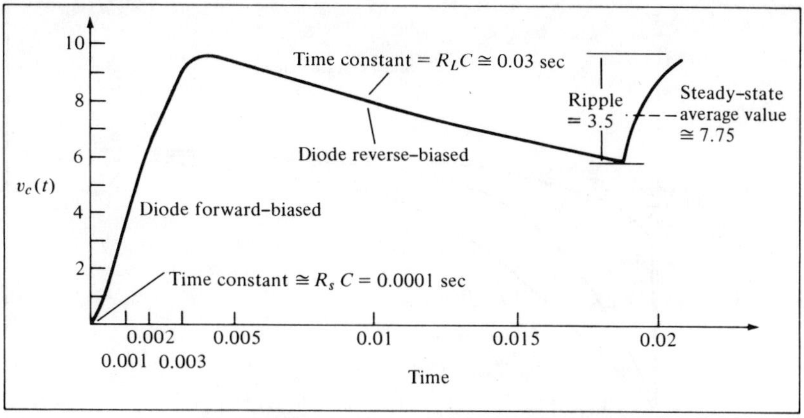

FIGURE 14.18 Response of rectifier-filter.

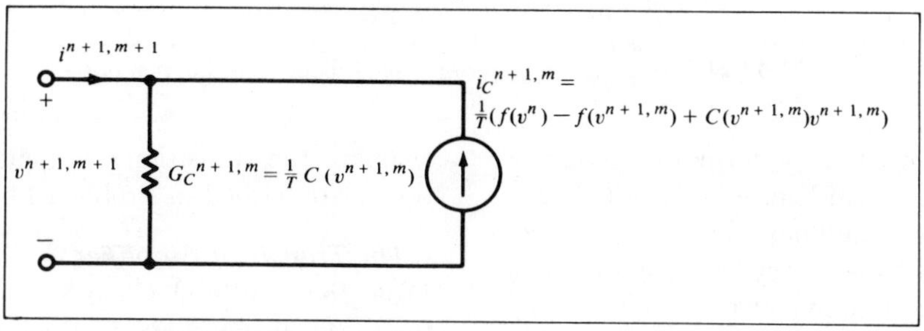

FIGURE 14.19 Companion model of nonlinear capacitor.

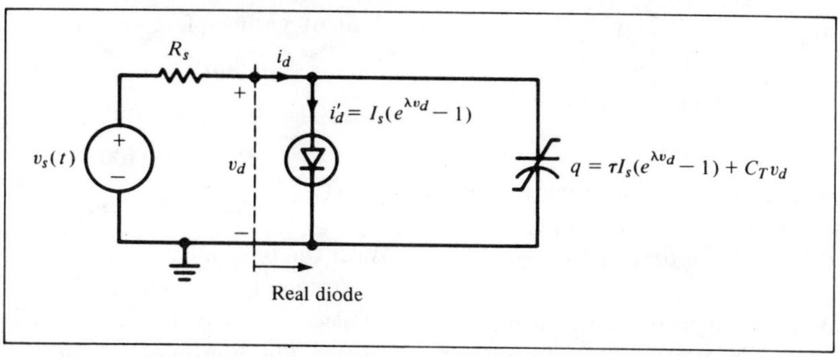

FIGURE 14.20 Dynamic model of simple diode circuit.

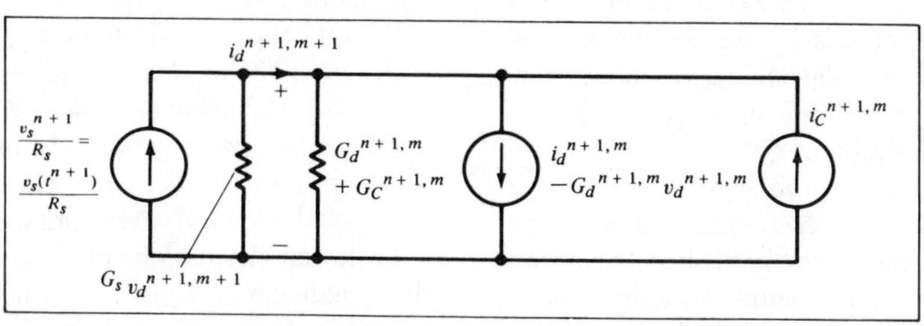

FIGURE 14.21 Companion model of diode circuit.

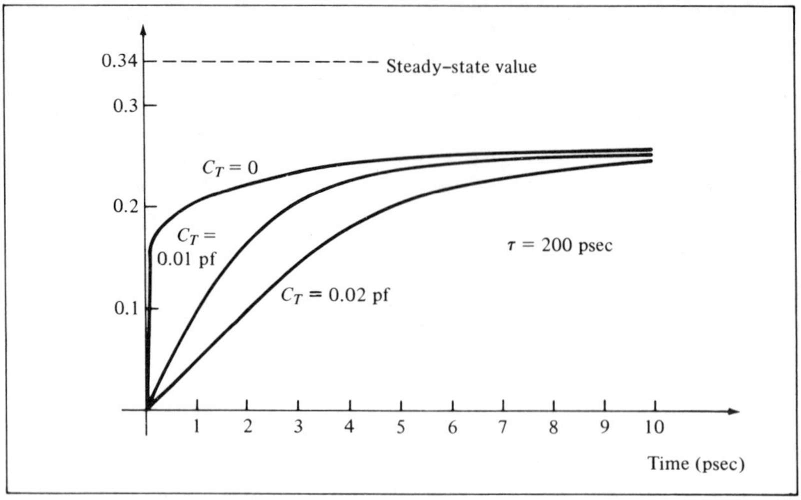

FIGURE 14.22 Initial part of step response of diode as a function of C_T.

Figure 5.19 is used to perform this dynamic analysis and is given in Figure 14.24 as TNET3. The main program and flow chart are identical to those of the TNET2 program of Figure 14.16.

We adopt the resistive element values used with the original INVRT program, together with

$$\tau = 200 \text{ psec} \qquad C_T = 0.1 \text{ pf}$$

If

$$v_s(t) = 5(1 - u(t)) \qquad (\text{all } t)$$

then $v_s(0^-) = 5$ (a logical **1**) and the output is $v_0(0^-) \cong 0$ (a logical **0**). The transistor is therefore initially saturated. For $t > 0$, $v_s(t) = 0$ and the transistor will be driven toward the cutoff state.

The transition from saturation to cutoff is shown in Figure 14.25. Note that after the input becomes zero, the transistor remains in saturation for 2 nsec; this is known as the storage time of the inverter. The transistor then enters the normal (forward) state, where $v_0(t)$ continues to rise. Finally, at $t = 5$ nsec, $v_0(t) = 5$ volts, and, externally, the switching action is complete. Internally the emitter diode may not become reverse-biased immediately, so that the transistor cutoff occurs sometime later. This investigation is left as a problem.

The TNET3 program may also be used to evaluate the transition from cutoff to saturation, as described in Problem 14.18.

The Transistor Amplifier

From the results of Chapter 3, we appreciate that a Thevenin equivalent can be devised for the base network of Figure 14.23(b). Using the element values of

$$R_b = 240{,}000 \qquad R_s = 20{,}000 \qquad v_c = 10$$

from Chapter 3, we find

$$R_{\text{eq}} = 18{,}400$$
$$v_{\text{eq}} = 0.77 + 0.923 v_s(t) \qquad (14.25)$$

With the base network reduced to a source and resistance, the model for the amplifier becomes identical in form to that for the transistor inverter. The nonlinear dynamic model is therefore again given in Figure 14.23(c).

We find the operating point by a dc analysis with TNET3 ($1/T = 0$) using the element values in Equation (14.25). We expect the amplifier to operate relatively linearly around this point, so we feel comfortable in carrying out a small-signal dynamic analysis.

From our discussions of Chapters 7 and 11, the small-signal dynamic equivalent is the linearized form of the above model evaluated at the dc operating point. This is described by

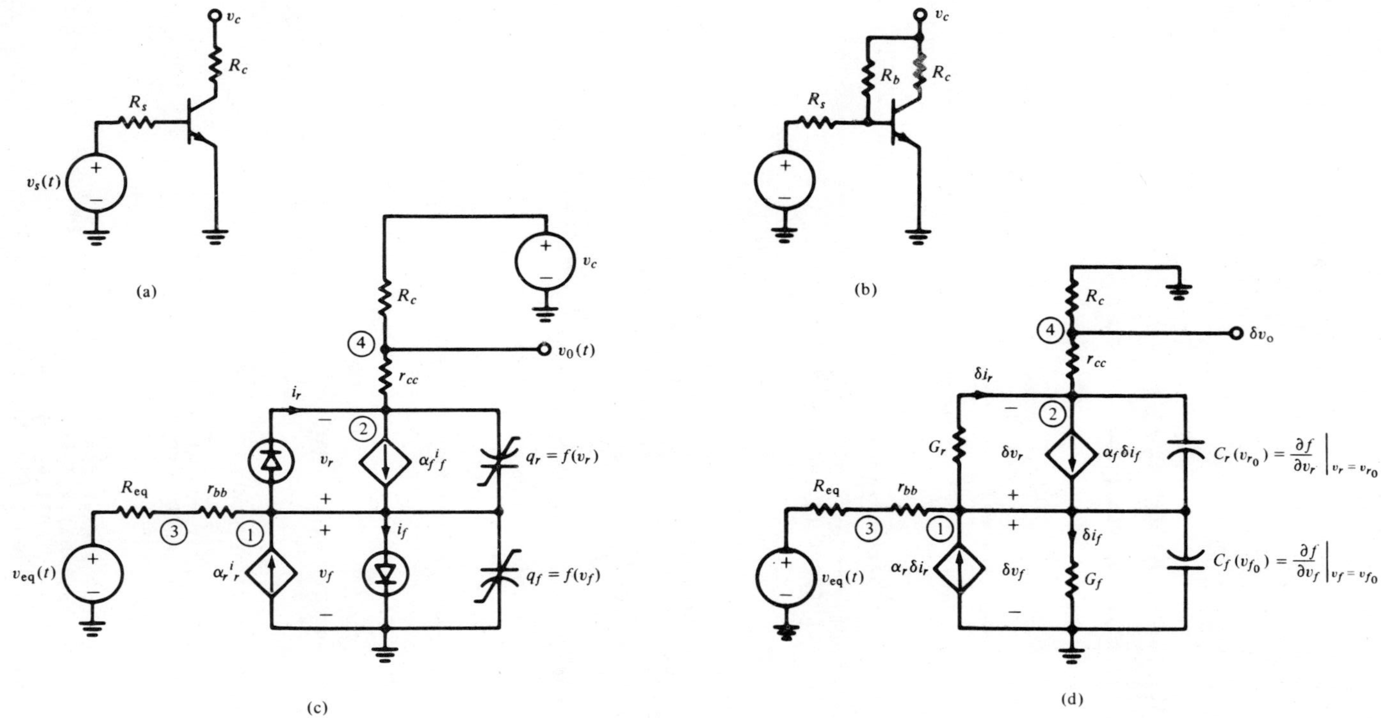

$$
\begin{bmatrix}
g_{bb} + G_f^*(1-\alpha_f) + G_r^*(1-\alpha_r) + \frac{1}{T}(C_r^* + C_f^*) & -G_r^*(1-\alpha_r) - \frac{1}{T}C & -g_{bb} & 0 \\
-G_r^* + \alpha_f G_f^* - \frac{1}{T}C_r^* & g_{cc} + G_r^* + \frac{1}{T}C_r^* & 0 & -g_{cc} \\
-g_{bb} & 0 & G_{eq} + g_{bb} & 0 \\
0 & -g_{cc} & 0 & G_c + g_{cc}
\end{bmatrix}
$$

(e)

$$
\begin{bmatrix}
g_{bb} + G_f(1-\alpha_f) + G_r(1-\alpha_r) + s(C_f + C_r) & -G_r(1-\alpha_r) - sC_r & -g_{bb} & 0 \\
-G_r + \alpha_f G_f - sC_r & g_{cc} + G_r + sC_r & 0 & -g_{cc} \\
-g_{bb} & 0 & G_{eq} + g_{bb} & 0 \\
0 & -g_{cc} & 0 & G_c + g_{cc}
\end{bmatrix}
$$

(f)

FIGURE 14.23 Comparison of amplifier, inverter analysis. (a) Inverter; (b) amplifier; (c) nonlinear, dynamic model; (d) small signal, dynamic; (e) node matrix of (a), as obtained from companion model; (f) node matrix of (b), as obtained from Laplace transform model.

simply the node matrix of the companion model discussed above, with $C_f(v_f)$, $C_r(v_r)$, $G_f(v_f)$, and $G_r(v_r)$ evaluated at the dc operating point. The only difference between these node matrices is that $1/T$ in the companion matrix is replaced by s in the Laplace-transformed matrix of the small-signal model.

For the problem at hand, a dc analysis using TNET3 at $v_{eq} = 0.77$ gives the values

$$v_0 = 4.4 \qquad G_f = 0.46\ \Omega \qquad G_r = 0$$

$$C_f = 9.25\text{ pf} \qquad C_r = 0.1\text{ pf}$$

```
C****  TRANSIENT ANALYSIS OF INVERTER (TNET3)
      IMPLICIT REAL*8(A-H,O-Z)
      REAL*8 G(10,10),I(10),V(10),VSAV(10)
      DIMENSION NODOUT(2,5)
      NDIM=10
    1 M=0
      N=0
C****  READ INPUT DATA
      CALL READ(NN,NODOUT,NOUT,T,TMAX)
      TIME=0.D0
      DO 6 J=1,NN
    6 V(J)=0.D0
C****  SET UP NODE EQUATIONS
    3 CALL NODEQ(G,V,I,VSAV,TIME,T,NN,NDIM)
C****  SOLVE NODE EQUATIONS
      CALL REDUCE(G,NN,I,NDIM)
C****  V(J)-I(J) IS THE INCREMENT IN THE JTH NODE VOLTAGE
      M=M+1
      ERR=0.D0
      DO 2 J=1,NN
      ERR=ERR+(V(J)-I(J))**2
    2 V(J)=I(J)
C****  TEST CONVERGENCE CRITERION
      IF(ERR.GT.1.D-6.AND.M.LT.50)GO TO 3
      IF(M.EQ.50)GO TO 8
C****  CALCULATE AND PRINT OUTPUT VOLTAGES
      CALL OUTPUT(NODOUT,NOUT,V,TIME)
      IF(TIME.GE.TMAX)STOP
C****  STORE NODE VOLTAGES FOR UPDATING CAPACITORS
      DO 5 J=1,NN
    5 VSAV(J)=V(J)
      M=0
      N=N+1
      TIME=TIME+T
      GO TO 3
    8 WRITE(6,9)
    9 FORMAT(' TOO MANY ITERATIONS')
      STOP
      END
      SUBROUTINE READ(NN,NODOUT,NOUT,T,TMAX)
      IMPLICIT REAL*8(A-H,O-Z)
      DIMENSION NODOUT(2,1)
      COMMON RCC,RBB,RS,RC,VC,ALFF,ALFR,VS,DVS,TAU,CT
      READ(5,1)RCC,RBB,RS,RC,VC,ALFF,ALFR,VS,DVS,TAU,CT
     &,T,TMAX
      READ(5,2)NN,NOUT,(NODOUT(1,J),NODOUT(2,J),J=1,NOUT)
      WRITE(6,3)RCC,RBB,RS,RC,VC,ALFF,ALFR,VS,DVS,TAU,CT
    6 RETURN
    1 FORMAT(6F10.0)
    2 FORMAT(20I2)
    3 FORMAT(' RCC =',1PD11.4,' RBB =',1PD11.4,' RS =',1PD11.4,
     &/' RC =',1PD11.4,' VC =',1PD11.4,' ALFF =',1PD11.4,
     &/' ALFR =',1PD11.4,' VS =',1PD11.4,' DVS =',1PD11.4,
     &/' TAU =',1PD11.4,' CT =',1PD11.4)
      END
```

FIGURE 14.24 Inverter-amplifier analysis program.

Evaluation of the nodal matrix using these linearized values gives the transfer function of the small-signal model as

$$\frac{\delta V_0}{\delta V_s}(s) = \frac{5.4 \times 10^{-6}(s - 0.447)}{(s + 0.00693)(s + 0.0000301)}$$

The associated step response is easily found to be

$$\delta v_0(t) = -0.05e^{-0.0069t} + 11.64e^{-3.01\times 10^{-5}t} - 11.59u(t) \quad (14.26)$$

The response $v_0 + \delta v_0(t)$ is plotted in Figure

```
      SUBROUTINE NODEQ(G,V,I,VSAV,TIME,T,NN,NDIM)
      IMPLICIT REAL*8 (A-H,O-Z)
      REAL*8 G(NDIM,1),I(1),VSAV(1),V(1),LAMDA
      COMMON RCC,RBB,RS,RC,VC,ALFF,ALFR,VS,DVS,TAU,CT
      DATA CSAT/1.D-9/,LAMDA/40.D0/
      DATA NOP/0/
C****  ZERO NODE MATRIX AND CURRENT VECTOR
      DO 1 K=1,NN
      I(K)=0.D0
      DO 1 J=1,NN
    1 G(J,K)=0.D0
      DXSAVF=CSAT*(DEXP(DMIN1(VSAV(1),.6D0)*LAMDA)-1.D0)
      DXSAVR=CSAT*(DEXP(DMIN1(VSAV(1)-VSAV(2),.6D0)*LAMDA)-1.D0)
      TINV=1.D0/T
      IF(TIME.EQ.0.D0)TINV=0.D0
      TAUT=TAU*TINV
      VSS=VS
      IF(TIME.NE.0.D0)VSS=VS+DVS
      GS=1.D0/RS
      GBB=1.D0/RBB
      GC=1.D0/RC
      GCC=1.D0/RCC
      VDF=DMIN1(V(1),.6D0)
      VDR=DMIN1(V(1)-V(2),.6D0)
      DEXF=CSAT*DEXP(LAMDA*VDF)
      DEXR=CSAT*DEXP(LAMDA*VDR)
      GDF=LAMDA*DEXF
      GDR=LAMDA*DEXR
      CURF=DEXF-CSAT-GDF*VDF
      CURR=DEXR-CSAT-GDR*VDR
      GCF=LAMDA*TAUT*DEXF
      GCR=LAMDA*TAUT*DEXR
      CURCF=TAUT*(DXSAVF-(DEXF-CSAT))+GCF*VDF
     &+VSAV(1)*CT*TINV
      CURCR=TAUT*(DXSAVR-(DEXR-CSAT))+GCR*VDR
     &+(VSAV(1)-VSAV(2))*CT*TINV
      GCF=GCF+CT*TINV
      GCR=GCR+CT*TINV
      IF(TIME.EQ.0.D0.OR.NQUIES.EQ.1)GO TO 2
      CF=GCF*T
      CR=GCR*T
      WRITE(6,3)GDF,GDR,CF,CR
      NQUIES=1
    2 G(1,1)=GBB+GDF+GDR-ALFF*GDF-ALFR*GDR+GCF+GCR
      G(1,2)=-GDR+ALFR*GDR-GCR
      G(1,3)=-GBB
      G(2,1)=-GDR+ALFF*GDF-GCR
      G(2,2)=GCC+GDR+GCR
      G(2,4)=-GCC
      G(3,1)=G(1,3)
      G(3,3)=GS+GBB
      G(4,2)=G(2,4)
      G(4,4)=GC+GCC
      I(1)=ALFR*CURR+ALFF*CURF-CURF-CURR+CURCF+CURCR
      I(2)=CURR-ALFF*CURF-CURCR
      I(3)=VSS*GS
      I(4)=VC*GC
      RETURN
```

Figure 14.24 (Continued)

```
    3 FORMAT(' AT DC OPERATING POINT'/' GF =',1PD11.4,' GR =',1PD11.4,
     &' CF =',1PD11.4,' CR =',1PD11.4)
      END
      SUBROUTINE OUTPUT(NODOUT,NOUT,V,TIME)
      IMPLICIT REAL*8 (A-H,O-Z)
      DATA INIT/0/
      DIMENSION NODOUT(2,1),V(1),VOUT(5)
      IF(INIT.NE.0)GO TO 3
      WRITE(6,4)(NODOUT(1,J),NODOUT(2,J),J=1,NOUT)
    3 DO 2 I=1,NOUT
      VOUT(I)=0.D0
      N1=NODOUT(1,I)
      N2=NODOUT(2,I)
      IF(N1.NE.0)VOUT(I)=VOUT(I)+V(N1)
      IF(N2.NE.0)VOUT(I)=VOUT(I)-V(N2)
    2 CONTINUE
      WRITE(6,1)TIME,(VOUT(I),I=1,NOUT)
      INIT=1
      RETURN
    1 FORMAT(6D12.3)
    4 FORMAT('     TIME      ',5(I2,' -',I2,' NODES'))
      END
      SUBROUTINE REDUCE(A,N,B,NDIM)
      IMPLICIT REAL*8(A-H,O-Z)
      DIMENSION A(NDIM,1),B(1)
C****  THIS SUBROUTINE SOLVES A*X=B BY GAUSS ELIMINATION
C****  N IS THE MATRIX SIZE; X IS RETURNED IN B; A IS DESTROYED
C****  CONVERT A TO UPPER TRIANGULAR FORM
      NM1=N-1
      IF(N.EQ.1)GO TO 8
      DO 1 J=1,NM1
      IJ=J+1
      DO 4 K=IJ,N
    4 A(J,K)=A(J,K)/A(J,J)
      B(J)=B(J)/A(J,J)
      DO 1 I=IJ,N
      ALPHA=A(I,J)
      DO 2 K=IJ,N
    2 A(I,K)=A(I,K)-ALPHA*A(J,K)
    1 B(I)=B(I)-ALPHA*B(J)
    8 B(N)=B(N)/A(N,N)
C****  PERFORM BACK SUBSTITUTION
      IF(N.LT.2)GO TO 7
      DO 3 J=2,N
      LAST=N-J+1
      LASTP1=LAST+1
      DO 3 I=LASTP1,N
    3 B(LAST)=B(LAST)-A(LAST,I)*B(I)
    7 RETURN
      END
```

(a)

```
10.         15.         500.        500.        5.          .98
.5          5.          -5.         200.        .1          300.
6000.
 4 3 1 2 1 0 4 0
```

(b)

Figure 14.24 (Continued)

14.26 together with the direct numerical solution of the nonlinear equations using the companion model. The difference is observed to be very small indeed, with a part of the error due to the imprecision of the approximation involved in the numerical solution. The conclusion is that small-signal analysis is quite adequate in evaluating the step responses of transistor amplifiers, even though the capacitors involved are quite nonlinear.

Summary

It might have been reasonable to expect that nonlinear dynamic analysis would be considered as a generalization of other types of analysis. The above example shows the simplicity of modifying the node matrix of the companion model to obtain the node matrix required for other types of analysis. These relationships are summarized in Table 14.3.

Those readers who are familiar with the auto-

TABLE 14.3

Relationships between Node Matrices

	Set	Analysis
Node matrix of companion model of nonlinear, dynamic model		Nonlinear, dynamic
	$\frac{1}{T} = 0$	Nonlinear, dc
	$\frac{1}{T} = s$	Small signal, pole-zero
	$\frac{1}{T} = j\omega$	Small-signal, sinusoidal steady state

matic equation formulation procedures of the DCAP program (see Instructors' Manual) will appreciate from Table 14.3 the conceptual simplicity of revising this program to perform all of these analyses. Programs developed on this principle have been used extensively in the design of amplifier, digital, and a variety of other types of circuits.

14.5 Conclusion

The computer-aided methods of this chapter are in many ways more general and powerful than those of Chapter 13. In Chapter 13, we examined only the solution of first-order differential equations of the form

$$\dot{\mathbf{x}} = \mathbf{f}(\mathbf{x}, t)$$

or

$$\dot{\mathbf{x}} - \mathbf{f}(\mathbf{x}, t) = 0 \qquad (14.27)$$

In contrast, the methods of this chapter are concerned with the solution of the more general problem

$$\mathbf{F}(\mathbf{x}, \dot{\mathbf{x}}, t) = \mathbf{0} \qquad (14.28)$$

Not only does Equation (14.28) include Equa-

```
1Ø.       15.       5ØØ.      5ØØ.      5.        .98
.5        5.        -5.       2ØØ.      .1        3ØØ.
6ØØØ.
 4 3 1 2 1 Ø 4 Ø
```

(a)

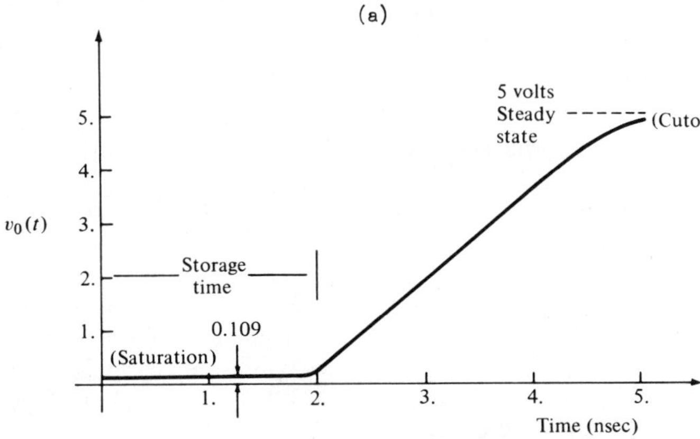

(b)

FIGURE 14.25 Solution of response of transistor switch. (a) Input data; (b) response.

```
10.        15.        18400.     5000.     10.        .98
.5         .77        .0923      200.      .1         10000.
100000.
 4 2 4 0 1 0            (a)
```

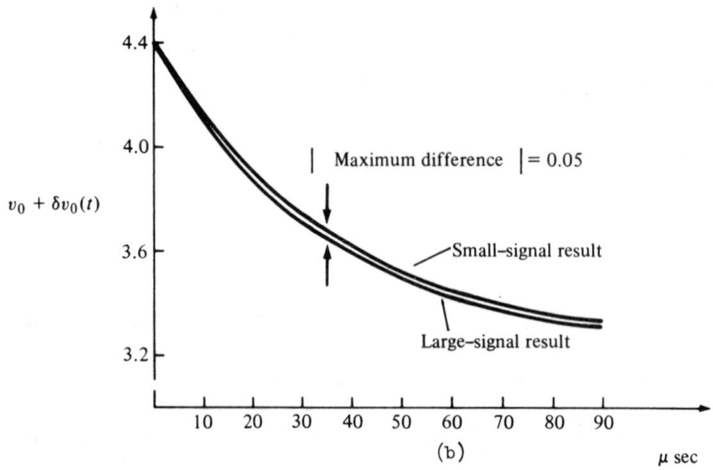

FIGURE 14.26 Solution of dynamic model of amplifier. (a) Input data; (b) response.

tion (14.27) as a special case, but it also includes the solution of algebraic equations of the form

$$\mathbf{F}(\mathbf{x}) = \mathbf{0}$$

as described in Section 5.6. In addition, it offers the possibility of simultaneous solution of algebraic and differential equations as illustrated in the diode switching and inverter problems of this chapter. Finally, it permits the equations to be assembled branch by branch on a nodal basis, which greatly simplifies the formulation.

We have not, however, discussed several important extensions of the methodology.

1. The backward Euler method is inaccurate relative to state-of-the-art integration methods. An example of a more accurate method is given in Problem 14.7; other methods are discussed in Reference 14.1.

2. Even an accurate integration formula should have provisions for controlling the integration step size T. The step size required to maintain a specified error can change dramatically during an integration, especially when the network is nonlinear and the small-signal natural frequencies change with the change in linearized element values. This problem is discussed in Reference 14.1.

In general, the reader should exercise care in attempting problems beyond those given in the text. It is not uncommon to encounter convergence problems or to require excessive computation times using the elementary numerical techniques of this chapter.

PROBLEMS

In developing programs for the following problems, use the TNET1, TNET2, and TNET3 programs as models for networks with more than one nonreference node. Wherever possible, use program variable names similar to the circuit notation (that is, program variable R1 represents the value of R_1). Time scaling may be used to assist in formating input.

14.1 (a)–(b) Develop discretized companion models of the networks of Figure P 14.1. Write the node equations of these models, and outline the solution process, including the updating procedure.

14.2 (a)–(b) For the networks of Figure P 14.1, let $R_1 = C_1 = R_2 = C_2 = 1$, $i_s(t) = 5u(t)$. Develop computer programs based on TNETI to find $v_0(t)$ for $0 \leqq t \leqq 3$ sec in 0.1 sec steps. Compare your solution with the analytic solution at $t = 1$ sec.

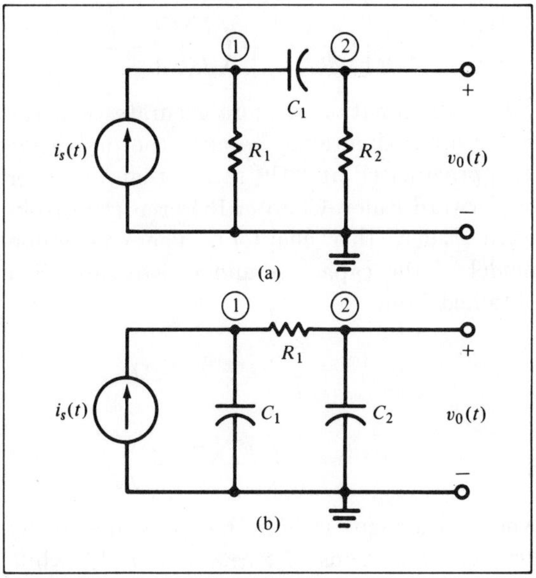

FIGURE P 14.1

14.3 (a)–(b) Develop discretized companion models of the networks of Figure P 14.2, write the *mesh* equations of these models, and outline the solution process including the updating procedure.

14.4 The voltage-source/capacitor combination of Figure P 4.3(a) is frequently encountered as an amplifier input because the capacitor, being an open circuit for dc analysis, prevents the signal source from loading the rest of the network (and vice versa) in a dc sense. (a) Show that a companion model suitable for

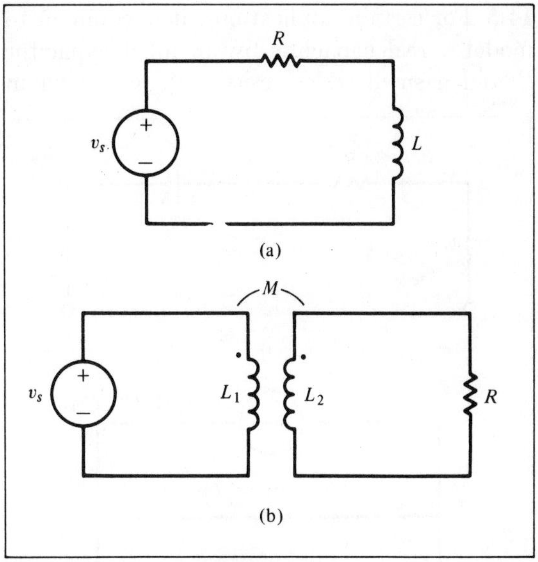

FIGURE P 14.2

node analysis is given in Figure P 14.3(b). (b) Show that as $T \rightarrow 0$, the companion model yields the signal-source equation $v^{n+1} = v_s^{n+1}$. This means that the capacitor appears to be a short for a sufficiently small time step.

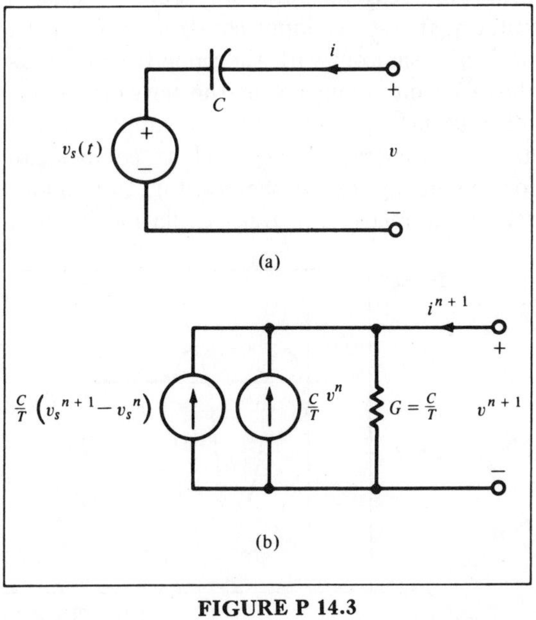

FIGURE P 14.3

14.5 For certain applications, it is common to model a real capacitor by an ideal capacitor C and a small series resistor R, as shown in

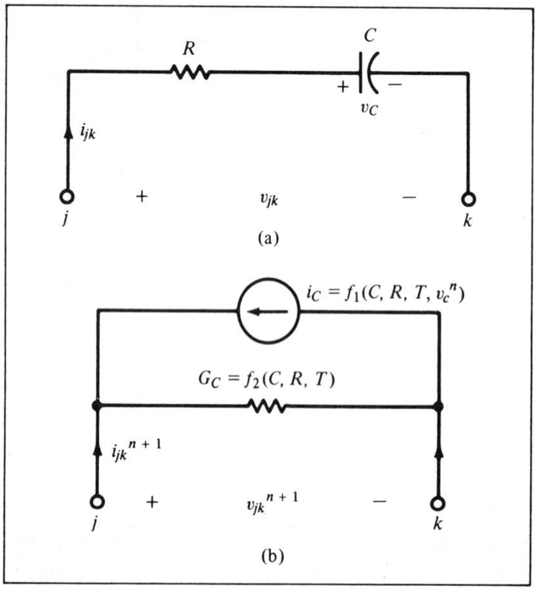

FIGURE P 14.4

Figure P 14.4. Determine a companion model in the form of Figure P 14.4(b).

14.6 The GCL network of Figure P 14.5 is initially at rest. An input constructed from the peak of a sinusoid with the same frequency as the resonant frequency of the network is applied as shown. The intent is to supply a small amount of energy at each cycle to compensate for the energy lost to the resistor, thus maintaining a relatively large oscillation with a small input current. (a) Construct a companion model and the corresponding set of discretized equations. Solve for v^n for $t = n\pi/25$, $n = 0, 1, \cdots, 300$. Use $A = .1$; assume zero-valued initial conditions. (b) Sketch the above results and note the tendency of the oscillation to approach a steady-state value. What is the distortion in the resulting oscillation? (See Chapter 3.) (c) If $A = 2$ above, what happens to the steady-state response?

14.7 An alternate approximation to the derivative is (reference 14.2)

$$\left[\frac{dv}{dt}\right]^{n+1/2} \triangleq \frac{1}{2}\left[\left(\frac{dv}{dt}\right)^{n+1} + \left(\frac{dv}{dt}\right)^{n}\right] = \frac{v^{n+1} - v^n}{T}$$

The approximation is more accurate than the Euler approximations, because the derivative is approximated at $t^{n+1/2}$ rather than at either t^n (forward Euler, Chapter 13) or at t^{n+1} (backward Euler, this chapter). The companion model of the capacitor and resistor are then obtained from

$$i_C{}^{n+1/2} = \frac{C}{T}\left(v_C{}^{n+1} - v_C{}^n\right)$$

$$i_R{}^{n+1/2} = \frac{1}{R}\left(\frac{v_R{}^{n+1} + v_R{}^n}{2}\right)$$

where all currents $i^{n+1/2}$ in the companion model are approximations of currents at $t^{n+1/2}$, while all voltages v^{n+1} are approximations of voltages at t^{n+1}. (a) Draw companion models of the capacitor and resistor. Note that the latter

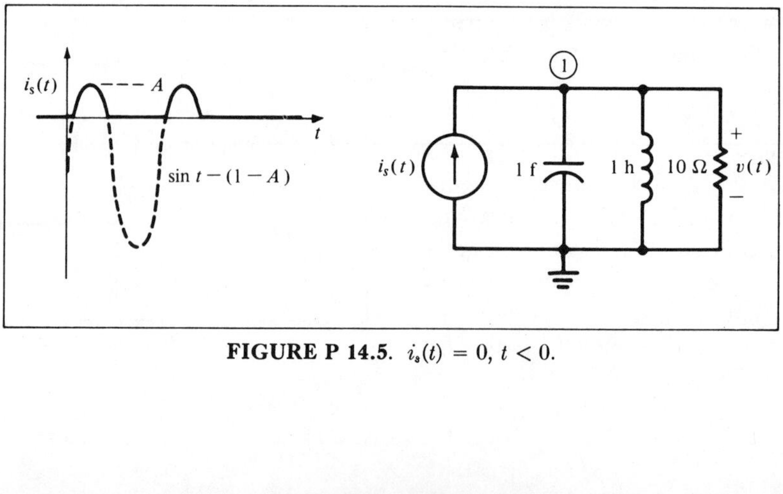

FIGURE P 14.5. $i_s(t) = 0$, $t < 0$.

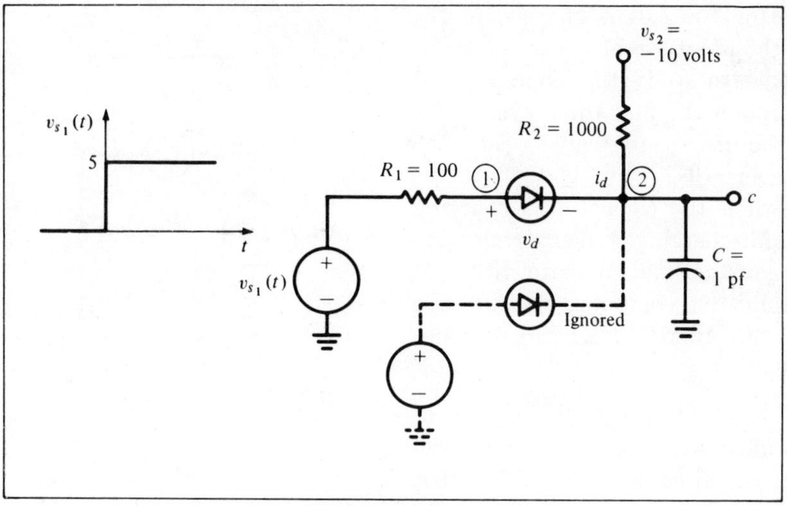

FIGURE P 14.6

model now involves a source. (b) Construct companion models of current and voltage sources. (c) Develop a companion model of the network of Figure 14.2 using the same approximation. Program the resultant algorithm and compare the errors with Table 14.1 at $t = 1$ for different step sizes. How does the error decrease with the step size?

14.8 You are asked to determine the transient response of the OR circuit of Figure P 14.6, assuming the diode model $i_d = 10^{-9}(e^{40v_d} - 1)$. (a) Estimate the *dc* voltage at *c* for $t < 0$. (b) Estimate the circuit time constant. (c) Determine the transient response at 40 time points for $0 \leq t \leq 3$ time constants. Use TNET2 as a model program.

14.9 The circuit of Figure P 14.7 (a) is a clamping circuit. The output is intended to look like the input, except that it is shifted downward so that its positive peaks are clamped to a near-zero value as shown. The circuit works by charging the capacitor on the positive peaks through the diode and the small resistor R_1; the rest of the time the diode is reverse-biased and C_1 is discharging slowly through the large resistance $R_1 + R_2$. The net effect is that C

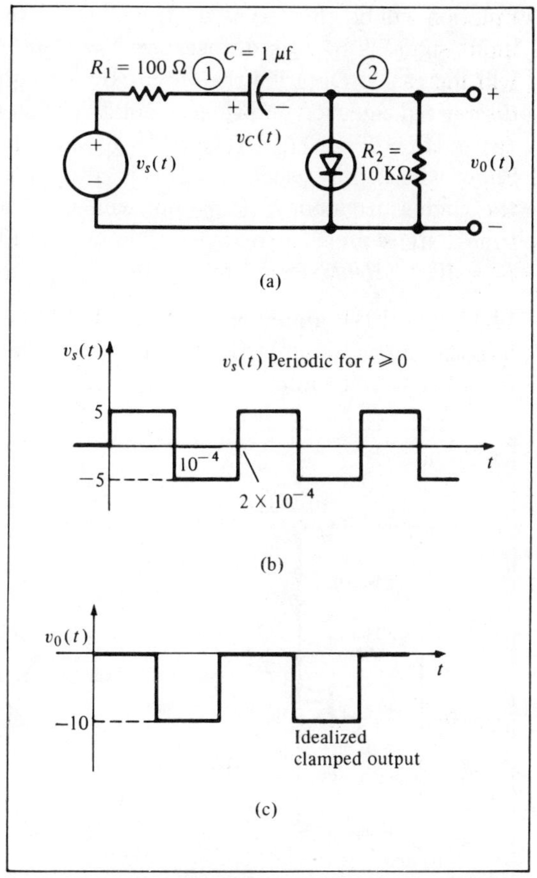

FIGURE P 14.7

acts like a battery once it is charged to the peak value of the input signal.

You are asked to study this charging and discharging through the first three cycles of a square wave input. In particular, calculate $v_c(t)$ and $v_0(t)$ every 10 μsec for $0 \leqq t \leqq 4 \times 10^{-4}$ sec. Indicate when the capacitor is charging and when it is discharging. Calculate the discharging time constant and compare with that determined from the computer results. Use $I_s = 10^{-9}$, $\lambda = 40$. Adopt TNET2 as a model program.

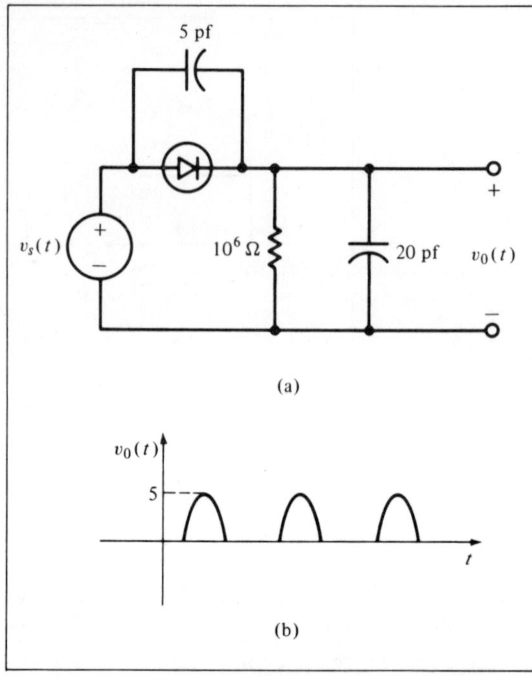

FIGURE P 14.8

14.10 Without the capacitors, the circuit of Figure P 14.8 ideally clips the negative peaks from the input signal as in Figure P 14.8(b). In practice, capacitances are likely to be present as shown; as a result, the clipping function will be affected as the frequency of the input signal is increased (because $i = C\, dv/dt$ will increase). Describe the circuit by a single discretized equation and then calculate $v_0(t)$ for (a) $v_s(t) = 5 \sin t$. (b) $v_s(t) = 5 \sin 10^4 t$. Show where the response shape is altered at the higher frequency. Use approximately 40 times steps over two cycles. Let $\lambda = 40$, $I_s = 10^{-9}$. (*Hint*: Scale time by 10^6).

14.11 The FET amplifier of Figure P 14.9(a) has the current-source signal input of Figure P 14.9(c). Determine $v_0(t)$ in 25 time steps for $0 \leqq t \leqq 2.5$ μsec. Check v_{gs} to insure that the FET remains in the pinch-off region.

14.12 Repeat Problem 14.11 for the voltage-source input of Figure P 14.9(d), in 25 steps for $0 \leqq t \leqq 0.5$ μsec. See Problem 14.4 for assistance. Explain the peak in the output.

14.13 A more realistic respresentation of the

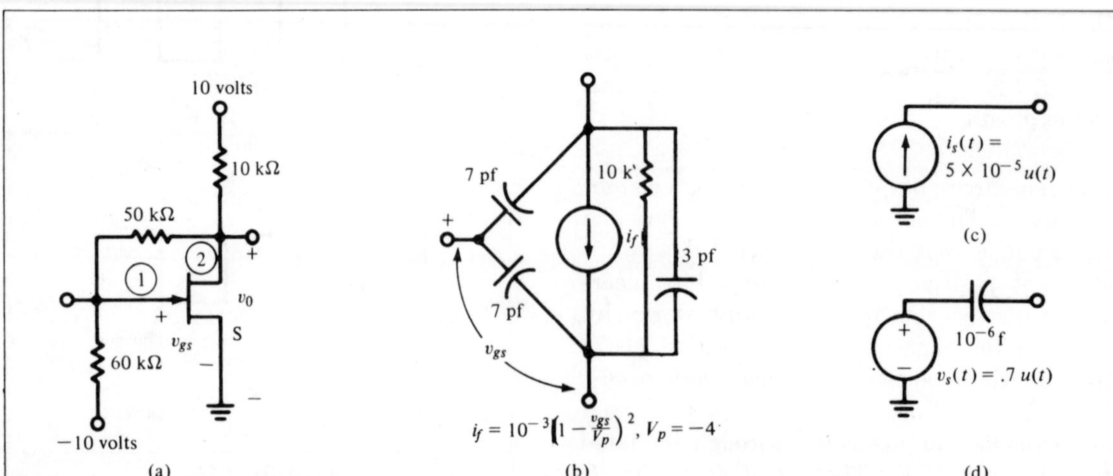

FIGURE P 14.9

internal physical phenomena associated with a diode than that used in Figure 14.20 is obtained by setting

$$q = \tau I_s(e^{\lambda v_d} - 1) + q_T(1 - v_d/V_0)^r$$

where $V_0 \cong 0.9$ volts and $r = 0.5$. Find the companion model of this nonlinear capacitor.

14.14 Develop a companion model for a nonlinear inductor described by $\lambda = f(i)$. Is your model more suitable for mesh or for node analysis?

14.15 The transient behavior of the diode in Figure 4.20 of the text can be estimated once the diode conducts sufficiently so that (1) $C_T v_d \ll \tau I_s(e^{\lambda v_d} - 1)$ and (2) the $v_s - R_s$ combination appears as an effective current source

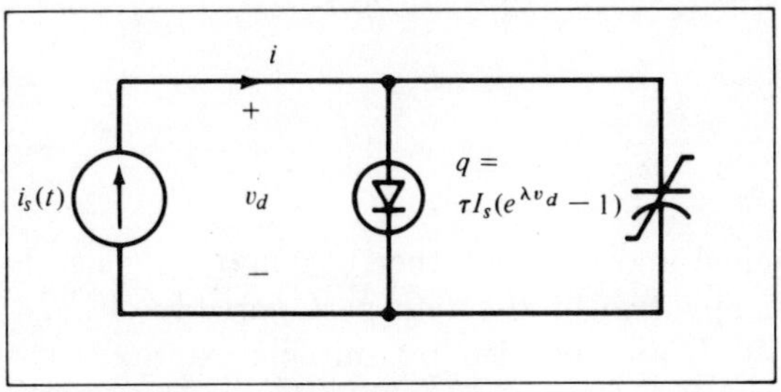

FIGURE P 14.10

of value $i_s = v_s/R_s$. (a) Show that for the model of Figure P 14.10,

$$i_s(t) = \frac{q(t)}{\tau} + \frac{dq(t)}{dt}$$

(b) Solve for $q(t)$ in terms of $q(0)$, assuming $i_s(t) = 10^{-3}u(t)$; what is the time constant? (c) Solve for $v_d(t)$ is terms of $q(t)$. (d) Sketch $q(t)$ and $v_d(t)$, beginning with $v_d(0) = 0.25$ (the approximate final value shown in Figure 14.23 is at $t = 10$ psec). Assume $\tau = 200$ psec, $\lambda = 40$, $I_s = 10^{-9}$. Next, sketch the composite response, including the initial response shown in Figure 14.22.

14.16 A common specification to indicate the speed of a transistor switching circuit is the *rise time* required for the step response to change from 10% to 90% of its final value. (a) What is the rise time of the response Figure 14.25? (b) Change R_s and R_C to 1000 Ω. Use TNET3 to evaluate the response. Integrate to 10 nsec in 0.3 nsec steps. (c) Compare the storage and rise times in (b) and (c).

14.17 With the transistor having the response in Figure 14.25 initially saturated, TNET3 gives the linearized element values as

$$G_f = 1.08 \text{ ℧} \qquad G_r = 0.667 \text{ ℧}$$

$$C_f = 216 \text{ pf} \qquad C_r = 133 \text{ pf}$$

(a) Evaluate the natural frequencies and the unit stop response of the network model in the saturated state. Compare the natural frequencies with the storage time of the circuit. Comment on the usefulness of the step response. (b) A formula often given for the storage time of such a circuit is

$$t_s = \frac{\alpha_f \tau (1 + \alpha_r)}{1 - \alpha_f \alpha_r} \ln (1 + I_b\beta/I_c)$$

where I_b and I_c are the values of i_b and i_c in the saturated state (see Figure 5.17). Check the accuracy of this formula by comparing with the results of Figure 14.25.

14.18 Determine the response for $t \geqq 0$ of the transistor inverter of Figure 14.23 to the input $v_s(t) = 5u(t)$. This input can be realized by setting $VS = 0$ and $DVS = 5$ in the input data of TNET3. It is suggested that 30 psec time steps be used for $0 \leqq t \leqq 1$ nsec.

14.19 Sketch the small-signal frequency response of the amplifier of Figure 14.23, with the values used in the input data of Figure 14.26.

REFERENCES AND ADDITIONAL READING

14.1 D. A. CALAHAN, *Computer-Aided Network Design.* New York: McGraw-Hill Book Company, Inc., 1972, Chapter 9.

14.2 OMAR WING, *Circuit Theory with Computer Methods*, New York: Holt, Rinehart and Winston, Inc., 1972, Chapter 9.

15

Two-Port Analysis

15.1 Introduction

The notions of an input and an output or an excitation and a response are natural to most physical devices and systems. For electrical networks, the input and output are likely to be located at different parts of the network. In this case, a one-port representation of the network performance is impossible; the port concept must be extended to include two pairs of terminals, called a *two-port,* where a current or voltage is applied at one pair of terminals and the current or voltage response is measured at another pair. Examples are shown in Figure 15.1. We have already described in Chapter 2 how ports are created in a network by "soldering iron" or "pliers" types of connection.

A general two-port representation is shown in Figure 15.2. Assuming that the network is linear, the currents and voltages can be written with capital letters because they represent phasor or transform variables. As with a one-port, we define a port by requiring the same current to flow in one port terminal and out the other terminal. In general, if any two of the four-port variables V_1, I_1, V_2, I_2 are specified by suitable external connections, the remaining two variables are determined.

In this chapter we examine various methods for relating these four variables and for using the resultant description in network analysis. We begin by an intensive examination of the case when both ports are excited by voltage sources.

15.2 A Case Study: The Short-Circuit Parameters

Definition

If we apply voltage sources at each port, the currents will be linear functions of these voltages:

$$I_1 = y_{11}V_1 + y_{12}V_2 \tag{15.1}$$

$$I_2 = y_{21}V_1 + y_{22}V_2 \tag{15.2}$$

The symbol y_{ij} is used for the coefficients of

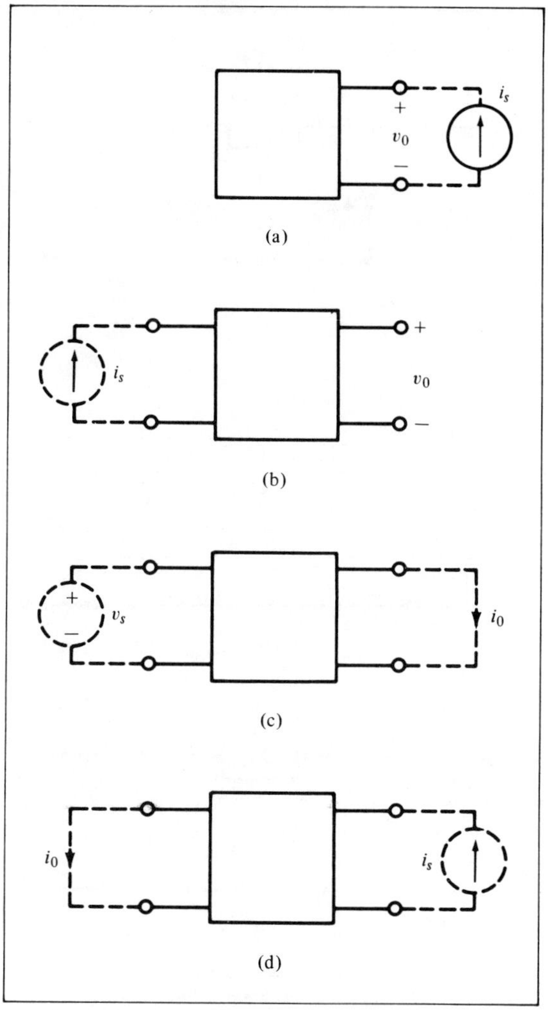

FIGURE 15.1 (a) One-port connection; (b)–(d) two-port connections.

these equations; because the coefficients relate voltage to current, they are admittances.

These y's are called the *short-circuit admittance* parameters of the two-port, for the following reasons. If we set $V_2 = 0$ in Equations (15.1)–(15.2), we note that

$$y_{11} = \left.\frac{I_1}{V_1}\right|_{V_2=0} \tag{15.3}$$

and

$$y_{21} = \left.\frac{I_2}{V_1}\right|_{V_2=0} \tag{15.4}$$

Similarly, if $V_1 = 0$, we have

$$y_{12} = \left.\frac{I_1}{V_2}\right|_{V_1=0} \tag{15.5}$$

and

$$y_{22} = \left.\frac{I_2}{V_2}\right|_{V_1=0}$$

All four y-parameters can therefore be determined by measurement or calculation of the variables *with one of the ports shorted*. This is illustrated in Figure 15.3.

Example 15.1

The y-parameters of the network of Figure 15.4(a) are to be found.

We proceed as follows. With port 2 shorted, the current at port 1 is calculated as shown

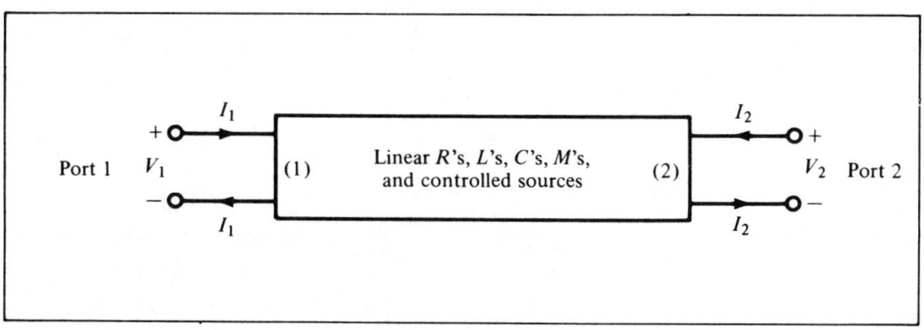

FIGURE 15.2 Nomenclature and reference directions for a linear two-port network.

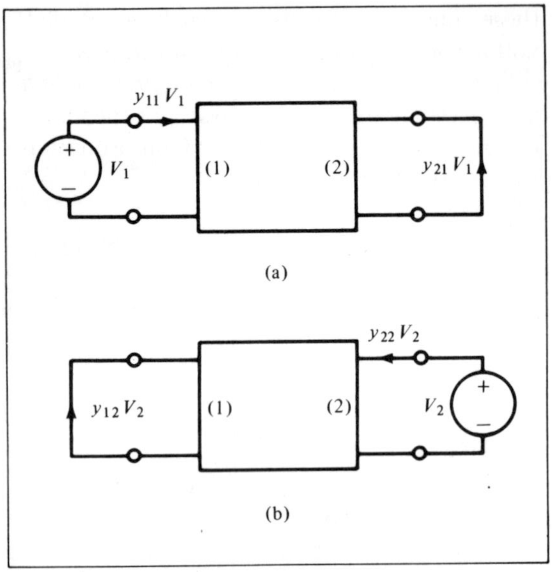

FIGURE 15.3 Physical interpretation of short-circuit admittance parameters. (a) y_{11} and y_{21}; (b) y_{12} and y_{22}.

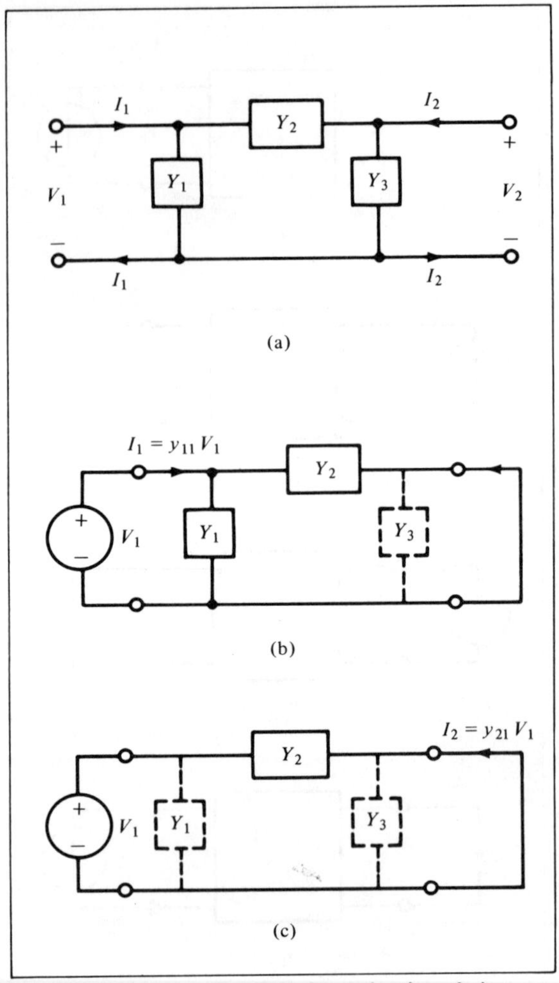

FIGURE 15.4 Example determination of y's.

in Figure 15.4(b). This calculation gives

$$I_1 = y_{11}V_1 = (Y_1 + Y_2)V_1$$

or

$$y_{11} = Y_1 + Y_2$$

The current flowing into the short at port 2 is seen from Figure 15.4(c) to be $-Y_2V_1$. The negative sign arises from the reference direction of I_2 flowing *into* the port. Therefore,

$$y_{21} = \left.\frac{I_2}{V_1}\right|_{V_2=0} = -Y_2$$

By shorting port 1 with a voltage applied at port 2, we can similarly obtain

$$y_{22} = Y_2 + Y_3 \qquad y_{12} = -Y_2$$

Example 15.2

We add a controlled source to the above network as shown in Figure 15.5, and determine the new y-parameters.

Inspection of this network shows that only the y_{21} calculation is changed from above. Now a voltage source V_1 causes an additional current of value Y_4V_1 to flow at the shorted port. Therefore,

$$y_{21} = \left.\frac{I_2}{V_1}\right|_{V_2=0} = \frac{-Y_2V_1 + Y_4V_1}{V_1}$$

$$= -Y_2 + Y_4$$

The parameters y_{11} and y_{22} are called *short-circuit driving point admittances* and y_{12} and y_{21} are *short-circuit transfer admittances*. They can represent response/excitation ratios useful in either sinusoidal or Laplace transform analysis.

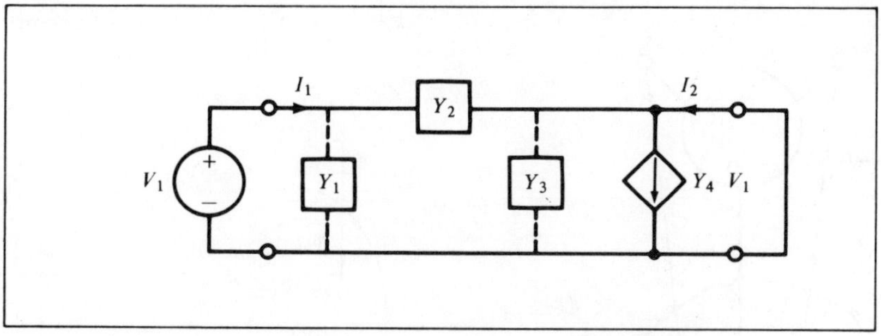

FIGURE 15.5 Network with controlled source.

Three-Terminal Two-Ports

The above examples illustrate a special but highly useful class of two ports: the three-terminal network. The general case is shown in Figure 15.6. One of the three terminals is designated a reference terminal (r); the port voltages are defined with respect to this terminal. The current flowing from this terminal is then $I_1 + I_2$, because the algebraic sum of currents entering the network must be zero.

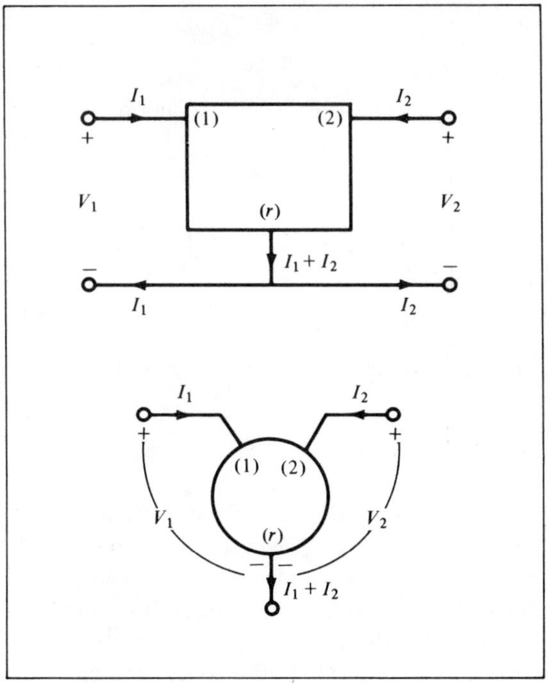

FIGURE 15.6 Three-terminal networks.

Three-terminal network descriptions are conceptually the simplest extensions of the two-terminal (branch) descriptions that we have used throughout the text. They are useful in describing many circuit components (such as transistors) which have three accessible nodes, as well as complicated networks that have an input node, an output node, and a common ground from input to output.

Analysis with Three-Terminal Networks

When three-terminal networks appear in a network analysis problem, we can perform the analysis using the same node, mesh, and loop analysis methods which we have already studied. In the following discussion, we will limit ourselves to node analysis, because this turns out to be most convenient to use with the y-parameters.

Once the y-parameters of a three-terminal network are found, we concentrate on the use of these parameters to represent the process of connection to other networks. A typical connection is shown in Figure 15.7, where a three-terminal network is connected to nodes i and j and the reference node of another network. If we examine the node equations of the interconnected networks, we find that the total currents flowing out of nodes i and j are altered by an amount

$$\begin{bmatrix} \Delta I_i \\ \Delta I_j \end{bmatrix} = \begin{bmatrix} y_{11} & y_{12} \\ y_{21} & y_{22} \end{bmatrix} \begin{bmatrix} V_i \\ V_j \end{bmatrix} \qquad (15.6)$$

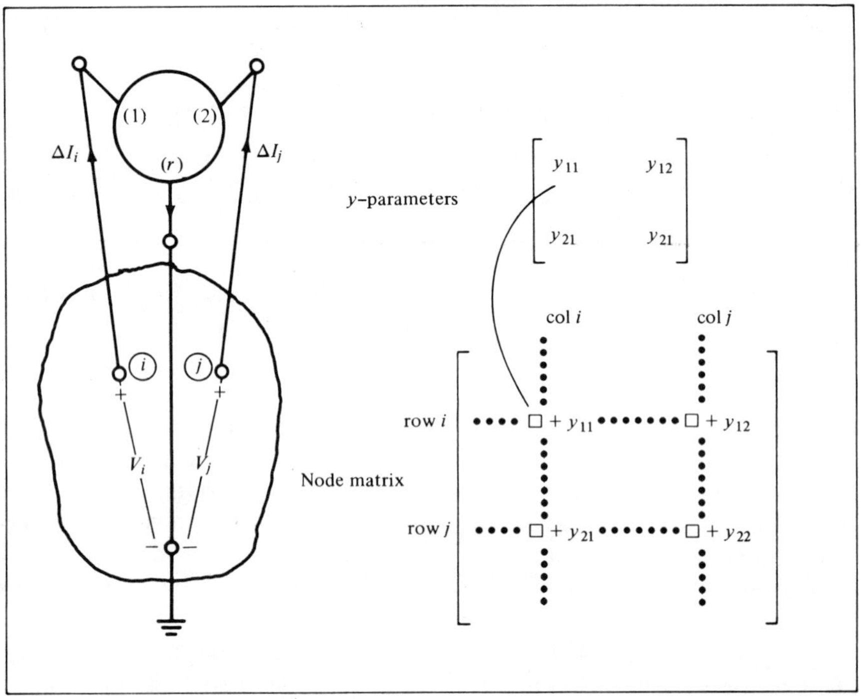

FIGURE 15.7 Representation of connection of three-terminal network (r connected to ground).

with the addition of the three-terminal network. These currents are represented by adding terms to the node equations in the i and j rows and columns as shown in the figure.

As an example of this procedure, consider the effect on the node matrix of adding an admittance Y across any two nodes i and j of a network. We know from Chapter 4 that this should result in Y appearing in the (i, i) and (j, j) positions and $-Y$ in the (i, j) and (j, i) positions of the node matrix. For our present study, we can view the admittance as forming the three-terminal network of Figure 15.8(b), with the port matrix

$$\begin{bmatrix} \Delta I_i \\ \Delta I_j \end{bmatrix} = \begin{bmatrix} Y & -Y \\ -Y & Y \end{bmatrix} \begin{bmatrix} V_i \\ V_j \end{bmatrix} \tag{15.7}$$

From Figure 15.7, we deduce the altered node matrix to be

$$\begin{array}{c} \\ \\ \text{row } i \\ \\ \text{row } j \\ \\ \end{array} \begin{array}{c} \begin{array}{ccccc} & \text{col } i & & \text{col } j & \end{array} \\ \left[\begin{array}{ccccc} & \vdots & & \vdots & \\ \cdots & \square + Y & \cdots & \square - Y & \cdots \\ & \vdots & & \vdots & \\ \cdots & \square - Y & \cdots & \square + Y & \cdots \\ & \vdots & & \vdots & \end{array}\right] \end{array}$$

A more general form of connection is illustrated in Figure 15.9, where the reference node r is connected to a node other than the ground node. Because connections have been made to three nodes, three node equations must now be altered. Noting that the port voltages are

$V_i - V_k$ and $V_j - V_k$, we can write

$$\text{Node } i: \quad \Delta I_i = y_{11}(V_i - V_k) + y_{12}(V_j - V_k) \tag{15.8}$$

$$\text{Node } j: \quad \Delta I_j = y_{21}(V_i - V_k) + y_{22}(V_j - V_k) \tag{15.9}$$

The current flowing *into* terminal (r) from node k is now $-\Delta I_i - \Delta I_j$, which we have found above. Therefore from Equations (15.8) and (15.9),

$$\begin{aligned}\text{Node } k: \quad & -\Delta I_i - \Delta I_j \\ &= (-y_{11} - y_{21})(V_i - V_k) \\ &+ (-y_{12} - y_{22})(V_j - V_k)\end{aligned} \tag{15.10}$$

Combining Equations (15.8) to (15.10) in matrix form gives

$$\begin{bmatrix} \Delta I_i \\ \Delta I_j \\ -\Delta I_i - \Delta I_j \end{bmatrix} = \begin{bmatrix} y_{11} & y_{12} & -y_{11} - y_{12} \\ y_{21} & y_{22} & -y_{21} - y_{22} \\ -y_{11} - y_{21} & -y_{12} - y_{22} & y_{11} + y_{12} + y_{21} + y_{22} \end{bmatrix} \begin{bmatrix} V_i \\ V_j \\ V_k \end{bmatrix}$$

This 3×3 port matrix is known as an *indefinite admittance matrix* (Reference 15.1). Once it is formed, the three terminals can be connected to any other network in any orientation. The node matrix is altered as shown in Figure 15.9, where now *nine* positions are affected. In the event that the reference node (r) is connected to ground, then the kth row and column are missing from the node matrix and the simpler case of Figure 15.7 results.

Because the addition of the indefinite matrix to the node matrix can be performed by inspection, the above is a highly useful generalization to three-terminal networks of the method of Chapter 4 for constructing the node matrix by inspection from branches (two-terminal networks).

Example 15.3

A three-terminal network with known y-parameters is connected to a network of admittances as shown in Figure 15.10. We are to find the voltage transfer function V_0/V_s in terms of the y-parameters and the admittances shown.

Converting the voltage source V_s and admittance Y_s to a Norton equivalent, the node equations *without* the three-terminal network attached are

$$\begin{bmatrix} Y_s & 0 & 0 \\ 0 & Y_L & 0 \\ 0 & 0 & Y_r \end{bmatrix} \begin{bmatrix} V_1 \\ V_2 \\ V_3 \end{bmatrix} = \begin{bmatrix} V_s Y_s \\ 0 \\ 0 \end{bmatrix} \tag{15.11}$$

With the network connected as shown, we identify the port node numbers as

$$i = 1 \qquad j = 2 \qquad k = 3$$

Following the model of Figure 15.9, we have the complete node matrix as

$$\begin{array}{c|ccc} & i = 1 & j = 2 & k = 3 \\ \hline i = 1 & Y_s + y_{11} & y_{12} & -y_{11} - y_{12} \\ j = 2 & y_{21} & Y_L + y_{22} & -y_{21} - y_{22} \\ k = 3 & -y_{11} - y_{21} & -y_{22} - y_{12} & Y_r + y_{11} + y_{12} + y_{21} + y_{22} \end{array} \tag{15.12}$$

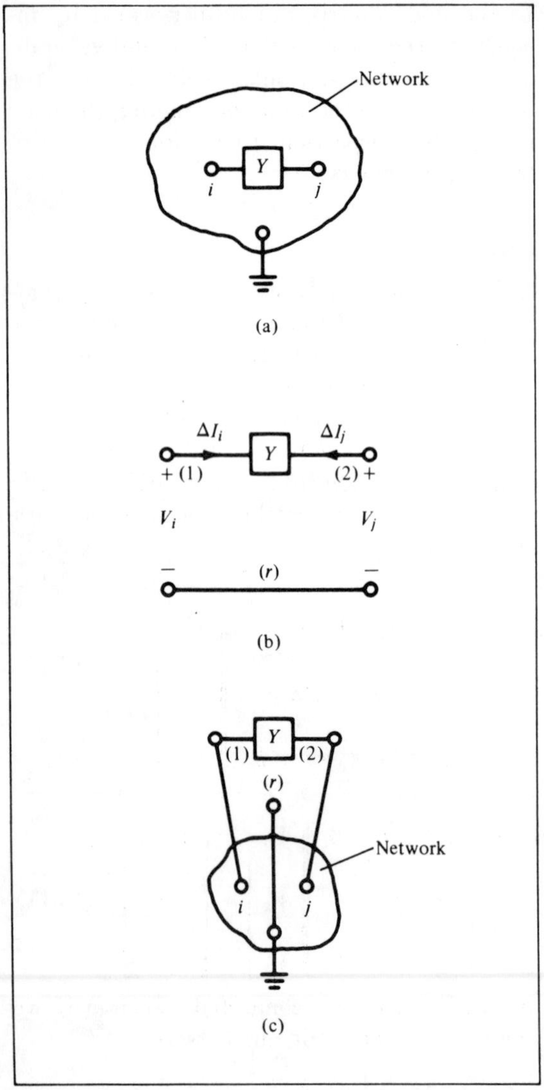

FIGURE 15.8 Network with simple branch attached.

We can now find the voltage ratio from

$$\frac{V_0}{V_sY_s} = \frac{V_2}{V_sY_s} = \frac{\Delta_{12}}{\Delta}$$

After considerable symbolic manipulation, we obtain

$$\frac{V_0}{V_s} = \frac{Y_s(\Delta_y - y_{21}Y_r)}{\Delta} \qquad (15.13)$$

where

$$\Delta = Y_sY_L(Y_r + y_{11} + y_{12} + y_{21} + y_{22})$$
$$+Y_sY_ry_{22} + Y_LY_ry_{11} + (Y_s + Y_L + Y_r)\,\Delta_y$$
$$+ y_{21}y_{12}(y_{12} - y_{21})$$

and $\Delta_y = y_{11}y_{22} - y_{12}y_{21}$ is the determinant of the short-circuit matrix.

This formula is useful for calculating the voltage transfer functions of three-terminal networks with sources and loads connected at one or both ports. This is such a frequent problem in network analysis that the formulas for several common configurations are shown in Table 15.1. For an example use of these formulas, see the application at the end of this chapter.

If the network in which the three-terminal network is imbedded is more complicated than that shown in Figure 15.10, we would probably resort to a computer-aided analysis using numerical values for the admittances and two-port parameters.

Determination of y-Parameters

The simple two-ports of Examples 15.1 to 15.2 do not bring out the principal conceptual and computational advantage of the two-port representation: To compact what can be a complicated internal network structure to a description involving only four functions or constants: y_{11}, y_{12}, y_{21}, and y_{22}. Determination of these parameters for a given network therefore involves an *elimination* of all internal solution variables (node voltages, loop currents, and so on) until we are left with only the port variables as solution variables.

To perform this contraction, we adopt general network analysis methods such as mesh and node analysis. Again we focus on the latter method for compatibility with computer-aided analysis.

Our general model is shown in Figure 15.11. In preparation for writing the node equations,

TABLE 15.1

Common Three-terminal Connections

Description	Model	Conditions	V_0/V_s
Unterminated	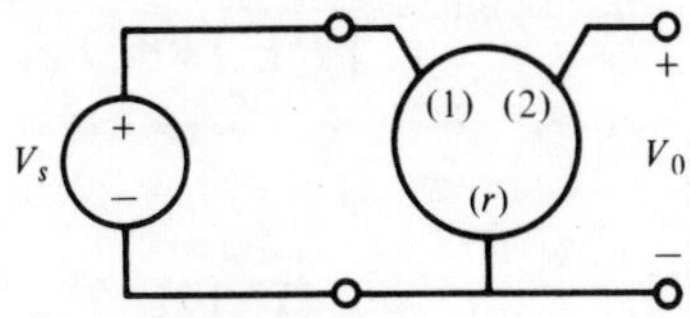	$Y_L = 0$; $Y_s = Y_r = \infty$	$\dfrac{-y_{21}}{y_{22}}$
Source resistance	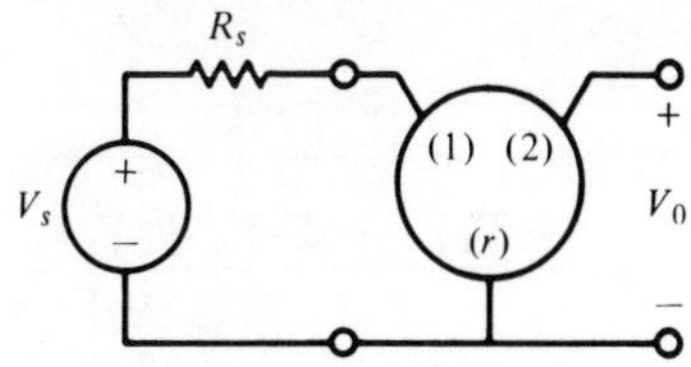	$Y_L = 0$; $Y_r = \infty$	$\dfrac{-y_{21}}{y_{22} + R_s\Delta_y}$
Load resistance	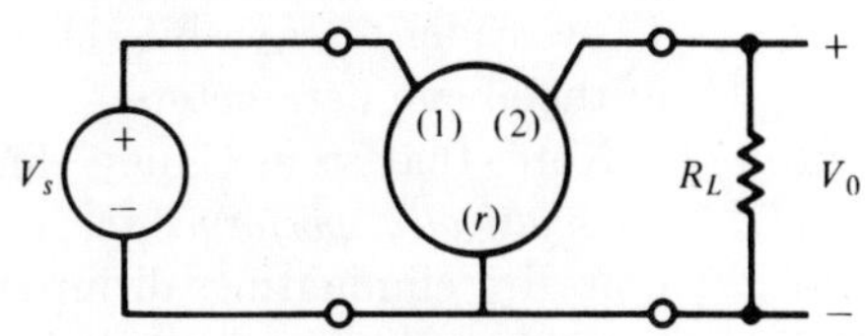	$Y_L = 1/R_L$; $Y_r = Y_s = \infty$	$\dfrac{-y_{21}}{y_{22} + 1/R_L}$

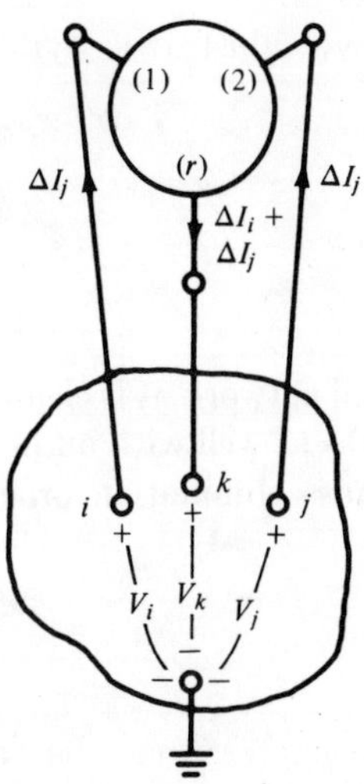

y-parameters

$$\begin{bmatrix} y_{11} & y_{12} & -y_{11}-y_{12} \\ y_{21} & y_{22} & -y_{21}-y_{22} \\ -y_{11}-y_{21} & -y_{12}-y_{22} & y_{11}+y_{12}+y_{21}+y_{22} \end{bmatrix}$$

Node matrix

$$\begin{array}{r} \\ \text{row } i \\ \text{row } j \\ \text{row } k \end{array} \begin{bmatrix} & \text{col } i & & \text{col } j & & \text{col } k & \\ \cdots & \square + y_{11} & \cdots & \square + y_{12} & \cdots & \square - y_{11} - y_{12} & \cdots \\ \cdots & \square + y_{21} & \cdots & \square + y_{22} & \cdots & \square - y_{21} - y_{22} & \cdots \\ \cdots & \square - y_{11} - y_{21} & \cdots & \square - y_{12} - y_{22} & \cdots & \square + y_{11} + y_{12} + y_{21} + y_{22} & \end{bmatrix}$$

FIGURE 15.9 General three-terminal connection.

the nodes are carefully numbered so that the two highest-numbered nodes correspond to the port terminals. For example, this numbering produces for the network of Figure 15.12 the node equations

$$\begin{bmatrix} Y_2+Y_3+Y_4 & -Y_2 & -Y_4 \\ -Y_2 & Y_1+Y_2 & 0 \\ -Y_4 & 0 & Y_4+Y_5 \end{bmatrix}\begin{bmatrix} V_1 \\ V_2 \\ V_3 \end{bmatrix} = \begin{bmatrix} 0 \\ \hat{I}_1 \\ \hat{I}_2 \end{bmatrix} \tag{15.14}$$

where port variables are indicated with a "∧" symbol, i.e., $V_2 = \hat{V}_1$ and $V_3 = \hat{V}_2$.

It is important to note that the current vector on the right-hand side will be zero-valued *except in the last two positions*, because the two-port cannot contain internal independent sources.

The contraction process begins by performing a Gauss elimination of the first row and column of the node matrix. For the above example, this involves reducing the first column below the diagonal to zero by adding multiples of the first row to the second and third rows.

$$\begin{bmatrix} Y_T & -Y_2 & -Y_4 \\ 0 & Y_1+Y_2-Y_2^2/Y_T & -Y_2Y_4/Y_T \\ 0 & -Y_2Y_4/Y_T & Y_4+Y_5-Y_4^2/Y_T \end{bmatrix}\begin{bmatrix} V_1 \\ V_2 \\ V_3 \end{bmatrix} = \begin{bmatrix} 0 \\ \hat{I}_1 \\ \hat{I}_2 \end{bmatrix}$$

where $Y_T = Y_2 + Y_3 + Y_4$. We are now left with two equations in two unknowns (V_2 and V_3). We can therefore write

$$\begin{bmatrix} Y_1+Y_2-Y_2^2/Y_T & -Y_2Y_4/Y_T \\ -Y_2Y_4/Y_T & Y_4+Y_5-Y_4^2/Y_T \end{bmatrix}\begin{bmatrix} V_2 \\ V_3 \end{bmatrix} = \begin{bmatrix} \hat{I}_1 \\ \hat{I}_2 \end{bmatrix}$$

But these are of the desired form

$$\begin{bmatrix} y_{11} & y_{12} \\ y_{21} & y_{22} \end{bmatrix}\begin{bmatrix} \hat{V}_1 \\ \hat{V}_2 \end{bmatrix} = \begin{bmatrix} \hat{I}_1 \\ \hat{I}_2 \end{bmatrix}$$

where

$$\hat{V}_1 = V_2$$

$$\hat{V}_2 = V_3$$

$$y_{11} = \frac{Y_1Y_2 + Y_1Y_3 + Y_1Y_4 + Y_2Y_3 + Y_2Y_4}{Y_2 + Y_3 + Y_4}$$

$$y_{12} = y_{21} = \frac{-Y_2Y_4}{Y_2 + Y_3 + Y_4}$$

$$y_{22} = \frac{Y_2Y_4 + Y_3Y_4 + Y_2Y_5 + Y_3Y_5 + Y_4Y_5}{Y_2 + Y_3 + Y_4}$$

(The reader may wish to check the dimensions of the above parameters.)

Note that the Gauss elimination process *automatically performs the contraction* we require by eliminating all internal node voltage variables. We simply halt this process before the last two node voltages (the terminal voltages) are eliminated, and the result is the short-circuit admittance matrix!

Although the above illustrates symbolic elimination with a small network, it is clear that (1) the procedure works as well with numerical values, and (2) a Gauss elimination program such as proposed in Chapter 5 can be easily modified to interrupt the forward elimination with $n - 2$ equations remaining. This is left as a problem.

Extension to (n + 1) Terminals

All of the above concepts and procedures extend to the case when a *multiport* network is connected at $(n + 1)$ terminals to a larger network (see Figure 15.13). This is especially true of the method for calculating the $n \times n$ short-circuit admittance matrix by Gauss elim-

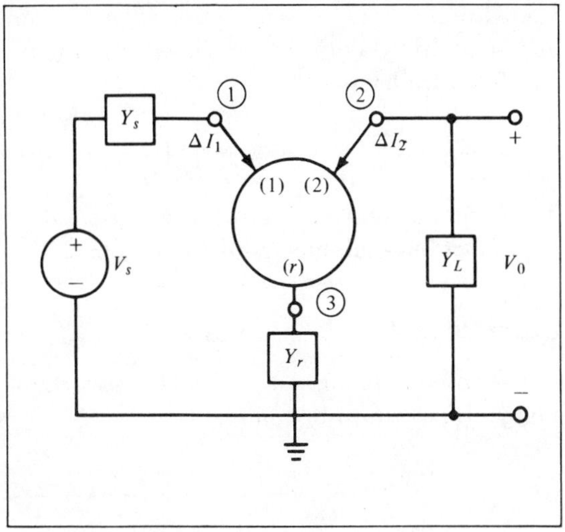

FIGURE 15.10 Example of three-terminal connection.

ination. The REDUCE subroutine is easily modified with an argument that indicates the number of ports, so that elimination is not carried beyond this point.

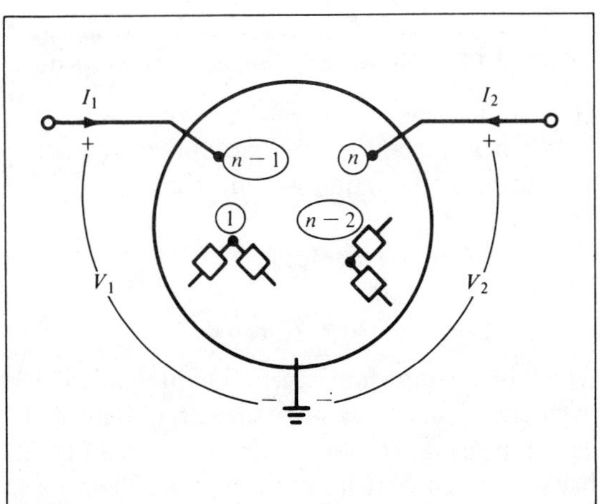

FIGURE 15.11 Model for determining y-parameters.

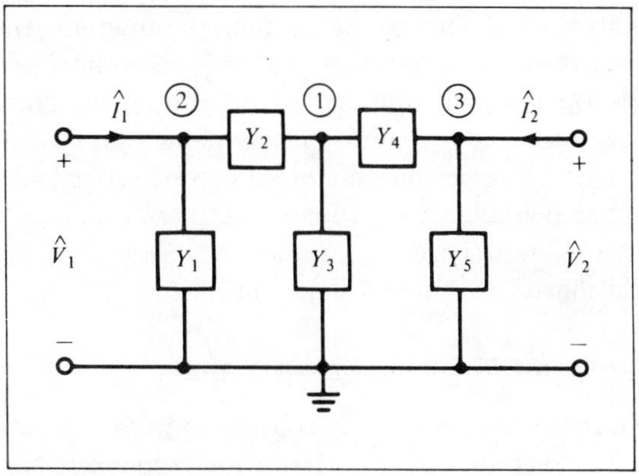

FIGURE 15.12 Example network for y-parameter calculation.

15.3 Other Parameter Sets

Introduction

The y-parameters of Section 15.2 have by far the greatest general utility for hand- and computer-aided analysis. However, in most cases, two-port parameters are determined by measurement; unfortunately, the y-parameters are not always the easiest to measure, especially because they require that both ports be short-

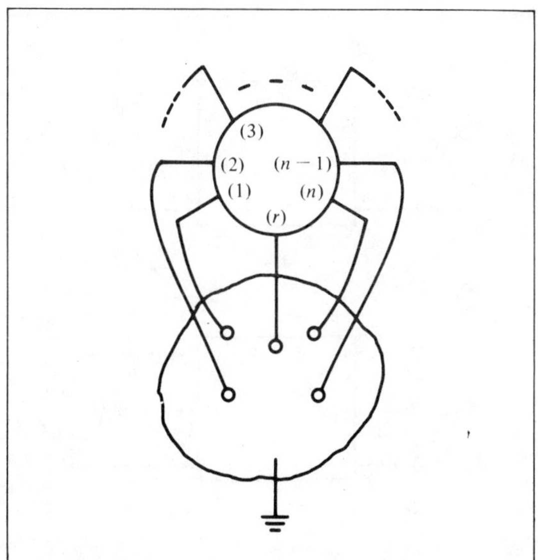

FIGURE 15.13 General $(n + 1)$-terminal connection.

circuited to determine all four parameters. In many cases, it happens that it is more natural to open-circuit one or both of the ports. This requirement is readily met by using one of the other two-port parameter sets to be described. It is possible, then, to convert these other parameters to the y-parameters using the relationships to be developed in Section 15.4.

Open-Circuit Impedance Parameters

Rather than choosing voltage sources for excitations and port currents for responses, we could just as well apply current sources at the ports. The port voltages could then be written

$$V_1 = z_{11}I_1 + z_{12}I_2 \qquad (15.15)$$

$$V_2 = z_{21}I_1 + z_{22}I_2 \qquad (15.16)$$

The z's are evidently *impedances*, because they relate complex-valued currents to complex-valued voltages.

These impedances are determined similarly to the y-parameters. In Equations (15.15) and (15.16), we set $I_2 = 0$ by open-circuiting port 2; then with a current source I_1 at port 1, the port voltages are used to determine

$$z_{11} = \left.\frac{V_1}{I_1}\right|_{I_2=0} \qquad z_{21} = \left.\frac{V_2}{I_1}\right|_{I_2=0} \qquad (15.17)$$

This calculation is shown in Figure 15.14(a). From applying the current source at port 2 and open-circuiting port 1 as shown in Figure 15.14(b), we have

$$z_{12} = \left.\frac{V_1}{I_2}\right|_{I_1=0} \qquad z_{22} = \left.\frac{V_2}{I_2}\right|_{I_1=0} \qquad (15.18)$$

These parameters are called the *open circuit impedance* or z-parameters of the two-port.

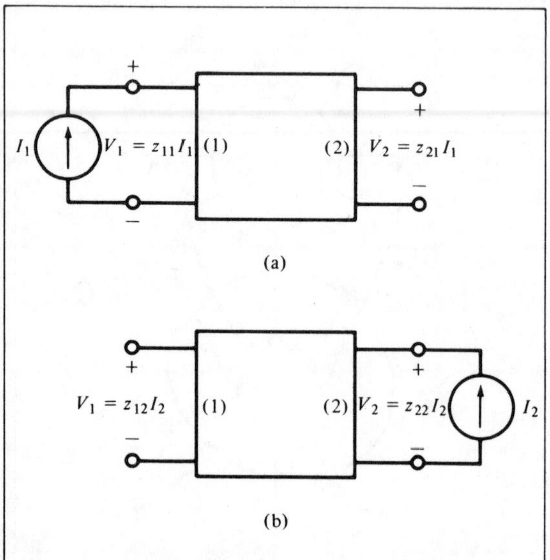

FIGURE 15.14 Physical interpretation of open-circuit impedance parameters. (a) z_{11} and z_{21}; (b) z_{12} and z_{22}.

Example 15.4

The two-port parameters of the network of Figure 15.15 are to be found.

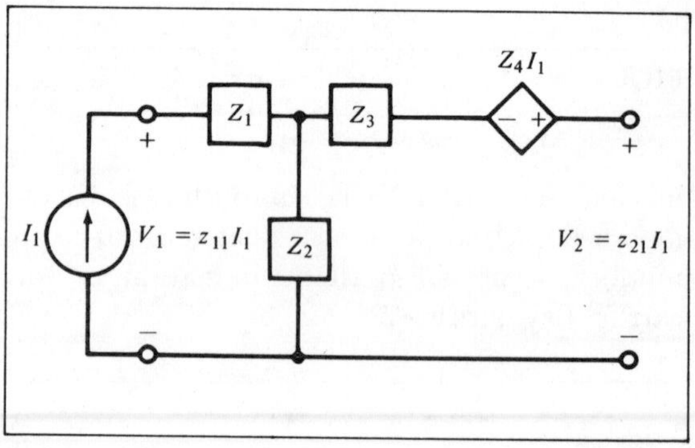

FIGURE 15.15 Network for z-parameter calculation.

When a current source is applied at port 1 as shown in the figure, we find that

$$z_{11} = \left.\frac{V_1}{I_1}\right|_{I_2=0} = Z_1 + Z_2$$

which is the impedance seen "looking in" port 1 with the second port open-circuited. The voltage appearing at port 2 does not include Z_3 because no current flows through it. Therefore, $V_2 = Z_2I_1 + Z_4I_1$ and

$$z_{21} = Z_2 + Z_4$$

The other z-parameters are found to be

$$z_{22} = Z_2 + Z_3 \qquad z_{12} = Z_2$$

Hybrid Parameters

Another useful parameter set is obtained by specifying I_1 and V_2 and solving for V_1 and I_2

$$\begin{bmatrix} V_1 \\ I_2 \end{bmatrix} = \begin{bmatrix} h_{11} & h_{12} \\ h_{21} & h_{22} \end{bmatrix} \begin{bmatrix} I_1 \\ V_2 \end{bmatrix} \tag{15.19}$$

The coefficients of this equation pair are called the *hybrid or h-parameters* of the two-port. Because of the mixed nature of the variables in this case, the various h-parameters are dimensionally different; h_{11} is an impedance, h_{22} is an admittance, and h_{12} and h_{21} are both dimensionless ratios.

Equation (15.19) suggests how the hybrid parameters are defined and measured, namely

$$h_{12} = \left.\frac{V_1}{V_2}\right|_{I_1=0}$$

$$h_{21} = \left.\frac{I_2}{I_1}\right|_{V_2=0}$$

These parameters involve both open- and short-circuit measurements.

Example 15.5

We are to find the h-parameters of the network of Figure 15.16.

From the defining equations for the h-parameters, we apply a current source I_1 and voltage source V_2 and determine V_1 and I_2. Using voltage and current divider formulas, we find

$$h_{11} = \left.\frac{V_1}{I_1}\right|_{V_2=0} = R + \frac{1}{sC + \frac{1}{sL}}$$

$$= \frac{s^2RLC + sL + R}{s^2LC + 1}$$

$$h_{12} = \left.\frac{V_1}{V_2}\right|_{I_1=0} = \frac{\frac{1}{sC}}{sL + \frac{1}{sC}}$$

$$= \frac{1}{s^2LC + 1}$$

$$h_{21} = \left.\frac{I_2}{I_1}\right|_{V_2=0} = \frac{-\frac{1}{sL}}{sC + \frac{1}{sL}}$$

$$= \frac{-1}{s^2LC + 1}$$

$$h_{22} = \left.\frac{I_2}{V_2}\right|_{I_1=0} = \frac{1}{sL + \frac{1}{sC}}$$

$$= \frac{sC}{s^2LC + 1}$$

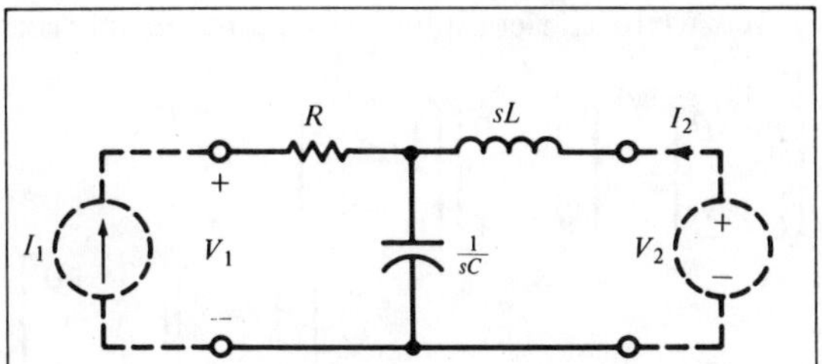

FIGURE 15.16 Network for h-parameter calculation.

ABCD Parameters

The fourth popular parameter set is obtained by expressing V_1 and I_1 as functions of V_2 and $-I_2$.

$$\begin{bmatrix} V_1 \\ I_1 \end{bmatrix} = \begin{bmatrix} A & B \\ C & D \end{bmatrix} \begin{bmatrix} V_2 \\ -I_2 \end{bmatrix} \tag{15.20}$$

The coefficients of this set of equations are called the *transmission* or *ABCD* parameters of the two-port. These parameters are particularly convenient when we are dealing with a cascade of two-ports as shown in Figure 15.17.

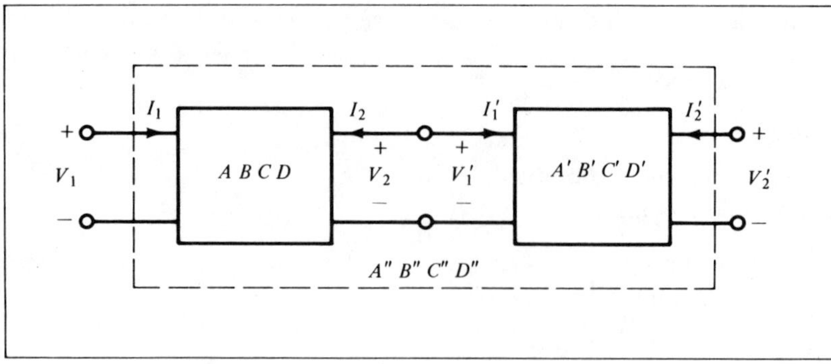

FIGURE 15.17 A cascade of two-port networks.

If the transmission parameters of both two-ports are known, the calculation of the parameters of the cascade is straightforward. In Figure 15.17 the first two-port is described by Equation (15.20); if the transmission parameters of the second two-port are $A'B'C'D'$, as indicated,

$$\begin{bmatrix} V_1' \\ I_1' \end{bmatrix} = \begin{bmatrix} A' & B' \\ C' & D' \end{bmatrix} \begin{bmatrix} V_2' \\ -I_2' \end{bmatrix} \qquad (15.21)$$

For the cascade connection, Kirchhoff's laws force the equalities

$$I_1' = -I_2$$

and

$$V_1' = V_2$$

so that Equation (15.21) can be written

$$\begin{bmatrix} V_2 \\ -I_2 \end{bmatrix} = \begin{bmatrix} A' & B' \\ C' & D' \end{bmatrix} \begin{bmatrix} V_2' \\ -I_2' \end{bmatrix} \qquad (15.22)$$

Substituting this into Equation (15.20) gives directly

$$\begin{bmatrix} V_1 \\ I_1 \end{bmatrix} = \begin{bmatrix} A & B \\ C & D \end{bmatrix} \begin{bmatrix} A' & B' \\ C' & D' \end{bmatrix} \begin{bmatrix} V_2' \\ -I_2' \end{bmatrix}$$

$$= \begin{bmatrix} A'' & B'' \\ C'' & D'' \end{bmatrix} \begin{bmatrix} V_2' \\ -I_2' \end{bmatrix} \qquad (15.23)$$

This shows that *the matrix of transmission parameters of the cascade is the product of the matrices of the individual two-ports*. For example, $A'' = AA' + BC'$, and so on.

Example 15.6

The $ABCD$ parameters of the ladder network of Figure 15.18 are to be found as the product of the $ABCD$ parameters of the subnetworks formed by the individual elements. The transfer function $V_2(s)/V_1(s)$ is also to be found.

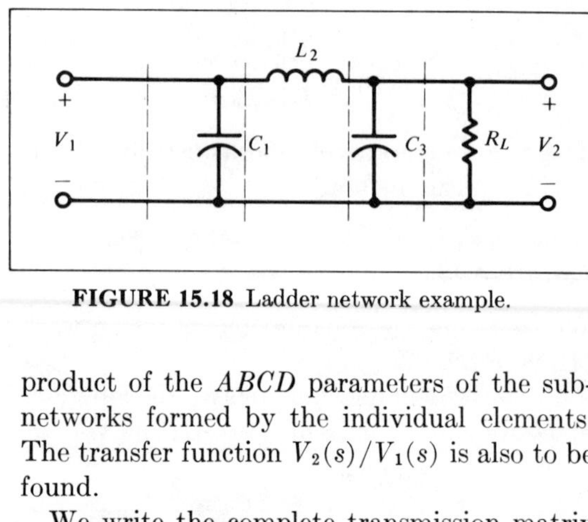

FIGURE 15.18 Ladder network example.

We write the complete transmission matrix as

$$\begin{bmatrix} A & B \\ C & D \end{bmatrix} = \begin{bmatrix} 1 & 0 \\ sC_1 & 1 \end{bmatrix} \begin{bmatrix} 1 & sL_2 \\ 0 & 1 \end{bmatrix}$$

$$\times \begin{bmatrix} 1 & 0 \\ sC_3 & 1 \end{bmatrix} \begin{bmatrix} 1 & 0 \\ \dfrac{1}{R_L} & 1 \end{bmatrix}$$

Therefore,

$$\left.\frac{V_2}{V_1}(s)\right|_{I_2=0} = \frac{1}{C}$$

$$= \frac{1}{G_L + s(C_1 + C_3) + s^2 G_L L_2 C_1 + s^3 L_2 C_1 C_3}$$

The reader is invited to compare the computational efficiency of the above calculation of V_2/V_1 versus calculation from the y-parameters using Gauss elimination.

15.4 Reciprocity

The two-port of Example 15.1 has the property that $y_{21} = y_{12}$. From the observation that

$$y_{21} = \left.\frac{I_2}{V_1}\right|_{I_2=0} \qquad y_{12} = \left.\frac{I_1}{V_2}\right|_{I_1=0}$$

we conclude the following: If a voltage source V_s exciting port 1 produces a current I_0 at port 2, the same voltage V_s applied at port 2

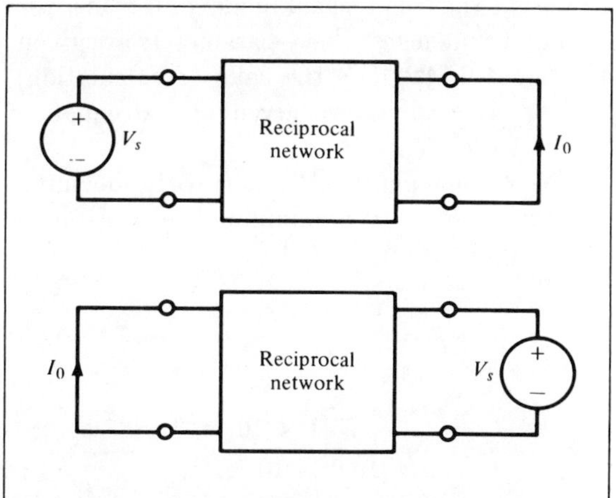

FIGURE 15.19 Meaning of reciprocity.

will also produce I_0 at port 1. Such a network, illustrated in Figure 15.19, is termed *reciprocal*.

We can usually identify a reciprocal network from the elements it contains: *A linear network of resistors, capacitors, inductors, and mutual inductance is always reciprocal.* The presence of a controlled source usually causes a network to become nonreciprocal.

For example, the network of Figure 15.20

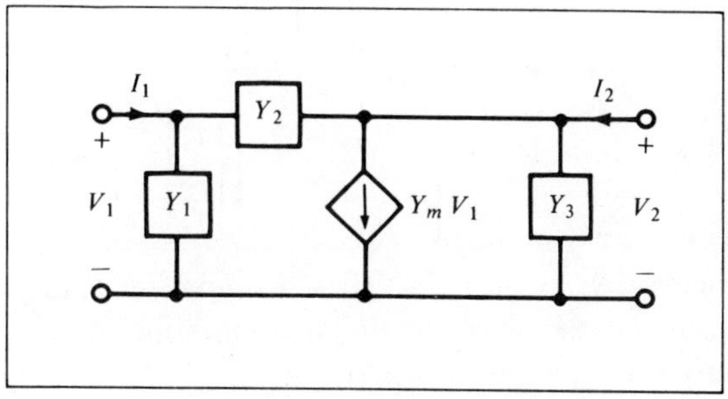

FIGURE 15.20 Nonreciprocal network.

has the node equations

$$\begin{bmatrix} Y_1 + Y_2 & -Y_2 \\ -Y_2 + Y_m & Y_2 + Y_3 \end{bmatrix} \begin{bmatrix} V_1 \\ V_2 \end{bmatrix} = \begin{bmatrix} I_1 \\ I_2 \end{bmatrix}$$

In this particular case, we can write

$$y_{12} = -Y_2 \qquad y_{21} = -Y_2 + Y_m$$

Clearly, if $Y_m = 0$ (that is, the controlled source is absent), the network is reciprocal; otherwise it is nonreciprocal.

15.5 Interrelations among Parameters

It is evident that the various two-port parameters are not independent of one another because any single parameter set is sufficient to completely characterize a two-port network. In using two-port parameters it is very convenient to be able to convert from one parameter set to another. For example, it is easy to measure the h-parameters of transistors, but for calculation of their performance in a network the y-parameters or the $ABCD$ parameters may be more convenient.

It is a simple matter to find the relationships among the various parameter sets. For example, the defining equation for the y-parameters is

$$\begin{bmatrix} y_{11} & y_{12} \\ y_{21} & y_{22} \end{bmatrix} \begin{bmatrix} V_1 \\ V_2 \end{bmatrix} = \begin{bmatrix} I_1 \\ I_2 \end{bmatrix}$$

so that

$$\begin{bmatrix} V_1 \\ V_2 \end{bmatrix} = \begin{bmatrix} y_{11} & y_{12} \\ y_{21} & y_{22} \end{bmatrix}^{-1} \begin{bmatrix} I_1 \\ I_2 \end{bmatrix}$$

$$= \frac{1}{\Delta_y} \begin{bmatrix} y_{22} & -y_{12} \\ -y_{21} & y_{11} \end{bmatrix} \begin{bmatrix} I_1 \\ I_2 \end{bmatrix} \qquad (15.24)$$

where $\Delta_y = y_{11}y_{22} - y_{12}y_{21}$. Comparing Equation (15.24) with the defining equation for the z-parameters, we can write

$$\begin{bmatrix} z_{11} & z_{12} \\ z_{21} & z_{22} \end{bmatrix} = \frac{1}{\Delta_y} \begin{bmatrix} y_{22} & -y_{12} \\ -y_{21} & y_{11} \end{bmatrix} \qquad (15.25)$$

so that $z_{11} = y_{22}/\Delta_y$ and so on.

Between some of the parameters the relationship may be even simpler. From Equation (15.19), when $V_2 = 0$,

$$V_1 = h_{11}I_1 \qquad (15.26)$$

and

$$I_2 = h_{21}I_1 \qquad (15.27)$$

Therefore,

$$h_{11} = \left.\frac{V_1}{I_1}\right|_{V_2=0} = \frac{1}{y_{11}} \qquad (15.28)$$

and

$$h_{21} = \left.\frac{I_2}{I_1}\right|_{V_2=0} = \frac{y_{21}}{y_{11}} \qquad (15.29)$$

All of the relationships among the four sets of parameters presented in the previous section are given in Table 15.2. All of the matrices on a single row of this table are equal. Reading across the (1,1) elements of the second row, for example, we see that

$$z_{11} = \frac{y_{22}}{\Delta_y} = \frac{\Delta_h}{h_{22}} = \frac{A}{C}$$

Conclusion

The two-port parameters given in Table 15.2 by no means exhaust the sets of relationships that can be written to relate the two-port currents and voltages. The interested reader can find more material on this subject both from a theoretical (References 15.3 and 15.4) and a practical (Reference 15.2) viewpoint.

APPLICATION

Frequency-Dependent Two-Port

It is very common to display frequency-response plots of network parameters, that is, the magnitude and phase response versus frequency. Consider the two-port of Figure 15.21 with a load resistance attached; this might represent the small-signal model of a transistor at high frequencies. The y-parameters are given in Figure 15.22. For the sake of calculation, the precise values are given at a frequency $f_0 = 2.15 \times 10^9$ hz.

The voltage gain V_0/V_s at f_0 with no source resistance ($R_s = 0$) and infinite load resistance ($G_L = 0$) is, from Table 15.1

$$\mu = \frac{V_0}{V_s} = \frac{-y_{21}}{y_{22}}$$

$$= \frac{-75.3 \times 10^{-3}e^{-j40°}}{16.0 \times 10^{-3}e^{j66°}}$$

$$= -4.7e^{-j106°}$$

That is, the two-port will amplify a sinusoidal voltage input by a factor of 4.7 at f_0. From the graph, it is clear that y_{22} increases with frequency and y_{21} decreases with frequency; therefore, $|\mu|$ will decrease with frequency.

If a load resistance is added, the voltage ratio

TABLE 15.2

Conversion Chart for Two-port Parameters

	$[y]$	$[z]$	$[h]$	$[T]$	Reciprocity Condition
$[y]$	$\begin{matrix} y_{11} & y_{12} \\ y_{21} & y_{22} \end{matrix}$	$\begin{matrix} \frac{z_{22}}{\Delta_z} & \frac{-z_{12}}{\Delta_z} \\ \frac{-z_{21}}{\Delta_z} & \frac{z_{11}}{\Delta_z} \end{matrix}$	$\begin{matrix} \frac{1}{h_{11}} & \frac{-h_{12}}{h_{11}} \\ \frac{h_{21}}{h_{11}} & \frac{\Delta_h}{h_{11}} \end{matrix}$	$\begin{matrix} \frac{D}{B} & \frac{-\Delta_T}{B} \\ \frac{-1}{B} & \frac{A}{B} \end{matrix}$	$y_{12} = y_{21}$
$[z]$	$\begin{matrix} \frac{y_{22}}{\Delta_y} & \frac{-y_{12}}{\Delta_y} \\ \frac{-y_{21}}{\Delta_y} & \frac{y_{11}}{\Delta_y} \end{matrix}$	$\begin{matrix} z_{11} & z_{12} \\ z_{21} & z_{22} \end{matrix}$	$\begin{matrix} \frac{\Delta_h}{h_{22}} & \frac{h_{12}}{h_{22}} \\ \frac{-h_{21}}{h_{22}} & \frac{1}{h_{22}} \end{matrix}$	$\begin{matrix} \frac{A}{C} & \frac{\Delta_T}{C} \\ \frac{1}{C} & \frac{D}{C} \end{matrix}$	$z_{12} = z_{21}$
$[h]$	$\begin{matrix} \frac{1}{y_{11}} & \frac{-y_{12}}{y_{11}} \\ \frac{y_{21}}{y_{11}} & \frac{\Delta_y}{y_{11}} \end{matrix}$	$\begin{matrix} \frac{\Delta_z}{z_{22}} & \frac{z_{12}}{z_{22}} \\ \frac{-z_{21}}{z_{22}} & \frac{1}{z_{22}} \end{matrix}$	$\begin{matrix} h_{11} & h_{12} \\ h_{21} & h_{22} \end{matrix}$	$\begin{matrix} \frac{B}{D} & \frac{\Delta_T}{D} \\ \frac{-1}{D} & \frac{C}{D} \end{matrix}$	$h_{12} = -h_{21}$
$[T]$	$\begin{matrix} \frac{-y_{22}}{y_{21}} & \frac{-1}{y_{21}} \\ \frac{-\Delta_y}{y_{21}} & \frac{-y_{11}}{y_{21}} \end{matrix}$	$\begin{matrix} \frac{z_{11}}{z_{21}} & \frac{\Delta_z}{z_{21}} \\ \frac{1}{z_{21}} & \frac{z_{22}}{z_{21}} \end{matrix}$	$\begin{matrix} \frac{-\Delta_h}{h_{21}} & \frac{-h_{11}}{h_{21}} \\ \frac{-h_{22}}{h_{21}} & \frac{-1}{h_{21}} \end{matrix}$	$\begin{matrix} A & B \\ C & D \end{matrix}$	$\Delta_T = 1$

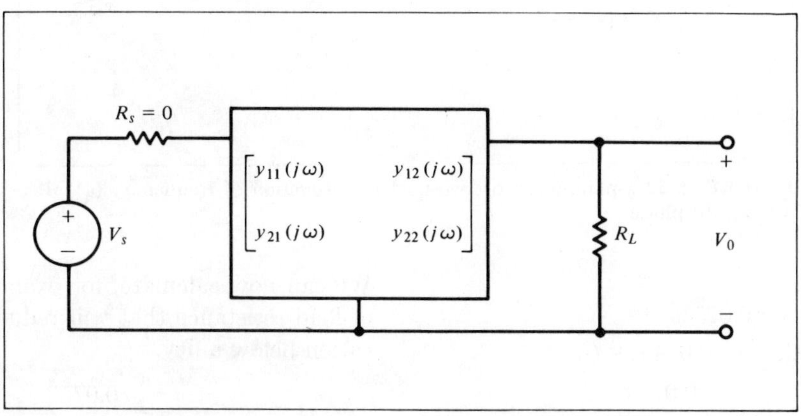

FIGURE 15.21 Two-port with load and source.

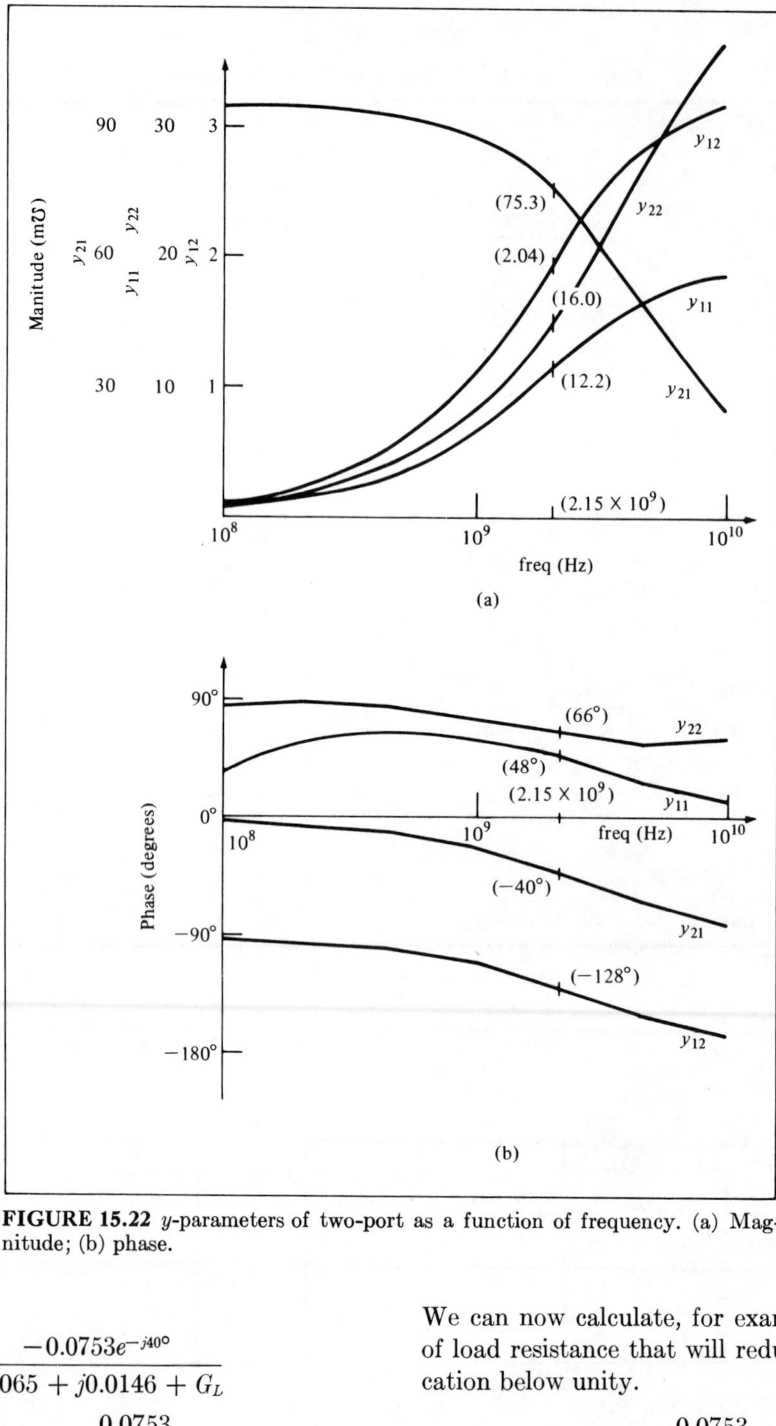

FIGURE 15.22 y-parameters of two-port as a function of frequency. (a) Magnitude; (b) phase.

becomes

$$\mu = \frac{-0.0753e^{-j40°}}{0.0065 + j0.0146 + G_L}$$

$$|\mu| = \frac{0.0753}{\sqrt{(0.0065 + G_L)^2 + (0.0146)^2}}$$

We can now calculate, for example, the value of load resistance that will reduce the amplification below unity.

$$1 \geqq \frac{0.0753}{\sqrt{(0.0065 + G_L)^2 + (0.0146)^2}}$$

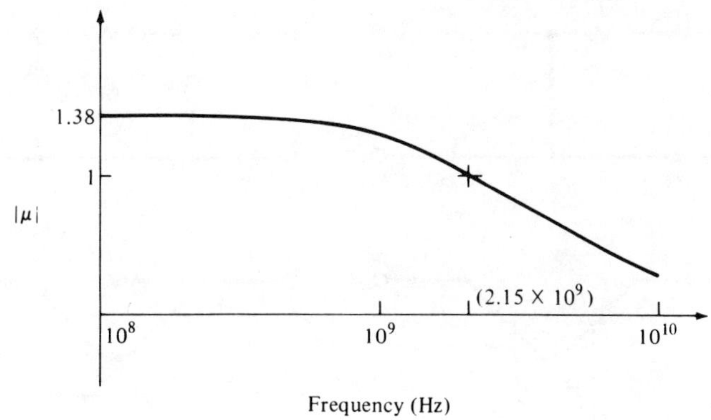

FIGURE 15.23 Magnitude response with load.

giving

$$G_L \geqq 0.068 \text{ ℧}$$

or

$$R_L \leqq 14.7 \ \Omega$$

With this value of load resistance added to the two-port, the magnitude response is plotted in Figure 15.23; we observe that μ does indeed pass through unity at $f_0 = 2.15 \times 10^9$ hz.

Several problems utilizing the above data are given in Problem 15.24.

PROBLEMS

15.1 (a)–(g) Determine the y-parameters of the networks shown in Figure P 15.1.

15.2 Given a set of y-parameters for a three-terminal network, we can always construct the equivalent network of the form of Figure P 15.2. Find Y_m as a function of the y-parameters. This representation is useful with the DCAP program of Chapter 5.

15.3 Given

$$y_{11} = 10 \qquad y_{12} = 0$$

$$y_{21} = 5 \qquad y_{22} = 20$$

(a) Can this be modeled by a network without controlled sources? Why not? (b) Sketch a network model of the network. Label carefully all V's and I's.

15.4 (a) The voltage-controlled current source of Figure P 15.3(a) is added to a network at three nodes ⓘ, ⓙ, ⓚ as shown. Obtain the two-port matrix for this element and show where $\pm g_m$ appears in the resultant node matrix, using the results of Figure 15.9. (b) Repeat (a) for the connection of Figure P 15.3(b).

15.5 The three-terminal network of Figure P 15.4 is added to the ladder network shown by connecting: (a) ⓐ to ②, ⓑ to ③, ⓒ to ground. (b) ⓐ to ④, ⓑ to ②, ⓒ to ground. (c) ⓐ to ③, ⓑ to ④, ⓒ to ⑤. (d) ⓐ to ①, ⓑ to ①, ⓒ to ②. Find the altered node matrix in each case, displaying the locations of the y-parameters of the three-terminal network.

15.6 Show that the y-parameters of two three-terminal networks connected in parallel (Figure P 15.5) add to give the y-parameters of the combined network (that is, $y_{11} = y_{11}' + y_{11}''$, and so on).

15.7 Find the complete Thevenin equivalent

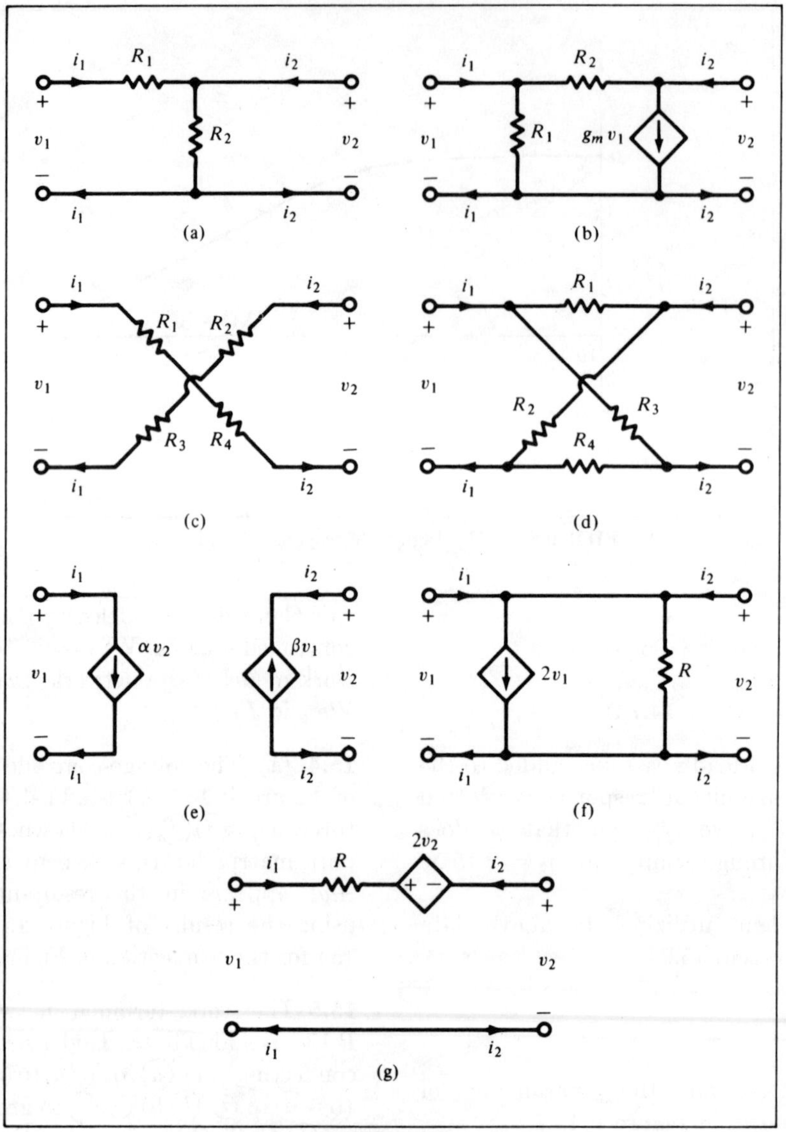

FIGURE P 15.1

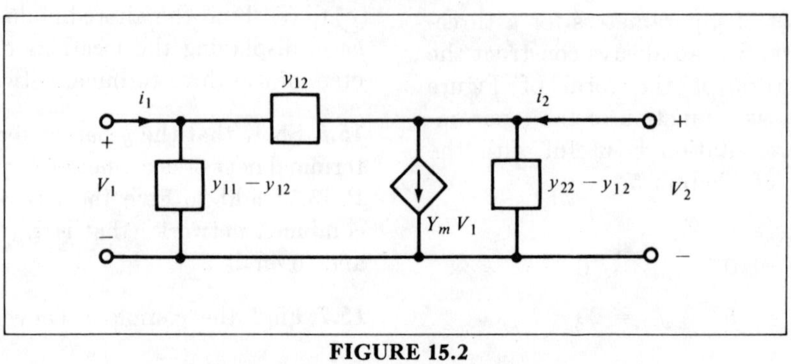

FIGURE 15.2

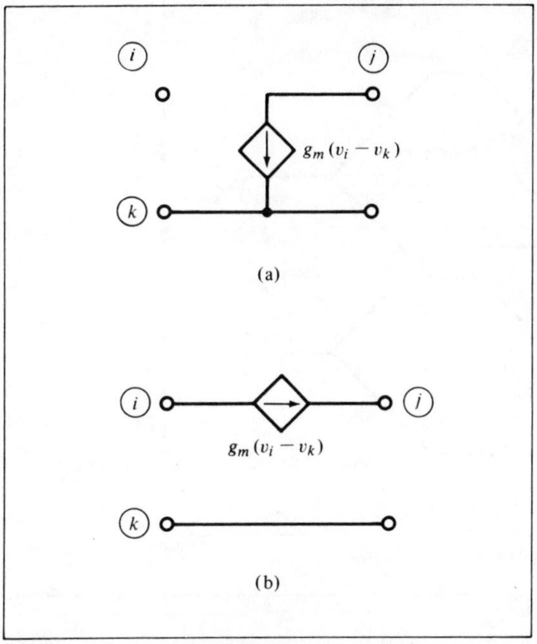

FIGURE P 15.3

of the network of Figure P 15.6 in terms of V_1 and the y-parameters.

15.8 Find the y-parameters of the gyrator of Figure P 15.7. Then insert admittance Y_0 across the "output" and show that the input *resistance* is Y_0. Use the formula

$$Y_{\text{in}} = y_{11} - \frac{y_{12}y_{21}}{y_{22} + Y_0}$$

15.9 If the y-parameters of the network N (Figure P 15.8) are y_{11}, y_{12}, y_{21}, y_{22}, what are the y-parameters of N'?

15.10 Show that the y-parameters of the network of Figure P 15.9 are given by

$$\begin{bmatrix} y_{11} & y_{12} \\ y_{21} & y_{21} \end{bmatrix} = \begin{bmatrix} 0.0192 & -1.9 \times 10^{-4} \\ -0.0188 & 2.85 \times 10^{-4} \end{bmatrix}$$

Use Gaussian elimination from the node equations.

15.11 The following problems relate to the transistor amplifier of Figure P 15.10. (a) Obtain the y-parameters of the amplifier (with $R_L = \infty$) from the node equations. Then find the voltage gain $\delta v_2/\delta v_1$ using the results of Table 15.1. (b) A load resistor R_L is added in series with a large capacitor so that the capacitor can be regarded as a short circuit for rapidly varying signals. Find the small-signal voltage

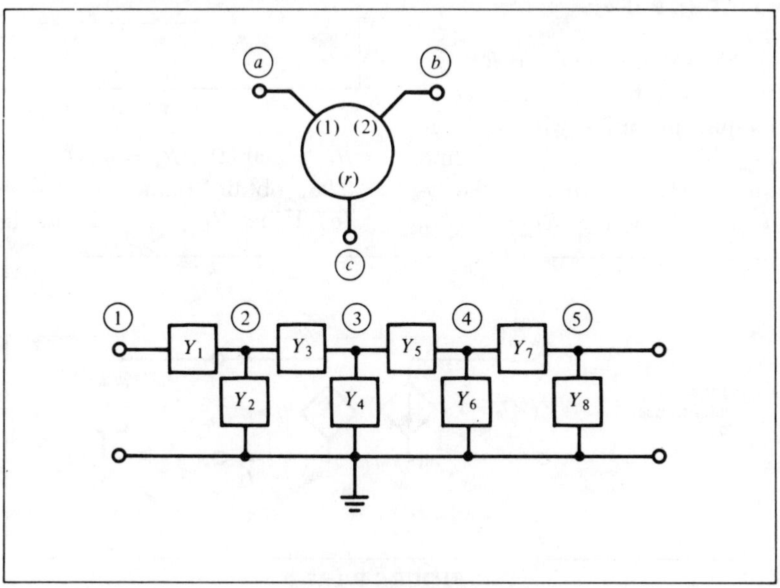

FIGURE P 15.4

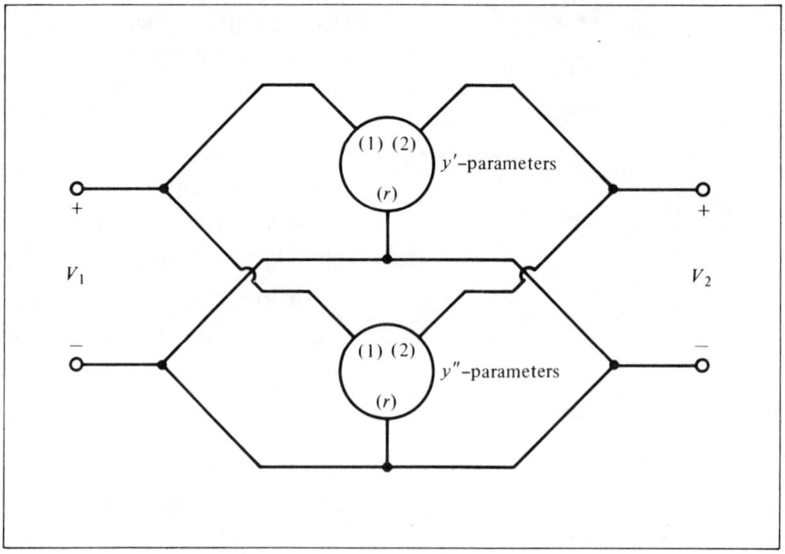

FIGURE P 15.5

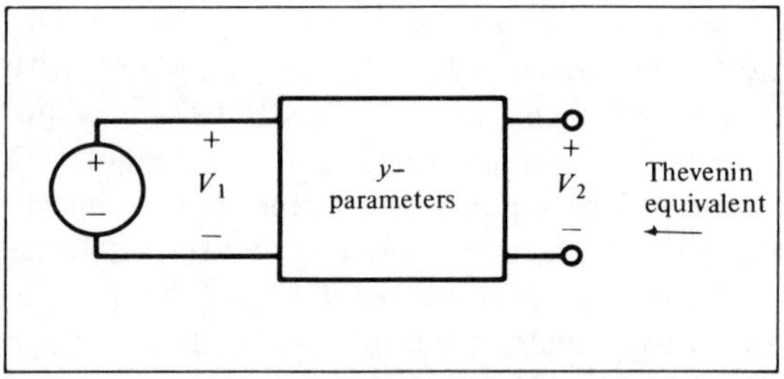

FIGURE P 15.6

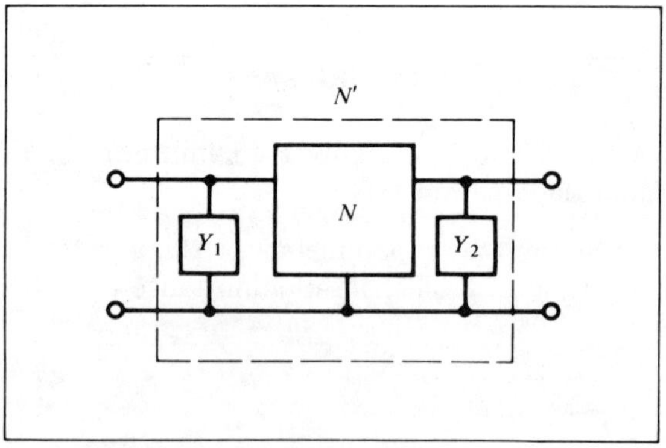

FIGURE P 15.8

gain $\delta v_2/\delta v_1$ from the y-parameters and R_L. (c) A voltage source with internal resistance is applied at the input in series with a large capacitor as in (b). With $R_L = \infty$, find the voltage gain $\delta v_0/\delta v_s$ in terms of the y-parameters and R_s. (d) Using $R_1 = 50\ k\Omega$, $R_C = 100\ k\Omega$, $R_2 = 4\ k\Omega$, $G_d = 0.02\ \mho$, $\alpha = 0.98$, obtain numerical values in part (a). (e) Using $R_L = 50\ k\Omega$ and the values of (d),

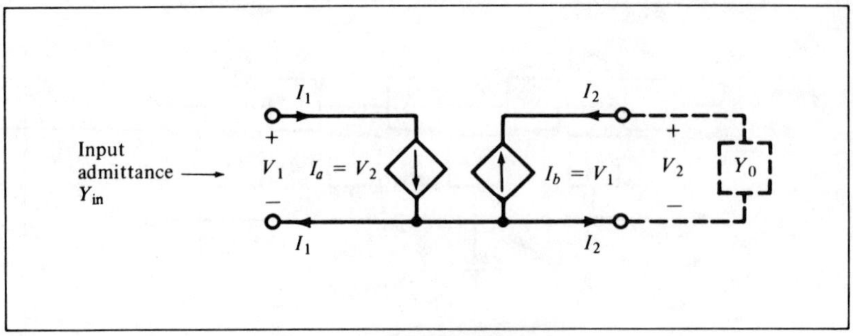

FIGURE P 15.7

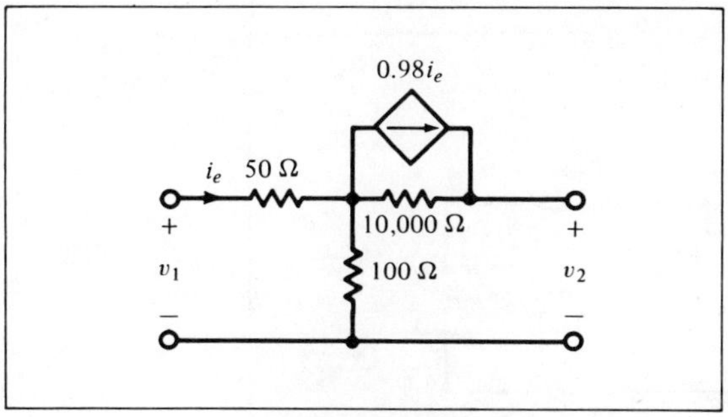

FIGURE P 15.9

find $\delta v_0/\delta v_1$ in (b). (f) Using $R_s = 1\ k\Omega$ and the values of (d), find $\delta v_0/\delta v_s$ in (c).

15.12 Transmission lines form an important class of distributed networks encountered in every branch of electrical engineering. A uniform line can be modeled by the Tee networks of Figure P 15.11. (a) Find the y-parameters of a line, simplifying expressions as much as possible. The constants associated with the line are 1. Z_0, its characteristic impedance. 2. γ, its propagation constant. 3. l, the length of the line. Both Z_0 and γ can be complex constants. (b) If the line is terminated in its characteristic impedance ($Z_L = Z_O$) at one end, find the impedance observed at the other end directly from the z- or y-parameters. Determine the voltage gain V_2/V_1 with such a load. A typical value for a coaxial line is $Z_O = 50$ (a real constant) and $\gamma = j\omega/(3 \times 10^8)$ meters^{-1} (m^{-1}). (c) If $\omega = \pi \times 10^8$ rad/sec and $l = 1$ m, what value of resistance $Z_L = R_L$ will cause a decrease of the voltage gain $|V_2/V_1|$ by a factor of 2 from its value at $R_L = \infty$? What value of Z_L would cause an infinite gain? Devise a network that will produce this infinite gain.

15.13 (a)–(g) Obtain the z-parameters of the networks of Figure P 15.1.

15.14 Obtain the y-parameters of the coupled inductances of Figure P 15.12 from the z-parameters. Use s-domain notation.

15.15 Determine the z-parameters of the two-port of Figure P 15.13 in terms of the node determinant and its cofactors.

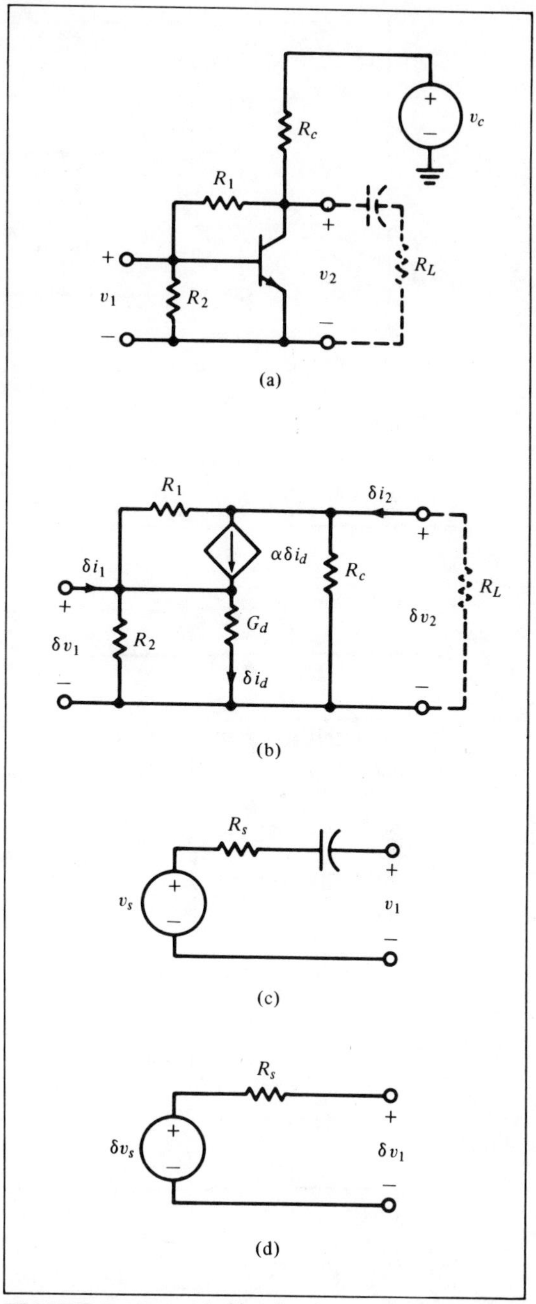

FIGURE P 15.10 (a) Circuit; (b) small-signal model.

15.16 Mesh analysis and Gauss elimination may be used to find the z-parameters in the same way that node analysis is used to determine the z-parameters. Find the z-param-

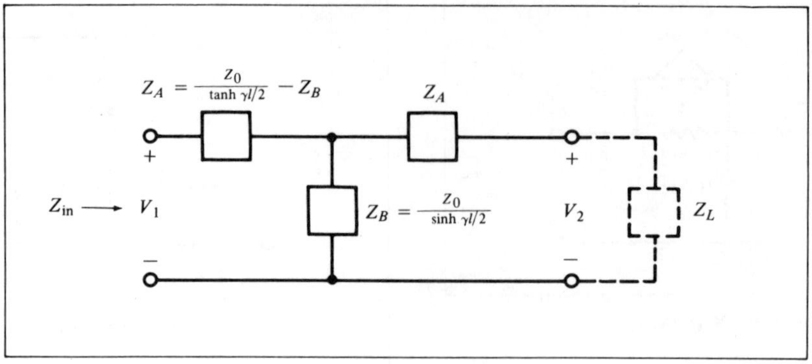

FIGURE P 15.11

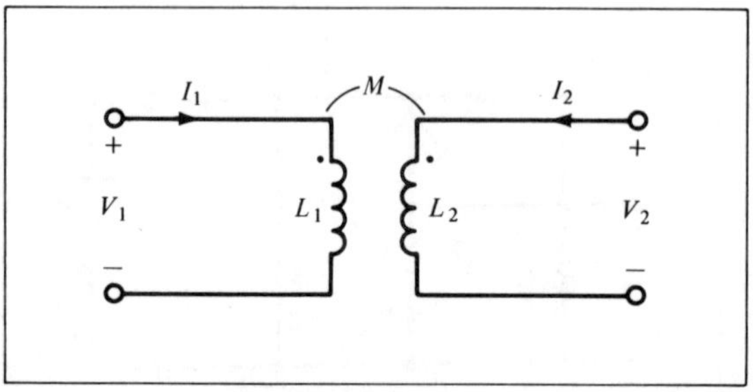

FIGURE P 15.12

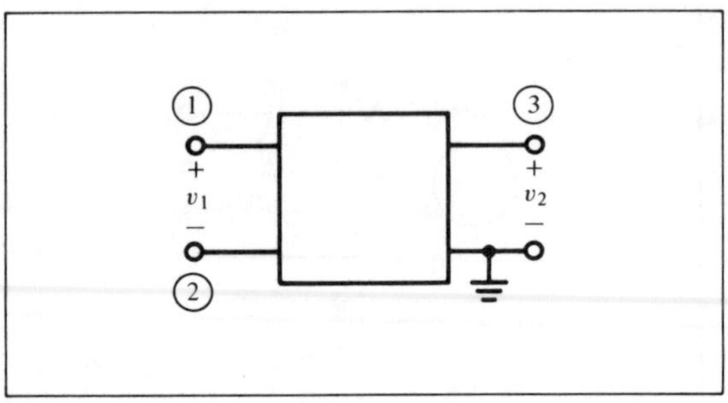

FIGURE P 15.13

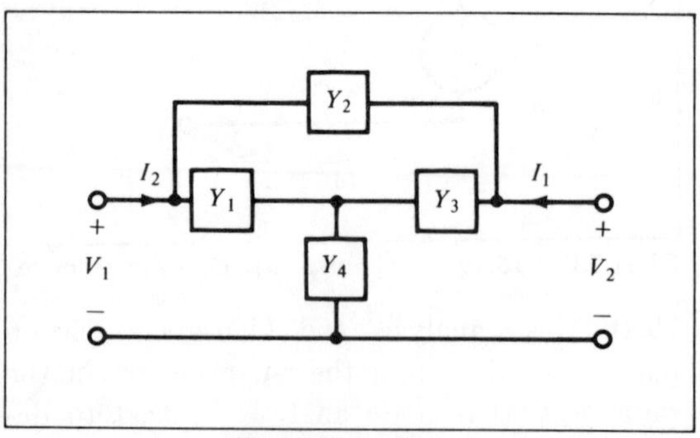

FIGURE P 15.14

eters of the network of Figure P 15.14 with this analogous method.

15.17 Two three-terminal networks are equivalent if they have the same two-port param-

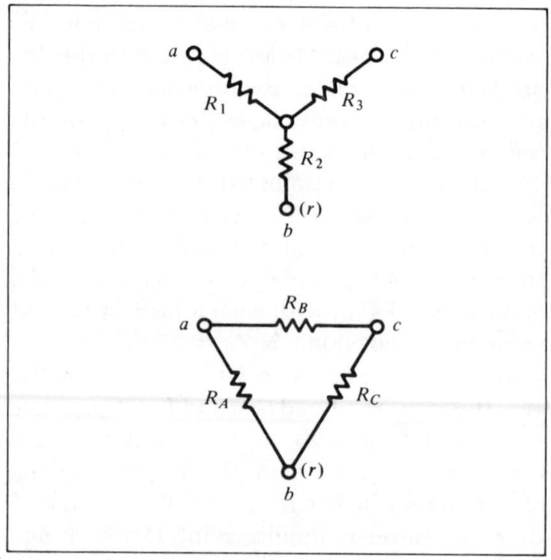

FIGURE P 15.15

eters. The Tee and Pi networks of Figure P 15.15 are equivalent if

$$R_1 = R_A R_B/(R_A + R_B + R_C)$$

$$R_2 = f_2(R_A, R_B, R_C)$$

$$R_3 = f_3(R_A, R_B, R_C)$$

Find f_2 and f_3. It is suggested that you find the

z-parameters of one of the networks and use Table 15.2 to relate the y- and z-parameters.

15.18 Use the most convenient set of parameters to solve the following problems. (a) If $I_1 = 3$ and $I_2 = 0$, we measure $V_1 = 5$ and $V_2 = -2$; if $I_1 = 0$ and $I_2 = 2$, we measure $V_1 = 6$ and $V_2 = 3$; find V_1 and V_2 for $I_1 = 5$, $I_2 = 6$. (b) If $V_1 = 2$ and $V_2 = 0$, $I_1 = -3$ and $I_2 = 1$; if $V_1 = 0$ and $V_2 = -1$, $I_1 = 6$ and $I_2 = 7$; find I_1 and I_2 for $V_1 = 1$ and $V_2 = 1$. (c) If $V_1 = 0$ and $I_2 = 3$, $I_1 = 5$ and $V_2 = 0$; if $V_1 = -3$ and $I_2 = 0$, $I_1 = -9$ and $V_2 = 6$; find I_1 and V_2 for $V_1 = 3$ and $I_2 = 7$. (d) If $V_1 = 1$ and $I_1 = 0$, $V_2 = 6$ and $I_2 = 5$; if $V_1 = 0$ and $I_1 = 10$, $V_2 = 5$ and $I_2 = 3$; find V_2 and I_2 for $V_1 = 1$ and $I_1 = -1$.

15.19 Two sets of measurements are made on a reciprocal network as illustrated in Figure P 15.16; find the y-parameters from these measurements. If the network were nonreciprocal, how would this alter your solution?

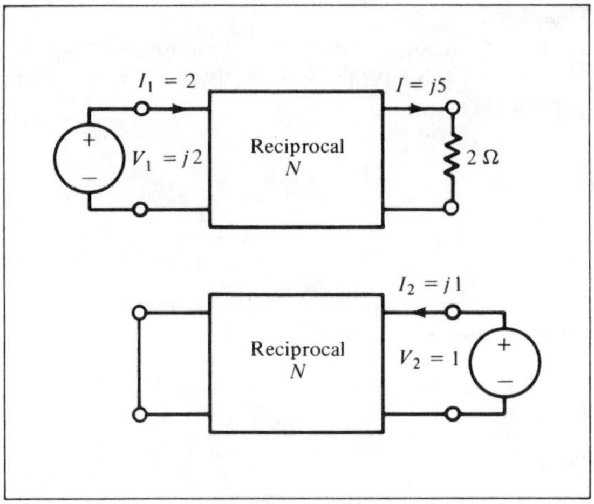

FIGURE P 15.16

15.20 Show that the transmission (ABCD) parameters of the Pi network of Figure P 15.17 are

$$A = 1 + Z_1/Z_B \qquad C = \frac{(Z_1 + Z_A + Z_B)}{Z_A Z_B}$$

$$B = Z_1 \qquad D = 1 + Z_1/Z_A$$

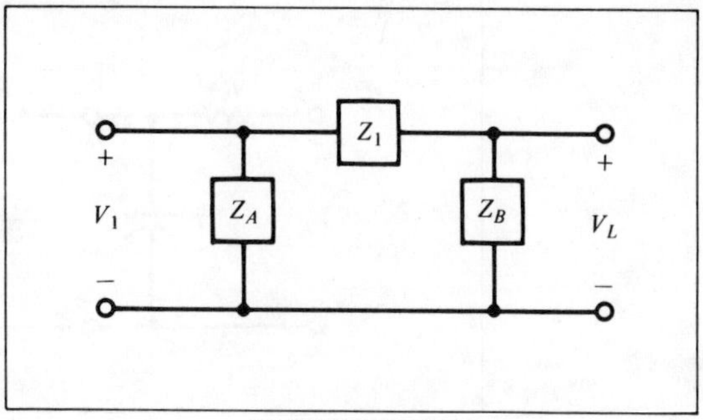

FIGURE P 15.17

15.21 Determine the $ABCD$ parameters of the network of Figure P 15.18.

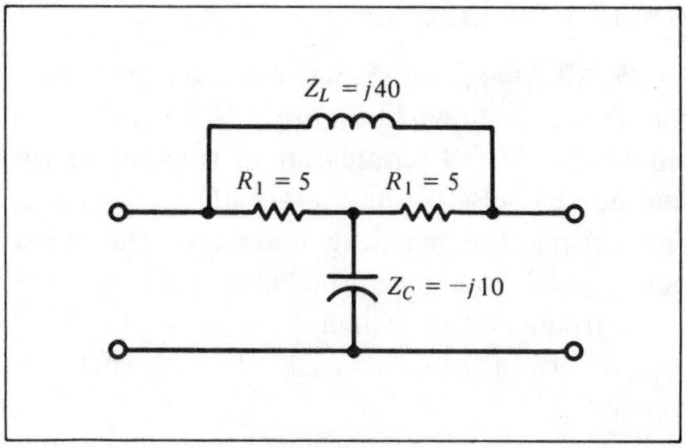

FIGURE P 15.18

15.22 (a) If a network is reciprocal, only three of the four two-port parameters are independent. For example for the z-parameters $z_{12} = z_{21}$ and for the y-parameters $y_{12} = y_{21}$. Derive the constraint that reciprocity places on the transmission ($ABCD$) parameters? (b) If a reciprocal network has transmission parameters A, B, C, and D, what are the transmission parameters when ports 1 and 2 are interchanged?

15.23 If a two-port network is reciprocal, show how its two-port parameters (any set will do) can be determined from driving-point measurements alone.

15.24 The application at the end of the chapter considers the response of the two-port in Figure 15.20 with a zero impedance (voltage) source. Assuming the load resistance remains infinite, what source resistance can be intro-

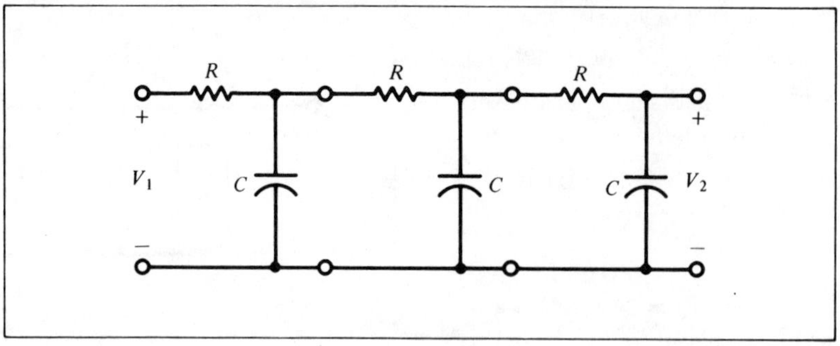

FIGURE P 15.19

duced without reducing the $|V_0/V_s|$ below 4.0 (use the y-parameters given in Figure 15.21) at 2.15×10^9 hz?

15.25 An RC ladder sometimes used in oscillator circuits is shown in Figure P 15.19: (a) Calculate the $ABCD$ parameters of this ladder by finding the $ABCD$ parameters of an L-section and raising the resulting matrix to the third power. (b) Calculate the sinusoidal steady-state frequency at which V_2 lags V_1 by 180° and determine the ratio $|V_2/V_1|$ at this frequency.

REFERENCES

15.1 M. E. VAN VALKENBURG, *Network Analysis.* Englewood Cliffs, N.J.: Prentice-Hall, Inc., 1964, (Second Edition), Chapter 11.

15.2 J. B. CRUZ AND M. E. VAN VALKENBURG, *Introductory Signals and Circuits.* Boston: Blaisdell Publishing Co., 1967, Chapter 14.

15.3 C. M. CLOSE, *The Analysis of Linear Circuits.* New York: Harcourt, Brace Javanovich, 1966, Chapter 7.

15.4 F. F. KUO, *Network Analysis and Synthesis.* New York: John Wiley & Sons, Inc., 1964, (Second Edition), Chapter 9.

16

Transformers; Analogs

16.1 Introduction

There are several additional topics that are often considered useful for study in an elementary network analysis text. These have little relation to one another, but are collected together in this chapter for organizational convenience.

16.2 Transformers

Introduction

The equations for a mutual inductance are given by

$$v_1 = L_1 \frac{di_1}{dt} + M \frac{di_2}{dt} \qquad (16.1)$$

$$v_2 = M \frac{di_1}{dt} + L_2 \frac{di_2}{dt} \qquad (16.2)$$

where L_1 and L_2 are self-inductances and M is the mutual inductance. We can be satisfied with this mathematical description from an analytical point of view; however, real *transformers* that utilize the mutual inductance effect are seldom modeled in this form. In this section, we examine a number of common transformer models and show how they are related to one another.

Coefficient of Coupling

The self- and mutual inductances are not independent in coupled coils. The self-inductance depends on flux linking the coil producing the flux while the mutual inductance is a measure of the flux linking another coil. The interrelationship between these inductances is expressed very simply by the fact that

$$\frac{M}{\sqrt{L_1 L_2}} = \text{Constant} \triangleq k \qquad (16.3)$$

where k is called the *coefficient of coupling* between the coils. It can be shown that k is simply the fraction of the flux from one coil that links the second coil. Physically, k can approach, but never exceed, unity. For all real coils $k < 1$, but it can be made as high as 0.998 by proper design.

We can attach a physical significance to the coefficient of coupling restriction by considering the network of Figure 16.1. In this network

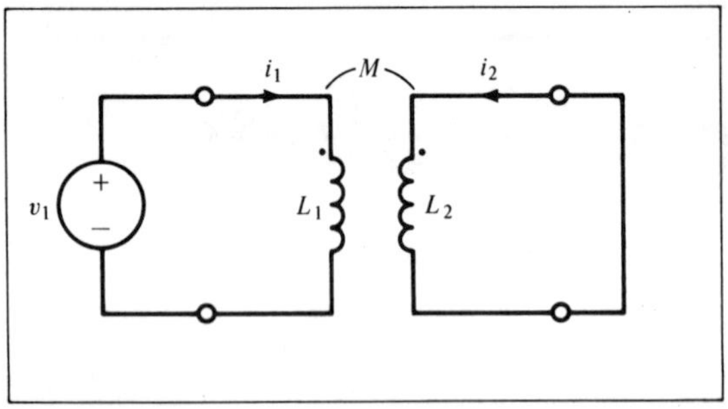

FIGURE 16.1 Coupled coils connected to a voltage source.

$v_2 = 0$, and therefore from Equation (16.2),

$$\frac{di_2}{dt} = -\frac{M}{L_2}\frac{di_1}{dt} \tag{16.4}$$

Substituting Equation (16.4) into (16.1) gives

$$v_1 = \left(L_1 - \frac{M^2}{L_2}\right)\frac{di_1}{dt} \tag{16.5}$$

which, using Equation (16.3), becomes

$$v_1 = L_1(1 - k^2)\frac{di_1}{dt} \tag{16.6}$$

This shows that the equivalent inductance of the network of Figure 16.1 is just $L_1(1 - k^2)$, and the energy stored in the coupled coils is therefore

$$w_m|_{v_2=0} = \tfrac{1}{2}L_1(1 - k^2)i_1^2 \tag{16.7}$$

Thus, the restriction $k < 1$ guarantees that the equivalent inductance, and hence stored energy, will not be negative.

Coupled Coil Equivalents

When a pair of coupled coils are connected with a common terminal as shown in Figure 16.2(a), an equivalent can be constructed using three two-terminal inductors. The inductors can be arranged in a Tee or a Pi geometry as shown in Figure 16.2(b) and (c).

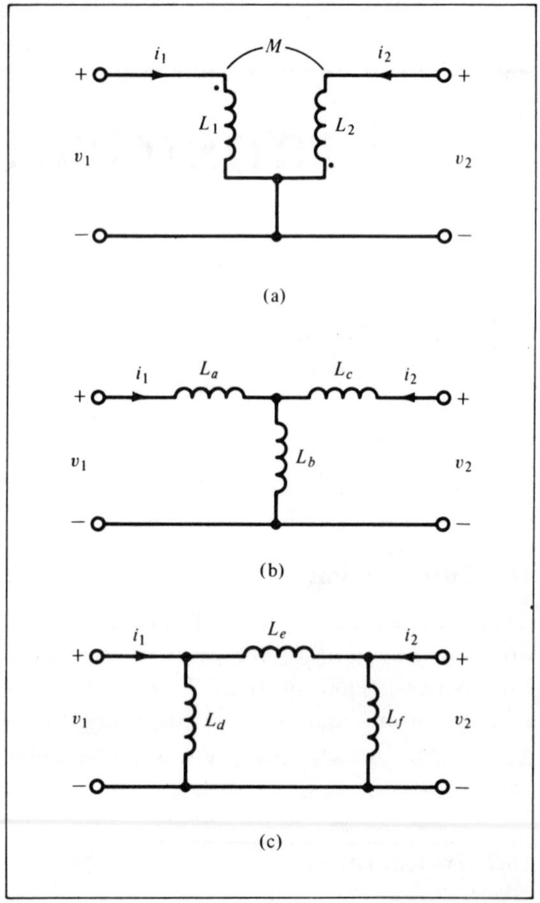

FIGURE 16.2 Coupled coils and two possible equivalents. (b) An inductive Tee; (c) an inductive Pi.

For the coupled coils the terminal voltages and currents are related by

$$v_1 = L_1\frac{di_1}{dt} - M\frac{di_2}{dt} \tag{16.8}$$

$$v_2 = -M\frac{di_1}{dt} + L_2\frac{di_2}{dt} \tag{16.9}$$

In the Tee circuit Kirchhoff's voltage law

around the $v_1 - L_a - L_b$ loop requires that

$$v_1 = L_a \frac{di_1}{dt} + L_b \frac{d(i_1 + i_2)}{dt}$$

$$= (L_a + L_b) \frac{di_1}{dt} + L_b \frac{di_2}{dt} \qquad (16.10)$$

Similarly, for the $L_b - L_c - v_2$ loop,

$$v_2 = L_b \frac{di_1}{dt} + (L_b + L_c) \frac{di_2}{dt} \qquad (16.11)$$

Equivalence between the two networks requires that Equations (16.10) and (16.11) be identical to Equations (16.8) and (16.9). This establishes the inductances needed for the Tee equivalent:

$$L_b = -M$$

$$L_a = L_1 + M \qquad (16.12)$$

and

$$L_c = L_2 + M$$

Note that if either dot in Figure 16.2(a) is shifted, the sign of M is changed in Equations (16.8), (16.9), and (16.12).

We will leave the development of the Pi circuit equivalence relation as an exercise.

Analysis with a Load

Consider the model of Figure 16.3 where a voltage source v_s is connected across the inductor L_1 and a load resistor R_L is placed across

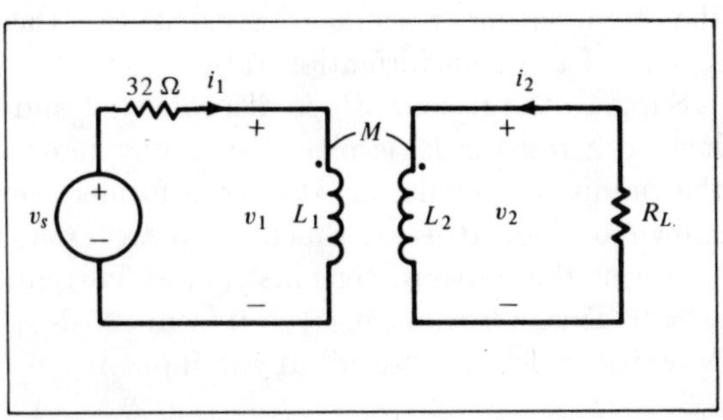

FIGURE 16.3 Coupled coils with a voltage excitation and a resistive load.

L_2. We would like to determine the relationship between the load variables (v_2 or i_2) and the source v_s.

For the coupled coils the voltages and currents are related by

$$v_1 = L_1 \frac{di_1}{dt} + M \frac{di_2}{dt} \qquad (16.13)$$

and

$$v_2 = M \frac{di_1}{dt} + L_2 \frac{di_2}{dt} \qquad (16.14)$$

In addition, by KVL,

$$v_1 = v_s \qquad (16.15)$$

and from Ohm's law,

$$v_2 = -R_L i_2 \qquad (16.16)$$

Combining Equations (16.16) and (16.14) gives

$$\frac{di_1}{dt} = \frac{-R_L i_2 - L_2 \frac{di_2}{dt}}{M}$$

which can be used to eliminate i_1 from Equation (16.13),

$$v_1 = v_s = L_1 \left[\frac{-R_L i_2 - L_2 \frac{di_2}{dt}}{M} \right] + M \frac{di_2}{dt}$$

$$= -\frac{L_1}{M} R_L i_2 - \frac{(L_1 L_2 - M^2)}{M} \frac{di_2}{dt} \qquad (16.17)$$

This first-order ordinary differential equation can be solved for i_2 once v_s and the inductance values are specified. However, we can arrive at some interesting results without solving Equation (16.17). If M is replaced by $k\sqrt{L_1 L_2}$, this equation can be written

$$v_1 = -\frac{1}{k} \sqrt{\frac{L_1}{L_2}} R_L i_2 - \sqrt{L_1 L_2} \left(\frac{1 - k^2}{k} \right) \frac{di_2}{dt} \qquad (16.18)$$

Now, suppose we idealize the coupled coils by letting $k \to 1$. Such a pair of coupled coils are said to be "perfectly coupled." As $k \to 1$, the second term in Equation (16.18) goes to zero leaving

$$v_1 = \sqrt{\frac{L_1}{L_2}}\,(-R_L i_2) = \sqrt{\frac{L_1}{L_2}}\,v_2 \quad (16.19)$$

Therefore, the *voltage ratio is reduced to the real constant* $\sqrt{L_1/L_2}$. This constant can be larger or smaller than unity depending on the relative sizes of the two self-inductances.

What about i_1 and i_2? From Equation (16.13),

$$\frac{di_1}{dt} = \frac{v_1}{L_1} - \frac{M}{L_1}\frac{di_2}{dt}$$

Integrating both sides and letting $k = 1$ (perfect coupling), this becomes

$$i_1(t) = \frac{1}{L_1}\int_0^t v_1(\xi)\,d\xi - \sqrt{\frac{L_2}{L_1}}\,i_2(t) + i_1(0) + \sqrt{\frac{L_2}{L_1}}\,i_2(0) \quad (16.20)$$

where the initial conditions insure that the equations are satisfied if $i_1(t)$ and $i_2(t)$ are constant. If, with $k = 1$, we let $L_1 \to \infty$ and $L_2 \to \infty$, but maintain fixed the ratio

$$\frac{L_2}{L_1} = n^2 \quad (16.21)$$

then with zero-valued initial conditions Equation (16.20) would reduce to

$$i_1 = -ni_2 \quad (16.22)$$

In this limit, the coupled inductors reduce to an *ideal transformer* described by

$$v_2 = nv_1 \quad (16.23)$$

$$i_2 = -\frac{1}{n}\,i_1 \quad (16.24)$$

The constant

$$n = \frac{v_2}{v_1} = -\frac{i_1}{i_2} = \sqrt{\frac{L_2}{L_1}}$$

is the *transformation ratio* of the ideal transformer. Physically, because the inductance L_i is proportional to the square of the turns ratio N_i [see Equation (7.38)], the quantity n is the turns ratio of the transformer. The assumptions leading to this idealized model are summarized in Figure 16.4(a).

The usual symbol for an ideal transformer is like the one for coupled coils. The difference lies in the fact that the only parameter of the ideal transformer is n. This similarity of symbols is unfortunate because it is misleading. It suggests inductive effects in the ideal transformer and *there are none!* In fact, the ideal transformer has no dynamics of its own. It is a static element because no time derivatives appear in Equations (16.23) and (16.24) which define this ideal element.

Let us explore this a little further by considering the circuit of Figure 16.4(b). In general, from (16.23) and (16.24) we know

$$\frac{v_1}{i_1} = \frac{v_2/n}{-ni_2} = \frac{1}{n^2}\,\frac{-v_2}{i_2}$$

With R_L connected as shown in Fig. 16.4(b)

$$\frac{v_2}{i_2} = -R_L$$

and therefore

$$\frac{v_1}{i_1} = \frac{R_L}{n^2}$$

The terminated ideal transformer "looks like" the terminating resistance divided by the square of the transformation ratio.

Suppose the resistor R_L is disconnected and instead a resistor R_i is placed in shunt across the input terminals of the transformer as shown in Fig. 16.4(c). Since no path exists between the output terminals, the current $i_2 = 0$. This in turn means $i_1 = 0$ from (16.24). Now the resistance "seen" at the input is

$$\frac{v_1}{i_1} = R_i$$

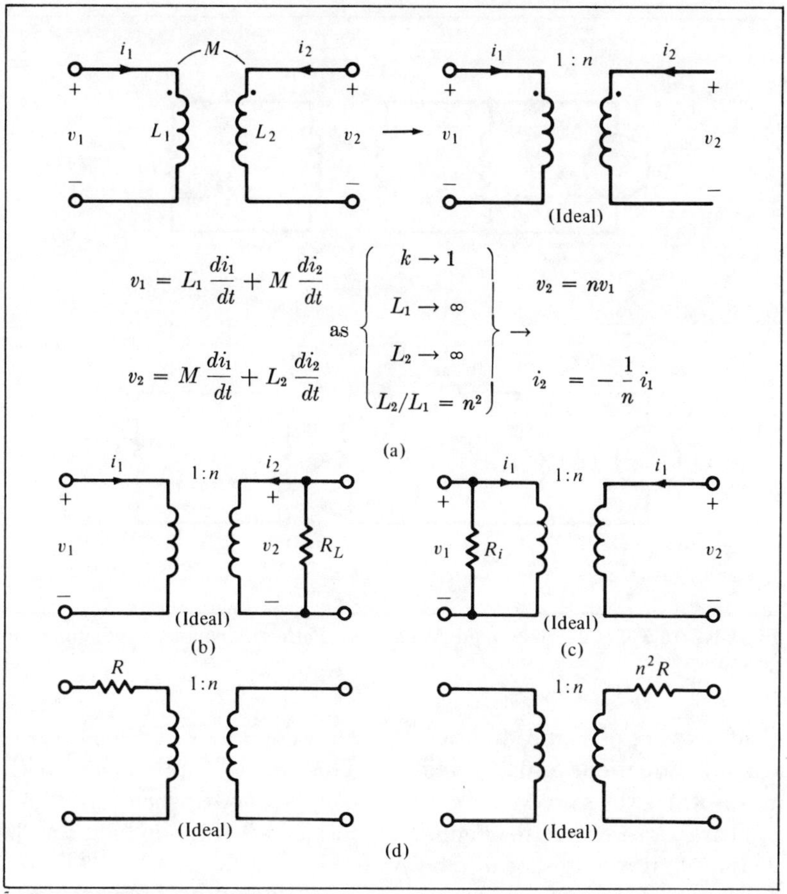

FIGURE 16.4 (a) Coupled coils reduced to an *ideal* transformer.

but $v_2 = nv_1$ as before. If we make

$$R_i = \frac{R_L}{n^2}$$

the two circuits in Fig. 16.4(b) and (c) are *completely equivalent electrically.* This means that there is no electrical measurement one can make at the indicated external terminals which will establish which circuit is, in fact, present! The reader should try calculating various parameters of the two circuits to establish the correctness of this equivalence. A similar equivalence for a series resistance is indicated in Fig. 16.4(d).

Complete Transformer Model

The ideal transformer can be used as part of a model of the original coupled inductors from which it was derived. Returning to Equation (16.20), if Equation (16.19) is used to eliminate i_2, we have (with zero-valued initial conditions)

$$i_1 = \frac{1}{L_1} \int_0^t v_1(\xi)\, d\xi + \frac{1}{R_L} \frac{L_2}{L_1} v_1 \qquad (16.25)$$

This indicates that when $k = 1$, the input current in Figure 16.3 consists of two components. (1) an inductive component due to L_1 and (2) a resistive component due to R_L transformed by an ideal transformer with a $1{:}\sqrt{L_2/L_1}$ transformation ratio. Equation (16.25) sug-

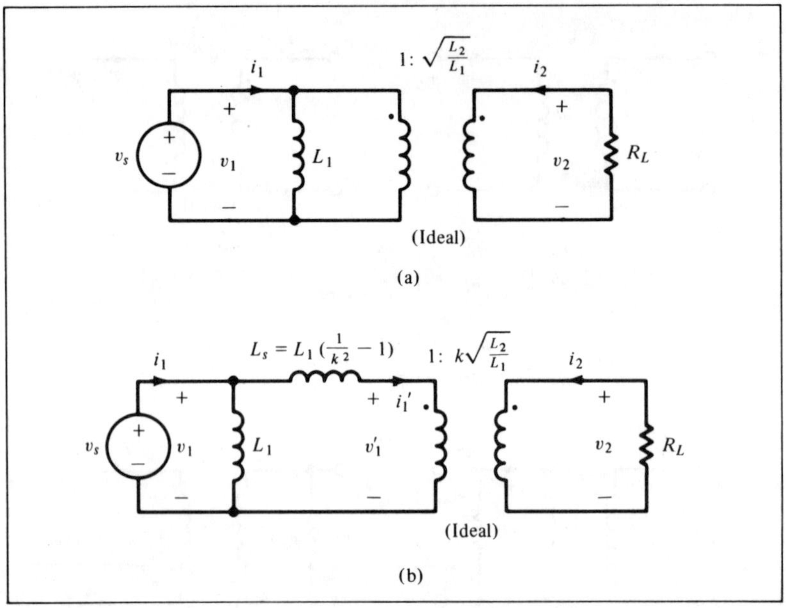

FIGURE 16.5 Transformer equivalents. (a) Perfectly coupled inductors; (b) general case.

gests the equivalent network of Figure 16.5(a). This network satisfies Equations (16.25) and (16.19) which means it has the same *input* and *forward transfer* characteristics as in Figure 16.3 when $k = 1$. Indeed, it can be shown to be a completely valid model for the perfectly coupled coil.

When $k \neq 1$, the equivalent network becomes more complicated, as shown in Figure 16.5(b). It is left as an exercise to show that this model satisfies the original mutual inductance equations of Equations (16.1) to (16.2). This model approaches both the perfectly coupled model when $k = 1$ and the ideal transformer when the inductances become infinite.

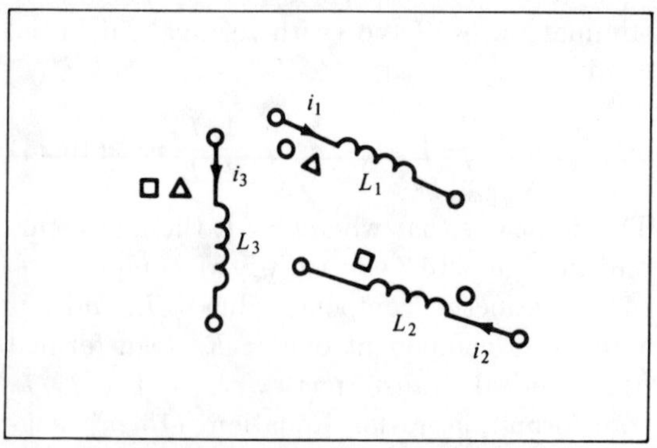

FIGURE 16.6 Three coupled coils.

Multiwinding Transformers

Although we have been considering two coupled coils, there is nothing which prevents our generalizing to any number of coupled coils. An example with three coupled coils is shown in Figure 16.6. Each coil will have a self-inductance and a mutual inductance to the other coils. The polarity of the mutuals are indicated by pairs of dots and similar symbols. Selecting an arbitrary set of positive current directions, we have in the general case

$$v_1 = L_1 \frac{di_1}{dt} + M_{12} \frac{di_2}{dt} + M_{13} \frac{di_3}{dt}$$

$$v_2 = M_{21} \frac{di_1}{dt} + L_2 \frac{di_2}{dt} + M_{23} \frac{di_3}{dt} \qquad (16.26)$$

$$v_3 = M_{31} \frac{di_1}{dt} + M_{32} \frac{di_2}{dt} + L_3 \frac{di_3}{dt}$$

For a multiwinding transformer, it is possible to estimate the mutual inductances from a knowledge of the turns ratios and coefficients of coupling. The L_j's and M_{ij}'s can be calculated from such estimates as follows.

Assume that the self-inductance of each winding is proportional to the square of the turns on that winding, so that for the jth winding

$$L_j = KN_j^2 \tag{16.27}$$

where K is independent of j. Therefore, if we are given the self-inductance (L_r) of one winding, and the turns ratios N_j/N_r with respect to the same winding, we calculate

$$L_j = \left(\frac{N_j}{N_r}\right)^2 L_r$$

Next, given the coupling coefficients k_{ij} between the ith and jth winding, we calculate

$$M_{ij} = k_{ij}\sqrt{L_i L_j}$$

which defines all the inductance values used in Equation (16.26). For example, it is common to know the primary inductance of a transformer used in power supplies ($\cong$10 h) and the voltage ratios between other windings and the primary winding. Because the voltage ratio is proportional to the turns ratio [see Equation (16.19)], this establishes the turns ratios needed in the above. A guess of all coupling coefficients (say 0.99) would then permit all L_j's and M_{ij}'s to be found.

16.3 Analogs

We have studied a variety of general analytic and computer-oriented methods for solution of electrical networks. Often we have referred to the application of these same techniques to other engineering systems that are *analogous* to electrical networks. The time has come to justify this pretension.

In the following sections, we study several such systems. In each case, we establish this analogy by showing that the system has (1) connection laws (topological), (2) solution variables, and (3) element (constitutive) equations similar to electrical networks. As we will see, each system will have two types of solution variables: (1) a *flow* or *through* variable, corresponding to electrical current $i(t)$ and (2) a *potential* or *across* variable corresponding to electrical voltage $v(t)$.

A word of caution before proceeding, however. We will study highly simplified model problems from other disciplines so as to make obvious the analogy involved. State-of-art analysis problems in these disciplines are of course quite different. In mechanical engineering, for example, we will study a simple one-dimensional problem; practical mechanical design problems are often three-dimensional requiring considerable insight by the analyst to formulate properly. Even so, practical considerations such as choice of loop or node analysis survive even the most advanced treatment. Interestingly, node analysis methods are almost universally preferred due to their generality and their ease of programming.

16.4 Mechanical System Analogies

Translational Analogies

The reader is probably familiar from elementary physics with the laws of motion of one-dimensional *translational* mechanical systems, that is, systems that move only along a line. The solution variables in such systems are

$x(t)$,	Displacement
$\nu(t) = \dot{x}(t)$,	Velocity
$\dot{\nu}(t) = \ddot{x}(t)$,	Acceleration
$f(t)$,	Force
$p(t)$,	Momentum

The mechanical elements we will study are mass, viscous friction, and stiffness. All quantities are assumed in the mks system.

The *mass element* is described by Newton's law as

$$f(t) = M\frac{d^2x(t)}{dt^2} = M\frac{d\nu(t)}{dt} \tag{16.28}$$

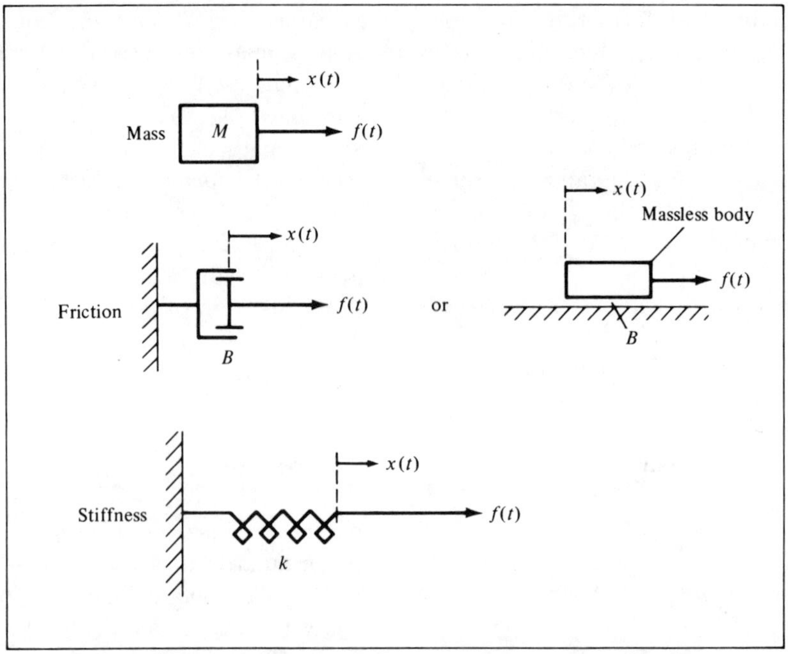

FIGURE 16.7 Symbols and references for translational elements.

where references for $x(t)$ and $\nu(t)$ are illustrated in Figure 16.7. The viscous friction element is defined by

$$f(t) = \frac{\eta A}{d}\nu(t)$$

$$= B\nu(t) \qquad (16.29)$$

This element models the force between moving massless bodies separated by a thin film of cross-sectional area A and thickness d. The constant η is a measure of the viscousity of the film. The so-called *damper* or *dashpot* symbol of Figure 16.7 represents this element. Usually, one of the bodies is stationary, so that the $f(t)$ of Equation (16.29) is the force exerted on the moving body.

The stiffness element is defined by

$$f(t) = Kx(t) \qquad (16.30)$$

where $x(t)$ is the displacement from a reference position. It is common to associate this equation with a spring, which produces a reaction force proportional to the distance it is stressed.

Excitations are commonly applied to mechanical systems by applying a force or specifying the displacement or velocity at a point.

A simple example of a mechanical system formed from the above elements is shown in Figure 16.8(a). A point mass M resting on an inclined plane is attached via a spring K to a fixed point on the plane, and viscous friction B is present between the plane and the mass. There are four forces acting in the x direction along the plane: the force of gravity

$$f_1 = -Mg\sin\theta$$

the frictional force

$$f_2 = B\nu$$

the inertial force

$$f_3 = M\frac{d\nu}{dt}$$

and the spring force, which can be written

$$f_4 = Kx$$

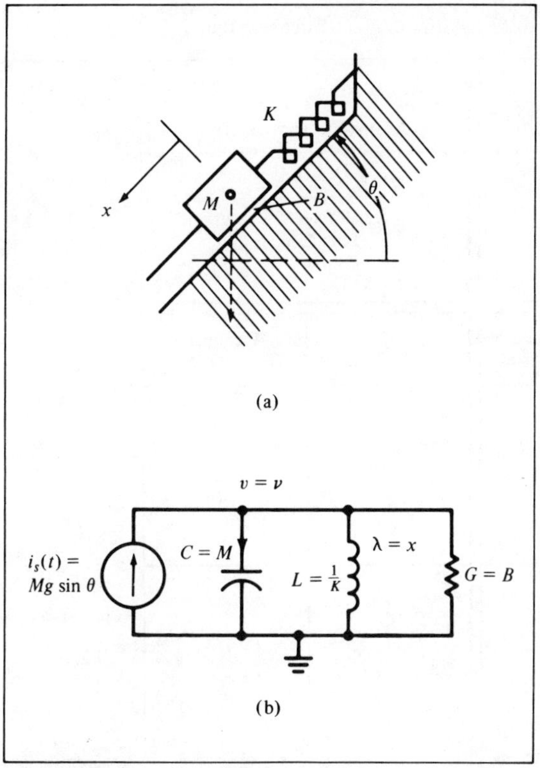

FIGURE 16.8 (a) Mechanical system; (b) electrical analog with i analogous to f.

if we assume the x-origin at the point where the spring is unstressed.

D'Alembert's principle, from elementary physics, requires

$$\Sigma f_i = 0 = M\frac{d\nu}{dt} + B\nu + Kx - Mg\sin\theta$$

or

$$M\frac{d^2x}{dt} + B\frac{dx}{dt} + Kx = Mg\sin\theta \quad (16.31)$$

Because θ is fixed, this equation is a second-order linear differential equation. If we make a proper substitution of electrical for mechanical quantities, we immediately recognize Equation (16.31) to be the node equation of a parallel GCL network. The substitution is indicated in Figure 16.8(b), yielding the equation

$$C\frac{d^2\lambda_L}{dt^2} + G\frac{d\lambda_L}{dt} + \frac{\lambda_L}{L} = i_s(t) \quad (16.32)$$

or, because $\lambda_L = Li_L$,

$$CL\frac{d^2i_L}{dt^2} + GL\frac{di_L}{dt} + i_L = i_s(t) \quad (16.33)$$

Therefore, solving the parallel GCL network is mathematically equivalent to solving the original mechanical problem.

The above relationship is based on two similarities of mechanical systems and electrical networks. First, D'Alembert's principal that requires forces to sum to zero at a point is identical in form to KCL which requires currents to sum to zero at a node. Further, the above associations of Figure 16.8 where velocity is replaced by voltage, hints that KVL would be analogous to a statement that "around a loop of mechanical parts, the velocities must sum to zero," a reasonable requirement. Therefore, we accept the analogies

$$f_{\text{trans}} \sim i_{\text{elec}}(\text{Flow variable})$$

$$\nu_{\text{trans}} \sim v_{\text{elec}}(\text{Potential variable})$$

$$\sum_{\text{at point}} f_{\text{trans}} = 0 \sim \sum_{\text{at node}} i_{\text{elec}} = 0$$

$$\sum_{\text{around loop}} \nu_{\text{trans}} = 0 \sim \sum_{\text{around loop}} v_{\text{elec}} = 0$$

where "$\sim$" means "analogous to."

This analogy is strengthened by observing that the individual mechanical elements of Equations (16.28) to (16.30) can be replaced by electrical equivalents on an *element-by-element* basis.

These analogies are summarized in Table 16.1. This feature allows us to convert from a mechanical system to an electrical network by inspection (or by a simple computer program) *without* writing the network equations. The following example illustrates this process.

TABLE 16.1 Electrical analogs of lumped, translational mechanical systems.

Translational Mechanical System			Analog Electrical Systems	
Variables: Force, Velocity, Displacement, Momentum			Current, Voltage, Flux Linkage, Charge	
Element Name	Symbol Ref. Direct.	Terminal Characteristic	Element	Terminal Characteristic
Mass	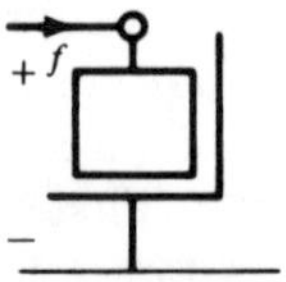	$p = Mv$ $f = \frac{dp}{dt}$	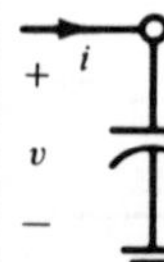	$q = Cv$ $i = \frac{dq}{dt}$
Spring	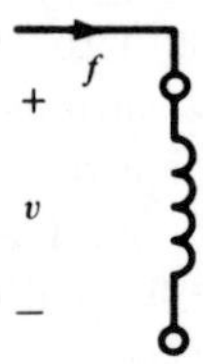	$f = Kx$ $v = \frac{dx}{dt}$	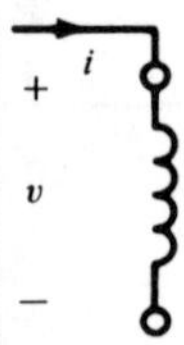	$i = \frac{1}{L}\lambda$ $v = \frac{d\lambda}{dt}$
Dash Pot	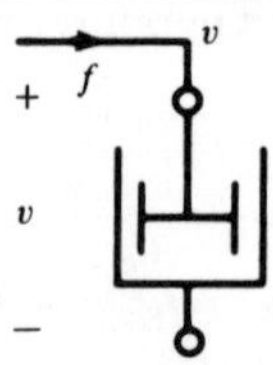	$f = Bv$	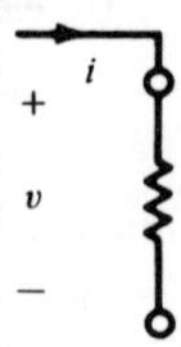	$i = Gv$
D'Almbert's Principle $\sum f = 0$			KCL $\sum i = 0$	

Example 16.1

The mechanical system of Figure 16.9(a) is excited by a velocity source $v_s(t)$ applied to the end of the spring K_1. We are to find the velocities of each mass as a function of time, assuming that the system is initiated in the zero state.

Each of the masses identifies a connection point in the system where $\sum f_i = 0$. The network analogy of Figure 16.9(b) has two corresponding nodes. Each of these electrical nodes has elements attached that correspond to the elements connected to the mass in the mechanical system. The mass itself is represented by a capacitor connected to ground. A grounded voltage source analogy to the velocity source completes the network model of the figure. If the effect of gravity is to be considered, then a force of value Mg is represented by constant current sources in the network model as shown.

The dc (static), transient, pole-zero, and sinusoidal steady-state (vibrational) response are all common quantities to calculate in mechanical systems. For example, the transfer function V_2/V_s in the above example is determined by node analysis of the electrical network to be

$$\frac{V_2}{V_s}(s) = \frac{L_3(sGL_2 + 1)}{s^4(C_1L_1C_2L_2L_3) + s^3(C_1 + C_2)L_1L_2L_3G + s^2(C_1L_1G(L_2 + L_3) + CL_3G(L_1 + L_2)) + s(GL_2L_3 + GL_1L_2) + L_1 + L_2 + L_3}$$

Similarly,

$$\frac{N_2}{N_s}(s) = \frac{K_1(sB + K_2)}{s^4(M_1M_2) + s^3B(M_1 + M_2) + s^2(M_1B(K_2 + K_3) + M_2B(K_1 + K_2)) + sB(K_1 + K_3) + K_1K_2 + K_1K_3 + K_2K_3}$$

If all masses M_i, stiffness constants K_j, and B have unit value, then the above becomes

$$\frac{N_2}{N_s} = H_1(s) = \frac{(s+1)}{s^4 + 2s^3 + 4s^2 + 2s + 3}$$

$$= \frac{s+1}{(s^2+1)(s^2+2s+3)}$$

The displacement $x_2(t)$, being the integral of the velocity $\nu_2(t)$, has the transfer function

$$\frac{X_2}{N_s} = \frac{H_1(s)}{s}$$

The above system has a natural frequency at $s = \pm j1$, giving rise to a steady-state oscillation of the form $A \cos(t + \phi)$.

Rotational Analogies

Most mechanical systems with which the electrical engineer becomes involved are rotational rather than translational. Electrical motors and generators are the principal example. Although we will not study such systems in detail, the following discussions will establish the existence of network analogies to such systems.

We recall from elementary physics that rotating bodies are also described by differential relationships. For example, two masses rotating on a flexible shaft subject to bearing friction are represented in Figure 16.10. We identify three types of elements in such a system; each element is described by equations relating an angular displacement θ, an angular velocity $\dot{\theta} = \omega$, an angular acceleration $\ddot{\theta} = \dot{\omega}$, and/or a torque τ. The elements are described as follows:

1. Inertial elements (J_1 and J_2) are described by

$$\tau = J\ddot{\theta} = J\dot{\omega}$$

2. Frictional elements (B) are described by

$$\tau = B\omega$$

3. Stiffness elements (K_1, K_2, and K_3) representing shaft flexibility, are described by

$$\tau = K\Delta\theta$$

where $\Delta\theta$ is the difference in angular displacements of the ends of the shaft.

The equations for the system are developed by identifying a point with each rotating mass. At this point we can write

$$\sum_{\text{at a point}} \tau_i = 0 \qquad \text{(Flow equation)}$$

Similarly, relative to a stationary reference, the

sum of angular velocities around any loop must be zero.

$$\sum_{\text{around loop}} \omega_i = 0 \qquad \text{(Potential equation)}$$

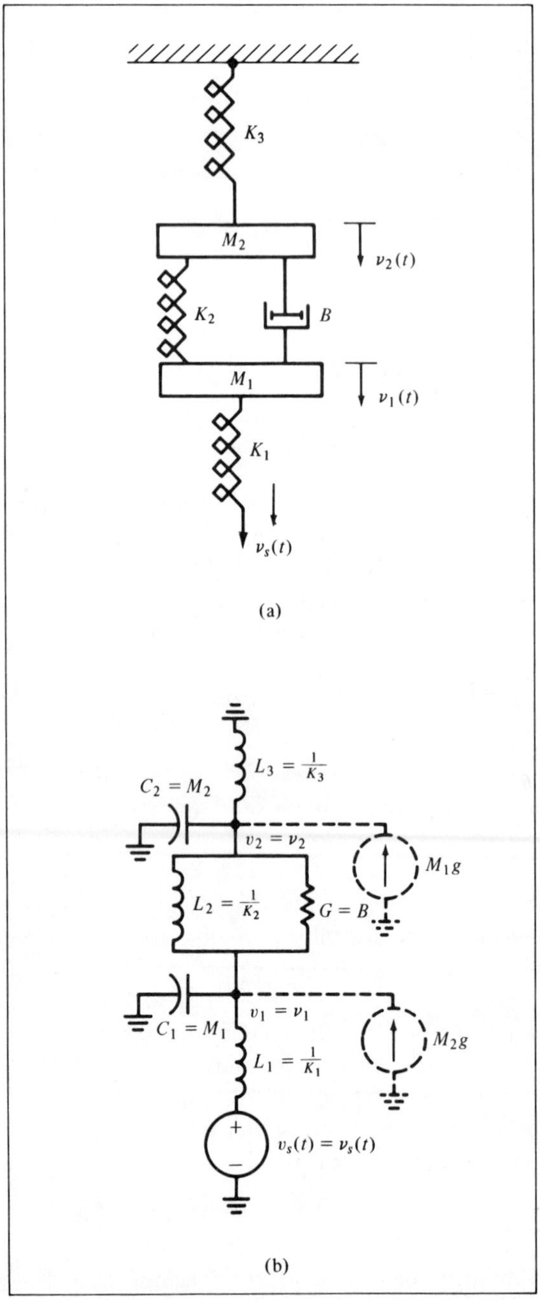

FIGURE 16.9 Mechanical and electrical analogies.

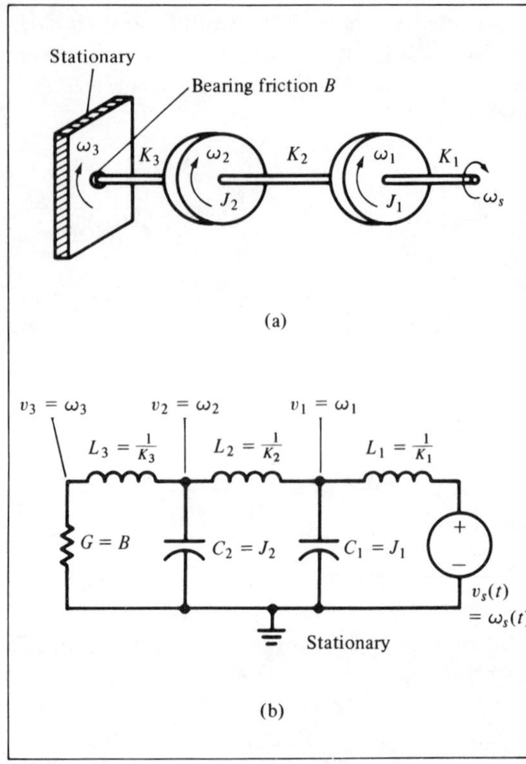

FIGURE 16.10 Mechanical and electrical analogies.

The electrical network analogy has nodes at the end of each connecting shaft. Attached to each node is a capacitor, resistor, inductor, voltage source, or current source, depending on whether an inertial load, a frictional loss, a flexible shaft, a velocity input, or a torque input (respectively) is connected to the corresponding point in the rotational system. This is exemplified in Figure 16.10 where the four node voltages represent the velocities at four shaft endings in the mechanical system. The complete set of analogous quantities are

$$\tau \sim i \qquad \omega \sim v \qquad \theta \sim \lambda$$

$$J \sim C \qquad B \sim G \qquad K \sim \frac{1}{L}$$

Idealized Transformation Elements

Just as there are electrical elements besides resistors, inductors, and capacitors, there are mechanical elements in addition to those we have studied. Many common mechanical elements such as pins, screws, and cams do not have useful electrical network analogies. However, there are two familiar elements that do

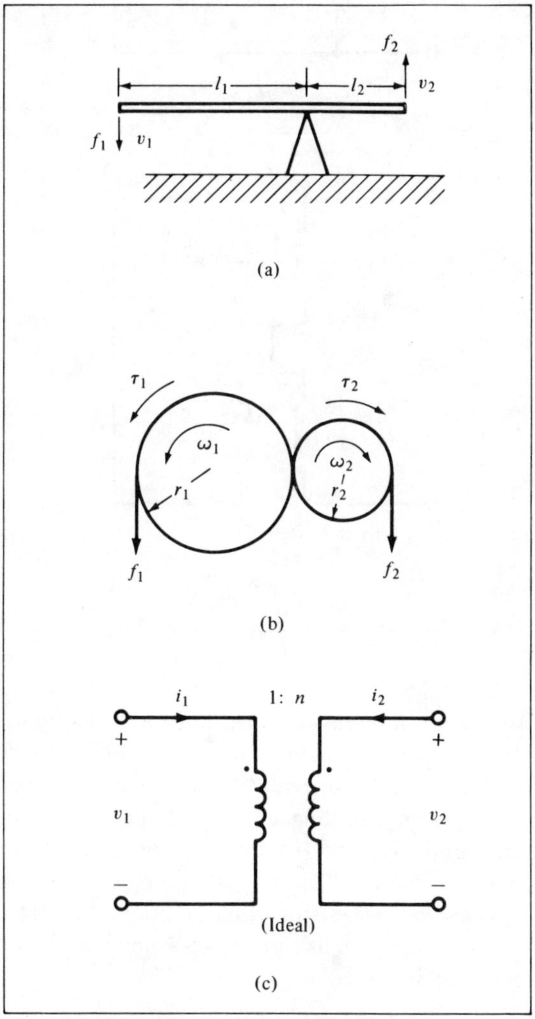

FIGURE 16.11 Idealized elements.

have an interesting network analogy: the *lever* and the *gear*.

The lever of Figure 16.11(a) is described by the torque equation

$$f_1 l_1 = -f_2 l_2$$

and the angular velocity equation

$$\frac{v_1}{l_1} = \frac{v_2}{l_2}$$

for small displacements from a horizontal position. The gear is similarly defined by the force equation

$$\frac{\tau_1}{r_1} = -\frac{\tau_2}{r_2}$$

and by a angular displacement equation

$$\omega_1 r_1 = \omega_2 r_2$$

Both of these are similar in form to the equations for an ideal transformer [Figure

TABLE 16.2

Equations of Transformation Elements

Electrical	$\frac{i_1}{i_2} = -n$	$\frac{v_1}{v_2} = \frac{1}{n}$
Lever	$\frac{f_1}{f_2} = -\frac{l_2}{l_1}$	$\frac{v_1}{v_2} = \frac{l_1}{l_2}$
Gear	$\frac{\tau_1}{\tau_2} = -\frac{r_1}{r_2}$	$\frac{\omega_1}{\omega_2} = \frac{r_2}{r_1}$

16.11(c)]. This analogy is formalized in Table 16.2 where the transformer turns ratio n is seen to be analogous to the ratios of the lever arm lengths and the gear radii.

PROBLEMS

16.1 (a)–(d) Find the impedances $Z(s)$ in Figure P 16.1.

16.2 An experimental method of determining whether two coils are inductively coupled is to measure the inductance at one pair of coil terminals as shown in Figure P 16.2. (a) Determine L_{oc} and L_{sc} in terms of the values L_1, L_2, and M. (b) Show how these two measured

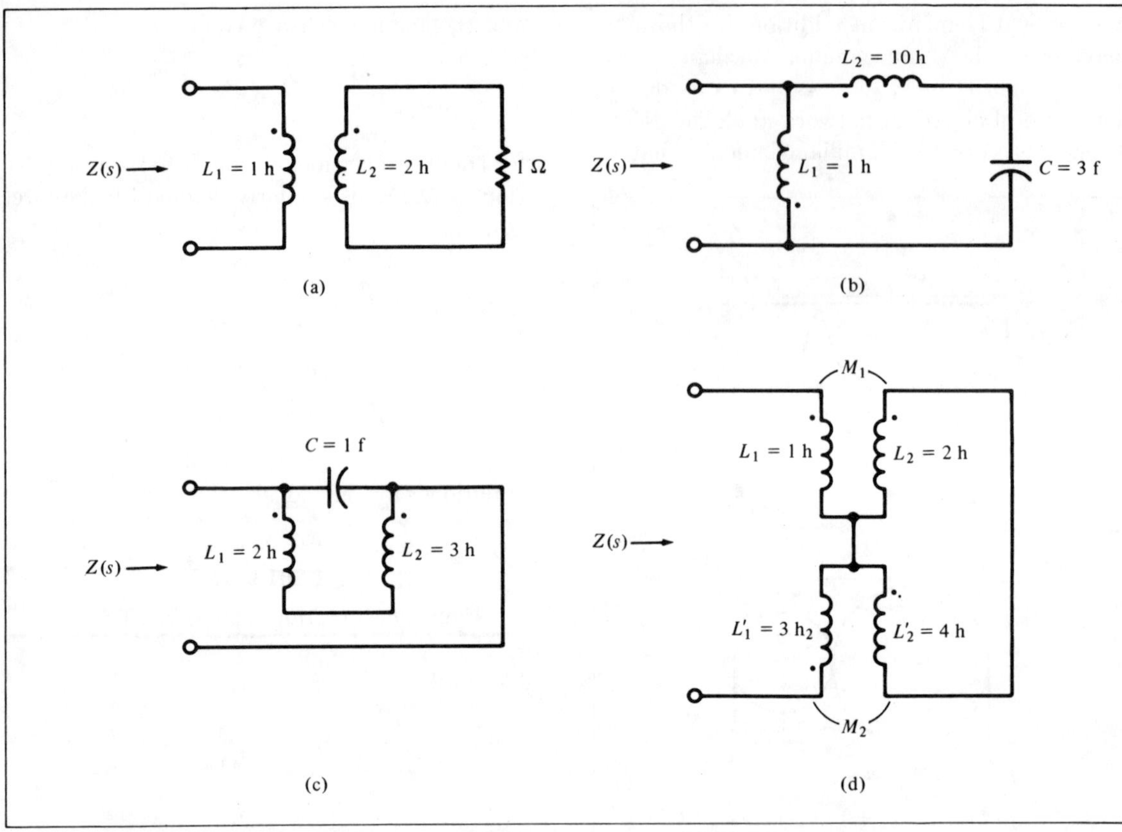

FIGURE P 16.1 (a) $k = 0.5$; (b) $k = 0.9$; (c) $k = 0.95$; (d) $k = 1$.

inductances can be used to calculate the coefficient of coupling.

16.3 The turns ratio V_1/V_2 of the perfectly-coupled transformer of Figure P 16.3 is 2:1. Find the value of L_1 that insures a voltage ratio $V_2(j\omega)/V_s(j\omega)$ having a phase shift less than a degree for $10 \leqq f \leqq 20{,}000$ hz.

16.4 From Equation (16.19), the voltage ratio is equal to the turns ratio of the windings. In Figure P 16.4, the voltages appearing at the windings of a two-winding transformer are shown. Find the differential equations of the transformer if $k = 1$.

16.5 (a) Show that the node equations of the capacitor network of Figure P 16.5 are similar to the mesh equations of a mutual inductance. Note that the element values may be negative and physically unrealizable. (b) Extend the concepts of coupling coefficient, perfect coupling, and idealness to this network. (c) Show that a resistor R_L across one pair of terminals appears as a resistor n^2R_L at the other terminals, where n is properly defined and the coupling is "ideal".

16.6 The frequency response of a double-terminated transformer is shown in Figure P 16.6. Using the transformer model of Figure 16.5, (a) find n from the midband gain, (b) find L_s and L_1 from the cutoff frequencies, (c) find L_2 and k.

16.7 The ideal transformer can be replaced by

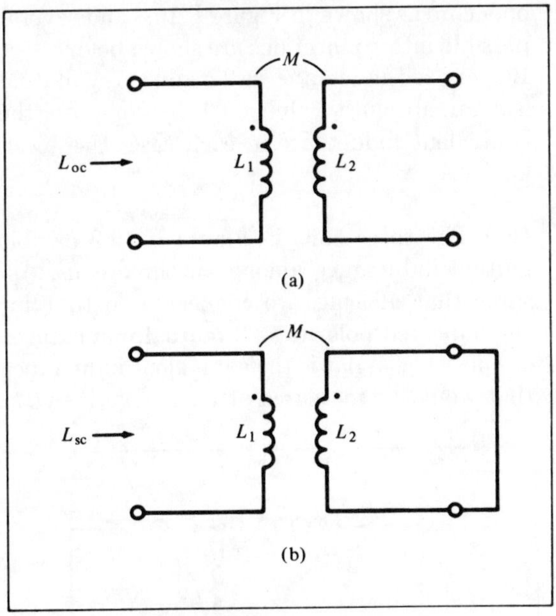

FIGURE P 16.2

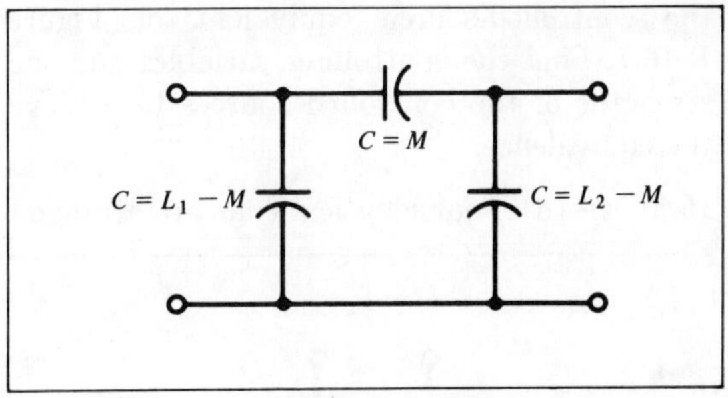

FIGURE P 16.5

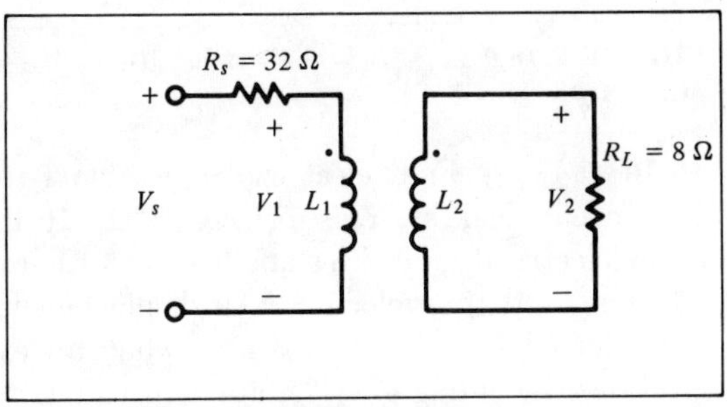

FIGURE P 16.3

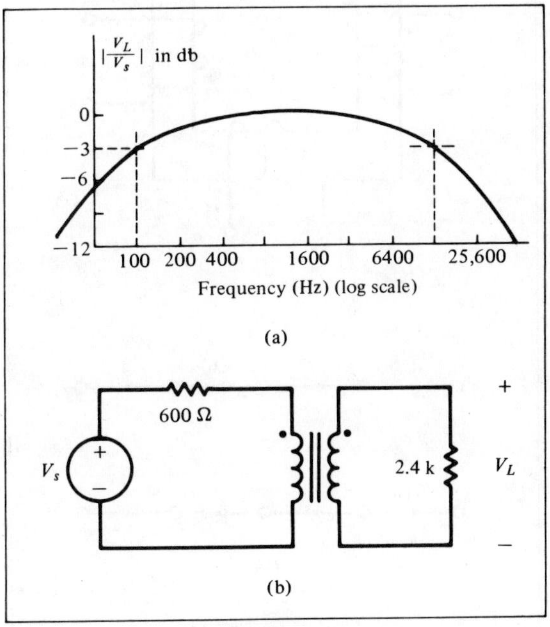

FIGURE P 16.6

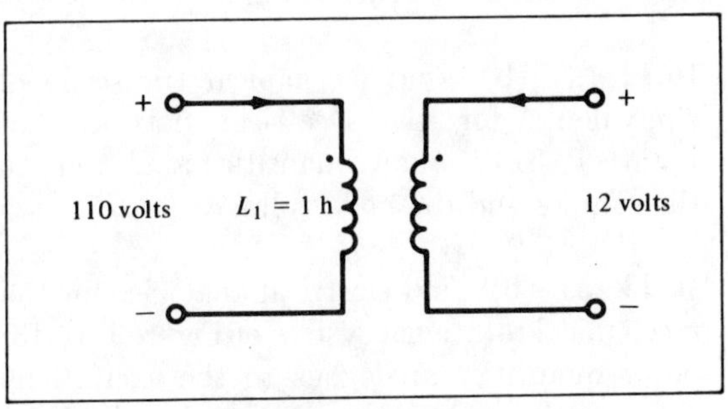

FIGURE P 16.4

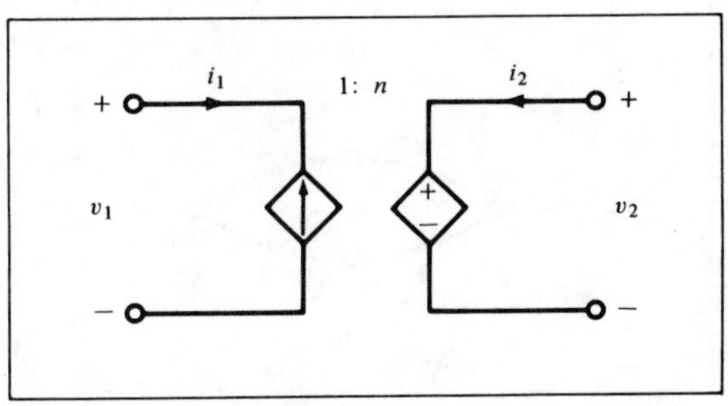

FIGURE P 16.7

the controlled-source equivalent of Figure P 16.7. Find the controlling variables and the strengths of the controlled sources to achieve the equivalence.

16.8 (a)–(d) Four identical coils are arranged on a core as shown in Figure P 16.8 and several possible interconnections are shown below:
If every self-inductance is 2 h and every mutual has an absolute value of 1 h, compute the equivalent inductance in each case. Use mesh analysis.

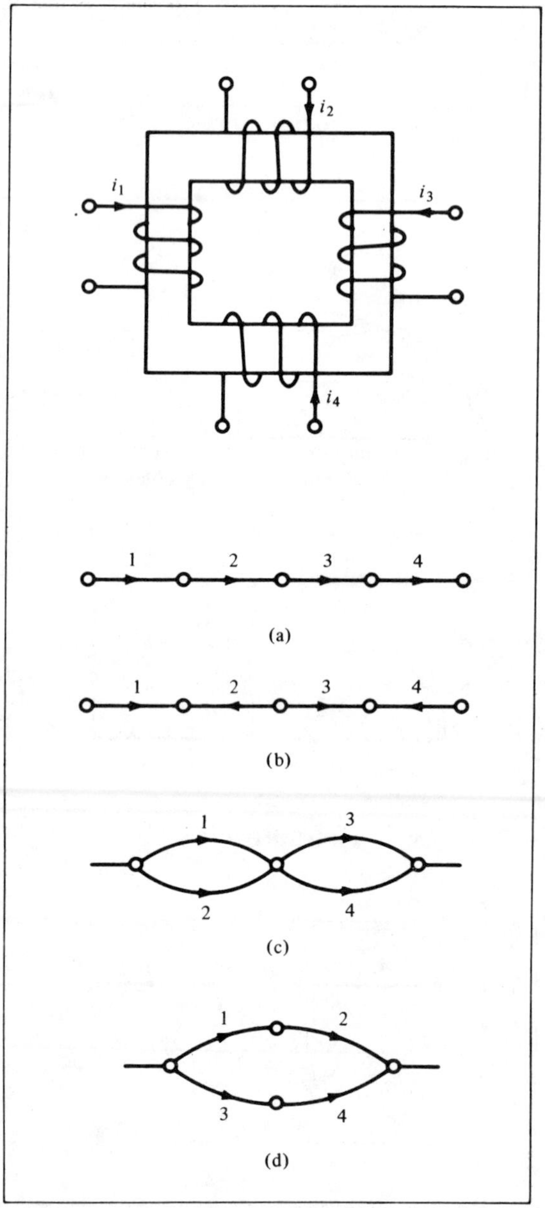

FIGURE P 16.8

16.9 The three coils in Figure P 16.9 exhibit mutual inductance among all three coils. Assume that all coils are connected so that for the indicated polarities all mutual inductances are positive. What is the equivalent inductance that would be measured at terminals $A - B$?

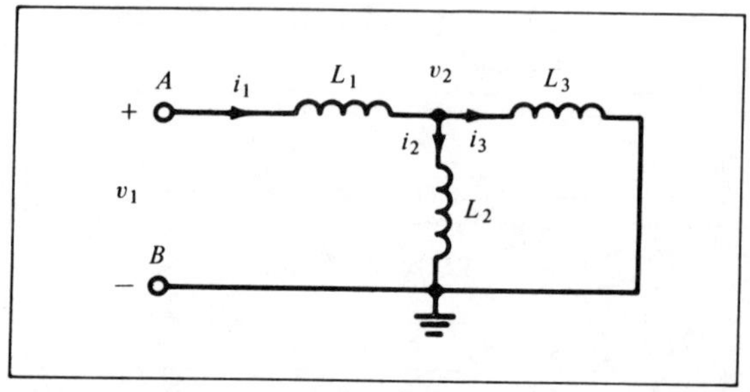

FIGURE P 16.9 $L_1 = L_2 = L_3 = 2$ h; $M_{12} = M_{13} = M_{23} = 1$ h.

16.10 (a)–(b) Find electrical equivalents for the mechanical systems of Figure P 16.10. Show electrical quantities analogous to the excitations and the velocities. All displacements are referred from their positions when all excitations, including gravity, are zero-valued.

16.11 (a)–(b) Find the natural frequencies of the mechanical systems of Figure P 16.11.

16.12 (a)–(b) Find mechanical translational equivalents for the electrical networks of Figure P 16.12. Show quantities analogous to the sources and the node voltages.

16.13 (a)–(b) Find electrical analogies for the mechanical rotational system of Figure P 16.13. Show quantities analogous to the excitations and angular velocities. All angular displace-

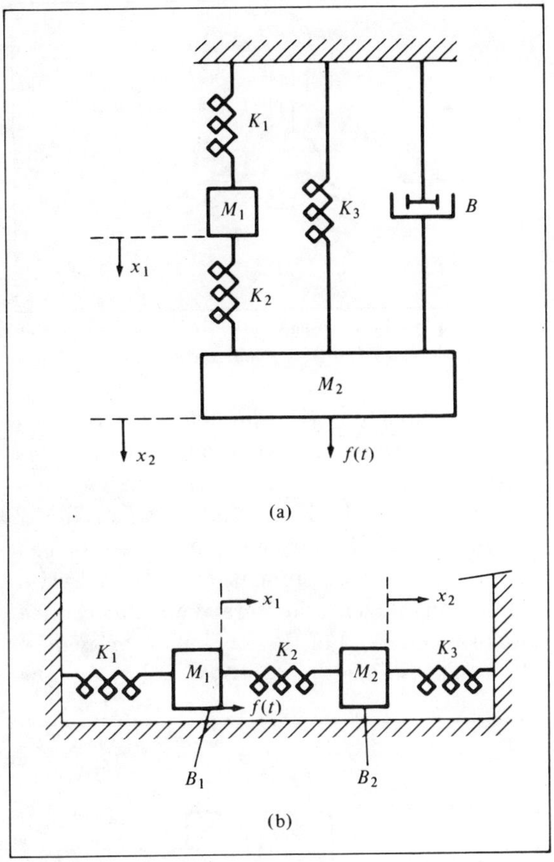

FIGURE P 16.10 (a) Gravity; (b) no gravity.

ments are referenced positive in the clockwise direction from their positions when excitations are zero-valued.

16.14 (a)–(d) Find rotational mechanical analogies to the electrical networks of Figure P 16.14. Show quantities analogous to the node voltages and to the sources.

16.15 One class of natural systems closely analogous to electrical networks are water and natural gas distribution systems. The analogous quantities are pressure (expressed in feet, that is, the number of feet of water which would produce an equivalent pressure) versus voltage and flow versus current. For incom-

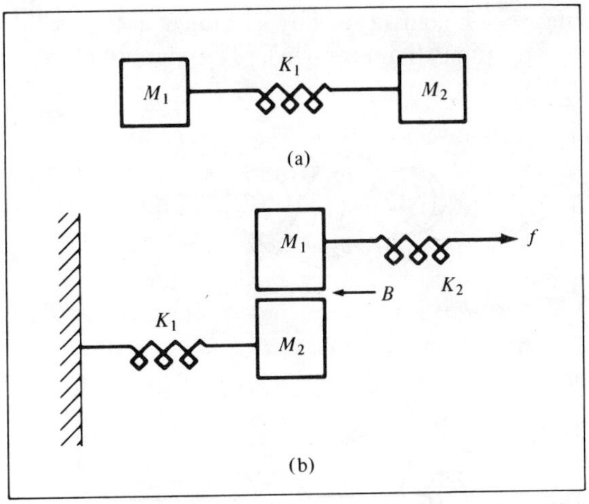

FIGURE P 16.11 (a) No gravity; (b) no gravity.

pressible flow, the constitutive equation relating h (Figure P 16.15) and q for a pipe are

$$h = \pm \frac{4.7L\,|\,q\,|^{1.85}}{C^{1.85}d^{4.87}}$$

where L is the length of pipe (ft), C is a rough-

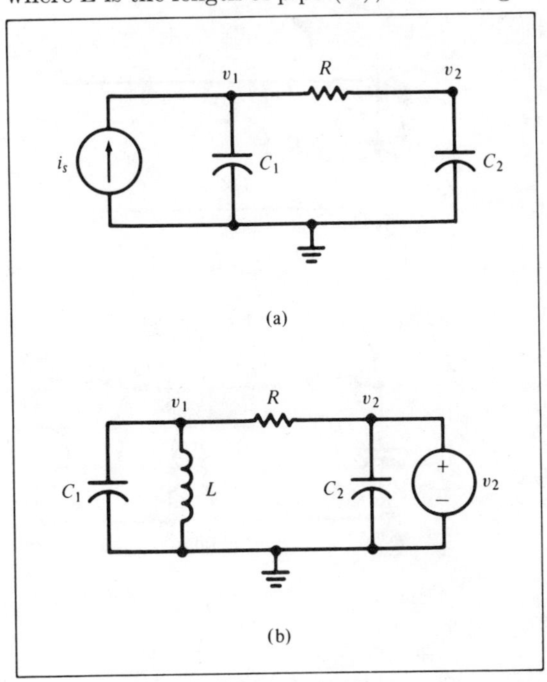

FIGURE P 16.12

ness coefficient (typically a value of 100), and d is the pipe diameter (in). The sign is chosen

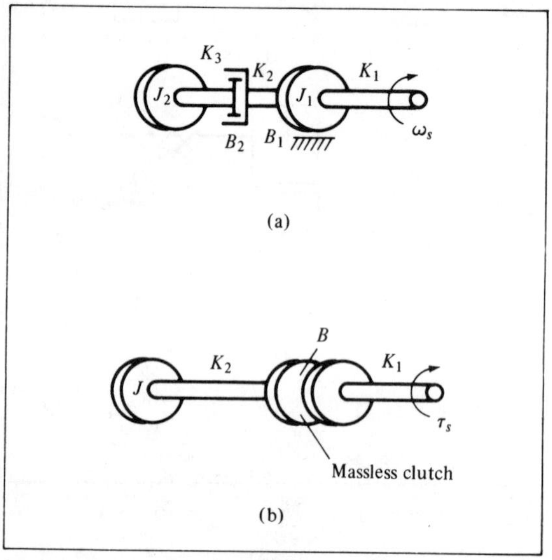

FIGURE P 16.13 (a) Velocity excitation; (b) torque excitation.

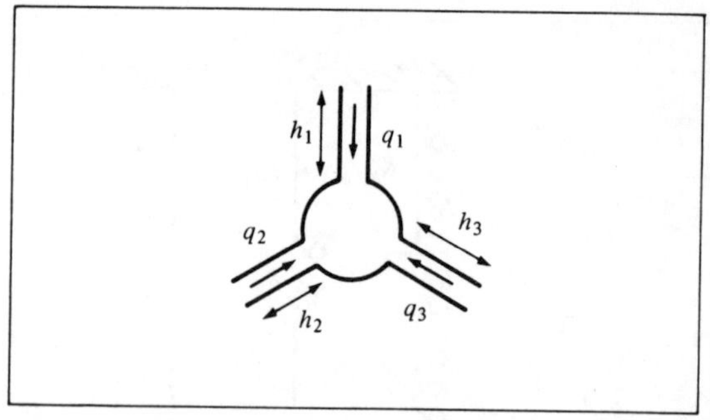

FIGURE P 16.15 h (head loss, ft) $\sim v$ (potential, volts); q (flow, gal/min) $\sim i$ (current, amps).

so that h and q have the same sign. Sources such as water towers are commonly modeled as pressure sources. The solution of such a system can proceed similarly to that of a non-linear electrical network: linearization of h–q or q–h characteristic, development of a branch-wise companion model, and iterative solution of a network composed of these companson models. Alter DNET in Chapter 5 to solve for the

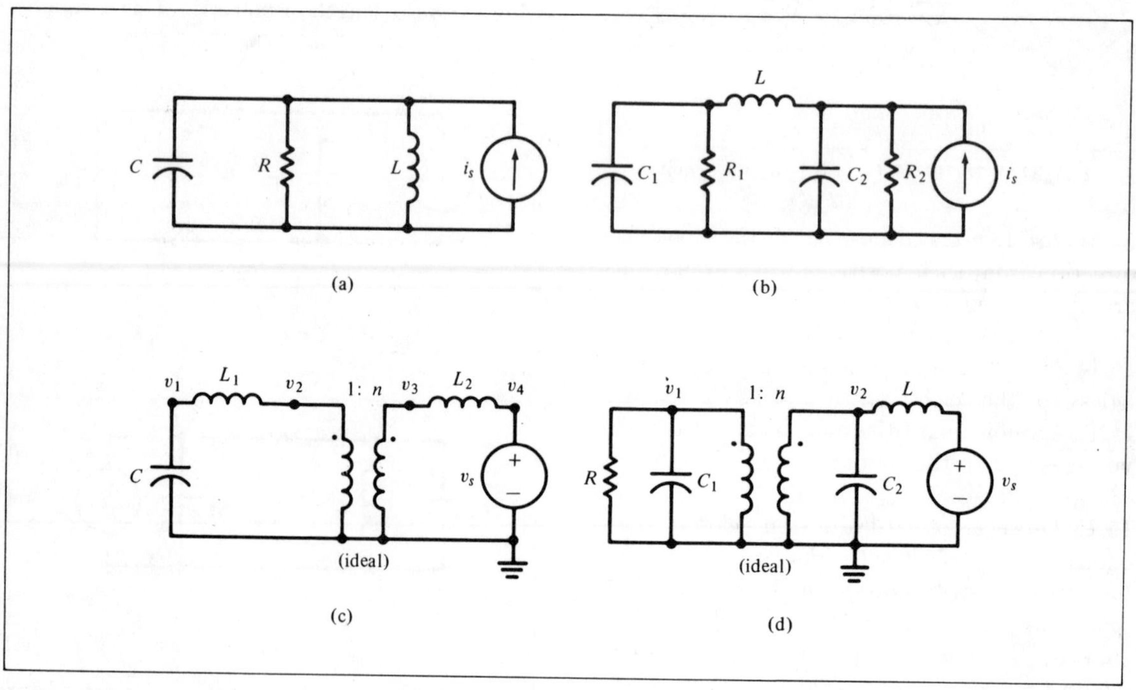

FIGURE P 16.14

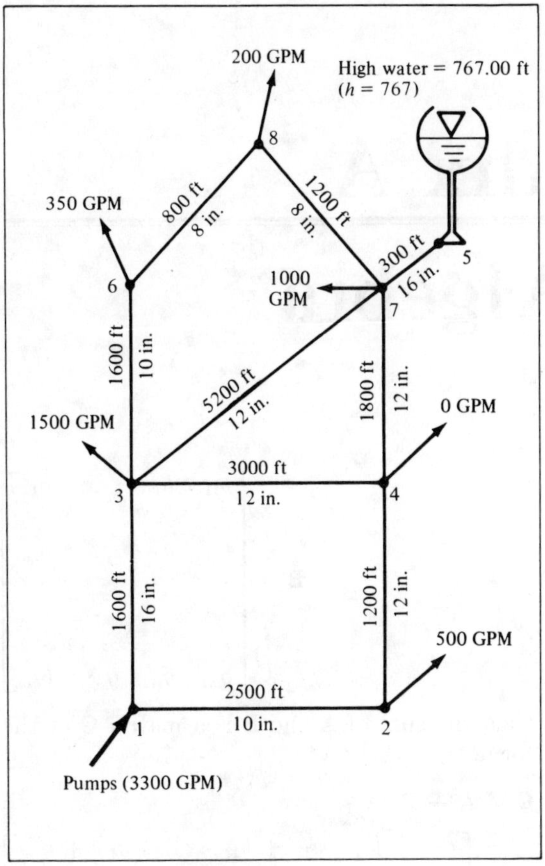

FIGURE P 16.16

flow and pressure drop in each pipe, using node analysis. Use initial nodal pressures near the value of the pressure source (747 ft).

REFERENCES AND ADDITIONAL READING

Transformers

16.1 W. H. HAYT, JR. AND J. E. KIMMERLY, *Engineering Circuit Analysis.* New York: McGraw-Hill Book Company, Inc., 1963, Chapter 16.

16.2 C. M. CLOSE, *The Analysis of Linear Circuits.* New York: Harcourt, Brace and World, Inc., 1966, Chapter 7.

Analogs

16.3 D. K. CHENG, *Analysis of Linear Systems.* Reading, Mass.: Addison-Wesley Publishing Company, Inc., 1959, Chapter 4.

16.4 C. M. CLOSE, *The Analysis of Linear Circuits.* New York: Harcourt, Brace and World, Inc., 1966, Chapter 11.

16.5 M. F. GARDNER AND J. L. BARNES, *Transients in Linear Systems*, New York; John Wiley & Sons, Inc., 1942, Chapter 2.

Appendix A
Matrix Algebra

Introduction

A matrix is a rectangular array of quantities ordered in rows and columns on which we define a set of rules (an algebra) that govern the arithmetic operations such as addition and multiplication performed on the matrix. An example is

$$\mathbf{A} = \begin{bmatrix} a_{11} & a_{12} & a_{13} \\ a_{21} & a_{22} & a_{23} \end{bmatrix}$$

This matrix, with two rows and three columns, is said to be of order 2×3. A matrix with only one column is called a *vector*.

Equality

If two matrices have the same order and if quantities in corresponding positions of the matrices are identical, the matrices are said to be equal.

Addition of Matrices

If

$$\mathbf{A} = \begin{bmatrix} a_{11} & a_{12} & \dots & a_{1n} \\ a_{21} & a_{22} & \dots & a_{2n} \\ \cdot & \cdot & \cdot & \cdot \\ \cdot & \cdot & \cdot & \cdot \\ \cdot & \cdot & \cdot & \cdot \\ a_{m1} & a_{m2} & \dots & a_{mn} \end{bmatrix}$$

$$\mathbf{B} = \begin{bmatrix} b_{11} & b_{12} & \dots & b_{1n} \\ b_{21} & b_{22} & \dots & b_{2n} \\ \cdot & \cdot & \cdot & \cdot \\ \cdot & \cdot & \cdot & \cdot \\ \cdot & \cdot & \cdot & \cdot \\ b_{m1} & b_{m2} & \dots & b_{mn} \end{bmatrix}$$

then the sum of $\mathbf{A}$ and $\mathbf{B}$ is a matrix $\mathbf{C}$ of the form

$$\mathbf{C} = \mathbf{A} + \mathbf{B}$$

$$= \begin{bmatrix} a_{11} + b_{11} & a_{12} + b_{12} & \dots & a_{1n} + b_{1n} \\ a_{21} + b_{21} & a_{22} + b_{22} & \dots & a_{2n} + b_{2n} \\ \cdot & \cdot & \cdot & \cdot \\ \cdot & \cdot & \cdot & \cdot \\ \cdot & \cdot & \cdot & \cdot \\ a_{m1} + b_{m1} & a_{m2} + b_{m2} & \dots & a_{mn} + b_{mn} \end{bmatrix}$$

That is, the sum of two $m \times n$ matrices is an $m \times n$ matrix with the (i, j)th entry being the sum of the (i, j)th entries of the two matrices; in compact form, this can be stated

$$c_{ij} = a_{ij} + b_{ij} \qquad (1 \leqq i \leqq m)$$
$$(1 \leqq j \leqq n)$$

Matrix Multiplication

If $\mathbf{A}$ has m rows and p columns and $\mathbf{B}$ has p rows and n columns, then the product $\mathbf{C} = \mathbf{AB}$

has the (i, j) entry

$$c_{ij} = \sum_{k=1}^{p} a_{ik}b_{kj} \qquad (1 \leqq i \leqq m)$$
$$(1 \leqq j \leqq n)$$

This product has a number of special characteristics.

1. $\mathbf{C}$ has the same number of rows as $\mathbf{A}$ and the same number of columns as $\mathbf{B}$.
2. This product is defined only if $\mathbf{A}$ has the same number of columns (p) as $\mathbf{B}$ has rows.
3. Even if both $\mathbf{AB}$ and $\mathbf{BA}$ are defined, in general $\mathbf{AB} \neq \mathbf{BA}$.

As a guide in multiplying matrices, note that the inner subscripts in the product $a_{ik}b_{kj}$ are equal. A typical calculation might be of the form

$$c_{23} = a_{21}b_{13} + a_{22}b_{23} + \cdots + a_{2p}b_{p3}$$

The most common form of multiplication is the product of a matrix $\mathbf{A}$ of order $m \times n$ and a vector of order $n \times 1$. For example,

$$\begin{bmatrix} 2 & 1 & 3 \\ 5 & -1 & -2 \end{bmatrix} \begin{bmatrix} 2 \\ 4 \\ 6 \end{bmatrix} = \begin{bmatrix} 26 \\ -6 \end{bmatrix}$$

where

$$26 = (2)(2) + (1)(4) + (3)(6)$$
$$-6 = (5)(2) + (-1)(4) + (-2)(6)$$

One product of special note involves the unit matrix defined by

$$\mathbf{I} \triangleq \begin{bmatrix} 1 & 0 & 0 & \ldots & 0 \\ 0 & 1 & 0 & \ldots & 0 \\ 0 & 0 & 1 & \ldots & 0 \\ \cdot & \cdot & \cdot & \cdot & \cdot \\ \cdot & \cdot & \cdot & \cdot & \cdot \\ \cdot & \cdot & \cdot & \cdot & \cdot \\ 0 & 0 & 0 & \ldots & 1 \end{bmatrix}$$

Here, $\mathbf{I}$ is a square $(n \times n)$ matrix. It is easily shown that $\mathbf{AI} = \mathbf{A}$, assuming that the order of $\mathbf{I}$ is chosen so that $\mathbf{AI}$ is defined. Similarly, $\mathbf{IA} = \mathbf{A}$.

Determinant of a Square Matrix

When $\mathbf{A}$ is an $n \times n$ (square) matrix, we can define the *determinant* $\Delta = |\mathbf{A}|$. The *value* of this determinant (also denoted as Δ) is defined as the sum of *all* products of the form

$$\pm a_{i_1j_1}a_{i_2j_2}a_{i_3j_3}\ldots a_{i_nj_n}$$

where none of the i's are taken from the same row and none of the j's are taken from the same column. The sign of the product is determined by rules to be studied shortly. There are n such terms, with each term involving n a_{ij}'s. The simplest examples are

$$\mathbf{A} = [a_{11}] \qquad \Delta = |\mathbf{A}| = a_{11}$$

$$\mathbf{A} = \begin{bmatrix} a_{11} & a_{12} \\ a_{21} & a_{22} \end{bmatrix} \qquad \Delta = |\mathbf{A}| = a_{11}a_{22} - a_{12}a_{21}$$

If in a determinant $\Delta = |\mathbf{A}|$ the kth row and the lth column are eliminated, the resulting determinant is called the $k - l$ *minor*. If the minor is multiplied by $(-1)^{k+l}$, the result is called the $k - l$ *cofactor* and is written Δ_{kl}. For example, associated with the determinant

$$\Delta = |\mathbf{A}| = \begin{vmatrix} a_{11} & a_{12} & a_{13} \\ a_{21} & a_{22} & a_{23} \\ a_{31} & a_{32} & a_{33} \end{vmatrix}$$

we could write

$$\Delta_{11} = (-1)^{1+1} \begin{vmatrix} a_{22} & a_{23} \\ a_{32} & a_{33} \end{vmatrix} = a_{22}a_{33} - a_{32}a_{23}$$

$$\Delta_{23} = (-1)^{2+3} \begin{vmatrix} a_{11} & a_{12} \\ a_{31} & a_{32} \end{vmatrix} = -a_{11}a_{32} + a_{31}a_{12}$$

Evaluation of Determinants

Beyond the simple 1×1 and 2×2 matrix determinant evaluations given above, this calculation is usually quite tedious. For the 3×3 case, the following technique is popular. We write the matrix

$$\mathbf{A} = \begin{bmatrix} a_{11} & a_{12} & a_{13} \\ a_{21} & a_{22} & a_{23} \\ a_{31} & a_{32} & a_{33} \end{bmatrix}$$

in the forms

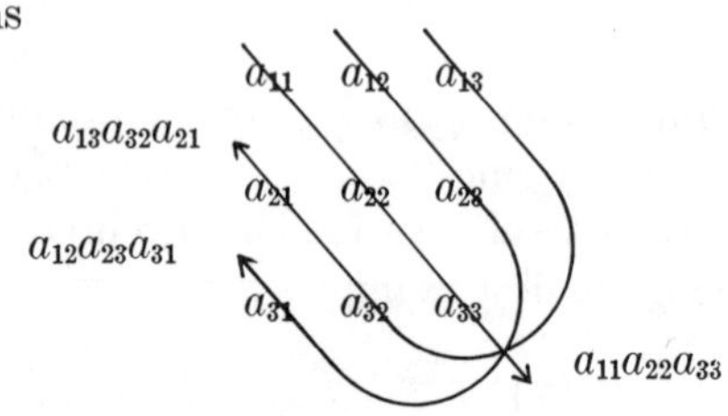

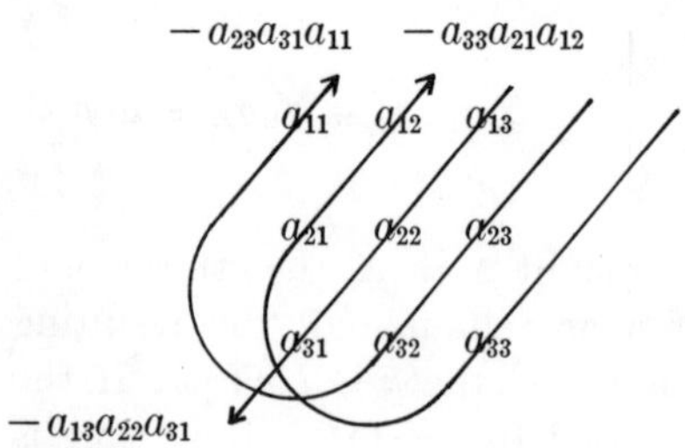

Then Δ is the sum of the positive and negative products shown, namely

$$\Delta = a_{11}a_{22}a_{33} + a_{13}a_{32}a_{21} + a_{12}a_{23}a_{31} - a_{13}a_{22}a_{31} - a_{23}a_{32}a_{11} - a_{33}a_{21}a_{12}$$

This procedure cannot conveniently be extended beyond $n > 3$. A more general method is the *Laplace expansion.* We begin by selecting a row *or* a column of the matrix on which to base the expansion. We next evaluate all cofactors for that row or column. Then if we have selected the jth column

$$\Delta = \sum_{k=1}^{n} a_{kj}\Delta_{kj}$$

and if we selected the ith row

$$\Delta = \sum_{k=1}^{n} a_{ik}\Delta_{ik}$$

For the general third-order case

$$[\mathbf{A}] = \begin{bmatrix} a_{11} & a_{12} & a_{13} \\ a_{21} & a_{22} & a_{23} \\ a_{31} & a_{32} & a_{33} \end{bmatrix}$$

if we expand along the first row ($i = 1$) we have

$$\Delta = a_{11}\Delta_{11} + a_{12}\Delta_{12} + a_{13}\Delta_{13}$$

$$= a_{11}\begin{vmatrix} a_{22} & a_{23} \\ a_{32} & a_{33} \end{vmatrix} - a_{12}\begin{vmatrix} a_{21} & a_{23} \\ a_{31} & a_{33} \end{vmatrix} + a_{13}\begin{vmatrix} a_{21} & a_{22} \\ a_{31} & a_{32} \end{vmatrix}$$

$$= a_{11}a_{22}a_{33} - a_{11}a_{23}a_{32} - a_{12}a_{21}a_{33} + a_{12}a_{23}a_{31} + a_{13}a_{21}a_{32} - a_{13}a_{22}a_{31}$$

This agrees with the previous evaluation for $n = 3$. The reader may wish to show that expanding along another row or column gives the same result. This method is valid for any n; however, the Δ_{ij} cofactors grow in size as n increases, so even this method becomes unwieldy for $n > 4$.

Matrix Inversion

The inverse of a matrix $\mathbf{A}$ is denoted $\mathbf{A}^{-1}$; it has the property that

$$\mathbf{A}^{-1}\mathbf{A} = \mathbf{A}\mathbf{A}^{-1} = \mathbf{I}$$

where $\mathbf{I}$ is the identity matrix.

The inverse has the form

$$\mathbf{A}^{-1} = \begin{bmatrix} \frac{\Delta_{11}}{\Delta} & \frac{\Delta_{21}}{\Delta} & \frac{\Delta_{31}}{\Delta} & \cdots & \frac{\Delta_{n1}}{\Delta} \\ \frac{\Delta_{12}}{\Delta} & \frac{\Delta_{22}}{\Delta} & \cdot & \cdots & \frac{\Delta_{n2}}{\Delta} \\ \frac{\Delta_{13}}{\Delta} & \cdot & \cdot & \cdots & \cdot \\ \cdot & \cdot & \cdot & \cdot & \cdot \\ \cdot & \cdot & \cdot & \cdot & \cdot \\ \cdot & \cdot & \cdot & \cdot & \cdot \\ \frac{\Delta_{1n}}{\Delta} & \frac{\Delta_{2n}}{\Delta} & \cdot & \cdots & \frac{\Delta_{nn}}{\Delta} \end{bmatrix}$$

$$= \frac{1}{\Delta} \begin{bmatrix} \Delta_{11} & \Delta_{21} & \Delta_{31} & \cdots & \Delta_{n1} \\ \Delta_{12} & \Delta_{22} & \cdot & \cdots & \Delta_{n2} \\ \Delta_{13} & \cdot & \cdot & \cdots & \cdot \\ \cdot & \cdot & \cdot & \cdots & \cdot \\ \cdot & \cdot & \cdot & \cdot & \cdot \\ \cdot & \cdot & \cdot & \cdot & \cdot \\ \Delta_{1n} & \Delta_{2n} & \cdot & \cdots & \Delta_{nn} \end{bmatrix}$$

It is especially important to note that (1) the cofactor Δ_{ji} appears in the (i, j) position of the inverse, and (2) the inverse does not exist if $\Delta = 0$. The matrix is said to be *singular* if $\Delta = 0$.

As a numerical example, we choose

$$\mathbf{A} = \begin{bmatrix} 1 & 2 & 3 \\ 1 & -2 & 1 \\ 2 & 3 & -1 \end{bmatrix}$$

We calculate

$$\begin{array}{lll} \Delta_{11} = -1 & \Delta_{12} = 3 & \Delta_{13} = 7 \\ \Delta_{21} = 11 & \Delta_{22} = -7 & \Delta_{23} = 1 \\ \Delta_{31} = 8 & \Delta_{32} = 2 & \Delta_{33} = -4 \end{array}$$

The determinant is found by the Laplace expansion down the first column

$$\Delta = (1)(-1) + (1)(11) + (2)(8) = 26$$

The inverse is then

$$\mathbf{A}^{-1} = \frac{1}{26} \begin{bmatrix} -1 & 11 & 8 \\ 3 & -7 & 2 \\ 7 & 1 & -4 \end{bmatrix}$$

The reader can verify that $\mathbf{AA}^{-1} = \mathbf{A}^{-1}\mathbf{A} = \mathbf{I}$ in this case.

Solution of Simultaneous Equations

A set of n simultaneous linear equations in n unknowns $x_1, x_2, \ldots, x_n$ can be written in either of the following forms

$$\begin{array}{c} a_{11}x_1 + a_{12}x_2 + \cdots + a_{1n}x_n = b_1 \\ a_{21}x_1 + a_{22}x_2 + \cdots + a_{2n}x_n = b_2 \\ \cdot \quad \cdot \quad \cdot \quad \cdot \\ \cdot \quad \cdot \quad \cdot \quad \cdot \\ \cdot \quad \cdot \quad \cdot \quad \cdot \\ a_{n1}x_1 + a_{n2}x_2 + \cdots + a_{nn}x_n = b_n \end{array}$$

$$\begin{bmatrix} a_{11} & a_{12} & \cdots & a_{1n} \\ a_{21} & a_{22} & \cdots & a_{2n} \\ \cdot & \cdot & \cdot & \cdot \\ \cdot & \cdot & \cdot & \cdot \\ \cdot & \cdot & \cdot & \cdot \\ a_{n1} & a_{n2} & \cdots & a_{nn} \end{bmatrix} \begin{bmatrix} x_1 \\ x_2 \\ \cdot \\ \cdot \\ \cdot \\ x_n \end{bmatrix} = \begin{bmatrix} b_1 \\ b_2 \\ \cdot \\ \cdot \\ \cdot \\ b_n \end{bmatrix}$$

where a_{ij} and b_i are constants. The latter has the compact form

$$\mathbf{Ax} = \mathbf{b}$$

One method of solution of equations of this form is Cramer's rule. The solution has the general form

$$x_r = \frac{\begin{vmatrix} a_{11} & a_{12} & \cdots & \overset{\substack{r\text{th column of }\mathbf{A}\\ \downarrow\ \text{replaced by }\mathbf{b}}}{b_1} & \cdots & a_{1n} \\ a_{21} & a_{22} & \cdots & b_2 & \cdots & a_{2n} \\ \cdot & \cdot & & \cdot & \cdot & \cdot \\ \cdot & \cdot & & \cdot & \cdot & \cdot \\ \cdot & \cdot & & \cdot & \cdot & \cdot \\ a_{n1} & a_{n2} & \cdots & b_r & & a_{nn} \end{vmatrix}}{\begin{vmatrix} a_{11} & a_{12} & a_{13} & \cdots & a_{1n} \\ a_{21} & a_{22} & a_{23} & \cdots & a_{2n} \\ \cdot & \cdot & \cdot & \cdot & \cdot \\ \cdot & \cdot & \cdot & \cdot & \cdot \\ \cdot & \cdot & \cdot & \cdot & \cdot \\ a_{n1} & a_{n2} & a_{n3} & \cdots & a_{nn} \end{vmatrix}}$$

That is, the rth solution variable can be found as the ratio of two determinants:

1. The denominator is the determinant of the matrix **A**.
2. The numerator is the determinant of an $n \times n$ matrix formed from **A** by replacing the rth column by the right-hand side vector **b**.

For example, the matrix equation

$$\begin{bmatrix} 1 & 2 & 3 \\ 1 & -2 & 1 \\ 2 & 3 & -1 \end{bmatrix} \begin{bmatrix} x_1 \\ x_2 \\ x_3 \end{bmatrix} = \begin{bmatrix} 1 \\ -1 \\ 2 \end{bmatrix}$$

can be solved for x_1 as

$$x_1 = \frac{\begin{bmatrix} \overset{\substack{\mathbf{b}\\ \downarrow}}{1} & 2 & 3 \\ -1 & -2 & 1 \\ 2 & 3 & -1 \end{bmatrix}}{\begin{bmatrix} 1 & 2 & 3 \\ 1 & -2 & 1 \\ 2 & 3 & -1 \end{bmatrix}} = \frac{4}{26}$$

An alternative solution is via the matrix inverse. Multiplying both sides of the equations by the inverse of **A** gives

$$\mathbf{A}^{-1}\mathbf{A}\mathbf{x} = \mathbf{A}^{-1}\mathbf{b}$$

Because $\mathbf{A}^{-1}\mathbf{A} = \mathbf{I}$ and $\mathbf{I}\mathbf{x} = \mathbf{x}$, we find

$$\mathbf{x} = \mathbf{A}^{-1}\mathbf{b}$$

As an example, consider the equations

$$\begin{bmatrix} 1 & 2 & 3 \\ 1 & -2 & 1 \\ 2 & 3 & -1 \end{bmatrix} \begin{bmatrix} x_1 \\ x_2 \\ x_3 \end{bmatrix} = \begin{bmatrix} 1 \\ -1 \\ 2 \end{bmatrix}$$

The inverse of the 3×3 matrix has already been found in a previous example. We therefore have

$$\begin{bmatrix} x_1 \\ x_2 \\ x_3 \end{bmatrix} = \frac{1}{26} \begin{bmatrix} -1 & 11 & 8 \\ 3 & -7 & 2 \\ 7 & 1 & -4 \end{bmatrix} \begin{bmatrix} 1 \\ -1 \\ 2 \end{bmatrix}$$

$$= \begin{bmatrix} \dfrac{4}{26} \\[2ex] \dfrac{14}{26} \\[2ex] \dfrac{-2}{26} \end{bmatrix}$$

This solution vector can be checked by substituting in the original equations.

Recalling the general form of the inverse, we can represent the solution symbolically as

$$\mathbf{x} = \mathbf{A}^{-1}\mathbf{b}$$

$$\begin{bmatrix} x_1 \\ x_2 \\ x_3 \\ \cdot \\ \cdot \\ \cdot \\ x_n \end{bmatrix} = \frac{1}{\Delta} \begin{bmatrix} \Delta_{11} & \Delta_{21} & \Delta_{31} & \cdots & \Delta_{n1} \\ \Delta_{12} & \Delta_{22} & \Delta_{32} & \cdots & \Delta_{n2} \\ \Delta_{13} & \Delta_{23} & \Delta_{33} & \cdots & \Delta_{n3} \\ \cdot & \cdot & \cdot & \cdot & \cdot \\ \cdot & \cdot & \cdot & \cdot & \cdot \\ \cdot & \cdot & \cdot & \cdot & \cdot \\ \Delta_{1n} & \Delta_{2n} & \Delta_{3n} & \cdots & \Delta_{nn} \end{bmatrix} \begin{bmatrix} b_1 \\ b_2 \\ b_3 \\ \cdot \\ \cdot \\ \cdot \\ b_n \end{bmatrix}$$

For any solution variable x_r, we can write simply

$$x_r = \frac{1}{\Delta} \sum_{i=1}^{n} \Delta_{ir} b_i$$

Appendix B

Complex Numbers

Definitions and Representations

A complex number is an ordered pair of real numbers with which we associate special arithmetic operations. We write this pair of real numbers σ and ω and the resultant complex number s as

$$s = \sigma + j\omega$$

When $j \triangleq \sqrt{-1}$, we can think of σ and ω as defining a point in the *complex plane* with the real axis corresponding to the axis of abscissas and the imaginary axis corresponding to the axis of ordinates (see Figure B.1). The sym-

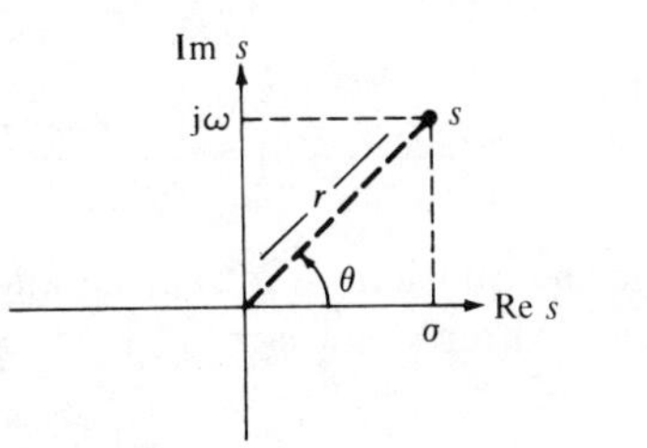

FIGURE B.1 $\sigma = r\cos\theta$; $\omega = r\sin\theta$; $r = \sqrt{\sigma^2 + \omega^2}$; $\theta = \tan^{-1}(\omega/\sigma)$.

bolism Re[s] and Im[s] denotes the real part and imaginary parts of s, respectively. From the above figure, it is clear that a positive value of σ will result in the point being in the right half of the plane, and a positive value of ω will mean the point is in the upper half-plane.

The above representation in terms of real and imaginary parts is known as the rectangular form. Other representations are the trigonometric form

$$s = r(\cos\phi + j\sin\theta)$$

so that $\sigma = r\cos\phi$ and $\omega = r\sin\theta$, and the polar form

$$s = re^{j\theta}$$

These two forms can be related as follows. For all finite values of the angle θ the functions $\sin\theta$ and $\cos\theta$ can be expressed by the power series

$$\sin\theta = \theta - \frac{\theta^3}{3!} + \frac{\theta^5}{5!} - \frac{\theta^7}{7!} + \cdots$$

$$\cos\theta = 1 - \frac{\theta^2}{2!} + \frac{\theta^4}{4!} - \frac{\theta^6}{6!} + \cdots$$

The exponential function can also be expanded in an absolutely convergent power series for all finite values of the argument; in particular

$$e^{j\theta} = 1 + (j\theta) + \frac{(j\theta)^2}{2!} + \frac{(j\theta)^3}{3!} + \frac{(j\theta)^4}{4!} + \cdots$$

$$= 1 + j\theta - \frac{\theta^2}{2!} - j\frac{\theta^3}{3!} + \frac{\theta^4}{4!} + \cdots$$

where $(j)^2 = -1$, $(j)^3 = -j$, and so on.

Comparing the above it is easy to see that

$$e^{j\theta} = \cos\theta + j\sin\theta$$

Equality

Two complex numbers are equal if their real and imaginary parts are equal.

Addition

The sum of two complex numbers $s_1 = \sigma_1 + j\omega_1$ and $s_2 = \sigma_2 + j\omega_2$ is a complex number $s_3 = \sigma_3 + j\omega_3$ given by

$$\begin{aligned} s_3 &= s_1 + s_2 \\ &= \sigma_1 + j\omega_1 + \sigma_2 + j\omega_2 \\ &= \sigma_1 + \sigma_2 + j(\omega_1 + \omega_2) \end{aligned}$$

so that $\sigma_3 = \sigma_1 + \sigma_2$, $\omega_3 = \omega_1 + \omega_2$.

Multiplication

In polar form, the multiplication of $s_1 = re^{j\theta_1}$ and $s_2 = r_2e^{j\theta_2}$ is defined as

$$\begin{aligned} s_3 &= s_1s_2 \\ &= r_1e^{j\theta_1}r_2e^{j\theta_2} \\ &= r_1r_2e^{j(\theta_1+\theta_2)} \end{aligned}$$

This operation is consistent with the real arithmetic rule that $e^ae^b = e^{a+b}$. It is sometimes necessary to carry out the multiplication in rectangular form.

$$\begin{aligned} s_3 &= r_1r_2e^{j(\theta_1+\theta_2)} \\ &= r_1r_2(\cos(\theta_1+\theta_2) + j\sin(\theta_1+\theta_2)) \\ &= r_1r_2(\cos\theta_1\cos\theta_2 - \sin\theta_1\sin\theta_2) \\ &\quad + jr_1r_2(\sin\theta_1\cos\theta_2 + \cos\theta_1\sin\theta_2) \\ &= (\sigma_1\sigma_2 - \omega_1\omega_2) + j(\sigma_2\omega_1 + \sigma_1\omega_2) \end{aligned}$$

This is precisely the same result that would be obtained from expanding

$$s_3 = (\sigma_1 + j\omega_1)(\sigma_2 + j\omega_2)$$

and using the fact that $j^2 = -1$.

Division

Division in polar form follows from the above multiplication rule.

$$\begin{aligned} \frac{s_1}{s_2} &= \frac{r_1e^{j\theta_1}}{r_2e^{j\theta_2}} = (r_1e^{j\theta_1})\left(\frac{1}{r_2}e^{-j\theta_2}\right) \\ &= \frac{r_1}{r_2}e^{j(\theta_1-\theta_2)} \end{aligned}$$

In rectangular form, we have

$$s_3 = \frac{s_1}{s_2} = \frac{(\sigma_1 + j\omega_1)}{(\sigma_2 + j\omega_2)}$$

We simplify this expression by use of the notion of the complex conjugate. The complex conjugate of $s = \sigma + j\omega$ is denoted s^* and defined as $s^* = \sigma - j\omega$; that is, the sign of the imaginary part is changed. If we multiply s_2 by $s_2{}^*$ we get

$$\begin{aligned} s_2s_2{}^* &= (\sigma_2 + j\omega_2)(\sigma_2 - j\omega_2) \\ &= \sigma_2{}^2 + \omega_2{}^2 + j(-\sigma_2\omega_2 + \sigma_2\omega_2) \\ &= \sigma_2{}^2 + \omega_2{}^2 \end{aligned}$$

which is a real number. If we now multiply *both* the numerator and denominator of s_1/s_2 by $s_2{}^*$, we preserve the value of the ratio and obtain

$$\begin{aligned} s_3 &= \frac{s_1s_2{}^*}{s_2s_2{}^*} = \frac{(\sigma_1 + j\omega_1)(\sigma_2 - j\omega_2)}{(\sigma_2 + j\omega_2)(\sigma_2 - j\omega_2)} \\ &= \frac{\sigma_1\sigma_2 + \omega_1\omega_2}{\sigma_2{}^2 + \omega_2{}^2} + j\,\frac{(-\sigma_1\omega_2 + \sigma_2\omega_1)}{\sigma_2{}^2 + \omega_2{}^2} \end{aligned}$$

This is the quotient s_3 in rectangular form.

As an example, let $s_1 = 3 + j4$, $s_2 = 1 + j1$; then

$$s_1s_2 = -1 + j7 = \sqrt{50}e^{j\tan^{-1}(7/-1)}$$

$$\frac{s_1}{s_2} = \frac{7}{2} + j\,\frac{1}{2} = \frac{\sqrt{50}}{2}\,e^{j\tan^{-1}(1/7)}$$

This example points out an important aspect of determining the angle θ in the polar representation; the sign of the real and imaginary parts must be preserved, because $\tan^{-1}(7/-1) \neq -\tan^{-1}(7/1)$.

Appendix C

Fourier Series and Transform

In this appendix we show how nonsinusoidal periodic signals can be expressed in terms of sinusoidal components of different frequencies. In general, any periodic signal that is single-valued and continuous except possibly for a finite number of finite discontinuities in the neighborhood of any point may be represented by the sum of sinusoids of different frequencies. That is, we can write such a function as

$$
\begin{aligned}
f(t) &= a_0 + a_1 \cos \omega t + b_1 \sin \omega t \\
&\quad + a_2 \cos 2\omega t + b_2 \sin 2\omega t \\
&\quad \vdots \\
&= \sum_{n=0}^{\infty} a_n \cos n\omega t + b_n \sin n\omega t \qquad \text{(C.1)}
\end{aligned}
$$

where $\omega = 2\pi/T$ and T is the period. This is termed a *Fourier series representation* of the function $f(t)$.

The a_n and b_n can be determined as follows. Multiplying both sides of Equation (C.1) by $\cos m\omega t$ and integrating over a period, we obtain

$$
\begin{aligned}
\int_0^T f(t) \cos m\omega t \, dt &= \int_0^T \left[\sum_{n=0}^{\infty} (a_n \cos n\omega t + b_n \sin n\omega t) \right] \cos m\omega t \, dt \\
&= \sum_{n=0}^{\infty} \left[a_n \int_0^T \cos m\omega t \cos n\omega t \, dt \right. \\
&\quad \left. + b_n \int_0^T \cos m\omega t \sin n\omega t \, dt \right] \qquad \text{(C.2)}
\end{aligned}
$$

Here the integral can be taken inside the summation due to the continuity requirements on $f(t)$. The integrals on the right-hand side can be evaluated from a table of definite integrals:

$$\int_0^T \cos m\omega t \cos n\omega t\, dt = 0 \qquad (m \neq n)$$

$$= \frac{T}{2} \qquad (m = n \neq 0)$$

$$= T \qquad (m = n = 0)$$

$$\int_0^T \cos m\omega t \sin n\omega t\, dt = 0 \qquad (\text{all } m \text{ and } n)$$

Matching terms on each side of Equation (C.2), we can write

$$a_n = \frac{2}{T}\int_0^T f(t) \cos n\omega t\, dt \qquad (n \neq 0)$$

In a completely analogous procedure, we can obtain b_n by multiplying Equation (C.1) by $\sin m\omega t$, integrating, and applying the identity

$$\int_0^T \sin m\omega t \sin n\omega t\, dt = 0 \qquad (m \neq n)$$

$$= \frac{T}{2} \qquad (m = n \neq 0)$$

$$= 0 \qquad (m \text{ or } n = 0)$$

This yields

$$b_n = \frac{2}{T}\int_0^T f(t) \sin n\omega t\, dt \qquad (n \neq 0)$$

For the case that $n = 0$, we obtain

$$b_0 = 0 \qquad a_0 = \frac{1}{T}\int_0^T f(t)\, dt$$

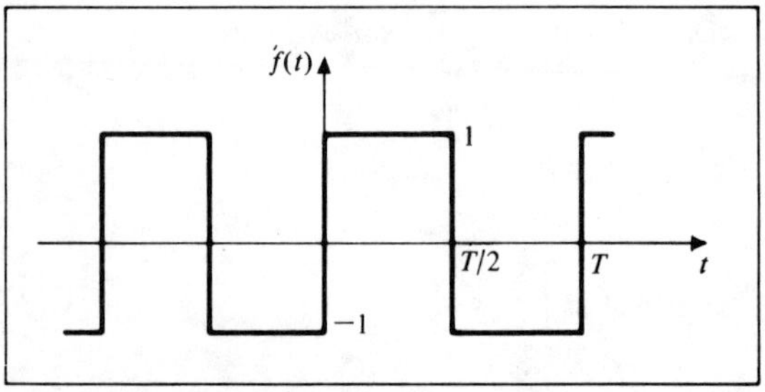

FIGURE C.1 Square wave signal.

Example

The Fourier series representation of the periodic signal of Figure C.1 is to be found.

We first determine the radian frequency to be $\omega = 2\pi/T$. Then we calculate

$$a_0 = \frac{1}{T}\int_0^T f(t)\, dt$$

$$= \frac{1}{T}\left[\int_0^{T/2} (1)\, dt + \int_{T/2}^T (-1)\, dt\right] = 0$$

$$a_n = \frac{2}{T}\int_0^T f(t) \cos n\omega t\, dt \qquad (n \neq 0)$$

$$= \frac{2}{T}\left[\int_0^{T/2} \cos n\omega t\, dt - \int_{T/2}^T \cos n\omega t\right]$$

$$= 0 \qquad \text{(C.3)}$$

$$b_n = \frac{2}{T}\int_0^T f(t) \sin n\omega t\, dt$$

$$= \frac{2}{T}\left[\int_0^{T/2} \sin n\omega t\, dt - \int_{T/2}^T \sin n\omega t\, dt\right]$$

$$= \frac{2}{T}\left[2\int_0^{T/2} \sin n\omega t\, dt\right]$$

$$= -\frac{2}{n\pi} \cos n\omega t \Big|_0^{T/2} \qquad \text{(C.4)}$$

$$= -\frac{2}{n\pi}(\cos n\pi - 1)$$

$$= 0 \qquad (n \text{ even})$$

$$= \frac{4}{n\pi} \qquad (n \text{ odd})$$

Therefore, we can represent $f(t)$ as (see Figure C.2)

$$f(t) = \sum_{n \text{ odd}} \frac{4}{n\pi} \sin n\omega t \qquad \left(\omega = \frac{2\pi}{T}\right)$$

It is useful to note that the symmetry of the above signal allowed the immediate simplifica-

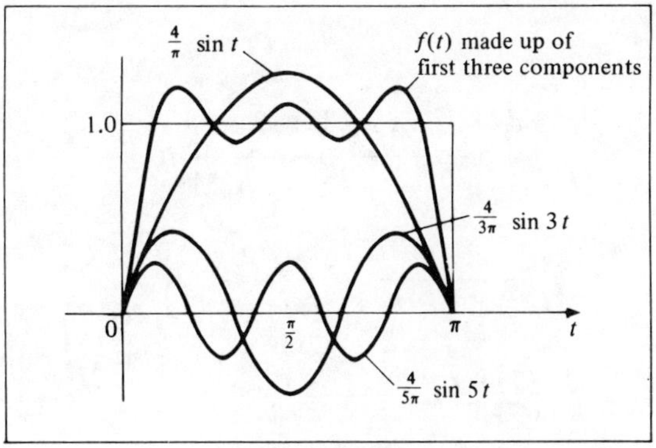

FIGURE C.2

tion of Equations (C.3) and (C.4). This occurred because the pairs of cosine and sine integrals in Equations (C.3) and (C.4) respectively were identical in value (except possibly for sign) over the integration interval. If the total signs of the integrals are the opposite, then cancellation occurs as in Equation (C.3); if the signs are the same, then we need only double one of the integrals as in Equation (C.4). Indeed, we can make the following general statement regarding the absence of the a_i or b_i terms which results from cancellation:

1. If $f(t)$ is an odd function, that is, $f(t) = -f(-t)$, then $a_n = 0$ and only sine terms will appear in the series; this is the phenomona observed above and is reasonable because both the function and its series will then both be odd functions.

2. If $f(t)$ is an even function, that is, $f(t) = f(-t)$, then $b_n = 0$ and only cosine terms will appear in the series; this will occur for the periodic functions in Figure C.3.

Exponential Form of Fourier Series

A comparison of the odd rectangular signal of Figure C.1 and the even rectangular signal of Figure C.3 (a) shows they differ only by a shift of $T/4$ sec or 90°. Yet their Fourier representations do not display this phase shift succinctly; if the time shift were $T/8$, then the function would be neither even nor odd and both a_n and b_n would be nonzero; we would be hard pressed to recognize this phase shift from an inspection of the values of a_n and b_n.

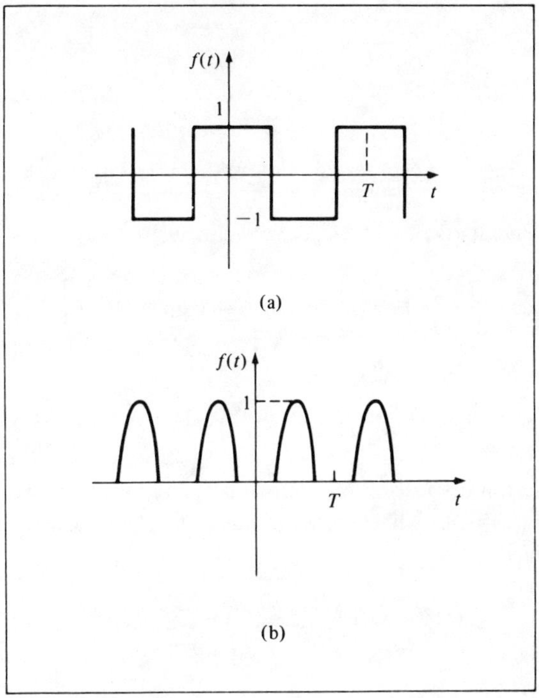

FIGURE C.3 Even functions of time.

A form more attuned to provide such insight can be obtained by employing the trigonometric identities

$$\cos\theta = \frac{e^{j\theta} + e^{-j\theta}}{2} \qquad \sin\theta = \frac{e^{j\theta} - e^{-j\theta}}{2j}$$

$$e^{j\theta} = \cos\theta + j\sin\theta$$

The series then becomes

$$f(t) = a_0 + \sum_{n=1}^{\infty}\left(\frac{a_n}{2}e^{jn\omega t} + \frac{a_n}{2}e^{-jn\omega t}\right.$$

$$\left. - j\frac{b_n}{2}e^{jn\omega t} + j\frac{b_n}{2}e^{-jn\omega t}\right)$$

$$= a_0 + \sum_{n=1}^{\infty}\left(\frac{a_n - jb_n}{2}e^{jn\omega t} + \frac{a_n + jb_n}{2}e^{-jn\omega t}\right)$$

$$= a_0 + \sum_{n=1}^{\infty} (c_n e^{jn\omega t} + c_n{}^* e^{-jn\omega t})$$

where

$$|c_n| = \sqrt{a_n{}^2 + b_n{}^2} \qquad \arg c_n = -\tan^{-1}\frac{b_n}{a_n}$$

Because $c_n = a_n - jb_n$, we can write

$$c_n = \frac{1}{2}\left[\frac{2}{T}\int_0^T f(t)(\cos n\omega t - j\sin n\omega t)\,dt\right]$$

$$= \frac{1}{T}\int_0^T f(t)e^{-jn\omega t}\,dt \qquad \text{(C.5)}$$

From inspection of Equation (C.5), we see that $c_n{}^* = c_{-n}$. The Fourier series of $f(t)$ can now be written

$$f(t) = a_0 + \sum_{n=1}^{\infty} c_n e^{jn\omega t} + \sum_{n=-1}^{-\infty} c_n e^{jn\omega t}$$

$$= \sum_{n=-\infty}^{\infty} c_n e^{jn\omega t}$$

where use has been made of the fact that

$$c_0 = \frac{1}{T}\int_0^T f(t)\,dt$$

$$= a_0$$

Example

The half-sinusoid of Figure C.4 is described by the complex Fourier coefficient

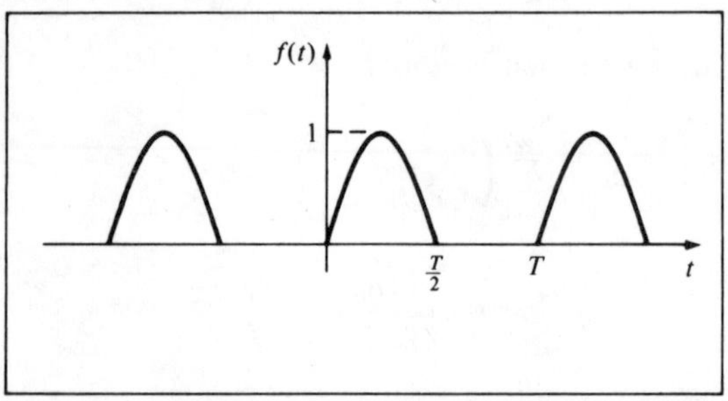

FIGURE C.4 Half sinusoid.

$$c_n = \frac{1}{T}\int_0^{T/2} \sin\omega t e^{-jn\omega t}\,dt$$

$$= \frac{1}{T}\int_0^{T/2}\left[\frac{e^{j\omega(1-n)t} - e^{-j\omega(1+n)t}}{2j}\right]dt$$

$$= \frac{1}{2jT}\left[\frac{e^{j\omega(1-n)t}}{j\omega(1-n)} + \frac{e^{-j\omega(1+n)t}}{j\omega(1+n)}\right]_0^{T/2}$$

$$= \frac{1}{2\pi}\left(\frac{1+e^{-jn\pi}}{1-n^2}\right) \qquad \text{(C.6)}$$

If the complex form of the series is desired for a half-sinusoid delayed by τ sec, then a typical term in the new series has the form

$$\hat{c}_n e^{-jn\omega t} = c_n e^{-jn\omega(t-\tau)}$$

$$= c_n e^{jn\omega\tau} e^{-jn\omega t}$$

so that the new Fourier coefficient is

$$\hat{c}_n = c_n e^{jn\omega\tau}$$

In particular, we note that

$$|\hat{c}_n| = |c_n||e^{jn\omega\tau}|$$

$$= |c_n|$$

Therefore, the coefficient of the nth term in the shifted and unshifted waveforms:

1. Are the same in magnitude, corresponding to $f(t)$ and $f(t-\tau)$ having the same magnitude.
2. Differ in phase (argument) by $n\omega\tau$ rad, corresponding to a shift of $n\tau$ sec in $f(t)$.

Circuit Analysis with Periodic Waveforms

The key to linear circuit analysis with general periodic waveforms is the relation between the c_n of the exponential form and the phasor representation of Chapter 10. We can write

$$c_n e^{jn\omega t} + c_n{}^* e^{-jn\omega t}$$

$$= 2|c_n|\left(\frac{e^{j(n\omega t+\phi_n)} + e^{-j(n\omega t+\phi_n)}}{2}\right)$$

$$= 2|c_n|\cos(n\omega t + \phi_n)$$

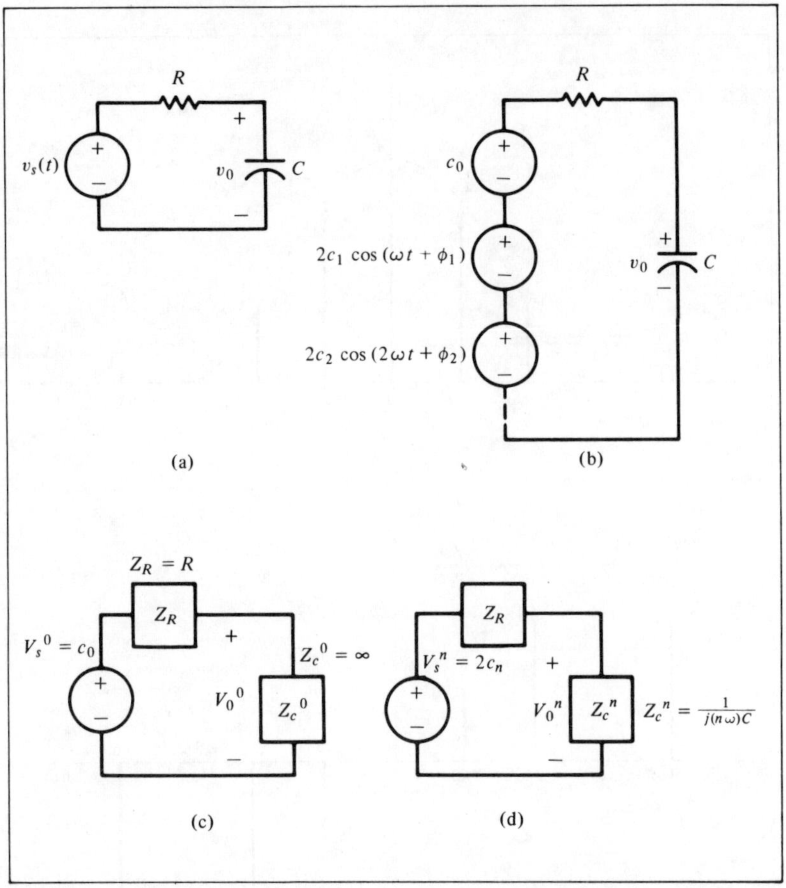

FIGURE C.5 Illustration of circuit analysis with periodic input. (a) Circuit with original signal; (b) circuit with expanded signal; (c) phasor model ($n = 0$); (d) phasor model ($n \neq 0$).

where $\phi_n = \arg c_n$. The phasor representation of $2|c_n| \cos(n\omega t + \phi_n)$ is $2|c_n|e^{j\phi_n} = 2c_n$. Therefore,

$$c_n = \tfrac{1}{2} \text{ (Phasor representation of } n\text{th component of series)}$$

If a voltage (current) source has a periodic waveform, this source may be replaced by a series (parallel) arrangement of sources in Figure C.5(b). Each of these, including the constant source of value c_0, can be represented in phasor form. From the superposition principle, these phasor sources can be applied to the phasor network model one at a time and the corresponding phasor components of the output determined. In Figure C.5, we note that the *impedances in each of these phasor networks will be different,* because each represents the original network at a different frequency ($n\omega$).

From Figure C.5(c)–(d), for example, we calculate the phasor response across the capacitor to be

$$V_0^n = \frac{V_s^n}{jn\omega RC + 1}$$

$$= \frac{2c_n}{jn\omega RC + 1} \qquad (n \neq 0)$$

$$= c_0 \qquad (n = 0)$$

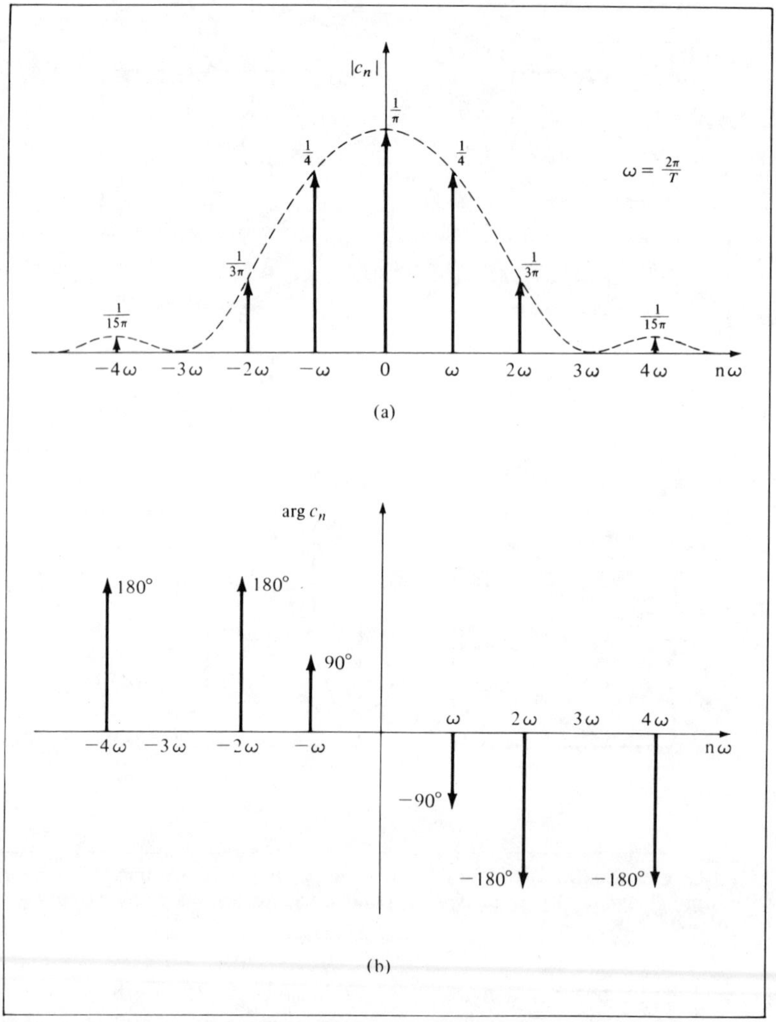

FIGURE C.6 Spectra of half sinusoid. (a) Amplitude; (b) phase.

and the total periodic response is

$$v_0(t) = c_0 + \sum_{n=1}^{\infty} |V_0{}^n| \cos(n\omega t + \phi_n)$$

where $\phi_n = \arg V_0{}^n$.

Spectral Representation

It is sometimes useful to display a plot of the magnitude and phase of c_n as a function of $n\omega$. For example, the Fourier coefficient given by Equation (C.6) is shown in Figure C.6. These plots are known as the *line or discrete spectra* of the signal. (They tell us the magnitudes and phases of each of the sinusoidal terms which constitute the signal.) In Figure C.6(a), for example, we would conclude that frequencies beyond $2\omega = 4\pi/T$ are relatively insignificant signal components, because their associated amplitudes are so small. Therefore,

$$f(t) \cong \sum_{n=-2}^{2} c_n e^{jn\omega t}$$

$$\cong \frac{1}{3\pi}\,(e^{j(-2\omega t+\pi)} + e^{j(2\omega t-\pi)}) + \frac{1}{4}\,(e^{j(-\omega t+\pi/2)}$$

$$+ e^{j(\omega t-\pi/2)}) + \frac{1}{\pi}\,e^{j(0)}$$

$$= \frac{2}{3\pi}\cos(2\omega t - \pi) + \frac{1}{2}\cos\left(\omega t - \frac{\pi}{2}\right) + \frac{1}{\pi} \qquad \text{(C.7)}$$

If the signal is changed in amplitude or shifted in time, these variations will appear in the amplitude and phase spectra of the signal.

Fourier Integral

Although our signal analysis with Fourier series has involved only periodic functions, we can rather simply extend the representation to aperiodic functions, provided we are willing to ignore certain mathematical rigor. This extension follows from the question "what happens to the Fourier series if $T \to \infty$." The period being infinite, the signal does not repeat itself and so is aperiodic (see Figure C.7).

We begin by repeating the exponential form of the Fourier series as

$$f(t) = \sum_{n=-\infty}^{\infty} c_n e^{jn\omega t} \qquad \text{(C.8)}$$

$$c_n = \frac{1}{T}\int_{-T/2}^{T/2} f(t)e^{-jn\omega t}\,dt \qquad \text{(C.9)}$$

When T approaches infinity, some changes in notation are necessary. First, because $\omega = 2\pi/T$, ω approaches zero, and n becomes meaningless as a counter. Second, the discrete variable $n\omega$, which identifies a frequency on the discrete (line) signal spectrum, becomes a continuous variable. The following changes in variable are useful to denote these effects:

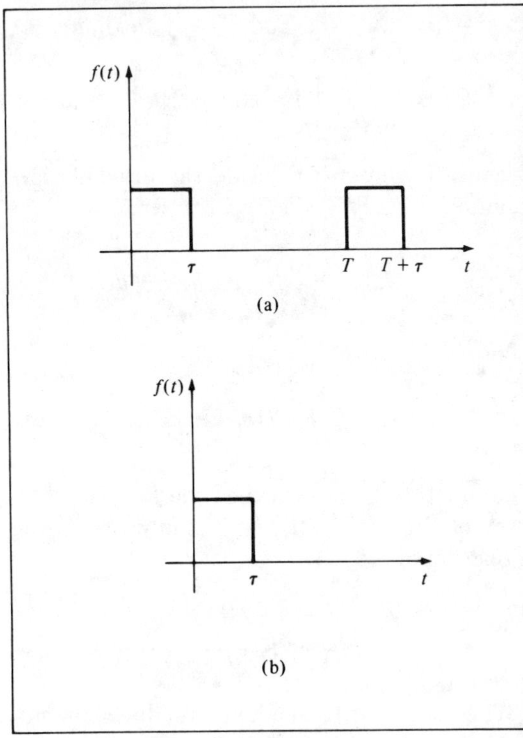

FIGURE C.7 Pulse representations. (a) Periodic (T large); (b) aperiodic (T infinite).

Angular frequency	$n\omega \to \omega$
Line spectrum spacing	$\omega \to \Delta\omega$
Period	$T \to 2\pi/\Delta\omega$

Equations (C.8) and (C.9) therefore become

$$f(t) = \sum_{\omega=-\infty}^{\infty} C_\omega e^{j\omega t}$$

$$(\omega = 0, \pm\Delta\omega, \pm 2\Delta\omega, \ldots) \qquad \text{(C.10)}$$

$$C_\omega = \frac{\Delta\omega}{2\pi}\int_{-T/2}^{T/2} f(t)e^{-j\omega t}\,dt \qquad \text{(C.11)}$$

Substituting Equation (C.11) into (C.10), we have

$$f(t) = \frac{1}{2\pi}\left[\sum_{\omega=-\infty}^{\infty}\int_{-T/2}^{T/2} f(t)e^{-j\omega t}\,dt\right]e^{j\omega t}\Delta\omega$$

As $T \to \infty$, $\Delta\omega \to d\omega$ and $\sum \to \int$ so that

$$f(t) = \frac{1}{2\pi}\int_{-\infty}^{\infty}\left[\int_{-\infty}^{\infty} f(t)e^{-j\omega t}\,dt\right]e^{j\omega t}\,d\omega$$

Because the quantity inside the brackets is a function of $j\omega$, $[F(j\omega)]$, we can write

$$f(t) = \frac{1}{2\pi}\int_{-\infty}^{\infty} F(j\omega)e^{j\omega t}\,d\omega$$

where

$$F(j\omega) \triangleq \int_{-\infty}^{\infty} f(t)e^{-j\omega t}\,dt$$

The function $F(j\omega)$ is called the *Fourier transform* of $f(t)$, and $f(t)$ is the *inverse* Fourier transform of $F(j\omega)$.

$$F(j\omega) = \mathfrak{F}[f(t)]$$
$$f(t) = \mathfrak{F}^{-1}[F(j\omega)]$$

The above has been a heuristic development, in that no care was exercised in converting the summation to an integral. In fact, $f(t)$ must satisfy a set of conditions known as the Dirichlet conditions. The interested reader may refer to Reference 11.4 for a more rigorous development.

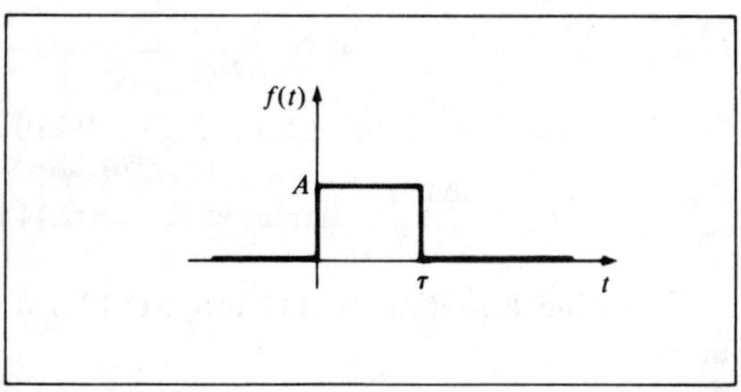

FIGURE C.8

Example

The Fourier transform of the pulse function (see Figure C.8)

$$f(t) = \begin{cases} A & 0 \leqq t \leqq \tau \\ 0 & \text{elsewhere} \end{cases}$$

is given by

$$F(j\omega) = \mathfrak{F}[f(t)] = \int_0^{\tau} Ae^{-j\omega t}\,dt$$

$$= \frac{A}{j\omega}(1 - e^{-j\omega\tau})$$

$$= \frac{2A}{\omega}\sin\left(\frac{\omega\tau}{2}\right)e^{-j\omega\tau/2}$$

The amplitude spectrum of this function is given by

$$|F(j\omega)| = \frac{2A}{\omega}\left|\sin\left(\frac{\omega\tau}{2}\right)\right|$$

$$= A\tau\left|\frac{\sin(\omega\tau/2)}{\omega\tau/2}\right|$$

This function of ω is illustrated in Figure C.9. It is clear that the largest components of the amplitude function lie in the interval $0 \leqq |\omega| \leqq 2\pi/\tau$. This range widens as the pulse duration τ is made shorter. We can conclude that the frequency bandwidth required to reproduce a rectangular pulse increases with the reciprocal of the pulse width.

Methods for the analysis of circuits with aperiodic inputs proceed analogously to the methods illustrated in Figure C.5 with periodic

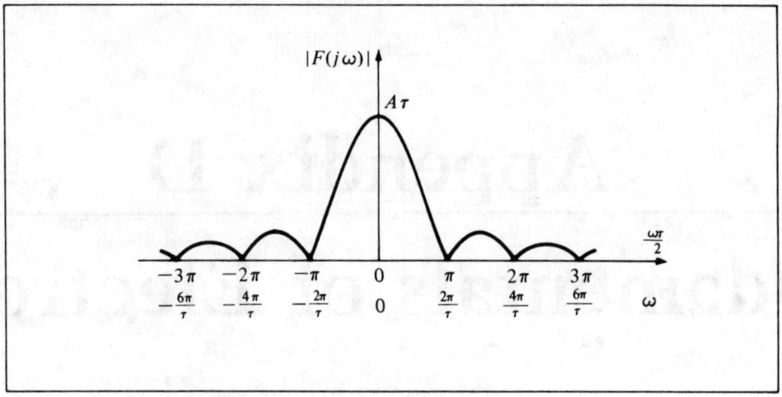

FIGURE C.9 Amplitude spectrum of $f(t)$.

excitations. For example, the pulse response of the RC network illustrated in that figure is

$$V_0(j\omega) = \frac{V_s(j\omega)}{j\omega RC + 1}$$

$$= \frac{2A}{\omega} \sin\left(\frac{\omega\tau}{2}\right) \frac{e^{-j\omega\tau/2}}{j\omega RC + 1}$$

from which the amplitude and phase spectra of the output function $v_0(t)$ can be found.

Appendix D

Fundamentals of Electronics

Introduction

This text is concerned principally with the analysis of electrical networks. Such a study requires an operational knowledge of mathematics to formulate and solve equations. We have reviewed certain of these mathematical techniques in previous appendices. However, the network models on which these equations are based are derived from *physical* laws. (See Figure D.1.) In this appendix, we survey in a qualitative way the physical processes on which often-used circuit models of devices such as diodes and transistors are based.

The basic building block of many electronic devices is the *pn* junction. We will spend a significant part of our discussion on this topic. With an understanding of *pn* junction operation, we will then discuss the principal physical processes involved in the *bipolar junction transistor* introduced in Chapter 3. Then we examine an alternative form of transistor known as the *field-effect transistor*; this device is used in problem assignments appearing throughout the text.

Electrical Classification of Materials

Based on their electrical conductivity, solids are divided into three classes. *Conductors* contain many free electrons and thus have a high *conductivity* σ illustrated in Figure D.2. Materials with few free electrons have low conductivity and are called *insulators*; those materials

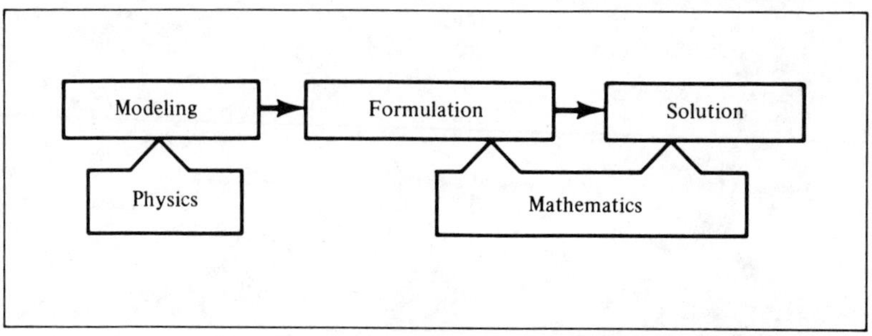

FIGURE D.1 Bases of circuit analysis.

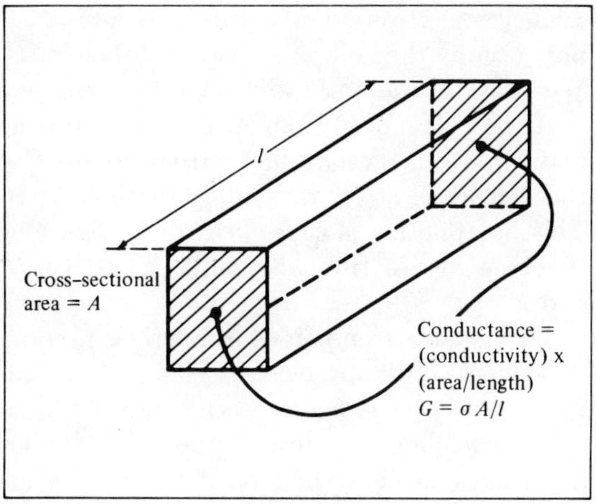

FIGURE D.2 Determination of conductance of bar of material.

with conductivity between the conductors and insulators are called *semiconductors*. It is the control of this intermediate level of conductivity by a signal that results in useful semiconductor device properties.

The semiconductors of special interest to us appear in Group IV of the periodic table of the elements, which is partially produced in Table D.1. The three columns (III, IV, and V) correspond to those elements that have 3, 4, and 5 electrons (respectively) in the *valence* energy band. These are called simply *valence* electrons.

Periodic Table of Groups III–V

Group III	Group IV	Group V
Boron	Carbon	Nitrogen
Aluminum	Silicon	Phosphorus
Gallium	Germanium	Arsenic
Indium	Tin	Antimony

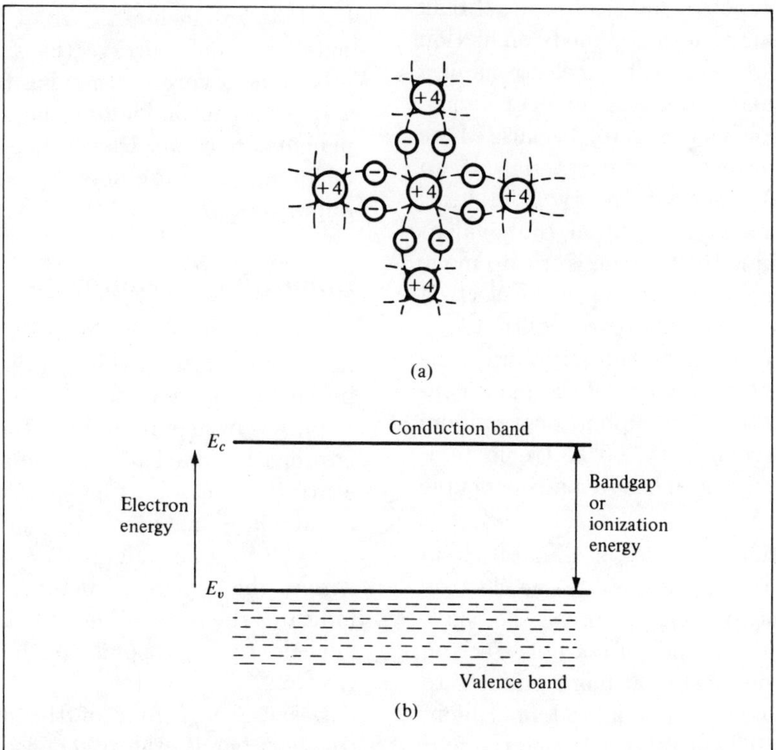

FIGURE D.3 Semiconductor crystal respresentations. (a) Pictorial; (b) energy diagram.

Electrical Properties of Pure Semiconductors

Of particular interest to us are germanium and silicon. These elements tend to form crystals in their pure or *intrinsic* state. The crystal structure results from each atom sharing an electron with each of its four neighboring atoms in *covalent* bonds. Figure D.3(a) is a two-dimensional pictorial representation of this crystal structure. Because the valence electrons and the associated positive charge in the nucleus are our primary concern, Figure D.3 shows only these charges. This figure, then, applies to either germanium or silicon.

These bonds are representations of the energy levels that exist in a crystal structure. As shown in Figure D.3(b), the valence electrons are held below an energy level E_v by the crystal structure. The process of conduction in a pure semiconductor requires an electron to break this bond or "jump" to a specified conduction energy level E_c. At 0°K, all the electrons in a semiconductor material are in covalent bonds; thus, the conductivity is zero, because there are no free electrons in the material. If an electron obtains sufficient energy from heat, light, or electromagnetic radiation, the covalent bond is broken, and the electron is free to move about in the semiconductor. When the electron leaves the covalent bond, a space, called a *hole*, is left in the covalent bond structure. The result is a *hole-electron pair*. These pairs continually regenerate and recombine on a random basis. On the average, at 300°K (room temperature) intrinsic silicon has 10^9 pairs per cubic centimeter.

The conduction process is illustrated in Figure D.4. In the figure there is one electron which has sufficient energy to break the bonds of its neighbor atoms and form a hole-electron pair. The electron located at point x is free to move about in the crystal in a random fashion. The hole associated with this hole-electron pair is shown at point A. Without a change in energy, the electron in the valence band (valence electron in a covalent bond) at B may occupy the hole at A, leaving a hole at B. We consider this to be the same hole and say that the hole moved from A to B. In the same manner, the hole may move from B to C by the valence electron at C filling the hole at B. This hole motion is continued in the figure as the hole moves through points A, B, C, D, and E.

The hole-electron pair continue their random movement until an electric field is applied. Then a drift of electrons occurs in one direction, depending on the direction of the field; this leaves an excess of broken covalent bonds (holes) which appear to be flowing in the opposite direction. These holes can be viewed as *carriers* of current, just as the electrons are. Therefore, a *drift* current (moving to the right, for example) produced by an electric field has two components: electron carriers (moving to the left) producing a current (moving to the right) and hole carriers (moving to the right) producing a current (moving to the right). In a pure semiconductor, these current components are equal. However, they can be made unequal, as we are about to show, by adding impurities.

Impurities in Semiconductors

The commercial refinement of silicon from its natural state in the earth's crust results in impurities of less than 1 part in 10^{10}. If an impurity from Group V of the periodic table is properly added to intrinsic silicon to a concentration of 1 part in 10^6, the crystal that results has one extra (valence) electron for every 10^6 atoms. Figure D.5 shows one atom within the crystal structure, where the $+5$ indicates the presence of an extra charge in the nucleus to correspond to the extra valence electron.

Because only four of the electrons are in covalent bonds, the other electron associated with this atom is loosely bound to its parent atom and is free to wander about in the solid;

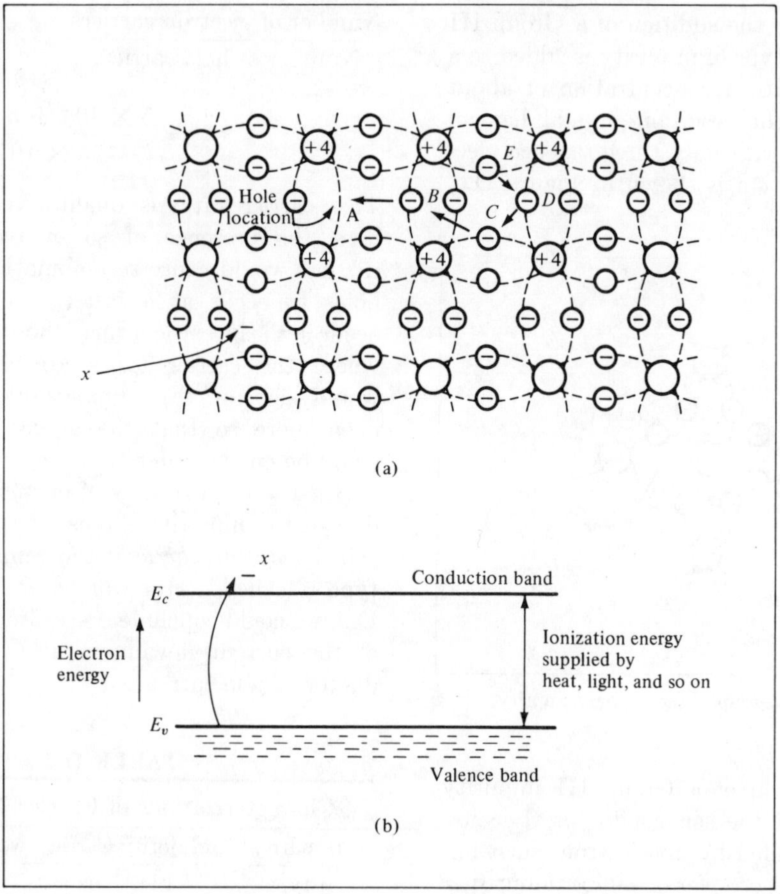

FIGURE D.4 Electron and hole conduction. (a) Pictoral representation; (b) energy diagram.

that is, it requires very little energy to elevate it from the valence energy band to the conduction energy band. Because of the extra electrons, a Group V impurity is also called a *donor* or *n*-type impurity. The number of such impurities required for large changes in conductivity is not great. For example, if a donor impurity is added in a concentration of 1 part in 100 million, the conductivity of germanium at 30°C is increased by a factor of 12.

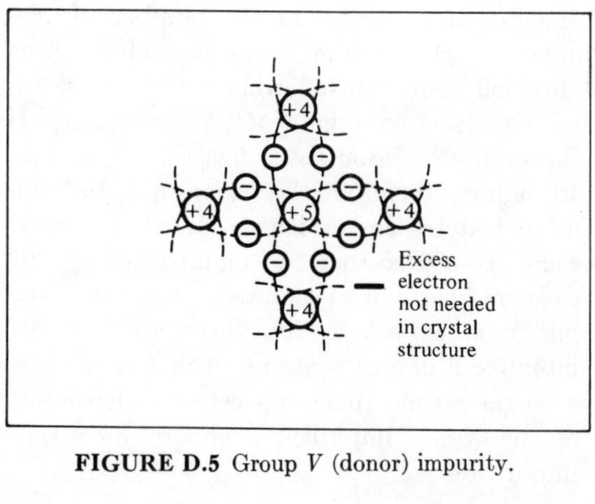

FIGURE D.5 Group *V* (donor) impurity.

Because conduction may also be accomplished by the motion of holes through the semiconductor, its electrical properties may also be altered by the addition of a Group III impurity. If this type of impurity is added to a semiconductor with a concentration of about 1 part in 10^{10}, the resultant crystal has an occasional atom with only three valence electrons. This situation is shown in Figure D.6.

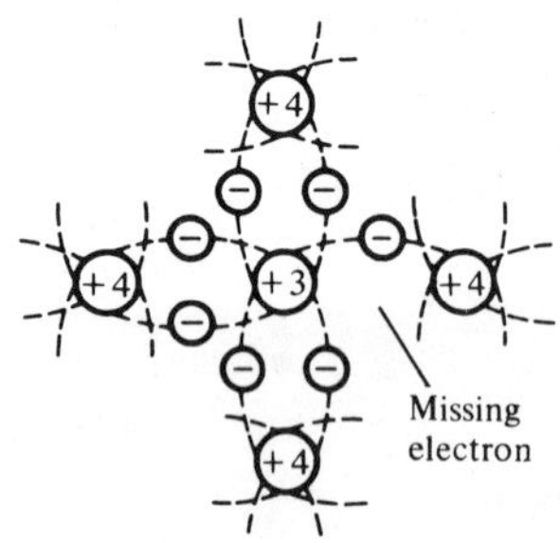

FIGURE D.6 Group *V* (acceptor) impurity.

Thus, the addition of a Group III impurity supplies holes to the semiconductor. Because a hole may be filled by an electron, such impurities are called *acceptor* or *p-type* impurities. When either *n*- or *p*-type impurities are intentionally added, the semiconductor is said to be *doped*.

An appreciation of the influence of a small amount of impurity on the number of free holes and electrons in a semiconductor can be obtained from the following calculations. Intrinsic silicon contains 5×10^{23} atoms/cm³. If the undesired impurity is less than 1 part in 10 billion, there are less than 5×10^{13} undesired atoms/cm³ which may contribute carriers. At 300°K there are approximately 10^{13} electron-hole pairs generated, and if the impurity is 1 part in 10^7, there are 5×10^{16} intentional impurity atoms, each contributing a single carrier (hole or electron), depending on the type of impurity. Therefore, for *n*-type impurities

$$\frac{\text{Number of electron carriers}}{\text{Number of hole carriers}} = \frac{5 \times 10^{16} + 5 \times 10^{13}}{5 \times 10^{-3}} \cong 10^3$$

This carrier ratio is qualitative at best. In fact, the presence of so many free *n*-type carriers would cause recombination of the few holes to occur at a faster rate than would occur in the intrinsic form of the semiconductor, where the carrier types are balanced. This would depress the number of hole carriers even more so that the actual carrier ratio could be on the order of 10^6 in this case.

Because the majority of current carriers are due to the impurity atoms, these carriers are called *majority carriers*; the remaining carrier type is called a *minority carrier*. From Table D.2, we could conclude, for example, that most of the current flowing in a *p*-type semiconductor is hole current.

TABLE D.2

Characterizations of Carrier Types

Impurity	Majority Carrier	Minority Carrier
n-type	Electrons	Holes
p-type	Holes	Electrons

One additional important aspect of carrier flow should be reviewed. Because each atom is electrically neutral, the total charge present in the semiconductor is always zero. The form of the charges is different, however; part of the charge is locked into the crystal structure in the form of ionized atoms that have taken on free electrons or holes; the other charge form is the free electrons or holes themselves. This distinction is important to all our device studies.

pn Junction

The most common method of obtaining a desired impurity level is illustrated in Figure

D.7. Here, an impurity—let us say a donor—in vapor form is passed over a silicon wafer that has an oxide layer over most of its surface. The oxide protects the part of the surface it covers; however, windows are carefully etched in this oxide layer to allow injection of the impurity into selective regions of the semiconductor. This process does not disturb the crystal structure of the silicon. If the silicon wafer already contains a small amount of impurity (p-type in Figure D.7), then the

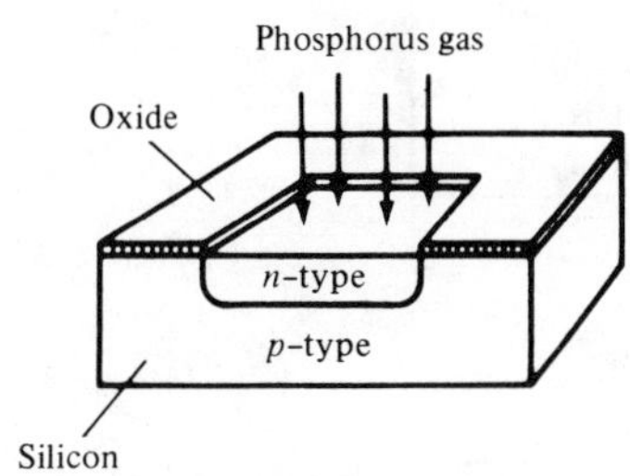

FIGURE D.7 Impurity injection process.

n-type impurity can be made more intense than the p-type in the sense that the majority carrier becomes the n-type. The intersection between these impurity levels within the semiconductor can be made quite abrupt, resulting in a *pn junction*. When electrical connections are made to the n- and p-type semiconductors, the resulting device is called a junction diode.

Although the crystal structure must be continuous at the junction, it is useful to imagine the diode being formed by bringing together p-type crystals and n-type crystals. The majority carrier [holes on the right and electrons on the left, as in Figure D.8(a)] are moving at random through the crystal. The holes tend to diffuse to the left in the direction of a lower concentration of holes, and electrons diffuse to the right toward a lower concentration of electrons. This creates two diffusion currents, both moving to the left (because electron current moves oppositely to electron flow). This process of carrier diffusion, however, sets up two opposing effects.

First, when an electron diffuses to the right to combine with a hole, it leaves the left side with an ionized atom with a unit positive charge. Similarly, a unit negative charge is left when a hole diffuses to the left. The result is a region *depleted* of free charge, containing immobile positive and negative ions in close proximity. From elementary physics, we realize that an electric field will appear between charges of opposite polarity, as in Figure D.8(b); this field inhibits further carrier diffusion. It is useful to represent the field in terms of the resultant voltage (potential) differences or "hill" between the right and left side of the junction, shown in Figure D.8(b) of height V_0. Thus, only a few holes can slide up the hill and a few electrons can slide down it, resulting in less diffusion current flowing to the left.

Second, an electron on the right that is beyond the depletion region does not experience a force. However, owing to the random carrier motion, some of the electrons arrive at the edge of the depletion region. An electron arriving at this point then experiences a force due to the electric field. This electron is swept up the potential hill, depicted in Figure D.8(c). By similar reasoning, the holes on the left that arrive at the edge of the depletion region are swept down the hill. The results are two so-called *drift* currents, both moving to the right; these cancel the reduced diffusion currents noted above.

Forward bias. If an external voltage v_d is applied across the diode, the junction potential changes to $V_0 - v_d$ [Figure D.9(b)]. If $v_d > 0$, we say that the diode is forward-biased. The number of majority carriers having sufficient energy to diffuse across the junction increases by orders of magnitude, and the diffusion current increases. Because the minority carrier

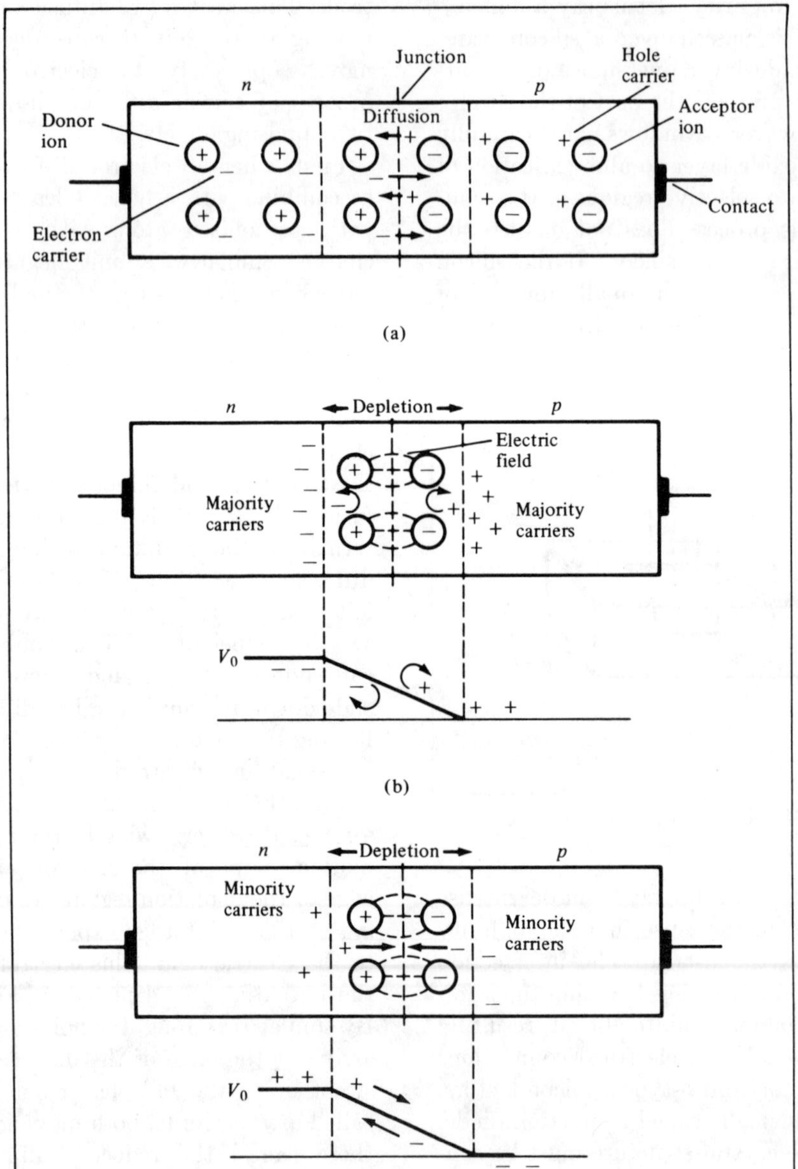

FIGURE D.8 Explanation of carrier motions in diode. (a) Free holes on p-side attracted to n-side; free electrons on n-side attracted to p-side; result: (1) two *diffusion* currents, and (2) immobile + ions on n-side and − ions on p-side defining a region *depleted* of free charges. (b) Ions set up an electric field in depletion region; result: opposing potential "hill" impedes further diffusion currents. (c) Thermally generated holes and electrons randomly arriving at edge of depletion region are swept across by aiding potential "hill"; result: two *drift* currents, cancelling diffusion currents in (a); net effect: no total current flow.

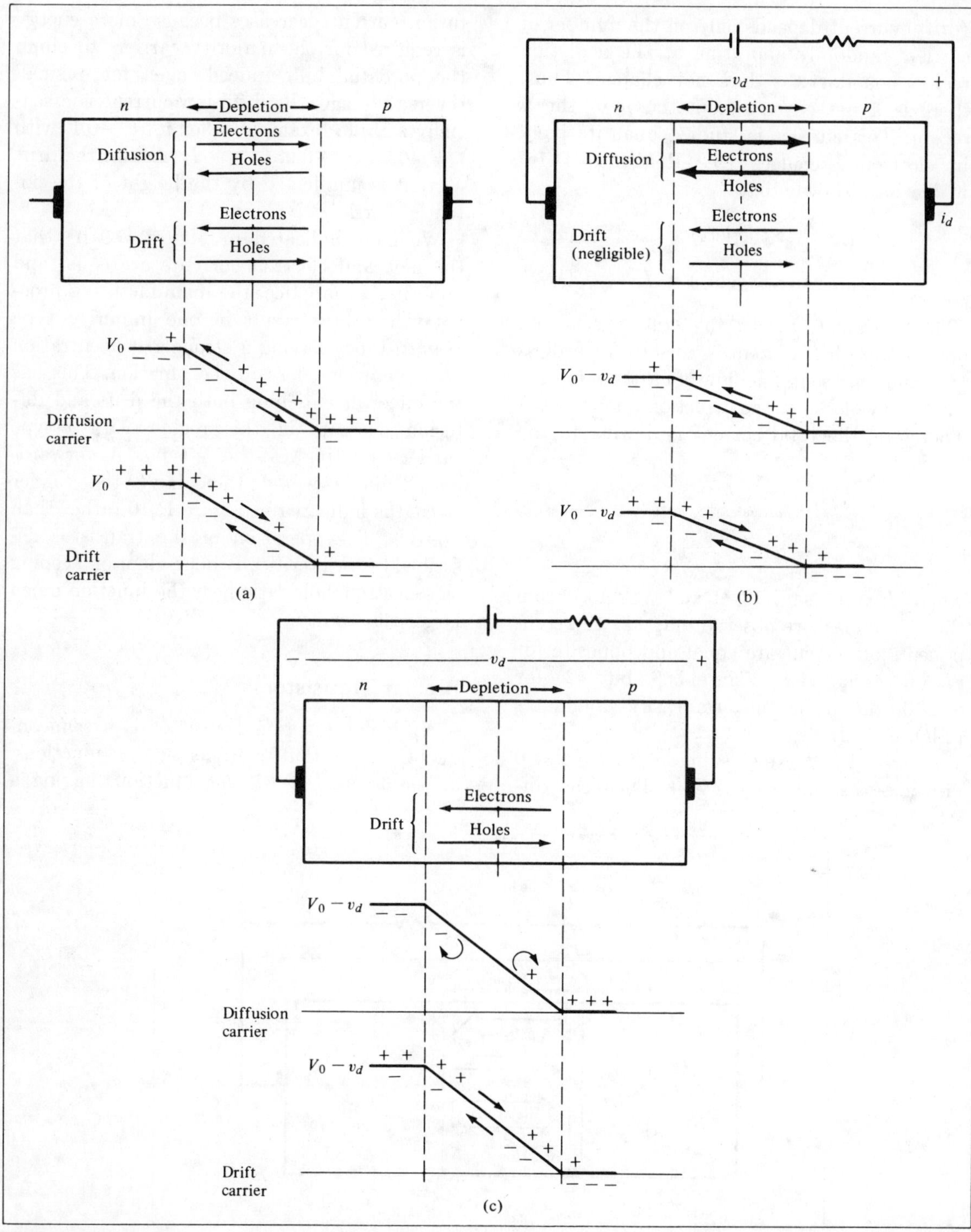

FIGURE D.9 *pn* junction diode bias study. (a) Carrier flows with no bias; drift and diffusion currents in equilibrium. (b) Potential "hill" lowered by forward-bias voltage; result: many diffusion carriers can surmount "hill," drift currents remain same. (c) Potential "hill" raised by reverse-bias voltage; result: no diffusion carriers can surmount hill, (drift) current depends only on the number of carriers randomly appearing at the depletion region boundary, it does not change. Thus, there is a net current from the p to the n region. This current is studied quantitatively in electronics courses (see Reference D.1), where it is shown that

$$i_{\text{diffusion}} \text{ (to the left)} = I_s e^{\lambda v_d} \quad \text{(D.1)}$$

$$i_{\text{drift}} \text{ (to the right)} = I_s \quad \text{(D.2)}$$

(The constant I_s is called the *saturation current*, because this drift current is constant regardless of v_d and represents the flow of *all* the minority carriers available at the depletion boundaries.) Therefore, the total current i_d flowing to the left is

$$\begin{aligned} i_d &= i_{\text{diffusion}} - i_{\text{drift}} \\ &= I_s(e^{\lambda v_d} - 1) \end{aligned}$$

where $I_s \ll 1$ (say, 10^{-9}) and $\lambda \cong 40$. From these formulas, we observe that (1) the components of current are equal and opposite for $v_d = 0$, as noted in Figure D.8, but (2) for $v_d > 0$, the diffusion term easily dominates [Figure D.9(b)].

Reverse bias. Now suppose that $v_d < 0$ [reverse-bias, Figure D.9(c)]. Then the diffusion current decreases because more energy is required for the majority carriers to climb the potential hill. Indeed, even for a small reverse voltage, the diffusion current is essentially zero; for example, when $v_d = -0.1$, with $\lambda = 40$, $i_d = -0.98 I_s \cong -I_s$. Again, the drift current is unaffected by the height of the potential barrier.

We have indicated in Figure D.9(b) that the hole and electron currents are equal and opposite. In practice, the manufacturing process will always result in one impurity type (say, n-type) having a stronger concentration of free carriers close to the junction. This has the effect of enriching both the drift and diffusion currents in this region with n-type carriers relative to the p-type. A forward-biased diode would then have the carrier strengths indicated in Figure D.10 rather than D.9(b). The operation of the transistor described below is critically dependent on keeping this electron/hole ratio near the junction much larger than unity.

Bipolar Transistors

A bipolar junction transistor is a semiconductor device that contains two pn junctions, an *emitter* and a *collector* junction, having a

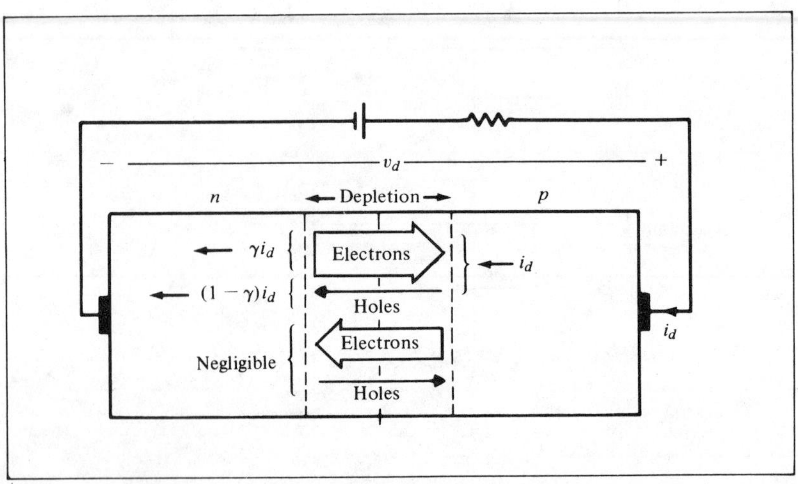

FIGURE D.10 Diode with heavily doped n region ($\gamma \cong 1$).

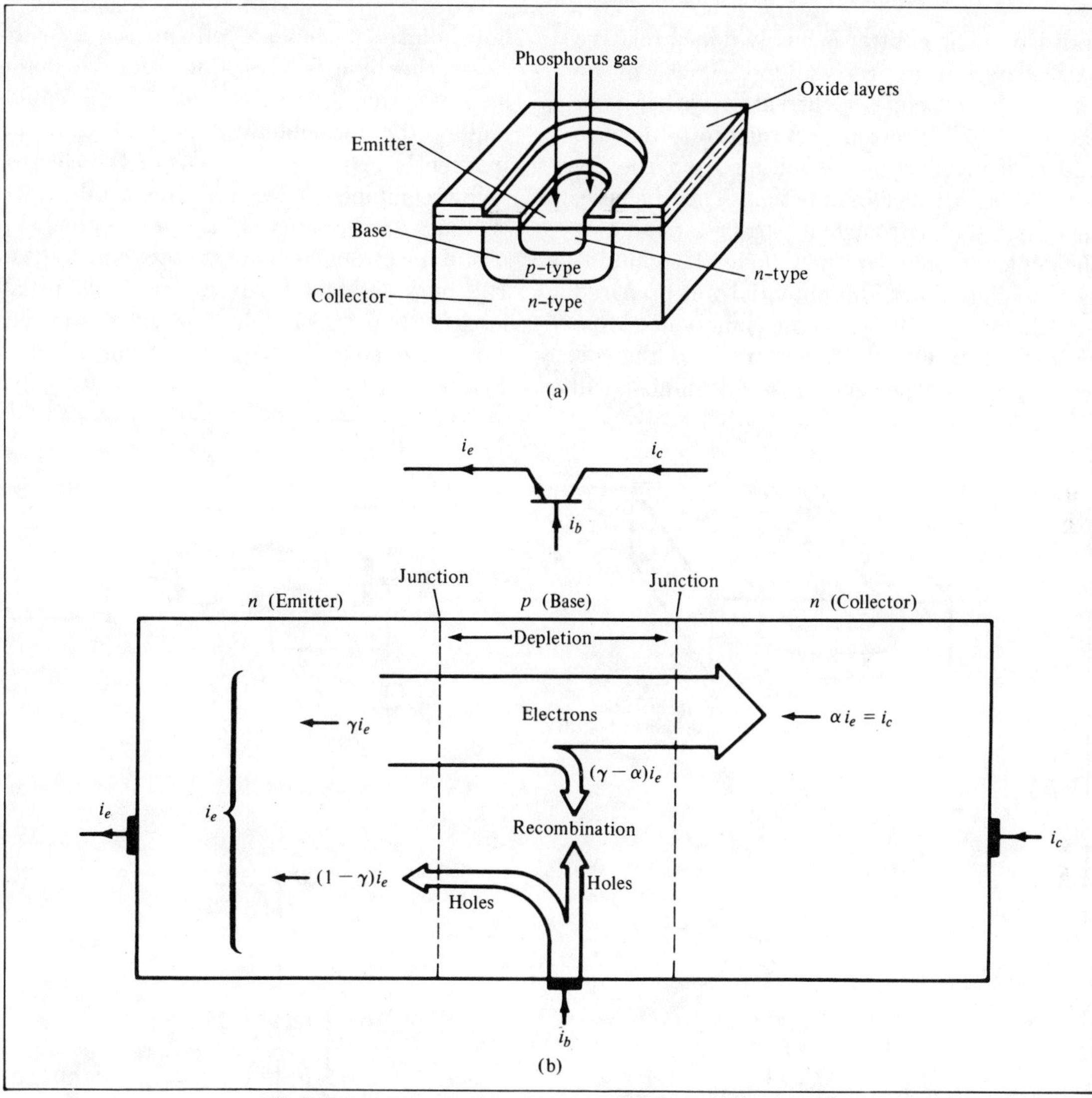

FIGURE D.11 *npn* Transistor construction and operation.

common region called the base. Figure D.11(a) shows a typical fabrication process for an *npn* transistor. First the collector-base junction is formed by injecting *p*-type carriers into an *n*-type semiconductor wafer. Then, *n*-type impurities are injected into the *p*-type base to form the emitter-base junction. The resulting base can be very thin (as small as 0.1 mm), which is critical to transistor operation. Of course, a *pnp* transistor can be made similarly with the opposite impurity types. Either type can serve a number of different functions, depending on whether the two junctions are forward- or reverse-biased (four possibilities, Table 5.1). We will study the case when the emitter junction is forward-biased and the collector junction reverse-biased. This is called the *forward* or *normal* region of operation.

Figure D.11(b) is a representation of the principal internal carrier flows within an *npn* transistor. The emitter is highly doped relative to the base, so that nearly all ($\gamma \cong 0.995$) of the forward-biased emitter current is via electron carriers. These electrons diffusing into the base find a depleted region resulting from the reversed-biased collector junction. They behave, then, like the drift current carriers arriving at the depletion boundary in a diode: They find it easy to slide down the potential hill prepared for them by the collector junction! Most emitter electrons will in fact arrive at the collector junction; however, a small number will find holes to recombine with in the base. Because the base is very thin and the doping level is lower in the base than in the emitter region, the recombination rate is small and frequently less than 1 percent of the electrons will recombine. Indeed, of the total emitter current i_e, the fraction α that both (1) is useful electron current at the emitter, and (2) survives the base transport, is usually greater than 0.99. This provides the vital controlled source in the transistor model of Figure 5.10(a).

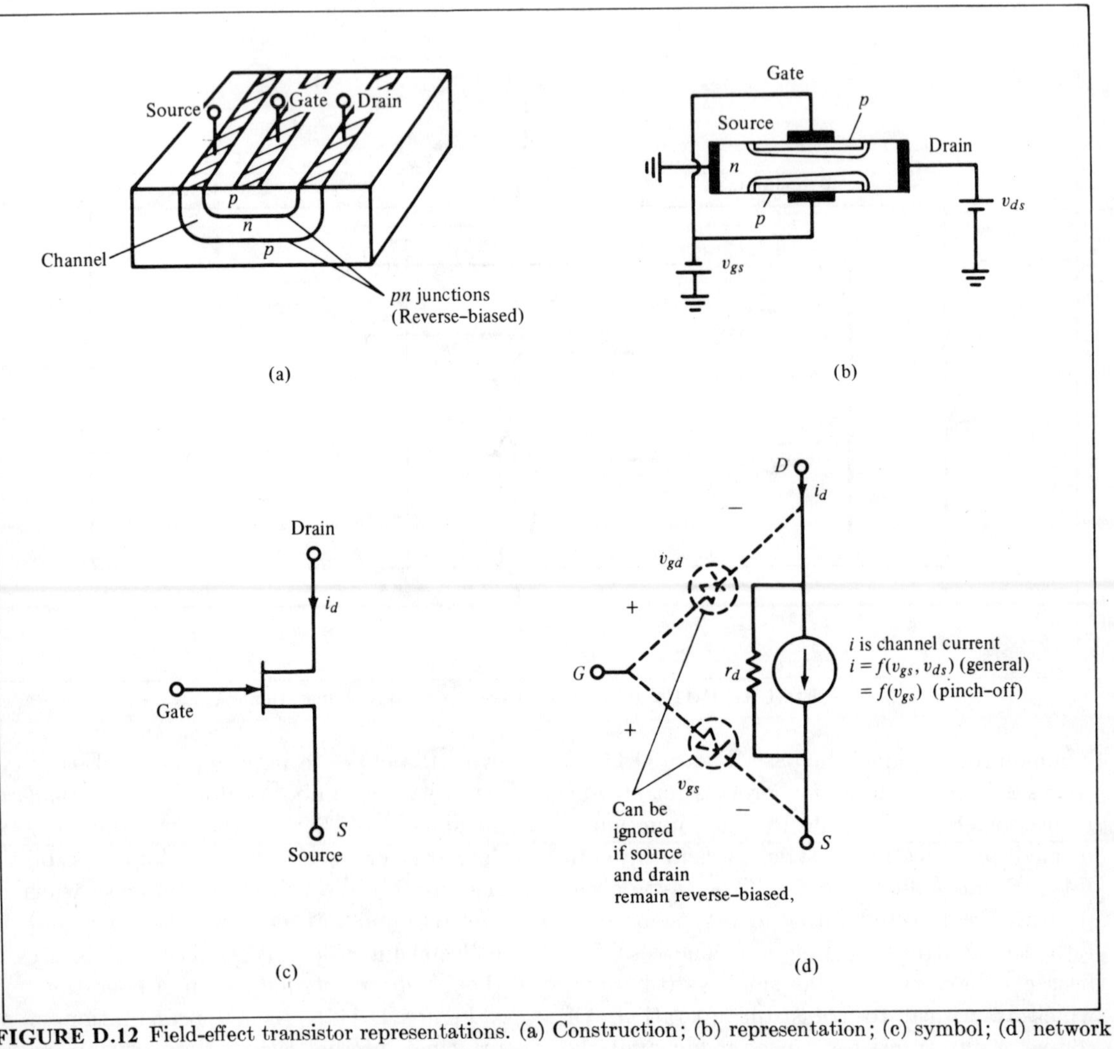

FIGURE D.12 Field-effect transistor representations. (a) Construction; (b) representation; (c) symbol; (d) network model.

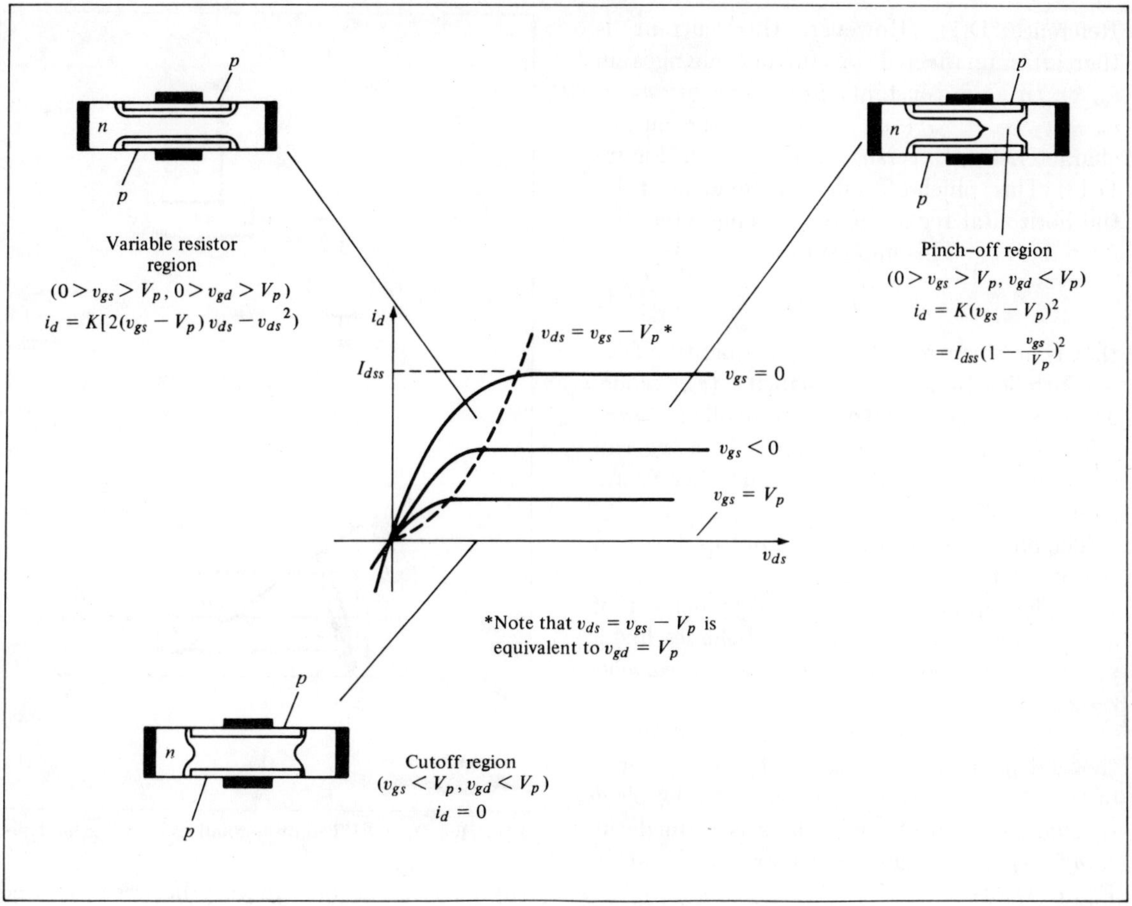

FIGURE D.13 Output characteristic of FET.

Field-Effect Transistor

The field-effect transistor (FET) is a device used to accomplish many of the same circuit functions as the bipolar transistor employed throughout this text: amplification, logic, memory, and so on. However, it operates on physical principles quite different from the bipolar transistor and so its model is described by different network equations.

Basically, a field-effect transistor operates by using the signal input to modify the conductivity of a channel of semiconductor material. One form of FET is shown in Figure D.12(a) and (b). Here, electrons flow from the *source* to the *drain* ends of a channel. Along the length of this channel, two depletion regions are maintained by reverse-biasing two elongated *pn* junctions. The *p*-side of the junction is called the *gate,* where the signal is applied. The channel width is modified by varying the reverse-bias voltage from gate to channel. A larger reverse-bias results in a larger depleted region. For a sufficiently negative drain voltage ($v_{gd} = V_p < 0$), the drain end of the channel will become completely depleted or "pinched-off." This does not imply that the channel current i_{ds} will become zero; rather, i_{ds} will still flow through this small depleted region. (See

Reference D.1). However, this current is thereafter unaffected by further changes in v_{gd} or (if v_{gd} is constant) in v_{ds}, because $v_{ds} = v_{dg} + v_{gs} = -v_{gd} + v_{gs}$. A plot of the output characteristic (i_{ds} versus v_{ds}) is shown in Figure D.13. This pinch-off effect is evidenced by the horizontal region of the *i-v* characteristics for a fixed v_{gs}. This implies that

$$i_{ds} = f(v_{gs}) \tag{D.3}$$

that is, in this region the output quantity (i_{ds}) is controlled by an input quantity (v_{gs}) alone. Thus, we have achieved a controlled source effect (albeit nonlinear), from which we can hope to obtain amplification and other useful circuit functions.

The limitations of the above *pinch-off* region of operation occur when:

1. The drain end ceases to pinch-off if $v_{gd} > V_p$, and the open channel behaves like a resistor of varying width; this is the *variable resistor region.*
2. The source end pinches off if $v_{gs} < V_p$, so that a depleted region extends the entire length of the channel; current flow will not take place through such an obstacle. This is termed the *cutoff region.* These regions are depicted in Figure D.13.

It can be shown (Reference D.1) that if $r_d = \infty$ the dependence of (D.3) in the pinch-off region of operation is given by

$$i_d = I_{dss}\left(1 - \frac{v_{gs}}{V_p}\right)^2 \tag{D. 4}$$

where v_{gs} and V_p are both negative quantities for an *n*-channel device. Therefore, the output current is not linearly related to the input voltage. The constant I_{dss} is the current that flows when $v_{gs} = 0$.

A simple FET amplifier configuration is shown in Figure D.14(a). The load-line construction of Figure D.14(b) displays the nonlinearity of Equation (D.4) by the unequal vertical spacing between three constant-gate-voltage characteristics. For example, if

$$v_{gs} = -1 + \sin t$$

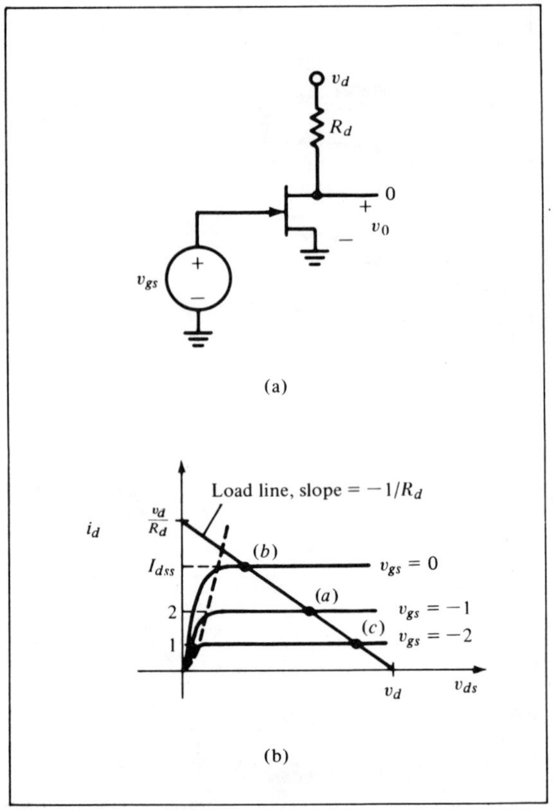

FIGURE D.14 FET amplifier and its load-line analysis.

the FET operation will vary between (b) and (c) around the dc solution at (a). Clearly, the output current i_{ds} and voltage v_0 will change appreciably more between (a) and (b) than between (a) and (c), producing a distortion. Analytically, if we let $v_{gs} = v_g + v_s(t)$, we can write

$$\begin{aligned} v_{ds} &= v_d - R_d i_{ds} \\ &= v_d - R_d I_{dss}\left(1 - \frac{v_g + v_s(t)}{V_p}\right)^2 \\ &= \underbrace{v_d - R_d I_d}_{\text{dc}} + \underbrace{2R_d I_{dss}\left(1 - \frac{v_g}{V_p}\right)\left(\frac{v_s(t)}{V_p}\right)}_{\text{Signal}} \\ &\qquad - \underbrace{\frac{R_d I_d v_s^2(t)}{V_p^2}}_{\text{Distortion}} \end{aligned}$$

where $I_d = I_{dss}(1 - v_g/V_p)^2$ is the dc value of i_{ds}. (The reader should fill in the details of the derivation.) From this expression we can find both the (1) the distortion term and (2) the linearized gain $(2R_dI_{dss}/V_p)[1 - (v_g/V_p)]$ for use in small-signal analysis.

The FET can be used to replace the bipolar transistor in digital applications as well. As with the bipolar transistor, the FET is driven to the limits of normal amplifier operation, where the output characteristic is highly nonlinear. In the case of a FET, this requires either

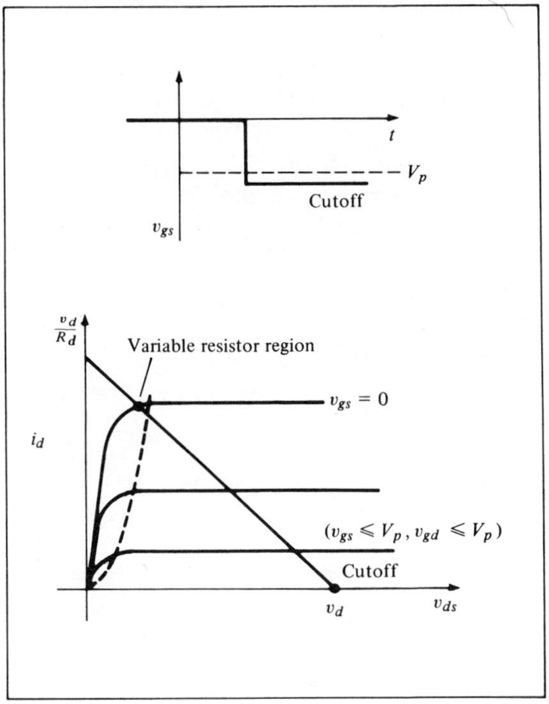

FIGURE D.15

(1) a large negative gate voltage to pinch off the gate-source part of the channel ($v_{gs} < V_p$), or (2) a negative gate voltage sufficiently close to zero to operate in or near the variable resistor region. In either case, the output voltage v_{ds} is essentially a constant, independent of small variations in v_{gs}, as required of a digital circuit.

FET amplifier and digital circuits are given as problems throughout the text. As indicated in Problem 14.12, the internal storage of charge in the FET can be represented for many purposes by linear capacitors between the three terminals. The static characteristics of the FET are available in the DCAP program of the Instructor's Manual (see Problem 5.14).

Other Types of FET's

The junction FET described above is only one type of FET used in modern electronic circuits. Another important class of FET's is the insulated-gate FET (IGFET) illustrated in Figure D.16. A metal control gate is separated from

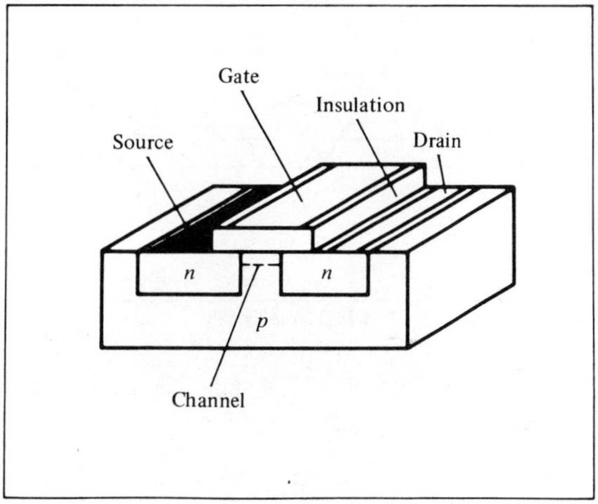

FIGURE D.16 IGFET construction.

the semiconductor channel by an insulating oxide layer.

The operation of the IGFET can be explained on an elementary level in terms of a charge-control concept. The metal control electrode (gate) acts as a charge storage element. A charge placed on the gate induces an equal but opposite charge in the semiconductor channel located beneath the gate. This induced charge can be used to control the conduction between the source and drain contacts either by depleting or enchancing the number of available carriers. In Figure D.16, for example, the application of a positive voltage to the gate attracts electrons to the channel region beneath

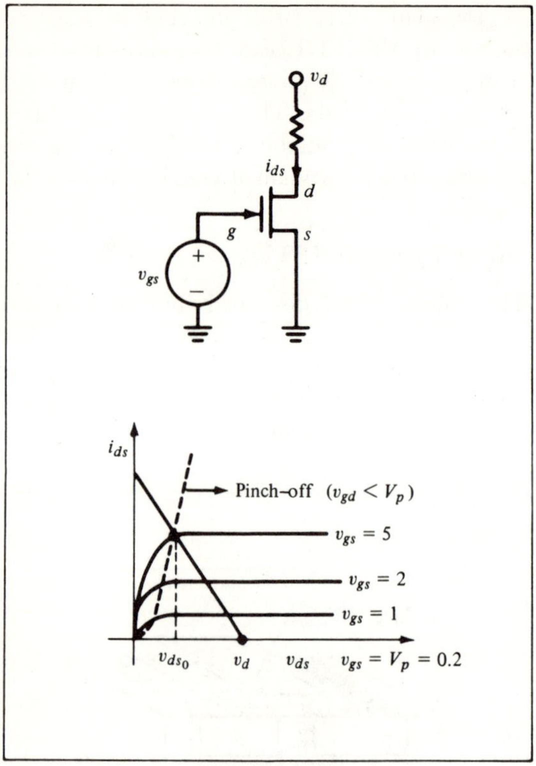

FIGURE D.17

the gate. If sufficient voltage is applied, the channel region changes from p-type to n-type. A conduction path is thus established between the n-type source and drain regions.

The IGFET equations are identical in form to the junction FET equations in the pinch-off, cutoff, and variable resistor regions. However, the IGFET has the very desirable property that the pinch-off voltage V_p can be either positive or negative. A digital circuit application of an IGFET with a positive pinch-off voltage is shown in Figure D.17.

When the input is a logical **0** ($v_{gs} = 0$), the IGFET is cutoff because $v_{gs} < V_p$; the output is then a logical **1** ($v_{ds} = v_d$). Conversely, when the input is a logical **1** ($v_{gs} = v_d$), the output is nearly a logical **0** ($v_{ds} = v_{ds_0} \cong 0$); the result is a logical inverter. Unlike a junction FET, both the input and output voltages can be positive, so that one IGFET can drive another IGFET directly. This makes the IGFET especially attractive for large digital circuit logic and memory arrays, where the number of electronic components is a critical factor.

REFERENCE

D.1 P. E. GRAY AND C. L. SEARLE, *Electronic Principles*. New York: John Wiley & Sons, 1969, Chapter 9.

Answers to Selected Problems

Chapter 2

2.2(a) Powers: $4\,\Omega: 2 \times 8 = 16$

$$3\,\Omega: \frac{14}{3} \times 14 = \frac{196}{3}$$

$$2\,\Omega: 2 \times 4 = 8$$

$$1\,\Omega: 2 \times 2 = 4$$

$$i_s: -\frac{20}{3} \times 14 = -\frac{280}{3}$$

(c) Powers: $1\,\Omega: 1 \times 1 = 1$

$$2\,\Omega: 2 \times 1 = 2$$

$$3\,\Omega: 3 \times 1 = 3$$

$$4\,\Omega: 8 \times 2 = 16$$

$$5\,\Omega: 11 \times \left(\frac{11}{5}\right) = \frac{121}{5}$$

$$6\,\Omega: \left(\frac{7}{10}\right) \times \left(\frac{21}{5}\right) = \frac{147}{50}$$

$$v_s: \left(\frac{117}{10}\right)\left(-\frac{21}{5}\right) = -\left(\frac{2457}{50}\right)$$

2.4 Left hand side of network: $i_{5\,\Omega} = -3$, $i_{200\Omega} = -4$, $i_{10\Omega} = 2$;
Right hand side of network: $i_{10\Omega} = 1$, $i_{5\Omega} = 6$,

2.6(a) $v_{R_2} = 2$, $i_{R_2} = 1.5$, $R_2 = 1.33$, $R_1 = 10$.

2.9(a) $15\,\Omega$

(c) $25\,\Omega$

2.10 $$\frac{\partial v_{R_2}}{\partial G_1} = \frac{\partial}{\partial G_1}\left(\frac{G_1 R_2 v_s}{1 + G_1 R_2}\right) = \frac{v_s R_2}{(1 + G_1 R_2)^2}$$

$$\hat{v}_{R_1} = \frac{R_2}{1 + G_1 R_2} \qquad v_{R_1} = \frac{v_s}{1 + G_1 R_2}$$

2.14(a)
$$i_a + i_d - i_e = 0 \quad ①$$
$$-i_a + i_b - i_c = 0 \quad ②$$
$$-i_b + i_e + i_f = 0 \quad ③$$
$$i_c - i_d - i_f = 0 \quad ④$$

(b) ① + ② = $i_b + i_d - i_c - i_e = 0$

(c) ① + ② + ③ = $-i_c + i_d + i_f = 0$

(d) Each current appears twice, once positive and once negative.

2.18(a) $$v_2 = \frac{1}{4}$$

$$v_1 = \frac{5}{8}$$

(b) $$v_o = \frac{11}{24},\ i_o = \frac{11}{24}$$

2.20 $v = -\frac{24}{13}$ in each case

2.24 (b) $i_{eq} = 3,\ R_{eq} = \frac{2}{3}$

(c) $i_{eq} = \frac{v_s + 3}{R} + 1,\ R_{eq} = R$

2.26 $v_s = \frac{3}{4}$

2.28 $639.3 < R < 667.7$

2.30 $i_{1\text{ amp}} = 1,\ i_{3\,\Omega} = -\frac{1}{2},\ i_{2\,\Omega} = -\frac{1}{2}$

2.33 (c) Power $= 0$

2.38 (a) $v_{eq} = -\frac{AR_f}{R_f + (1 + A)R_s}$ $\quad R_{eq} = 0$

Chapter 3

3.3 $v_A = 0,\ v_B = 1 \Rightarrow v_C = 1$

(a) OR circuit

A	B	C
0	0	0
0	1	1
1	0	1
1	1	1

3.7 (b) When $v_A = 5$ and $v_B = 0$, diodes D_1 and D_2 conduct, and diodes D_3, D_4, and D_5 do not conduct.

(d) All diodes conduct; $v_1 = v_2 = v_3 = 4.85$

3.13 (a) Same load line, new operating point is (5.23, 0.95)
$v_{s_{max}} = 0.428,\ v_{s_{min}} = -0.388.$

(d) $v_{s_{max}} = 0.62,\ v_{s_{min}} = -0.4$; new operating point is (3.15, 1.37).

3.15 (a) $v_{eq} = \frac{(i_s R_{b_1} R_{b_2} + v_c R_{b_2})}{(R_{b_1} + R_{b_2})}$

$R_{eq} = \frac{R_{b_1} R_{b_2}}{(R_{b_1} + R_{b_2})}$

3.18 $R_{eq} = \frac{(R_b(R_{bc} + R_c) - \alpha R_{bc} R_b)}{(R_b + R_{bc} + R_c)}$

3.20 $v_{eq} = \frac{(i_s R_b R_{bc} + v_c R_b)}{(R_b + R_{bc})}$

3.23 Distortion $= 0.638$ percent

3.24 (d) $v_c^{m+1} = \frac{R_s f^m - T^m v_c^m R_s + v_s R_f}{R_s + R_f - T^m R_s}$

Chapter 4

4.2 (a) $v_o = v_4 = \frac{v_s G_1 \Delta_{14}}{\Delta}$

(c) $v_o = v_1 - v_2$

$= \frac{v_s G_4 (\Delta_{11} - \Delta_{31} + \Delta_{12} - \Delta_{32})}{\Delta}$

4.4 (a) $v_o = v_2 = -\frac{4}{3}$

(b) $v_o = \frac{i_s}{12}$

4.6 (a) $v_o = \frac{52}{79}$

(b) $i_1 = -\frac{9}{4} \Rightarrow v_o = 0$

4.8 (a) $v_o = \frac{\Delta_{12} + \alpha\Delta_{22} - \alpha\Delta_{32}}{\Delta} i_s$

(c) $v_o = \frac{-2\Delta_{15} + 2\Delta_{25} + 2\Delta_{45}}{\Delta}$

4.11 $\frac{v_3}{v_2} = \frac{\Delta_{13}}{\Delta_{12}}$

4.17 Five loop equations are required.

4.18 (b) Defining meshes in the obvious way:

$(R_1 + R_2 + R_4)i_1 - R_2 i_2 = v_1 - v_2$

$-R_2 i_1 + (R_2 + R_3 + R_5)i_2 = v_2 - v_3$

Chapter 5

5.5(a) $v_1 = -321.4,\ v_2 = 35.57$

5.7 $v_{eq} = 0.9694 v_s;\ R_{eq} = 30.53$

5.11(a) $R_d = \dfrac{1}{\lambda I_s}\left(\dfrac{1}{\dfrac{i_d}{I_s} + 1}\right)$

5.12(b) $v^{m+1} = \tilde{v}^m + \widehat{R}^m i^{m+1}$

$$\tilde{v}^m = \frac{R_o i^m}{[1 - \alpha\beta(i^m)^2] - \widehat{R}^m i^m}$$

5.14(a) $i_d^{m+1} = i^m + G_{gs}{}^m v_{gs}{}^{m+1}$

Where

$$G_{gs}{}^m = \frac{2I_{dss}}{V_p{}^2}(v_{gs}{}^m - V_p)$$

$$i^m = I_{dss}\left(1 - \frac{v_{gs}{}^m}{V_p}\right)^2 - G_{gs}{}^m v_{gs}{}^m$$

5.19 $r_{cc} \leq 43.5$

5.21 $R_e < 20$

5.26 $\hat{x} - x^{m+1} = \dfrac{F''(\xi)}{2F'(x^m)}(\hat{x} - x^m)^2$

Chapter 6

6.2 $v_o = \dfrac{23}{3}$ volts

6.4 $v_o = R_1(i_2 - i_1)$

6.6 $R_{eq} = 0.0625$ ohms; $v_{eq} = 0.4375$ volts

6.8 $v_{eq} = v_c;\ R_{eq} = R_c + R_e + R_b(1 - \alpha)$

6.10 If $\lambda = 40$ volts^{-1}, $G_d = 0.04$ ohms per milliampere

6.16 $\delta v_o = 0.733;\ \delta v_s = v_{eq};\ R_{eq} = 120.4$ ohms

6.18 $i_c = \left(a_1 - a_2\dfrac{b_1}{b_2}\right)v_c$

$$+ \left(a_3 - a_2\frac{b_3}{b_2}\right)\delta i_s + \frac{a_2}{b_2}v_{ce}$$

6.20(a) $v_{gs} = -2.71$ volts;

(c) $\dfrac{\delta v_o}{\delta v_s} = 0.802$

Chapter 7

7.2 (a)

v	C_{eq}
0.0000	0.5
0.2035	0.4744
0.4328	0.3886
0.7744	0.1889

(b) $C_2 = 0.275$ Farads

7.4 $i_1 = 1.875$ amps; $i_2 = 3.125$ amps; flux linkages in 2 h. inductor = 10 volt-secs; $L_{total} = 3.875$ henry; 48.4 joules.

7.6(a) $v_c = t - 1$ volts, $1 < t < 2$; $v_c = 1$ volt, $2 < t < 3$; $v_c = 2t - 5$, $3 < t < 4$

7.8 (a) -0.4 amps per second

(b) 0.4 volt-secs

(c) 0.08 joules

(d) -0.16 watts

7.10 0.368 volts; 0.0677 joules.

7.14(a) $L_{total} = L_o\left(1 - \dfrac{i^2}{2} + \dfrac{2}{1 + i^2}\right)$

7.16(b) 1.2×10^{-3} joules

7.18(a) $\dfrac{di_2}{dt} + 1.786 i_2 = -2.143 v_s(t)$

Chapter 8

8.1(b) $r(t) = \dfrac{-E}{\alpha}(1 - e^{\alpha t}) \qquad t \geq 0$

8.3(b) $v_c(t) = \dfrac{1}{3}(1 - e^{-3t/4}),$

$$v_o(t) = \frac{1}{3}e^{-3t/4} \qquad t \geq 0$$

8.5(a) $t_D = -RC \ln\left(1 - \dfrac{V_o}{IR}\right)$

8.7 (a) At $t = 0^-$, $i(t) = 1$

(b) $i(t) = -e^{-t/2} \qquad t \geq 0$

8.9 $v_c(t) = \frac{10}{3}(1 - e^{-3t/2}) \qquad t \geq 0$

8.11 $t_{\text{rise}} = 2.3$ seconds

8.13 (c) $K = v_c(0) - \dfrac{V_s}{\left(C\gamma + \frac{1}{R}\right)}$

8.15 (c) $i_o(0^-) = -1$

8.17 Treat as single inductor; $W_L = 0.045$ joules.

8.19 $V_s = 3.68$

8.21 $v_c(t) = \sum_{m=0,1,\ldots}^{\infty} \left(\frac{2}{RC} e^{-(t-2m)/RC} u(t - 2m) - \frac{1}{RC} e^{-(t-2m-1.5)/RC} u(t - 2m - 1.5)\right)$

Chapter 9

9.1 (a) $s_1, s_2 = -500, -2000$

9.2 $s_1, s_2 = -10^6 \pm j\sqrt{3} \times 10^6$

9.5 $M = 0$

9.7 $i_L(t) = 0.032\, e^{-20t} \cos(45.8t - 23.6°)$

9.9 $R = 50\,\Omega$, $v_c(t) = 0.202(e^{-4.98t} - e^{-0.02t})$

9.11 $v_{c_2}(t) = 10 - 10.96\, e^{-0.121t} + 0.96\, e^{-1.38t} \quad t \geq 0$

9.13 $v_{c_2}(t) = -9(\cos t - 1) \qquad t \geq 0$

9.15 $s(t) = [1 + 1.29e^{-20t} \times \cos(24.5t + 140.8°)]u(t)$

9.19 (a) $i_L(t) = (1 + 1.081\, e^{-4.39\times 10^8 t} - 2.081 \times e^{-2.28\times 10^8 t})u(t) + 2 \sum_{m=1,2,\ldots}^{\infty} (-1)^m \times [1 + 1.081\, e^{-4.39\times 10^8(t-m)} - 2.081\, e^{-2.28\times 10^8(t-m)}]u(t - m)$

9.21 $r(t) = \frac{1}{2}[\cos t + \sin t - e^{-t}]$

9.25 Middle piece: $r(t) = -t^2 + 4t - 2$

Chapter 10

10.1 (a) $V_{\text{ave}} = \dfrac{V_m}{\pi}$, $V_{\text{eff}} = \dfrac{V_m}{2}$

(c) $V_{\text{ave}} = 0$, $V_{\text{eff}} = \dfrac{V_m}{2}$

10.2 (b) $P_{\text{ave}} = 17.5$ watts

10.4 (a) $5\, e^{j53°}$;

(d) $10.2\, e^{-j101.3°}$

10.5 (a) $2.6 + j1.5$;

(c) $7.12 - j3.63$

10.6 (b) $5\, e^{j30°}$

(d) $10e^{-j45°}$

10.7 (c) V lags I

(e) I lags V

10.12 $I = 8.06 \cos(2t - 60.5°)$

10.14 (b) $V_{\text{eq}} = 1.26\, e^{-j116.6°}$

$Z_{\text{eq}} = 1.78\, e^{-j26.6°}$

10.17 $R_s = \dfrac{R_p}{(1 + \omega^2 R_p^2 C_p^2)}$

10.20 $L = 2.98 \times 10^{-4}$

10.22 (a) $C = 2.01 \times 10^{-5}$

10.24 (a) $P_{\max} = 5 \times 10^{-5}$

10.27 $L' = 31.8 \times 10^{-6}$

Chapter 11

11.1 (c) $F(s) = \dfrac{8}{(s + 1000)} - \dfrac{5}{(s + 2000)}$

(g) $F(s) = \dfrac{s}{(s^2 - 1)}$

11.2 (b) $(2s + 3)V(s) = 4 + \dfrac{1}{(s + 3)}$

11.3 (c) $f(t) = \left(\frac{4900}{199} e^{-t} + \left(\frac{15000}{199}\right) e^{-200t}\right) u(t)$

11.4(d) $F(s) = \dfrac{(1 - e^{-2s})}{s^2}$

11.6(c) $f(\infty) = 0$; (e) $f(\infty)$ not defined

11.8(a) $v_L(t) = (13.33\, e^{-2000t} - 3.33\, e^{-500t})u(t)$

11.10 $v_{c_2}(t) = (0.78\, e^{-1.245\times 10^6 t} - 0.065\, e^{-1.486\times 10^7 t})u(t)$

11.13 $v_1 = -2$

11.15(b) $h_{ss}(t) = \dfrac{1}{6}\cos 2t$

11.17 First network:

$$v_o(t) = \sqrt{2}\, e^{-0.707t} \sin(0.707t),$$

$$\left.\frac{dv_o}{dt}\right|_{t=0} = 1$$

11.19(c) $\delta v_o(t) = (0.16 \times 10^{-4}\, e^{-10^8 t} - 0.133\, e^{-0.604\times 10^7 t} + 0.13)u(t)$

11.21(b) $\dfrac{\delta V_o}{\delta V_s} = \dfrac{-2.2s}{(s + 600)}$

Chapter 12

12.2 $h(t) = \dfrac{1}{99}\left(\sin t + \dfrac{1}{100}\sin 10t\right)$

12.4 (a) $R = 14.43$ ohms

(b) $R = 4.16$ ohms

(c) $R = 73.6$ ohms

12.6 $\dfrac{I_o}{I_s}(s) = \dfrac{s^2 \dfrac{M}{L_2}}{s^2 + \dfrac{R}{L_2}s + \dfrac{1}{L_2 C}}$;

$$\Delta\omega = \frac{R}{L_2}\text{ rad/sec; } Q = \frac{\omega_0 L_2}{R}$$

12.8 $\alpha = 2$

12.10 $s_p = -\dfrac{G}{2C} \pm j\sqrt{\dfrac{1}{LC(1 \pm k)}}$; $k > 0.05$;

$C = 20$ f, $L = 0.05$ h

$M = .0025$ h

12.14(b) $H_b(s) = \dfrac{4 \times 10^{10}(s + 1000)}{(s + 4000)(s + 10000)}$

12.18(b) $\omega_2 = \dfrac{g_m}{C_1}$, $A = 20\log_{10}(g_m R)$

(c) $v_o(t) = -g_m R[1 - e^{-t/R(C_1+C_2)}] + \dfrac{C_1}{C_1 + C_2} e^{-t/R(C_1+C_2)}$

12.20(a) $H(j1) = -1300$

Chapter 13

13.1 (b) $\dfrac{L di_L}{dt} = -R_1 i_L - v_c + v_s$

$$\frac{C dv_c}{dt} = i_L - \frac{1}{R_2 + R_3} v_c + \frac{R_3}{R_2 + R_3} i_s$$

$$v_o = \frac{R_3}{R_2 + R_3} v_c + \frac{R_2 R_3}{R_2 + R_3} i_s$$

(e) $\dfrac{C_2 dv_{c_2}}{dt} = -i_L - \dfrac{(1 + g_m R_1)}{R_1} v_{c1} - \left(\dfrac{1 + g_m R_1}{R_1} + \dfrac{1}{R_2}\right) v_{c2}$

[Two (2) other state equations]

13.3 $v_c(t) = 5.16\, e^{-218.75t} \cos(121t + 90°)$

13.5 Let $C(v_j) = 3k^3 v_j^2 = C_j$

$$\frac{dv_1}{dt} = -\frac{v_1}{R_1 C_1} - \frac{v_2}{R_1 C_1} + \frac{v_s}{R_1 C_1}$$

$$\frac{dv_2}{dt} = -\frac{v_1}{R_1 C_2} - \frac{v_2}{C_2}\left(\frac{1}{R_1} + \frac{1}{R_2}\right) + \frac{v_s}{C_2 R_1}$$

13.7(c) $k = 25.5025$

13.11(c) $e^{At} = \begin{bmatrix} 0.5\, e^{-3t} + 0.5\, e^{t} & ? \\ ? & 0.5\, e^{-3t} + 0.5\, e^{t} \end{bmatrix}$

13.12(a) Output equation:

$$y(t) = 1 + 0.1705\, e^{-3.62t} - 1.1495\, e^{-1.38t} - 0.615 x_1(0)\, e^{-3.62t} + 1.615 x_1(0)\, e^{-1.38t} + x_2(0)\, e^{-3.62t} + x_2(0)\, e^{-1.38t}$$

Chapter 14

14.2 Node equations:

$$\left(\frac{C_1}{T} + G_1\right) v_1^{n+1} - G_1 v_2^{n+1} = i_s + \frac{C_1 v_1 n}{T}$$

$$-G_1 v_1^{n+1} + \left(\frac{C_2}{T} + G_1 + G_2\right) v_2^{n+1} = \frac{C_2 v_2^n}{T}$$

14.8 Node equations (m superscripts ignored):

$$(G_1 + G_d) v_1^{n+1} - G_d v_2^{n+1}$$
$$= -i_d^{n+1} + v_{s_1} G_1$$
$$-G_d v_1^{n+1} + (G_2 + G_d) v_2^{n+1}$$
$$= i_d^{n+1} + v_{s_2}^{n+1} G_2 + \frac{C v_2^{n+1}}{T}$$

14.10 $v_o^{n+1,m+1}$

$$= \frac{i_d^{n+1,m} + \dfrac{C_2}{T} v_o^n + v_s^{n+1}\left(G_d^{n+1,m} + \dfrac{C_1}{T}\right)}{G_d^{n+1,m} + \dfrac{C_1}{T} + \dfrac{C_2}{T} + 10^{-6}}$$

14.15 (b) $q(t) = [q(0)\, e^{-t/\tau} + 10^{-3}\tau(1 - e^{-t/\tau})] \times u(t)$

14.17 (a) Natural frequencies $= -1.51 \times 10^9, -8.57 \times 10^9$

Chapter 15

15.1 (d) $y_{11} = \dfrac{(G_1 + G_3)(G_2 + G_4)}{G_1 + G_2 + G_3 + G_4}$

$y_{12} = \dfrac{G_2(G_3 + G_4) + G_4(G_1 + G_3)}{G_1 + G_2 + G_3 + G_4}$

15.1 (g) $y_{11} = G,\ y_{12} = -3G,\ y_{21} = -G,\ y_{22} = 3G$

15.4 (a) $y_{11} = y_{22} = y_{12} = 0,\ y_{21} = g_m$

15.6 $y_{11} = y_{11}' + y_{11}''$

15.8 $Y_{\text{in}} = \dfrac{1}{Y_0}$

15.11 (a) $\dfrac{\delta v_2}{\delta v_1} = \dfrac{(G_1 - G_d)}{(G_1 + G_c)}$

(d) $\dfrac{\delta v_2}{\delta v_1} = -667$

15.12 (b) $Z_{\text{in}} = Z_0$

15.13 $z_{21} = z_{12} = \dfrac{(R_2R_3 - R_1R_4)}{(R_1 + R_2 + R_3 + R_4)}$

15.17 $R_3 = \dfrac{R_B R_C}{(R_A + R_B + R_C)}$

15.20 $A = 1 + \dfrac{Z_1}{Z_B},\ C = \dfrac{1}{Z_s} + \dfrac{1}{Z_B} + \dfrac{Z_1}{(Z_A Z_B)}$

15.24 $R_s \cong 6.1\ \text{k}\Omega$

15.25 $\left.\dfrac{V_2}{V_1}\right|_{s=j\sqrt{6}/RC} = -0.0286$

Chapter 16

16.1 (a) $Z(s) = \dfrac{3}{4}\,\dfrac{s\left(s + \frac{2}{3}\right)}{\left(s + \frac{1}{2}\right)}$

(d) $Z(s) = \dfrac{s[(L_1 + L_1')(L_2 + L_2') - (M - M_2)^2]}{(L_2 + L_2')}$

16.3 $L_1 > 0.96$

16.6 (a) $n = k\sqrt{\dfrac{L_2}{L_1}}$

(c) $L_2 = 1.972,\ M = 0.956$

16.8 (a) $L_{\text{eq}} = 20$

(c) $L_{\text{eq}} = 5$

16.11 (a) $F_{\text{nat}} = \sqrt{\dfrac{K_1(M_1 + M_2)}{M_1 M_2}}$

Index